FORENSIC SCIENCE

An Introduction

Saferstein | Roy

Pearson

Cover art: PeopleImages.com—Yuri A/Shutterstock; stylephotographs/123rf.com; LEDOMSTOCK/Shutterstock; Looker_Studio/Shutterstock; pogonici/123rf.com; lightwise/123rf.com

Library of Congress Cataloging-in-Publication Data

Cataloging-in-Publication Data is available on file at the Library of Congress.

2 2023

Pearson

ISBN-10: 0-13-804620-4

ISBN-13: 978-0-13-804620-0

For my husband, Sean, and my children, Milo and Stella.
Everything I do in this world, I do to make your lives better.

-Tiffany Roy

Contents

Chapter 7

Forensic Toxicology 250

Chapter 8

The Microscope 294

Chapter 9

Forensic Serology 324

Chapter 12

Trace Evidence I: Hairs and Fibers . 454

Chapter 13

Trace Evidence II: Metals, Paint, and Soil. 494

Chapter 14

Forensic Aspects of Fire and Explosion Investigation . 532

Chapter 15

Forensic Investigation of Explosions 562

Chapter 16

Forensic Biometrics: Fingerprints and Facial Recognition 584

Chapter 17

Firearms, Tool Marks, and Other Impressions 626

Preface

The level of sophistication that forensic science has brought to criminal investigations is awesome. But one cannot lose sight of the fact that, once all the drama of a forensic science case is put aside, what remains is an academic subject emphasizing science and technology. It is to this end that this second edition of *Forensic Science: An Introduction* is dedicated.

Only the most relevant scientific and technological concepts are presented to the reader, so that the subject is not watered down with superfluous discussions that are of no real significance to current forensic science practices. It is the author's belief that, by learning in an interactive environment using the Internet, the reader will be a more motivated and active participant in the learning process. This text is accompanied by MyLab Criminal Justice with Pearson eText and offers book-specific learning objectives, virtual laboratory exercises, Web resources, practice tests, and more to aid student learning and comprehension.

One of the constants of forensic science is how frequently its applications become front-page news. Whether the story is sniper shootings or the tragic consequences of the terrorist attacks of 9/11/01, forensic science is at the forefront of the public response. In order to merge theory with practice, a significant number of actual forensic Case Files are included in the text. The intent is for all the case illustrations to capture the interest of the reader and to move forensic science from the domain of the abstract into the real world of criminal investigation.

Within and at the end of each chapter, the student will encounter Quick Reviews and a Chapter Summary that recap all of the major points of the chapter. The end-of-chapter summary is followed by review questions, as well as application and critical thinking exercises designed to have the reader further explore the chapter's content and its significance. Most chapters also include Laboratory Experiments, which have students apply the Next Generation Science Standards to a crime-scene activity. In some chapters, virtual crime scene exercises enable the reader to move through various types of crime scenes while identifying and collecting physical evidence.

New to This Edition Highlights

Information throughout the text has been updated and many new figures have been added to illustrate concepts discussed in the chapters. Numerous case files have been added to select chapters to illustrate how forensic technology has been applied to solving crimes of notoriety.

- Chapter 1 has new cases added, including Making a Murderer: Steven Avery
- Chapter 2 contains additional language and clarification regarding sketching, photographing, and taking notes at a crime scene.
- Chapter 3, "Physical Evidence" has been updated to introduce the reader to the new FBI Next Generation Identification System (NGI), created as a repository for biometric information.
- Chapter 4 has been updated to reflect the evolution of forensic anthropology, including new terms and updated references.
- Chapter 5 has been updated with new and more recent references
- Chapter 6 has been updated with new cases, including El Chapo, and changes in drug laws
- Chapter 7 has been updated with new and more recent references
- Chapter 8 has been updated with new cases, included Baby Doe, and advancements in stable isotope testing and palynology
- Chapter 9 has been updated with new and more recent references

- Chapter 10, "DNA: The Indispensable Tool" has been revised to reflect newly implemented technology, like forensic genetic genealogy, the method used to catch the Golden State Killer.
- Chapter 11 has been updated with all new photographs of bloodstains, as well as changes to impact spatter terminology to reflect the latest standards in the field
- Chapter 12 has been updated with new cases, including Casey Anthony and Wayne Williams, and updates to the section on estimating ancestry based on microscopic hair analysis
- Chapter 13 has been updated with new and more recent references
- Chapter 14, "Forensic Aspects of Fire Investigation," has been expanded to cover the discussion of the deviation of fire from normal behavior and how it impacts burn pattern interpretations at fire scenes
- Chapter 15 has been updated with new cases, including the Beirut Explosion
- Chapter 16 has been expanded to include the application of iris and facial biometrics used to create biometric databases
- Chapter 17 has been updated with new cases, including the case against Aaron Hernandez
- Chapter 18 has been updated with new cases, include The Life and Deaths of Robert Durst
- Chapter 19 has been updated with new and more recent references
- Chapter 20 has been updated with new cases, including the FBI v. Apple and the Michelle Carter case, as well as a discussion of new technology, like the StingRay device
- Chapter 21 has been updated with new and more recent references

New to MyLab Criminal Justice

The **Forensic Science Virtual Lab** exercises are intended to give the student a first-hand look at the types of tests and examinations performed in the crime lab. Using 360-degree photography, microscope imagery and explanatory videos, these laboratory exercises will bring the content to life and give students an opportunity to experience a day in the life of a forensic scientist. These Virtual Lab exercises are located within MyLab Criminal Justice. A complete list of the labs are provided on the next page.

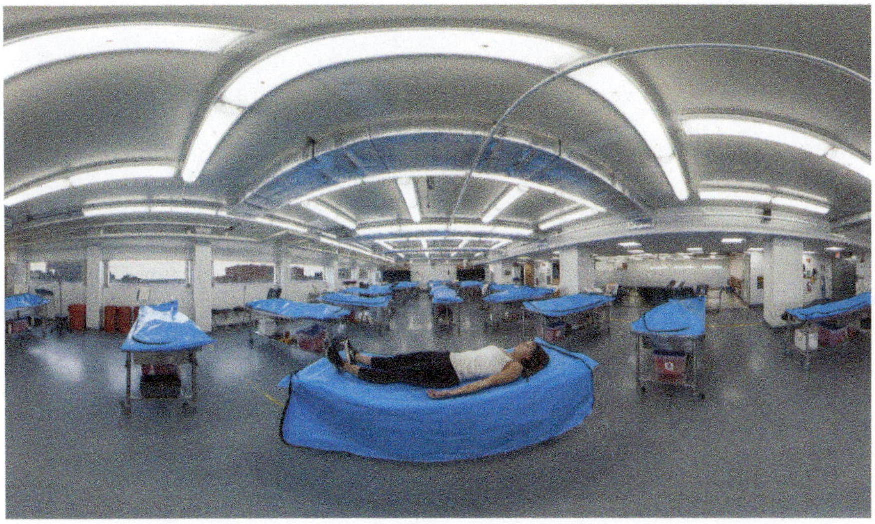

360-degree photographs bring the content to life.

Forensic Science Virtual Labs

- 1.1: Tour of a Crime Lab
- 1.2: Services at the Crime Lab
- 2.1: Processing the Crime Scene
- 2.2: Packaging Physical Evidence
- 3.1: Evidence with the Strongest Link
- 3.2: Forensic Databases
- 4.1: Medical Examiner's Office
- 4.2: Shooting Victim
- 4.3: Forensic Anthropology
- 6.1: Drug Evidence
- 7.1: Analyzing Mass Spectra
- 8.1: Using the Microscope
- 9.1: Blood Agglutination
- 9.2: Immunochromatographic Cards
- 9.3: Blood Testing
- 9.4: Brentamine Test for Acid Phosphatase
- 10.1: DNA Testing
- 10.2: DNA Profile Comparison
- 11.1: Estimating Angle of Impact
- 11.2: Bloodstain Classification
- 12.1: Analyzing Hair Specimens
- 12.2: Analyzing Fiber Specimens
- 14.1: Analyzing Ignitable Liquids
- 16.1: Fingerprints
- 16.2: IrisCode
- 16.3: Facial Recognition
- 17.1: Serial Number Restoration
- 17.2: Footwear Impressions
- 18.1: Document Examination
- 19.1: Computer Forensics
- 20.1: StingRay Devices
- 20.2: Geolocation

Acknowledgments

I would like to express my most sincere thanks to Dr. Richard Saferstein and his family for the opportunity to work on this text. It is an honor and privilege. I will work hard to ensure it is up to the high standard set by Richard in his lifetime. I am most appreciative of my former student, Karlee Jonas, for her assistance with text revision on this edition.

I would like to thank the subject matter experts who reviewed this text content and recommended updates, including Amanda Anderson, Lynn English High School; Erin Miles, Agoura High School; Carlos Rivera, Medical Arts High School; Linda Stefaniak, Allentown High School; and Bryan Whitmire, Buna High School. I would like to thank Andrew Donofrio and Dr. Peter Stephenson for their previous contributions to this book on the subject of mobile and digital forensics. Thank you to everyone who assisted Richard in making this book a success by providing photos and materials including Anita Wonder, Robert J. Phillips, Norman H. Reeves, Jeffrey C. Kercheval, Robert Thompson, Roger Ely, Jose R. Almirall, Michael Malone, Ronald Welsh, Ken Radwill, David Pauly, Jan Johnson, Natalie Borgan, Dr. Barbara Needell, Robin D. Williams, Peter Diaczuk, and Jacqueline E. Joseph. I'm appreciative of the contributions, reviews, and comments that Dr. Claus Speth, Dr. Mark Taff, Dr. Elizabeth Laposata, Thomas P. Mauriello, and Michelle D. Miranda provided during the preparation of Chapter 4, "Death Investigation."

I am grateful to the law enforcement agencies, governmental agencies, private individuals, and equipment manufacturers cited in the text for contributing their photographs and illustrations.

Finally, I would like to thank my family, friends and colleagues for all they have done to support me in my academic and professional pursuits. Any successes I achieve in my lifetime, I owe to the family and friends who love and support me through hard times and celebrate with me in the good.

Tiffany Roy, JD, MSFS.

About the Authors

Richard Saferstein, Ph.D., retired in 1991 after serving twenty-one years as the Chief Forensic Scientist of the New Jersey State Police Laboratory, one of the largest crime laboratories in the United States. He currently acts as a consultant for attorneys and the media in the area of forensic science. During the O. J. Simpson criminal trial, Dr. Saferstein provided extensive commentary on forensic aspects of the case for the *Rivera Live* show, the E! television network, ABC radio, and various radio talk shows. Dr. Saferstein holds degrees from the City College of New York and earned his doctorate degree in chemistry in 1970 from the City University of New York. From 1972 to 1991, he taught an introductory forensic science course in the criminal justice programs at The College of New Jersey and Ocean County College. These teaching experiences played an influential role in Dr. Saferstein's authorship in 1977 of the widely used introductory textbook *Criminalistics: An Introduction to Forensic Science*, currently in its eleventh edition. Saferstein's basic philosophy in writing *Criminalistics* is to make forensic science understandable and meaningful to the nonscience reader while giving the reader an appreciation for the scientific principles that underlie the subject.

Dr. Saferstein has authored or co-authored more than forty-four technical papers covering a variety of forensic topics. Dr. Saferstein has authored *Basic Laboratory Exercises for Forensic Science* (Prentice Hall, 2011) and co-authored *Lab Manual for Criminalistics* (Prentice Hall, 2015). He has also edited two editions of the widely used professional reference books *Forensic Science Handbook*, Volume 1 (Prentice Hall, 2002), *Forensic Science Handbook*, Volume 2 (Prentice Hall, 2005), and *Forensic Science Handbook*, Volume 3 (Prentice Hall, 2009). Dr. Saferstein is a member of the American Chemical Society, the American Academy of Forensic Sciences, the Canadian Society of Forensic Scientists, the International Association for Identification, the Northeastern Association of Forensic Scientists, and the Society of Forensic Toxicologists.

In 2006, Dr. Saferstein received the American Academy of Forensic Sciences Paul L. Kirk award for distinguished service and contributions to the field of criminalistics.

Tiffany Roy, MSFS, JD, is a Forensic DNA expert with over fourteen years of forensic biology experience in both public and private laboratories in the United States. She has processed thousands of DNA samples and thousands of cases over the course of her career. She has provided expert witness testimony in more than one hundred cases in state, federal, and international courts.

She instructs undergraduates at Palm Beach Atlantic University, University of Maryland Global Campus, and Southern New Hampshire University Global Campus. She currently acts as a consultant for attorneys and the media in the area of forensic biology through her firm, ForensicAid, LLC.

Roy holds degrees from Syracuse University, Massachusetts School of Law, and University of Florida in the areas of Biology, Law, and Forensic Science. Her teaching, legal writing, and testimonial experience help her to take complex scientific concepts and make them easily understandable for the nonscientist. Roy assisted Dr. Saferstein in completing *Forensic Science: From the Crime Scene to the Crime Lab 4e* (Pearson 2019) and has authored the text *Criminalistics: An Introduction to Forensic Science 13e* (2021) and *The Complete Guide to the American Board of Criminalistics Molecular Biology Examination* (CRC Press 2020).

Roy is a member of the American Academy of Forensic Sciences, the Northeastern Association of Forensic Scientists, and the Massachusetts Board of Bar Examiners. She is certified in the area of Molecular Biology by the American Board of Criminalistics. Aside from her teaching, writing, and consulting, Roy also assists with international capacity building initiatives, providing subject matter expertise for trainings for criminal justice stakeholders in the Middle East and Africa.

Welcome...

to the exciting second edition of *Forensic Science: An Introduction*. Richard Saferstein and Tiffany Roy have carefully adapted and updated their classic *Criminalistics: An Introduction to Forensic Science* text to create a comprehensive program designed specifically for high school students and teachers.

Accessible Text and Motivational 4-Color Presentation

The layout and design make learning forensic science even more motivating and exciting.

Students live in a visual world, and the functional use of full color conveys forensic science to today's students. Over 150 full-color photos and illustrations motivate students to read.

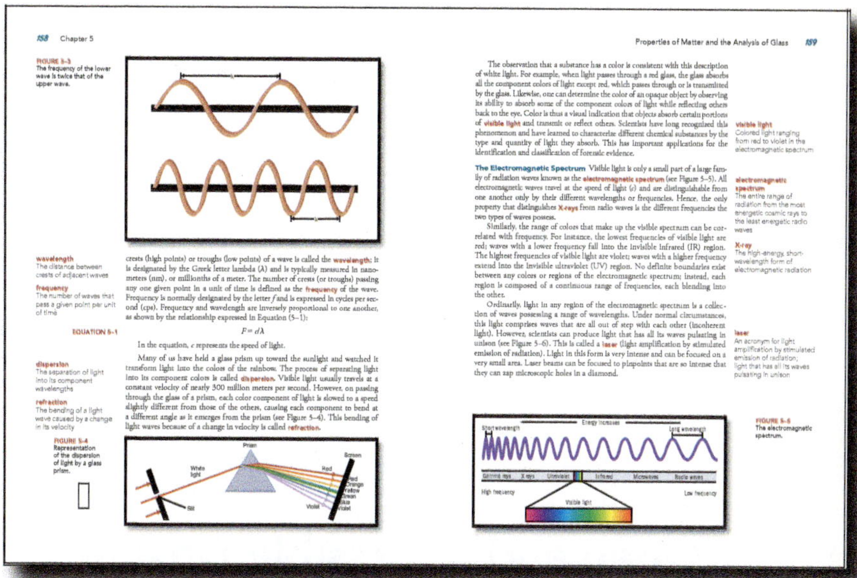

Chapter Openers

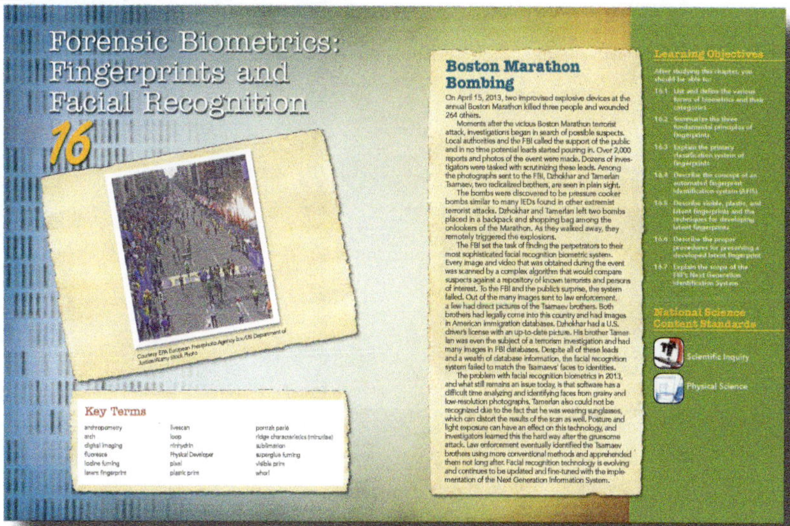

Each chapter opens with a real-life case study and stunning visual that captures students' interest and brings content to life.

Learning Objectives help students focus on the key takeaways for that chapter.

National Science Education Standards align with the chapter content and highlight the multidisciplinary nature of forensic science.

Dimensional Illustrations

The full-color art program helps students better understand key forensics concepts.

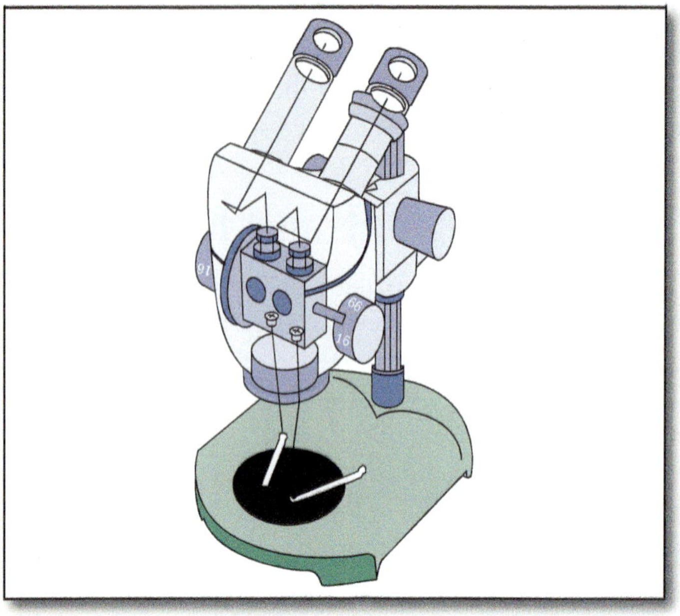

Open and Accessible Design

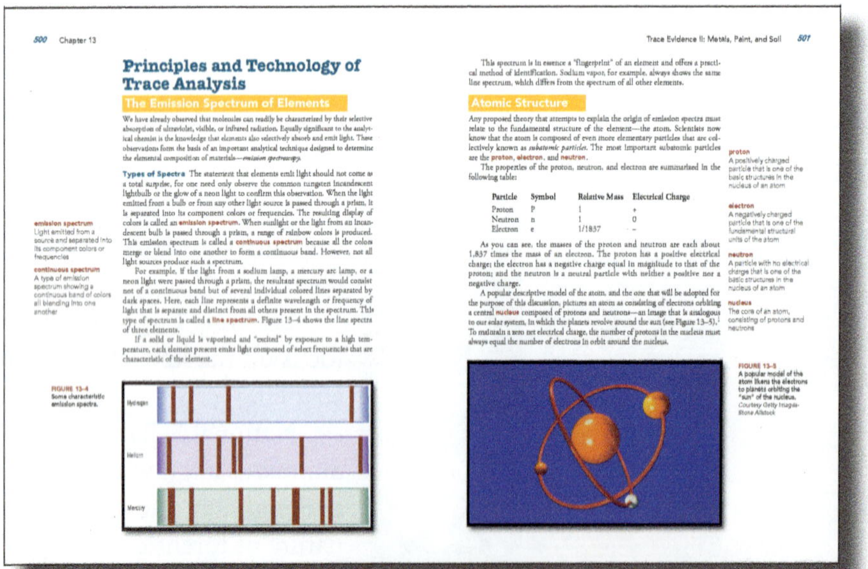

500 Chapter 13

Principles and Technology of Trace Analysis

The Emission Spectrum of Elements

We have already observed that molecules can readily be characterized by their selective absorption of ultraviolet, visible, or infrared radiation. Equally significant to the analytical chemist is the knowledge that elements also selectively absorb and emit light. These observations form the basis of an important analytical technique designed to determine the elemental composition of materials—*emission spectroscopy*.

Types of Spectra The statement that elements emit light should not come as a total surprise, for one need only observe the common tungsten incandescent lightbulb or the glow of a neon light to confirm this observation. When the light emitted from a bulb or from any other light source is passed through a prism, it is separated into its component colors or frequencies. The resulting display of colors is called an **emission spectrum**. When sunlight or the light from an incandescent bulb is passed through a prism, a range of rainbow colors is produced. This emission spectrum is called a **continuous spectrum** because all the colors merge or blend into one another to form a continuous band. However, not all light sources produce such a spectrum.

emission spectrum Light emitted from a source and separated into its component colors or frequencies

continuous spectrum A type of emission spectrum showing a continuous band of colors all blending into one another

For example, if the light from a sodium lamp, a mercury arc lamp, or a neon light were passed through a prism, the resultant spectrum would consist not of a continuous band but of several individual colored lines separated by dark spaces. Here, each line represents a definite wavelength or frequency of light that is separate and distinct from all others present in the spectrum. This type of spectrum is called a **line spectrum**. Figure 13–4 shows the line spectra of three elements.

If a solid or liquid is vaporized and "excited" by exposure to a high temperature, each element present emits light composed of select frequencies that are characteristic of the element.

FIGURE 13–4 Some characteristic emission spectra.

Trace Evidence II: Metals, Paint, and Soil 501

This spectrum is in essence a "fingerprint" of an element and offers a practical method of identification. Sodium vapor, for example, always shows the same line spectrum, which differs from the spectrum of all other elements.

Atomic Structure

Any proposed theory that attempts to explain the origin of emission spectra must relate to the fundamental structure of the element—the atom. Scientists now know that the atom is composed of even more elementary particles that are collectively known as *subatomic particles*. The most important subatomic particles are the **proton**, **electron**, and **neutron**.

proton A positively charged particle that is one of the basic structures in the nucleus of an atom

electron A negatively charged particle that is one of the fundamental structural units of the atom

The properties of the proton, neutron, and electron are summarized in the following table:

Particle	Symbol	Relative Mass	Electrical Charge
Proton	P	1	+
Neutron	n	1	0
Electron	e	1/1837	–

As you can see, the masses of the proton and neutron are each about 1,837 times the mass of an electron. The proton has a positive electrical charge; the electron has a negative charge equal in magnitude to that of the proton; and the neutron is a neutral particle with neither a positive nor a negative charge.

neutron A particle with no electrical charge that is one of the basic structures in the nucleus of an atom

A popular descriptive model of the atom, and the one that will be adopted for the purpose of this discussion, pictures an atom as consisting of electrons orbiting a central **nucleus** composed of protons and neutrons—an image that is analogous to our solar system, in which the planets revolve around the sun (see Figure 13–5).[1] To maintain a zero net electrical charge, the number of protons in the nucleus must always equal the number of electrons in orbit around the nucleus.

nucleus The core of an atom, consisting of protons and neutrons

FIGURE 13–5 A popular model of the atom likens the electrons to planets orbiting the "sun" of the nucleus. *Courtesy Getty Images-Stone Allstock*

Design elements bring the course content to life and provide visual cues to guide student reading.

Key Terms

Forensic-specific vocabulary is highlighted in the text and defined in the margins.

Engaging Case Files

Chapter 8

Identifying Baby Doe

The body of a young girl was found on the shore of Deer Island outside Boston by a woman walking her dog (Figure 8–19). The child soon came to be known as "Baby Doe" by the local news media.

FIGURE 8–19

Two weeks later, police released a computer-generated composite image of the girl along with photos of the polkadot leggings and zebra-print blanket that were discovered with her body. In the days after the images were released, tips poured in as millions viewed the photos. Baby Doe's image was placed on billboards in 50 locations throughout the state. In the next three months, investigators followed up 210 leads but were unsuccessful in identifying her.

Investigators began considering the possibility that she washed up on the shore of Deer Island from a boat, or was transported to the area from somewhere else. In order to determine if they were focusing their search in the right place, they explored two forensic examination types which have the potential to indicate geographic origin: Palynology and Stable Isotope Analysis.

To perform the pollen analysis, investigators took vacuum samples from the blanket and pants, and forensic scientists were able to retrieve enough pollen to complete a standard pollen count (Figure 8–20). The pollen taxa encountered were expected to be found in the broad-leafed forests of the Northeastern United States.

FIGURE 8–20

The Massachusetts State Police also sent samples to IsoForensics, a lab in Salt Lake City, Utah. Using stable isotopes in the girl's hair and teeth, analysts could examine the chemical characteristics she encountered throughout her life.

"What you're eating, what you're drinking, that's recorded in your tissues immediately," IsoForensics President Lesley Chesson said. "You're recording the isotopes of your drinking water into your hair, into your bones." Researchers use that information to paint a general picture of the person's life; where they lived, where they have been recently. An isotope analysis showed the girl spent time in New England before her death.

Efforts of law enforcement paid off when a tip led to the identity of Baby Doe in September 2015. The girl was positively identified as Bella Bond, 2 ½ years old, from Dorchester, MA. As a result of the tip, police subsequently charged Michael P. McCarthy, the child's mother's boyfriend, with her murder. He was convicted and sentenced to life in prison.

Linked to the chapter material, the Case File feature boxes provide students with quick and pertinent facts about real forensic cases.

Quick Labs

Inquiry is at the heart of science, and it's no exception here. In-text Quick Labs are hands-on activities that allow students to apply and experience key forensic concepts.

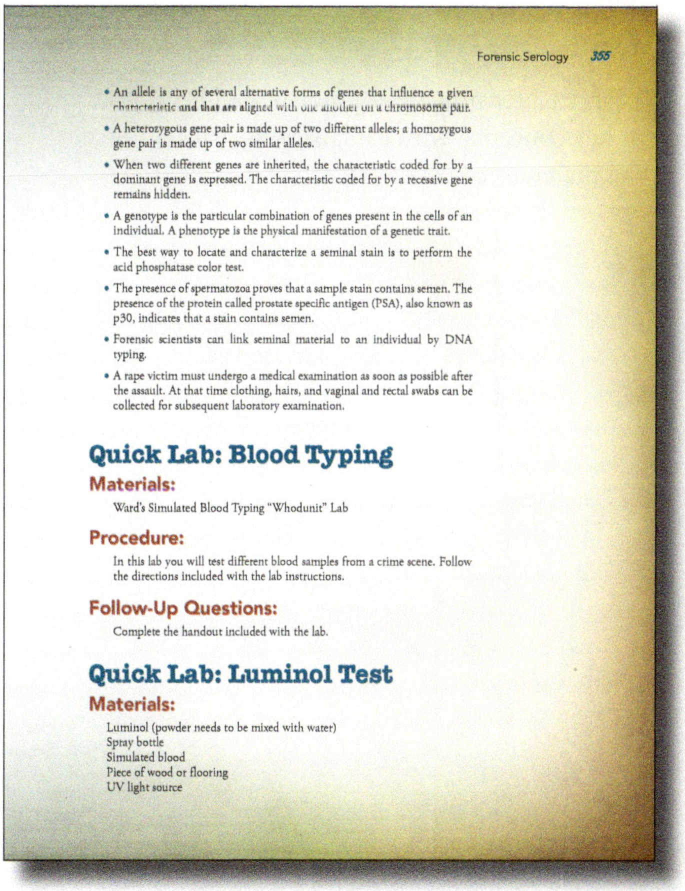
Forensic Serology 355

- An allele is any of several alternative forms of genes that influence a given characteristic and that are aligned with one another on a chromosome pair.
- A heterozygous gene pair is made up of two different alleles; a homozygous gene pair is made up of two similar alleles.
- When two different genes are inherited, the characteristic coded for by a dominant gene is expressed. The characteristic coded for by a recessive gene remains hidden.
- A genotype is the particular combination of genes present in the cells of an individual. A phenotype is the physical manifestation of a genetic trait.
- The best way to locate and characterize a seminal stain is to perform the acid phosphatase color test.
- The presence of spermatozoa proves that a sample stain contains semen. The presence of the protein called prostate specific antigen (PSA), also known as p30, indicates that a stain contains semen.
- Forensic scientists can link seminal material to an individual by DNA typing.
- A rape victim must undergo a medical examination as soon as possible after the assault. At that time clothing, hairs, and vaginal and rectal swabs can be collected for subsequent laboratory examination.

Quick Lab: Blood Typing

Materials:

Ward's Simulated Blood Typing "Whodunit" Lab

Procedure:

In this lab you will test different blood samples from a crime scene. Follow the directions included with the lab instructions.

Follow-Up Questions:

Complete the handout included with the lab.

Quick Lab: Luminol Test

Materials:

Luminol (powder needs to be mixed with water)
Spray bottle
Simulated blood
Piece of wood or flooring
UV light source

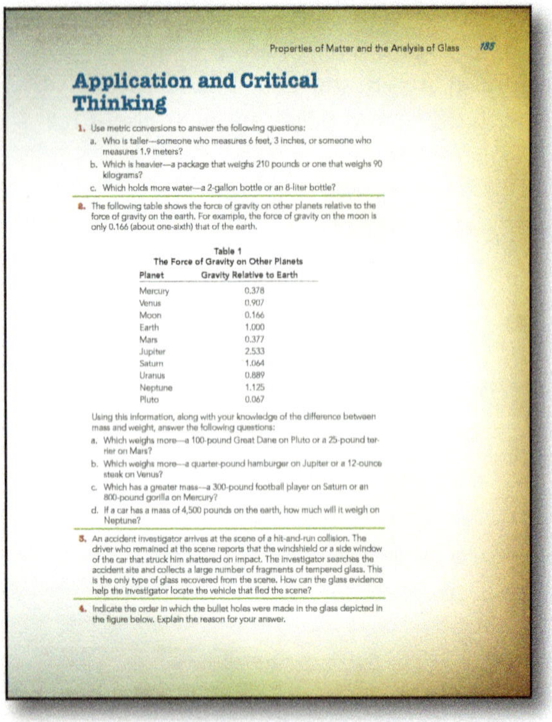

Application and Critical Thinking

1. Use metric conversions to answer the following questions:
   a. Who is taller—someone who measures 6 feet, 3 inches, or someone who measures 1.9 meters?
   b. Which is heavier—a package that weighs 210 pounds or one that weighs 90 kilograms?
   c. Which holds more water—a 2-gallon bottle or an 8-liter bottle?

2. The following table shows the force of gravity on other planets relative to the force of gravity on the earth. For example, the force of gravity on the moon is only 0.166 (about one-sixth) that of the earth.

Table 1
The Force of Gravity on Other Planets

Planet	Gravity Relative to Earth
Mercury	0.378
Venus	0.907
Moon	0.166
Earth	1.000
Mars	0.377
Jupiter	2.533
Saturn	1.064
Uranus	0.889
Neptune	1.125
Pluto	0.067

Using this information, along with your knowledge of the difference between mass and weight, answer the following questions:
   a. Which weighs more—a 100-pound Great Dane on Pluto or a 25-pound terrier on Mars?
   b. Which weighs more—a quarter-pound hamburger on Jupiter or a 12-ounce steak on Venus?
   c. Which has a greater mass—a 300-pound football player on Saturn or an 800-pound gorilla on Mercury?
   d. If a car has a mass of 4,500 pounds on the earth, how much will it weigh on Neptune?

3. An accident investigator arrives at the scene of a hit-and-run collision. The driver who remained at the scene reports that the windshield or a side window of the car that struck him shattered on impact. The investigator searches the accident site and collects a large number of fragments of tempered glass. This is the only type of glass recovered from the scene. How can the glass evidence help the investigator locate the vehicle that fled the scene?

4. Indicate the order in which the bullet holes were made in the glass depicted in the figure below. Explain the reason for your answer.

Application and Critical Thinking

Each chapter contains many activities designed to encourage application of critical thinking skills as they pertain to everyday life.

Chapter Review and Assessment

Each chapter provides a point-by-point summary of key concepts, with explanations that reinforce the materials covered.

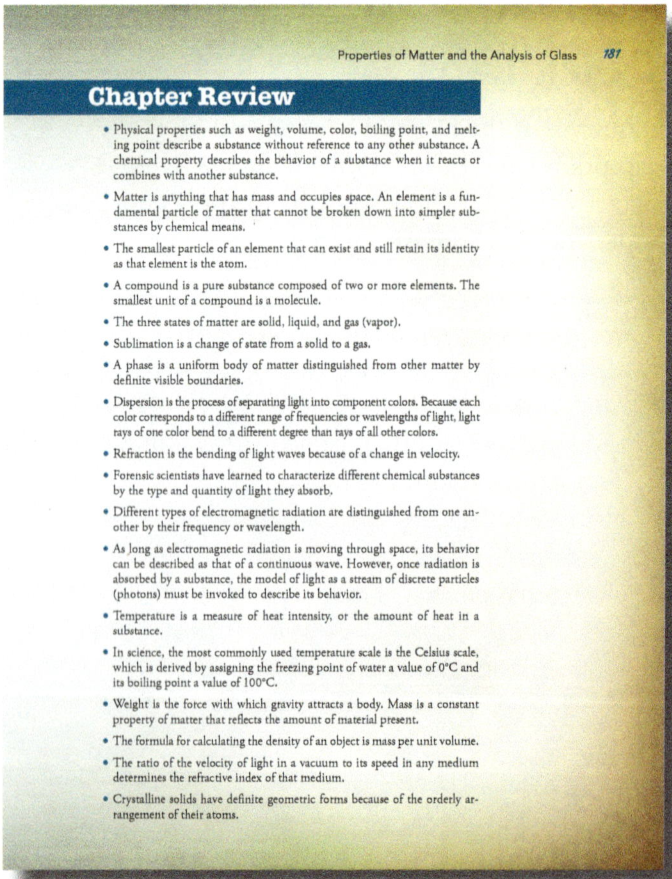

Chapter Review

- Physical properties such as weight, volume, color, boiling point, and melting point describe a substance without reference to any other substance. A chemical property describes the behavior of a substance when it reacts or combines with another substance.
- Matter is anything that has mass and occupies space. An element is a fundamental particle of matter that cannot be broken down into simpler substances by chemical means.
- The smallest particle of an element that can exist and still retain its identity as that element is the atom.
- A compound is a pure substance composed of two or more elements. The smallest unit of a compound is a molecule.
- The three states of matter are solid, liquid, and gas (vapor).
- Sublimation is a change of state from a solid to a gas.
- A phase is a uniform body of matter distinguished from other matter by definite visible boundaries.
- Dispersion is the process of separating light into component colors. Because each color corresponds to a different range of frequencies or wavelengths of light, light rays of one color bend to a different degree than rays of all other colors.
- Refraction is the bending of light waves because of a change in velocity.
- Forensic scientists have learned to characterize different chemical substances by the type and quantity of light they absorb.
- Different types of electromagnetic radiation are distinguished from one another by their frequency or wavelength.
- As long as electromagnetic radiation is moving through space, its behavior can be described as that of a continuous wave. However, once radiation is absorbed by a substance, the model of light as a stream of discrete particles (photons) must be invoked to describe its behavior.
- Temperature is a measure of heat intensity, or the amount of heat in a substance.
- In science, the most commonly used temperature scale is the Celsius scale, which is derived by assigning the freezing point of water a value of 0°C and its boiling point a value of 100°C.
- Weight is the force with which gravity attracts a body. Mass is a constant property of matter that reflects the amount of material present.
- The formula for calculating the density of an object is mass per unit volume.
- The ratio of the velocity of light in a vacuum to its speed in any medium determines the refractive index of that medium.
- Crystalline solids have definite geometric forms because of the orderly arrangement of their atoms.

Next Generation Science Standards* Overview

The Next Generation Science Standards (NGSS) provide an important opportunity to improve not only science education but also student achievement. Based on the Framework for K–12 Science Education, the NGSS are intended to reflect a new vision for American science Education.

The forensic science course, being an integrated science, is not intended to directly address specific NGSS expectations. However, it incorporates the science and engineering practices and crosscutting concepts from the Framework for K–12 Science Education, which are the foundation for the NGSS standards.

The Framework identifies seven crosscutting concepts and eight science and engineering practices. The seven crosscutting concepts bridge disciplinary boundaries, uniting core ideas throughout the fields of science and engineering. The seven crosscutting concepts are as follows.

1. **Patterns**—Observed patterns of forms and events guide organization and classification, and they prompt questions about relationships and the factors that influence them.
2. **Cause and effect: Mechanism and explanation**—Events have causes, sometimes simple, sometimes multifaceted. A major activity of science is investigating and explaining causal relationships and the mechanisms by which they are mediated. Such mechanisms can then be tested across given contexts and used to predict and explain events in new contexts.
3. **Scale, proportion, and quantity**—In considering phenomena, it is critical to recognize what is relevant at different measures of size, time, and energy and to recognize how changes in scale, proportion, or quantity affect a system's structure or performance.
4. **Systems and system models**—Defining the system under study—specifying its boundaries and making explicit a model of that system—provides tools for understanding and testing ideas that are applicable throughout science and engineering.
5. **Energy and matter: Flows, cycles, and conservation**—Tracking fluxes of energy and matter into, out of, and within systems helps one understand the systems' possibilities and limitations.
6. **Structure and function**—The way in which an object or living thing is shaped and its substructure determine many of its properties and functions.
7. **Stability and change**—For natural and built systems alike, conditions of stability and determinants of rates of change or evolution of a system are critical elements of study.

The eight practices of science and engineering identified as essential for all students to learn are listed below:

1. Asking questions (for science) and defining problems (for engineering)
2. Developing and using models
3. Planning and carrying out investigations
4. Analyzing and interpreting data
5. Using mathematics and computational thinking
6. Constructing explanations (for science) and designing solutions (for engineering)
7. Engaging in argument from evidence
8. Obtaining, evaluating, and communicating information

*Next Generation Science Standards is a registered trademark of Achieve. Neither Achieve nor the lead states and partners that developed the Next Generation Science Standards was involved in the production of, and does not endorse, this product.

Handbook of Forensic Services—FBI

The *Handbook of Forensic Services* provides guidance and procedures for the safe and efficient methods of collecting, preserving, packaging, and shipping evidence, and describes the forensic examinations performed by the FBI's Laboratory Division and Operational Technology Division.

The contents of the Handbook are to be found by the reader on either the iPhone app entitled "FBI Handbook" or the Android app entitled "Handbook of Forensic Services." The handbook can also be found online:

https://www.fbi.gov/file-repository/handbook-of-forensic-services-pdf.pdf/view

Introduction

1

GL Archive/Alamy Stock Photo

Key Terms

expert witness
Locard's exchange principle
scientific method

Steven Avery: Making a Murderer

The case of Steven Avery captured America's attention when it was featured in the Netflix documentary *Making a Murderer*. The documentary detailed the case of Avery, who was wrongfully convicted in 1985 of sexual assault and attempted murder. After serving 18 years of a 20-year sentence, he was exonerated by DNA testing and released.

In 2003, Avery filed a $36 million lawsuit against Manitowoc County, its former sheriff, and its former district attorney for wrongful conviction and imprisonment. In November 2005, with his civil suit still pending, he was arrested for the murder of Wisconsin photographer Teresa Halbach.

Teresa Halbach was last seen on October 31, 2005. Her last known appointment was a meeting with Avery, at his home on the grounds of Avery's Auto Salvage, to photograph his sister's minivan that he was offering for sale on Autotrader.com. During the investigation into her disappearance, Halbach's vehicle was found partially concealed in the salvage yard, and bloodstains recovered from its interior matched Avery's DNA. Investigators later identified charred bone fragments found in a burn pit near Avery's home as belonging to Halbach's. A Manitowoc deputy found the key to Halbach's vehicle in Avery's bedroom.

Avery's attorneys assert that there was a conflict of interest in the participation of the Manitowoc Sheriff's Department and suggested evidence tampering. Avery was arrested and charged with Halbach's murder, kidnapping, sexual assault, and mutilation of her corpse on November 11, 2005. Although Manitowoc County ceded control of the murder investigation to the neighboring Calumet County Sheriff's Department because of Avery's lawsuit, Manitowoc sheriff's deputies participated in repeated searches of Avery's trailer, garage, and property, supervised by Calumet County officers. The case serves as a prime example of the kinds of legal questions that can be raised if proper consideration isn't given to crime scene search and recovery procedures.

Learning Objectives

After studying this chapter, you should be able to:

1.1 Distinguish between forensic science and criminalistics

1.2 Describe the organization and services of a typical comprehensive crime laboratory in the criminal justice system

1.3 Explain how physical evidence is analyzed and presented in the courtroom by the forensic scientist, and how admissibility of evidence is determined in the courtroom

1.4 Explain the role and responsibilities of the expert witness and what specialized forensic services, aside from the crime laboratory, are generally available to law enforcement personnel

1.5 Learn where to search for information about forensic science on the Internet

National Science Content Standards

Scientific Inquiry

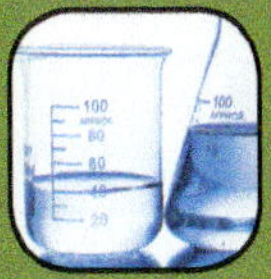
Physical Science

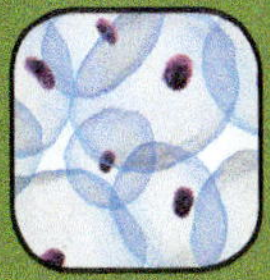
Life Science

Definition and Scope of Forensic Science

Forensic science in its broadest definition is the application of science to law. As our society has grown more complex, it has become more dependent on rules of law to regulate the activities of its members. Forensic science applies the knowledge and technology of science to the definition and enforcement of such laws.

Each year, as government finds it increasingly necessary to regulate the activities that most intimately influence our daily lives, science merges more closely with civil and criminal law. Consider, for example, the laws and agencies that regulate the quality of our food, the nature and potency of drugs, the extent of automobile emissions, the kind of fuel oil we burn, the purity of our drinking water, and the pesticides we use on our crops and plants. It would be difficult to conceive of any food and drug regulation or environmental protection act that could be effectively monitored and enforced without the assistance of scientific technology and the skill of the scientific community.

Laws are continually being broadened and revised to counter the alarming increase in crime rates. In response to public concern, law enforcement agencies have expanded their patrol and investigative functions, hoping to stem the rising tide of crime. At the same time they are looking more to the scientific community for advice and technical support for their efforts. Can the technology that put astronauts on the moon, split the atom, and eradicated most dreaded diseases be enlisted in this critical battle? Unfortunately, science cannot offer final and authoritative solutions to problems that stem from a maze of social and psychological factors. However, as the contents of this book will attest, science occupies an important and unique role in the criminal justice system—a role that relates to the scientist's ability to supply accurate and objective information that reflects the events that have occurred at a crime. A good deal of work remains to be done if the full potential of science as applied to criminal investigations is to be realized.

Considering the vast array of civil and criminal laws that regulate society, forensic science, in its broadest sense, has become so comprehensive a subject as to make a meaningful introductory textbook treatment of its role and techniques most difficult, if not overwhelming. For this reason, we must find practical limits that narrow the scope of the subject. Fortunately, common usage provides us with such a limited definition: **Forensic science is the application of science to the criminal and civil laws that are enforced by police agencies in a criminal justice system.** *Forensic science* is an umbrella term encompassing a myriad of professions that use their skills to help law enforcement officials conduct their investigations.

The diversity of professions practicing forensic science is illustrated by the 11 sections of the American Academy of Forensic Science, the largest forensic science organization in the world:

1. **Criminalistics**
2. **Digital and Multimedia Sciences**
3. **Engineering Science**
4. **General**
5. **Jurisprudence**

6. **Odontology**
7. **Pathology/Biology**
8. **Physical Anthropology**
9. **Psychiatry/Behavioral Sciences**
10. **Questioned Documents**
11. **Toxicology**

Even this list of professions is not exclusive. It does not encompass skills such as fingerprint examination, firearm and tool mark examination, and photography. Obviously, to author a book covering all of the major activities of forensic science as they apply to the enforcement of criminal and civil laws by police agencies would be a major undertaking.

Thus, this book will further restrict itself to discussions of the subjects of chemistry, biology, physics, geology, and computer technology, which are useful for determining the evidential value of crime-scene and related evidence. Forensic psychology, anthropology, and odontology also encompass important and relevant areas of knowledge and practice in law enforcement, each being an integral part of the total forensic science service that is provided to any up-to-date criminal justice system. However, these subjects go beyond the intended scope of this book, and except for brief discussions, along with pointing the reader to relevant websites, the reader is referred elsewhere for discussions of their applications and techniques. Instead, this book focuses on the services of what has popularly become known as the crime laboratory, where the principles and techniques of the physical and natural sciences are practiced and applied to the analysis of crime-scene evidence.

For many, the term criminalistics seems more descriptive than forensic science for describing the services of a crime laboratory. The two terms will be used interchangeably in this text. Regardless of title—criminalist or forensic scientist—the trend of events has made the scientist in the crime laboratory an active participant in the criminal justice system.

Prime-time television shows like *CSI: Crime Scene Investigation* have greatly increased the public's awareness of the use of science in criminal and civil investigations. However, by simplifying scientific procedures to fit into the available airtime, these shows have created unrealistic expectations of forensic science skills within both the public and the legal community. In these shows, members of the CSI team collect evidence at the crime scene, process all evidence, question witnesses, interrogate suspects, carry out search warrants, and testify in court. In the real world, these tasks are almost always delegated to different people in different parts of the criminal justice system. Procedures that could take days, weeks, months, or years in reality appear on these shows to take mere minutes. This false image is especially relevant to the public's high interest in and expectations for DNA evidence.

The dramatization of forensic science on television has led the public to believe that every crime scene will yield forensic evidence and produces unrealistic expectations that a prosecutor's case should always be bolstered and supported by forensic evidence. This phenomenon is known as the "CSI effect." Some jurists have come to believe that this phenomenon ultimately detracts from the search for truth and justice in the courtroom.

History and Development of Forensic Science

Forensic science owes its origins first to the individuals who developed the principles and techniques needed to identify or compare physical evidence and second to those who recognized the need to merge these principles into a coherent discipline that could be practically applied to a criminal justice system.

The roots of forensic science reach back many centuries, and history records a number of instances in which individuals used close observation of evidence and applied basic scientific principles to solve crimes. Not until relatively recently, however, did forensic science take on the more careful and systematic approach that characterizes the modern discipline. Only by looking at the historical development and current advancement of forensic science from its humble beginnings to where we are today can we get a true understanding of how far we have come.

FIGURE 1–1
Crime Scene, NYC, Early 20th Century. *Courtesy Science History Images/ Alamy Stock Photo*

Early Developments

One of the earliest records of applying forensics to solve criminal cases comes from third-century China. A manuscript titled *Yi Yu Ji* (*A Collection of Criminal Cases*) reports how a coroner solved a case in which a woman was suspected of murdering her husband and burning the body, then claiming that he died in an accidental fire. Noticing that the husband's corpse had no ashes in its mouth, the coroner performed an experiment to test the woman's story. He burned two pigs—one alive and one dead—and then checked for ashes inside the mouth of each. He found ashes in the mouth of the pig that was alive before it was burned, but none in the mouth of the pig that was dead beforehand. The coroner thus concluded that the husband, too, was dead before his body was burned. Confronted with this evidence, the woman admitted her guilt. The Chinese were also among the first to recognize the potential of fingerprints as a means of identification.

Although cases such as that of the Chinese coroner are noteworthy, this kind of scientific approach to criminal investigation was for many years the exception rather than the rule. Limited knowledge of anatomy and pathology hampered the development of forensic science until the late 17th and early 18th centuries. For example, the first recorded notes about fingerprint characteristics were prepared in 1686 by Marcello Malpighi, a professor of anatomy at the University of Bologna in Italy. Malpighi, however, did not acknowledge the value of fingerprints as a method of identification. The first scientific paper about the nature of fingerprints did not appear until more than a century later, but that work also did not recognize their potential as a form of identification.

Initial Scientific Advances

As physicians gained a greater understanding of the workings of the body, the first scientific treatises on forensic science began to appear, such as the 1798 work *A Treatise on Forensic Medicine and Public Health* by the French physician François-Emanuel Fodéré. Breakthroughs in chemistry at this time also helped forensic science take significant strides forward. In 1775, the Swedish chemist Carl Wilhelm Scheele devised the first successful test for detecting the poison arsenic in corpses. By 1806, the German chemist Valentin Ross had discovered a more precise method for detecting small amounts of arsenic in the walls of a victim's stomach. The most significant early figure in this area was Mathieu Orfila, a Spaniard who is considered the father of forensic toxicology. In 1814, Orfila published the first scientific treatise on the detection of poisons and their effects on animals. This treatise established forensic toxicology as a legitimate scientific endeavor.

The mid-1800s saw a spate of advances in several scientific disciplines that furthered the field of forensic science. In 1828, William Nichol invented the polarizing microscope. Eleven years later, Henri-Louis Bayard formulated the first procedures for microscopic detection of sperm. Other developments during this time included the first microcrystalline test for hemoglobin (1853) and the first presumptive test for blood (1863). Such tests soon found practical applications in criminal trials. Toxicological evidence at trial was first used in 1839, when a Scottish chemist named James Marsh testified on the detection of arsenic in a victim's body. During the 1850s and 1860s, the new science of photography was also used in forensics, recording images of prisoners and crime scenes.

FIGURE 1–2 Mathieu Orfila.
The Print Collector Heritage Images/Newscom

Late 19th-Century Progress

By the late 19th century, public officials were beginning to apply knowledge from virtually all scientific disciplines to the study of crime. Anthropology and morphology (the study of the structure of living organisms) were applied to the first system of personal identification, devised by the French scientist Alphonse Bertillon in 1879. Bertillon's system, which he dubbed anthropometry, was a systematic procedure that involved taking a series of body measurements as a means of distinguishing one individual from another. For nearly two decades, this system was considered the most accurate method of personal identification, before being replaced by fingerprinting in the early 1900s. Bertillon's early efforts earned him the distinction of being known as the father of criminal identification.

Bertillon's anthropometry, however, would soon be supplanted by the more reliable method of identification by fingerprinting. Two years before the publication of Bertillon's system, the U.S. microscopist Thomas Taylor suggested that fingerprints could be used as a means of identification, but his ideas were not immediately followed up. Three years later, the Scottish physician Henry Faulds made a similar assertion in a paper published in the journal *Nature*. However, the Englishman Francis Henry Galton undertook the first definitive study of fingerprints and developed a methodology of classifying them for filing. In 1892, Galton published a book titled *Finger Prints*, which contained the first statistical proof supporting the uniqueness of his method of personal identification. His work went on to describe the basic principles that form the present system of identification by fingerprints. Sir Edward Richard Henry contributed to forensic science by developing a fingerprint identification system while he was stationed in India. Henry's method created a primary fingerprint classification system with 1,024 classifications.

The first treatise describing the application of scientific disciplines to the field of criminal investigation was written by Hans Gross in 1893. Gross, a public prosecutor and judge in Graz, Austria, spent many years studying and developing principles of criminal investigation. In his classic book, *Handbuch für Untersuchungsrichter als System der Kriminalistik* (later published in English under the title *Criminal Investigation*), he detailed the assistance that investigators could expect from the fields of microscopy, chemistry, physics, mineralogy, zoology, botany, anthropometry, and fingerprinting. He later introduced the forensic journal *Archiv für Kriminal Anthropologie und Kriminalistik*, which still reports improved methods of scientific crime detection.

Ironically, the best-known figure in 19th-century forensics was not a real person, but a fictional character, the legendary detective Sherlock Holmes. Many people today believe that Holmes's creator, Sir Arthur Conan Doyle, had a considerable influence on popularizing scientific crime-detection methods. In adventures with his partner and biographer, Dr. John Watson, Holmes first applied the newly developing principles of serology (the study of blood and bodily fluids), fingerprinting, firearms identification, and questioned-document examination long before their value was recognized and accepted by real-life criminal investigators. Holmes's feats excited the imagination of an emerging generation of forensic scientists and criminal investigators. Even in the first Sherlock Holmes novel, *A Study in Scarlet*, published in 1887, we find examples of Doyle's uncanny ability to describe scientific methods of detection years before they were actually discovered and implemented.

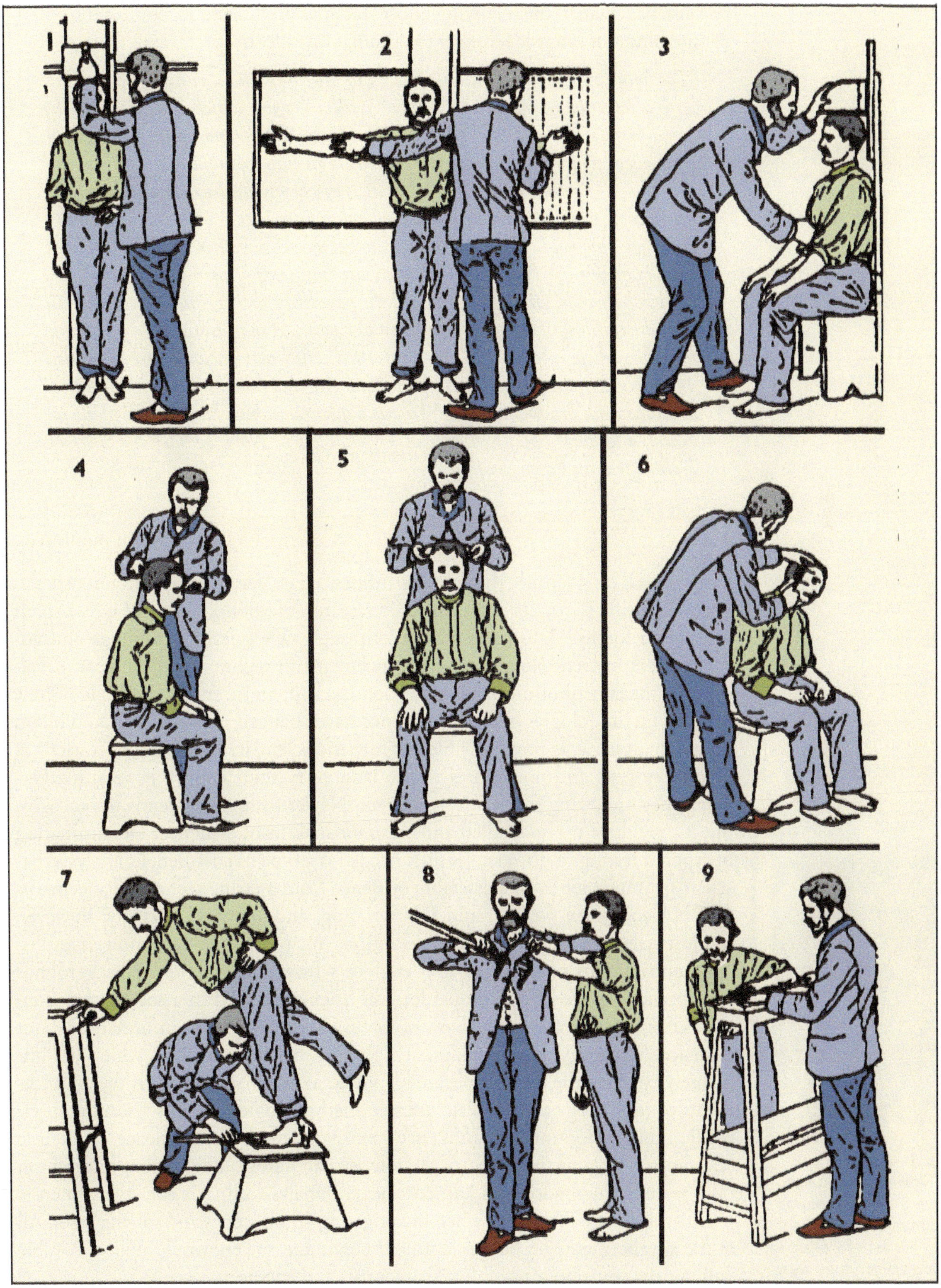

FIGURE 1–3 Bertillon's system of bodily measurements as used for the identification of an individual.
Courtesy Sirchie Finger Print Laboratories, Inc., Youngsville, N.C., www.sirchie.com

For instance, here in the following quote, Holmes probes and recognizes the potential usefulness of forensic serology to criminal investigation.

> *"I've found it. I've found it," he shouted to my companion, running towards us with a test tube in his hand. "I have found a reagent which is precipitated by hemoglobin and by nothing else. . . . Why, man, it is the most practical medico-legal discovery for years. Don't you see that it gives us an infallible test for blood stains? . . . The old guaiacum test was very clumsy and uncertain. So is the microscopic examination for blood corpuscles. The latter is valueless if the stains are a few hours old. Now, this appears to act as well whether the blood is old or new. Had this test been invented, there are hundreds of men now walking the earth who would long ago have paid the penalty of their crimes. . . . Criminal cases are continually hinging upon that one point. A man is suspected of a crime months perhaps after it has been committed. His linen or clothes are examined and brownish stains discovered upon them. Are they blood stains, or rust stains, or fruit stains, or what are they? That is a question which has puzzled many an expert, and why? Because there was no reliable test. Now we have the Sherlock Holmes test, and there will no longer be any difficulty."*

20th-Century Breakthroughs

Past research on scientific thought has influenced current practices in forensic science. Scientific thoughts about forensic research methodologies and cost-benefit analysis in forensic labs have changed through new scientific findings and advancements in technology. These advancements in technology have been developed by the contributions of diverse scientists and engineers. An example of how research methodologies in forensic science have changed include DNA and blood typing analysis. It is now possible for forensic scientists to conduct a variety of laboratory tests and procedures in the laboratory that identify genetic markers when they are trying to identify a suspect. Newer technology tends to be costly, but it provides a more detailed and accurate analysis of evidence. Determination of types of testing to do versus results needed is a type of cost-benefit analysis that scientists implement when assessing evidence from a crime scene. Advancements in DNA collection swabs, rapid DNA testing, and analysis are costly; however, forensic scientists might choose these options due to their reliable and fast results.

FIGURE 1–4 Sir Arthur Conan Doyle's legendary detective Sherlock Holmes applied many of the principles of modern forensic science long before they were adopted widely by police. *Courtesy Paul Chauncey/Alamy Stock Photo*

Contributions of scientists and engineers from many different backgrounds and specialties have helped propel forensic science forward in a society that is always changing. Locard set the basis of forensic science with his information about evidence. Watson, Crick, Franklin, and Wilkins discovered DNA, which has immensely changed the way people are identified at crime scenes. With the advancements of technology and forensic research methodologies, current research on scientific thought continues to influence forensic science. The influence of forensic science goes beyond the scope of the forensic laboratory. The current research on both research methodologies and cost-benefit analysis impacts society as forensic scientists improve their skills when investigating crime scenes and solving cases. All of the advancements on both scientific thought and society would not be possible without the contributions of diverse scientists and engineers.

The pace of technological change quickened considerably in the 20th century, and with it the rate of advancement in the field of forensic science. In 1901, Dr.

Karl Landsteiner discovered that blood can be grouped into different categories, now recognized as the blood types A, B, AB, and O. The possibility that blood grouping could be useful in identifying an individual intrigued Dr. Leone Lattes, a professor at the Institute of Forensic Medicine at the University of Turin in Italy. In 1915, Lattes devised a relatively simple procedure for determining the blood group of a dried bloodstain, a technique that he immediately applied to criminal investigations.

At around the same time, Albert S. Osborn was conducting pioneering work in document examination. In 1910, Osborn wrote the first significant text in this field, *Questioned Documents*. This book is still considered a primary reference for document examiners. Osborn's development of the fundamental principles of document examination was responsible for the acceptance of documents as scientific evidence by the courts.

One of the most important contributors to the field in the early 20th century was the Frenchman Edmond Locard. Although Hans Gross was a pioneer advocate of the use of the scientific method in criminal investigation, Locard first demonstrated how the principles enunciated by Gross could be incorporated within a workable crime laboratory. Locard's formal education was in both medicine and law. In 1910, he persuaded the Lyons police department to give him two attic rooms and two assistants to start a police laboratory. During Locard's first years of work, the only available instruments were a microscope and a rudimentary spectrometer. However, his enthusiasm quickly overcame the technical and monetary deficiencies he encountered. From these modest beginnings, Locard's research and accomplishments became known throughout the world by forensic scientists and criminal investigators. Eventually he became the founder and director of the Institute of Criminalistics at the University of Lyons; this quickly developed into a leading international center for study and research in forensic science.

Locard asserted that when two objects come into contact with each other, a cross-transfer of materials occurs (**Locard's exchange principle**). He strongly believed that every criminal can be connected to a crime by dust particles carried from the crime scene. This concept was reinforced by a series of successful and well-publicized investigations. In one case, presented with counterfeit coins and the names of three suspects, Locard urged the police to bring the suspects' clothing to his laboratory. On careful examination, he located small metallic particles in all the garments. Chemical analysis revealed that the particles and coins were composed of exactly the same metallic elements. Confronted with this evidence, the suspects were arrested and soon confessed to the crime. After World War I, Locard's successes served as an impetus for the formation of police laboratories in Vienna, Berlin, Sweden, Finland, and Holland.

Locard's exchange principle
When two objects come into contact with one another, there is exchange of materials between them

The microscope came into widespread use in forensic science during the 20th century, and its applications grew dramatically. Perhaps the leading figure in the field of microscopy was Dr. Walter C. McCrone. During his lifetime, McCrone became the world's preeminent microscopist. Through his books, journal publications, and research institute, he was a tireless advocate for applying microscopy to analytical problems, particularly forensic science cases. McCrone's exceptional communication skills made him a much-sought-after instructor, and he educated thousands of forensic scientists throughout the world in the application of microscopic techniques. Dr. McCrone used microscopy,

FIGURE 1–5
Edmond Locard.
©Collection Roger-Viollet/The Image Works

often in conjunction with other analytical methodologies, to examine evidence in thousands of criminal and civil cases throughout a long and illustrious career.

Another trailblazer in forensic applications of microscopy was U.S. Army Colonel Calvin Goddard, who refined the techniques of firearms examination by using the comparison microscope. Goddard's work allowed investigators to determine whether a particular gun has fired a bullet by comparing the bullet with one that has been test-fired from the suspect's weapon. His expertise established the comparison microscope as the indispensable tool of the modern firearms examiner.

Modern Scientific Advances

Since the mid-20th century, a revolution in computer technology has made possible a quantum leap forward in human knowledge. The resulting explosion of scientific advances has dramatically impacted the field of forensic science by introducing a wide array of sophisticated techniques for analyzing evidence related to a crime. Procedures such as chromatography, spectrophotometry, and electrophoresis (all discussed in later chapters) allow the modern forensic scientist to determine with astounding accuracy the identity of a suspect substance, and to connect even tiny fragments of evidence to a particular person and place.

The most significant modern advance in forensic science undoubtedly has been the discovery and refinement of DNA typing in the late 20th and early 21st centuries. Sir Alec Jeffreys developed the first DNA profiling test in 1984, and two years later he applied it for the first time to solve a crime by identifying Colin Pitchfork as the murderer of two young English girls. The same case also marked the first time DNA profiling established the innocence of a criminal suspect. Made possible by scientific breakthroughs in the 1950s and 1960s, DNA typing offers law enforcement officials a powerful tool for establishing the precise identity of a suspect, even when only a small amount of physical evidence is available. Combined with the modern analytical tools mentioned earlier, DNA typing has revolutionized the practice of forensic science.

Another significant recent development in forensics is the establishment of computerized databases on physical evidence such as fingerprints, markings on bullets and shell casings, and DNA. These databases have proven to be invaluable, enabling law enforcement officials to compare evidence found at crime scenes to records of thousands of pieces of similar information. This has significantly reduced the time required to analyze evidence and increased the accuracy of the work done by police and forensic investigators.

FIGURE 1–6
Sir Alec Jeffreys.
Homer Sykes/Alamy Stock Photo

Although this brief narrative is by no means a complete summary of historical advances in forensics, it provides an idea of the progress made in the field by dedicated scientists and law enforcement personnel. Even Sherlock Holmes probably couldn't have imagined the lengths to which science today is applied in the service of criminal investigation.

Scientists and engineers from diverse backgrounds have contributed to developing new techniques and technology, allowing for new ways to analyze a greater variety of evidence than in the past. Access to these new methods has advanced scientific thought and expanded society's expectations of what type of evidence could and should be included as part of a forensic investigation. In the past, analysts were limited in the type of analysis that could be accomplished. With all of the new innovations, a challenge for forensic scientists is to determine what tests are going to provide the most crucial information while also keeping the cost of the procedures in mind. Since there are a variety of tests that can be done for a single piece of evidence, forensic scientists are selective based on current advancements in research methodologies. This type of cost-benefit analysis is critical to the effectiveness of a forensic laboratory.

Quick Review

- Forensic science is the application of science to criminal and civil laws that are enforced by police agencies in a criminal justice system.
- The first system of personal identification was called anthropometry. It distinguished one individual from another based on a series of body measurements.
- Forensic science owes its origins to individuals such as Bertillon, Galton, Lattes, Goddard, Osborn, and Locard, who developed the principles and techniques needed to identify or compare physical evidence.
- Locard's exchange principle states that when two objects come into contact with each other, there is exchange of materials between them. This cross-transfer of materials can connect a criminal suspect to their victim.

Crime Laboratories

The steady advance of forensic science technologies during the 20th century led to the establishment of the first facilities specifically dedicated to forensic analysis of criminal evidence. These crime laboratories are now the centers for both forensic investigation of ongoing criminal cases and research into new techniques and procedures to aid investigators in the future.

History of Crime Labs in the United States

The oldest forensic laboratory in the United States is that of the Los Angeles Police Department, created in 1923 by August Vollmer, a police chief from Berkeley, California. In the 1930s, Vollmer headed the first U.S. university institute for criminology and criminalistics at the University of California at Berkeley. However, this institute lacked any official status in the university until 1948, when a school of criminology was formed. The famous criminalist Paul Kirk was selected to head its criminalistics department. Many graduates of this school have gone on to develop forensic laboratories in other parts of the state and country.

In 1932, the Federal Bureau of Investigation (FBI), under the directorship of J. Edgar Hoover, organized a national laboratory that offered forensic services to all law enforcement agencies in the country. During its formative stages, Hoover consulted extensively with business executives, manufacturers, and scientists whose knowledge and experience guided the new facility through its infancy. The FBI Laboratory is now the world's largest forensic laboratory, performing more than one million examinations every year. Its accomplishments have earned it worldwide recognition, and its structure and organization have served as a model for forensic laboratories formed at the state and local levels in the United States as well as in other countries. Furthermore, the opening of the FBI's Forensic Science Research and Training Center in 1981 gave the United States, for the first time, a facility dedicated to conducting research to develop new and reliable scientific methods that can be applied to forensic science. This facility is also used to train crime laboratory personnel in the latest forensic science techniques and methods.

(a)

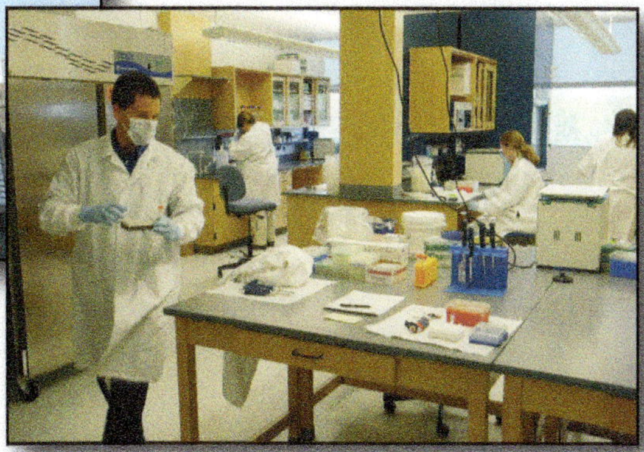

(b)

FIGURE 1–7
(a) Exterior and (b) interior views of the FBI crime laboratory in Quantico, Virginia. *Charles Dharapak/ AP/Shutterstock*

Despite the existence of the FBI Laboratory, the United States has no national system of forensic laboratories. Instead, many local law enforcement jurisdictions—city, county, and state—around the country each operate their own independent crime labs. California, for example, has numerous federal, state, county, and city crime laboratories, many of which operate independently. However, in 1972 the California Department of Justice created a network of integrated state-operated crime laboratories consisting of regional and satellite facilities. An informal exchange of information and expertise occurs within California's criminalist community through a regional professional society, the California Association of Criminalists. This organization was the forerunner of a number of regional organizations that have developed throughout the United States to foster cooperation among the nation's growing community of criminalists.

Organization of a Crime Laboratory

The development of crime laboratories in the United States has been characterized by rapid growth accompanied by a lack of national and regional planning and coordination. Approximately four hundred public crime laboratories operate at various levels of government—federal, state, county, and municipal. The size and diversity of crime laboratories make it impossible to select any one model that best describes a typical crime laboratory. Although most of these facilities function as part of a police department, others operate under the direction of the prosecutor's or district attorney's office; some work with the laboratories of the medical examiner or coroner. Far fewer are affiliated with universities or exist as independent agencies in government. Laboratory staff sizes range from one person to more than a hundred, and their services may be diverse or specialized, depending on the responsibilities of the agency that houses the laboratory.

The Growth of Crime Laboratories Crime laboratories have mostly been organized by agencies that either foresaw their potential application to criminal investigation or were pressed by the increasing demands of casework. Several reasons explain the unparalleled growth of crime laboratories during the past 35 years. Supreme Court decisions in the 1960s were responsible for greater police emphasis on securing scientifically evaluated evidence. The requirement to advise criminal suspects of their constitutional rights and their right of immediate access to counsel has all but eliminated confessions as a routine investigative tool. Successful prosecution of criminal cases requires a thorough and professional police investigation, frequently incorporating the skills of forensic science experts. Modern technology has provided forensic scientists with many new skills and techniques to meet the challenges accompanying their increased participation in the criminal justice system.

Coinciding with changing judicial requirements has been the staggering increase in crime rates in the United States over the past 40 years. This factor alone would probably have accounted for the increased use of crime laboratory services by police agencies, but only a small percentage of police investigations generate evidence requiring scientific examination. There is, however, one important exception to this observation: drug-related arrests. All illicit-drug seizures must be sent to a forensic laboratory for confirmatory chemical analysis before the case can be adjudicated. Since the mid-1960s, drug abuse has accelerated to nearly uncontrollable levels and has resulted in crime laboratories being inundated with drug specimens. Current estimates indicate that nearly half of all requests for examination of forensic evidence deal with abused drugs.

A more recent impetus leading to the growth and maturation of crime laboratories has been the advent of DNA profiling. Since the early 1990s, this technology has progressed to the point at which traces of blood, semen stains, hair, and saliva residues left behind on stamps, cups, bitemarks, and so on have made possible the individualization or near-individualization of biological evidence. To meet the demands of DNA technology, crime labs have expanded staff and in many cases modernized their physical plants. The labor-intensive demands and sophisticated requirements of the technology have affected the structure of the forensic laboratory as has no other technology in the past 50 years. Likewise, DNA profiling has become the dominant factor in explaining how the general public perceives the workings and capabilities of the modern crime laboratory. In coming years thousands of forensic scientists will be added to the rolls of both public and private forensic laboratories to process crime-scene evidence for DNA and to acquire DNA profiles, as mandated by state laws, from the hundreds of thousands of individuals convicted of crimes. This endeavor has already added many new scientists to the field and will eventually more than double the number of scientists employed by forensic laboratories in the United States.

FIGURE 1–8 A forensic scientist performing DNA analysis.
Mauro Fermariello/Science Source

A major problem facing the forensic DNA community is the substantial backlog of unanalyzed DNA samples from crime scenes. The number of unanalyzed casework DNA samples reported by state and national agencies is more than 169,000 in 2017[1]. The estimated number of untested convicted offender samples is 800,000 in 2017. In an attempt to eliminate the backlog of convicted offender or arrestee samples to be analyzed and entered into the Combined DNA Index System (CODIS), the federal government has initiated funding for in-house analysis of samples at the crime laboratory or outsourcing samples to private laboratories for analysis.

In 2008, California began collecting DNA samples from all people arrested on suspicion of a felony, not waiting until a person is convicted. The state's database, with approximately one million DNA profiles, is already the third largest in the world, behind those maintained by the United Kingdom and the FBI. Currently, 30 states and the federal government have laws permitting DNA collection from arrestees.

Crime Laboratories in the United States Historically, a federal system of government, combined with a desire to retain local control, has produced a variety of independent laboratories in the United States, precluding the creation of a national system. Crime laboratories to a large extent mirror the fragmented law enforcement structure that exists on the national, state, and local levels. The federal government has no single law enforcement or investigative agency with unlimited jurisdiction.

Four major federal crime laboratories have been created to help investigate and enforce criminal laws that extend beyond the jurisdictional boundaries of state and local forces. The FBI (Department of Justice) maintains the largest crime laboratory in the world. An ultramodern facility housing the FBI's forensic science services is located in Quantico, Virginia. Its expertise and technology support its broad investigative powers. The Drug Enforcement Administration laboratories (Department of Justice) analyze drugs seized in violation of federal laws regulating the production, sale, and transportation of drugs. The laboratories of the Bureau of Alcohol, Tobacco, Firearms and Explosives (Department of Justice) analyze alcoholic beverages and documents relating to alcohol and firearm excise tax law enforcement and examine weapons, explosive devices, and related evidence to enforce the Gun Control Act of 1968 and the Organized Crime Control Act of 1970. The U.S. Postal Inspection Service maintains laboratories concerned with criminal investigations relating to the postal service. Each of these federal facilities offers its expertise to any local agency that requests assistance in relevant investigative matters.

Most state governments maintain a crime laboratory to service state and local law enforcement agencies that do not have ready access to a laboratory. Some states, such as Alabama, California, Illinois, Michigan, New Jersey, Texas, Washington, Oregon, Virginia, and Florida, have developed a comprehensive statewide system of regional or satellite laboratories. These operate under the direction of a central facility and provide forensic services to most areas of the state. The concept of a regional laboratory operating as part of a statewide system has increased the accessibility of many local law enforcement agencies to a crime laboratory, while minimizing duplication of services and ensuring maximum interlaboratory cooperation through the sharing of expertise and equipment.

Local laboratories provide services to county and municipal agencies. Generally, these facilities operate independently of the state crime laboratory and are financed directly by local government. However, as costs have risen, some counties have combined resources and created multicounty laboratories to service their jurisdictions. Many of the larger cities in the United States maintain their own crime laboratories, usually under the direction of the local police department. Frequently, high population and high crime rates combine to make a municipal facility, such as that of New York City, the largest crime laboratory in the state.

Crime Laboratories Abroad Like the United States, most countries in the world have created and now maintain forensic facilities. In contrast to the American system of independent local laboratories, Great Britain has developed a national system of regional laboratories under the direction of the government's Home Office. In the early 1990s, the British Home Office reorganized the country's forensic laboratories into the Forensic Science Service and instituted a system in which police agencies are charged a fee for services rendered by the laboratory. The fee-for-service concept encouraged the creation of a number of private laboratories that provide services to both police and criminal defense attorneys. One such organization is LGC. In 2010, the British government announced the closure of the Forensic Science Service, citing financial losses. The laboratories closed in 2012, and forensic work in England and Wales is now contracted out to the private sector. Since privatization, LGC has grown to be the largest forensic science provider in the United Kingdom, employing more than seven hundred forensic scientists servicing both police agencies and the private sector.

In Canada, forensic services are provided by three government-funded institutes: (1) six Royal Canadian Mounted Police regional laboratories, (2) the Centre of Forensic Sciences in Toronto, and (3) the Institute of Legal Medicine and Police Science in Montreal. The Royal Canadian Mounted Police opened its first laboratory in Regina, Saskatchewan, in 1937. Altogether, more than a hundred countries throughout the world have at least one laboratory facility offering services in the field of forensic science.

Services of the Crime Laboratory

Bearing in mind the independent development of crime laboratories in the United States, the wide variation in total services offered in different communities is not surprising. There are many reasons for this, including (1) variations in local laws, (2) the different capabilities and functions of the organization to which a laboratory is attached, and (3) budgetary and staffing limitations.

In recent years, many local crime laboratories have been created solely to process drug specimens. Often these facilities were staffed with few personnel and operated under limited budgets. Although many have expanded their forensic services, some still primarily perform drug analyses. However, even among crime laboratories providing services beyond drug identification, the diversity and quality of services rendered varies significantly. For the purposes of this text, I have arbitrarily designated the following units as those that should constitute a "fullservice" crime laboratory.

Basic Services Provided by Full-Service Crime Laboratories

Physical science unit. The physical science unit applies principles and techniques of chemistry, physics, and geology to the identification and comparison of crime-scene evidence. It is staffed by criminalists who have the expertise to use chemical tests and modern analytical instrumentation to examine items as diverse as drugs, glass, paint, explosives, and soil. In a laboratory that has a staff large enough to permit specialization, the responsibilities of this unit may be further subdivided into drug identification, soil and mineral analysis, and examination of a variety of trace physical evidence.

Biology unit. The biology unit is staffed with biologists and biochemists who identify and perform DNA profiling on dried bloodstains and other body fluids, compare hairs and fibers, and identify and compare botanical materials such as wood and plants.

Firearms unit. The firearms unit examines firearms, discharged bullets, cartridge cases, shotgun shells, and ammunition of all types. Garments and other objects are also examined to detect firearms discharge residues and to approximate the distance from a target at which a weapon was fired. The basic principles of firearms examination are also applied here to the comparison of marks made by tools.

FIGURE 1–9
A forensic analyst examining a firearm.
Mediacolor's/Alamy Stock Photo

Document examination unit. The document examination unit studies the handwriting and typewriting on questioned documents to ascertain authenticity and/or source. Related responsibilities include analyzing paper and ink and examining indented writings (the term usually applied to the partially visible depressions appearing on a sheet of paper underneath the one on which the visible writing appears), obliterations, erasures, and burned or charred documents.

Photography unit. A complete photographic laboratory examines and records physical evidence. Its procedures may require the use of highly specialized photographic techniques and equipment, such as digital imaging, infrared, ultraviolet, and X-ray photography, to make invisible information visible to the naked eye. This unit also prepares photographic exhibits for courtroom presentation.

Optional Services Provided by Full-Service Crime Laboratories

Toxicology unit. The toxicology group examines body fluids and organs to determine the presence or absence of drugs and poisons. Frequently, such functions are shared with or may be the sole responsibility of a separate laboratory facility placed under the direction of the medical examiner's or coroner's office. In most jurisdictions, field instruments such as the Intoxilyzer are used to determine the alcoholic consumption of individuals. Often the toxicology section also trains operators and maintains and services these instruments.

Latent fingerprint unit. The latent fingerprint unit processes and examines evidence for latent fingerprints when they are submitted in conjunction with other laboratory examinations.

Polygraph unit. The polygraph, or lie detector, has come to be recognized as an essential tool of the criminal investigator rather than the forensic scientist. However, during the formative years of polygraph technology, many police agencies incorporated this unit into the laboratory's administrative structure, where it sometimes remains today. In any case, its functions are handled by people trained in the techniques of criminal investigation and interrogation.

Voiceprint analysis unit. In cases involving telephoned threats or tape-recorded messages, investigators may require the skills of the voiceprint analysis unit to tie the voice to a particular suspect. To this end, a good deal of casework has been performed with the sound spectrograph, an instrument that transforms speech into a visual graphic display called a voiceprint. The validity of this technique as a means of personal identification rests on the premise that the sound patterns produced in speech are unique to the individual and that the voiceprint displays this uniqueness.

Crime-scene investigation unit. The concept of incorporating crime-scene evidence collection into the total forensic science service is slowly gaining recognition in the United States. This unit dispatches specially trained personnel (civilian and/or police) to the crime scene to collect and preserve physical evidence that will later be processed at the crime laboratory.

Whatever the organizational structure of a forensic science laboratory may be, specialization must not impede the overall coordination of services demanded by today's criminal investigator. Laboratory administrators need to keep open the lines of communication between analysts (civilian and uniform), crime-scene investigators, and police personnel. Inevitably, forensic investigations require the skills of many individuals. One notoriously high-profile investigation illustrates this process—the search for the source of the anthrax letters mailed shortly after September 11, 2001. Figure 1–10 shows one of the letters and illustrates the multitude of skills required in the investigation—skills possessed by forensic chemists and biologists, fingerprint examiners, and forensic document examiners.

Other Forensic Science Services Even though this textbook is devoted to describing the services normally provided by a crime laboratory, the field of forensic science is by no means limited to the areas covered in this book. A number of specialized forensic science services outside the crime laboratory are routinely available to law enforcement personnel. These services are important aids to a criminal investigation and require the involvement of individuals who have highly specialized skills.

FIGURE 1–10
An envelope containing anthrax spores along with an anonymous letter was sent to the office of Senator Tom Daschle shortly after the terrorist attacks of September 11, 2001. A variety of forensic skills were used to examine the envelope and letter. Also, bar codes placed on the front and back of the envelope by mail-sorting machines contain address information and information about where the envelope was first processed.
Getty Images News/ Getty Images

Three specialized forensic services—forensic pathology, forensic anthropology, and forensic entomology—are frequently employed at a murder scene and will be discussed at greater length when we examine crime-scene procedures in

Chapter 2. Other services, such as those discussed next, are used in a wide variety of criminal investigations.

Medicolegal death investigators. Medicolegal death investigations are performed by trained medicolegal death investigators who are employed by the coroner or Medical examiner to investigate deaths occurring in their jurisdiction that are violent, suspicious, unexplained or unexpected.

Forensic psychiatry. Forensic psychiatry is a specialized area that examines the relationship between human behavior and legal proceedings. Forensic psychiatrists are retained for both civil and criminal litigations. In civil cases, they typically perform tasks such as determining whether an individual is competent to make decisions about preparing a will, settling property, or refusing medical treatment. In criminal cases, forensic psychologists evaluate behavioral disorders and determine whether defendants are competent to stand trial. Forensic psychiatrists also examine behavior patterns of criminals as an aid in developing a suspect's behavioral profile.

Forensic criminology. Forensic criminology uses principles from both sociology and psychology to study why a crime occurs, what can be done to prevent future crimes, and the effects of these crimes on society. A criminologist will look at psychological, environmental, social and hereditary causes of crimes.

Forensic odontology. Practitioners of forensic odontology help identify victims when the body is left in an unrecognizable state. Teeth are composed of enamel, the hardest substance in the body. Because of enamel's resilience, the teeth outlast tissues and organs as decomposition begins. The characteristics of teeth, their alignment, and the overall structure of the mouth provide individual evidence for identifying a specific person. With the use of dental records such as X-rays and dental casts or even a photograph of the person's smile, a set of dental remains can be compared to a suspected victim (see Figure 1–11).

Historically, forensic odontologists have also interpreted human bitemarks as a means of identifying a potential perpetrator with a mark left on a victim. Bitemark comparison evidence was first admitted in courts in the United States in 1975 and were soon admissible in all 50 states. In light of recent scientific advancements, namely DNA testing, it has been demonstrated that there are serious flaws in the assertions made by forensic dentists during a bitemark comparison. Since the advent of DNA, a number of cases involving trial level bitemark identifications have been overturned following post-conviction DNA testing. In 2009, the National

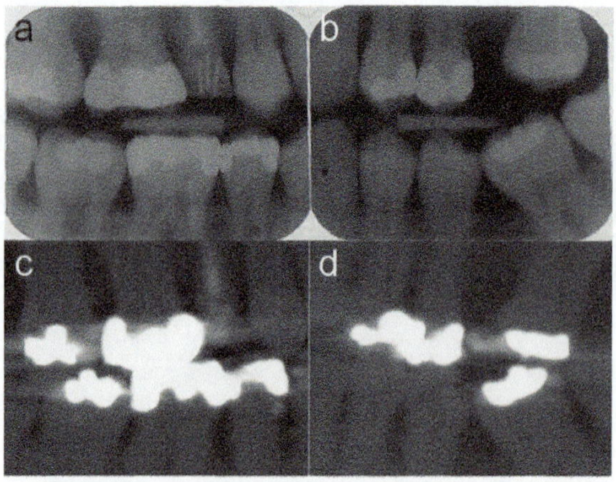

FIGURE 1–11
Images a) and c) and b) and d) are post mortem and antemortem dental
x-rays taken from the same person. The images were used to positively identify the individual using dental records.
Source: (a-c) Bowers, C. M., "Review of a Forensic Pseudoscience: Identification of Criminals from Bitemark Patterns," Journal of Forensic and Legal Medicine 61 (2019): 34–39. doi:10.1016/j. jflm.2018.11.001; (d) Reprinted from Wood RE. Forensic aspects of maxillofacial radiology. Forensic Sci Int 2006;159S:S47–S55, with permission from Elsevier.

Ted Bundy, Serial Killer

The name Ted Bundy is synonymous with the term serial killer. This handsome, gregarious, and worldly onetime law student is believed to be responsible for 40 murders between 1964 and 1978. His reign of terror stretched from the Pacific Northwest down to California and into Utah, Idaho, and Colorado, finally ending in Florida. His victims were typically young women, usually murdered with a blunt instrument or by strangulation, and sexually assaulted before and after death.

First convicted in Utah in 1976 on a charge of kidnapping, Bundy managed to escape after his extradition to Colorado on a murder charge. Ultimately, Bundy found his way to the Tallahassee area of Florida. There he unleashed mayhem, killing two women at a Florida State University sorority house and then murdering a 12-year-old girl three weeks later. Fortunately, future victims were spared when Bundy was arrested while driving a stolen vehicle. As police investigated the sorority murders, they noted that one victim, who had been beaten over the head with a log, raped, and strangled, also had bitemarks on her left buttock and breast.

Supremely confident that he could beat the sorority murder charges, the arrogant Bundy insisted on acting as his own attorney. His unfounded optimism was shattered in the courtroom when a forensic odontologist matched the bitemark on the victim's buttock to Bundy's front teeth. Bundy was ultimately executed in 1989.

ASSOCIATED PRESS/AP Images

Case Files

Academy of Sciences took a hard look at the scientific underpinnings of forensic bitemark identification and found glaring gaps in the scientific research.

There are two assumptions on which the practice of human bitemark comparison rests. First, that human dentition is substantially unique such that a person can be identified by marks made by their teeth. Second, that human skin is an accurate impression material for the individual marks made by the teeth, if they exist. In 2016, the U.S. American Board of Forensic Odontology published a rejection of past practices of using bitemarks for human identification. Still, some experts practicing in this area disagreed with criticisms that there was a lack of scientific evidence to support the "assumptions and assertions made by forensic dentists during bitemark comparisons."[2]

The controversy highlights an important struggle where law and science intersect. What evidence will be admitted in a criminal trial is often governed by legal precedent. Once that legal precedent is firmly established, as with human bitemark comparisons, it is hard to change what has been admissible in courts for the past 40 years. As more scientific evidence mounts rejecting the use of bitemark comparison as a means of human identification, scientists are becoming more skeptical of its use in courtrooms, but many prosecutors and judges are proceeding undeterred.

On February 12, 2016, the Texas Forensic Science Commission published a report on bitemark comparison in response to a complaint filed by the National Innocence Project on behalf of Steven Mark Chaney. The Commission concluded the following: (1) at the current time, the overwhelming majority of existing research does not support the contention that bitemark comparison can be performed reliably and accurately from examiner to examiner due to the subjective nature of the analysis and (2) in addition to the foundational scientific and research issues, there are significant quality control and infrastructure differences between forensic odontology and other patterned and impression disciplines performed in accredited laboratories. The Commission recommended bitemark comparison evidence *not be admitted in criminal cases in Texas* unless and until the following are established: (1) criteria for identifying when a patterned injury constitutes a human bitemark, (2) criteria for identifying when a human bitemark was made by an adult versus a child, and (3) rigorous and appropriately validated proficiency testing.

Forensic engineering. Forensic engineers are concerned with failure analysis, accident reconstruction, and causes and origins of fires or explosions. Forensic engineers develop and use models to find solutions to an engineering problem. These models can represent different phenomena, systems, or processes at a crime scene. For example, forensic engineers may design models to help investigate the collapse of a building. If findings indicate that there is possible criminal involvement, an investigation would follow. Forensic engineers use engineering designs to answer questions such as these: How did an accident or structural failure occur? Were the parties involved responsible? If so, how were they responsible? Accident scenes are examined, photographs are reviewed, and any mechanical objects involved are inspected.

Forensic computer and digital analysis. Forensic computer science is a new and fast-growing field that involves identifying, collecting, preserving, and examining information derived from computers and other digital devices, such as cell phones. Law enforcement aspects of this work normally involve recovering deleted or overwritten data from a computer's hard drive and tracking hacking activities within a compromised system. This field of forensic computer analysis will be addressed in detail in Chapter 19.

Quick Review

- The development of crime laboratories in the United States has been characterized by rapid growth accompanied by a lack of national and regional planning and coordination.
- Four major reasons for the increase in the number of crime laboratories in the United States since the 1960s are as follows: (1) the fact that the requirement to advise criminal suspects of their constitutional rights and their right of immediate access to counsel has all but eliminated confessions as a routine investigative tool, (2) the staggering increase in crime rates in the United States, (3) the fact that all illicit-drug seizures must be sent to a forensic laboratory for confirmatory chemical analysis before the case can be adjudicated in court, and (4) the advent of DNA profiling.
- The technical support provided by crime laboratories can be assigned to five basic services: the physical science unit, the biology unit, the firearms unit, the document examination unit, and the photography unit.
- Some crime laboratories offer optional services such as toxicology, fingerprint analysis, polygraph administration, voiceprint analysis, and crime-scene investigation.
- Special forensic science services available to the law enforcement community include forensic pathology, forensic anthropology, forensic entomology, forensic psychiatry, forensic odontology, forensic engineering, and forensic computer and digital analysis.

The Functions of the Forensic Scientist

Although a forensic scientist relies primarily on scientific knowledge and skill, only half of the job is performed in the laboratory. The other half takes place in the courtroom, where the ultimate significance of the evidence is determined. The forensic scientist must not only analyze physical evidence but also persuade a jury to accept the conclusions derived from that analysis.

Analyzing Physical Evidence

First and foremost, the forensic scientist must be skilled in applying the principles and techniques of the physical and natural sciences to analyze the many types of physical evidence that may be recovered during a criminal investigation. Scientists often seek answers to questions about our natural world based on information from text, models, phenomena, or investigations. When analyzing evidence, a forensic scientist must first ask questions. For example, the questions could be based on observations made at a crime scene. These questions lead forensic scientists to analyze and interpret various types of data. Of the three major avenues available to police investigators for assistance in solving a crime—confessions, eyewitness accounts by victims or witnesses, and evaluation of physical evidence retrieved from the crime scene—only physical evidence is free of inherent error or bias.

Criminal cases are replete with examples of individuals who were incorrectly charged with and convicted of committing a crime because of faulty memories or lapses in judgment. For example, investigators may be led astray during their preliminary evaluation of the events and circumstances surrounding the commission of a crime. These errors may be compounded by misleading eyewitness statements and inappropriate confessions. These same concerns don't apply to physical evidence.

scientific method
A process that uses strict guidelines to ensure careful and systematic collection, organization, and analysis of information

scientific hypothesis
An idea proposing an initial explanation to a problem, fact or situation observed

theory
An explanation of phenomena, an observable event, that has been tested repeatedly using the scientific method to make observations, measurements, and evaluating results

scientific law
An explanation of the reason why something happens

What about physical evidence allows investigators to sort out facts as they are and not what one wishes they were? Forensic scientists must continually ask questions, such as, "how does this piece of evidence connect to the crime?" Making observations is essential throughout an investigation to define the problem of making connections between a suspect and the crime scene. The hallmark of physical evidence is that it must undergo scientific inquiry. Science derives its integrity from adherence to strict guidelines that ensure careful and systematic collection, organization, and analysis of information—a process known as the **scientific method**. The underlying principles of the scientific method provide a safety net to ensure that the outcome of an investigation is not tainted by human emotion or compromised by distorting, belittling, or ignoring contrary evidence. In order to identify scientific ideas, principles, and theories, scientists use the scientific method to develop explanations or propose solutions. Forensic scientists utilize the scientific method when observing and processing a crime scene.

The scientific method, also referred to as experimental design, begins by formulating a question worthy of investigation, such as who committed a particular crime. The investigator next formulates a **scientific hypothesis**, a reasonable explanation proposed to answer the question. What follows is the basic foundation of scientific inquiry—the testing of the hypothesis through experimentation. The testing process must be thorough and recognized by other scientists as valid, resulting in the formation of a **theory**. With further experimentation and analysis, theories can be developed into **scientific law**. Scientists and investigators must accept the experimental findings even when they wish they were different. Finally, when the hypothesis is validated by experimentation, it becomes suitable as scientific evidence, appropriate for use in a criminal investigation and ultimately available for admission in a court of law.

When analyzing physical evidence from a crime scene, scientists need to develop explanations and propose solutions from the data they obtain. Scientists obtain evidence at the crime scene itself, which is then processed in forensic laboratories. Qualitative and quantitative data are analyzed to develop explanations and propose solutions. For example, forensic scientists might collect a shirt at the crime scene. This shirt is then analyzed for hair and fibers. The hair on the shirt might provide DNA, which can then be processed in the lab. This data can lead forensic scientists to different individuals who might have been at the crime scene. These explanations are consistent with scientific ideas, scientific principles, and scientific theories.

FIGURE 1–12
An individual undergoing a polygraph test.
Sandy Schaeffer/Mai/The LIFE Images Collection/Getty Images

Forensic scientists also use models to help them develop explanations and propose solutions for evidence found at a crime scene. A forensic scientist might use computer models to help recreate images from a crime scene. Engineers might also use computer models to help recreate activity that occurred at a crime scene. These models are consistent with scientific ideas, scientific principles, and scientific theories.

Determining admissibility of evidence Court systems personnel such as judges, prosecution attorneys, and defense attorneys all play different roles and have different responsibilities within the criminal justice system. The judge will determine what evidence is admissible in court and which expert witnesses will be allowed to provide testimony. Prosecution attorneys are responsible for developing the case against an individual charged with a crime. The burden of proof of guilt rests on the prosecution. Defense attorneys represent the defendant, who is the person charged with a crime. It is the goal of the defense to introduce reasonable doubt their client committed the crime and to remind the jury that all individuals are presumed innocent until proven guilty. In rejecting the scientific validity of the lie detector (polygraph), the District of Columbia Circuit Court in 1923 set forth what has since become a standard guideline for determining the judicial admissibility of scientific examinations. In *Frye v. United States*,[3] the court ruled that in order to be admitted as evidence at trial, the questioned procedure, technique, or principles must be "generally accepted" by a meaningful segment of the relevant scientific community. In practice, this approach requires the proponent of a scientific test to present to the court a collection of experts who can testify that the scientific issue before the court is generally accepted by the relevant members of the scientific community. Furthermore, in determining whether a novel technique meets criteria associated with "general acceptance," courts have frequently taken note of books and papers written on the subject, as well as prior judicial decisions relating to the reliability and general acceptance of the technique. In recent years, many observers have questioned whether this approach is sufficiently flexible to deal with new scientific issues that may not have gained widespread support within the scientific community.

The Federal Rules of Evidence offer an alternative to the *Frye* standard, one that some courts believe espouses a more flexible standard for admitting scientific evidence. Part of the Federal Rules of Evidence governs the admissibility of all evidence, including expert testimony, in federal courts, and many states have adopted codes similar to those of the Federal Rules. Specifically, Rule 702 of the Federal Rules of Evidence sets a different standard from "general acceptance" for admissibility of expert testimony. Under this standard, a witness "qualified as an expert by knowledge, skill, experience, training, or education" may offer expert testimony on a scientific or technical matter if "(1) the testimony is based upon sufficient facts or data, (2) the testimony is the product of reliable principles and methods, and (3) the witness has applied the principles and methods reliably to the facts of the case."

In a landmark ruling in the 1993 case of *Daubert v. Merrell Dow Pharmaceuticals, Inc.*,[4] the U.S. Supreme Court asserted that "general acceptance," or the *Frye* standard, is not an absolute prerequisite to the admissibility of scientific evidence under the Federal Rules of Evidence. According to the Court, the Rules of Evidence—especially Rule 702—assign to the trial judge the task of ensuring that an expert's testimony rests on a reliable foundation and is relevant to the case. Although this ruling applies only to federal courts, many state courts are expected to use this decision as a guideline in setting standards for the admissibility of scientific evidence.

Judging Scientific evidence In *Daubert*, the Court advocates that trial judges assume the ultimate responsibility for acting as a "gatekeeper" in judging the admissibility and reliability of scientific evidence presented in their courts. The Court offered some guidelines as to how a judge can gauge the veracity of scientific evidence, emphasizing that the inquiry should be flexible. Suggested areas of inquiry include the following:

1. **Whether the scientific technique or theory can be (and has been) tested**
2. **Whether the technique or theory has been subject to peer review and publication**
3. **The technique's potential rate of error**
4. **Existence and maintenance of standards controlling the technique's operation**
5. **Whether the scientific theory or method has attracted widespread acceptance within a relevant scientific community**

Some legal experts have expressed concern that abandoning *Frye's* general-acceptance test will result in the introduction of absurd and irrational pseudoscientific claims in the courtroom. The Supreme Court rejected these concerns, pointing out the inherent strengths of the American judicial process in identifying unreliable evidence:

> *In this regard the respondent seems to us to be overly pessimistic about the capabilities of the jury and of the adversary system generally. Vigorous cross-examination, presentation of contrary evidence, and careful instruction on the burden of proof are the traditional and appropriate means of attacking shaky but admissible evidence.*

In a 1999 decision, *Kumho Tire Co., Ltd. v. Carmichael,*[5] the Court unanimously ruled that the "gatekeeping" role of the trial judge applied not only to scientific testimony, but to all expert testimony:

> *We conclude that Daubert's general holding—setting forth the trial judge's general "gatekeeping" obligation—applies not only to testimony based on "scientific" knowledge, but also to testimony based on "technical" and "other specialized" knowledge. . . . We also conclude that a trial court may consider one or more of the more specific factors that Daubert mentioned when doing so will help determine that testimony's reliability. But, as the Court stated in Daubert, the test of reliability is "flexible," and Daubert's list of specific factors neither necessarily nor exclusively applies to all experts in every case.*

A leading case that exemplifies the type of flexibility and wide discretion that the *Daubert* ruling apparently gives trial judges in matters of scientific inquiry is *Coppolino v. State.*[6] Here a medical examiner testified to his finding that the victim had died of an overdose of a drug known as succinylcholine chloride. This drug had never before been detected in the human body. The medical examiner's findings were dependent on a toxicological report that identified an abnormally high concentration of succinic acid, a breakdown product of the drug, in the victim's body.

Case Files

Dr. Coppolino's Deadly House Calls

A frantic late-night telephone call brought a local physician to the Florida home of Drs. Carl and Carmela Coppolino. The physician arrived to find Carmela beyond help. Carmela Coppolino's body, unexamined by anyone, was buried in her family's plot in her home state of New Jersey.

A little more than a month later, Carl married a moneyed socialite, Mary Gibson. News of Carl's marriage infuriated Marjorie Farber, a former New Jersey neighbor of Dr. Coppolino who had been a having an affair with the good doctor. Soon Marjorie had an interesting story to recount to investigators. Her husband's death two years before, although ruled to be from natural causes, had actually been murder! Carl, an anesthesiologist, had given Marjorie a syringe containing some medication and told her to inject her husband, William, while he was sleeping. Ultimately, Marjorie claimed, she was unable to inject the full dose and called Carl, who finished the job by suffocating William with a pillow.

Marjorie Farber's astonishing story was supported in part by Carl's recent increase in his wife's life insurance. Carmela's $65,000 policy, along with his new wife's fortune, would keep Dr. Coppolino in high society for the rest of his life. Based on this information, authorities in New Jersey and Florida now obtained exhumation orders for both William Farber and Carmela Coppolino. After examination of both bodies, Dr. Coppolino was charged with the murders of William and Carmela.

Officials decided to try Dr. Coppolino first in New Jersey for the murder of William Farber. The Farber autopsy did not reveal any evidence of poisoning, but seemed to show strong evidence of strangulation. The absence of toxicological findings left the jury to deliberate the conflicting medical expert testimony versus the sensational story told by a scorned and embittered woman. In the end, Dr. Coppolino was acquitted.

The Florida trial presented another chance to bring Carl Coppolino to justice. Recalling Dr. Coppolino's career as an anesthesiologist, it was theorized that Coppolino had exploited his access to the many potent drugs used during surgery to commit these murders, specifically an injectable paralytic agent called succinylcholine chloride.

After having Carmela's body exhumed, it was found that Carmela had been injected in her left buttock shortly before her death. Ultimately, a completely novel procedure for detecting succinylcholine chloride was devised. Elevated levels of succinic acid were found in Carmela's brain, which proved that she had received a large dose of the paralytic drug shortly before her death. This evidence, along with the finding of the same drug residues in the injection site on her buttock, was presented in the Florida murder trial of Carl Coppolino, who was convicted of second-degree murder.

The defense argued that this test for the presence of succinylcholine chloride was new and the absence of corroborative experimental data by other scientists meant that it had not yet gained general acceptance in the toxicology profession. The court, in rejecting this argument, recognized the necessity for devising new scientific tests to solve the special problems that are continually arising in the forensic laboratory. It emphasized, however, that although these tests may be new and unique, they are admissible only if they are based on scientifically valid principles and techniques: "The tests by which the medical examiner sought to determine whether death was caused by succinylcholine chloride were novel and devised specifically for this case. This does not render the evidence inadmissible. Society need not tolerate homicide until there develops a body of medical literature about some particular lethal agent."

Providing Expert Witness Testimony

Because the results of their work may ultimately be a factor in determining a person's guilt or innocence, forensic scientists may be required to testify about their methods and conclusions at a trial or hearing. In the criminal justice system, trial courts have broad discretion in accepting an individual as an **expert witness** on any particular subject. Generally, if a witness can establish to the satisfaction of a trial judge that they possess a particular skill or have knowledge in a trade or profession that will aid the court in determining the truth of the matter at issue, that individual will be accepted as an expert witness. Depending on the subject area in question, the court will usually consider knowledge acquired through experience, training, education, or a combination as sufficient grounds for qualification as an expert witness.

expert witness
An individual whom the court determines to possess knowledge relevant to the trial that is not expected of the average layperson

In court, an expert witness may be asked questions intended to demonstrate his or her ability and competence pertaining to the matter at hand. Competency may be established by having the witness cite educational degrees, participation in special courses, membership in professional societies, and any professional articles or books published. Also important is the number of years of occupational experience the witness has in areas related to the matter before the court.

Unfortunately, few schools confer degrees in forensic science. Most chemists, biologists, geologists, and physicists prepare themselves for careers in forensic

FIGURE 1–13
An expert witness testifying in court.
Taylor Jones/ZUMA Press/Newscom

science by combining training under an experienced examiner with independent study. Of course, formal education in the physical sciences provides a firm foundation for learning and understanding the principles and techniques of forensic science. Nevertheless, for the most part, courts must rely on training and years of experience as a measurement of the knowledge and ability of the expert.

Before the judge rules on the witness's qualifications, the opposing attorney may cross-examine the witness and point out weaknesses in background and knowledge. Most courts are reluctant to disqualify an individual as an expert even when presented with someone whose background is only remotely associated with the issue at hand. The question of what credentials are suitable for qualification as an expert is ambiguous and highly subjective and one that the courts wisely try to avoid.

The weight that a judge or jury assigns to "expert" testimony in subsequent deliberations is, however, quite another matter. Undoubtedly, education and experience have considerable bearing on the value assigned to the expert's opinions. Just as important may be his or her demeanor and ability to explain scientific data and conclusions clearly, concisely, and logically to a judge and jury composed of nonscientists. The problem of sorting out the strengths and weaknesses of expert testimony falls to prosecution and defense counsel.

The ordinary or lay witness must testify on events or observations that arise from personal knowledge. This testimony must be factual and, with few exceptions, cannot contain the personal opinions of the witness. On the other hand, the expert witness is called on to evaluate evidence when the court lacks the expertise to do so. This expert then expresses an opinion as to the significance of the findings. The views expressed are accepted only as representing the expert's opinion and may later be accepted or ignored in jury deliberations.

The expert cannot render any view with absolute certainty. At best, the expert may only be able to offer an opinion based on a reasonable scientific certainty derived from training and experience. Obviously, the expert is expected to defend vigorously the techniques and conclusions of the analysis, but at the same time must not be reluctant to discuss impartially any findings that could minimize the significance of the analysis. The forensic scientist should not be an advocate of one party's cause, but only an advocate of truth. An adversary system of justice must give the prosecutor and defense ample opportunity to offer expert opinions and to argue the merits of such testimony. Ultimately, the duty of the judge or jury is to weigh the pros and cons of all the information presented in deciding guilt or innocence.

The necessity for the forensic scientist to appear in court has been imposed on the criminal justice system by a 2009 U.S. Supreme Court Case, *Melendez-Diaz v. Massachusetts.*[7] The *Melendez-Diaz* decision addressed the practice of using evidence affidavits or laboratory certificates in lieu of in-person testimony by forensic analysts. In its reasoning, the Court relied on a previous ruling, *Crawford v. Washington,*[8] where it explored the meaning of the Confrontation Clause of the Sixth Amendment. In the *Crawford* case, a recorded statement by a spouse was used against her husband in his prosecution. Crawford argued that this was a violation of his right to confront witnesses against him under the Sixth Amendment, and the Court agreed. Using the same logic in *Melendez-Diaz,* the Court reasoned that introducing forensic science evidence via an affidavit or a certificate denied a defendant the opportunity to cross-examine the analyst. In 2011, the Supreme Court reaffirmed the *Melendez-Diaz* decision in the case of *Bullcoming v. New Mexico*[9] by rejecting a substitute expert witness in lieu of the original analyst:

> *The question presented is whether the Confrontation Clause permits the prosecution to introduce a forensic laboratory report containing a testimonial certification—made for the purpose of proving a particular fact through the in-court testimony of a scientist who did not sign the certification or perform or observe the test reported in the certification. We hold that surrogate testimony of that order does not meet the constitutional requirement. The accused's right is to be confronted with the analyst who made the certification, unless that analyst is unavailable at trial, and the accused had an opportunity, pretrial, to cross examine that particular scientist.*

Furnishing Training in the Proper Recognition, Collection, and Preservation of Physical Evidence

The competence of a laboratory staff and the sophistication of its analytical equipment have little or no value if relevant evidence cannot be properly recognized, collected, and preserved at the site of a crime. For this reason, the forensic staff must have responsibilities that will influence the conduct of the crime-scene investigation.

The most direct and effective response to this problem has been to dispatch specially trained evidence-collection technicians to the crime scene. A growing number of crime laboratories and the police agencies they service keep trained "evidence technicians" on 24-hour call to help criminal investigators retrieve evidence. These technicians are trained by the laboratory staff to recognize and gather pertinent physical evidence at the crime scene. They are assigned to the laboratory full time for continued exposure to forensic techniques and procedures. They have at their disposal all the proper tools and supplies for proper collection and packaging of evidence for future scientific examination.

Unfortunately, many police forces still have not adopted this approach. Often a patrol officer or detective collects the evidence. The individual's effectiveness in this role depends on the extent of his or her training and working relationship with the laboratory. For maximum use of the skills of the crime laboratory, training of the crime-scene investigator must go beyond superficial classroom lectures to involve extensive personal contact with the forensic scientist. Each must become aware of the other's problems, techniques, and limitations.

The training of police officers in evidence collection and their familiarization with the capabilities of a crime laboratory should not be restricted to a select group of personnel on the force. Every officer engaged in fieldwork, whether it be traffic, patrol, investigation, or juvenile control, often must process evidence for laboratory examination. Obviously, it would be difficult and time consuming to give everyone the in-depth training and attention that a qualified criminal investigator requires. However, familiarity with crime laboratory services and capabilities can be gained through periodic lectures, laboratory tours, and dissemination of manuals prepared by the laboratory staff that outline the proper methods for collecting and submitting physical evidence to the laboratory.

Professional Standards, Employability Skills, and Ethics

Professionalism is the ability to show respect to everyone around you while you perform your responsibilities as best you can. By displaying professional standards, you increase your chances of obtaining and maintaining a job. Some of these professional standards and employability skills to exemplify include: good attendance, punctuality, meeting deadlines, and working towards personal and team goals daily.

Ethical conduct is essential for working in law enforcement and forensic science. Ethical behaviors include being fair and honest in academic, personal, and interpersonal relationships. In forensic science, this includes accurately representing ones qualifications, maintaining the integrity of all items of evidence, true and accurate representation of any data collected and analyzed, impartiality during examination of evidence and when testifying in court, as well as reporting colleagues who violate the ethical code.

Some common violations to ethical standards that forensic scientists can face include misrepresentation of qualifications or credentials, pressures to testify for or against a defendant, omission of unfavorable information, falsification of data or notes, and biased examination of evidence.

Bias in Forensic Science

In the pursuit of justice, forensic evidence is a critical component of most court proceedings. Forensic scientists play a pivotal role in carefully examining and analyzing each item of evidence collected to determine its value to the case. It is for this reason a forensic scientist must be aware of and make every attempt to avoid potential bias. Bias can occur when contextual information about the case, irrelevant to the analysis, is shared with forensic experts. Conclusions can be overstated or even be incorrect when forensic experts are given specific details about a suspect's criminal record, ethnicity, confessions, eyewitness identifications, or other evidential connections rather than providing impartial conclusions based on what is actually supported by the evidence. Bias can also occur when forensic experts work from a suspect and try to connect evidence to them rather than using the evidence to lead the decision-making process. Another type of bias that can occur is confirmation bias.

Confirmation bias occurs when an individual gives more emphasis to evidence that supports their existing belief and excludes evidence that would reject their belief. An expert testimony may be disqualified if their conclusion is swayed based on these existing beliefs. Confirmation bias is more likely to occur if a forensic expert is told the outcome of previous analysis for an item of evidence. If they come to a different conclusion, they may question their own outcome as incorrect.

Bias can also occur during the collection of evidence at a crime scene and forensic analysis of the evidence. The crime scene is the first stop for the collection of forensic evidence. After evidence is collected, it is further analyzed in various capacities. Scientists will analyze evidence for fingerprints, biological fluids, and hair and fibers. The forensic analysis of the evidence is subject to human interpretation. Bias during both evidence collection and forensic analysis can be influenced by outside factors, such as experience in the field at similar crime scenes; eyewitness reports; or personal information about suspects or victims. The effect of these biases can impact the proper collection of evidence or full analysis of evidence found at a crime scene.

Quick Review

- A forensic scientist must be skilled in applying the principles and techniques of the physical and natural sciences to analyzing evidence that may be recovered during a criminal investigation.
- The cases *Frye v. United States* and *Daubert v. Merrell Dow Pharmaceuticals, Inc.*, set guidelines for determining the admissibility of scientific evidence into the courtroom.
- An expert witness evaluates evidence based on specialized training and experience.
- Forensic scientists participate in training law enforcement personnel in the proper recognition, collection, and preservation of physical evidence.
- Forensic scientists must carefully examine each item of evidence to determine its value to the case.
- Professionalism is the ability to show respect to everyone around you while you perform your responsibilities as best you can.
- Ethical conduct is essential for forensic science professionals in all aspects of their work.
- A forensic scientist must work to avoid ethical standard violations in all aspects of their job.
- Bias can occur when contextual information about a case is shared with forensic experts.

Exploring Forensic Science on the Internet

There are no limits to the amount or type of information that can be found on the Internet. The fields of law enforcement and forensic science have not been left behind by advancing computer technology. Extensive information about forensic science is available on the Internet. The types of web pages range from simple explanations of the various fields of forensics to intricate details of crime-scene reconstruction. People can also find information on which colleges offer degree programs in forensics or pages posted by law enforcement agencies that detail their activities, as well as possible employment opportunities.

General Forensics Sites

Reddy's Forensic Page (www.forensicpage.com) is a valuable starting point. This site is a collection of forensic web pages in categories such as new links in forensics; general forensic information sources; associations, colleges, and societies; literature and journals; forensic laboratories; general web pages; forensic-related mailing lists and newsgroups; universities; conferences; and various forensic fields of expertise.

Another website offering a multitude of information related to forensic science is Zeno's Forensic Site (http://www.forensic.to/links/pages/). Here users can find links related to forensic education and expert consultation, as well as a wealth of information concerning specific fields of forensic science.

A comprehensive and useful website for those interested in law enforcement is Officer.com (www.officer.com). This comprehensive collection of criminal justice resources is organized into easy-to-read subdirectories that relate to topics such as law enforcement agencies, police association and organization sites, criminal justice organizations, law research pages, and police mailing-list directories.

An Introduction to Forensic Firearm Identification (www.firearmsid.com/) This website contains an extensive collection of information relating to the identification of firearms. An individual can explore in detail how to examine bullets, cartridge cases, and clothing for gunshot residues and suspect shooters' hands for primer residues. Information on the latest technology involving the automated firearms search system NIBIN can also be found on this site.

Carpenter's Forensic Science Resources (www.tncrimlaw.com/forensic/) This site provides a bibliography involving forensic evidence. For example, the user can find references about DNA, fingerprints, hairs, fibers, and questioned documents as they relate to crime scenes and assist investigations. This website is an excellent place to start a research project in forensic science.

Crime Scene Investigator Network (www.crime-scene-investigator.net/index.html) For those who are interested in learning the process of crime-scene investigation, this site provides detailed guidelines and information regarding crime-scene response and the collection and preservation of evidence. For example, information concerning the packaging and analysis of bloodstains, seminal fluids, hairs, fibers, paint, glass, firearms, documents, and fingerprints can be found through this website. It explains the importance of inspecting the crime scene and the impact forensic evidence has on the investigation.

Crimes and Clues (www.crimeandclues.com/) Users interested in learning about the forensic aspects of fingerprinting will find this to be a useful and informative website. The site covers the history of fingerprints, as well as subjects pertaining to the development of latent fingerprints. The user will also find links to other websites covering a variety of subjects pertaining to crime scene investigation, documentation of the crime scene, and expert testimony.

Questioned-Document Examination (www.qdewill.com/) This basic, informative web page answers frequently asked questions concerning document examination, explains the application of typical document examinations, and details the basic facts and theory of handwriting and signatures. There are also links to noted document examination cases that present the user with real-life applications of forensic document examination.

Chapter Review

- Forensic science is the application of science to criminal and civil laws that are enforced by police agencies in a criminal justice system.
- The first system of personal identification was called anthropometry. It distinguished one individual from another based on a series of body measurements.
- Forensic science owes its origins to individuals such as Bertillon, Galton, Lattes, Goddard, Osborn, and Locard, who developed the principles and techniques needed to identify or compare physical evidence.
- Locard's exchange principle states that when two objects come into contact with each other, there is exchange of materials between them. This cross-transfer of materials can connect a criminal suspect to his or her victim.
- The development of crime laboratories in the United States has been characterized by rapid growth accompanied by a lack of national and regional planning and coordination.
- Four major reasons for the increase in the number of crime laboratories in the United States since the 1960s are as follows: (1) the fact that the requirement to advise criminal suspects of their constitutional rights and their right of immediate access to counsel has all but eliminated confessions as a routine investigative tool, (2) the staggering increase in crime rates in the United States, (3) the fact that all illicit-drug seizures must be sent to a forensic laboratory for confirmatory chemical analysis before the case can be adjudicated in court, and (4) the advent of DNA profiling.
- The technical support provided by crime laboratories can be assigned to five basic services: the physical science unit, the biology unit, the firearms unit, the document examination unit, and the photography unit.
- Some crime laboratories offer optional services such as toxicology, fingerprint analysis, polygraph administration, voiceprint analysis, and crime-scene investigation.
- Special forensic science services available to the law enforcement community include forensic pathology, forensic anthropology, forensic entomology, forensic psychiatry, forensic odontology, forensic engineering, and forensic computer and digital analysis.
- A forensic scientist must be skilled in applying the principles and techniques of the physical and natural sciences to analyzing evidence that may be recovered during a criminal investigation.
- The cases *Frye v. United States* and *Daubert v. Merrell Dow Pharmaceuticals, Inc.*, set guidelines for determining the admissibility of scientific evidence into the courtroom.
- An expert witness evaluates evidence based on specialized training and experience.
- Forensic scientists participate in training law enforcement personnel in the proper recognition, collection, and preservation of physical evidence.

Review Questions

1. All the following are basic services provided by "full-service crime laboratories" except the:
   a. biology unit.
   b. firearms unit.
   c. document examination unit.
   d. toxicology unit.

2. The standard guideline for determining the judicial admissibility of scientific examination stemmed from which court ruling?
   a. *Mincey v. Arizona*
   b. *Kumho Tire Co., Ltd. v. Carmichael*
   c. *Frye v. United States*
   d. *Coppolino v. State*

3. The scientist who maintained that when a criminal came in contact with an object or person a cross-transfer of evidence would occur was
   a. Hans Gross.
   b. Walter McCrone.
   c. Edmond Locard.
   d. Albert Osborn.

4. Which individual contributed knowledge to the forensic characterization of blood?
   a. Albert Osborn
   b. Calvin Goddard
   c. Alec Jeffries
   d. Thomas Taylor

5. The basic functions of a forensic scientist include:
   a. furnishing training on the proper collection of physical evidence.
   b. analysis of physical evidence.
   c. providing expert testimony.
   d. all of the above.

6. True or False: The first forensic laboratory in the United States was created in 1923.

7. True or False: The United States is one of only a handful of countries in the world that have created and now maintain forensic facilities.

8. True or False: In 1993 the U.S. Supreme Court decided that the *Frye v. United States* precedent is an absolute prerequisite to the admissibility of scientific evidence.

9. True or False: Only individuals who have accredited professional credentials from recognized university or medical programs are allowed to testify in court as an expert witness.

10. True or False: Strict guidelines that ensure careful and systematic collection, organization, and analysis of information is a process known as the scientific method.

11. Define forensic science.

12. What was the name of the first system of personal identification? What criteria did it use to distinguish individuals?

13. Explain Francis Henry Galton's major contribution to the development and advancement of forensic science. How did this advancement improve forensic science?

14. Who is known as "the father of forensic toxicology" and why?

15. Explain two major contributions to forensic science made by Hans Gross. Why are they considered important to forensic science?

16. With what area of forensic investigation are Karl Landsteiner and Louis Lattes associated?

17. Who was the first person to apply the principles of forensic science to a working crime laboratory? What crime was he investigating?

18. What is Locard's exchange principle?

19. With what instrument did Dr. Walter C. McCrone make significant contributions to forensic science?

20. List four major reasons for the increase in the number of crime laboratories in the United States since the 1960s.

21. List four government agencies that offer forensic services at the federal level.

22. The current system of crime laboratories in the United States can best be described as

a. centralized.

b. regional.

c. decentralized.

d. national.

23. List three advantages of having regional crime laboratories operate as part of a statewide system. Are there any disadvantages to this setup?

24. How does the organization of Great Britain's forensic laboratories differ from that of the United States?

25. Which unit examines body fluids and organs for drugs and poisons?

26. Which unit examines and compares tool marks?

27. What part of the body do forensic odontologists use to identify a victim? Why is this body part particularly useful as a source of identification?

28. Describe the criteria for admissibility of scientific evidence as laid out in *Frye v. United States*.

29. In its decision in *Daubert v. Merrell Dow Pharmaceuticals, Inc.*, whom did the U.S. Supreme Court charge with ensuring that an expert's testimony rests on a reliable foundation and is relevant to the case?

30. In the criminal justice system, what is an expert witness? Identify and explain knowledge of terminology and procedures in your explanation.

31. What is the main difference between the testimony given by an expert witness and that given by a lay witness?

32. How can you distinguish between scientific hypotheses, theories, and laws? Write a brief written response.

33. What are the procedures used by the criminal justice system to determine if a witness qualifies as an expert witness?

34. Working with a partner, use logical thinking to research and discuss the effect of bias on the various aspects of a criminal investigation, such as evidence collection, forensic analysis, and expert testimony. How can you use critical-thinking skills to determine if an expert witness testimony has been falsified? Write a brief written response.

35. Evaluate the two different methodologies for problem solving: experimental designs and engineering designs. In what type of crime scenes would each method be most effectively used?

Learn the Vocabulary

Use the information from this chapter along with your prior knowledge and experience to learn the key terms. Choose from the following activities:

1. Make flashcards with a term on the front and its definition on the back. Use the flashcards to quiz a partner on correctly pronouncing and defining the terms.

2. Make an audio or video recording of each term and its definition.

3. Work with a partner to write a sentence for each term. Read the sentences out loud to a partner or share them with the class.

4. Write a list of new terms and language you have encountered in this chapter. Reuse the terms to write sentences and then read the sentences out loud to a partner. Reuse the terms again by writing a short speech. The speech might explain the terms or teach a lesson using the terms. Give the speech to a partner or to the class.

Application and Critical Thinking

1. Most crime labs in the United States are funded and operated by the government and provide services free to police and prosecutors. Great Britain, however, uses a quasi-governmental agency that charges fees for its services and keeps any profits it makes. Suggest potential strengths and weaknesses of each system. In a brief essay, relate the impact of past and current research on scientific thought and society regarding the cost-benefit analysis of each system.

2. Police investigating an apparent suicide collect the following items at the scene: a note purportedly written by the victim, a revolver bearing very faint fingerprints, and traces of skin and blood under the victim's fingernails. What units of the crime laboratory will examine each piece of evidence?

3. List at least three advantages of having an evidence-collection unit process a crime scene instead of a patrol officer or detective.

4. What legal issue was raised on appeal by the defense in Carl Coppolino's Florida murder trial? What court ruling is most relevant to the decision to reject the appeal? Explain your answer.

5. Create a timeline relating the impact of past and current research on scientific thought and society.
   a. Cover the historical development of the various forensic disciplines, including research methodologies.
   b. Analyze how these disciplines have evolved by describing the current roles, functions, and responsibilities of each discipline.

6. Diverse scientists and engineers have impacted both past and current research on scientific thought and society. Research and relate one significant historical and one significant modern contribution made to the development and advancement of forensic science. In a brief essay, explain who the person was and what specifically they contributed.

7. Identify and write a summary describing the professional and ethical standards and employability skills required of a forensic science professional. Working with a partner, brainstorm a list of professional and ethical standards and employability skills. For each item, identify one way you can demonstrate this professional standard or employability skill.

8. What are some of the ethical dilemmas a forensic scientist might face? How can these be avoided?

9. Compare and contrast the admissibility of expert witness testimony in terms of the Frye Standard and the Daubert Standard under federal rules of evidence. Be sure to identify and explain proper expert witness testimony terminology and procedures in your comparison.

10. Choose one of the diverse scientists or engineers who have contributed to the development of modern forensic science to research. You will need to:
    a. Identify the impact this individual has had on modern forensic science.
    b. Describe the research methodologies used by this person, and how the methods have changed.
    c. Relate the impact of this individual on scientific thought on the area of specialty.

    How has society's view of forensic science changed as a result of this individual?

11. In a brief written response, describe how cost-benefit analysis impacts research.

12. A Timeline of Forensic Science

The following images depict different types of evidence or techniques for analyzing evidence. Place the images in order pertaining to the time in history (least recent to most recent) at which that evidence or technique was first introduced. Do this using the letter assigned to each image.

Eduard Skorov/123RF

(A)

Touch Imaging Ltd./Alamy Stock Photo

(B)

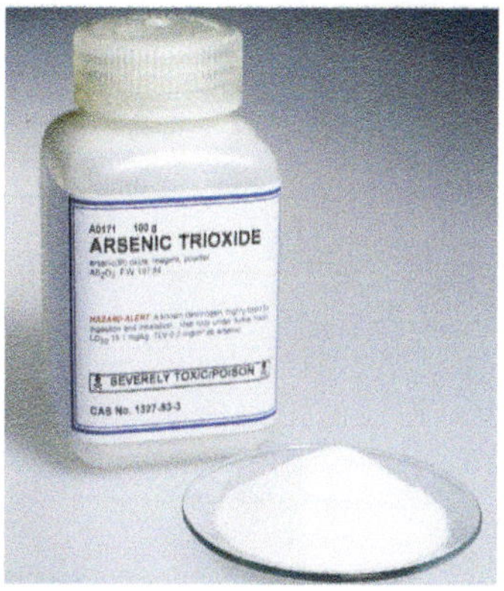

Richard Megna/Fundamental Photographs

(C)

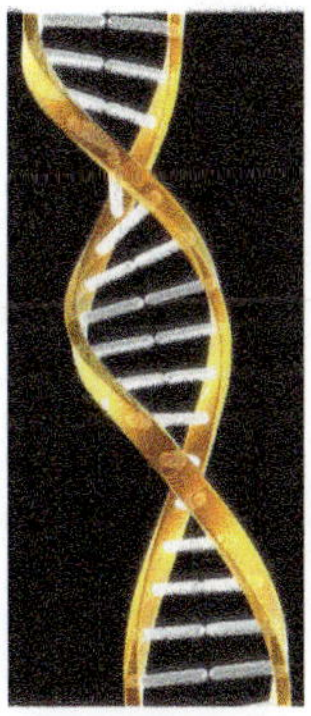

James Steidl/ Shutterstock

(D)

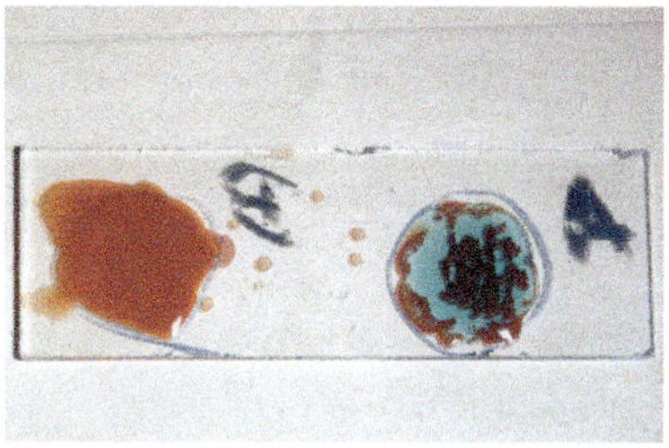

George Whitely/Science Source

(E)

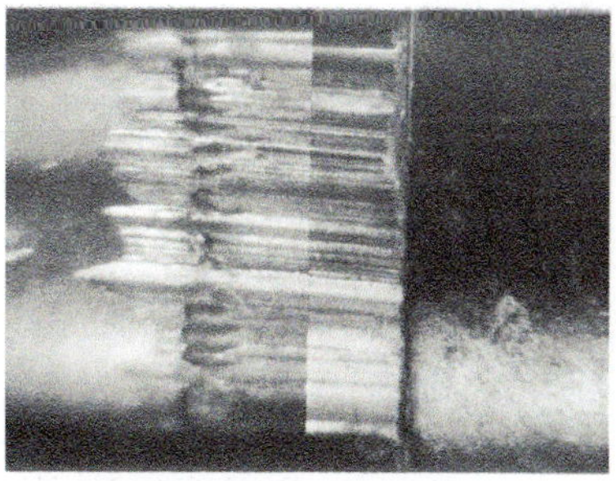

Orlando/Stringer/Hulton Archive/ Getty Images

(F)

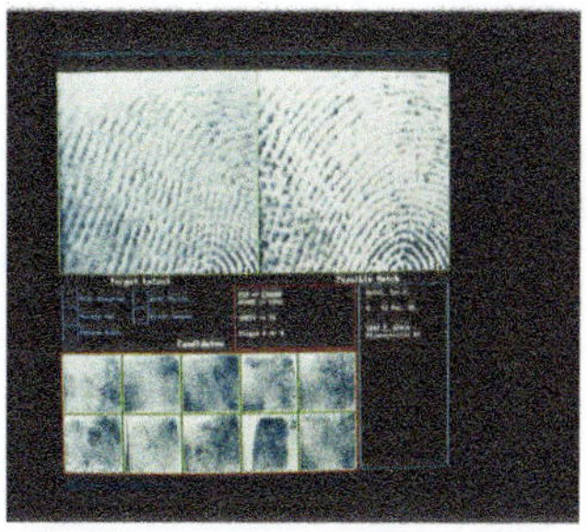

Alan Abramowitz/The Image Bank/Getty Images

(G)

13. Evidence Processing at the Crime Laboratory

You are the evidence technician at the front desk of the state crime lab. You receive the following items of evidence (see page 43) to check-in on a very busy day. You must indicate which unit each piece of evidence must be sent to for analysis. Organize your data in a chart to communicate which unit to send each item of evidence. Your crime lab has a criminalistics (physical science) unit, a drug unit, a biology unit, a firearms unit, a document examination unit, a toxicology unit, a latent fingerprinting unit, an anthropology unit, and a forensic computer and digital analysis unit.

A.

B.

C.

D.

E.

F.

G.

H.

I.

J.

K.

L.

M.

14. Choose a specialty within forensic science to research. Identify past and current methods used within this specialty. Relate the impact this specialty has had on scientific thought and society both in the past and modern day. Be sure to include the research methodologies, how the cost is related to the benefits obtained, and the contributions that have been made by the diverse scientists and engineers working within the specialty chosen. Create a presentation on your specialty to share with your class.

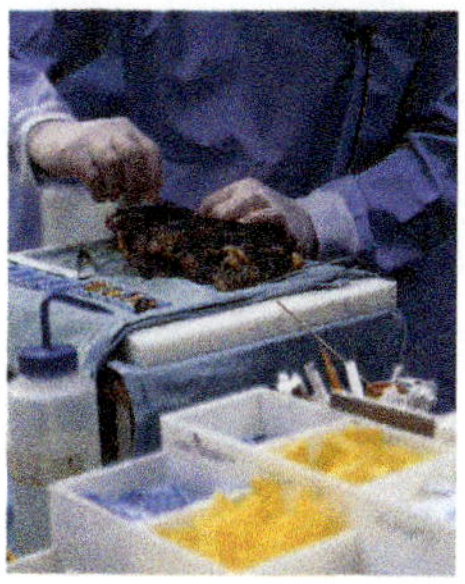

Keith Brofsky/Photodisc/ Getty Images

(A)

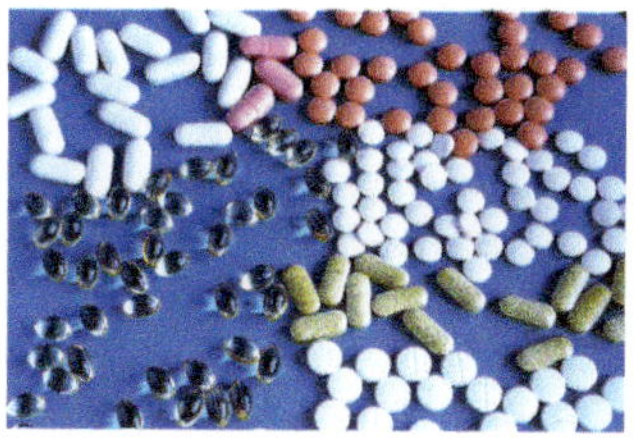

Michael P. Gadomski/Science Source

(B)

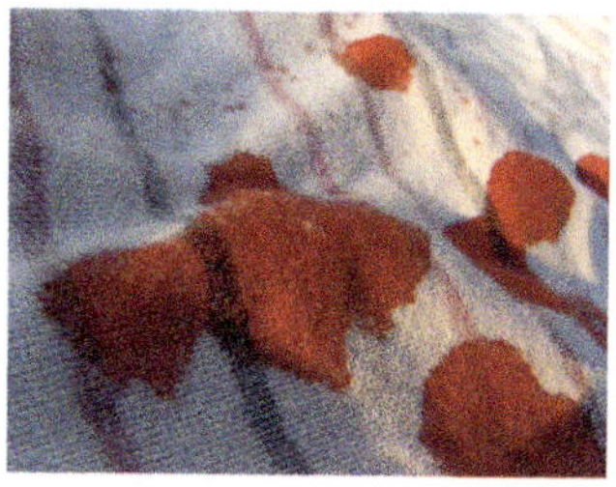

Saivann/ Shutterstock

(C)

Haemin Rapp/ Shutterstock

(D)

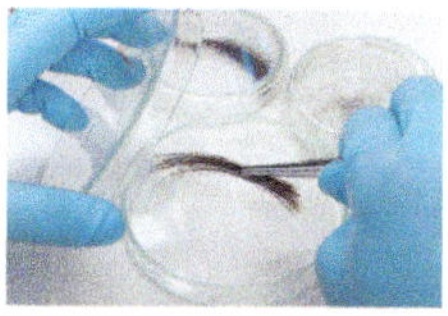

Tonhom1009/ Shutterstock

(E)

Jack Hollingsworth/Photodisc/ Getty Images

(F)

Horacio Villalobos/ Corbis Historical/Getty Images

(G)

Creative Touch Imaging Ltd./Alamy Stock Photo

(H)

Hugh Threlfall/Alamy Stock Photo

(I)

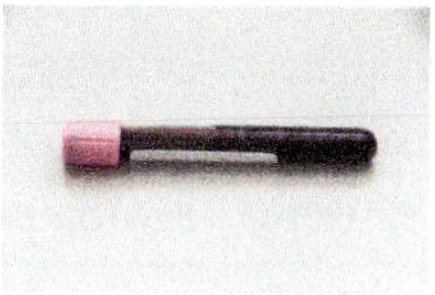

Tonhom1009/ Shutterstock

(J)

Tetyana Tsvihun/123RF

(K)

Olga Yastremska/123RF

(L)

Gasparij/123RF

(M)

Laboratory Experiments

This activity requires the use of the following practices of science and engineering:

- Analyzing and interpreting data
- Constructing explanations
- Asking questions
- Identifing problems
- Safely conducting investigations using the appropriate tools and equipment

This activity consists of the following crosscutting concepts:

- **Patterns**—The observed pattern of the medulla in hair helps to distinguish the hair's origin.
- **Cause and Effect**—The Locard's Exchange Principle is based on the idea that every contact leaves a trace. The presence or absence of hair evidence can be used to determine if a person was present in a given location.

Before you begin, read the scenario and steps out loud with a partner. Ask for clarification as necessary to make sure you understand the experiment.

The Scenario

Bridgette's parents were waiting for her when she got home at 2 a.m. 2 hours after her curfew. They asked her where she'd been and she said at her friend Michelle's house working on her English project. Michelle and her family are animal lovers and have lots of pets. Her parents were concerned that she might be lying and was actually at her new boyfriend's house (who has no pets), whose parents are out of town. They were very upset but wanted to discuss things in the morning. After Bridgette got ready for bed, her father went into her room and took the shirt she had been wearing that evening. Her father has brought the t-shirt to your forensics class to determine if she was telling the truth. The shirt was freshly laundered prior to that evening and Bridgette's family does not have any pets. Complete the following procedure to answer that question.

1. Clean the laboratory bench top with soap and water and then cover it with butcher paper, plastic sheeting, or a trash bag to inhibit contamination.
2. Lay the T-shirt flat on the bench top.
3. Using a hand lens and forceps, scan the shirt for any hairs, and collect them using the forceps. Place the hairs you find inside small envelopes or use a clean sheet of paper to make druggist folds. Take special care around the collar area.
4. Turn the shirt over and repeat this process.
5. Examine the reference samples under the microscope and sketch.
6. Examine your hairs under the light microscope up to 100x magnification, and sketch. You must examine five samples. Group together hairs that look the same.
7. Try to identify the hair based on the reference samples provided.
8. First define the problem you are trying to solve based on observations. Then, use your observations to formulate five questions you would ask Bridgette based on the evidence you found on the shirt. Write a paragraph analyzing the evidence you collected to determine if Bridgette was at her friend's house.

9. Read your summary to a partner or to the class. Be prepared to answer questions about your observations and summary.

Endnotes

1. US Government Accountability Office Congressional Report, March 2019 found here: https://www.gao.gov/assets/700/697768.pdf
2. Bowers, C. M., "Review of a Forensic Pseudo science: Identification of Criminals from Bitemark Patterns," *Journal of Forensic and Legal Medicine* 61 (2019): 34–39. doi:10.1016/j.jflm.2018.11.001.
3. 293 Fed. 1013 (D.C. Cir. 1923).
4. 509 U.S. 579 (1993).
5. 526 U.S 137 (1999)., Kumho Tire Co., Ltd. v. Carmichael
6. 223 So. 2d 68 (Fla. App. 1968), app. dismissed, 234 So. 2d (Fla. 1969), cert. denied, 399 U.S. 927 (1970).
7. 557 U.S. 305 (2009).
8. 541 U.S. 36 (2004).
9. 131 S. Ct. 2705 (2011)., Bullcoming v. New Mexico

The Crime Scene

2

VINCENZO PINTO/AFP/Getty Images

Key Terms

buccal swab
chain of custody
finished sketch
International System of Units (SI)
physical evidence
qualitative data
quantitative data
rough sketch
standard/reference sample
substrate control
United States customary units

Amanda Knox: A Flawed Case of Murder

In September 2007, Amanda Knox moved to Perugia, Italy, as a foreign exchange student attending language classes at the University for Foreigners. Knox shared an upstairs flat in a cottage with Meredith Kercher and two other women. Within weeks of her arrival, Knox became romantically linked to an Italian student, Raffaele Sollecito, and began spending nights at his home.

On November 1, Kercher was brutally murdered. Early that afternoon, Kercher's naked body was found inside her bedroom covered by a bedspread soaked in blood and with stab wounds to her throat. As the investigation proceeded, police matched fingerprints found in Kercher's bedroom to Rudy Guede, whom Knox and Kercher had met weeks earlier when he was playing guitar on the downstairs floor of their cottage flat. Guede was subsequently arrested and charged with the Kercher murder. His DNA was later found at the crime scene, on and inside Kercher's body. The prosecution also charged Knox and Sollecito with murder and sexual assault.

According to the prosecution's theory, Knox was part of a satanic ritual sex game that went out of control. The Knox case became the subject of intense media scrutiny, focusing in part on Knox's alleged romantic escapades. The prosecution alleged that the murder weapon was a kitchen knife found in Sollecito's kitchen that had Kercher's DNA on the blade and Knox's DNA on the handle. The DNA on the handle could have arisen from Knox's handling of the knife while cooking for Sollecito. Experts questioned the veracity of the DNA protocols conducted on the knife blade. Expert analysis concluded that the knife wounds were inconsistent with the knife recovered from Sollecito's residence. Guede was convicted of Kercher's murder. Knox was convicted of slander, sexual violence, and murder and sentenced to twenty six years in prison. Four years later her conviction was set aside by the Italian supreme court and a new trial was ordered. In 2015, Italy's highest court—the Supreme Court of Cassation—definitively exonerated Knox and Sollecito.

Learning Objectives

After studying this chapter, you should be able to:

2.1 Describe the various measures taken while securing, recording, and searching the crime scene

2.2 Describe proper techniques for packaging common types of physical evidence

2.3 Explain the concept of chain of custody

2.4 Relate what steps are typically required to maintain appropriate health and safety standards at the crime scene

2.5 Understand the implications of relevant U.S. Supreme Court decisions in conducting a crime scene investigation

National Science Content Standards

Scientific Inquiry

Physical Science

History and Nature of Science

Physical Evidence and the Crime Scene

physical evidence
Any object that can establish that a crime has or has not been committed or can link a crime and its victim or its perpetrator

qualitative data
Any data that is described using words and cannot be counted or measured

quantitative data
Any data that is numerical and can be counted or measured

International System of Units (SI)
Often referred to as the metric system, a measurement system based on the standards units of meter, kilogram, second, kelvin, ampere, mole, and candela

United States customary units
System of measurements used in the U.S. based off the English system using inches, feet, yards, and miles for lengths and ounces, pound, and ton for weights, fluid ounce, cup, pint, quart, and gallon for volume, and degrees Fahrenheit for temperature

As automobiles run on gasoline, crime laboratories "run" on **physical evidence**. Physical evidence encompasses any and all objects that can establish that a crime has or has not been committed or can link a crime and its victim or its perpetrator. But if physical evidence is to be used effectively to aid the investigator, its presence first must be recognized at the crime scene. If all the natural and commercial objects within a reasonable distance of a crime were gathered so that the scientist could uncover significant clues from them, the deluge of material would quickly immobilize the laboratory facility. Physical evidence achieves its optimum value in criminal investigations only when its collection is performed with a selectivity governed by the collector's thorough knowledge of the crime laboratory's techniques, capabilities, and limitations.

Forthcoming chapters discuss methods and techniques available to forensic scientists to evaluate physical evidence. Although current technology has given the crime laboratory capabilities far exceeding those of past decades, these advances are no excuse for complacency on the part of criminal investigators. Crime laboratories do not solve crimes; only a thorough and competent investigation conducted by professional police officers will enhance the chances for a successful criminal investigation. Forensic science is, and will continue to be, an important element of the total investigative process, but it is only one aspect of an endeavor that must be a team effort. The investigator who believes the crime laboratory to be a solution for carelessness or ineptness is in for a rude awakening.

Forensic science begins at the crime scene. If the investigator cannot recognize physical evidence or cannot properly preserve it for laboratory examination, no amount of sophisticated laboratory instrumentation or technical expertise can salvage the situation. The know-how for conducting a proper crime-scene search for physical evidence is within the grasp of any police department, regardless of its size. With proper training, police agencies can ensure competent performance at crime scenes. In many jurisdictions, police agencies have delegated this task to a specialized team of technicians. However, the techniques of crime-scene investigation are easy to master and certainly lie within the bounds of comprehension of the average police officer.

Not all crime scenes require retrieval of physical evidence, and limited resources and personnel have forced many police agencies to restrict their efforts in this area to crimes of a more serious nature. Once the commitment is made to process a crime site for physical evidence, however, certain fundamental practices must be followed.

It is essential that forensic workers are able to collect both **qualitative** and **quantitative data** with accuracy and precision so that it can be used as evidence. They must also use mathematical calculations to assess quantitative relationships in data to discover any relationships or correlations. Forensic workers need to understand how to use the **International System of Units (SI)** and **United states customary units** as they collect both qualitative and quantitative data as evidence. Qualitative and quantitative data should be organized using appropriate methods to communicate the data. Quantitative data is usually communicated in the form of graphs, tables, and charts. An example of quantitative data is a table with measurements of the amount of drugs/toxins found during an autopsy. Qualitative data is communicated in reports, tables, and images. An example of qualitative data is a report documenting the description of the drugs/powders found at the crime scene.

Scientific practices are applied when trying to answer questions about the natural world. Forensic scientists ask many questions before, during, and after investigations.

These scientific practices should be applied when a scientist plans and conducts descriptive investigations. For example, the collection of evidence at a crime scene. Scientific practices should also be applied when a scientist plans and conducts comparative investigations. For example, the analysis of various tire treads at the edge of a forest crime scene. Lastly, forensic scientists plan and conduct experimental investigations while using scientific practices. For example, a blood spatter analysis and recreation will help an analyst determine how the crime occurred. Forensic scientists also use engineering practices. Engineering practices are used to determine the cause of an accident or crime by reconstructing the sequence of events objectively and neutrally. Scientific practices are focused on when trying to explain a natural phenomena or answer a question, while engineering practices formulate a problem solved through design. Both practices require asking questions; defining the problem; developing and then using models to aid in understanding; planning and then carrying out an investigation; analyzing and interpreting the data collected; the use of mathematical and computational thought processes; formulating explanations and design solutions; and argumentation of the evidence to identify solutions.

It is the responsibility of forensic scientists to analyze, evaluate, and critique scientific explanations when they process crime scenes or solve crimes. Empirical evidence, experimental testing, observational testing, and logical reasoning are all used by forensic scientists to establish the truth. A forensic scientist collects empirical evidence at a crime scene. Forensic scientists use physical evidence collected to analyze scientific explanations at a crime scene. This evidence must then be evaluated for use in solving the crime. In addition to empirical evidence, forensic scientists use observational testing to help analyze and evaluate scientific explanations. Forensic scientists always use observational skills when entering a crime scene to process it. Some evidence collected at a crime scene, such as blood and fibers, is tested through experimental testing in a forensic lab. Blood samples can be tested in a lab to determine whose DNA is present in the evidence. Another part of forensic science involves using logical reasoning. Forensic scientists use their skills and the collection of physical evidence to work through logical explanations of possible scenarios surrounding a crime.

Securing the Crime Scene

The first officer to arrive at the scene of a crime is responsible for taking steps to preserve and protect the area to the greatest extent possible. The officer should not let their guard down; the scene should always be treated as though the crime were still occurring until it is proved otherwise. Arriving officers should immediately ascertain that the perpetrator is no longer in the immediate area of the crime scene and is not a threat to anyone at or near the crime scene. Special note should be taken of any vehicles or people leaving the scene. Of course, first priority should be given to obtaining medical assistance for individuals in need of it. If medical assistance is needed, the officer should direct medical workers to approach the body by an indirect route to minimize the possibility of disturbing evidence. This pathway should later be used by investigative personnel for the same reason. The first responding officer must quickly evaluate the victim's condition before the victim is taken to a medical facility. The officer must also record any statements made by the victim and instruct the emergency medical personnel to record any statements the victim makes on the way to the hospital. This information should later be included in notes.

FIGURE 2–1 The first investigators to arrive must secure the crime scene and establish the crime-scene perimeter. *Courtesy Sirchie Finger Print Laboratories, Inc., Youngsville, N.C., www.sirchie.com*

The officer should call for any backup or investigative personnel required and, as soon as possible, detain all potential suspects or witnesses still at the scene. It is essential that the crime scene specialists are able to work as a team and collaborate with outside experts and agencies as necessary. The officer must identify all individuals at the scene, including bystanders and medical personnel. At the same time, the officer should exclude all unauthorized personnel from the scene. This includes family and friends of the victim, who should be shown as much compassion as possible.

The first responder(s) are responsible for establishing the boundaries of the scene to be protected. The boundaries should encompass the center of the scene where the crime occurred, any paths of entry or exit, and any areas where evidence may have been discarded or moved. For indoor scenes this may include anything from a single room to an entire house and yard. The center of the crime scene is usually apparent, and a sufficient area around this spot should be closed off. The boundaries of an outdoor crime scene are more difficult to determine and can span miles, especially if a vehicle is involved.

The officer should initially denote the boundaries of the scene using crime-scene tape, ropes, or traffic cones (see Figure 2–1). As additional officers arrive, investigators should immediately take measures to isolate the area around the taped-off section. Police barricades, along with the strategic positioning of guards, will prevent unauthorized access to the area. Only investigative personnel assigned to the scene should be admitted. The responding officers must keep an accurate log of who enters and exits the scene and the time at which they do so.

Sometimes the exclusion of unauthorized personnel proves more difficult than expected. Violent crimes are especially susceptible to attention from higher level police officials and members of the press, as well as by emotionally charged neighbors and curiosity seekers. Every individual who enters the scene has the potential to destroy physical evidence, even if by unintentional carelessness. To exercise proper control over the crime scene, the officer protecting it must have the authority to exclude everyone, including fellow police officers not directly involved in processing the site or in conducting the investigation. Seasoned criminal investigators are always prepared to relate horror stories about crime scenes where physical evidence was rendered totally valueless by hordes of people who trampled through the site. Securing and isolating the crime scene are critical steps in an investigation, the accomplishment of which is the mark of a trained and professional crime-scene investigative team.

It is worth noting that personnel should never do anything while at the crime scene—including smoking, eating, drinking, or littering—that might alter the scene. No aspects of the scene, including a body at a death scene, should be moved or disturbed unless they pose a serious threat to investigating officers or bystanders. This means that no one should open or close faucets or flush toilets at the scene. Also, officers should avoid altering temperature conditions at the scene by adjusting windows, doors, or the heat or air-conditioning.

Recording the Scene

Once the scene has been secured, a lead investigator starts evaluating the area. First, the investigator determines the boundaries of the scene and then establishes the perpetrator's path of entry and exit. Logic dictates that obvious items of crime-scene evidence will first come to the attention of the crime-scene investigator. These items must be documented and photographed. The investigator then proceeds with an initial walk-through of the scene to gain an overview of the

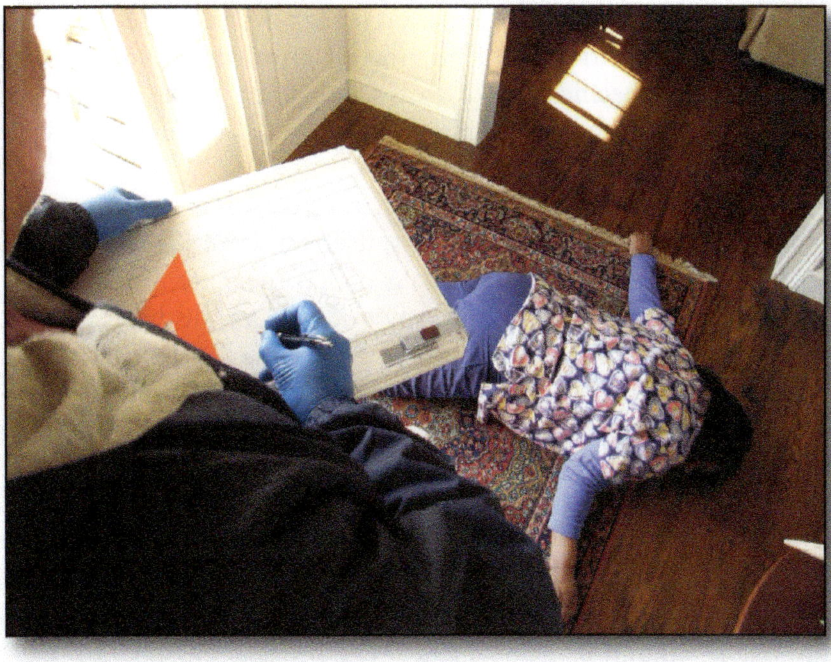

FIGURE 2–2 Sketching a victim at the crime scene to show the victim's relation to the crime. *Courtesy Evident, Union Hall, VA 24176-4025, www.evidentcrimescene.com*

situation and develop a strategy for systematically examining and documenting the entire crime scene.

Investigators have only a limited amount of time to work a crime scene in its untouched state. They must not lose the opportunity to observe the scene and activities in its original condition and begin documenting the crime scene examination with a detailed, technical written record. Such records not only will prove useful during the subsequent investigation but also must be presented at a trial in order to document the condition of the crime scene and the location of physical evidence. Notes, photographs, and sketches are the three methods of crime scene recording (see Figure 2–2).

Notes The note-taking process begins with the call to a crime-scene investigator to report to a scene. The first notes should identify the person who contacted the investigator and record the time of the contact and any preliminary information disclosed, including the case number. When the lead investigator arrives, the note taker should record the date and time of arrival, who is present, and the identities of any other personnel who are being contacted.

If additional personnel are contacted, their names, titles, and times of arrival should be recorded. A crime scene should be off-limits to any people who are not directly involved with the processing. Investigators must keep very precise records of personnel movements in and out of the scene, beginning with the movements of the first responding officer. It is also important to record the tasks assigned to each member of a team, as well as the beginning and ending times of the processing of the scene.

Before the scene is sketched, photographed, or searched, the lead investigator carries out the initial walk-through. During this walk-through, the investigator should take notes on many aspects of the crime scene in its original condition. Notes taken by an investigator should be uniform in format for all cases. The notes should be in ink (preferably black or blue) and written in a bound notebook. Most important, notes should be written at the time of the crime-scene investigation, not left to record from memory at a later time. At this time, the investigator may need to interview the first responding officer.

The officer should supply information on any events at the crime scene that the officer or others witnessed. When the walk-through is complete, the lead investigator assigns specific tasks or areas to members of the crime-scene team. The notes should also record these assignments, as well as the times at which each task was started and completed.

Once a search for evidence has taken place, the team members mark the location of all evidence using evidence markers, referred to as evidence tents, that often include a metric scale and photo crosshairs on a corner cutout on the flat surface. The investigator should note whether any evidence was disturbed by emergency medical personnel, a suspect, or investigative personnel. Before the team collects items of evidence, the investigator must fully describe each item in their notes. The person who collects a piece of evidence should note who found it, where it was located, how it was packaged, who packaged it, and when it was packaged. The notes should also mention whether the evidence underwent any field tests or processing.

If a victim is present at a homicide scene, the investigator should observe and record the state of the body before the medical examiner or coroner moves it. The notes should describe the victim's appearance and record the position of the body and any visible wounds or blood spatter. The investigator should make note of any identifying features or marks on the body, such as tattoos. The investigator should also make a list of objects found on the body, such as a wallet or jewelry, before collecting those items. Moving the body may reveal previously unseen injuries and physical evidence that the investigator should record. The notes should indicate when the medical examiner or coroner moved the body and whether the victim was moved or affected in some way by emergency medical procedures. Any preliminary identification of a victim or suspect should be recorded.

Audio recording of notes at a scene can be advantageous because detailed notes can be spoken much faster than they can be written. This may also leave the investigators' hands free to carry out other tasks while recording the notes. However, care must be taken to avoid embarrassing conversation in recordings that will be used as evidence in a trial. Digital voice recordings are easily uploaded to a computer, but they must be copied to a disk or flash drive to produce a hard copy. Another method of recording notes is by narrating a digital video of the crime scene. This has the advantage of combining note taking with photography. However, at some point the recording must be transcribed into a written document.

The note taker must keep in mind that this written record may be their only way of refreshing their memory months, perhaps years, after a crime has been processed. The notes must be sufficiently detailed to anticipate this need.

Photography The most important prerequisite for photographing a crime scene is for it to be unaltered. Unless injured people are involved, objects must not be moved until they have been photographed from all necessary angles. If objects are removed, positions changed, or items added, the photographs may not be admissible as evidence at a trial, and their intended value will be lost. If evidence has been moved or removed before photography, the fact should be noted in the report, but the evidence should not be reintroduced into the scene in order to take photographs.

Forensic photography documents and preserves the original crime scene intact for future investigation and shows the layout of the scene, the position and dimensions of evidence to be collected, and the physical relationship of objects at the scene to one another. Photographs taken from many angles can show possible lines of sight of victims, suspects, and witnesses. Photography is also important for documenting biological evidence in its original condition because this kind of evidence is often altered during testing. A photography report, which is a qualitative data collection method, organizes this collection of photographs.

In order to be admissible as evidence within the justice system, photographs must meet the following criteria: (1) they must represent the crime scene as found, without manipulation or bias; (2) any re-creations of scenes or manipulation must be documented; (3) they must have bearing on the case in question; and (4) they must not engage in emotional manipulation of the jury.

Photographs cannot stand alone, however, they are complementary to notes and sketches and are an important part of the forensic documentation process. This process involves the collection, the organization, and the investigation of visual information through digital photography, digital videography, note taking, manual sketching, **computer-aided drafting (CAD)**, and three-dimensional imaging systems. Documenting a crime scene by using CAD or three-dimensional imaging systems allows forensic scientists to use models to represent the phenomena of crime scenes, aiding in an engineering approach to finding a solution to what happened during the commission of the crime. These methods can be used to develop a type of model that can be used to represent the phenomena, systems, processes, or possible solutions to engineering problems within the crime scene. Investigators should be aware of both the advantages as well as the limitations of these models in relation to their size, scale, properties, and materials. For example, an advantage of using photographs is you can document everything from the entire scene to an individual item of evidence. However, when taking photographs, it is hard to know the true size of an item. To combat this issue, place a ruler next to the item and be sure to include the ruler in the photo. To address issues with the scale of a crime scene, photos are taken at close, mid, and long range. Properties and materials of the scene are documented through notes and sketches.

Computer-aided drafting (CAD)
The use of computers (or workstations) to aid in the creation, modification, analysis, or optimization of a drawing

Crime-scene photographs have great value in their ability to show the layout of the scene, the position of evidence to be collected, and the relation of objects at the scene to one another. Photographs taken from many angles can show possible lines of sight of victims, suspects, or witnesses. An accurate description of the scene must be available to investigators for future analysis.

Currently digital photography is the preferred approach to crime-scene photography. A digital photograph is made when a light-sensitive microchip inside a digital camera is exposed to light coming from an object or scene. A digital camera captures light on each of millions of tiny picture elements called "pixels." The light is recorded on each pixel as a specific electric charge using a charged coupled device (CCD) or complementary metal oxide semiconductor (CMOS). The camera reads this charge number as image information, and then stores the image as a file on a memory card.

The number of pixels used to capture light is directly related to the resolution of the picture. Resolution is defined as the minimum distance that must separate two objects in order for them to be viewed as distinct objects. The lower the distance needed, the greater the resolution of the photograph.

Photographs of increasingly higher resolution show more and more detail and sharpness. The greater the number of pixels featured on the digital camera, the better the resolution will be. Because the number of pixels on a digital camera is

FIGURE 2–3 Individual images (top) are shown before being electronically stitched together into a single panoramic image (bottom). Individual photographs should be taken with about a 30 percent overlap. *Courtesy Imaging Forensics, Fountain Valley, Calif., www.imagingforensics.com*

in the millions, it is usually referred to in terms of megapixels. A camera that has four million pixels is a four-megapixel camera. A standard four-megapixel camera can create a clear image on a photographic print of up to 8 by 10 inches. As the number of megapixels increases, the clarity increases, allowing photographers to create bigger prints. Crime-scene photographers usually use cameras that feature as many as twelve megapixels or more. The Scientific Working Group on Image Technology (SWGIT) recommends using a single-lens reflex (SLR) camera with at least eight-megapixel resolution. Digital photography of crime scenes has many advantages. Digital imaging technologies can increase the pace of investigation, aid in the provision of evidence, as well as the identification of suspects. Investigators can observe the images immediately after taking them to ensure that important photographs are clear and show the best possible detail. As technology advances, digital photography provides other advantages over traditional film. For example, photographers can electronically stitch together individual images of a crime scene that were captured with a digital camera to create a nearly three dimensional panoramic view of the crime scene (see Figure 2–3). With the aid of a computer, any area of the scene captured digitally can be enhanced and examined in fine detail.

However, the effectiveness and utility of such technologies depend on the quality of image and the reliability of the process. The very nature of digital images opens digital photography up to important criticisms within forensic science casework. Because the photographs are digital, they can be easily manipulated by using computer software. This manipulation goes beyond traditional photograph enhancement such as adjusting brightness and contrast or color balancing. Computer software allows a person to crop a photo, remove repeating patterns such as window screens, superimpose images, and alter photos significantly in many other ways. Because the primary function of crime-scene photography is to provide an accurate depiction, this is a major concern, and care must be taken to maintain an original version of the image without modification. The nature of digital images, however, opens digital photography to important criticisms within forensic science casework. Because the photographs are digital, they can be easily manipulated by using computer software. This manipulation goes beyond traditional photograph enhancement such as adjusting brightness and contrast or color balancing. Because the main function of crime-scene photography is to provide an accurate depiction, this is a major concern. To ensure that their digital images are admissible, many jurisdictions set guidelines for determining the circumstances under which digital photography may be used and establish and enforce strict protocols for image security and chain of custody.

Photographic Procedures. Each crime scene should be photographed as completely as possible. This means that the crime scene should include the area in which the crime took place and all adjacent areas where important acts occurred immediately before or after the commission of the crime.

Overview Photographs The first pictures the photographer takes are overview photographs of the entire scene and surrounding area, including points of exit and entry. These photographs provide information about the general area where the crime took place, the specific scene of the crime, about the relation of items of evidence to each other and to fixed features of the crime scene, and about the actual items of evidence. Overview photos are generally shot from a natural perspective (although aerial shots are also possible).

Overview photos begin from a distance and gradually zone in on the specific crime scene. At times, crime-scene photographs may find it difficult or impossible to take external overview photos at an active crime scene. Overview photos should be taken from the outside borders of the scene from various angles. If the crime took place indoors, the entire room should be photographed to show each wall area. Rooms adjacent to the actual crime site must be similarly photographed. If the crime scene includes a body, photographs must show the body's position and location relative to the entire scene. When taking overview photographs, the crime-scene photographer should include at least one object in multiple overview photographs to help visually piece the scene together. This object is called a "visual tag." Although one generally should avoid having individuals present in photographs, it may be helpful to photograph witnesses in the locations from which they viewed the crime. This can help the jury determine what a witness could or could not have seen from these locations. One may also wish to photograph the sight lines of suspects and victims. If a camera boom is available, the photographer should take pictures from overhead.

Medium-Range Photographs The next set of pictures should be medium range photographs that show the layout of smaller significant areas of the crime scene. Medium-range shots should be taken with evidence markers in place to show the spatial relationships between and among pieces of evidence in greater detail than in the overview photographs. A key medium-range photograph standard to all crime scenes is one that includes the "center" of the scene. In violent crimes, this usually includes the site where the victim was found and the surrounding area.

Close-Up Photographs Close-up photographs, taken last, show the greatest detail of individual objects or evidence. Macro lenses should be used when taking close-ups in order to fill the frame. A supplemental lens called an "extension tube" as well as close-up filter sets for magnification may also be used. At a minimum, there are four photographs required at a crime scene: an overview photograph, a medium-range photograph, a close-up photograph, and a close-up photograph with an ABFO scale. The ABFO scale was originally designed by the American Board of Forensic Odontology is used to photograph bite marks identified on a victim's body. These photographs create an adequate visual record of the position and appearance of an item of evidence at a crime scene.

Sketches. Once the crime-scene investigator has taken sufficient notes and photographs, the investigator sketches the scene. The sketch serves many important functions in the legal investigation of a crime. If done correctly, a sketch can clearly show the layout of an indoor or outdoor crime scene and the relationship in space of all the items and features significant to the investigation. Sketches are especially important to illustrate the location of collected evidence. A sketch can clarify objects and features already described in notes or shown in photographs. Sketches can also show measurements over long distances and the topography of outdoor scenes. Possible paths of entry, exit, and movement through the scene may be speculated from a good sketch. Additionally, the state of the scene illustrated by the sketch may help to demonstrate the feasibility of a witness's testimony. To be effective, a sketch must be clear enough to be used in reconstruction by other investigative personnel and to illustrate aspects of the crime scene to a jury.

rough sketch
A sketch, drawn at the crime scene, that contains an accurate depiction of the dimensions of the scene and shows the location of all objects having a bearing on the case an accurate depiction of the dimensions of the scene and showing the location of all objects having a bearing on the case.

The Rough Sketch The investigator may have neither the skill nor the time to make a polished sketch of the scene. However, this is not required during the early phase of the investigation. What is necessary is a **rough sketch**, either physical or digital, containing an accurate depiction of the dimensions of the scene and showing the location of all objects having a bearing on the case. It shows all recovered items of physical evidence, as well as other important features of the crime scene (see Figure 2–4). Coordinates, or measurements, are made from fixed points by choosing two non-movable items, such as doors or walls, from which to measure from the evidence to that fixed point to triangulate the item position within the scene.

The following tools are required to create the sketch:

- Graph paper or drawing paper
- Two measuring tapes
- Clipboard or drawing surface
- Pencils
- Straightedge L-ruler
- Erasers
- Compass or GPS device
- Optional tools including drawing compasses, protractors (half- or full-circle), architect scales, French curves, drafting triangles, and rolling measuring devices.

All sketches must include the following features:

- The title block contains the case number, the agency number (if applicable), the name and title of the artist, the location of the scene, and the date and time at which the sketch was created. It may also state the victim's name, the names of any suspects, or the type of crime. The title block should appear in the lower-right corner of the sketch paper.

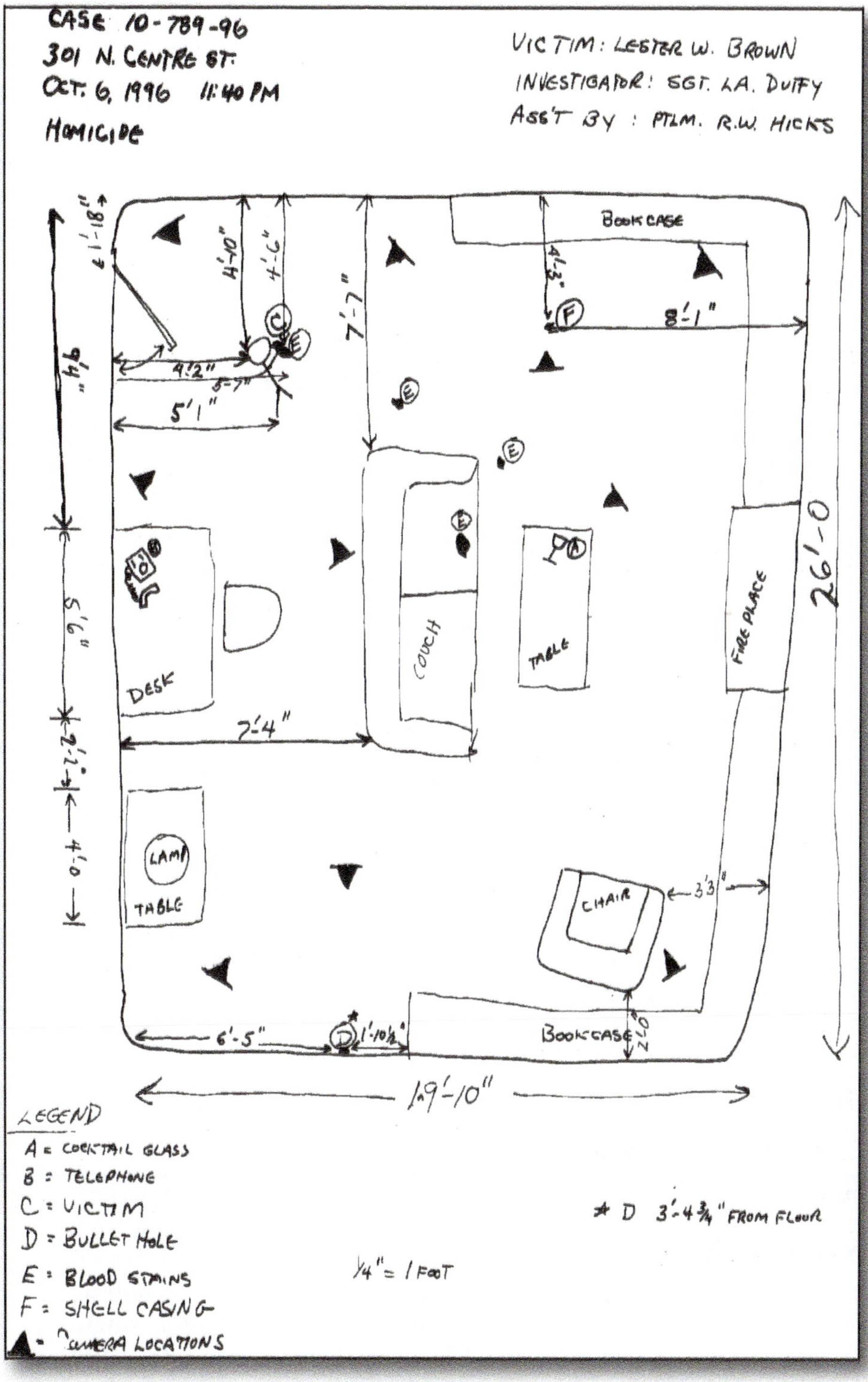

FIGURE 2–4 Rough-sketch diagram of a crime scene. *Courtesy Sirchie Finger Print Laboratories, Inc., Youngsville, N.C., www.sirchie.com*

- The heading is a description of why the sketch was created.
- The legend-key should contain the key to the identity and dimensions of objects or evidence and may be represented by symbols, letters, or numbers.
- The compass directions are shown by including an arrow to denote north in relation to the scene.
- The body of the sketch contains the drawing itself and all dimensions and objects located within it.

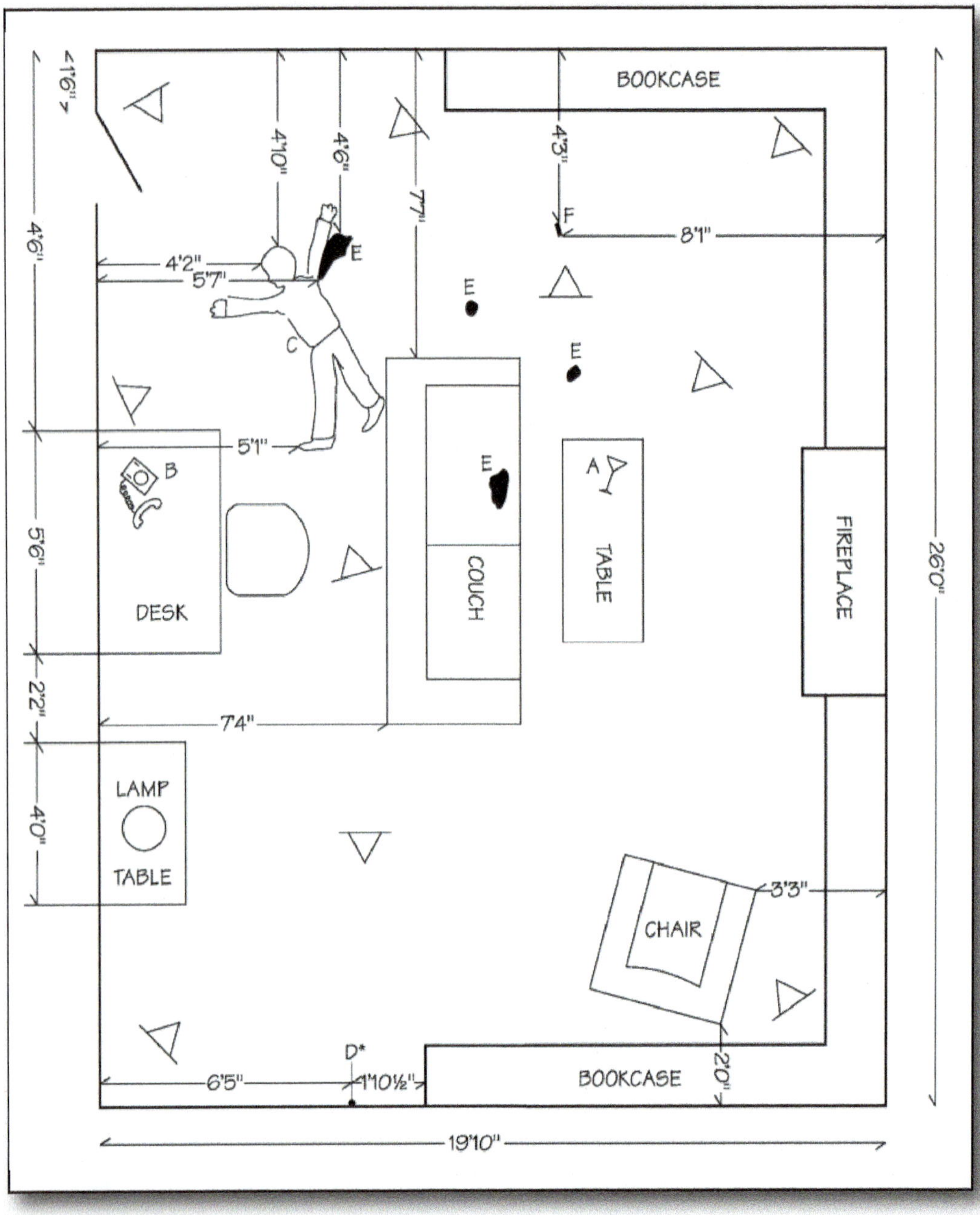

FIGURE 2–5 Finished sketch of a crime scene.
Courtesy Sirchie Finger Print Laboratories, Inc., Youngsville, N.C., www.sirchie.com

finished sketch
A precise rendering of the crime scene, usually drawn to scale

The Finished Sketch Unlike the rough sketch, the finished sketch (see Figure 2–5) is constructed with care and concern for aesthetic appearance. The finished sketch must be drawn to scale. The legend should contain the scale of proportion (e.g., 1/20 = 19). When the finished sketch is completed, it must reflect information contained within the rough sketch to be admissible evidence in a courtroom. An indoor overview sketch shows the floors of one or more rooms, doors and windows, and any evidence or objects on the floor. An outdoor overview sketch is like an aerial view of a small or large outdoor scene. A final sketch can be done by an investigator or a hired professional. It can be done by hand in pen or by computer.

The process of computer-aided drafting (CAD) has become the standard method for reconstructing crime scenes from rough sketches (see Figure 2–6). The software, ranging from simple, low-cost programs to complex, expensive

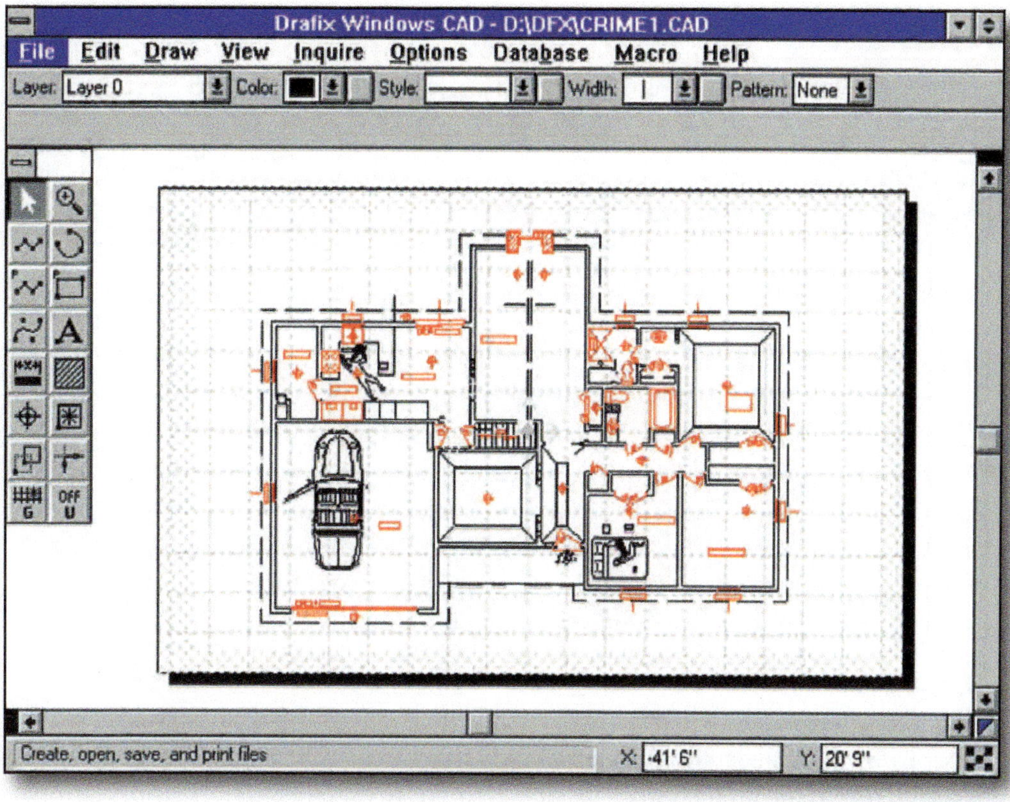

FIGURE 2–6 Construction of a crime scene diagram with the aid of a computer-aided drafting program.
Courtesy Sirchie Finger Print Laboratories, Inc., Youngsville, N.C., www.sirchie.com

ones, contains predrawn objects, such as intersections, roadways, buildings, and rooms, onto which information can be added.

A generous symbol library provides a variety of images that can be used to add intricate details, such as blood spatters, to a crime-scene sketch. Equipped with a zoom function, computerized sketching programs can enlarge a specific area for a more detailed picture. These sketches may also be able to show bullet trajectories in scenes where a gun was involved. CAD programs allow users to select a scale size so that the final product can be produced in a size suitable for courtroom presentation.

Three-dimensional CAD sketches can also be created. These sketches show the nature of the crime scene from many angles. This can be helpful in crime reconstruction and in the trial setting. Some can also be animated to show how the events suggested in the reconstruction would lead to the final state of the scene.

Quick Review

- Physical evidence includes any and all objects that can establish that a crime has been committed or can link the crime and its victim or its perpetrator.
- Forensic science begins at the crime scene, where investigators must recognize and properly preserve evidence for laboratory examination.
- The first officer to arrive must secure the crime scene.
- Investigators record the crime scene by using photographs, sketches, and notes and make a preliminary examination of the scene as the perpetrator left it.

Searching the Crime Scene

Investigators must establish a center of operations, or command center, at the scene. Here, members of the investigative team receive their assignments, store their equipment, and meet to discuss aspects of the case. The command center must be located outside the taped-off boundary of the scene and contain the basic equipment needed to photograph, sketch, process, and collect evidence. An equipped crime-scene vehicle usually serves the purpose well. If multiple scenes are involved, the command center should also be a center for communicating with investigators at the other scenes. At the command center, the lead investigator assigns tasks after the initial walk-through. Basic tasks include locating possible evidence, assessing the evidence, processing evidence (e.g., dusting for fingerprints and casting footprints or tire impressions), and photographing and sketching the scene. The tasks should be carried out in this exact order to properly process the scene. The number of personnel assigned to each task depends on the scene and the discretion of the lead investigator. In some cases, a single crime-scene investigator might be required to handle all these tasks.

Considerations in Searching the Crime Scene How one conducts a crimescene search will depend on the locale and size of the area, as well as on the actions of the suspect(s) and victim(s) at the scene. When possible, one person should supervise and coordinate the collection of evidence. Without proper control, the search may be conducted in an atmosphere of confusion with needless duplication of effort. Evidence collectors may subdivide the scene into segments and use search patterns to search each segment individually, or the search may start at some outer point and gradually move toward the center of the scene in a circular fashion (see Figure 2–7). The areas searched must include all probable points of entry and exit used by the criminals.

What to search for will be determined by the particular circumstances of the crime. Obviously, the skill of crime-scene investigators at recognizing evidence and searching relevant locations is paramount to successful processing of the crime scene. Although training will impart general knowledge for conducting a proper crime-scene investigation, ultimately the investigator must rely on the experience gained from numerous investigations to formulate a successful strategy for recovering relevant physical evidence at crime scenes. For example, in a homicide case, the search will center on the weapon and any type of evidence left as a result of contact between the victim and the assailant. The cross-transfer of evidence, such as hairs, fibers, and blood, between individuals involved in the crime is particularly useful for linking suspects to the crime scene and for corroborating events that transpired during the commission of the crime. During the investigation of a burglary, efforts will be made to locate tool marks at the point of entry. In most crimes, a thorough and systematic search for latent fingerprints is required. The systematic search of a crime scene takes place at different levels. The primary systematic search is used to search the scene of the crime looking for relevant items of evidence. The secondary systematic search occurs when investigators look through photographs, sketches, witnesses, or relatives that could provide more information relevant to the case.

Vehicle searches must be carefully planned and systematically carried out. The nature of the case determines how detailed the search must be. In hit-and-run cases, the outside and undercarriage of the car must be examined with care. Particular

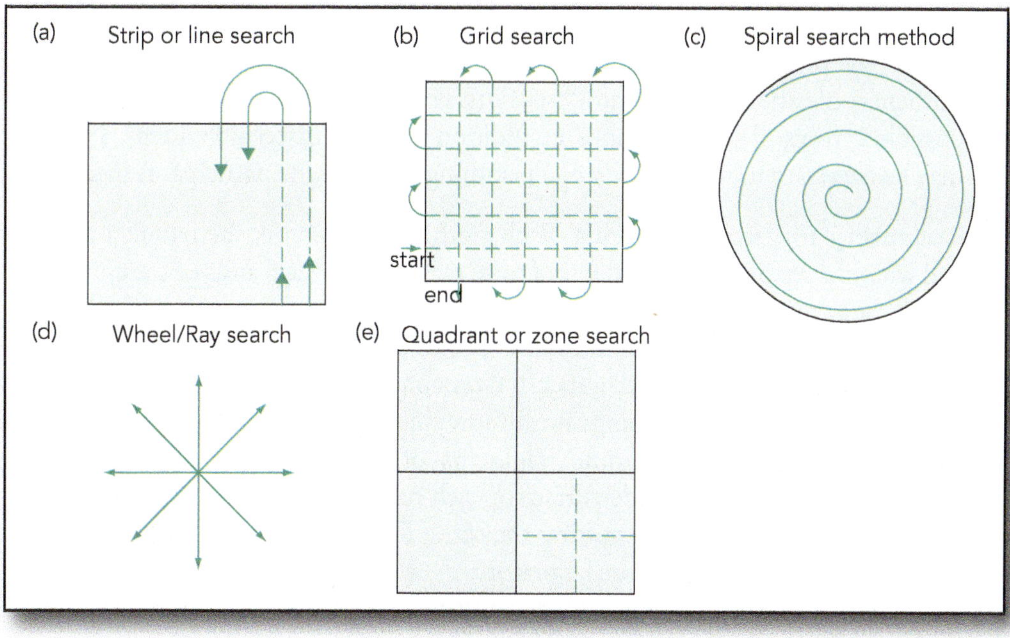

FIGURE 2-7 Several typical examples of crime-scene search patterns. The pattern selected normally depends on the size and locale of the scene and the number of collectors participating in the search.

attention is paid to looking for any evidence resulting from a cross-transfer of evidence between the car and the victim—including blood, tissue, hair, fibers, and fabric impressions. Traces of paint or broken glass may be located on the victim. In cases of homicide, burglary, kidnapping, and so on, all areas of the vehicle, inside and outside, are searched with equal care for physical evidence.

Considerations in Searching a Potential Suspect

When it is necessary to search a person in connection to a crime, law enforcement personnel must follow the guidelines set forth by the 4th amendment. The 4th amendment protects an individual against unreasonable searches and seizures. Law enforcement must show probable cause to a judge in order to obtain a warrant that will describe the particular place to be searched, and the specific persons, property, and/or evidence to be seized.

Some exceptions to a search warrant include exigent situations such as a threat to the lives or safety of others, or there is a risk of immediate destruction or removal of evidence from the scene; items in plain view; consented search; search as part of a lawful arrest; and stop-and-frisk, also known as a Terry stop, in which police may stop a person if there is reasonable suspicion of a crime and conduct a brief pat-down search to ensure safety.

Collecting and Packaging Physical Evidence

Physical evidence can be anything from massive objects to microscopic traces. It is essential that proper techniques for collecting, packaging, and preserving physical evidence are properly documented, including chain of custody, throughout the process. Often, many items of evidence are obvious in their presence, but others may be detected only through examination in the crime laboratory. For example, minute traces of blood may be discovered on garments only after a thorough search in the laboratory, or the presence of hairs and fibers may be revealed in vacuum sweepings or on garments only after close laboratory scrutiny. For this reason, it is important to collect possible carriers of trace evidence in addition to more discernible items. Hence, it may be necessary to take custody of all clothing worn by the participants in a crime.

Collecting Physical Evidence Each clothing item should be handled carefully and wrapped separately to avoid loss of trace materials. Critical areas of the crime scene should be vacuumed and the sweepings submitted to the laboratory for analysis. The sweepings from different areas must be collected and packaged separately. A portable vacuum cleaner equipped with a special filter attachment is suitable for this purpose. Additionally, fingernail scrapings from individuals who were in contact with other individuals may contain minute fragments of evidence capable of linking the assailant and victim. The undersurface of each nail is best scraped with a dull object such as a toothpick to avoid cutting the skin. These scrapings will be subjected to microscopic examination in the laboratory.

The search for physical evidence must extend beyond the crime scene to the autopsy room of a deceased victim. Here, the medical examiner or coroner carefully examines the victim to establish a cause and manner of death. Tissues and organs are routinely retained for pathological and toxicological examination. At the same time, arrangements must be made between the examiner and investigator to secure a variety of items that may be obtainable from the body for laboratory examination (see pg. 109).

In recent years, many police departments have gone to the expense of purchasing and equipping "mobile crime laboratories" (see Figure 2–8) for their evidence technicians. However, the term *mobile crime laboratory* is a misnomer. These vehicles carry the necessary supplies to protect the crime scene; photograph, collect, and package physical evidence; and perform latent print development. They are not designed to carry out the functions of a chemical laboratory. *Crime-scene search vehicle* would be a more appropriate but perhaps less dramatic name for such a vehicle.

Handling Evidence Investigators must handle and process physical evidence in a way that prevents any change from taking place between the time the evidence is removed from the crime scene and the time it is received by the crime laboratory. Changes can arise through contamination, breakage, evaporation, accidental scratching or bending, or improper or careless packaging. The use of latex gloves or disposable forceps when touching evidence often can prevent such problems. Any equipment that is not disposable should be cleaned and/or sanitized between collecting each piece of evidence. Evidence should remain unmoved until investigators have documented its location and appearance in notes, sketches, and photographs.

Evidence best maintains its integrity when kept in its original condition as found at the crime site. Whenever possible, one should submit evidence to the laboratory intact. The investigator normally should not remove blood, hairs, fibers, soil particles, and other types of trace evidence from garments, weapons, or other articles that bear them. Instead, the investigator should send the entire object to the laboratory for processing.

Of course, if evidence is adhering to an object in a precarious manner, good judgment dictates removing and packaging the item. Use common sense when handling evidence adhering to a large structure, such as a door, wall, or floor;

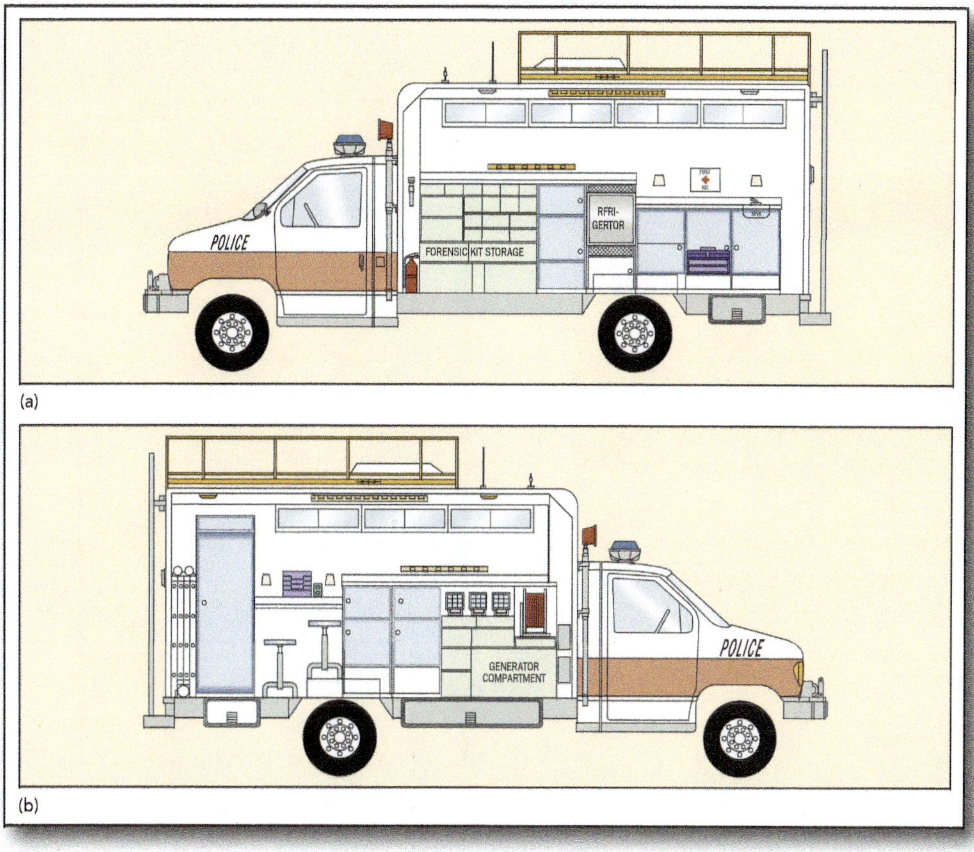

FIGURE 2–8 Inside view of a mobile crime-scene van: (a) driver's side and (b) passenger's side. *Courtesy Sirchie Finger Print Laboratories, Inc., Youngsville, N.C., www.sirchie.com*

remove the specimen with a forceps or other appropriate tool. In the case of a bloodstain, one may either scrape the stain off the surface, transfer the stain to a moistened swab, or cut out the area of the object bearing the stain.

Packaging and Preserving Evidence The well-prepared evidence collector arrives at a crime scene with a large assortment of packaging materials and tools, ready to encounter any type of situation. Forceps and similar tools may be used to pick up small items. Unbreakable plastic pill bottles with pressure lids are excellent containers for hairs, glass, fibers, and various other kinds of small or trace evidence. Alternatively, manila envelopes, screw-cap glass vials, sealable plastic bags, or metal pillboxes are adequate containers for most trace evidence encountered at crime sites (see Figure 2–9). Charred debris recovered from the scene of a suspicious fire must be sealed in an airtight container to prevent the evaporation of volatile petroleum residues. New paint cans or tightly sealed jars are recommended in such situations (see Figure 2–10). Ordinary envelopes should not be used as evidence containers because powders and fine particles will leak out of their corners. Instead, small amounts of trace evidence can be conveniently packaged in a carefully folded paper, using what is known as a "druggist fold" (see Figure 2–11). This consists of folding one end of the paper over by one-third, then folding the other end (one-third) over that, and repeating the process from the other two sides. After folding the paper in this manner, tuck the outside two edges into each other to produce a closed container that keeps the specimen from falling out.

Place each different item or similar items collected at different locations in separate containers. Packaging evidence separately prevents damage through contact and prevents cross-contamination. Each package should be sealed with tamper evident tape.

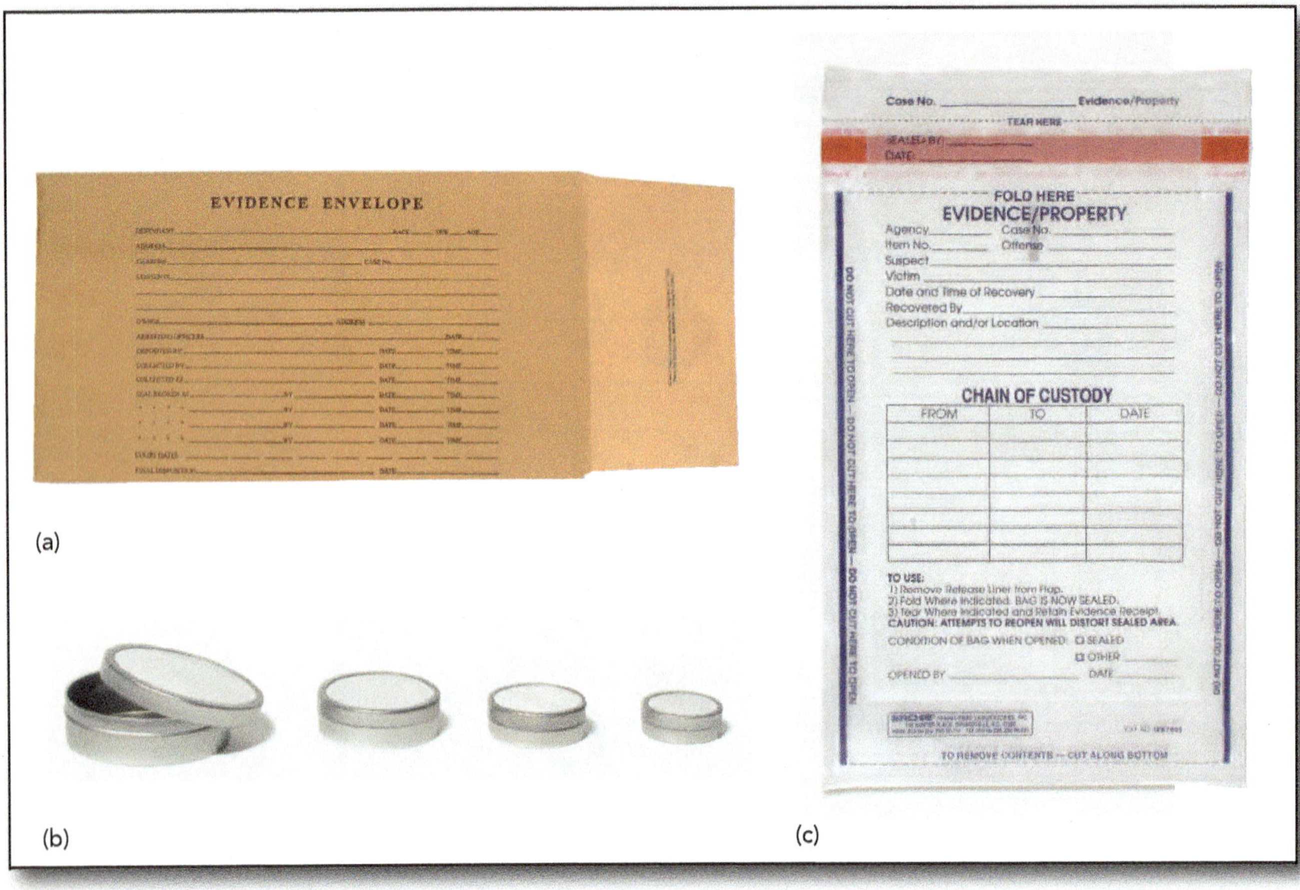

FIGURE 2–9 (a) Manila evidence envelope, (b) metal pillboxes, (c) sealable plastic evidence bag. *Courtesy Sirchie Finger Print Laboratories, Inc., Youngsville, N.C., www.sirchie.com*

FIGURE 2–10 Air-tight metal cans used to package arson evidence. *Courtesy Sirchie Finger Print Laboratories, Inc., Youngsville, N.C., www.sirchie.com*

FIGURE 2–11 A druggist fold is used to package paint transfer evidence. *Courtesy Sirchie Finger Print Laboratories, Inc., Youngsville, N.C., www.sirchie.com*

Biological Materials Use only disposable tools to collect biological materials for packaging. If biological materials such as blood are stored in airtight containers, the accumulation of moisture may encourage the growth of mold, which can destroy their evidential value. In these instances, wrapping paper, manila envelopes, or paper bags are the recommended packaging materials (see Figure 2–12). As a matter of routine, all items possibly containing biological fluid evidence should be airdried and placed individually in separate paper bags to ensure constant circulation of air around them. This will prevent the formation of mold and mildew.

Paper packaging is easily written on, but seals may not be sturdy. Finally, place a red biohazard sticker on both the secured evidence bag and the property receipt to ensure that all handlers will be aware the item is contaminated with biological fluids, such as blood, saliva, or semen (see Table 2–1).

DNA Evidence The advent of DNA analysis brought one of the most significant recent advances in crime-scene investigation. This technique is valuable for making it possible to identify suspects through detecting and analyzing minute quantities of DNA deposited on evidence as a result of contact with saliva, sweat, or skin cells. The search for DNA evidence should include any and all objects with which the suspect or victim may have come into bodily contact. Likely sources of DNA evidence include stamps and envelopes that have been licked, a cup or can that has touched a person's lips, chewing gum, the sweatband of a hat, and a bed sheet containing dead skin cells.

One key concern during the collection of a DNA-containing specimen is contamination. Contamination—in this case, introducing foreign DNA—can

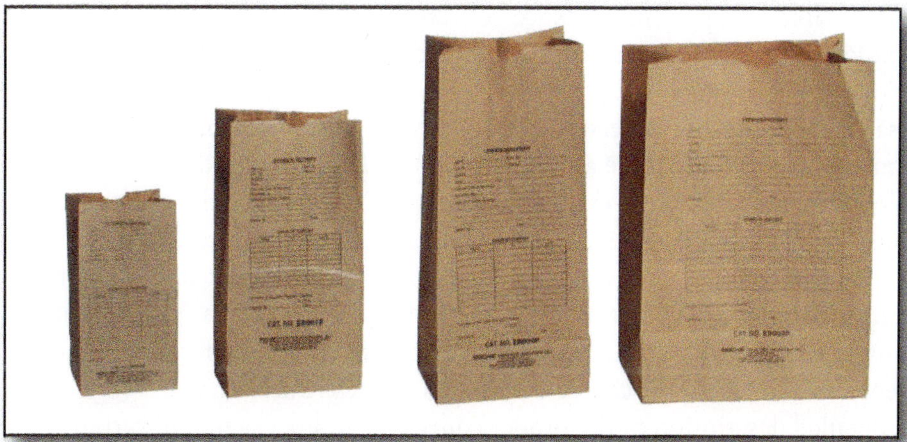

FIGURE 2–12 Paper bags are recommended evidence containers for objects suspected of containing blood and semen stains. Each object should be packaged in a separate bag. *Courtesy Sirchie Finger Print Laboratories, Inc., Youngsville, N.C., www.sirchie.com*

Table 2–1

Best Practices in Biological Evidence Packaging

Containers

- Use paper bags, manila envelopes, cardboard boxes, and similar porous materials for all biological evidence.
- Use butcher paper or art paper for wrapping evidence, for padding in the evidence container, and/or as a general drop cloth to collect trace evidence.
- Package evidence and seal the container to protect it from loss, cross-transfer, contamination, and/or deleterious change.
- For security purposes, seal the package in such a manner that opening it causes obvious damage or alteration to the container or its seal.

Item Packaging

- Package each item separately; avoid commingling items to prevent cross-contamination.
- Use a biohazard label to indicate that a potential biohazard is present.
- Plastic bags are not preferred for storage because of the possibility of bacterial growth or mold.
- If drying wet evidence is not possible, place the evidence in an impermeable, nonporous container and place the container in a refrigerator that maintains a temperature of 2–8°C (approximately 35–46°F) and that is located away from direct sunlight until the evidence can be air-dried or submitted to the laboratory.
- Seal each package with evidence tape or other seals, such as heat seals and gum seals; if possible, do not use staples. Mark across the seal with the sealer's identification or initials and the date.

Reprinted in part from *The Biological Evidence Preservation Handbook: Best Practices for Evidence Handlers*, http://nvlpubs.nist.gov/nistpubs/ir/2013/NIST.IR.7928.pdf.

occur from coughing or sneezing onto evidence during the collection process. Transfer of DNA can also occur when items of evidence are incorrectly placed in contact with each other during packaging. To prevent contamination, the evidence collector must wear a face mask and lab coat, use disposable latex gloves, and work with disposable forceps. The evidence collector must also take into consideration that biological materials, such as dried blood, should be considered potentially infectious. It's recommended practice that the biological evidence collector wear disposable coveralls, shoe covers, and eye protection as an extra precaution to avoid contaminating DNA evidence and being exposed to infectious diseases.

Blood analysis has great evidential value when it allows the investigator to demonstrate a transfer between a victim and a suspect. For this reason, all clothing from both the victim and suspect should be collected and sent to the laboratory for examination, even when the presence of blood on a garment does not appear obvious to the investigator. Laboratory search procedures are far more revealing and sensitive than any that can be conducted at the crime scene.

A detailed description of the proper collection and packaging of various types of physical evidence will be discussed in forthcoming chapters.

Case Files

Scott Peterson: A Case Of Circumstantial Evidence

On the surface, Scott Peterson and his wife, Laci, appeared to live a happy and contented lifestyle in Modesto, California. The 30-year-old Peterson and his 27-yearold former college sweetheart, a substitute teacher, were expecting their first child in about one month when Laci suddenly disappeared. Scott Peterson told investigators that he had last seen his wife on December 24, 2002, at 9:30 a.m. when he left home for a fishing trip off San Francisco Bay.

In April 2003, Laci's decomposed remains washed ashore not far from where Scott said he had gone fishing on the day she vanished. Peterson claimed that Laci was dressed in a white top and black pants when he last saw her, but when her body was found she was wearing khaki pants. Scott's sister recalled that Laci was wearing khaki pants the night before her disappearance.

When questioned, Peterson claimed that he had gone fishing for sturgeon or striped bass. However, the police investigation revealed that he failed to bring the appropriate fishing rod and lines to catch such fish. Further revelations surfaced when it became known that Scott was having an affair with another woman. A search of Scott's warehouse led to the recovery of a black hair on a pair of pliers resting in Scott's boat. A mitochondrial DNA profile of the hair was consistent with Laci's DNA. Scott Peterson was charged with murder and convicted and currently awaits his fate on death row.

From Modesto Police Department Handout

Maintaining the Chain of Custody

chain of custody
A list of all people who came into possession of an item of evidence

In the criminal justice system it is important that investigators establish continuity of possession, or the **chain of custody**. This must be established and documented whenever evidence is presented in court as an exhibit. This means that every person who handled or examined the evidence must be accounted for. At each transfer of evidence, the chain of custody must be updated to describe the condition of the evidence, who received the evidence, dates and times of transfer, who delivered the evidence, and the final disposition of the evidence. Failure to substantiate the evidence's chain of custody may lead to serious questions regarding the authenticity and integrity of the evidence and the examinations of it. Adhering to standard procedures and using proper terminology in documenting the location of evidence, marking it for identification, and properly completing evidence submission forms for laboratory analysis is the best guarantee that the evidence will withstand inquiries of what happened to it from the time of its finding to its presentation in court.

All items of physical evidence should be carefully packaged and marked upon their retrieval at crime sites. This should be done with the utmost care to avoid destroying their evidential value or restricting the number and kind of examinations to which the criminalist may subject them. If possible, the evidence itself should be marked for identification. Normally, the collector's initials and the date of collection are inscribed directly on the article. However, if the evidence collector is unsure of the necessity of marking the item itself, or of where to mark it, it is best to omit this step. Where appropriate, the evidence is to be tagged for identification.

Once an evidence container is selected for the evidence, whether a box, bag, vial, or can, it also must be marked for identification. A minimum record would show the collector's initials, location of the evidence, and date of collection. If the evidence is turned over to another individual for care or delivery to the laboratory, this transfer must be recorded in notes and other appropriate forms. In fact, every individual who possesses the evidence must maintain a written record of its acquisition and disposition. Frequently, all of the individuals involved in the collection and transportation of the evidence may be requested to testify in court. Thus, to avoid confusion and to retain complete control of the evidence at all times, the chain of custody should be kept to a minimum.

Obtaining Standard/Reference Samples

standard/reference sample
Physical evidence whose origin is known, such as blood or hair from a suspect, that can be compared to crime-scene evidence

The examination of evidence, whether soil, blood, glass, hair, fibers, and so on, often requires comparison with a known **standard/reference sample**. Although most investigators have little difficulty recognizing and collecting relevant crime-scene evidence, few seem aware of the necessity and importance of providing the crime lab with a thorough sampling of standard/reference materials. Such materials may be obtained from the victim, a suspect, or other known sources. For instance, investigation of a hit-and-run incident may require the removal of standard/reference paint from a suspect vehicle. This will permit its comparison to paint recovered at the scene.

The presence of standard/reference samples greatly facilitates the work of the forensic scientist. For example, hair found at a crime scene will be of optimum value only when compared to standard/reference hairs removed from the suspect and victim. Likewise, bloodstained evidence must be accompanied by a wholeblood or **buccal swab** standard/reference sample obtained from all relevant crime-scene participants. The quality and quantity of standard/reference specimens often determine the evidential value of crime-scene evidence, and these standard/reference specimens must be treated with equal care.

Some types of evidence must also be accompanied by the collection of **substrate controls**. These are materials close to areas where physical evidence has been deposited. For example, substrate controls are normally collected at arson scenes. If an investigator suspects that a particular surface has been exposed to gasoline or some other accelerant, the investigator should also collect a piece of the same surface material that is believed not to have been exposed to the accelerant. At the laboratory, the substrate control is tested to ensure that the surface on which the accelerant was deposited does not interfere with testing procedures. Another common example of a substrate control is a material on which a bloodstain has been deposited. Unstained areas close to the stain may be sampled to determine whether this material can interfere with the interpretation of laboratory results. Thorough collection and proper packaging of standard/reference specimens and substrate controls are the mark of a skilled investigator.

buccal swab
A swab of the inner portion of the cheek, performed to collect cells for use in determining the DNA profile of an individual

substrate control
Uncontaminated surface material close to an area where physical evidence has been deposited; used to ensure that the surface on which a sample has been deposited does not interfere with laboratory tests

Submitting Evidence

Evidence is usually submitted to the laboratory either by personal delivery or by mail shipment. The method of transmittal is determined by the distance the submitting agency must travel to the laboratory and the urgency of the case. If the evidence is delivered personally, the deliverer should be familiar with the case, to facilitate any discussions between laboratory personnel and the deliverer concerning specific aspects of the case.

If desired, most evidence can be conveniently shipped by mail. However, postal regulations restrict the shipment of certain chemicals and live ammunition and prohibit the mailing of explosives. In such situations, the laboratory must be consulted to determine the disposition of these substances. Care must also be exercised in the packaging of evidence in order to prevent breakage or other accidental destruction while it is in transit to the laboratory.

Most laboratories require that an evidence submission form accompany all evidence submitted (see Figure 2–13). This form must be properly completed. Its information will enable the laboratory analyst to make an intelligent and complete examination of the evidence. Particular attention should be paid to providing the laboratory with a brief description of the case history. The description should be sufficient to provide task relevant information, though attempts should be made to sanitize the narrative of any information that could potentially bias the examiner or analysis. This information will allow the examiner to analyze the specimens in a logical sequence and make the proper comparisons, and it will also facilitate the search for trace quantities of evidence.

The particular kind of examination requested for each type of evidence should be delineated. However, the analyst will not be bound to adhere strictly to the specific tests requested by the investigator. During the examination new evidence may be uncovered, and as a result the complexity of the case may change. Furthermore, the analyst may find the initial requests incomplete or not totally relevant to the case. Finally, a list of items submitted for examination must be included on the evidence submission form. Each item is to be packaged separately and assigned a number or letter, which should be listed in an orderly and logical sequence on the form. Once evidence is received, it will undergo the requested tests and examinations. When a piece of evidence has been fully examined and tested, it must be submitted to long-term storage. The storage area and containers should be secure to prevent tampering and maintain the chain of custody.

Quick Review

- The search pattern selected at a crime scene depends on the size and locale of the scene and the number of collectors participating in the search.
- Many items of evidence may be detected only through examination at the crime laboratory. For this reason, it is important to collect possible carriers of trace evidence, such as clothing, vacuum sweepings, and fingernail scrapings, in addition to more discernible items.
- Each item of physical evidence collected at a crime scene must be placed in a separate appropriate container to prevent damage through contact or cross-contamination.
- Investigators must maintain the chain of custody, a record for denoting the location of the evidence.
- Proper standard/reference samples, such as hairs, blood, and fibers, must be collected at the crime scene and from appropriate subjects for comparison purposes in the laboratory.
- The 4th amendment protects an individual's rights against unlawful search and seizure. Law enforcement must show probable cause to obtain a warrant to search. The warrant describes the particular place, person to be searched and/or things to be seized.
- There are some exceptions when a search can occur without a warrant. These include exigent situations, plain view, consent, and searches during a lawful arrest.

Ensuring Crime-Scene Safety

The increasing spread of AIDS and hepatitis B has sensitized the law enforcement community to the potential health hazards at crime scenes. Law enforcement officers have an extremely small chance of contracting AIDS or hepatitis at the crime scene. Both diseases are normally transmitted by the exchange of body fluids, such as blood, semen, and vaginal and cervical secretions; intravenous drug needles and syringes; and transfusion of infected blood products. However, the presence of blood and semen at crime scenes presents the investigator with biological specimens of unknown origin; the investigator has no way of gauging what health hazards they may contain. Therefore, caution and protection must be used at all times.

CRIME:				COUNTY OF:			
Homicide; Aggravated Sexual Assault				Mercer			
VICTIM:	Age	Sex	Race	SUSPECT:	Age	Sex	Race
Jane Doe (v)	25	F	C	John Doe(s)	30	M	C

SUBMITTING AGENCY: (Address)
Trenton P.D. Central Evidence 225 N. Clinton Avenue, Trenton, N.J. 08609

FORWARD REPLIES TO: (Name)	(Address)	Telephone Number:
Capt. John Smith, Chief of Detectives	Same as above	609 555-5555

INVESTIGATED BY:	DELIVERED BY: (Signature of Person Delivering Evidence)
Detective John Jones	

BRIEF HISTORY OF CASE: (Include Date and Location, if Applicable)
On March 1st, 2001 the victim was found dead in her bedroom. Victim was partially clothed and autopsy revealed that victim was stabbed numerous times and medical examiner stated that there was evidence of sexual assault. Victim was last seen drinking with suspect at a local tavern. Suspect was arrested and item # 23, a folding knife, was found in his pocket.

* Note* Suspect is HIV+

EXAMINATION REQUESTED ON SPECIMENS LISTED BELOW:

Examine items 2, 5, 8-11, 13-15, 18 and 19 for seminal material and compare to controls #'s 12 and 26. Examine #1, 2, 4, 7,13, 14, 15, 18, 20, 21, 22 and 23 for trace evidence transfer. Examine #3 for saliva from suspect. Examine #6 for skin, blood and trace evidence. Examine #'s 20-23 for transfer blood evidence and compare to control #'s 12 and 26.

Item #	* Code	LIST OF SPECIMENS *SOURCE OF EVIDENCE CODE (V-Victim, S-Suspect, SC-Scene)
1	V	Debris collection
2	V	Clothing, white panties
3	V	Dried Secretions/bite marks
4	V	Head Hair Combings
5	V	Oral Specimens
6	V	Fingernail Specimens
7	V	Pubic Hair Combings
8	V	External Genital Specimen
9	V	Vaginal Specimens
10	V	Cervical Specimens

(LABORATORY USE ONLY) DRUG ☐ TRACE ☐ BIO/CHEM ☐ TOX ☐ ABC ☐ EQUINE ☐ BALLISTICS ☐

FOR ADDITIONAL INFORMATION USE FORM 631A AND ATTACH Page 1 of 2 pages

FIGURE 2–13 An example of a properly completed evidence submission form.
Pearson Education Inc

Fortunately, inoculation can easily prevent hepatitis B infection in most people. Furthermore, the federal Occupational Safety and Health Administration (OSHA) requires that law enforcement agencies offer hepatitis B vaccinations to all officers who may have contact with body fluids while on the job, at no expense to the officer.

Each crime scene is unique and carries with it its own collection of hazards. Fortunately, a number of options are available to crime-scene investigators for dealing with the potential hazards that crime scenes present. More frequently than not, once the scene is secured and isolated, it should become apparent that the locale may not contain urgent safety concerns, as in burglaries and car thefts. Nevertheless, routine safety practices must be enforced and appropriate tools and safety equipment must be used. The approved safety equipment and practices are outlined for use in the laboratory, field investigation, or classroom in the Texas Education Agency-approved safety standards. This includes donning latex or nitrite gloves. The latter provides better protection from chemicals. Gloves offer protection from inadvertent contact with blood or other biological materials. They also prevent accidental deposition of fingerprints on objects the scene investigators may touch. Gloves must be changed frequently; in fact, a new pair of gloves must be worn for each item of evidence handled by the investigator. When removed, the gloves must be disposed of in a biohazard bag.

Protective footwear is an important component of the crime-scene investigator's garb. Shoes must be covered with rubber booties when moving about indoors. The investigator should routinely wear shoes or boots that provide good traction and ample support. Inexpensive shoes are recommended, as the investigator must be prepared to dispose of them if they become contaminated with unknown liquids or chemicals.

A basic concern of the crime scene is eye protection. It's appropriate to wear eyeglasses or goggles at crime scenes. However, if concerns exist about encountering splashing liquids, a face shield should be donned to maximize eye and face protection.

Crime scenes that contain the greatest risks to health and safety typically entail exposure to potentially life-threatening biological hazards. These scenes call for maximum respiratory, eye, and skin protection. A wide variety of respiratory masks are available. They include single and double filter masks. Tyvek protective suits are a good option for keeping biohazards off the skin. These suits allow the wearer to move about with ease and flexibility.

When processing and collecting evidence at a crime scene, personnel should be alert to sharp objects, knives, hypodermic syringes, razor blades, and similar items. If such sharp objects are encountered and must be recovered as evidence, the items should be placed in a puncture-resistant container and properly labeled.

When potentially infectious materials are present at a crime scene, personnel should maintain a red biohazard plastic bag for the disposal of contaminated gloves, clothing, masks, pencils, wrapping paper, and so on. On departure from the scene, the biohazard bag must be taped shut and transported to an approved biohazardous waste pickup site.

Quick Review

- Law enforcement officers and crime-scene technicians at a crime scene must use caution and protect themselves at all times from contracting AIDS or hepatitis. Bodily fluids must always be treated as though they were infectious.
- Crime-scene technicians most often use dust particle masks at routine crime scenes. They are considered the most common type of respiratory protection. These masks are considered to be disposable and should be discarded after one use.
- It is recommended that personnel always wear latex gloves and possibly wear chemical-resistant clothing, Tyvek-type shoe covers, a particle mask/respirator, goggles, and possibly face shields when potentially infectious material is present. Gloves should be changed often while processing the scene.
- When processing and collecting evidence at a crime scene, personnel should be alert to sharp objects, knives, hypodermic syringes, razor blades, and similar items.
- Eating, drinking, smoking, and chewing gum are prohibited at the immediate crime scene.

Chapter Review

- Physical evidence includes any and all objects that can establish that a crime has been committed or can link the crime and its victim or its perpetrator.
- Forensic science begins at the crime scene, where investigators must recognize and properly preserve evidence for laboratory examination.
- The first officer to arrive must secure the crime scene.
- Investigators record the crime scene by using photographs, sketches, and notes and make a preliminary examination of the scene as the perpetrator left it.
- The search pattern selected at a crime scene depends on the size and locale of the scene and the number of collectors participating in the search.
- Many items of evidence may be detected only through examination at the crime laboratory. For this reason, it is important to collect possible carriers of trace evidence, such as clothing, vacuum sweepings, and fingernail scrapings, in addition to more discernible items.
- Each item of physical evidence collected at a crime scene must be placed in a separate appropriate container to prevent damage through contact or cross-contamination.
- Investigators must maintain the chain of custody, a record for denoting the location of the evidence.
- Proper standard/reference samples, such as hairs, blood, and fibers, must be collected at the crime scene and from appropriate subjects for comparison purposes in the laboratory.
- Law enforcement officers and crime-scene technicians at a crime scene must use caution and protect themselves at all times from contracting AIDS or hepatitis. Bodily fluids must always be treated as though they were infectious.
- Crime-scene technicians most often use dust particle masks at routine crime scenes. They are considered the most common type of respiratory protection. These masks are considered to be disposable and should be discarded after one use.
- It is recommended that personnel always wear latex gloves and possibly wear chemical-resistant clothing, Tyvek-type shoe covers, a particle mask/respirator, goggles, and possibly face shields when potentially infectious material is present. Gloves should be changed often while processing the scene.
- When processing and collecting evidence at a crime scene, personnel should be alert to sharp objects, knives, hypodermic syringes, razor blades, and similar items.
- Eating, drinking, smoking, and chewing gum are prohibited at the immediate crime scene.

Quick Lab: Crime-Scene Sketch

Materials:

Graph paper
Measuring tape/meter stick
Crime scene tape
Evidence packaging
Notepad
Mock crime scene
L rulers
Tamper evident tape
Metric rulers
Evidence tents
ABFO Scales

Procedure:

You have been introduced to the appropriate steps to process a crime scene. An important part of this process is surveying the scene, taking diligent notes, and creating a sketch of the scene. With a partner or small group, use crime scene tape to indicate the boundaries of your scene. Mark any items of evidence using evidence tents, create a sketch of the scene presented to you and keep notes of what evidence you find. In your sketch, provide an accurate depiction of the entire scene with dimension measurements as well as location measurements for all pieces of physical evidence. Take photographs of the crime scene using the ABFO Scales.

Follow-Up Questions:

1. **Why is it important to take diligent notes when processing the crime scene?**
2. **What is the chain of custody?**
3. **Why do we sketch the crime scene as well as take photographs of it?**
4. **Write a scenario to describe the events that could have occurred in the crime scene you sketched. Use your forensics vocabulary along with basic, clear, language in a meaningful way to describe what could have happened. Be sure to address the following in your scenario: How would you use engineering practices to design a solution and determine whether the scene was an accident or a crime? How would you apply scientific practices to plan and conduct a descriptive, comparative, and experimental investigation to determine the outcome of the case? Read your scenario out loud to a partner or to the class, and answer any questions they have to explain your sketch and the crime scene.**
5. **Choose a crime-scene search pattern for your crime scene. Create a drawing to show how a forensic scientist would search your crime scene using this pattern. Explain why you chose it.**
6. **Create a descriptive scenario to describe the events that could have occurred in the crime scene you sketched.**

Review Questions

1. All of the following are items to be collected from a deceased's body and sent to the forensic laboratory except
   a. head and pubic hairs.
   b. fingernail scrapings.
   c. blood.
   d. all of the above.

2. Which of the following is not an allowance made to justify a warrantless search?
   a. to catch an escaping suspect
   b. the existence of emergency circumstances
   c. a search made by consent of the parties involved
   d. a search made incident to a lawful arrest

3. The most important prerequisite for photographing a crime scene is
   a. to place items in indirect light so that nuances can be picked up in the photograph.
   b. to place rulers or other items in the photographs to show scale.
   c. that the crime scene is in an unaltered condition.
   d. that the photographer start with close-ups of any important object and back up to get the full scale.

4. Uncontaminated surface material close to an area where physical evidence has been deposited is
   a. a standard sample.
   b. a reference sample.
   c. a substrate control.
   d. ground material.

5. True or False: In order for physical evidence to be used effectively for aiding an investigator, it must first be photographed and tagged as evidence and turned over to the officer in charge of the crime scene.

6. True or False: Before taking photographs of the crime scene, the investigator first draws a rough sketch, containing an accurate depiction of the dimensions of the scene and showing the location of all objects.

7. True or False: If bloodstained materials are stored in airtight containers, such as pill bottles or vials, the accumulation of moisture may encourage the growth of mold, which can destroy the evidential value of blood.

8. True or False: The examination of evidence requires comparison with a substrate control to ensure the evidentiary value of the crime-scene evidence.

9. True or False: Charred debris recovered from the scene of a suspicious fire may be sealed in a plastic evidence container.

10. What is physical evidence? Give three examples of physical evidence that may be found at a crime scene.

11. What is the first step in processing a crime scene?

12. Why is it important to exclude onlookers from a crime scene?

13. Name three methods for recording a crime scene and list one unique advantage of each.

14. What is the most important prerequisite for video recording a crime scene?

15. Name and describe the two kinds of sketches prepared for a crime scene.

16. What information must be included in written technical notes made at the crime scene?

17. List at least four pieces of evidence from a murder scene that must be collected and sent to the forensic laboratory. Why must these items be collected at the crime scene instead of later?

18. What is the best way to maintain the integrity of physical evidence?

19. Describe the best way for an investigator to handle trace evidence found on articles collected at the crime scene. Name two exceptions to this general rule.

20. Describe a druggist's fold and explain why it is a better way to store trace evidence than an ordinary mailing envelope.

21. Why should all items of evidence be placed in separate containers?

22. Name one type of evidence that should be stored in an airtight container and one type that should not be stored in such a container. Explain why each type of evidence should be stored (or not stored) in this way.

23. In the criminal justice system, why is it important to maintain chain of custody? Employing appropriate terminology, identify each procedure pertaining to the chain of custody and, in writing, explain how it's important.

24. What is a substrate control and how is it used by the crime-scene investigator?

25. Discuss how you could develop a plan to use a model of a crime scene to address each of the following engineering problems:
    a. A solution for a car malfunction at the scene of a car accident
    b. The process of a collision that caused a building to collapse

    Write a short summary of your findings.

Learn the Vocabulary

Use the information from this chapter along with your prior knowledge and experience to learn the key terms. Choose from the following activities:

1. Make flashcards with a term on the front and its definition on the back. Use the flashcards to quiz a partner on correctly pronouncing and defining the terms.

2. Make an audio or video recording of each term and its definition.

3. Work with a partner to write a sentence for each term. Read the sentences out loud to a partner or share them with the class.

4. Write a list of new terms and language you have encountered in this chapter. Reuse the terms to write sentences and then read the sentences out loud to a partner. Reuse the terms again by writing a short speech. The speech might explain the terms or teach a lesson using the terms. Give the speech to a partner or to the class.

Application and Critical Thinking

1. Give at least three examples of how evidence may be destroyed by onlookers at a crime scene.

2. What important elements are missing from the following crime-scene sketch?

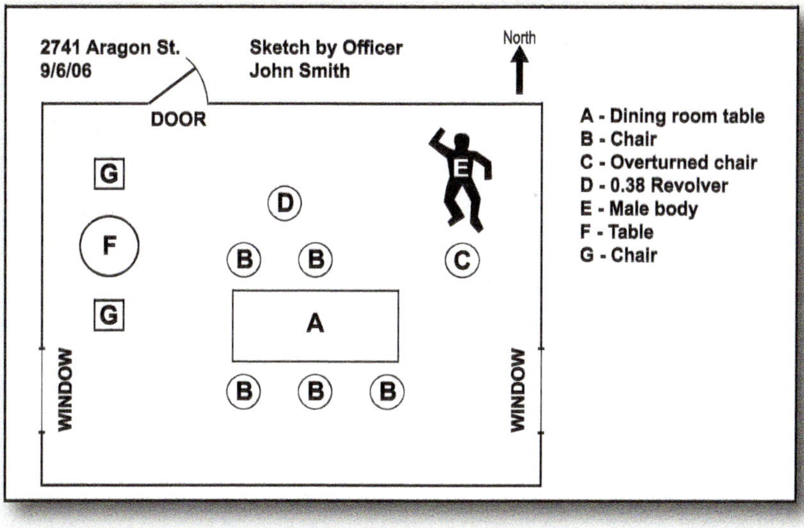

3. An investigator at a murder scene notes signs of a prolonged struggle between the attacker and victim. Name at least three types of physical evidence for which the investigator would likely collect standard/reference samples and explain why the investigator would collect them.

4. Explain the procedures, using correct terminology employed by the criminal justice system, that must be followed to maintain the chain of custody for items of evidence. Create a visual, such as a poster or slide presentation, that identifies each step in the process. Be sure to use correct terminology

Laboratory Experiments

This activity requires the use of the following practices of science and engineering:

- Planning and carrying out investigations
- Obtaining, evaluating, and communicating information
- Using appropriate safety equipment and practices during laboratory/classroom and field investigations as outlined in Texas Education Agency-approved safety standards

This activity consists of the following crosscutting concepts:

- **Patterns**—the students will use search patterns to organize the discovery of evidence in crime scene.
- **Scale, Proportion and Quantity**—students will use scale drawing to represent a crime scene.

Before you begin, read the scenario and steps out loud with a partner. Ask for clarification as necessary to make sure you understand the experiment.

The Scenario

You and your team are called to the scene of a robbery at the 200 block of South Main St. at 3:24 AM on Sunday, December 27. The business owner was alerted by a monitored alarm system that someone had entered the building. Instead of immediately calling the police, the owner went to his business on his own to investigate what happened. Upon arrival, he noticed a broken window and the rear door open. The owner decided to call the police. It is your team's responsibility to search and document the crime scene. You will be given 15 minutes in the crime scene to search and make a rough sketch. Generate a finished sketch of the entire crime scene. Next, organize the qualitative data you collected and communicate this data by making a table or report (paragraph). Then organize the quantitative data you collected and communicate this data by making a table. Finally, present to the class what your team found and what conclusions you made from your initial search of the scene. The information should be communicated in an appropriate format, such as an oral report.

1. Use the knowledge you have gained in this chapter to demonstrate as a team how to process the crime scene described in the scenario. As part of your analysis, explain the role and task of each member of your team—including any outside experts and agencies that you will collaborate with—and describe their role in processing and examining the crime scene. As part of a team, your group will need to:
   A. Develop a detailed, technical written record documenting, based on observations and activities, what was done as part of the crime scene examination.
   B. Determine which search pattern would be best to locate evidence for the type of scene you are investigating. Utilize this method to conduct a primary search followed up by a secondary search pattern to ensure no physical evidence was overlooked. Be sure to document all evidence located in your detailed notes.
   C. Document the crime scene and all items of evidence using appropriate photographic or audiovisual equipment.

D. Generate a physical or digital crime scene sketch, including coordinates or measurements from fixed points, compass directions, scale of proportion, legend-key, heading, and title block.

E. Demonstrate proper techniques as you collect, package, and preserve any physical evidence found at the scene. Document each step in the process. Maintain and document chain of custody for each item of evidence collected.

2. Working with a partner describe and discuss, using details and specifics, the elements of criminal law that guide search and seizure of persons, property, and evidence. Working with a partner, discuss your findings. Write a brief summary of the elements discussed.

3. What are the advantages and limitations of creating a model (sketch) of the crime scene? What are ways that you can identify these advantages and limitations? How are size, scale, properties, and materials relevant?

Physical Evidence

3

Andrew Francis Wallace/Toronto Star/Getty Images

Key Terms

class characteristics
comparison
identification
individual characteristics
product rule
Rapid DNA

Bruce McArthur: A Mountain of Physical Evidence

Over the course of a decade, gay men were vanishing off the streets of Toronto, Canada. The last in the string of unexplained disappearances was Andrew Kinsman, who went missing on January 26, 2017. As part of the investigation, police combed through his belongings for information on his whereabouts the day he went missing. On his calendar, there was just one entry—the name "Bruce."

Police were able to pull surveillance footage of Kinsman getting into a red 2004 Dodge Caravan on the day he disappeared. When investigators cross-referenced those pieces of information, there was only one person named "Bruce" who owned a red 2004 Dodge Caravan.

Police began to monitor Bruce McArthur, a 66-year-old landscaper from Toronto. After weeks of surveillance, officers observed a young man entering McArthur's apartment. Fearing for the man's safety, police decided to make entry. Once inside, they discovered the young man tied up but otherwise unharmed. It was at this time that Bruce McArthur was taken into custody on suspicion of murder. The subsequent search yielded a trove of physical evidence that linked McArthur to the killings of eight men with ties to Toronto's Gay Village. A search of his van yielded blood that matched the DNA profiles of some of his victims. Police found a black duffel bag containing duct tape, a surgical glove, rope, zip ties, a black bungee cord, and syringes in McArthur's bedroom. He shaved his victims' heads and beards before burying them and kept hair from the men in bags. He buried the remains of his dismembered victims in planter pots at the home of a landscaping client.

In the face of the "unprecedented amount of real, forensic, digital and documentary evidence" described by the prosecutor in the case, Bruce McArthur pled guilty to eight counts of murder in a Toronto courtroom on January 29, 2019. During the sentencing hearing, the prosecutor walked through each piece of physical evidence amassed by investigators, stating: "For years, members of the LGBTQ community in Toronto believed they were being targeted by a killer and they were right." McArthur was sentenced to life in prison without the possibility of parole for his crimes.

Learning Objectives

After studying this chapter, you should be able to:

3.1 Review the common types of physical evidence encountered at crime scenes

3.2 Summarize the significance of physical evidence and the steps involved in its analysis

3.3 Explain the function of national databases available to forensic scientists

National Science Content Standards

Scientific Inquiry

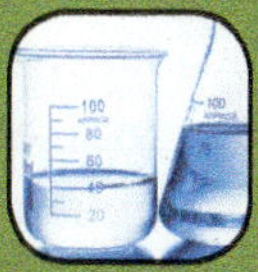
Physical Science

Science in Personal and Social Perspective

History and Nature of Science

It would be impossible to list all the objects that could conceivably be important to a crime; every crime scene obviously has to be treated on an individual basis, having its own peculiar history, circumstances, and problems. It is practical, however, to list items whose scientific examination is likely to yield significant results in ascertaining the nature and circumstances of a crime. The investigator who is thoroughly familiar with the recognition, collection, and analysis of these items, as well as with laboratory procedures and capabilities, can make logical decisions when the uncommon and unexpected are encountered at the crime scene. Just as important, a qualified evidence collector cannot rely on collection procedures memorized from a pamphlet but must be able to make innovative, on-the-spot decisions at the crime scene.

Common Types of Physical Evidence

Blood, semen, and saliva All suspected blood, semen, or saliva—liquid or dried, animal or human—present in a form that suggests a relation to the offense or the people involved in a crime. This category includes blood or semen dried onto fabrics or other objects, as well as cigarette butts that may contain saliva residues. These substances are subjected to serological and biochemical analysis to determine their identity and possible origin.

Documents Any handwriting and typewriting submitted so that authenticity or source can be determined. Related items include paper, ink, indented writings, obliterations, and burned or charred documents.

Drugs Any substance seized in violation of laws regulating the sale, manufacture, distribution, and use of drugs.

Explosives Any device containing an explosive charge, as well as all objects removed from the scene of an explosion that are suspected to contain the residues of an explosive.

Fibers Any natural or synthetic fiber whose transfer may be useful in establishing a relationship between objects and/or people.

Fingerprints All prints of this nature, hidden (latent) and visible.

Firearms and ammunition Any firearm, as well as discharged or intact ammunition, suspected of being involved in a criminal offense.

Glass Any glass particle or fragment that may have been transferred to a person or object involved in a crime. Windowpanes containing holes made by a bullet or other projectile are included in this category.

Hair Any animal or human hair present that could link a person with a crime.

Impressions Tire markings, shoe prints, depressions in soft soils, and all other forms of tracks. Glove and other fabric impressions, as well as bite marks in skin or foodstuffs, are also included.

Organs and physiological fluids Body organs and fluids are submitted for analysis to detect the possible existence of drugs and poisons. This category includes blood to be analyzed for the presence of alcohol and other drugs.

Paint Any paint, liquid or dried, that may have been transferred from the surface of one object to another during the commission of a crime. A common example is the transfer of paint from one vehicle to another during an automobile collision.

Petroleum products Any petroleum product removed from a suspect or recovered from a crime scene. The most common examples are gasoline residues removed from the scene of an arson, or grease and oil stains whose presence may suggest involvement in a crime.

Plastic bags A disposable polyethylene bag such as a garbage bag may be evidential in a homicide or drug case. Examinations are conducted to associate a bag with a similar bag in the possession of a suspect.

Plastic, rubber, and other polymers Remnants of these manufactured materials recovered at crime scenes may be linked to objects recovered in the possession of a criminal suspect.

Powder residues Any item suspected of containing powder residues resulting from the discharge of a firearm (see Figure 3–1).

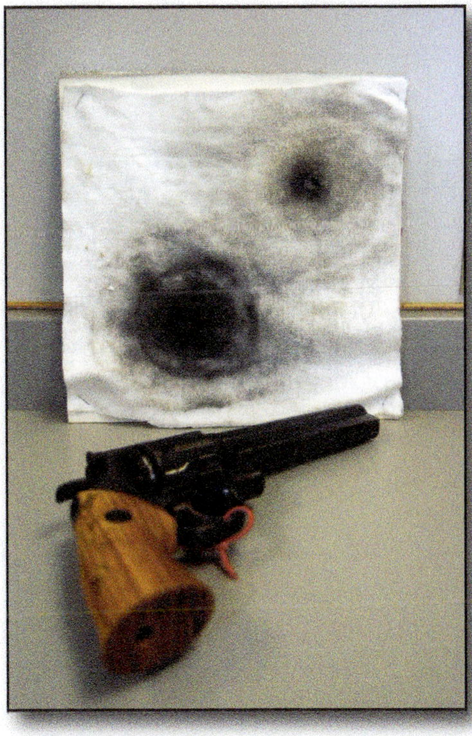

FIGURE 3–1 The gun is fired at a set distance from the target, and the gunpowder left on the target is compared to powder stains found on a victim's clothing. The density and shape of the powder stains vary with the distance the gun was fired. *Mikael Karlsson/Alamy Stock Photo*

Serial numbers This category includes all stolen property submitted to the laboratory for the restoration of erased identification numbers.

Soil and minerals All items containing soil or minerals that could link a person or object to a particular location. Common examples are soil embedded in shoes and insulation found on garments.

Toolmarks This category includes any objects suspected of containing the impression of another object that served as a tool in a crime. For example, a screwdriver or crowbar could produce tool marks by being impressed into or scraped along a surface of a wall.

Vehicle lights Examination of vehicle headlights and taillights is normally conducted to determine whether a light was on or off at the time of impact.

Wood and other vegetative matter Any fragments of wood, sawdust, shavings, or vegetative matter discovered on clothing, shoes, or tools that could link a person or object to a crime location.

Quick Review

- Biological crime-scene evidence includes blood, saliva, semen, DNA, hair, organs, and physiological fluids.
- Impression crime-scene evidence includes tire markings, shoe prints, depressions in soft soils, all other forms of tracks, glove and other fabric impressions, tool marks, and bite marks.
- Manufactured items considered common items of crime-scene evidence include firearms, ammunition, fibers, paint, glass, petroleum products, plastic bags, rubber, polymers, and vehicle headlights.

The Examination of Physical Evidence

Forensic scientists have to analyze, evaluate, and critique scientific explanations by using empirical evidence, observational testing, and logical reasoning. Physical evidence is usually examined by a forensic scientist for identification or comparison.

Identification

identification
The process of determining a substance's physical or chemical identity

Identification has as its purpose the determination of the physical or chemical identity of a substance with as near absolute certainty as existing analytical techniques will permit. For example, the crime laboratory is frequently asked to identify the chemical composition of preparations that may contain illicit drugs such as heroin, cocaine, or barbiturates. It may be asked to identify gasoline in residues recovered from the debris of a fire, or it may have to identify the nature of explosive residues—for example, dynamite or TNT. Also, the identification of blood, semen, hair, or wood would, as a matter of routine, include a determination of species origin. For example, did a bloodstain originate from a human as opposed to a dog or cat? Each of these requests requires the analysis and ultimate identification of a specific physical or chemical substance to the exclusion of all other possible substances.

The process of identification first requires the adoption of testing procedures that give characteristic results for specific standard materials. Once these test results have been established, they may be permanently recorded and used repeatedly to prove the identity of suspect materials. For example, to ascertain that a particular suspect powder is heroin, the test results on the powder must be identical to those that have been previously obtained from a known heroin sample.

Second, identification requires that the number and type of tests needed to identify a substance be sufficient to exclude all other substances. This means that the examiner must devise a specific analytical scheme that will eliminate all but one substance from consideration. Hence, if the examiner concludes that a white powder contains heroin, the test results must have been comprehensive enough to have excluded all other drugs—or, for that matter, all other substances—from consideration.

Simple rules cannot be devised for defining what constitutes a thorough and foolproof analytical scheme. Each type of evidence obviously requires different tests, and each test has a different degree of specificity. Thus, one substance could conceivably be identified by one test, whereas another may require the combination of five or six different tests to arrive at an identification. In a science in which the practitioner has little or no control over the quality and quantity of the specimens received, a standard series of tests cannot encompass all possible problems and pitfalls. So the forensic scientist must determine at what point the analysis can be concluded and the criteria for positive identification satisfied; for this, the forensic scientist must rely on knowledge gained through education and experience. Ultimately, the conclusion will have to be substantiated beyond any reasonable doubt in a court of law.

Comparison

A **comparison** analysis subjects a suspect specimen and a standard/reference specimen to the same tests and examinations in order to determine whether they have a common origin. For example, the forensic scientist may place a suspect at a particular location by noting the similarities of a hair found at the crime scene to hairs removed from a suspect's head (Figure 3–2). Or a paint chip found on a hit-and-run victim's garment may be compared with paint removed from a vehicle suspected of being involved in the incident.

comparison
The process of ascertaining whether two or more objects have a common origin

The forensic comparison is actually a two-step procedure. First, combinations of select properties are chosen from the suspect and the standard/reference specimen for comparison. The question of which and how many properties are selected obviously depends on the type of materials being examined. (This subject

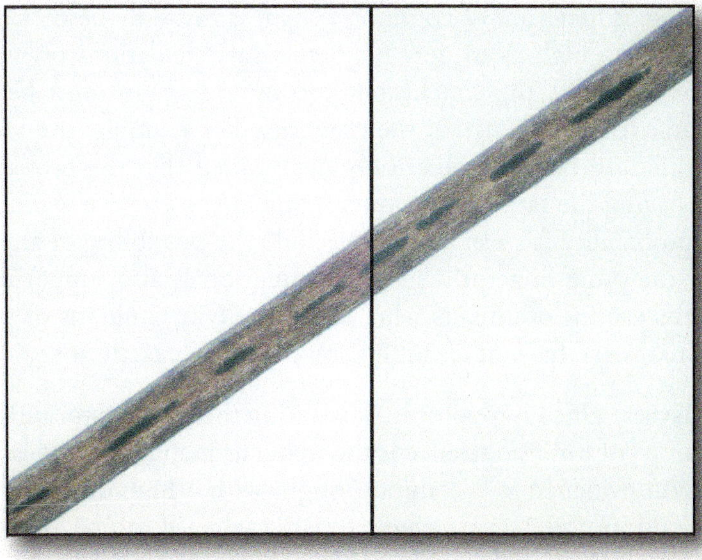

FIGURE 3–2
Side-by-side comparison hairs. *Courtesy Chris Palenik, Microtrace, Elgin, IL., www.microtracescientific.com*

will receive a good deal of discussion in forthcoming chapters.) The overriding consideration must be the ultimate evidential value of the conclusion.

This brings us to the second objective. Once the examination has been completed, the forensic scientist must draw a conclusion about the origins of the specimens. Do they or do they not come from the same source? Certainly if one or more of the properties selected for comparison do not agree, the analyst will conclude that the specimens are not the same and hence could not have originated from the same source. Suppose, on the other hand, that all the properties do compare and the specimens, as far as the examiner can determine, are indistinguishable. Does it logically follow that they come from the same source? Not necessarily so.

To comprehend the evidential value of a comparison, one must appreciate the role that probability has in ascertaining the origins of two or more specimens. Simply defined, probability is the frequency of occurrence of an event. If a coin is flipped 100 times, in theory we can expect heads to come up 50 times. Hence, the probability of the event (heads) occurring is 50 in 100. In other words, probability defines the odds that a certain event will occur. When making a comparison and analyzing data, forensic professionals must be able to identify significant statistical features, patterns, sources of error, and limitations to the methods being used. When analyzing data, it is important to identify results that are reliable and not due to chance, meaning they are statistically significant. Patterns in the data must be recognized to understand if the data is true or skewed in one direction. Sources of error, such as procedural, instrumental, environmental, or human, must be avoided. There also must be an understanding of the limitations to the conclusions that can be made based on the data results.

individual characteristics
Properties of evidence that can be attributed to a common source with an extremely high degree of certainty

Individual Characteristics Evidence that can be associated with a common source with an extremely high degree of probability is said to possess **individual characteristics**. Examples of such associations are the matching ridge characteristics of two fingerprints, the comparison of random striations (markings) on bullets or tool marks, the comparison of irregular and random wear patterns in tire or footwear impressions, the comparison of handwriting characteristics, the fitting together of the irregular edges of broken objects in the manner of a jigsaw puzzle (see Figure 3–3), or matching sequentially made plastic bags by striation marks running across the bags (see Figure 3–4).

In all of these cases, it is not possible to state with mathematical exactness the probability that the specimens are of common origin; it can only be concluded that this probability is so high as to defy mathematical calculations or human comprehension. Furthermore, the conclusion of common origin must be substantiated by the practical experience of the examiner. For example, the French scientist Victor Balthazard has mathematically determined that the probability of two individuals having the same fingerprints is one in 1×10^{60}, or 1 followed by 60 zeros. This probability is so small as to exclude the possibility of any two individuals having the same fingerprints. This contention is also supported by the experience of fingerprint examiners who, after classifying millions of prints over the past hundred years, have never found any two to be exactly alike.

class characteristics
Properties of evidence that can be associated only with a group and never with a single source

Class Characteristics One disappointment awaiting the investigator unfamiliar with the limitations of forensic science is the frequent inability of the laboratory to relate physical evidence to a common origin with a high degree of certainty. Evidence is said to possess **class characteristics** when it can be associated

FIGURE 3–3 The body of a woman was found with evidence of a stablike wound in the neck. A pathologist found a knife blade tip in the wound in the neck. The knife blade tip was compared with the broken blade of a knife found in the trousers pocket of the accused. A close examination reveals the fit of the indentations on the edges and individual characteristics of stria from the sharpening process. *Courtesy Peter Diaczuk, John Jay College, City University of New York*

FIGURE 3–4 The bound body of a young woman was recovered from a river. Her head was covered with a black polyethylene trash bag (shown on the right). Among the items recovered from one of several suspects was a black polyethylene trash bag (shown on the left). A side-by-side comparison of the two bags' extrusion marks and pigment bands showed them to be consecutively manufactured. This information allowed investigators to focus their attention on one suspect, who ultimately was convicted of the homicide. *Courtesy Richard Saferstein, Ph.D*

only with a group and never with a single source. Here again, probability is a determining factor. For example, if we compare two single-layer automobile paint chips of a similar color, their chance of originating from the same car is not nearly as great as when we compare two paint chips with seven similar layers of paint, not all of which were part of the car's original color. The former will have class characteristics and could be associated at best only with one car model (which may number in the thousands), whereas the latter may be judged to have individual characteristics and to have a high probability of originating from one specific car. Blood offers another good example of evidence that can have class characteristics. For example, suppose that two blood specimens are compared and both are found to be of human origin, type A. The frequency of occurrence in the population of type A blood is 26 percent—hardly offering a basis for establishing the common origin of the stains. However, if other blood factors are also determined and are found to compare, the probability that the two blood

product rule
A formula for determining how frequently a certain combination of characteristics occurs in a population; the product rule states that one must first determine the probability of each characteristic occurring separately, then multiply together the frequencies of all independently occurring characteristics; the result is the overall frequency of occurrence for that particular combination of characteristics

specimens originated from a common source increases. Thus, if one uses a series of blood factors that occur independently of each other, then one can apply the **product rule** to calculate the overall frequency of occurrence of the blood in a population. In this case, the product rule states that multiplying together the frequency of each factor present in a blood sample will determine how common blood containing that combination of factors is in the general population.

For example, in the O. J. Simpson murder case, a bloodstain located at the crime scene was found to contain a number of factors that compared to O. J.'s blood:

Blood Factors	Frequency
A	26%
EsD	85%
PGM 2+2–	2%

The product of all the frequencies shown in the table determines the probability that any one individual possesses such a combination of blood factors. In this instance, applying the product rule, 0.26 × 0.85 × 0.02 equals 0.0044. Thus, only 0.44 percent, or less than 1 in 200 people, would be expected to have this particular combination of blood factors. These bloodstain factors did not match either of the two victims, Nicole Brown Simpson or Ronald Goldman, thus eliminating them as possible sources of the blood. Although the forensic scientist has still not individualized the bloodstains to one person (in this case, O. J. Simpson), data have been provided that will permit investigators and the courts to better assess the evidential value of the crime-scene stain.

As we will learn in Chapter 10, the product rule is used to determine the frequency of occurrence of DNA profiles typically determined from blood and other biological materials. Importantly, modern DNA technology provides enough factors to allow an analyst to individualize blood, semen, and other biological materials to a single person. Figure 3–5 shows a computer generated image of DNA superimposed on a fingerprint representing two of the most frequently found individualized items of evidence at crime scenes.

Quick Review

- Two methods used by forensic scientists when examining physical evidence are identification and comparison.
- Identification is the process of determining a substance's chemical or physical identity to the exclusion of all other substances; for example, drugs, explosives, petroleum products, blood, semen, and hair species.
- A comparison analysis determines whether a suspect specimen and a standard/reference specimen have a common origin.
- Evidence that can be associated with a common source with an extremely high degree of probability is said to possess individual characteristics.
- Evidence associated with only a group is said to have class characteristics.
- The overall frequency of occurrence of an event can be obtained by multiplying together the frequencies of all independently occurring instances related to that event. This is known as the product rule.

Assessing the Significance of Physical Evidence

One of the current weaknesses of forensic science is the inability of the examiner to assign exact or even approximate probability values to the comparison of most class physical evidence. When analyzing evidence, a forensic scientist must identify the problem, ask questions, and define problems about different pieces of evidence from an investigation based on observations. For example, what is the probability that a nylon fiber originated from a particular sweater or that a hair came from a particular person's head or that a paint chip came from a car suspected to have been involved in a hit-and-run accident? Few statistical data are available from which to derive this information, and in a society that is increasingly dependent on mass-produced products, the gathering of such data is becoming an increasingly elusive goal.

One of the primary endeavors of forensic scientists must be to create and update statistical databases for evaluating the significance of class physical evidence. Of course, when such information—for example, the population frequency of blood factors—is available, it is used; but for the most part, the forensic scientist must rely on personal experience when interpreting the significance of class physical evidence.

People who are unfamiliar with the realities of modern criminalistics are often disappointed to learn that most items of physical evidence retrieved at crime scenes cannot be linked definitively to a single person or object. Although investigators always try to uncover physical evidence with individual characteristics—such as fingerprints, tool marks, and bullets—the chances of finding class physical evidence are far greater. To deny or belittle the value of such evidence is to reject the potential role that criminalistics can play in a criminal investigation.

FIGURE 3–5 A computer-generated image of DNA superimposed on a fingerprint representing two of the most frequently found individualized items of evidence at crime scenes. *Alfred Pasieka / Science Source*

Often, most of the evidence gathered is subjective in nature, prone to human error and bias. The extent to which cognitive biases may influence decision-making in forensic science is an important question with implications for training and practice. There have been a number of studies that support the idea of susceptibility of forensic science practitioners to various types of confirmation bias. As a result, it has been recommended that forensic laboratories implement procedures designed to reduce access to unnecessary information, use of multiple comparison samples rather than a single suspect exemplar, and replication of results by analysts blinded to previous results.

The believability of eyewitness accounts, confessions, and informant testimony can all be disputed, maligned, and subjected to severe attack and skepticism in the courtroom. Under these circumstances, errors in human judgment are often magnified to detract from the credibility of the witness.

Assessing the Value of Physical Evidence

The value of class physical evidence lies in its ability to corroborate events with data in a manner that is, as nearly as possible, free of human error and bias. It is the thread that binds together other investigative findings that are more dependent on human judgments and, therefore, more prone to human failings. The fact that scientists have not yet learned to individualize many kinds of physical evidence means that criminal investigators should not abdicate or falter in their pursuit of all investigative leads. However, the ability of scientists to achieve a high degree of success in evaluating class physical evidence means that criminal investigators can pursue their work with a much greater chance of success.

Admittedly, defining the significance of an item of class evidence in exact mathematical terms is usually a difficult if not impossible goal. Although class evidence is by its nature not unique, meaningful items of physical evidence, such as those listed at the beginning of this chapter, are extremely diverse in our environment. Select, for example, a colored fiber from an article of clothing and try to locate the exact same color on the clothing of random individuals you meet, or select a car color and try to match it to other automobiles you see on local streets. Furthermore, keep in mind that a forensic comparison goes beyond a mere color comparison and involves examining and comparing a variety of chemical and/or physical properties (Figure 3–6). The chances are low of encountering two indistinguishable items of physical evidence at a crime scene that actually originated from different sources. Obviously, given these circumstances, only objects that exhibit significant diversity are appropriate for classification as physical evidence.

In the same way, when one is dealing with more than one type of class evidence, their collective presence may lead to an extremely high certainty that they originated from the same source. As the number of different objects linking an individual to a crime increases, the probability of involvement increases dramatically. A classic example of this situation can be found in the evidence presented at the trial of Wayne Williams. Wayne Williams was charged with the murders of two individuals in the Atlanta, Georgia, metropolitan area; he was also linked to the murders of ten other boys and young men. An essential element of the state's case involved the association of Williams with the victims through a variety of fiber evidence. Twenty-eight different types of fibers linked Williams to the murder victims, evidence that the forensic examiner characterized as "overwhelming."

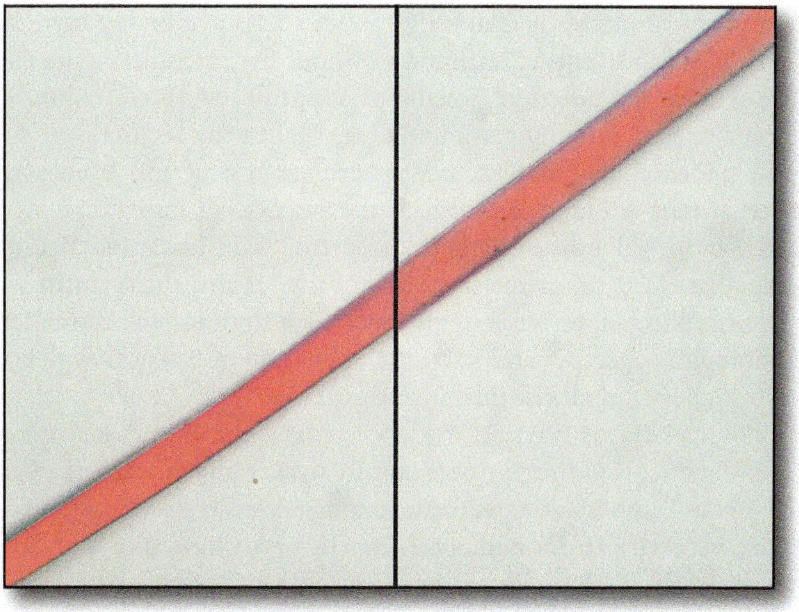

FIGURE 3–6 Side-by-side comparison of fibers. *Courtesy Chris Palenik, Microtrace, Elgin, IL., www.microtracescientific.com*

Before moving beyond the subject of the nature and values that contrast individual and class evidence, one major concern must be addressed. In 2009, the report entitled *Strengthening Forensic Science in the United States: A Path Forward* was issued by the National Research Council (NRC) of the National Academy of Sciences.[1] The report addressed an overarching concern that has permeated the foundation of forensic evidence; that is, whether certain items of class or individual evidence is worthy of admission in a court of law. Putting aside concerns about the lack of a statistical basis to define probabilities to support the comparative significance of many types of physical evidence, the NRC report decries the fact that many forensic determinations involve subjective evaluations, whose correctness is not readily verifiable and whose accuracy is highly dependent on the experience and training of the examiner. While the NRC report encourages research to put forensic science on a more objective footing, this goal will be painstakingly slow. Alternatively, crime laboratories have been encouraged to implement quality assurance measures to measure the competency of their examiners; these include peer review, proficiency testing, analyst certification, and periodic external audits.

Cautions and Limitations in Dealing with Physical Evidence

In further evaluating the contribution of physical evidence, one cannot overlook one important reality in the courtroom: the weight or significance accorded physical evidence is a determination left entirely to the trier of fact, usually a jury of laypeople. Given the high esteem in which scientists are generally held by society and the infallible image created for forensic science by books and television, scientifically evaluated evidence often takes on an aura of special reliability and trustworthiness in the courtroom. Often physical evidence, whether individual or class, is accorded great weight during jury deliberations and becomes a primary factor in reinforcing or overcoming lingering doubts about guilt or innocence.

In fact, a number of jurists have already cautioned against giving carte blanche approval to admitting scientific testimony without first considering its relevance to the case. Given the potential weight of scientific evidence, failure to take proper safeguards may unfairly prejudice a case against the accused.

Physical evidence may also exclude or exonerate a person from suspicion. For instance, if type A blood is linked to the suspect, all individuals who have types B, AB, and O blood can be eliminated from consideration. Because it is not possible to assess at the crime scene what value, if any, the scientist will find in the evidence collected, or what significance such findings will ultimately have to a jury, thorough collection and scientific evaluation of physical evidence must become a routine part of all criminal investigations.

Just when an item of physical evidence crosses the line that distinguishes class from individual is a difficult question to answer and is often the source of heated debate and honest disagreement among forensic scientists. How many striations are necessary to individualize a mark to a single tool and no other? How many color layers individualize a paint chip to a single car? How many ridge characteristics individualize a fingerprint, and how many handwriting characteristics tie a person to a signature? These questions defy simple answers. The task of the forensic scientist is to find as many characteristics as possible to compare one substance with another. The significance attached to the findings is decided by the quality and composition of the evidence, the case history, and the examiner's experience. Ultimately, the conclusion can range from mere speculation to near certainty.

There are practical limits to the properties and characteristics the forensic scientist can select for comparison. Carried to the extreme, no two things in this world are alike in every detail. Modern analytical techniques have become so sophisticated and sensitive that the criminalist must define the limits of natural variation among materials when interpreting the data gathered from a comparative analysis. For example, we will learn in Chapter 5 that two properties, density and refractive index, are best suited for comparing two pieces of glass. But the latest techniques that have been developed to measure these properties are so sensitive that they can even distinguish glass originating from a single pane of glass. Certainly this goes beyond the desires of a criminalist trying to determine only whether two glass particles originated from the same window. Similarly, if the surface of a paint chip is magnified 1,600 times with a powerful scanning electron microscope, it is apparent that the fine details that are revealed could not be duplicated in any other paint chip. Under these circumstances, no two paint chips, even those coming from the same surface, could ever compare in the true sense of the word. Therefore, practicality dictates that such examinations be conducted at a less revealing, but more meaningful, magnification (see Figure 3–7).

Distinguishing evidential variations from natural variations is not always an easy task. Learning how to use the microscope and all the other modern instruments in a crime laboratory properly is one thing; gaining the proficiency needed to interpret the observations and data is another. As new crime laboratories are created and others expand to meet the requirements of the law enforcement community, many individuals are starting new careers in forensic science. They must be cautioned that merely reading relevant textbooks and journals is no substitute for experience in this most practical of sciences.

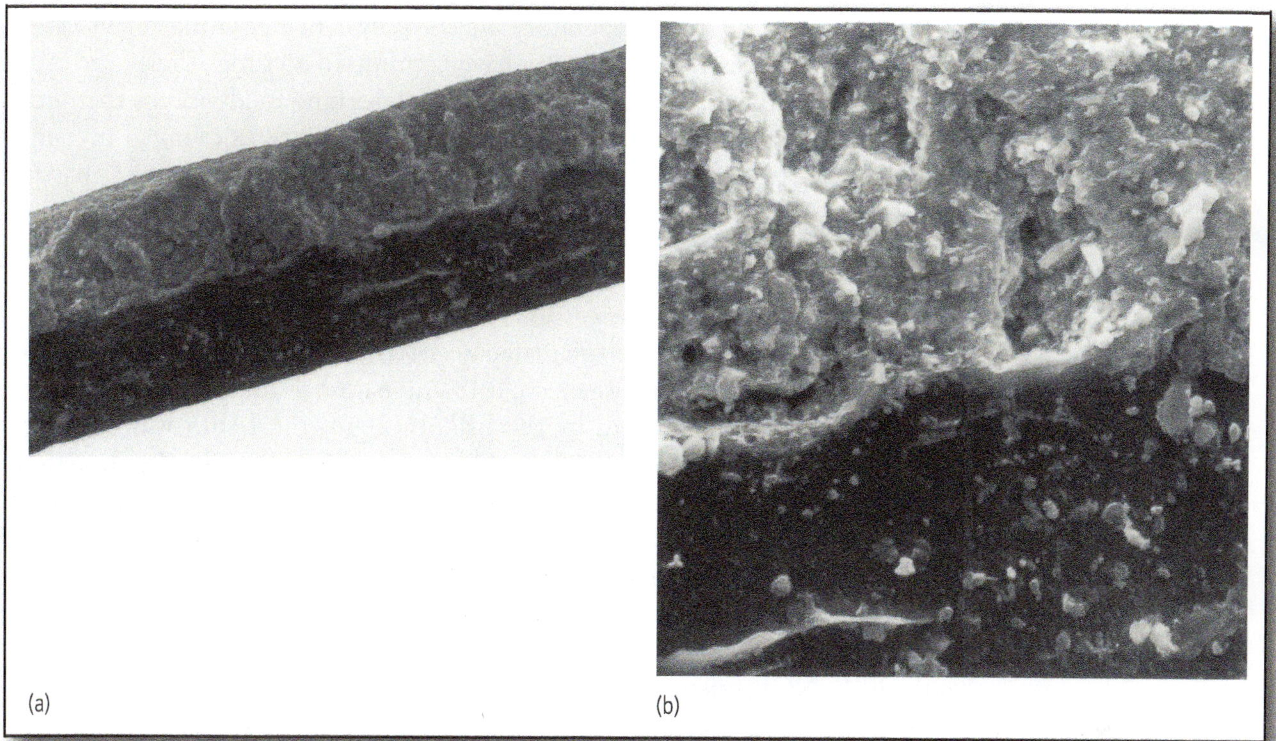

FIGURE 3–7 (a) Two-layer paint chip magnified 244 times with a scanning electron microscope. (b) The same paint chip viewed at a magnification of 1,600 times. *Courtesy Richard Saferstein, Ph.D.*

Quick Review

- The value of class physical evidence lies in its ability to corroborate events with data in a manner that is, as nearly as possible, free of human error and bias.
- As the number of different objects linking an individual to a crime scene increases, so does the likelihood of that individual's involvement with the crime.
- A person may be exonerated or excluded from suspicion if physical evidence collected at a crime scene is found to be different from standard/reference samples collected from that subject.

Forensic Databases

In a criminal investigation, the ultimate contribution a criminalist can make is to link a suspect to a crime through comparative analyses. This comparison defines the unique role of the criminalist in a criminal investigation. Of course, a one-to-one comparison requires a suspect. Little or nothing of evidential value can be accomplished if crime-scene investigators acquire fingerprints, hairs, fibers, paint, blood, and semen without the ability to link these items to a suspect. In this respect, computer technology has dramatically altered the role of the crime laboratory in the investigative process.

No longer is the crime laboratory a passive bystander waiting for investigators to uncover clues about who may have committed a crime. Today, the crime laboratory is on the forefront of the investigation seeking to identify perpetrators. This dramatic enhancement of the role of forensic science in criminal investigation has come about through the creation of computerized databases that not only link all 50 states, but tie together police agencies throughout the world.

Fingerprint Databases

The premier model of all forensic database systems is the Integrated Automated Fingerprint Identification System (IAFIS), a national fingerprint and criminal history system maintained by the FBI. In 2014, the IAFIS was effectively replaced and integrated into the Next Generation Identification (NGI) system. IAFIS, which first became operational in 1999, contains fingerprints and access to corresponding criminal history information for nearly 75 million subjects (or 750 million fingerprint images), which are submitted voluntarily to the FBI by state, local, and federal law enforcement agencies. In the United States, each state has its own Automated Fingerprint Identification System (AFIS), which is linked to the FBI's NGI system (see Figure 3–8).

A crime-scene fingerprint or latent fingerprint is a dramatic find for the criminal investigator. Once the quality of the print has been deemed suitable for the NGI system search, the latent-print examiner creates a digital image of the print with either a digital camera or a scanner. Next, the examiner, with the aid of a coder, marks points on the print to guide the computerized search. The print is then electronically submitted to the NGI system, and within minutes the search is completed against all fingerprint images in the NGI system; the examiner may receive a list of potential candidates and their corresponding fingerprints for comparison and verification (see Figure 3–9).

FIGURE 3–8 A forensic scientist using the IAFIS database.
Ulrich Baumgarten/Getty Images

Case Files

Gerald Wallace

In 1975, police found Gerald Wallace's body on his living room couch. He had been savagely beaten, his hands bound with an electric cord. Detectives searched his ransacked house, cataloging every piece of evidence they could find. None of it led to the murderer. They had no witnesses. Sixteen years after the fact, a lone fingerprint, lifted from a cigarette pack found in Wallace's house and kept for 16 years in the police files, was entered into the Pennsylvania State Police AFIS database. Within minutes, it hit a match. That print, police say, gave investigators the identity of a man who had been at the house the night of the murder. Police talked to him. He led them to other witnesses, who led them to the man police ultimately charged with the murder of Gerald Wallace.

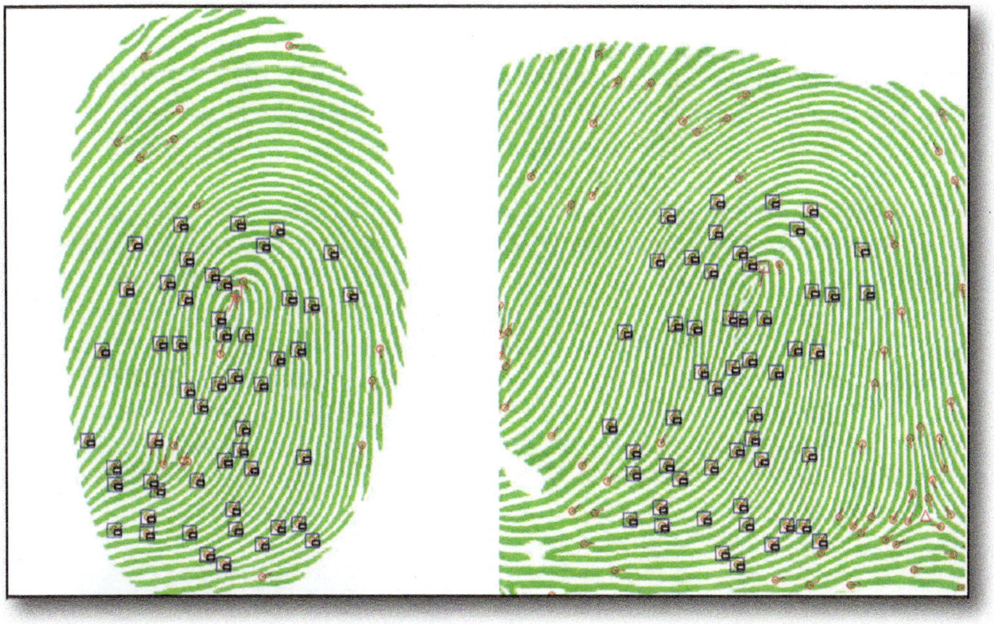

Latent Print   File Print

FIGURE 3–9 The computerized search of a fingerprint database first requires that selected ridge characteristics be designated by a coder. The positions of these ridge characteristics serve as a basis for comparing the print against fingerprints. *Courtesy Sirchie Finger Print Laboratories, Inc., Youngsville, N.C., www.sirchie.com*

Many countries throughout the world have created national automated fingerprint identification systems that are comparable to the FBI's model. For example, the United Kingdom has also recently created an integrated system that links police and justice agencies called IDENT1. Its primary capabilities include fingerprint and palm print analysis, print search capabilities with access to international databases, verification of the identities of arrested persons, and information sharing capabilities between state, local, and federal agencies throughout England, Scotland, and Wales.

DNA Databases

In 1998, the FBI's Combined DNA Index System (CODIS) became fully operational (see Figure 3–10). CODIS enables federal, state, and local crime laboratories to electronically exchange and compare DNA profiles, thereby linking crimes to each other and to convicted offenders. All 50 states have enacted legislation to establish a data bank containing DNA profiles of individuals convicted of felony sexual offenses (and other crimes, depending on each state's statute). CODIS creates investigative leads from three indexes: the forensic, offender, and arrestee indices. The forensic index currently contains about 915,000 DNA profiles from unsolved crime-scene evidence. Based on a match, police in multiple jurisdictions can identify serial crimes, allowing coordination of investigations and sharing of leads developed independently. The offender index contains the profiles of more than 13.6 million convicted individuals. The FBI has joined numerous states that collect DNA samples from those awaiting trial and will collect DNA from detained immigrants.

This information will be entered into the arrestee index database, presently at 3.4 million.[2] Ultimately, the success of the CODIS program is measured by the crimes it helps to solve. To this end, CODIS has produced more than 451,000 hits assisting in more than 440,000 investigations.

Constitutional issues regarding the appropriateness of collecting DNA from arrestees not convicted of any crime, but who nevertheless were the subject of a CODIS search against DNA collected from unsolved crimes, was decided in the case of *Maryland* v. *King*.[3]

When officers make an arrest supported by probable cause to hold for a serious offense and bring the suspect to a station to be detained in custody, taking and analyzing a cheek swab of the arrestee's DNA is, like fingerprinting and photographing, a legitimate police booking procedure that is reasonable under the Fourth Amendment.

FIGURE 3–10 A forensic scientist using the CODIS database. *Mikael Karlsson/Alamy Stock Photo*

Case Files

The Center City Rapist

Fort Collins, Colorado, and Philadelphia, Pennsylvania, are separated by nearly 1,800 miles, but in 2001 they were tragically linked though DNA. Troy Graves left the Philadelphia area in 1999, joined the Air Force, and settled down with his wife in Colorado. A frenzied string of eight sexual assaults around the Colorado University campus set off a manhunt that ultimately resulted in the arrest of Graves. However, his DNA profile inextricably identified him as Philadelphia's notorious "Center City rapist." This assailant attacked four women in 1997 and brutally murdered Shannon Schieber, a Wharton School graduate student, in 1998. His last known attack in Philadelphia was the rape of an 18-year-old student in August 1999, shortly before he left the city. In 2002 Graves was returned to Philadelphia, where he was sentenced to life in prison without parole.

With the Supreme Court sanctioning the collection of cheek or buccal swabs from arrestees, the necessity for the analysis of a swab as close to the time of arrest as possible becomes apparent. The term **Rapid DNA** has become part of the lingo of forensic science and describes approaches for rapidly obtaining a DNA profile from a buccal swab. A number of compact instruments are already commercially available and others are being developed. These allow for the development of a DNA profile from a buccal swab in less than 90 minutes. Experts envision that rapid DNA devices will take their place alongside fingerprinting units for the routine processing of arrestees. Recently, the FBI permitted profiles collected by Rapid DNA into the CODIS database (see Figure 3–11).

Rapid DNA
A process for developing DNA profiles from a buccal swab or a swab containing evidential DNA in 90 minutes or less that are compatible with a CODIS search

Unfortunately, hundreds of thousands of samples are backlogged, still awaiting DNA analysis and entry into the offender index. Law enforcement agencies search this index against DNA profiles recovered from biological evidence found at unsolved crime scenes. This approach has proven to be tremendously successful in identifying perpetrators because most crimes involving biological evidence are committed by repeat offenders.

Several countries throughout the world have initiated national DNA data banks. The United Kingdom's *National DNA Database*, established in 1995, was the world's first national database. Currently it holds more than 6 million profiles, and DNA can be taken for entry into the database from anyone arrested for an offense likely to involve a prison term. The National DNA Data Bank, housed in Ottawa, Canada, contains more than 379,000 DNA profiles from convicted individuals and has assisted in more than 54,000 cases, including more than 3,500 murders and more than 5,900 sexual assaults.

Case Files

Aztec Gold Metallic Hit and Run

A 53-year-old man was walking his dog in the early morning hours. He was struck and killed by an unknown vehicle and later found lying in the roadway. No witnesses were present, and the police had no leads regarding the suspect vehicle. A gold metallic painted plastic fragment recovered from the scene and the victim's clothing were submitted to the Virginia Department of Forensic Science for analysis.

The victim's clothing was scraped, and several minute gold metallic paint particles were recovered. Most of these particles contained only topcoats, whereas one minute particle contained two primer layers and a limited amount of colorcoat. The color of the primer surface layer was similar to that typically associated with some Fords. Subsequent spectral searches in the Paint Data Query (PDQ) database indicated that the paint most likely originated from a 1990 or newer Ford.

The most discriminating aspect of this paint was the unusual-looking gold metallic topcoat color. A search of automotive repaint books yielded only one color that closely matched the paint recovered in the case. The color, Aztec Gold Metallic, was determined to have been used only on 1997 Ford Mustangs.

The results of the examination were relayed via telephone to the investigating detective. The investigating detective quickly determined that only 11,000 1997 Ford Mustangs were produced in Aztec Gold Metallic. Only two of these vehicles were registered, and had been previously stopped, in the jurisdiction of the offense. Ninety minutes after the make, model, and year information was relayed to the investigator, he called back to say he had located a suspect vehicle. Molding from the vehicle and known paint samples were submitted for comparison. Subsequent laboratory comparisons showed that the painted plastic piece recovered from the scene could be physically fitted together with the molding, and paint recovered from the victim's clothing was consistent with paint samples taken from the suspect vehicle.

Source: Brenda Christy, Virginia Department of Forensic Science

Case Files

NIBIN Links Handgun to Suspects

After a series of armed robberies in which suspects fired shots, the sheriff's office of Broward County, Florida, entered the cartridge casings from the crime scenes into NIBIN. Through NIBIN, four of the armed robberies were linked to the same .40-caliber handgun. A short time later, sheriff's deputies noticed suspicious activity around a local business. When they attempted to interview the suspects, the suspects fled in a vehicle. During the chase, the suspects attempted to dispose of a handgun; deputies recovered the gun after making the arrests. The gun was testfired and the resulting evidence entered into NIBIN, which indicated a possible link between this handgun and the four previous armed robberies. Firearms examiners confirmed the link by examining the original evidence. The suspects were arrested and charged with four prior armed robbery offenses.

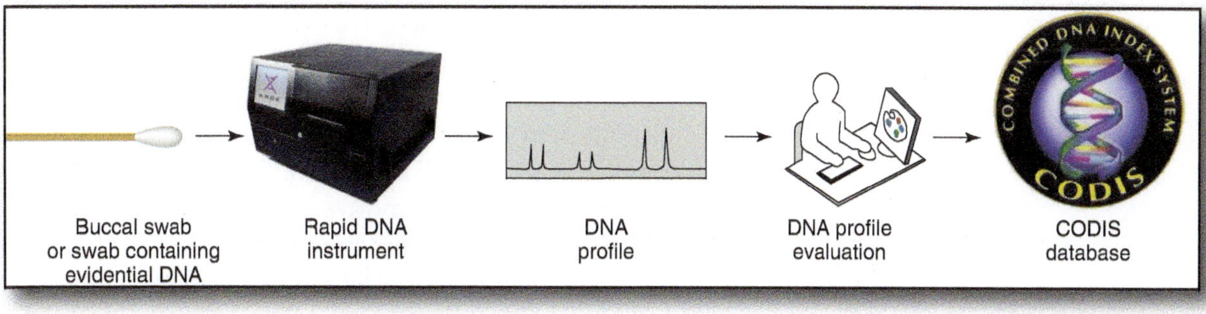

FIGURE 3–11 From a swab, a rapid DNA instrument is used to create a DNA profile that is compatible with the CODIS database. *ANDE; Federal Bureau of Investigation*

Genealogy Databases

An emerging use of forensic DNA profiles from crime scene samples involves searching unknown profiles through genealogy databases, like GEDmatch, to identify close relatives. These databases contain samples that are processed by commercial genealogy companies and uploaded by private citizens. The availability of commercial DNA testing kits, from companies like 23andMe and Ancestry.com, is credited with increasing the size and popularity of these genealogy databases.

The primary purpose of these databases is to assist professional and amateur genealogists in identifying potential relatives, but because they contain DNA profiles from a wide cross section of the population, they have proved to be extremely useful in cases where investigators have DNA profiles but have not yet matched it to a source. Searching the databases is akin to a familial search, where a link can be established to a particular family lineage. Once a family lineage has been identified, the family tree can then be further examined to determine if there are individuals that fit the description of the potential person of interest.

Other Databases

The *National Integrated Ballistics Information Network* (NIBIN), maintained by the Bureau of Alcohol, Tobacco, Firearms, and Explosives, allows firearms analysts to acquire, digitize, and compare markings made by a firearm on bullets and cartridge casings recovered from crime scenes. Since the program's inception in 1999, NIBIN partners have processed approximately 99,000 NIBIN leads and 110,000 NIBIN hits. Approximately 16 million images in the network include 3.3 million pieces of evidence. The heart of NIBIN is the *Integrated Ballistic Identification System* (IBIS), comprising a microscope and a computer unit that can capture an image of a bullet or cartridge casing. The images are then forwarded to a regional server, where they are stored and correlated against other images in the regional database. IBIS does not positively match bullets or casings fired from the same weapon; this must be done by a firearms examiner. IBIS does, however, facilitate the work of the firearms examiner by producing a short list of candidates for the examiner to manually compare.

The *International Forensic Automotive Paint Data Query* (PDQ) database contains chemical and color information pertaining to original automotive paints. This database, developed and maintained by the Forensic Laboratory Services of the Royal Canadian Mounted Police (RCMP), contains information about make, model, year, and assembly plant on more than 21,000 samples corresponding to over 85,000 layers of paint. Contributors to the PDQ include the RCMP and forensic laboratories in Ontario and Quebec, as well as 40 U.S. forensic laboratories and police agencies in 21 other countries. Accredited users of PDQ are required to submit 60 new automotive paint samples per year for addition to the database. The PDQ database has found its greatest utility in the investigation of hit-and-runs by providing police with possible make, model, and year information to aid in the search for the unknown vehicle.

The previously described databases are maintained and controlled by government agencies. There is one exception: a commercially available computer retrieval system for comparing and identifying crime-scene shoe prints known as *SICAR* (shoeprint image capture and retrieval).[4] SICAR's pattern-coding system enables an analyst to create a simple description of a shoe print by assigning codes to individual pattern features (see Figure 3–12). Shoe print images can be entered into SICAR by either a scanner or a digital camera. This product has a comprehensive shoe sole database (Solemate) that includes more than 39,000 footwear entries providing investigators with a means for linking a crime-scene footwear impression to a particular shoe manufacturer.

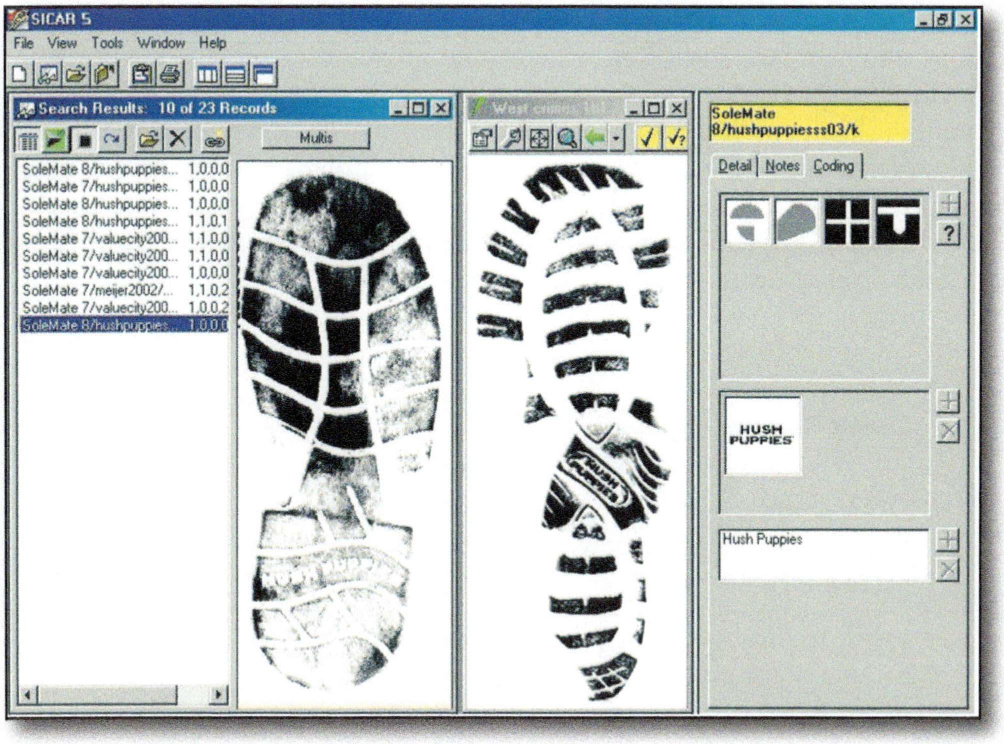

FIGURE 3–12
The crime-scene footwear print on the right is being searched against 8,000 sole patterns to determine its make and model. *Courtesy Foster & Freeman Limited, Worcestershire, U.K., www.fosterfreeman.co.uk*

It has been estimated that there are approximately 40,000 unidentified human remains that have been buried or cremated by law enforcement authorities in the United States. In a typical year, medical examiners and coroners handle approximately 4,000 unidentified human decedent cases, 1,000 of which remain unidentified after one year.

The National Missing and Unidentified Persons System (NamUs) was created in 2007 as a national centralized repository and resource center for missing persons and unidentified decedent records. NamUs is a free online system that can be searched by medical examiners, coroners, law enforcement officials, and the general public from all over the country in hopes of resolving these cases. NamUs is comprised of three databases, all of which are open to the general public. The total number of cases reported to NamUs is nearly 16,000, with nearly 10,000 open cases.

The Missing Persons Database contains information about missing persons that can be entered by anyone; however, before a person appears as a case on NamUs, there must be verification by law enforcement. When a new missing persons case is entered into NamUs, the system automatically performs cross-matching with links to state clearinghouses, medical examiners' and coroners' offices, law enforcement agencies, and victim assistance groups to check potential matches between cases.

The Unidentified Persons Database contains information entered by medical examiners and coroners. Unidentified persons are people who have died and whose bodies have not been identified. Anyone can search this database using characteristics such as sex, ancestry, tattoos and other distinctive body features, as well as dental information. However, sensitive case data is restricted and can be viewed only by select agencies.

The Unclaimed Persons Database contains information about deceased persons who have been identified by name, but for whom no next of kin or family member has been identified or located to claim the body for burial or other disposition. Only medical examiners and coroners may enter cases in this database. However, the database is searchable by the public using a missing person's name and year of birth.

In 2011, the NamUs database was awarded to the University of North Texas Health Science Center for system management and ongoing development.

Quick Review

- The creation of computerized databases for fingerprints, criminal histories, DNA profiles, markings on bullets and cartridges, automotive paints, and shoe prints has dramatically enhanced the role of forensic science in criminal investigation.
- IAFIS is the Integrated Automated Fingerprint Identification System, a national fingerprint and criminal history database maintained by the FBI. IAFIS allows criminal investigators to compare fingerprints at a crime scene to an index of 680 million known prints. CODIS is the FBI's Combined DNA Index System. It enables federal, state, and local crime laboratories to electronically exchange and compare DNA profiles, linking crimes to each other and to convicted offenders.

Chapter Review

- Biological crime-scene evidence includes blood, saliva, semen, DNA, hair, organs, and physiological fluids.
- Impression crime-scene evidence includes tire markings, shoe prints, depressions in soft soils, all other forms of tracks, glove and other fabric impressions, tool marks, and bite marks.
- Manufactured items considered common items of crime-scene evidence include firearms, ammunition, fibers, paint, glass, petroleum products, plastic bags, rubber, polymers, and vehicle headlights.
- Two methods used by forensic scientists when examining physical evidence are identification and comparison.
- Identification is the process of determining a substance's chemical or physical identity to the exclusion of all other substances; for example, drugs, explosives, petroleum products, blood, semen, and hair species.
- A comparison analysis determines whether a suspect specimen and a standard/reference specimen have a common origin.
- Evidence that can be associated with a common source with an extremely high degree of probability is said to possess individual characteristics.
- Evidence associated with only a group is said to have class characteristics.
- The overall frequency of occurrence of an event can be obtained by multiplying together the frequencies of all independently occurring instances related to that event. This is known as the product rule.
- The value of class physical evidence lies in its ability to corroborate events with data in a manner that is, as nearly as possible, free of human error and bias.
- As the number of different objects linking an individual to a crime scene increases, so does the likelihood of that individual's involvement with the crime.
- A person may be exonerated or excluded from suspicion if physical evidence collected at a crime scene is found to be different from standard/reference samples collected from that subject.
- The creation of computerized databases for fingerprints, criminal histories, DNA profiles, markings on bullets and cartridges, automotive paints, and shoe prints has dramatically enhanced the role of forensic science in criminal investigation.
- IAFIS is the Integrated Automated Fingerprint Identification System, a national fingerprint and criminal history database maintained by the FBI. IAFIS allows criminal investigators to compare fingerprints at a crime scene to an index of 680 million known prints. CODIS is the FBI's Combined DNA Index System. It enables federal, state, and local crime laboratories to electronically exchange and compare DNA profiles, linking crimes to each other and to convicted offenders.

Review Questions

1. When evidence can be associated only with a group and never with a single source, that evidence is said to possess
   a. class categories.
   b. class characteristics.
   c. group categories.
   d. group characteristics.

2. The examination of physical evidence by a forensic scientist is usually undertaken for
   a. identification or comparison purposes.
   b. proving an suspect's guilt in a courtroom.
   c. proving a suspect's innocence in a courtroom.
   d. assisting law enforcement in the apprehension of an offender.

3. Properties of evidence that can be attributed to a common source with an extremely high degree of certainty is/are
   a. comparison characteristics.
   b. referent characteristics.
   c. class characteristics.
   d. individual characteristics.

4. Most items of physical evidence retrieved at crime scenes
   a. can be definitively linked to a single person or object.
   b. can be used in court with little challenge to their scientific validity.
   c. can be manipulated by forensic experts to "tell" the story that they wish to tell.
   d. cannot be definitively linked to a single person or object.

5. The weight or significance accorded physical evidence in a courtroom is left to
   a. the expert witnesses or criminalists who are testifying.
   b. the laboratory analysts who are interpreting the results of their tests.
   c. the trier of facts, usually a jury of laypeople.
   d. the prosecutor, who instructs the jury as to how much weight to assign to an article.

6. True or False: Physical evidence is usually examined by a forensic scientist to determine the identity of the offender of a crime.

7. True or False: Technology surrounding the forensic sciences has evolved to the point that an examiner can assign approximate probability values to the comparison of most class physical evidence.

8. True or False: Generally speaking, the chances of an investigator finding physical evidence possessing individual characteristics is far greater than finding class physical evidence.

9. True or False: A database applicable to DNA profiling is CODIS.

10. True or False: The weight or significance accorded physical evidence during court proceedings is dictated to the jury by the judge during jury instructions.

11. List five examples of each of the following types of physical evidence: biological materials, impressions, and manufactured items.

12. What is the purpose of identification? What are the two main requirements for identification?

13. List three reasons why it is not possible to define a simple analytical scheme that can be applied to all types of evidence.

14. What is a comparison analysis? What are the two steps in comparison?

15. Define *individual characteristics* and give two examples.

16. Define *class characteristics* and give two examples.

17. State the product rule and explain how it applies to the comparison of physical evidence.

18. What is the greatest weakness of class physical evidence? What is the main reason for this weakness?

19. What is the greatest value of class physical evidence?

20. State two reasons why it is important to consider the relevance of scientific evidence before allowing it to be introduced into a criminal case.

21. Why does a forensic scientist often opt not to use the most sensitive analytical techniques when comparing suspect substances?

22. What is a forensic database and how are such databases useful to criminalists?

23. Name three forensic databases and describe the type of information stored in each.

24. Describe the difference between the forensic and offender indexes in the CODIS database.

Learn the Vocabulary

Use the information from this chapter along with your prior knowledge and experience to learn the key terms. Choose from the following activities:

1. Make flashcards with a term on the front and its definition on the back. Use the flashcards to quiz a partner on correctly pronouncing and defining the terms.

2. Make an audio or video recording of each term and its definition.

3. Work with a partner to write a sentence for each term. Read the sentences out loud to a partner or share them with the class.

4. Write a list of new terms and language you have encountered in this chapter. Reuse the terms to write sentences and then read the sentences out loud to a partner. Reuse the terms again by writing a short speech. The speech might explain the terms or teach a lesson using the terms. Give the speech to a partner or to the class.

Application and Critical Thinking

1. Arrange the following tasks in order from the one that would require the least extensive testing procedure to the one that would require the most extensive. Explain your answer.
   a. determining whether an unknown substance contains an illicit drug
   b. determining the composition of an unknown substance
   c. determining whether an unknown substance contains heroin

2. The following are three possible combinations of DNA characteristics that may be found in an individual's genetic profile. Using the probability rule, rank each of these combinations of DNA characteristics from most common to least common. The number after each characteristic indicates its percentage distribution in the population.
   a. FGA 24,24 (3.6%), TH01 6,8 (8.1%), and D16S539 11, 12 (8.9%)
   b. vWA 14,19 (6.2%), D21S11 30,30 (3.9%), and D13S317 12,12 (8.5%)
   c. CSF1PO 9,10 (11.2%), D18S51 14,17 (2.8%), and D8S1179 17,18 (6.7%)

3. For each of the following pieces of evidence, indicate whether the item is more likely to possess class or individual characteristics and explain your answers.
   a. an impression from a new automobile tire
   b. a fingerprint
   c. a spent bullet cartridge
   d. a mass-produced synthetic fiber
   e. pieces of a shredded document
   f. commercial potting soil
   g. skin and hair scrapings
   h. fragments of a multilayer custom automobile paint

4. Which of the forensic databases described in the text contain information that relates primarily to evidence exhibiting class characteristics? Which ones contain information that relates primarily to evidence exhibiting individual characteristics? Explain your answers.

5. Conduct a survey on a topic of your choice to gather data that can be analyzed to identify patterns, significant statistical features, sources of error, and limitations. Organize your data visually in a chart or graph. Present your findings to the class.

Laboratory Experiments

This activity requires the use of the following practices of science and engineering:

- Planning and carrying out investigations
- Analyzing and interpreting data
- Constructing explanations
- Engaging in argument from evidence
- Obtaining, evaluating, and communicating information

This activity consists of the following crosscutting concept:

- **Patterns**—This figures prominently in the practice of analyzing and interpreting data. Patterns are discernible in the fiber identification, fingerprints, DNA, tire treads, and shoe prints.

Before you begin, read the scenario and steps out loud with a partner. Ask for clarification as necessary to make sure you understand the experiment.

The Scenario

You respond to a homicide at a two-story townhouse on June 5, 2014 at 6 PM. As you pull up to the location, the first officer on scene gives you the following information:

> They received a call from the victim's husband, Steve Walters, who has been out of town for the last week on business. He said when he arrived home he noticed the front door was locked but not dead bolted. When he walked in he noticed the house ransacked and blood on the floor. He yelled for his wife, Angela, and began to run upstairs when he saw her on the floor face down not moving. He attempted to revive her but she was already dead. He then called 911. When the officers arrived they looked inside and cleared the house for any suspects. They pulled back and called you to investigate the homicide. Your initial observation of the victim is that she appears to have been killed by blunt force trauma to the head. The victim's hands and feet were bound with duct tape. You suspect that this may have been a robbery that got out of hand.

PART I

Below is a list of evidence found at the crime scene. Applying scientific practices to plan your comparative investigation, identify each piece of evidence as individual vs. class:

A. Gold metallic fiber caught in the duct tape

B. Shoe prints on the kitchen tile floor

C. Six *good* fingerprints (combined) from the refrigerator (3), back doorknob (2), and toilet handle (1)

D. Reddish oily substance on victims face

E. Ski mask found on master bedroom floor

F. Blood smears on stair carpet

G. Tire impressions in alley behind house

H. Blood droplets on Angela's cell phone found in driveway

I. Scrapings under victim's finger nails

PART II

Applying scientific practices to plan your experimental investigation, determine what information you would process each piece of evidence for.

PART III

A. What follow-up investigation would you do if the following results came back from your evidence processing, as well as your investigation?

B. Gold fiber is a fine nylon blend commonly used in stitching or embroidering. Shoe prints come back to a Rough House brand work boot size 10. The wear pattern suggests that the owner walks on the outside of their feet.

C. Fingerprints not belonging to the victim cannot be matched in AFIS.

D. Reddish substance on victim's face is standard transmission oil.

E. No information is obtained from the ski mask.

F. Blood smears are same blood type as victim; DNA match to victim.

G. Tire tread pattern comes back to a Michelin Cross Terrain tire size 235/70R16 1045 FRD, most commonly used in SUVs.

H. Blood droplets on Angela's cell phone do not match victim or husband. However, there is a CODIS match to an unknown suspect in three other burglaries and one sexual assault. Angela's cell phone records show last three calls at 1:20, 1:21, and 1:23 to 602-888-5555 on June 5, 2014.

I. There is skin tissue under three of the victim's fingernails on her right hand. There is a CODIS match to an unknown suspect in three other burglaries and one sexual assault.

PART IV

During your investigation you discover the following:

A. The husband's rare coin collection is missing.

B. The victim's wallet and credit cards are missing.

C. The husband returned to Phoenix two days early from his business trip to see his girlfriend, April Meadows.

PART V

1. Your investigation and the evidence lead you to infer that the victim's husband, his girlfriend April, and the owner of phone number 602-888-5555 (Michelle Rivera) are suspects. What questions would you ask of each person during your interview to support your case?
2. During your interviews you discover that Michelle has a brother, Fabian, who occasionally lives with her and drives a Chevy Blazer. What follow-up investigation would you do? And what questions would you ask him if you were to interview him?
3. What is your hypothesis as to what happened to the victim?

Endnotes

1. National Research Council (U.S.). (2009). Strengthening forensic science in the United States: A path forward.
2. *Collecting DNA at Arrest: Policies, Practices, and Implications*, https://www.ncjrs.gov/pdffiles1/nij/grants/242812.pdf.
3. Maryland v. King, 569 U.S. 435 (2013)
4. Foster & Freeman Limited, Worcestershire, U.K., www.fosterfreeman.co.uk.

Death Investigation

4

Jacob Langston/Orlando Sentinel/MCT/Alamy Stock Photo

Key Terms

- algor mortis
- autopsy
- cause of death
- forensic anthropology
- forensic entomology
- forensic pathologist
- livor mortis
- manner of death
- petechiae
- postmortem interval (PMI)
- rigor mortis

The George Zimmerman Case: Self-defense or Murder?

The setting was Sanford, Florida, when during a rainy night Trayvon Martin, a 17-year-old African American high school student, was returning to the Twin Lakes housing complex after walking to a local convenience store to purchase candy and a beverage. George Zimmerman, a local neighborhood watch volunteer, noticed Martin walking around the community and proceeded to call 911 dispatch. As Martin realized Zimmerman was following him, he began to run with Zimmerman in pursuit behind him. The dispatcher, still on the phone with Zimmerman, can be heard telling him that he should not follow Martin. After the call is disconnected, a violent struggle ensued and Martin suffered a single gunshot wound to his chest. Police reported to the scene not long after, finding Martin dead on the ground, and Zimmerman bloodied and bruised with a broken nose.

At trial, George Zimmerman faced charges of manslaughter and murder in the second degree. The trial engendered intense media discussions revolving around racial bias that had not been seen since the days of the O.J. Simpson trial. The prosecution claimed that Zimmerman was an overzealous watch volunteer when he chased after Martin and provoked a confrontation. Zimmerman's defense countered these accusations by citing Florida's "Stand Your Ground" law, and suggested that Martin's death was a product of self-defense. According to Zimmerman's testimony, Martin had him pinned down on the sidewalk, and was punching and beating his head. During the altercation Zimmerman's gun was exposed and as Martin reached for the weapon, Zimmerman grabbed and fired his weapon killing Martin. During the trial, expert witness Anthony Di Maio, a forensic pathologist, claimed that the angle of the gunshot wound and the area in which Martin was shot proved to be consistent with Martin straddling Zimmerman while beating his face. Di Maio went one step further by explaining the significance of a "powder tattooing" on Martin's body. Di Maio's testimony revealed that the gunshot powder residue found on Trayvon Martin was evidence that he was shot at close range, further supporting Zimmerman's account

Zimmerman was found not guilty of all charges and released amid cries of social and legal injustice and racial prejudice.

Learning Objectives

After studying this chapter, you should be able to:

4.1 Describe the role of a forensic pathologist

4.2 Describe autopsy

4.3 Describe common causes of death

4.4 List various categories associated with the manner of death

4.5 Describe chemical and physical changes helpful for estimating time of death

4.6 Discuss the role of the forensic anthropologist in death investigation

4.7 Describe the role of the forensic entomologist in death investigation

National Science Content Standards

Scientific Inquiry

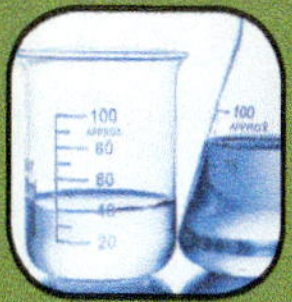

Physical Science

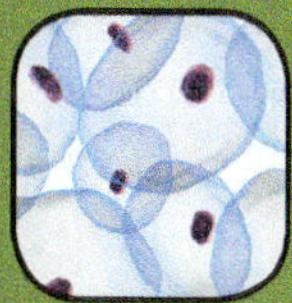

Life Science

Role of the Forensic Pathologist

forensic pathologists Investigative personnel, typically medical examiners or coroners, who investigate the cause, manner, and time of death of a victim in a crime; can also be a physician who has been trained to conduct autopsies

Few investigations bring with them the intense focus of community interest and news media coverage as that of a suspicious death. Generally, **forensic pathologists** associated with the medical examiner's or coroner's office are responsible for determining the cause of an undetermined or unexpected death. These officers coordinate their response with that of law enforcement in the ensuing investigation. The titles *coroner* and *medical examiner* are often used interchangeably, but there are significant differences in their job descriptions. In the United States, there's a mix of state medical examiner systems, county medical examiner offices, and county coroner systems. The coroner is an elected official and may or may not possess a medical degree. The term *coroner* dates back hundreds of years to the rule of King Richard I of England (1189–1199), who created the office of the coroner to collect money and personal possessions from people who had died. The medical examiner, on the other hand, is almost always an appointed official and is usually a physician who generally is a board-certified forensic pathologist and is responsible for certifying the manner and the cause of a death.

The tasks of examining the case for the cause and manner of death and recording the results on a death certificate are the responsibilities of both offices. However, although both the coroner's office and the medical examiner's office are charged with investigating suspicious deaths, only the pathologist is trained to perform an autopsy. Ideally, the coroner or medical examiner's office should be staffed with physicians who are board certified in forensic pathology and should charge them with determining the cause of death by autopsy. The cause-of-death determination, however, involves not just an autopsy but also the history of death, witness statements, relevant medical records, and any scene investigation, all of which constitute the surrounding circumstances of death.

From a practical point of view, it is often not feasible for the forensic pathologist to personally solicit information regarding the circumstances surrounding a death or to respond in person to every death scene. Thus, the gathering of vital information and the scene investigation can be delegated to trained coroner/medical examiner investigators who, when a crime scene is involved, coordinate their efforts with the those of crime-scene and criminal investigators. The forensic pathologist's work is also aided by the skills of specialists including forensic anthropologists, forensic entomologists, and forensic odontologists. Working as a team, forensic professionals must be able to communicate their explanations and solutions to the question of what happened, both individually and collaboratively, in a variety of settings and formats. For example, in the lab, a forensic pathologist might communicate individually with an investigator to discuss or clarify findings in an autopsy report. A forensic scientist might communicate, in the field, collaboratively with forensic investigators on a team to process a crime scene to gather evidence efficiently and effectively. They may be required to engage in respectful scientific argumentation, using applied scientific explanations and empirical evidence, if all members of the team are not in agreement to the proposed solutions.

Scene Investigation

With regard to any scene investigation, protection of the overall scene and the body are of paramount importance, as is the ultimate removal of the body in a medically acceptable manner. The death investigation involves documenting and photographing the undisturbed scene; collecting relevant physical evidence; attempting to determine time of death, which must be done in a timely fashion at the scene; and, among other things, ascertaining premortem locations of the body and whether any postmortem movement of the body occurred. Examples of observations that can be made of the body at the scene include bruises along the upper lip, which may be evidence of smothering; a black eye limited to the eyelids, which implies an injury from inside the head; or bleeding from the ear, which implies a basal skull fracture.

A critical phase of the death investigation will be a preliminary reconstruction of events that preceded the onset of death, so all significant details of the scene must be recorded. Blood spatter and blood flow patterns must be documented. Blood should be sampled for testing in case some of the blood was cast off by a perpetrator. Any tire marks or shoe prints must be documented. Fingerprints must be processed and collected. Of particular importance is the search for any evidence discarded, dropped, or cast off by a perpetrator. When a weapon is involved, there must be a concerted effort to locate and recover the suspect weapon. In the case of firearm deaths, fired bullets or casings must be found and their locations documented. In such firearm deaths, before the body is moved or clothing is removed, blood spatter directionality and trace evidence (such as hairs) on the hands must be documented. Paper bags then should be placed over the hands and secured around the wrist or arm (paper prevents moisture condensation) to preserve any additional evidence.

Photographs must always be taken before the scene is altered in any way (except from lifesaving efforts). This includes moving the body or anything on the body, such as clothing or jewelry. A particularly violent scene can carry with it a large amount of blood and disorder. Blood may be found at different locations throughout the scene. This could prove to be important in shaping the events that led to the final outcome; it may be possible to determine the initial location of the injury, as well as victim and assailant movements throughout the course of events. Initially it may be difficult to properly infer the source of the wounds and the order in which they were received at the scene. Photographs then will play a very large role when reconstructing the events later. As always, photographs should be taken with a scale, always first overall, then at medium range, then close up. The photographer must also be careful not to get caught up in capturing the injuries exclusively. Negative findings can also be significant. This means photographs should also be taken of areas on the body where injuries are not apparent.

Quick Review

- Forensic pathologists associated with the medical examiner's or coroner's office are responsible for determining the cause of an undetermined or unexpected death.
- Although both the coroner's office and the medical examiner's office are charged with investigating suspicious deaths, only the pathologist is trained to perform an autopsy. The tasks of examining a body for the cause and manner of death and recording the results in the death certificate are the responsibilities of both offices.
- Protection of the body and the overall scene is of paramount importance, as is the ultimate removal of the body in a medically acceptable manner.

Protection of the body and the overall scene is of paramount importance, as is the ultimate removal of the body in a medically acceptable manner. Often the initial phase of the investigation will focus on determining the identity of the deceased, often called the *decedent*. Although this task may be relatively simple to accomplish through a visual examination, complications can arise. Body decomposition and the existence of extensive trauma can complicate the identification. This may necessitate the application of more sophisticated technology, such as DNA, fingerprinting, dental examination, and facial reconstruction.

The Autopsy

autopsy
A surgical procedure performed by a pathologist on a dead body to ascertain—from the body, organs, and bodily fluids—the cause of death

An **autopsy**, in its broadest definition, is simply the examination of a body after death (i.e., a postmortem examination). The autopsy can be further described as one of two types: a clinical/hospital autopsy or a forensic/medicolegal autopsy. The clinical/hospital autopsy focuses on the internal organ findings and medical conditions. Its purpose is to confirm the clinical diagnoses, the presence and extent of disease, any medical conditions that were overlooked, and the appropriateness and outcome of therapy. In contrast, the goal of a forensic/medicolegal autopsy is to determine the cause of death and confirm the manner of death, often to be used in criminal proceedings. The forensic autopsy usually emphasizes external and internal findings while developing meaningful forensic correlations between sustained injuries and the crime scene (see Figures 4–1 and 4–2).

As part of a forensic autopsy, a corpse is subjected to an intense physical analysis just like any other piece of evidence. Internal organs are weighed and dissected to search for signs of injury or disease that could have resulted in the death. Toxicological tests are performed on urine and blood, microscopic examinations are performed on tissue samples, and DNA is analyzed. All of the results are compiled and analyzed to establish cause and manner of death in the autopsy report.

All the steps of the forensic autopsy must be carefully documented and photographed. The documentation should include date, time, place, by whom the

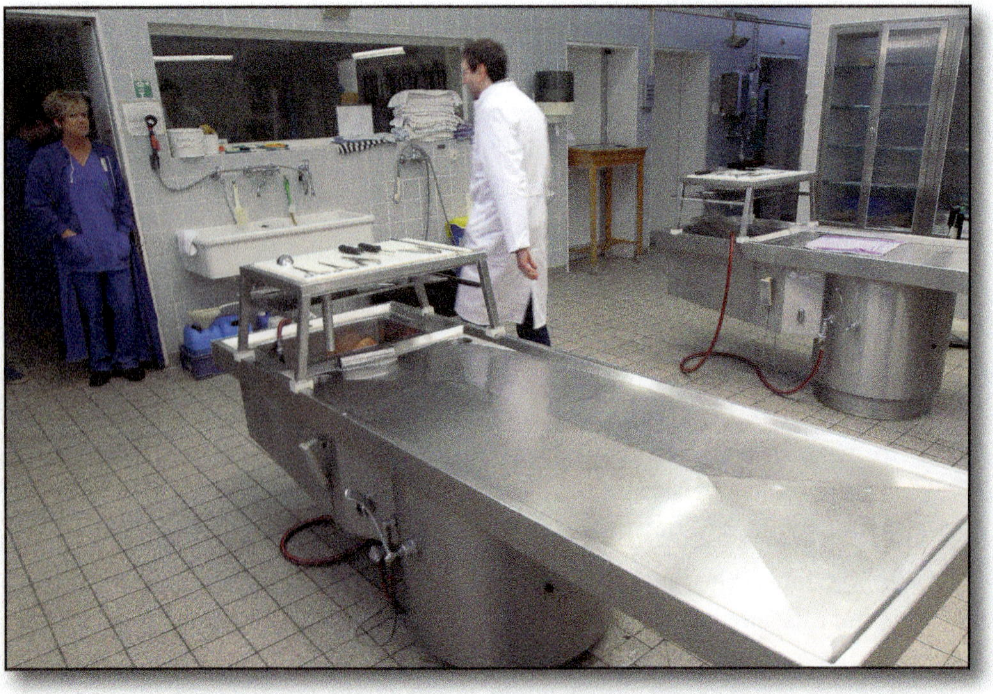

FIGURE 4–1 An autopsy suite. *Robert Seeberg/Agencja Fotograficzna Caro/Alamy Stock Photo*

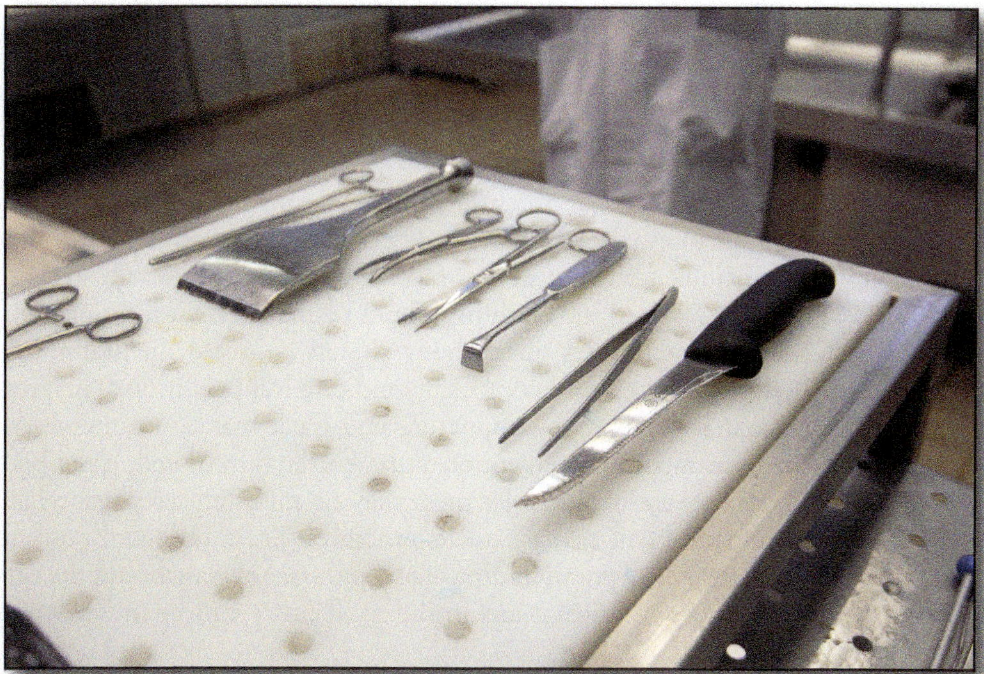

FIGURE 4–2 Tools used for an autopsy. *Robert Seeberg/Agencja Fotograficzna Caro/Alamy Stock Photo*

autopsy was performed, and who attended the autopsy. Photographs of the injuries, complete with a scale, and descriptions of each photograph's location are important when correlating external wounds with internal damage. Negative photographs—photographs of uninjured parts of the body—are also important. The autopsy report and photographs are so important because, once the body is buried, no further evidence can be collected and no additional findings can occur.

Evidence from the Autopsy The search for physical evidence must extend beyond the crime scene to the autopsy room of a deceased victim. Here, the medical examiner or pathologist carefully examines the victim to establish the cause and manner of death. As a matter of routine, tissues and organs are retained for pathological and toxicological examination. At the same time, arrangements must be made between the examiner and investigator to secure a variety of items that may be obtainable from the body for laboratory examination. The following are among the items to be collected and sent to the forensic laboratory:

- Victim's clothing
- Fingernail scrapings or clippings
- Combings from head and pubic areas
- Buccal swab (for DNA typing purposes)
- Vaginal, anal, and oral swabs (in sex-related crimes)
- Bullets recovered from the body
- Swabs of body areas, such as breasts or penis, suspected of being in contact with DNA arising from touching or saliva
- Hand swabs from shooting victims (for gunshot residue analysis)

These items of evidence should be properly packaged and labeled like all other evidence. Once the body is buried, efforts at obtaining these items may prove difficult or futile. Furthermore, a lengthy time delay in obtaining many of these items will diminish or destroy their forensic value.

External Examination The forensic autopsy consists of an external examination and an internal examination. The first steps taken for the external examination include a broad overview of the condition of the body and the clothing. Obvious damage to the clothing should be matched up to injuries on the body. General characteristics of the body should be noted, including sex, height, weight, approximate age, color of hair, and physical condition. The presence of tattoos and scars, as well as puncture and track marks, are noted. All evidence of apparent medical intervention must be carefully noted, described, and photographed because occasionally these may be misinterpreted, especially chest tube insertions and emergency cardiac punctures. The mouth and nose are examined for the presence of vomit and/or blood and trace evidence, and the ears are examined for blood. Any irritations in the nasal cavity can be indicative of drug sniffing.

Often, paper bags are placed over the hands at the crime scene until it is time to examine them. This prevents contamination and possible loss of trace evidence, such as hairs and fibers. This preservation of evidence can play an important role in identifying a suspect. A victim will sometimes have skin and DNA under their fingernails from fighting with the assailant.

The external examination also consists of classifying the injuries. This includes distinguishing between different types of wounds, such as a stab wound versus a gunshot wound. The injuries that are examined may include abrasions, contusions, lacerations, and sharp-injury wounds. Hemorrhages in the eyelids (petechiae) are also essential to note, as they can indicate strangulation. Attention is also paid to the genitalia, especially in cases where sexual abuse is suspected. In these cases, vaginal, oral, and rectal samples are taken.

The discharge from a firearm will produce characteristic markings on the skin. This discharge is a combination of soot and gunpowder. It will leave markings called *stippling* or *tattooing* around the bullet hole. The stippling can be analyzed in terms of its span and density in order to approximate the range of fire. The range of fire may prove to be the most important factor in distinguishing a homicide from a suicide.

X-ray examinations can be very useful in the autopsy process. They are most commonly performed in gunshot wound cases and stab wound cases. Even if the bullet, knife, or other piercing weapon is recovered outside the body, an X-ray will identify any fragments still inside the body. An X-ray will also help determine the path of the projectile or sharp utensil. X-rays can also be very helpful in cases where the victim was beaten, especially situations in which the victim is a child: an X-ray can show past bone fractures and a possible pattern of abuse.

Internal Examination The dissection of the human body generally entails the removal of all internal organs through a Y-shaped incision beginning at the top of each shoulder and extending down to the pubic bone. Performing the internal examination entails weighing, dissecting, and sectioning each organ of the body. When required and in accordance with jurisdictional rules, microscopic examination of the sectioned organs is conducted, which can help in determining the cause of death. For example, microscopic examination of lungs and liver can confirm chronic intravenous drug abuse. Examination of the cranium requires cutting an incision from behind one ear to the other, peeling the scalp upward and backward, and sawing the skull in a circular cut; then the skull cap is removed to reveal the brain, as shown in Figure 4–3.

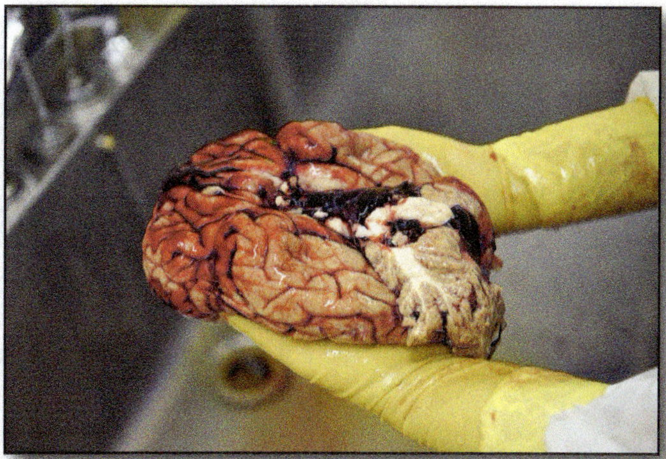

FIGURE 4–3 A brain during autopsy. *Courtesy SPL/Science Source*

Special care is taken to identify any preexisting conditions or malformations in the organs that might have contributed to the death of the victim. Pulmonary edema (fluid accumulation in the lungs) is frequently found in victims of chronic cocaine and amphetamine abuse. Heart malformations may cause suspicious death in an otherwise healthy individual.

Special attention is paid to the digestive tract if poisoning is suspected. The stomach can show partially digested or dissolved pills. Chemical analyses can also be carried out to show signs of poisoning. The amount of pills or tablets in the stomach can aid in the determination of manner of death as well. It is not always a sure sign, but typically it is unlikely that a person will accidentally swallow a large number of pills. This would suggest suicide rather than an accidental overdose. Stomach contents may reveal the deceased's last meal. The extent of digestion can help with determining the time of death.

Toxicology The internal examination is also where toxicological specimens are taken. These include samples of blood, stomach content, bile, and urine. All bile in the gallbladder and all stomach content are collected. In addition to these, brain matter, liver, and vitreous humor are also gathered. These specimens can play especially large roles in cases where poisoning or drug abuse is suspected.

Blood is often tested to determine the presence and levels of alcohol and drugs. Blood should be taken from areas of the body where there is the least chance of contamination. Blood should never be collected from body cavities, where it may be contaminated from adjacent structures. Many changes occur in the body after death, and these changes can alter the drugs present in the system at the time of death. This can make interpreting how much of a drug was present, if any at all, a very challenging task. Some drugs redistribute or reenter the blood after death and thus may complicate the interpretation of postmortem blood levels of these drugs. This phenomenon is known as *postmortem redistribution*. For this reason, it is best to collect blood at distant areas of the body to allow the toxicologist to compare the agreement of the drug concentrations found. The ideal location to retrieve the blood is internally, directly from the inferior vena cava (the large vein inside the lower abdominal region, which receives its blood from the femoral veins) using a syringe. Where postmortem redistribution of drugs may have occurred, blood should also be collected at autopsy from the superior venous system directly above the heart.

For illicit as well as legal substances, it is necessary to know what levels are indicative of therapeutic use and what levels indicate toxicity of a given substance. Much information regarding therapeutic versus toxic drug levels has been published. This data can help pathologists and toxicologists ascertain the cause of death. Most drug-related deaths are quite apparent from the blood concentrations of alcohol and/or a drug found in the postmortem toxicological report. (Note that depressant drugs will act in concert with alcohol.) However, in some cases of drug-induced death, drug levels may not always provide evidence. Cocaine is a prime example of this. Cocaine-induced sudden death is an event with an incubation period. Structural alterations of the cardiovascular system are required, and such alterations take months, or perhaps years, of chronic cocaine use. In these individuals, death and toxicity may occur after the use of even a trivial amount of the drug.

Unlike drug analyses, general testing for poisons is not a routine procedure carried out by the pathologist. However, if a specific poison is suspected, a particular test must be performed. A body that displays a cherry-red discoloration often leads a pathologist to suspect carbon monoxide poisoning. The pathologist would then perform a toxicological test of the blood. Poisoning by cyanide could also produce a pinkish discoloration. Often, cyanide toxicity will show additional signs, such as a distinct smell of burnt almonds. Corrosion around the lips of a victim may lead to a suspicion of ingesting an acid or alkaline substance.

Quick Review

- An autopsy, in its broadest definition, is simply the examination of a body after death.
- The forensic autopsy consists of an external examination and an internal examination.
- The first steps taken for the external examination include a broad overview of the condition of the body and the clothing.
- The external examination also consists of classifying the injuries. This includes distinguishing between different types of wounds, such as a stab wound versus a gunshot wound.
- The dissection of the human body generally entails the removal of all internal organs through a Y-shaped incision beginning at the top of each shoulder and extending down to the pubic bone.
- The internal examination entails weighing, dissecting, and sectioning each organ of the body.
- Blood is often routinely tested to determine the presence and levels of alcohol and drugs.
- Some drugs redistribute or reenter the blood after death and thus may complicate the interpretation of postmortem blood levels of these drugs.

Case Files

Harold Shipman, Dr. Death

Ray Bradbury/Alamy Stock Photo

Kathleen Grundy's sudden death in 1998 was shocking news to her daughter, Angela Woodruff. Mrs. Grundy, an 81-year-old widow, was believed to be in good health when her physician, Dr. Harold Shipman, visited her a few hours before her demise. Some hours later, when friends came to her home to check on her whereabouts, they found Mrs. Grundy lying on a sofa fully dressed and dead.

Dr. Shipman pronounced her dead and informed her daughter that an autopsy was not necessary. A few days later, Mrs. Woodruff was surprised to learn that a will had surfaced leaving all of Mrs. Grundy's money to Dr. Shipman. The will was immediately recognized as a forgery and led to the exhumation of Mrs. Grundy's body. A toxicological analysis of the remains revealed a lethal quantity of morphine.

In retrospect, there was good reason to suspect that Dr. Shipman was capable of foul play. In the 1970s, he was asked to leave a medical practice because of a drug abuse problem and charges that he obtained drugs by forgery and deception. However, Dr. Shipman was quickly back to practicing medicine. By 1998, local undertakers became suspicious at the number of his patients who were dying. What is more, they all seemed to be elderly women who were found sitting in a chair or lying fully clothed on a bed.

As police investigated, the horror of Dr. Shipman's deeds became apparent. One clinical audit estimated that Dr. Shipman killed at least 236 of his patients over a 24-year period. Most of the deaths were attributed to fatal doses of heroin or morphine. Toxicological analysis on seven exhumed bodies clearly showed significant quantities of morphine. Convicted of murder, Dr. Shipman hanged himself in his jail cell in 2004.

Cause of Death

cause of death Identifies the injury or disease that led to the chain of events resulting in death

A primary objective of the autopsy is to determine the cause of death. The **cause of death** is that which initiates the series of events ending in death. The most important determination in a violent death is the character of the injury that started the chain of events that resulted in death. However, if the sequence of events leading to death is sufficiently prolonged, then the decedent may actually suffer from adverse medical conditions brought about by the initial injury and then die as a result of those conditions. In that case, it will be up to the forensic pathologist to determine that the original injury inflicted on the victim was the underlying cause of death. Scientists determine the cause of death from an autopsy report obtained through resources. Some of the more common causes of death are discussed here.

Blunt-Force Injury A blunt-force injury is caused by a nonsharpened object such a bat or pipe. A blunt-force injury can abrade, or scrape, tissue. If tissue is crushed by a blunt force to the point of causing skin to overstretch, a laceration will form, characterized by the skin splitting and tearing. Lacerations exhibit abrasions around the open wound, tissue bridging within the open wound, and torn or disturbed tissue beneath the skin surrounding the open portion of the wound. Blunt-force injury can also crush tissue. This will cause bleeding from tiny ruptured blood vessels within and beneath the skin, known as a contusion, or bruise (see Figure 4–4). Much has been written about determining the age of bruises, but forensic pathologists have become keenly aware that attempting to "age" bruises based on color and changes in color over time is fraught with difficulty, and contusions must be interpreted with great care and reserve. Some contusions only become visible externally over time, and frequently, bruises will not be visible externally but become eminently visible internally within soft tissues (e.g., in the abdomen and on the back, arms, and legs).

A contusion can sometimes exhibit the pattern of the weapon used. For example, if a person wearing a ring strikes another person, the ring may imprint its pattern onto the skin. A person who stomps on another may leave the impression of their shoe heel. Over time, however, the bruise will lose its original shape and pattern and undergo color changes. Some objects will produce a characteristic bruised perimeter and a white center.

The outward appearance of the injuries does not always coincide with the injuries sustained inside the body. This is something the pathologist must keep in mind when examining blunt-force injuries. A single blow to certain parts of the body can cause instantaneous death with little visible damage. Likewise, a blow to the head can cause a concussion that can be instantly fatal.

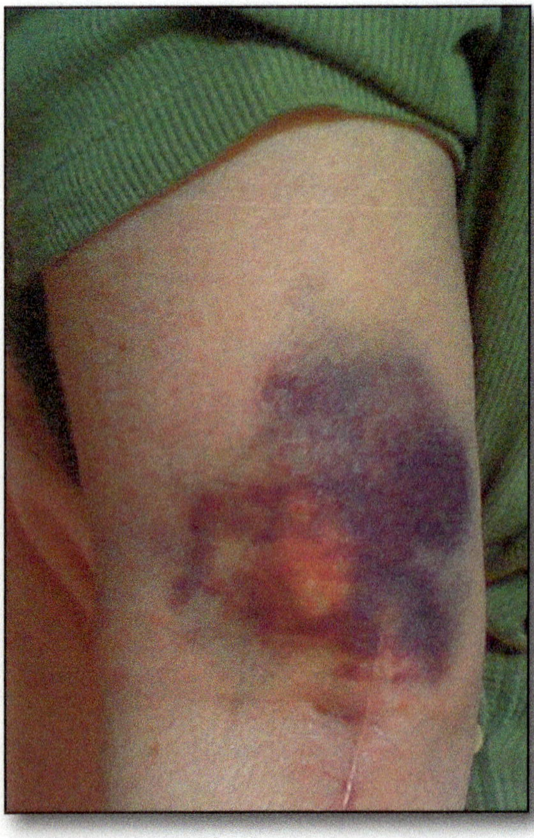

FIGURE 4–4 Bruising (contusions) on the skin.

Sharp-Force Injuries Sharp-force injuries occur from weapons with sharp edges, such as knives or blades. These weapons are capable of cutting or stabbing. A *cut* is formed when the weapon produces an injury that is longer than it is deep. In contrast, a *stab* is deeper than its length. As shown in Figure 4–5, the tissue associated with these types of wounds is not crushed or torn but sliced.

A scene that involves a sharp-force injury is usually especially bloody and unruly. Blood may be found at different locations throughout the scene. Again, this information may make it possible to determine the initial location of the injury as well as where the body was moved throughout the course of events. Particularly important in sharp-force cases is to examine the victim for defensive wounds. A victim's forearm that exhibits wounds may indicate defense wounds. These occur when the victim attempts to fight off the attacker or block assaults. Though defense wounds are more typical on the outer forearms, they can also be evident on the lower extremities if the victim tries to protect themself by kicking. A lack of any defense wounds can lead a pathologist to conclude that the victim was either unconscious or somehow tied up during the assault.

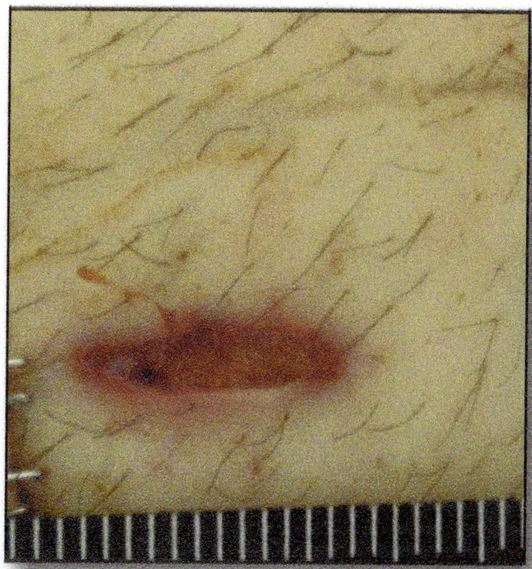

FIGURE 4–5 A stab wound. *Elizabeth Laposata*

Asphyxia Asphyxia encompasses a variety of conditions that involve interference with the intake of oxygen. For example, death at a fire scene is caused primarily by the extremely toxic gas carbon monoxide. When carbon monoxide is present, hemoglobin, the protein in red blood cells that transports oxygen, will bind to the carbon monoxide instead of oxygen. This is carbon monoxide poisoning, and this deadly complex of hemoglobin and carbon monoxide is known as carboxyhemoglobin. Bound up with carbon monoxide, the hemoglobin is prevented from transporting oxygen throughout the body, causing asphyxia. High levels of carbon monoxide in the blood will cause death. Low levels of carbon monoxide can cause a victim to become disoriented and lose consciousness.

petechiae
Pinpoint hemorrhaging often observed in the white area of the victim's eyes; often observed in strangulation cases

Carbon monoxide will not continue to build up in the body after death. The levels found in a fire victim then can be used to determine whether the individual was breathing at the time of the fire. The presence of soot is another indicator that the victim was alive during the fire. These black particles are often seen in the airway of fire victims who inhaled smoke before death. During the autopsy, soot can be observed, especially in the larynx and trachea and even in the lungs. Sometimes the victim will actually swallow the soot. In these cases, traces can be found in the esophagus and the lining of the stomach.

The ultimate cause of a death from hanging is typically the cessation of blood flow to or from the brain. Victims of hangings may show signs of petechiae on the eyelids, along with a swollen and a blue/purplish appearance of the face. **Petechiae** are very small and are caused by blood having escaped into the tissues as a result of capillaries bursting (see Figure 4–6). Although petechiae are witnessed in hanging cases, they are more common in strangulation deaths. Typically the hyoid bone (the bone on which the tongue rests) and thyroid cartilage (located below the hyoid) are not fractured in cases of hanging. A break of the thyroid cartilage is common, however, in manual strangulation cases.

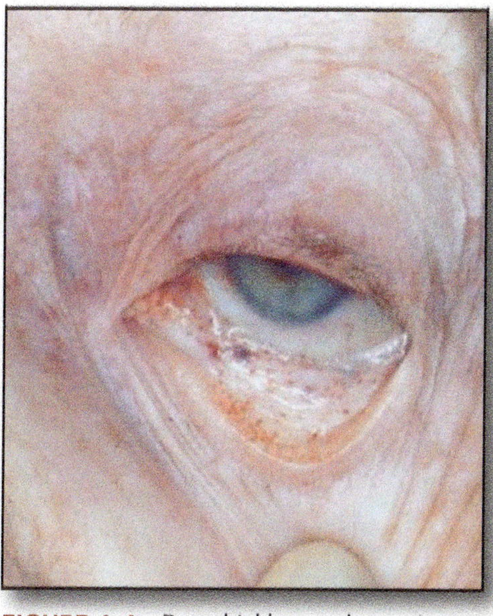

FIGURE 4–6 Petechial hemorrhages in a victim's eye. *Elizabeth Laposata*

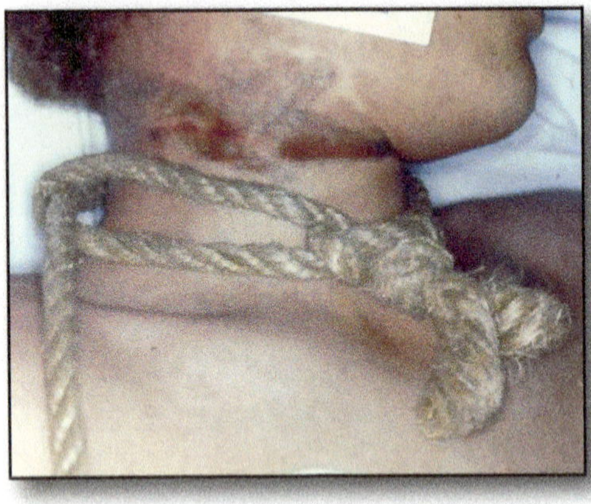

FIGURE 4–7 A ligature pattern on a neck with corresponding ligature. *Elizabeth Laposata*

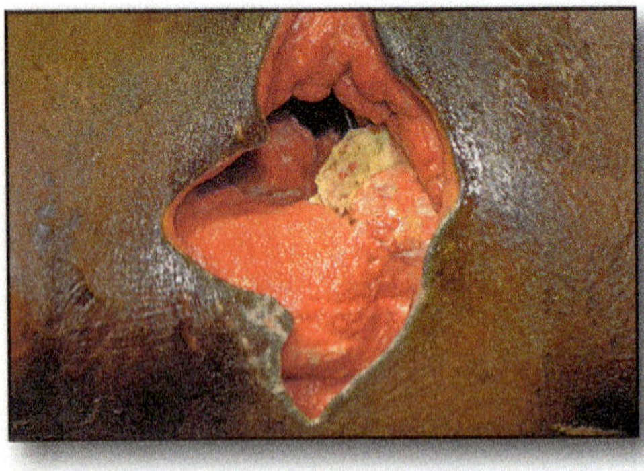

FIGURE 4–8 A contact gunshot wound to the temple of a suicide victim. *Medicimage/UIG Universal Images Group/Newscom*

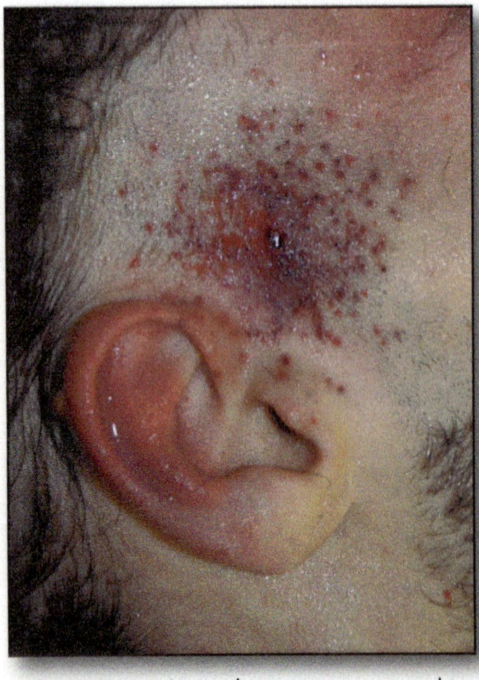

FIGURE 4–9 A gunshot entrance wound to the head from a weapon fired several inches from the target. *Elizabeth Laposata*

In hangings it is vitally important to document exactly how the victim was initially found and the position of the encircling noose, as shown in Figure 4–7. The type of knot used may strongly support the notion that another person was involved in the hanging. This means that the knot should always be preserved for later examination. Either the noose should be slipped off the victim's head intact, or the noose should be cut distant from the knot. Defense wounds are common on strangulation victims. Often the marks found on the neck of a victim are the victim's own, made in the attempt to loosen whatever was constricting their neck. Even in cases of hanging by suicide, there can be defensive wounds on the neck.

Smothering can occur by various materials that block the mouth, nose, and internal airway. Pillows or a hand can inhibit breathing. Gags that are used to silence a victim can be sucked into the airway and block oxygen flow. Typically a death by smothering is homicidal in nature. Accidental smothering usually occurs only in infants or in cases where a victim is trapped under an obstruction.

Gunshot Wounds When evaluating a gunshot wound, the estimated range of fire is one of the most important characteristics to analyze (see Figure 4–8). The appearance of the wound can help in estimating whether the firearm used to inflict the wound was discharged while in contact with the victim's body or from a distance of only inches to many feet away (Figure 4–9). The investigator will compare powder residue distribution around the wound to test fires collected from the inflicting firearm to make this estimate. Obviously if the firearm was fired at a distance of several feet, suicide is a highly unlikely cause of death because the wound could not have been self-inflicted. Gunpowder residue on the victim's hand is a possible indicator of suicide, but this is not always the case. Evidence of contact shots, that is, shots fired with the gun held against the body of the victim, typically indicates that the death was not an accident. The autopsy must include a determination of the path or "wound track" of the projectile. The wound track is determined by observing the wound from the outside of the body, following the track of the projectile through the body, and documenting its terminus. The pathologist will recover any and all projectiles from the body, carefully protecting the forensic markings. The autopsy of gunshot victims should include several facts in addition to the general autopsy facts; scene investigation and the results of toxicological and serological analyses are important. All findings regarding the bullet wounds should be noted, as well as descriptions of the clothing. The police report with a thorough description of the scene is also important.

A gunshot wound may not necessarily explain why a victim died. A person who sustains a gunshot wound can bleed to death in a matter of minutes or up to several hours. Infection can also be a contributory cause of death, especially in cases where the victim was shot in the abdomen; the victim might live several days but eventually succumb to infection. In cases where the victim was shot in the head but survives in a comatose state, pneumonia often develops. These intervening factors are considered contributory causes of death, but the gunshot wound is still considered the underlying cause of death.

Substance Abuse Drug abuse continues to be an enormous problem in the United States. Drug enforcement is a multibillion-dollar industry. Many of the abused drugs in the country are illegal, but not all are. Deaths as a result of substance abuse are common cases that a forensic pathologist must face. Because drug abuse is so common, the forensic pathologist will routinely test for the presence of drugs in nearly all investigations, and routine tests are available for many commonly abused drugs. As technology has improved, many drugs can be detected at very low levels. These factors have helped considerably in making substance abuse testing easier and less expensive.

Drug abuse can directly cause death, or it can cause complications that can serve as a contributing factor to death. An abuser can misuse a drug or a number of drugs for years, accumulating detrimental effects in that time. Death as a result of those effects is typically labeled a natural death by the pathologist. Drugs can also alter a person's judgment and psychomotor skills to the point that a fatal accident occurs. Drugs are also often at the source of acts of violence that result in death.

Manner of Death

The **manner of death** relates to the circumstances that led to the fatal result and is the culmination of the complete investigation, including the determination of cause of death. The manner of death is determined from an autopsy report obtained through resources. The certification of the circumstances and manner of death is the responsibility of the coroner's and medical examiner's offices. The manner in which death occurred is classified in death certifications as one of five categories: *homicide, suicide, accidental, natural,* or *undetermined.*

manner of death A determination made by a forensic pathologist of the cause of death. Five broad categories are homicide, suicide, accidental, natural, and undetermined

Homicide Although there is no universal agreement on its definition, generally the term *homicide,* as certified by coroners' and medical examiners' offices, is defined as a nonaccidental death resulting from grossly negligent, reckless, or intentional actions of another person. Both the cause and manner of death, as certified by the coroner's/medical examiner's offices, can become the subject of expert debate during any subsequent judicial proceedings. However, this does not result in a revision of the death certification unless there has been negligence on the part of the certifying offices.

If the pathologist was unable to go to the scene, the pathologist should receive adequate information detailing the conditions of the scene from coroner/medical examiner investigators and law enforcement personnel. This information should include how the body was discovered as well as when and where. It is also an important first step for investigators to make note of the algor mortis, livor mortis, and/or rigor mortis of the body at the scene. These will help determine time of death.

Suicide Suicide is the result of an individual taking their own life with lethal intention. For a determination of suicide, it must be demonstrated that the individual carried out the act alone. If there is any doubt about the intentions of the victim, the death is not classified as a suicide; the death is ruled as an accident or even as undetermined. The most common methods of suicide include self-inflicted gunshot wounds, hanging, and drug overdosing. Although drug abuse is deliberately committed by a victim, it is not considered suicide unless it was clearly intended as a lethal act.

Various challenges are associated with discriminating suicide from an accident or even homicide. The victim's personal history, including their psychiatric history, becomes relevant. Suicidal threats or past attempts would give obvious evidence of a suicide as opposed to an accident. In all cases of suspected suicide, a thorough search of the victim's possessions should be made to locate a suicide note.

Multiple gunshot wounds might lead one to suspect homicide. However, a person who is committed to ending their own life may take several shots if the wounds are not instantly fatal. It is imperative to confirm that it is physically possible that the victim could inflict the wounds. There are a few areas of the body that strongly point toward homicide. These are areas that are not easily accessible to the victim's own reach. For example, anywhere on the back of a victim is difficult and sometimes impossible for the victim to have shot by their own hand. This is especially true if the wound was made in the back of the head. For suicides, the most common shot is to the temple of the head. The mouth, forehead, and chest are also common.

Also, if the wound was immediately incapacitating, the weapon should be present. Blood spatter analysis should be consistent with the proposed order of events. All victims involved in gunshot cases should have their hands swabbed for gunshot residue.

Accidental In all deaths that are ruled accidental, there must not be intent to cause harm through gross negligence on the part of a perpetrator or the victim. Traffic accidents make up a large percentage of accidental deaths, followed by drug overdoses and drownings. The surviving driver may have vehicular homicide charges brought against them especially if the driver is determined to have been driving under the influence of drugs or alcohol. In this case, the official manner of death certified on the death certificate in many jurisdictions would be *vehicular homicide*.

All cases that have the possibility of being a ruled an accident should have toxicological analyses carried out. The presence of drugs and/or alcohol in the victim's system can potentially affect the determination. Also, the pathologist should be aware that some events might be disguised as accidents to cover up a homicide or suicide. For example, bodies recovered from a house fire might show evidence that the victims were dead before the fire started. This evidence might include a lack of soot in the victims' airways or no indication of elevated levels of carbon monoxide. This scenario, although not common, illustrates how the autopsy and scene can apparently not correlate with each other. No matter how obvious a scene may appear, the two should always correspond with one another. Cases of electrocution are generally ruled as accidents, but this may be difficult to prove. High-voltage electrocutions will usually leave burns on the body. Low-voltage electrocutions, however, may show few or no signs of trauma. The scene then becomes crucial in ascertaining the events surrounding the death.

The determination of manner of death in drownings (accidental, suicidal, or homicidal), falls (accidental, pushed, or deliberate), and asphyxiations can be exceedingly difficult, and therefore the investigation into all of its components becomes much more important than the autopsy.

Natural Causes The differentiation between the categories of manner of death can be difficult to make. The distinction between natural and accidental deaths can pose challenges. The classification of natural death includes disease and continual environmental abuse. This abuse can encompass various events, such as chronic drug and alcohol abuse or longtime exposure to natural toxins or asbestos. Again, although drug abuse is deliberately committed by the victim, a death caused by drug use is not considered suicide unless it is clear that drugs were taken as an intentionally lethal act. Acute ethanol intoxication can be ruled as either natural or accidental depending on the circumstances. If the victim suffers from chronic alcoholism, the death is ruled to be natural. If the victim is a teenager experimenting with alcohol for the first time, the death is ruled an accident.

Undetermined A death is ruled to be undetermined only when a rational classification cannot be established. This can happen when the mechanism that caused the death cannot be determined by a physical finding at the autopsy or because of the absence of meaningful findings in the subsequent toxicological and microscopic examinations.

Quick Review

- The manner in which death occurred is classified in death certificates as one of five categories: homicide, suicide, accidental, natural, or undetermined.
- Homicide is generally defined as a nonaccidental death resulting from grossly negligent, reckless, or intentional actions of another person.
- Suicide is the result of an individual taking their own life with lethal intention. Although drug abuse is deliberately committed by a victim, it is not considered the cause of suicide unless it was clearly intended as a lethal act.
- In all deaths that are ruled accidental, there must not be intent to cause harm through gross negligence on the part of a perpetrator or the victim. Traffic accidents make up a large percentage of accidental deaths, followed by drug overdoses and drownings.
- The classification of natural death includes disease and continual environmental abuse. This abuse can encompass various events, such as chronic drug and alcohol abuse or longtime exposure to natural toxins or asbestos.
- An undetermined cause of death arises when the cause of death cannot be determined by a physical finding at the autopsy or because of the absence of meaningful findings in the subsequent toxicological and microscopic examinations.

Estimating Time of Death

A pathologist can never give an exact time of death. However, there are many characteristics that the examiner can analyze in order to arrive at an approximate time of death. Some features can give a very probable time of death, but others are extremely variable. Witnesses can serve to reconstruct the events leading up to the death and the incidents that occurred after the death, along with the times when they occurred, but a single witness's account alone is not enough to make an accurate determination. The chemical and physical changes that occur after death must also be examined.

algor mortis
A process that occurs after death in which the body temperature continually cools until it reaches the ambient or room temperature

Algor Mortis After death the body undergoes a process in which it continually adjusts to equalize with the environmental temperature. This process is known as **algor mortis**. An algor mortis determination must be performed at the scene as early as possible. The first step is to determine as best as possible what the environmental temperatures may have been prior to discovering the body. Then the environmental temperature and the bilateral axillary and/or ear canal temperatures are recorded at the crime scene (rectal temperatures are usually too disruptive at the scene). The cooling rate of a typical body can be used to estimate the time of death. At average ambient temperatures of 70–72°F, the body loses heat at a rate of approximately of 1–1.5°F per hour until the body reaches the ambient or room temperature. However, the rate of heat loss is influenced by factors such as ambient temperature, the size of the body, and the victim's clothing. Because of such factors, this method can only approximate the amount of time that has elapsed since death.

livor mortis
A medical condition that occurs after death and results in the settling of blood in areas of the body closest to the ground

Livor Mortis Another condition that begins when circulation ceases is **livor mortis**. When the human heart stops pumping, the blood begins to settle in the parts of the body closest to the ground. As shown in Figure 4–10, the skin becomes a bluish-purple color in these areas. The onset of this condition begins 20 minutes to 3 hours after death and under average conditions continues for up to 16 hours after death, at which point all lividity, or coloring, is fixed. Initially, lividity can be pressed out of the vessels when the skin is pressed, that is, lividity can be "blanched." With time, coloring becomes "fixed" in the vessels, beginning in the most dependent (lowest) areas and progressing to the least dependent areas, and then finally no blanching can be elicited anywhere. In any case, levels of lividity are tested at the scene with regard to whether it is completely fixed, blanches when subjected to light pressure, or blanches when subjected to significant pressure. A range of time of death can be estimated if at least some of the lividity is still blanching. However, the environmental temperature and the rate of body temperature decline (i.e., algor mortis) directly affect the rate of fixation of lividity and therefore must be taken into account when attempting to estimate time of death from lividity.

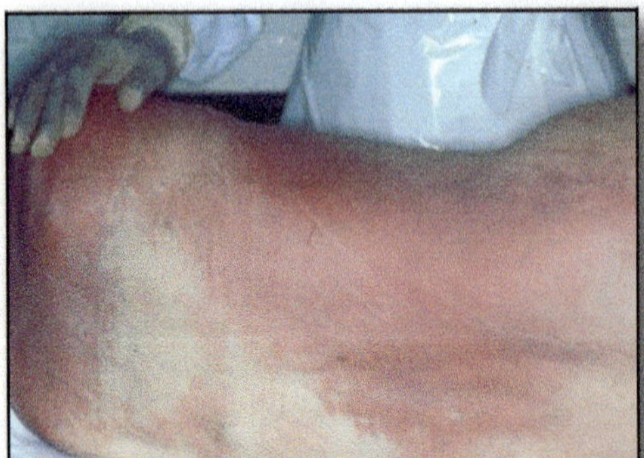

FIGURE 4–10 Livor mortis. *Elizabeth Laposata*

Different lividity patterns in a body may indicate that the body was moved after death but before livor mortis had fully fixed. The skin does not become discolored in areas where the body is restricted by either clothing or an object pressing against the body.

This information can be useful in determining whether the victim's position was changed after death. Livor that is a deep purple is often seen in cases where the victim suffered asphyxia or heart failure.

Rigor Mortis Immediately following death, a chemical change occurs in the muscles that causes them to become rigid, as shown in Figure 4–11. This condition, **rigor mortis**, evolves over the first 24 hours under average temperature and body conditions. This rigidity subsides as time goes on, however, and disappears after about 36 hours under average conditions. Rigor will develop in the position that the body was in at the time of death, essentially freezing the body in that pose. Discovering a body in a position that defies gravity is a likely indicator that the body was moved after death.

rigor mortis
A medical condition that occurs after death and results in the stiffening of muscle mass. The rigidity of the body begins within 24 hours of death and disappears within 36 hours of death

Although rigor mortis can roughly indicate a time of death, there are factors that can alter this determination. An environment that is hot can speed up the process significantly. Conditions that affected the body before death, such as exercise or physical activity, can also speed up the process. Because rigor mortis occurs as a result of the muscles stiffening, individuals with decreased muscle mass may not develop rigor completely. Examples of these individuals may be infants or elderly or obese persons.

Potassium Eye Levels Another approach helpful for estimating the time of death is to determine potassium levels in the decedent's ocular fluid, that is, the fluid within the eye, also known as the *vitreous humor*. It is important to draw a clean, bloodless vitreous sample from one eye with a syringe as soon as possible at the scene, then draw a second sample from the other eye an hour or two later. After death, cells within the inner surface of the eyeball release potassium into the ocular fluid. By analyzing the amount of potassium present at various intervals after death, the forensic pathologist can determine the rate at which potassium is released into the vitreous humor and use it to approximate the time of death. However, the rate of potassium release also is dependent on ambient temperatures.

Stomach Contents Special attention must be paid to the digestive tract. The identification of food items in the stomach may help determine the location of the decedent prior to death (during their last meal). The quantity,

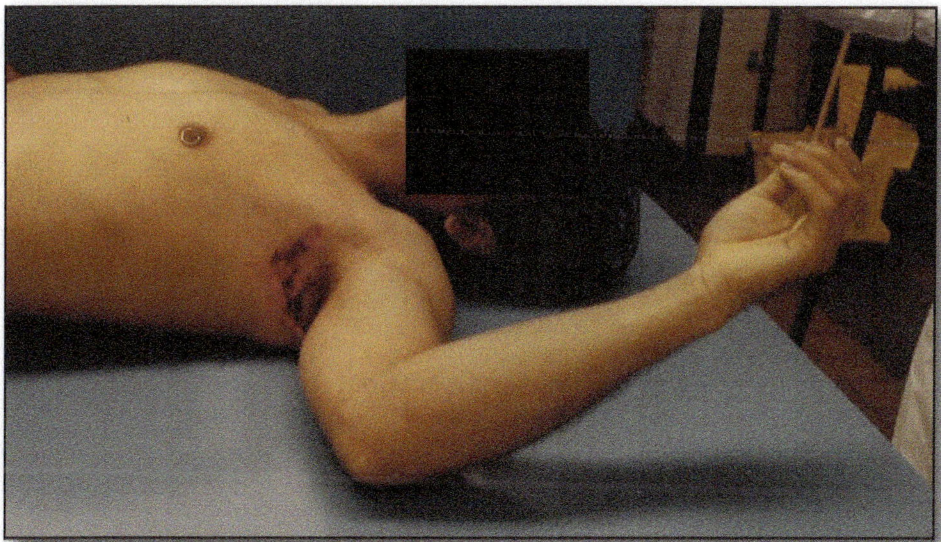

FIGURE 4–11 Rigor mortis in the arm of a decedent.
Elizabeth Laposata

consistency, and color of bile, and the degree of digestion of food in the stomach and its passage into the small intestine can help determine the time of death. The stomach also can contain partially digested or dissolved pills. Chemical analyses can be carried out to identify and analyze substances found in the stomach. These can aid in the determination of cause and manner of death.

Decomposition Once decomposition has set in, the preceding methods of determining time of death are no longer of any use. After death, two decomposition processes take place: autolysis and putrefaction. *Autolysis* is fundamentally self-digestion by cells' own enzymes, and its rate varies from organ to organ depending on the mechanism of death, the enzyme content of the respective organs, the position of the body, and environmental factors. *Putrefaction* is decomposition carried out by microorganisms such as bacteria. Putrefaction is accompanied by bloating, discoloration, and a foul smell caused by accumulating gases. Again, the rate of putrefaction is dependent on the mechanism of death (for example, congestive respiratory versus sudden cardiac death) allowing bacteria to spread from the bowel, presence or absence of infection, environmental temperatures and humidity, degree of obesity, extent of clothing, and so on. Green discoloration often begins in the abdomen. Darker green or purple discoloration follows on the face. The skin begins to blister with gas and then peel (called *slippage*). The skin of the hands and feet can actually detach and come off the body like a glove. This stage is also accompanied by bloating, which causes the eyes to bulge and the tongue to protrude. The chest and extremities will then turn a green/purple discoloration and bloat.

In the postmortem period of decomposition, a waxy substance called *adipocere* may form. Adipocere adds a white or gray wax-like consistency to fatty tissues in the face and extremities that can take on a yellow to tan color. Typically, adipocere takes about three months to develop.

Quick Review

- After death, the body undergoes a process known as algor mortis in which it will continually adjust to equalize with the environmental temperature.
- Another condition beginning when circulation ceases is livor mortis. When the human heart stops pumping, the blood begins to settle in the parts of the body closest to the ground. The skin becomes a bluish-purple color in these areas.
- Immediately following death, a chemical change known as rigor mortis occurs in the muscles, causing them to become rigid.
- Another approach helpful for estimating the time of death is to measure potassium levels in the ocular fluid.
- The identification of food items in the stomach may help to determine the location of the decedent prior to death, during their last meal.

Role of the Forensic Anthropologist

The discipline of **forensic anthropology** is concerned primarily with the recovery, examination, and identification of human skeletal remains. Skeletal remains are remarkably durable in various environmental contexts, and can withstand burial over decades or centuries. Because of their resistance to decomposition, bones can provide a multitude of individual characteristics long after a victim's death. An examination of bones can help anthropologists estimate a victim's age, sex, population affinity, and height. There are 206 total bones in the human body. The major bones of the human skeletal system include the skull, spine, arms, hands, legs, feet, chest, and pelvis. Additionally, pathologies related to disease processes, as well as traumatic injuries, can be observed from the skeleton. Time since death may also be estimated from skeletal remains.

forensic anthropology The examination of human skeletal remains from medico-legal contexts using principles of skeletal biology

Human and Non-human Bones

When bones are identified at a potential scene, one of the first questions should be, "Is it bone?" followed closely by, "Is it human?" To answer this question, anthropologists compare composition and structure of human and non-human bones to make a determination. Mammalian bones all share similar morphological features between species. The biggest difference is often bone size. Some animals, like deer, have bones that are similar in size to an adult human. Some animals, like bears and raccoon, have paws that can be mistaken for a child's hand. Humans are bipedal, walk on two legs, so have a broad and shallow pelvis. Most other mammals are quadrupedal, walk on four legs, and have a pelvis that is long and narrow. Vertebrae in humans are large, broad and have a flat vertebral body while other mammals have a small vertebral body that is often convex on one side and concave on the other. Animal bones have thicker cortical bone than humans. The composition of human bones tends to be thinner than non-human bones. Human bones are very porous, whereas non-human bones are not.

Recovering and Processing Remains

Thorough documentation is required throughout the processes of recovery and examination of human remains. A site where human remains are found must be treated as both a recovery scene and crime scene. When civilians encounter human remains, they should contact law enforcement to secure the scene. The scene should be secured as soon as possible to prevent any further alteration of the environment. The scene should then be searched to locate all bones, if they are scattered, and any other items of evidence such as footwear impressions or discarded items. Many tools can be useful when searching for evidence at a site, including aerial photography, metal detectors, ground-penetrating radar, infrared photography, apparatuses that detect the gases produced by biological decomposition, and so-called cadaver dogs that detect the odors caused by biological decomposition. The collection methods for bone evidence include to first make a scene sketch and take images of the location of each bone. Bone evidence should be carefully removed with gloved hands or forceps. Bone evidence should then be packaged in a container that is labeled. The chain of custody should be maintained. Bone evidence should then be preserved in a freezer until observed and analyzed in the laboratory. All items that are found must be tagged, photographed, sketched, and documented in notes. Once all bones and other evidence are found, a scene sketch should be made to show the exact location of each item (preferably using Global Positioning System [GPS] coordinates) and the spatial relationship of all evidence. Once the skeletal remains have been recovered, processed, and preserved, they can then be transferred to an anthropologist for examination.

Estimating Victim Characteristics The sex of the decedent can be determined by the size and shape (morphology) of various skeletal features, especially those of the pelvis and skull. Female pelvic bones tend to form a wider, more circular opening than that in a male pelvis because of the potential for childbearing. The female sacrum (flat bone above the tailbone) is wider and shorter (see Figure 4–12[a]) than a male's; the length and width of the male sacrum are

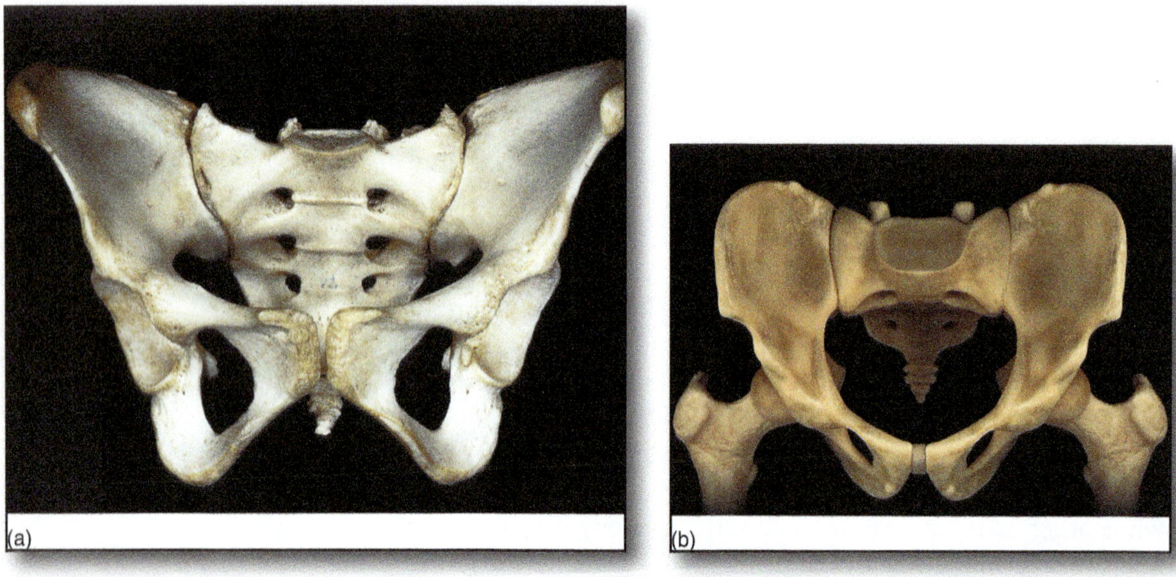

FIGURE 4–12 (a) Frontal view of female pelvis and sacrum. This view shows the wide, circular nature of the pelvic opening and the short, wide nature of the sacrum. (b) Human male pelvis. This view shows the narrow pelvic opening and long, narrow sacrum. *VideoSurgery/Science Source; Science Picture Co/Alamy Stock Photo*

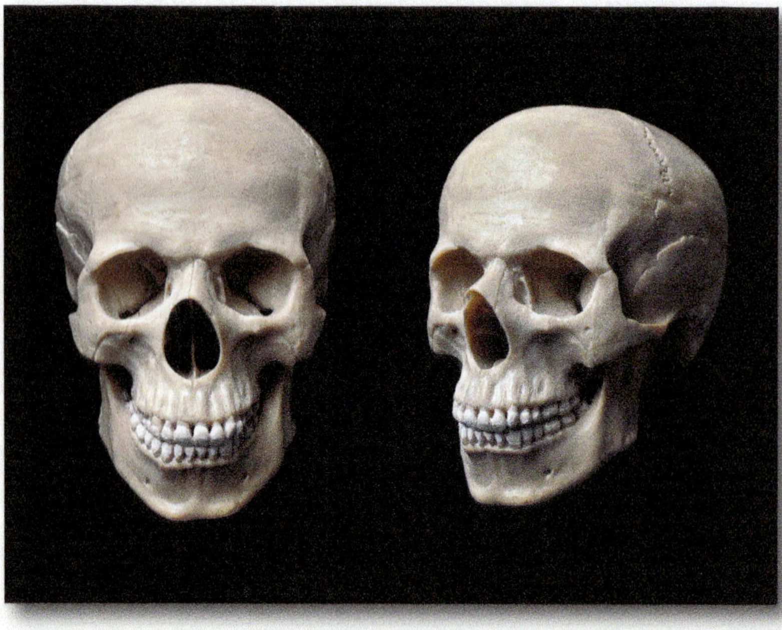

FIGURE 4–13 Male (left) and female (right) human skulls showing male skull's larger size. *Croisy/Shutterstock*

Table 4–1
Summary of Skeletal Features by Sex

	Female	*Male*
Skull	Gracile	Robust
Frontal bossing	Diminished	Pronounced
Brow bone	Diminished	Pronounced
Mastoid process	Diminished	Pronounced
Mandible (jaw) angle	Obtuse (>90 degrees)	Approximately right (90 degrees)
Pelvis opening	Wide, large	Narrow, smaller
Sacrum	Short, wide, straighter	Approximately equal width/length, turned inward
Subpubic angle	Approximately right (90 degrees)	Acute (<90 degrees)
Femur	Angled inward from pelvis	Relatively straight from pelvis

roughly equal (see Figure 4–12[b]). The angle formed at the bottom of the pelvis (subpubic angle) is approximately a right angle (90 degrees) in females, but it is acute (less than 90 degrees) in males. This is generally described as a U shape in females, and a V shape in males. In general, male skulls are larger in overall size than those of females. A male cranium tends to have a more pronounced mastoid process (a bony protrusion behind the jaw) and glabella region of the frontal bone (brow bone) than a female cranium (see Figure 4–13). See Table 4–1 for a summary of the differing features of female and male skeletons from head to toe. These are typical cases; not all skeletons may display the given characteristics to clearly indicate the sex of the decedent.

The method for determining the age of a decedent varies depending on the victim's growth stage. For infants and toddlers, age can be estimated by the length of the long bones (e.g., femur and humerus) when compared to a known growth curve. Different elements of the skull also fuse together at different stages during early development, and the appearance of fused or divided elements can be used to estimate the age of bones still in early developmental stages (see Figure 4–14). In infant skeletons through the late teenage years, formation of teeth can be used in age determination; this is based on the fact that permanent teeth start to form at birth. If the skeletal remains belong to a child, the age of the decedent may be determined by observing the fusion or lack of fusion of epiphyseal regions of bones such as those of the mandible (i.e., lower jaw), fingers, wrist, long bones, and clavicle (see Figure 4–15). The average age at which each of these regions fuses, while population-specific, is known and can be compared against the state

FIGURE 4–14 A fetal skull showing the separated bones of the skull before they have had a chance to fuse. *Martin Shields/Alamy Stock Photo*

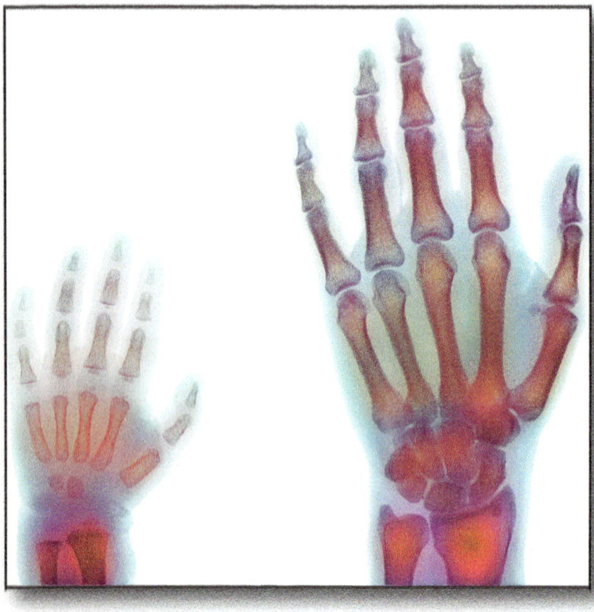

FIGURE 4–15 Colored X-rays of healthy human hands at 3 years (left) and at 20 years. Bones display in red, and flesh is in blue. The child's hand has areas of cartilage in the joints between the finger bones (i.e., epiphyseal areas), where bone growth and fusion will occur. In the adult hand, all the bones are present, and the joints have closed. *Courtesy SPL/Science Source*

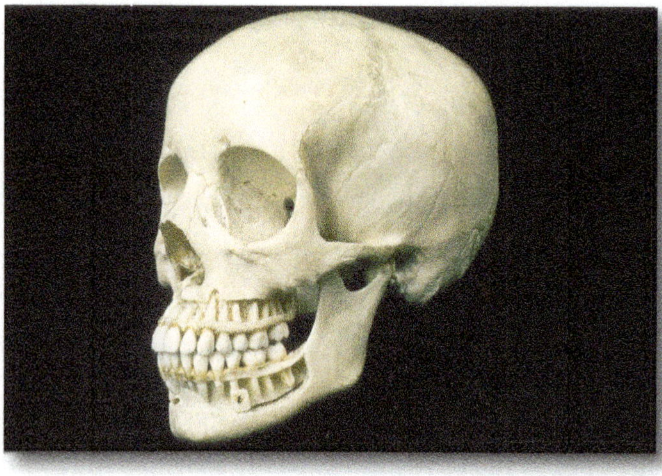

FIGURE 4–16 The skull of an adult. *John Watney/Science History Images/Alamy Stock Photo*

of the remains to provide a range of possible ages for the decedent. A child's cranium may also be identified by its smaller size and the presence of developing teeth as contrasted with the skull of an adult. After age 21, age is estimated by the level of change the surfaces of the bones have undergone, especially in areas of common wear such as the pubic symphysis. The pubic symphyseal face shown in Figure 4–17 is a raised platform that slowly changes over the years from a furrowed surface to a pitted, well-defined area. Age estimates can also be derived from sternal ribs, cranial sutures, and histology.

Although the categorization "race" has come under scrutiny and is difficult to define, forensic anthropologists have historically used broad classes to characterize the likely (but not definite) population affinity of skeletal remains. The population affinity of the decedent can be assessed by the appearance of various

Table 4–2
Summary of Skeletal Characteristics Indicating Population Affinity

	Eye Orbits	*Nasal Aperture*	*Incisors*	*Cranium Frontal Plane*
European	Oval	Long, narrow	Smooth	Flat
Asian	Circular	Small, rounded	Shoveled interior	Flat or projected outward
African	Square	Wide	Smooth	Projected outward

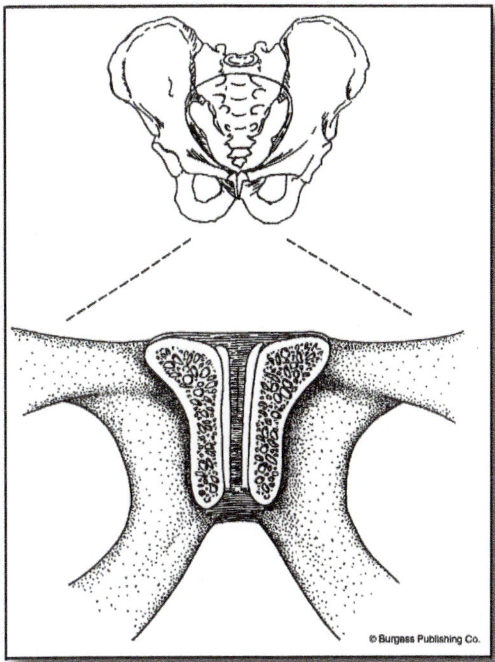

FIGURE 4–17 The symphysis pubis shown magnified beneath human pelvic bones.
Pearson Education Custom Publishing

cranial features (and post-cranial features, to an extent) on the skeletal remains. For example, eye orbits tend to be circular in Asian skeletons, oval in European skeletons, and square in African skeletons. The frontal plane of the cranium may also vary. The frontal plane of Asian craniums may be flat or projected outward, that of European craniums is flat, and that of African craniums is projected outward. The nasal cavity tends to be small and rounded in Asians, long and narrow in Europeans, and wide in Africans. Skeletal remains of decedents of Asian population affinity, including those of Native American descent, also tend to have "scooped-out" or shovel-shaped incisor teeth. See Table 4–2 for a summary of the differing features of skeletons that can indicate population affinity. These are typical cases; not all skeletons may display the given characteristics to indicate the population affinity of the decedent.

The height of the victim when alive can be estimated by measuring the long bones of the skeleton, especially in the lower limbs. Even partial bones can yield useful results. However, meaningful stature calculations from known equations must be based on the determined sex and ancestry of the remains. See Table 4-3 for examples of equations used to calculate the height of the decedent from skeletal remains. These equations should yield estimations within 5 cm of actual height.

Table 4–3
Equations for Height Calculation from Skeletal Remains

	European	*African*	*Unknown Population Affinity*
Female	Height (cm) = femur length (cm) × 2.47 + 54.10	Height (cm) = femur length (cm) × 2.28 + 59.76	Height (cm) = femur length (cm) × 3.01 + 32.52
	Height (cm) = humerus length (cm) × 3.36 + 57.97	Height (cm) = humerus length (cm) × 3.08 + 64.67	Height (cm) = humerus length (cm) × 4.62 + 19.00
Male	Height (cm) = femur length (cm) × 2.32 + 65.53	Height (cm) = femur length (cm) × 2.10 + 72.22	Height (cm) = femur length (cm) × 2.71 + 45.86
	Height (cm) = humerus length (cm) × 2.89 + 78.10	Height (cm) = humerus length (cm) × 2.88 + 75.48	Height (cm) = humerus length (cm) × 4.62 + 19.00

Other Contributions of Forensic Anthropology

A forensic anthropologist may assist in the creation of facial reconstructions to help identify skeletal remains. **Forensic anthropology** is concerned primarily with the identification and examination of human skeletal remains. Historically, facial approximation clay was placed and shaped over the victim's actual cranium, and it took into account the decedent's estimated age, population affinity, and sex. Today, this technique is performed digitally with computer software (see Figure 4–18). With the help of this technique, a composite of the victim can be drawn and advertised in an attempt to identify the victim.

forensic anthropology Ancestry assessment represents a major component of forensic anthropological analysis of recovered human remains. Interpretations of ancestry, together with other aspects of the biological profile, can help narrow the search of missing persons and contribute to eventual positive identification. Such information can prove useful to authorities involved in the identification and investigative process since many lists of missing persons have a reference to this parameter. Recent research has strengthened available methodologies involving metric, non-metric morphological as well as chemical and genetic approaches.

Forensic anthropologists are also helpful in identifying victims of a mass disaster such as a plane crash. When such a tragedy occurs, forensic anthropologists can help identify victims using the collection of bone fragments. Usually, the identification of the remains will depend on medical records, especially dental records of the individuals. When skeletal remains containing a skull with teeth are found, investigators can compare the skeletal remains to dental records of individuals that have been reported as missing for possible identification. Medical procedures such as dentures, x-rays, and implants are recorded in a living person's medical charts and can be used as additional pieces of evidence to connect to skeletal remains. However, definite identification of remains can be made only by analyzing the decedent's DNA profile, fingerprints, or medical records. Recovered remains may still contain some soft tissue material, such as the tissue of the hand, which may yield a DNA profile for identification purposes. If the tissue is dried out, it may be possible to rehydrate it to recover fingerprints also.

Case Files

Identifying a Serial Killer's Victims

The worst serial killer in the United States calmly admitted his guilt as he led investigators to a crawl space under his house. There, John Wayne Gacy had buried 28 young men, after brutally raping and murdering them in cold blood. Because no forms of identification were found with the bodies, the police were forced to examine missing-person reports for leads. However, these boys and men were so alike in age, population affinity, and stature that police were unable to individually identify most of the victims. Clyde Snow, the world-renowned forensic anthropologist from Oklahoma, was asked to help the investigators make these difficult identifications.

Snow began by making a 35-point examination of each skull for comparison to known individuals. By examining each skeleton, he made sure each bone was correctly attributed to an individual. This was crucial to later efforts because some of the victims had been buried on top of older graves, mingling their remains. Once Snow was sure all the bones were sorted properly, he began his in-depth study. Long bones such as the femur (thigh bone) were used to estimate each individual's height. This helped narrow the search in the attempt to match the victims with the descriptions of missing people.

After narrowing the list of missing people to those fitting the general description, investigators consulted missing persons' hospital and dental records. Evidence of injury, illness, or surgery and other unique skeletal defects of the victims were matched to information in the records to make identifications. Snow also pointed out features that gave useful clues to the victim's behavior and medical history. For example, he discovered that one of Gacy's victims had a healed fracture on his left arm, and that his left scapula (shoulder blade) and arm bore the telltale signs of a left-handed individual. These details were matched to a missing-person report, and another young victim was identified.

For the most difficult cases, Snow called in the help of forensic sculptor and facial reconstructionist Betty Pat Gatliff. She used clay and depth markers to put the "flesh" back on the faces of these forgotten boys in the hopes that someone would recognize them after the photographs of the reconstructed faces were released to the media. Her efforts were successful, but investigators found some families unwilling to accept the idea that their loved one was among Gacy's victims. Even with Gatliff's help, nine of Gacy's victims remain unidentified.

The scope of forensic anthropologists' work is broad and includes contributions in a variety of areas, such as the following:

- Human/non-human identification
- Time since death estimation
- Native American Graves Protection and Repatriation Act (NAGPRA) assistance
- Genocide reconciliation
- Teaching
- Museum work

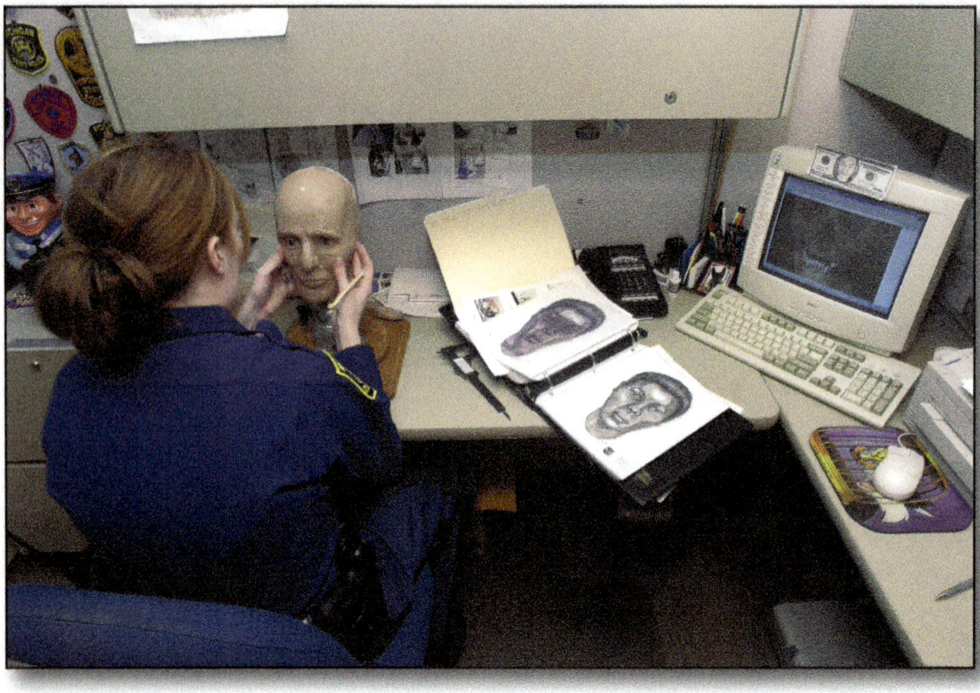

FIGURE 4–18 Trooper Sarah Krebs, a Michigan State Police forensic artist, works on a three-dimensional facial reconstruction from an unidentified human skull at Richmond Post in Richmond, MI. *Paul Sancya/AP Images*

Quick Review

- Forensic anthropology is concerned primarily with the identification and examination of human skeletal remains.
- The sex of the decedent can be determined by the size and shape of various skeletal features, especially those of the pelvis and skull, or cranium.
- The height of the victim when alive can be estimated by measuring the long bones of the skeleton, especially those in the lower limbs.

Role of the Forensic Entomologist

forensic entomology
The study of insect matter, growth patterns, and succession of arrival at a crime scene to determine the time since death

The study of insects and their relation to a criminal investigation is known as **forensic entomology**. In practice, forensic entomology is commonly used to estimate the time of death when the circumstances surrounding the crime are unknown. This determination can be carried out by observing the stage of development of maggots or insects' sequence of arrival.

Determining Time of Death

After decomposition begins, necrophilious insects, or insects that feed on dead tissue, are the first to infest the body, usually within 24 hours. The most common and important of these is the blowfly, recognized by its green or blue color. Blowfly eggs are laid in human remains and ultimately hatch into maggots, or fly larvae,

Case Files

The Danielle Van Dam Murder Case

Sometime during the night of February 1, 2002, 7-year-old Danielle Van Dam disappeared from her bedroom in the Sabre Springs suburb of San Diego, California. On February 27, three and a half weeks later, searchers found her naked body in a trash-covered lot about 25 miles from her home. Because of the high degree of decomposition of the girl's remains, the medical examiner could not pinpoint the exact time of the girl's death. Her neighbor, 50-year-old engineer David Westerfield, was accused of kidnapping Danielle, killing her, and dumping her body in the desert. During the subsequent investigation, Danielle's blood was found on Westerfield's clothes, her fingerprints and blood were found in his RV, and child pornography was found on his home computer.

The actual time of the 7-year-old's death became a central issue during the murder trial. Westerfield had been under constant police surveillance since February 4. Any suggestion that Danielle was placed at the dump site after that date would have eliminated him as a suspect. Conflicting expert testimony was elicited from forensic entomologists who were called on to estimate when the body was dumped.

The forensic entomologist who went to the dump site, witnessed the autopsy, and collected and analyzed insects from both locations estimated that Danielle died between February 16 and 18. A forensic entomologist and a forensic anthropologist both called to testify on behalf of the prosecution noted that the very hot, very dry weather at the dump site might have mummified Danielle's body almost immediately, thus causing a delay in the flies colonizing the body.

The jurors convicted Westerfield of the kidnapping and murder of Danielle Van Dam, and a San Diego judge sentenced David Westerfield to death. Danielle Van Dam's parents filed and settled a wrongful death suit against Westerfield requiring his automotive and homeowner's insurance carriers to pay the Van Dams an undisclosed amount, reported to be between $400,000 and $1 million.

FIGURE 4–19 A scanning electron micrograph of two-hour-old blowfly maggots.
Dr Jeremy Burgess/Science Source

that consume human organs and tissues (see Figure 4–19). Typically, a single blowfly can lay up to 2,000 eggs during its lifetime. The resulting larvae gather and feed as a "maggot mass" on the decomposing remains. Forensic entomologists can approximate how long a body has been left exposed by examining the stage of development of the fly larvae. This kind of determination is best for a timeline of hours to approximately one month because the blowfly goes through the stages of its life cycle at a known sequence and in known time intervals that span this period. By determining the most developed stage of fly found on the body, entomologists can approximate the **postmortem interval (PMI)**, or the time that has elapsed since death (see Figure 4–20). Newly emerged flies are of important forensic interest, as they indicate that an entire blowfly cycle has been completed on the decomposing body. Likewise, empty pupal cases indicate that a fly has completed its entire life cycle on the body. Flies known as cheese skippers are primarily found on human corpses in the later stages of decomposition, long after the blowflies have left the corpse.

postmortem interval (PMI)
The length of time that has elapsed since a person has died; if the time is not known, a number of medical or scientific techniques may be used to estimate it

Time determinations based on the blowfly cycle are not always straightforward, however. The time required for each stage of development is affected by environmental influences such as geographical location, climate, weather conditions, and the presence of drugs. For example, cold temperatures hinder the development of fly eggs into adult flies. The forensic entomologist must consider these conditions when estimating the PMI.

Information about the arrival of other species of insects may also help determine the PMI. The sequence of arrival of these groups depends mostly on the body's natural decomposition process. Predator insects generally arrive and prey on the necrophilious insects. Several kinds of beetles will be found, either feeding directly on the corpse's tissues or as predators feeding on blowfly eggs and maggots present on the corpse. Next, omnivore insects arrive at the body. These insects feed on the body, on other insects, and on any surrounding vegetation. Ants and wasps are an example of omnivore insects. Last comes the arrival of indigenous insects, such as spiders, whose presence on or near the body is coincidental as they move about their environment.

Other Contributions of Forensic Entomology

Entomological evidence can also provide other pertinent information. In general, insects first colonize the body's naturally moist orifices. However, if open wounds are present, they will colonize there first. Although the decomposition processes may conceal wounds, colonization away from natural orifices may indicate the locations of wounds on the body. If maggots are found extensively on the hands and forearms, for example, this suggests the presence of defensive wounds on the victim. Insects that have fed on the body may also have accumulated any drugs present in the flesh, and analyzing these insects can yield the identity of these drugs.

If resources allow, all insect evidence should be carefully collected by a forensic entomology expert. When this is not possible, collection should be carried

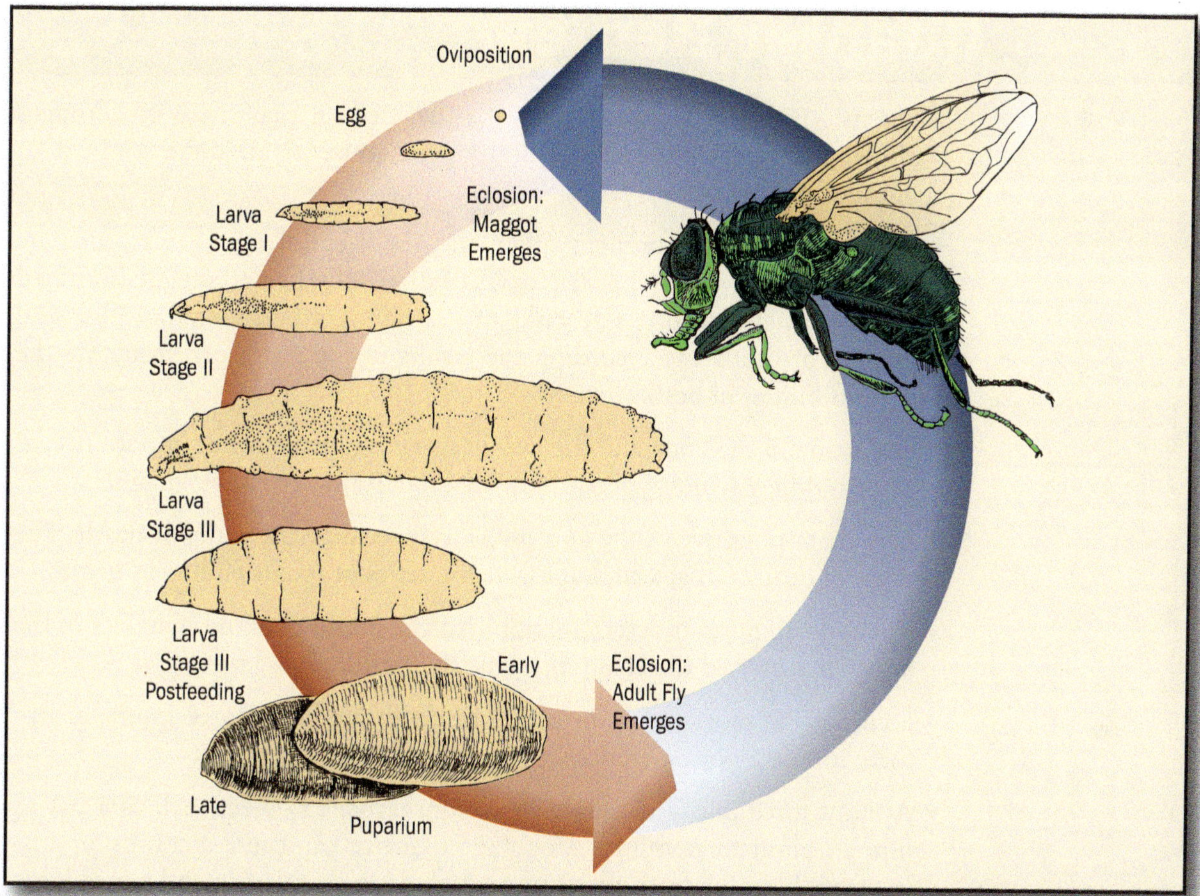

FIGURE 4–20 Typical blowfly life cycle from egg deposition to adult fly emergence. This cycle is representative of any one of nearly ninety species of blowflies in North America. *Based on E.P. Catts, Entomology and Death: A Procedural Guide. 1990, 2008.*

out by an investigator with experience in death investigation. The entire body and the area where insect evidence was found must be photographed and documented before collection. Insect specimens should be taken from each area on the body where they are found and labeled to show where they were collected from.

Quick Review

- Forensic entomologists can approximate how long a body has been left exposed by examining the stage of development of fly larvae on the body.
- Information about the arrival of other species of insects may also help determine the postmortem interval. The sequence of arrival of these groups depends mostly on the body's natural decomposition process.
- In general, insects first colonize the body's naturally moist orifices. However, if open wounds are present, they will colonize there first.
- Mammalian bones share similar morphology. Animal bones can be mistaken for human bones by the untrained eye. Humans are bipedal with a broad, shallow pelvis. Most other mammals are quadrupedal with a long, narrow pelvis. Animal bones have thicker cortical bone than humans.

Chapter Review

- Forensic pathologists associated with the medical examiner's or coroner's office are responsible for determining the cause of an undetermined or unexpected death.
- Although both the coroner's office and the medical examiner's office are charged with investigating suspicious deaths, only the pathologist is trained to perform an autopsy. The tasks of examining a body for the cause and manner of death and recording the results in the death certificate are the responsibilities of both offices.
- Protection of the body and the overall scene is of paramount importance, as is the ultimate removal of the body in a medically acceptable manner.
- A primary objective of the autopsy is to determine the cause of death. The cause of death is that which initiates the series of events ending in death.
- The most important determination in a violent death is the character of the injury that started the chain of events that resulted in death.
- Some of the more common causes of death are blunt-force injury, sharp-force injury, asphyxia, gunshot wound, and substance abuse.
- A blunt-force injury is caused by a nonsharpened object such as a bat or pipe. A blunt-force injury can abrade tissue or can cause a contusion arising from bleeding from tiny ruptured blood vessels within and beneath the skin.
- Sharp-force injuries occur from weapons with sharp edges, such as knives or blades.
- Asphyxia encompasses a variety of conditions that involve interference with the intake of oxygen. For example, death at a fire scene is caused primarily by the extremely toxic gas, carbon monoxide.
- Gunshot wounds originate from projectiles fired by a firearm. The distance a weapon was fired from a target is one of the most important factors in characterizing a gunshot wound.
- Because drug abuse is so common, a forensic pathologist will routinely order toxicological tests for the presence of drugs in nearly all autopsies.
- An autopsy, in its broadest definition, is simply the examination of a body after death.
- The forensic autopsy consists of an external examination and an internal examination.
- The first steps taken for the external examination include a broad overview of the condition of the body and the clothing.
- The external examination also consists of classifying the injuries. This includes distinguishing between different types of wounds, such as a stab wound versus a gunshot wound.

- The dissection of the human body generally entails the removal of all internal organs through a Y-shaped incision beginning at the top of each shoulder and extending down to the pubic bone.
- The internal examination entails weighing, dissecting, and sectioning each organ of the body.
- Blood is often routinely tested to determine the presence and levels of alcohol and drugs.
- Some drugs redistribute or reenter the blood after death and thus may complicate the interpretation of postmortem blood levels of these drugs.
- The manner in which death occurred is classified in death certificates as one of five categories: homicide, suicide, accidental, natural, or undetermined.
- Homicide is generally defined as a nonaccidental death resulting from grossly negligent, reckless, or intentional actions of another person.
- Suicide is the result of an individual taking their own life with lethal intention. Although drug abuse is deliberately committed by a victim, it is not considered the cause of suicide unless it was clearly intended as a lethal act.
- In all deaths that are ruled accidental, there must not be intent to cause harm through gross negligence on the part of a perpetrator or the victim. Traffic accidents make up a large percentage of accidental deaths, followed by drug overdoses and drownings.
- The classification of natural death includes disease and continual environmental abuse. This abuse can encompass various events, such as chronic drug and alcohol abuse or longtime exposure to natural toxins or asbestos.
- An undetermined cause of death arises when the cause of death cannot be determined by a physical finding at the autopsy or because of the absence of meaningful findings in the subsequent toxicological and microscopic examinations.
- After death, the body undergoes a process known as algor mortis in which it will continually adjust to equalize with the environmental temperature.
- Another condition beginning when circulation ceases is livor mortis. When the human heart stops pumping, the blood begins to settle in the parts of the body closest to the ground. The skin becomes a bluish-purple color in these areas.
- Immediately following death, a chemical change known as rigor mortis occurs in the muscles, causing them to become rigid.
- Another approach helpful for estimating the time of death is to measure potassium levels in the ocular fluid.
- The identification of food items in the stomach may help to determine the location of the decedent prior to death, during their last meal.
- Forensic anthropology is concerned primarily with the identification and examination of human skeletal remains.

- The sex of the decedent can be estimated by the size and shape of various skeletal features, especially those of the pelvis and skull, or cranium.
- The height of the victim when alive can be estimated by measuring the long bones of the skeleton, especially those in the lower limbs.
- Forensic entomologists can approximate how long a body has been left exposed by examining the stage of development of fly larvae on the body.
- Information about the arrival of other species of insects may also help determine the postmortem interval. The sequence of arrival of these groups depends mostly on the body's natural decomposition process.
- In general, insects first colonize the body's naturally moist orifices. However, if open wounds are present, they will colonize there first.

Review Questions

1. The titles of __________ and __________ are often used interchangeably, but there are significant differences in their job descriptions.
2. True or False: The medical examiner is an elected official and is not required to possess a medical degree.
3. Although both a coroner and a forensic pathologist are charged with investigating a suspicious death, only the __________ is trained to perform an autopsy.
4. True or False: If it appears that a victim did not shoot themself or anyone else, the victim's hands should not be swabbed.
5. The primary objective of the autopsy is to determine the __________.
6. True or False: The manner of death is defined as that which initiates the series of events ending in death.
7. A(n) __________ -force injury can abrade and crush tissue.
8. True or False: The outward appearance of the injuries will always match the injuries sustained inside the body.
9. Wounds on a victim's forearm may be __________ wounds.
10. True or False: A lack of any defense wounds can lead a pathologist to believe that the victim was either unconscious or somehow tied up during the assault.
11. Asphyxia encompasses a variety of conditions that involve interference with the intake of __________.
12. True or False: Death at a fire scene is primarily caused by the extremely toxic gas carbon monoxide.
13. The protein in red blood cells that transports oxygen is known as __________.
14. True or False: High levels of carbon monoxide must be present for a victim to become disoriented and lose consciousness.
15. True or False: Carbon monoxide will continue to build up in the body after death.

16. Carbon monoxide levels and the presence of soot can be used to determine whether the individual was ________ at the time of the fire.

17. Victims of hangings often show signs of __________ on the eyelids, cheeks, and forehead.

18. Petechiae are caused by the escaping of blood into the tissue as a result of __________ bursting.

19. True or False: Petechiae are more common in hangings than strangulation deaths.

20. True or False: Typically the hyoid bone and thyroid cartilage are not fractured in hanging cases.

21. True or False: For gunshot victims, the cause of death can be listed as a gunshot wound.

22. True or False: Because drug abuse is so common, the forensic pathologist will routinely test for the presence of drugs in nearly all investigations.

23. A(n) in its broadest definition __________ is simply the examination of a body after death.

24. True or False: There are two types of autopsies: a forensic/medicolegal autopsy and a clinical/hospital autopsy.

25. The autopsy consists of a(n) __________ examination and a(n) __________ examination.

26. The discharge from a firearm will produce characteristic markings on the skin known as __________.

27. True or False: X-ray examinations are most commonly performed in gunshot wound cases and stab wound cases.

28. Pulmonary __________, or fluid accumulation in the lungs, is frequently found in victims of chronic cocaine and amphetamine abuse.

29. True or False: The liver can contain partially digested or dissolved pills.

30. True or False: The ideal location to take a blood sample is from the heart.

31. __________ is the redistribution of drugs after death.

32. True or False: General testing for poisons is not a routine procedure carried out by the pathologist.

33. A body that displays a cherry-red discoloration often leads a pathologist to suspect poisoning by __________.

34. True or False: A pathologist can often give an exact time of death.

35. The process of the body's continually decreasing in temperature after death until it reaches the environmental temperature is known as __________.

36. The process of the blood settling in parts of the body closest to the ground after death is known as __________.

37. True or False: Different lividity patterns on a body may indicate that the body was moved after death but before livor mortis had fully fixed.

38. Levels of __________ in the ocular fluid can help indicate the time of death.

39. After death, two decomposition processes take place: __________ and __________.

40. The female bone structure differs from the male structure within the __________ area because of a woman's child-bearing capabilities.

41. True or False: A definite identification of remains cannot be made through the analysis of the decedent's DNA profile, fingerprints, or medical records.

42. True or False: A site where human remains are found must be treated as a crime scene, and the site and surrounding area should be secured, searched, and carefully processed.

43. The field of __________ takes advantage of the durable nature of bones over a long period of time to examine and identify human skeletal remains through a multitude of individual characteristics.

44. The study of insects and their relation to a criminal investigation __________, known as, is commonly used to estimate the time of death when the circumstances surrounding the crime are unknown.

45. By determining the oldest stage of fly found on the body and taking environmental factors into consideration, entomologists can approximate the __________ interval.

46. True or False: Another method to determine PMI is by observing the schedule of arrival of different insects species on the body.

47. Physical and chemical changes that occur following death include all of the following except
   a. livor mortis.
   b. biogor mortis.
   c. algor mortis.
   d. rigor mortis.

Learn the Vocabulary

Use the information from this chapter along with your prior knowledge and experience to learn the key terms. Choose from the following activities:

1. Make flashcards with a term on the front and its definition on the back. Use the flashcards to quiz a partner on correctly pronouncing and defining the terms.
2. Make an audio or video recording of each term and its definition.
3. Work with a partner to write a sentence for each term. Read the sentences out loud to a partner or share them with the class.
4. Write a list of new terms and language you have encountered in this chapter. Reuse the terms to write sentences and then read the sentences out loud to a partner. Reuse the terms again by writing a short speech. The speech might explain the terms or teach a lesson using the terms. Give the speech to a partner or to the class.

Application and Critical Thinking

1. Rigor mortis, livor mortis, and algor mortis are all used to help determine time of death. However, each method has its limitations. Explain the principles of each method and how how they apply to a deceased person. Describe at least one condition that would render each method unsuitable or inaccurate for determining time of death.

2. What kind of forensic expert would most likely be asked to help identify human remains in each of the following conditions?

a. A body that has been decomposing for a day or two

b. Fragmentary remains of a few arm bones and part of a jaw

c. A skeleton that is missing its skull

3. Identify a reasonable manner of death for each of the following situations:

a. A contact wound to the back of the head

b. An elevated carboxyhemoglobin blood level in a fire victim

c. A fractured hyoid bone

d. Death by overdose of a first-time user of alcohol

e. A gunshot wound to the chest from a distance of 3 feet

f. Sudden death of a young chronic user of cocaine

4. In cooperation with the medical examiner or coroner, evidence retrieved from a deceased victim and sent to the crime lab should include which items?

5. The following images depict the sequence of events at the site of a decomposing body. Place the arrival events in order of occurrence from earliest to latest.

Andrey Pavlov/Shutterstock

(a)

Xpixel/Shutterstock

(b)

irin-k/Shutterstock

(c)

Volker Steger/Science Source

(d)

Goldfinch4ever/123RF

(e)

Dorling Kindersley ltd/Alamy Stock Photo

(f)

6. Using a skeletal diagram of a human, identify the major bones of the human skeletal system used for human identification.

7. If possible, examine sample human and non-human bones, or refer to illustrations, to compare the composition and structure. In a short written response, describe the composition and structure of human and non-human bones.

8. In a brief written response, describe the collection and preservation methods used for bone evidence.

9. Explain in a brief written response how human remains are identified through dental records.

10. Use a case study or simulated autopsy report obtained through resources to determine cause and manner of death. Write a summary of your findings.

11. In a brief written response, describe how a forensic entomologist would determine the approximate time of death using entomology.

12. In a brief written response, describe how a forensic pathologist can differentiate between the types of wound patterns. Provide specific examples for each of the following types of wounds: lacerations and blunt force trauma resulting from stabbings, bludgeoning, gunshots, and strangulations.

Laboratory Experiments

This activity requires the use of the following practices of science and engineering:

- Planning and carrying out investigations
- Analyzing and interpreting data
- Using mathematics and computational thinking
- Constructing explanations
- Obtaining, evaluating, and communicating information
- Using appropriate safety equipment and practices during laboratory/classroom and field investigations as outlined in Texas Education Agency-approved safety standards

This activity consists of the following crosscutting concepts:

- **Patterns**—Evident in the relationship between skeletal remains and height, gender, ancestry and age.
- **Structure and function**—Shape of the pelvis allows for gender identification.

Before you begin, read the scenario and steps out loud with a partner. Ask for clarification as necessary to make sure you understand the experiment.

The Scenario

Two human skeletons were found on Mud Island in Memphis, Tennessee, on June 30 at 3:42 PM by a local woman out for a run. The remains appear to be intact, with all pieces of the skeletons present. Several people have gone missing in the Memphis area recently. Identification of the remains will initially be based on the skeletal features. The skeletons have features shown in the accompanying tables and images. Approximate and explain the sex, ancestry, age, and height of both individuals based on the characteristics of these human skeletons. Apply scientific practices to conduct comparative and descriptive investigations to differentiate between the two human skeletons. Working collaboratively with a partner, communicate your explanations and solutions in a paragraph describing how the characteristics of the human skeletal system can be used to determine biological sex and approximate range of age and height. Justify your decisions based on the data given.

Be prepared to communicate your findings, both individually and collaboratively, and be able to explain your reasoning to your classmates. If any of your classmates have reached different conclusions, be prepared to respectfully debate your findings supporting your opinions with scientific argumentation and empirical evidence. Explain the experiment to someone outside of the class.

Cranium	
Size	Medium
Forehead	Rounded, projected outward Mastoid
process	Absent
Jaw	Angle = 110 degrees
Teeth	All permanent
Sagittal suture	Not fused
Coronal suture	Not fused
Eye orbits	Squared
Nasal cavity	Large, wide
Incisors	Smooth
Pelvis	
Opening	*See figure*
Sacrum	*See figure*
Subpubic angle	90–100 degrees
Long Bones	
Femur	Fully fused, 44.1 cm long
Clavicle	Fully fused

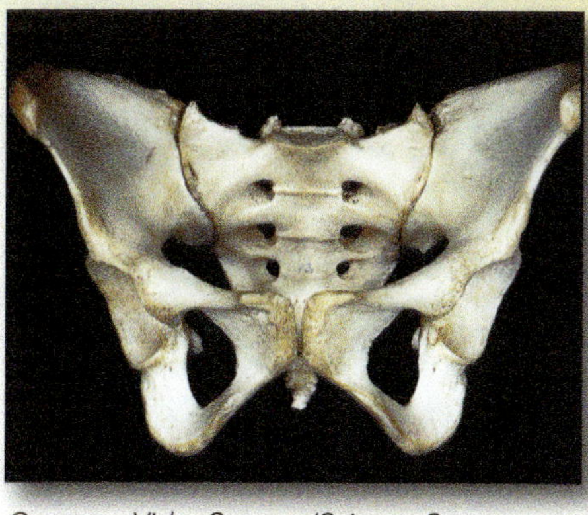

Courtesy VideoSurgery/Science Source

Cranium	
Size	Large
Forehead	Receding
Mastoid process	Large
Jaw	Close to 90 degrees
Teeth	All permanent
Sagittal suture	Fused
Coronal suture	Not fused
Eye orbits	Oval
Nasal cavity	Long, narrow
Incisors	Smooth
Pelvis	
Opening	*See figure*
Sacrum	*See figure*
Subpubic angle	<90 degrees
Long Bones	
Femur	Fully fused, 48.5 cm long
Clavicle	Fully fused

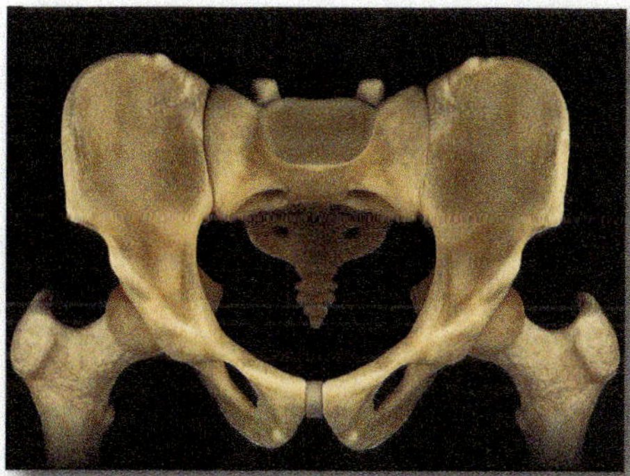

Courtesy Science Picture Co/Alamy Stock Photo

Properties of Matter and the Analysis of Glass

5

Courtesy New York Daily News/Getty Images

Key Terms

amorphous solid
atom
Becke line
birefringence
Celsius scale
chemical property
compound
concentric fracture
crystalline solid
density
dispersion
electromagnetic spectrum
element
Fahrenheit scale
frequency
gas
intensive property
laminated glass
laser
liquid
mass
matter
periodic table
phase
photon
physical property
physical state
radial fracture
refraction
refractive index
solid
sublimation
tempered glass
visible light
wavelength
weight
X-ray

The Lindbergh Baby Case

On the evening of March 1, 1932, a kidnapper crept up his homemade ladder and stole the baby of Charles and Anne Lindbergh directly from the second-floor nursery of their house in Hopewell, New Jersey. The only evidence of his coming was a ransom note, the ladder, a chisel, and the tragic absence of the infant. A couple of months later, though the $50,000 ransom had been paid, the baby turned up dead in the woods a mile away. There was no additional sign of the killer. Fortunately, when finally studied by wood technologist Arthur Koehler, the abandoned ladder yielded some important investigative clues.

By studying the types of wood used and the cutter marks on the wood, Koehler ascertained where the materials might have come from and what specific equipment was used to create them. Koehler traced the wood from a South Carolina mill to a lumberyard in the Bronx, New York. Unfortunately the trail went cold, as the lumberyard did not keep sales records of purchases. The break in the case came in 1934, when Bruno Richard Hauptmann paid for gasoline with a bill that matched a serial number on the ransom money. Koehler showed that microscopic markings on the wood were made by a tool in Hauptmann's possession. Ultimately, handwriting analysis of the ransom note clearly showed it to be written by Hauptmann.

Learning Objectives

After studying this chapter, you should be able to:

5.1 Discuss the composition of matter

5.2 Explain how density and refractive index of glass is measured and utilized for forensic characterization

5.3 Explain the nature of light as a wave and as a particle

5.4 Explain forensic methods for comparing glass fragments

5.5 Explain how glass fractures reveal information related to the force and direction of an impact

5.6 Describe the proper collection of glass evidence

National Science Content Standards

Scientific Inquiry

Properties of Matter

The forensic scientist must constantly determine the properties that impart distinguishing characteristics to matter, giving it a unique identity. The continuing search for distinctive properties ends only when the scientist has completely individualized a substance to one correct source. Properties are the identifying characteristics of substances. In this and succeeding chapters, we will examine properties that are most useful for characterizing soil, glass, and other physical evidence. However, before we begin, we can simplify our understanding of the nature of properties by classifying them into two broad categories: physical and chemical.

physical property A property that describes the behavior of a substance without reference to any other substance

chemical property A property that describes the behavior of a substance when it combines with another substance

Physical properties describe a substance without reference to any other substance. For example, weight, volume, color, boiling point, and melting point are typical physical properties that can be measured for a particular substance without altering the material's composition through a chemical reaction; they are associated only with the physical existence of that substance. A **chemical property** describes the behavior of a substance when it reacts or combines with another substance. For example, when wood burns, it chemically combines with oxygen in the air to form new substances; this transformation describes a chemical property of wood. In the crime laboratory, a routine procedure for determining the presence of heroin in a suspect specimen is to react it with a chemical reagent known as the Marquis reagent, which turns purple in the presence of heroin. This color transformation becomes a chemical property of heroin and provides a convenient test for its identification (see Figure 5–1).

Which physical and chemical properties the forensic scientist ultimately chooses to observe and measure depends on the type of material that is being examined. Logic requires, however, that if the property can be assigned a numerical value, it must relate to a standard system of measurement accepted throughout the scientific community. That standard system is called the metric system.

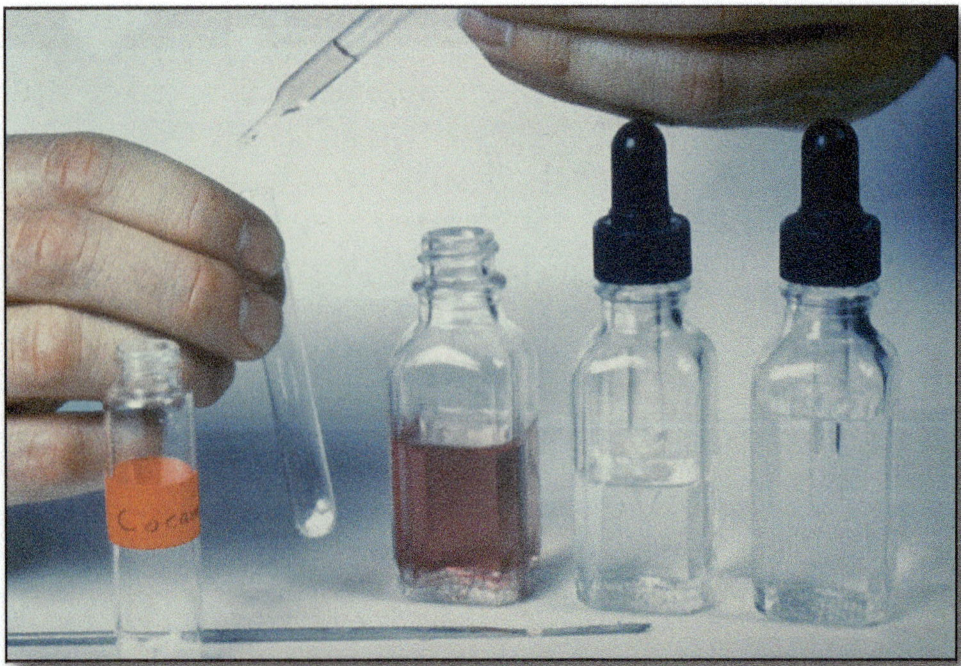

FIGURE 5–1 The color change associated with a drug color test represents an example of a chemical property that can be used to identify the drug of interest. *Courtesy Drug Enforcement Administration*

Inside the Science

The Metric System

Although scientists, including forensic scientists, throughout the world have been using the metric system of measurement for more than a century, the United States still uses the cumbersome "English system" to express length in inches, feet, or yards; weight in ounces or pounds; and volume in pints or quarts. The inherent difficulty of this system is that no simple numerical relationship exists between the various units of measurement. For example, to convert inches to feet one must know that 1 foot is equal to 12 inches; the conversion of ounces to pounds requires the knowledge that 16 ounces is equivalent to 1 pound.

In 1791, the French Academy of Science devised the simple system of measurement known as the metric system. This system uses a simple decimal relationship so that a unit of length, volume, or mass can be converted into a subunit by simply multiplying or dividing by a multiple of 10—for example, 10, 100, or 1,000. Even though the United States has not yet adopted the metric system, its system of currency is decimal and, hence, is analogous to the metric system. The basic unit of currency is the dollar. A dollar is divided into 10 equal units called dimes, and each dime is further divided into 10 equal units of cents.

The metric system has basic units of measurement for length, mass, and volume: the meter, gram, and liter, respectively. These three basic units can be converted into subunits that are decimal multiples of the basic unit by simply attaching a prefix to the unit name. The following are common prefixes and their equivalent decimal value:

Prefix	Equivalent Value
deci-	1/10 or 0.1
centi-	1/100 or 0.01
milli-	1/1,000 or 0.001
micro-	1/1,000,000 or 0.000001
nano-	1/1,000,000,000 or 0.000000001
kilo-	1,000
mega-	1,000,000

Hence, 1/10 or 0.1 gram (g) is the same as a decigram (dg), 1/100 or 0.01 meter is equal to a centimeter (cm), and 1/1000 liter (0.001) is a milliliter (mL). A metric conversion is carried out simply by moving the decimal point to the right or left and inserting the proper prefix to show the direction and number of places that the decimal point has been moved. For example, if the weight of a powder is 0.0165 gram, it may be more convenient to multiply this value by 100 and express it as 1.65 centigrams or by 1,000 to show it as its equivalent value of 16.5 milligrams. Similarly, an object that weighs 264,450 grams may be expressed as 264.45 kilograms simply by dividing it by 1,000. It is important to remember that in any of these conversions, the value of the measurement has not changed; 0.0165 gram is still equivalent to 1.65 centigrams, just as one dollar is still equal to 100 cents. We have simply adjusted the position of the decimal and shown the extent of the adjustment with a prefix.

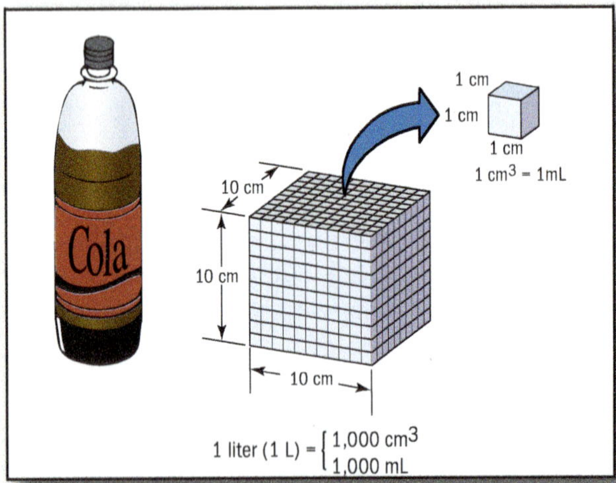

FIGURE 1 Volume equivalencies in the metric system.

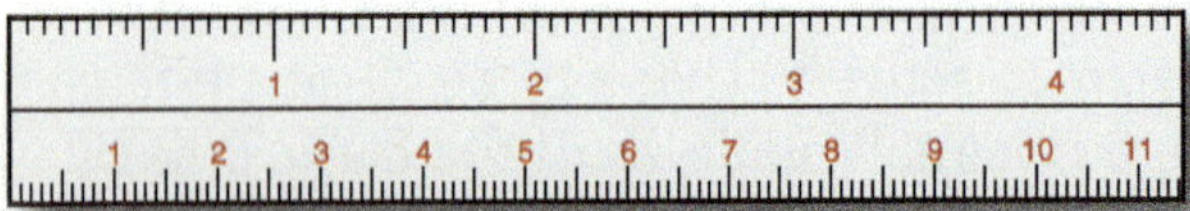

FIGURE 2 Comparison of the metric and English systems of length measurement; 2.54 centimeters = 1 inch.

One interesting aspect of the metric system is that volume can be defined in terms of length. A liter by definition is the volume of a cube with sides of length 10 centimeters. One liter is therefore equivalent to a volume of (10 cm)^3, or 1,000 cubic centimeters (cc). Thus, 1/1,000 liter or 1 milliliter (mL) is equal to 1 cubic centimeter (cc) (see Figure 1). Scientists commonly use the subunits mL and cc interchangeably to express volume.

At times, it may be necessary to convert units from the metric system into the English system, or vice versa (see Figure 2). To accomplish this, we must consult references that list English units and their metric equivalents. Some of the more useful equivalents follow:

1 inch = 2.54 centimeters	1 liter = 1.06 quarts
1 meter = 39.37 inches	1 kilogram = 2.2 pounds
1 pound = 453.6 grams	

The general mathematical procedures for converting from one system to another can be illustrated by converting 12 inches into centimeters. To change inches into centimeters, we need to know that there are 2.54 centimeters per inch. Hence, if we multiply 12 inches by 2.54 centimeters per inch (12 in. × 2.54 cm/in.), the unit of inches will cancel out, leaving the product 30.48 cm. Similarly, converting grams to pounds, 227 grams is equivalent to 227 g × 1 lb/453.6 g or 0.5 lb. These are just a few examples of how mathematical calculations can be used to assess quantitative relationships between different types of data.

The Nature of Matter

Before we can apply physical properties, as well as chemical properties, to the identification and comparison of evidence, we need to gain an insight into the composition of matter. Beginning with knowledge of the fundamental building block of all substances—the element—we will extend our discussion to compounds.

Elements and Compounds

Matter is anything that has mass and occupies space. As we examine the world that surrounds us and consider the countless variety of materials that we encounter, we must consider one of humankind's most remarkable accomplishments: the discovery of the concept of the atom to explain the composition of all matter. This search had its earliest contribution from the ancient Greek philosophers, who suggested air, water, fire, and earth as matter's fundamental building blocks. It culminated with the development of the atomic theory and the discovery of matter's simplest identity, the element.

matter
Anything that has mass and occupies space

An **element** is a fundamental particle of matter that cannot be broken down into simpler substances by chemical means. Elements provide the building blocks from which all matter is composed. At present, 118 elements have been identified (see Table 5–1); of these, 89 occur naturally on the earth, and the remainder have been created in the laboratory.

element
A fundamental particle of matter that cannot be broken down into simpler substances by chemical means

Table 5–1

List of Elements with Their Symbols and Atomic Masses

Element	*Symbol*	*Atomic Mass*[a] *(amu)*	*Element*	*Symbol*	*Atomic Mass*[a] *(amu)*
Actinium	Ac	(227)	Copernicium	Cp	(285)
Aluminum	Al	26.9815	Copper	Cu	63.546
Americium	Am	(243)	Curium	Cm	(247)
Antimony	Sb	121.75	Darmstadtium	Ds	(271)
Argon	Ar	39.948	Dubnium	Db	(260)
Arsenic	As	74.9216	Dysprosium	Dy	162.50
Astatine	At	(210)	Einsteinium	Es	(254)
Barium	Ba	137.34	Erbium	Er	167.26
Berkelium	Bk	(247)	Europium	Eu	151.96
Beryllium	Be	9.01218	Fermium	Fm	(253)
Bismuth	Bi	208.9806	Fluorine	F	18.9984
Bohrium	Bh	(262)	Francium	Fr	(223)
Boron	B	10.81	Gadolinium	Gd	157.25
Bromine	Br	79.904	Gallium	Ga	69.72
Cadmium	Cd	112.40	Germanium	Ge	72.59
Calcium	Ca	40.08	Gold	Au	196.9665
Californium	Cf	(251)	Hafnium	Hf	178.49
Carbon	C	12.011	Hassium	Hs	(265)
Cerium	Ce	140.12	Helium	He	4.00260
Cesium	Cs	132.9055	Holmium	Ho	164.9303
Chlorine	Cl	35.453	Hydrogen	H	1.0080
Chromium	Cr	51.996	Indium	In	114.82
Cobalt	Co	58.9332	Iodine	I	126.9045

Table 5–1 *(continued)*

List of Elements with Their Symbols and Atomic Masses

Element	*Symbol*	*Atomic Mass*[a] *(amu)*	*Element*	*Symbol*	*Atomic Mass*[a] *(amu)*
Iridium	Ir	192.22	Roentgenium	Rg	(272)
Iron	Fe	55.847	Rubidium	Rb	85.4678
Krypton	Kr	83.80	Ruthenium	Ru	101.07
Lanthanum	La	138.9055	Rutherfordium	Rf	(257)
Lawrencium	Lr	(257)	Samarium	Sm	105.4
Lead	Pb	207.2	Scandium	Sc	44.9559
Lithium	Li	6.941	Seaborgium	Sg	(263)
Lutetium	Lu	174.97	Selenium	Se	78.96
Magnesium	Mg	24.305	Silicon	Si	28.086
Manganese	Mn	54.9380	Silver	Ag	107.868
Meitnerium	Mt	(266)	Sodium	Na	22.9898
Mendelevium	Md	(256)	Strontium	Sr	87.62
Mercury	Hg	200.59	Sulfur	S	32.06
Molybdenum	Mo	95.94	Tantalum	Ta	180.9479
Neodymium	Nd	144.24	Technetium	Tc	98.9062
Neon	Ne	20.179	Thallium	Tl	204.37
Neptunium	Np	237.0482	Terbium	Tb	158.9254
Nickel	Ni	58.71	Tellurium	Te	127.60
Niobium	Nb	92.9064	Thorium	Th	232.0381
Nitrogen	N	14.0067	Thulium	Tm	168.9342
Nobelium	No	(254)	Tin	Sn	118.69
Osmium	Os	190.2	Titanium	Ti	47.90
Oxygen	O	15.9994	Tungsten	W	183.85
Palladium	Pd	106.4	Ununhexium	Lv	(292)
Phosphorus	P	30.9738	Ununoctium	Uuo	(294)
Platinum	Pt	195.09	Ununpentium	Uup	(288)
Plutonium	Pu	(244)	Ununquadium	Fl	(289)
Polonium	Po	(209)	Ununseptium	Uus	(unknown)
Potassium	K	39.102	Ununtrium	Uut	(284)
Praseodymium	Pr	140.9077	Uranium	U	238.029
Promethium	Pm	(145)	Vanadium	V	50.9414
Protactinium	Pa	231.0359	Xenon	Xe	131.3
Radium	Ra	226.0254	Ytterbium	Yb	173.04
Radon	Rn	(222)	Yttrium	Y	88.9059
Rhenium	Re	186.2	Zinc	Zn	65.57
Rhodium	Rh	102.9055	Zirconium	Zr	91.22

[a]Based on the assigned relative atomic mass of C = exactly 12; parentheses denote the mass number of the isotope with the longest half-life.

Figure 5–2 shows the **periodic table**, a chart of all the known elements arranged in a systematic fashion. This table is most useful to chemists because it arranges elements with similar chemical properties in the same vertical or horizontal rows. Vertical rows of elements in the table are called groups or families; horizontal rows are called series. Elements in the same group have similar properties (for example, atomic mass or density) and exhibit a clear trend in properties as one moves down the group. As a result, groups are considered the most important way of classifying the elements. In some parts of the table, however, similarities in properties of the elements in the same row makes the series a more useful method for classification.

periodic table
A chart of all the known elements arranged in a systematic fashion

For convenience, chemists have chosen letter symbols to represent the elements. Many of these symbols come from the first letter of the element's English name—for example, carbon (C), hydrogen (H), and oxygen (O). Others are twoletter abbreviations of the English name—calcium (Ca) and zinc (Zn). Some symbols are derived from the first letters of Latin or Greek names. Thus, the symbol for silver, Ag, comes from the Latin name argentum; copper, Cu, from the Latin cuprum; and helium, He, from the Greek name helios.

The smallest particle of an element that can exist and still retain its identity as that element is the **atom**. When we write the symbol C we mean one atom of carbon; the chemical symbol for carbon dioxide, CO_2, signifies one atom of carbon combined with two atoms of oxygen. When two or more elements are combined, as with carbon dioxide, a new substance is created, different

atom
The smallest unit of an element that can exist and still retain its identity as that element

Period \ Group	IA	IIA	IIIB	IVB	VB	VIB	VIIB	VIII			IB	IIB	IIIA	IVA	VA	VIA	VIIA	0
1	1 H																	2 He
2	3 Li	4 Be											5 B	6 C	7 N	8 O	9 F	10 Ne
3	11 Na	12 Mg											13 Al	14 Si	15 P	16 S	17 Cl	18 Ar
4	19 K	20 Ca	21 Sc	22 Ti	23 V	24 Cr	25 Mn	26 Fe	27 Co	28 Ni	29 Cu	30 Zn	31 Ga	32 Ge	33 As	34 Se	35 Br	36 Kr
5	37 Rb	38 Sr	39 Y	40 Zr	41 Nb	42 Mo	43 Tc	44 Ru	45 Rh	46 Pd	47 Ag	48 Cd	49 In	50 Sn	51 Sb	52 Tc	53 I	54 Xe
6	55 Cs	56 Ba	57 La a	72 Hf	73 Ta	74 W	75 Re	76 Os	77 Ir	78 Pt	79 Au	80 Hg	81 Tl	82 Pb	83 Bi	84 Po	85 At	86 Rn
7	87 Fr	88 Ra	89 Ac b	104 Rf	105 Db	106 Sg	107 Bh	108 Hs	109 Mt	110 Ds	111 Rg	112 Cn	113 Uut	114 Fl	115 Uup	116 Lv	117 Uus	118 Uuo

[a]Lanthanide series	58 Ce	59 Pr	60 Nd	61 Pm	62 Sm	63 Eu	64 Gd	65 Tb	66 Dy	67 Ho	68 Er	69 Tm	70 Yb	71 Lu
[b]Actinide series	90 Th	91 Pa	92 U	93 Np	94 Pu	95 Am	96 Cm	97 Bk	98 Cf	99 Es	100 Fm	101 Md	102 No	103 Lr

FIGURE 5–2 The periodic table.

in its physical and chemical properties from its elemental components. This new material is called a **compound**. Compounds contain at least two elements. Considering that there are 89 natural elements, it is easy to imagine the large number of possible elemental combinations that may form compounds. Not surprisingly, more than 16 million known compounds have already been identified.

compound
A pure substance composed of two or more elements

Just as the atom is the basic unit of an element, the molecule is the smallest unit of a compound. Thus, a molecule of carbon dioxide is represented by the symbol CO_2, and a molecule of table salt is symbolized by NaCl, representing the combination of one atom of the element sodium (Na) with one atom of the element chlorine (Cl).

States of Matter

As we look around us and view the materials that make up the earth, it becomes an awesome task even to attempt to estimate the number of different kinds of matter. A much more logical approach is to classify matter according to the physical form it takes. These forms are called **physical states**. There are three such states: solid, liquid, and gas (vapor).

physical state
The physical form taken by matter: solid, liquid, or gas

solid
A state of matter in which molecules are held closely together by strong attractive forces

liquid
A state of matter in which the attractive forces are strong enough to allow molecules to be in contact with one another but too weak to hold them rigidly in place

gas (vapor)
A state of matter in which the attractive forces between molecules are weak enough to permit them to move with complete freedom

sublimation
A change of state from a solid directly into a gas

In the **solid** state, molecules of matter are held closely together by strong attractive forces. This tightly packed molecular arrangement makes solid matter rigid and gives it a definite shape and volume. In a **liquid**, the attractive forces between molecules are weaker. The molecules are in contact with one another, but are not held as rigidly in place. Like a solid, a liquid also occupies a specific volume, but its fluidity causes it to take the shape of the container in which it resides. The attractive forces between molecules in a **gas (vapor)** are much weaker still, allowing them to move freely. Thus, gaseous matter has neither a definite shape nor volume, and it will completely fill any container into which it is placed.

Changes of State Substances can change from one state to another. For example, as water is heated, it is converted from a liquid form into a vapor. At a high enough temperature (100°C), water boils and rapidly changes into steam. Similarly, at 0°C, water solidifies or freezes into ice. Under certain conditions, some solids can be converted directly into a gaseous state. For instance, a piece of dry ice (solid carbon dioxide) left standing at room temperature quickly forms carbon dioxide vapor and disappears. This change of state from a solid to a gas is called **sublimation**.

In each of these examples, no new chemical species are formed; matter simply changes from one physical state to another. Water, whether in the form of liquid, ice, or steam, remains chemically H_2O. Simply, what has been altered are the attractive forces between the water molecules. In a solid, these forces are very strong, and the molecules are held closely together in a rigid state. In a liquid, the attractive forces are not as strong, and the molecules have more mobility. Finally, in the vapor state, appreciable attractive forces no longer exist between the molecules; thus, they may move in any direction at will.

Phases Chemists are forever combining different substances, no matter whether they are in the solid, liquid, or gaseous states, hoping to create new and useful products. Not all attempts at mixing matter are productive. For instance, oil spills demonstrate that oil and water do not mix. Whenever substances can be distinguished by a visible boundary, different **phases** are said to exist.

phase
A uniform substance separated from other substances by definite visible boundaries

Oil floating on water is an example of a two-phase system. The oil and water each constitute a separate liquid phase, clearly distinct from each other. Similarly, when sugar is first added to water, it will not dissolve, and two distinctly different phases exist: the solid sugar and the liquid water. However, after stirring, all the sugar dissolves, leaving just one liquid phase. As we shall see, forensic scientists use the existence of different phases to identify and classify evidence found at a crime scene.

Quick Review

- Physical properties such as weight, volume, color, boiling point, and melting point describe a substance without reference to any other substance. A chemical property describes the behavior of a substance when it reacts or combines with another substance.
- Matter is anything that has mass and occupies space. An element is a fundamental particle of matter that cannot be broken down into simpler substances by chemical means.
- The smallest particle of an element that can exist and still retain its identity as that element is the atom.
- A compound is a pure substance composed of two or more elements. The smallest unit of a compound is a molecule.
- The three states of matter are solid, liquid, and gas (vapor).
- Sublimation is a change of state from a solid to a gas.
- A phase is a uniform body of matter distinguished from other matter by definite visible boundaries.

Theory of Light

As with matter, knowledge of the nature and behavior of light is fundamental to understanding physical properties important to the examination of forensic evidence. Two simple models explain light's behavior. The first model describes light as a continuous wave; the second depicts it as a stream of discrete energy particles. Together, these two very different descriptions explain all of the observed properties of light, but by itself, no one model can explain all the facets of the behavior of light.

Light as a Wave

The wave concept depicts light as having an up-and-down motion of a continuous wave, as shown in Figure 5–3. Such a wave can be characterized by two distinct properties: wavelength and frequency. The distance between two consecutive

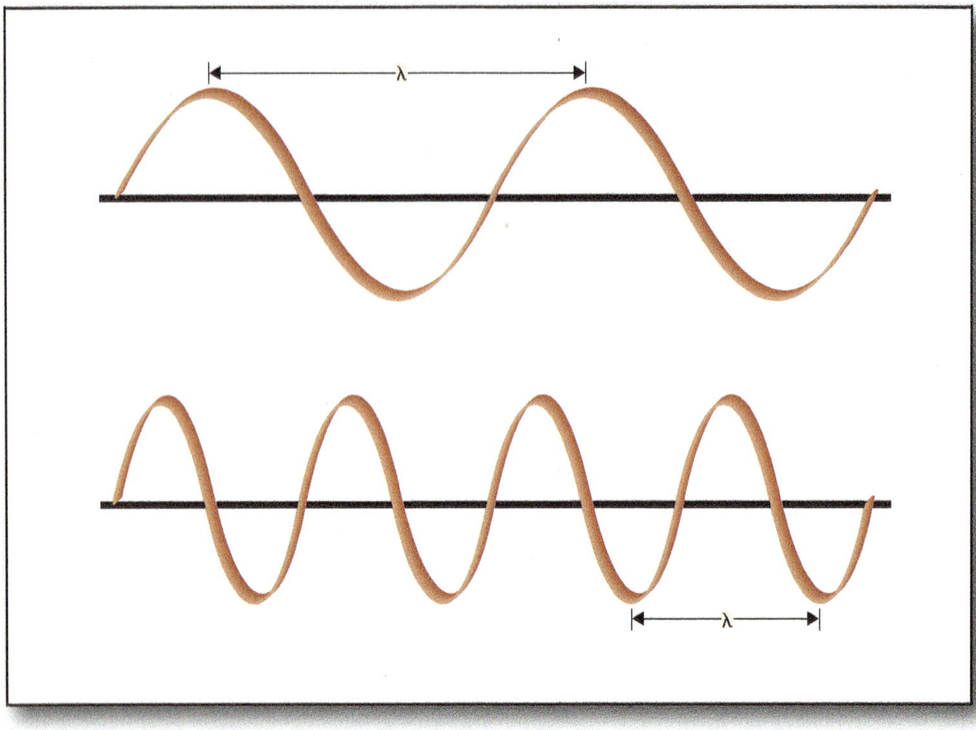

FIGURE 5–3
The frequency of the lower wave is twice that of the upper wave.

wavelength
The distance between crests of adjacent waves

frequency
The number of waves that pass a given point per unit of time

crests (high points) or troughs (low points) of a wave is called the **wavelength**; it is designated by the Greek letter lambda (l) and is typically measured in nanometers (nm), or millionths of a meter. The number of crests (or troughs) passing any one given point in a unit of time is defined as the **frequency** of the wave. Frequency is normally designated by the letter *f* and is expressed in cycles per second (cps). Frequency and wavelength are inversely proportional to one another, as shown by the relationship expressed in Equation (5–1):

EQUATION 5–1

$$F = c/l$$

In the equation, *c* represents the speed of light.

dispersion
The separation of light into its component wavelengths

refraction
The bending of a light wave caused by a change in its velocity

Many of us have held a glass prism up toward the sunlight and watched it transform light into the colors of the rainbow. The process of separating light into its component colors is called **dispersion**. Visible light usually travels at a constant velocity of nearly 300 million meters per second. However, on passing through the glass of a prism, each color component of light is slowed to a speed slightly different from those of the others, causing each component to bend at a different angle as it emerges from the prism (see Figure 5–4). This bending of light waves because of a change in velocity is called **refraction**.

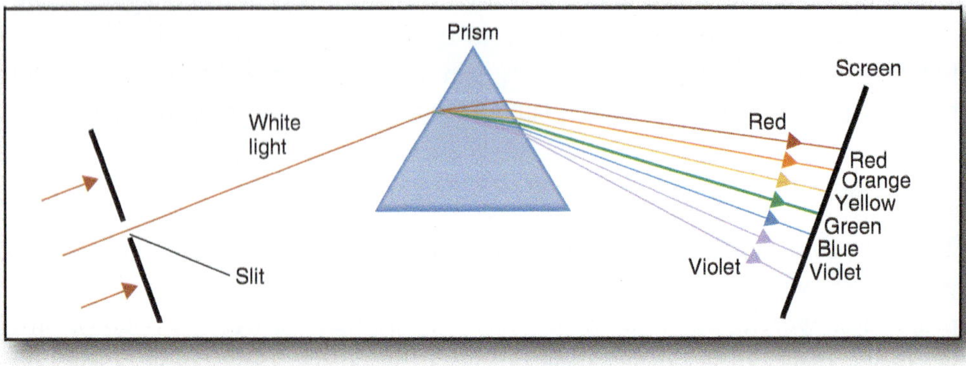

FIGURE 5–4
Representation of the dispersion of light by a glass prism.

The observation that a substance has a color is consistent with this description of white light. For example, when light passes through a red glass, the glass absorbs all the component colors of light except red, which passes through or is transmitted by the glass. Likewise, one can determine the color of an opaque object by observing its ability to absorb some of the component colors of light while reflecting others back to the eye. Color is thus a visual indication that objects absorb certain portions of **visible light** and transmit or reflect others. Scientists have long recognized this phenomenon and have learned to characterize different chemical substances by the type and quantity of light they absorb. This has important applications for the identification and classification of forensic evidence.

visible light
Colored light ranging from red to violet in the electromagnetic spectrum

The Electromagnetic Spectrum Visible light is only a small part of a large family of radiation waves known as the **electromagnetic spectrum** (see Figure 5–5). All electromagnetic waves travel at the speed of light (*c*) and are distinguishable from one another only by their different wavelengths or frequencies. Hence, the only property that distinguishes **X-rays** from radio waves is the different frequencies the two types of waves possess.

electromagnetic spectrum
The entire range of radiation from the most energetic cosmic rays to the least energetic radio waves

X-ray
The high-energy, short-wavelength form of electromagnetic radiation

Similarly, the range of colors that make up the visible spectrum can be correlated with frequency. For instance, the lowest frequencies of visible light are red; waves with a lower frequency fall into the invisible infrared (IR) region. The highest frequencies of visible light are violet; waves with a higher frequency extend into the invisible ultraviolet (UV) region. No definite boundaries exist between any colors or regions of the electromagnetic spectrum; instead, each region is composed of a continuous range of frequencies, each blending into the other.

Ordinarily, light in any region of the electromagnetic spectrum is a collection of waves possessing a range of wavelengths. Under normal circumstances, this light comprises waves that are all out of step with each other (incoherent light). However, scientists can produce light that has all its waves pulsating in unison (see Figure 5–6). This is called a **laser** (light amplification by stimulated emission of radiation). Light in this form is very intense and can be focused on a very small area. Laser beams can be focused to pinpoints that are so intense that they can zap microscopic holes in a diamond.

laser
An acronym for light amplification by stimulated emission of radiation; light that has all its waves pulsating in unison

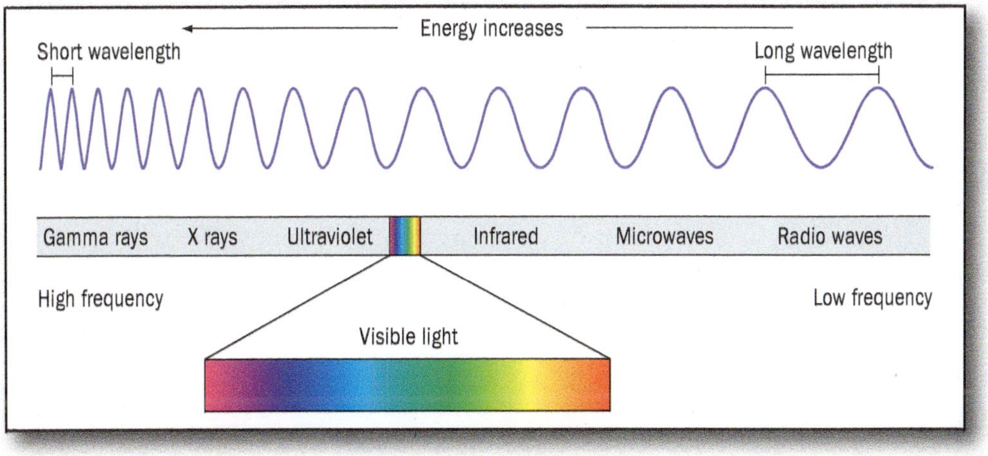

FIGURE 5–5
The electromagnetic spectrum.

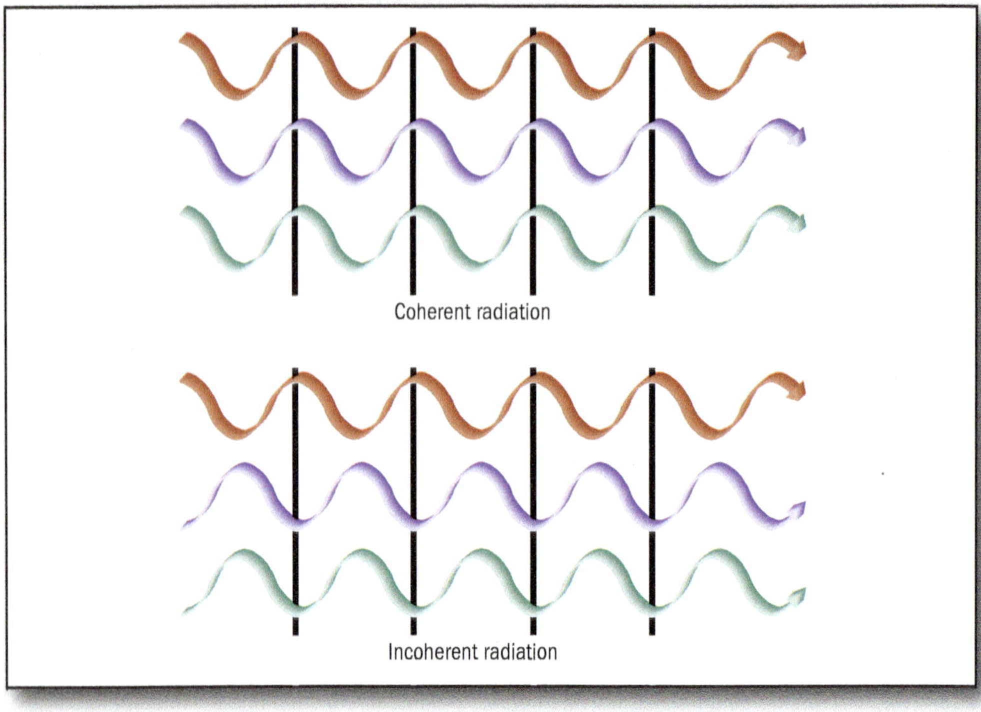

FIGURE 5–6
Coherent and incoherent radiation.

Light as a Particle

photon
A discrete particle of electromagnetic radiation

As long as electromagnetic radiation is moving through space, its behavior can be described as that of a continuous wave. However, once radiation is absorbed by a substance, the model of light as a stream of discrete particles must be invoked to describe its behavior. Here, light is depicted as consisting of energy particles that are known as **photons**. Each photon has a definite amount of energy associated with its behavior. This energy is related to the frequency of light, as shown by Equation (5–2):

EQUATION 5–2

$$E = hf$$

where E specifies the energy of the photon, f is the frequency of radiation, and h is a universal constant called Planck's constant. As shown by Equation (5–2), the energy of a photon is directly proportional to its frequency. Therefore, the photons of ultraviolet light will be more energetic than the photons of visible or infrared light, and exposure to the more energetic photons of X-rays presents more danger to human health than exposure to the photons of radio waves.

Just as a substance can absorb visible light to produce color, many of the invisible radiations of the electromagnetic spectrum are likewise absorbed. This absorption phenomenon is the basis for spectrophotometry, an analytical technique that measures the quantity of radiation that a particular material absorbs as a function of wavelength or frequency. We will examine spectrophotometry in more detail when we discuss the forensic analysis of drugs in Chapter 6.

Quick Review

- Dispersion is the process of separating light into component colors. Because each color corresponds to a different range of frequencies or wavelengths of light, light rays of one color bend to a different degree than rays of all other colors.
- Refraction is the bending of light waves because of a change in velocity.
- Forensic scientists have learned to characterize different chemical substances by the type and quantity of light they absorb.
- Different types of electromagnetic radiation are distinguished from one another by their frequency or wavelength.
- As long as electromagnetic radiation is moving through space, its behavior can be described as that of a continuous wave. However, once radiation is absorbed by a substance, the model of light as a stream of discrete particles (photons) must be invoked to describe its behavior.

Physical Properties of Matter

Temperature

Determining the physical properties of any material often requires measuring its temperature. For instance, the temperatures at which a substance melts or boils are readily determinable characteristics that will help identify it. Temperature is a measure of heat intensity, or the amount of heat in a substance. Temperature is usually measured by placing a thermometer in contact with a substance. The familiar mercury-in-glass thermometer functions because mercury expands more than glass when heated and contracts more than glass when cooled. Thus, the length of the mercury column in the glass tube measures the surrounding environment's temperature.

The construction of a temperature scale requires two reference points and a choice of units. The reference points most conveniently chosen are the freezing point and boiling point of water. The two most common temperature scales used are the **Fahrenheit** and **Celsius** (formerly called centigrade) **scales** (see Figure 5–7).

Fahrenheit scale
The temperature scale that defines the melting point of ice as 32° and the boiling point of water as 212°, with 180 equal divisions or degrees between

Celsius scale
The temperature scale that defines the melting point of ice as 0° and the boiling point of water as 100°, with 100 equal divisions or degrees between

The Fahrenheit scale is based on the assignment of a value of 32°F to the freezing point of water and a value of 212°F to its boiling point. The difference between the two points is evenly divided into 180 units. Thus, a degree Fahrenheit is 1/180 of the temperature change between the freezing point and boiling point of water. The Celsius scale is derived by assigning the freezing point of water a value of 0°C and its boiling point a value of 100°C. A degree Celsius is thus 1/100 of the temperature change between the two reference points. Scientists in most countries use the Celsius scale to measure temperature.

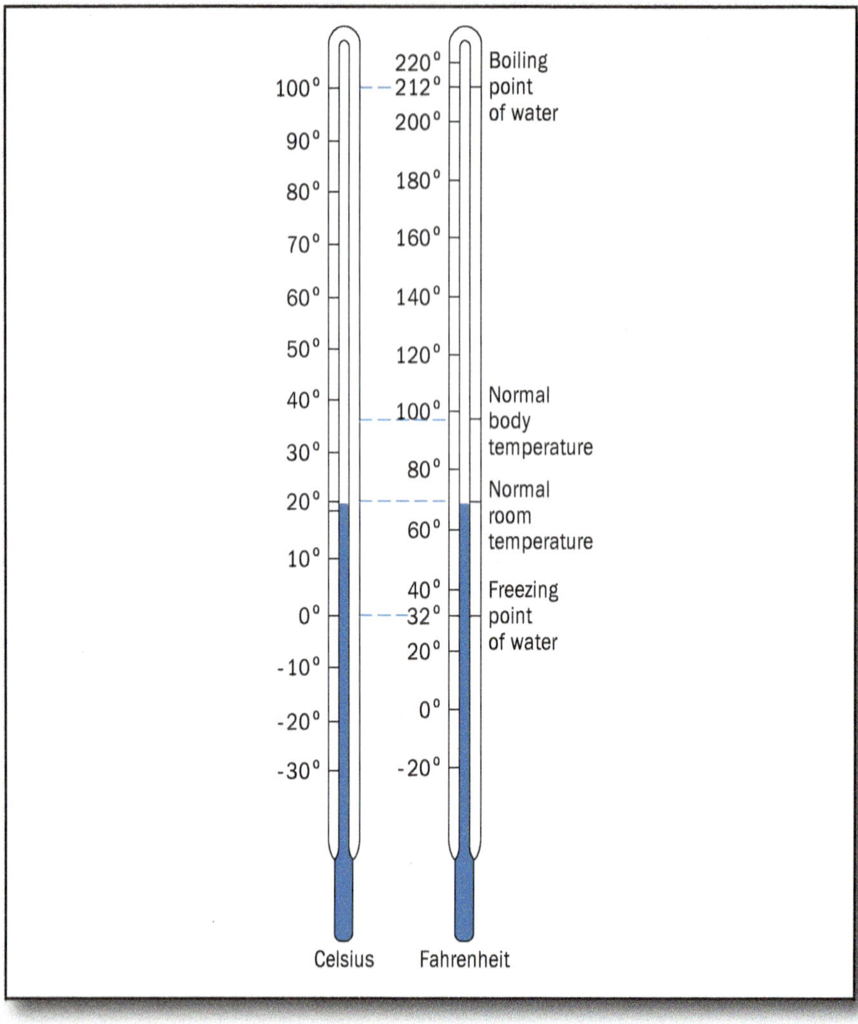

FIGURE 5–7 Comparison of the Celsius and Fahrenheit temperature scales.

Weight and Mass

weight The force with which gravity attracts a body

mass A constant property that refers to the amount of matter an object contains

The force with which gravity attracts a body is called **weight**. If your weight is 180 pounds, this means that the earth's gravity is pulling you down with a force of 180 pounds; on the moon, where the force of gravity is one-sixth that of the earth, your weight would be 30 pounds.

Mass differs from weight because it refers to the amount of matter an object contains and is independent of its location on the earth or any other place in the universe. The mathematical relationship between weight (w) and mass (m) is shown in Equation (5–3), where g is the acceleration imparted to a body by the force of gravity.

EQUATION 5–3

$$w = mg$$

The weight of a body is seen to be directly proportional to its mass; hence, a large mass weighs more than a small mass.

In the metric system, the mass of an object is always specified, rather than its weight. The basic unit of mass is the gram. An object that has a mass of 40 grams on the earth will have a mass of 40 grams anywhere else in the universe. Normally, however, the terms mass and weight are used interchangeably, and we often speak of the weight of an object when we really mean its mass.

The mass of an object is determined by comparing it against the known mass of standard objects. The comparison is confusingly called weighing, and the

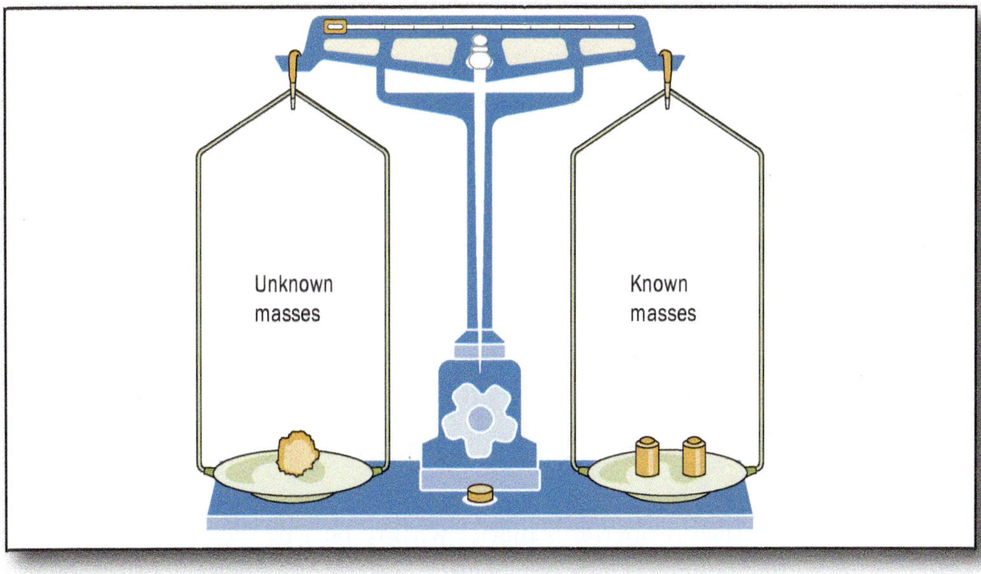

FIGURE 5–8
The measurement of mass.

standard objects are called weights (masses would be a more correct term). The comparison is performed on a balance. The simplest type of balance for weighing is the equal-arm balance shown in Figure 5–8. The object to be weighed is placed on the left pan, and the standard weights are placed on the right pan; when the pointer between the two pans is at the center mark, the total mass on the right pan is equal to the mass of the object on the left pan.

The modern laboratory has progressed beyond the simple equal-arm balance, and either the top-loading balance or the single-pan analytical balance (see Figure 5–9) is now likely to be used. The choice depends on the accuracy required and the amount of material being weighed. Each works on the same counterbalancing principle as the simple equal-arm balance, but the second pan, the one on which the standard weights are placed, is hidden from view within the balance's housing. Once the object whose

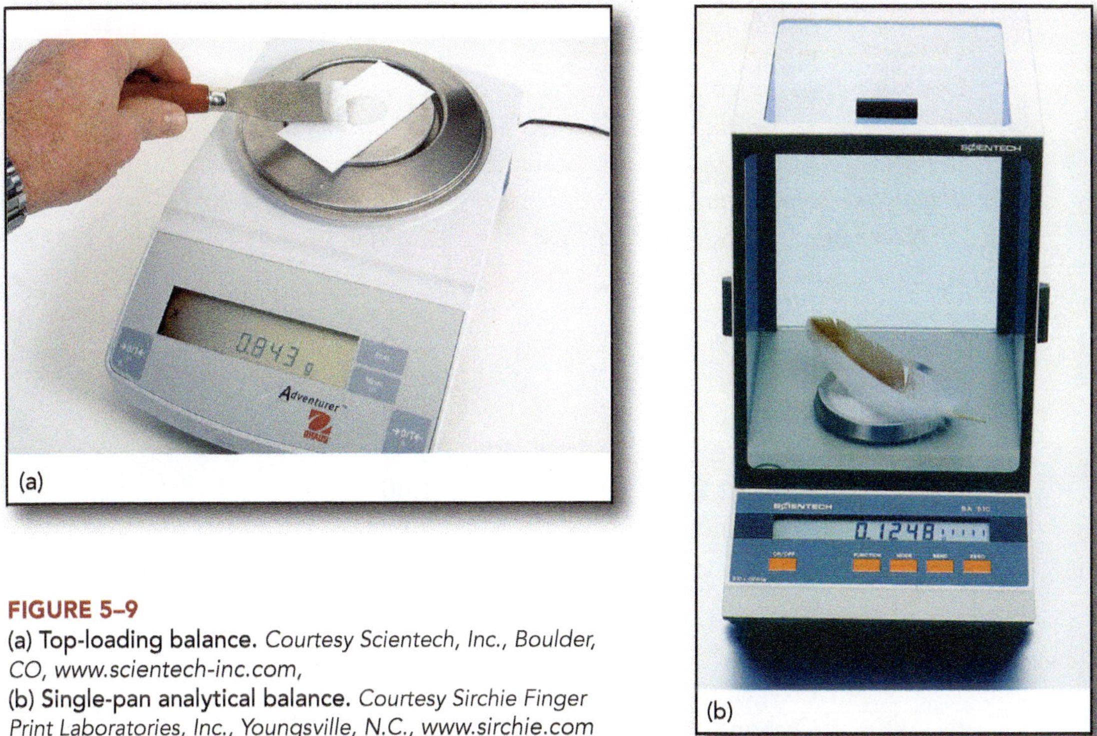

FIGURE 5–9
(a) **Top-loading balance.** *Courtesy Scientech, Inc., Boulder, CO, www.scientech-inc.com,*
(b) **Single-pan analytical balance.** *Courtesy Sirchie Finger Print Laboratories, Inc., Youngsville, N.C., www.sirchie.com*

weight is to be determined is placed on the visible pan, the operator selects the proper standard weights (also contained with the housing) by manually turning a set of knobs located on the front side of the balance. At the point of balance, the weights selected are automatically recorded on digital and optical readout scales. Other versions of single-pan balances rely on an electromagnetic field to generate a current to balance the force pressing down on the pan from the sample being weighed. When the scale is properly calibrated, the amount of current needed to keep the pan balanced is used to determine the weight of the sample. The strength of the current is converted to a digitized signal for a readout. The top-loading balance can accurately weigh an object to the nearest 1 milligram or 0.001 gram; the analytical balance is even more accurate, weighing to the nearest tenth of a milligram or 0.0001 gram.

Density

density
A physical property of matter that is equivalent to the mass per unit volume of a substance

An important physical property of matter with respect to the analysis of certain kinds of physical evidence is **density**. Density is defined as mass per unit volume (see Equation (5–4)).

EQUATION 5–4

$$\text{Density} = \frac{\text{mass}}{\text{volume}}$$

intensive property
A property that is not dependent on the size of an object

Density is an **intensive property** of matter—that is, it remains the same regardless of the size of an object; thus, it is a characteristic property of a substance and can be used in identification. Solids tend to be more dense than liquids, and liquids more dense than gases. The densities of some common substances are shown in Table 5–2.

Table 5–2
Densities of Select Materials (at 20°C Unless Otherwise Stated)

Substance	Density (g/mL)
Solids	
Silver	10.5
Lead	11.5
Iron	7.8
Aluminum	2.7
Window glass	2.47–2.54
Ice (0°C)	0.92
Liquids	
Mercury	13.6
Benzene	0.88
Ethyl alcohol	0.79
Gasoline	0.69
Water at 4°C	1.00
Water at 20°C	0.998
Gases	
Air (0°C)	0.0013
Chlorine (0°C)	0.0032
Oxygen (0°C)	0.0014
Carbon dioxide (0°C)	0.0020

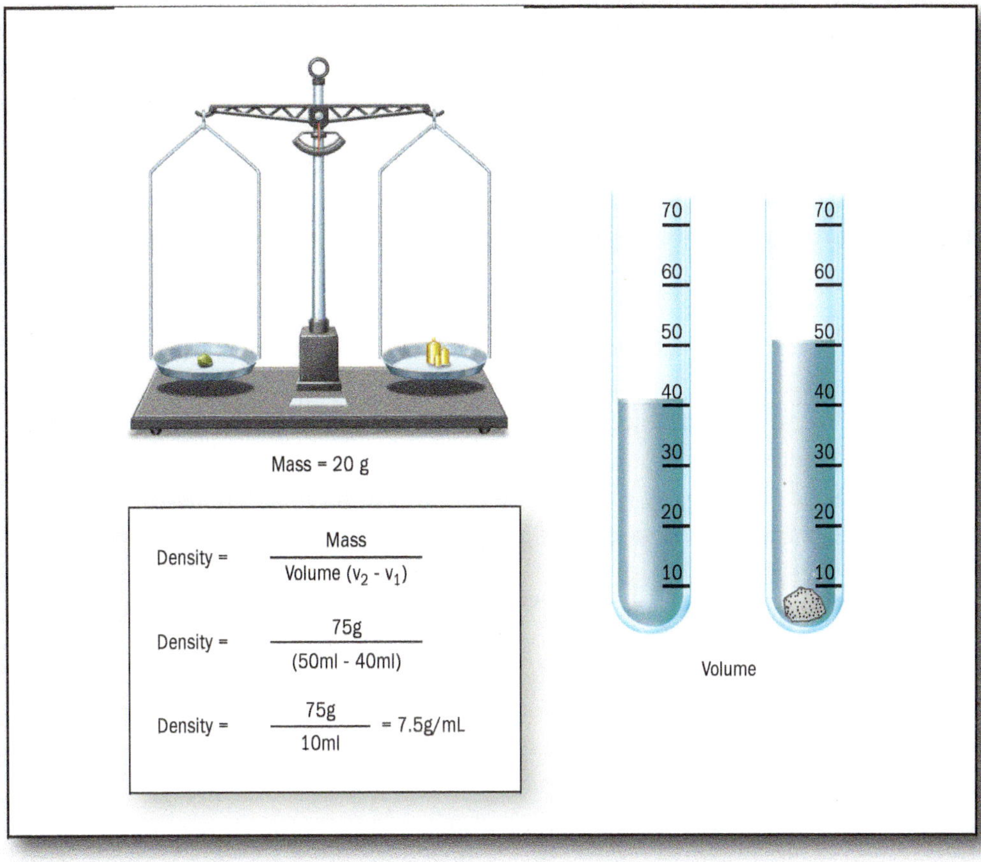

FIGURE 5–10
A simple procedure for determining the density of a solid is to first weigh it and then measure its volume by noting the volume of water it displaces.

A simple procedure for determining the density of a solid is illustrated in Figure 5–10. First, the solid is weighed on a balance against known standard gram weights to determine its mass. The solid's volume is then determined from the volume of water it displaces. This is easily measured by filling a cylinder with a known volume of water (V_1), adding the object, and measuring the new water level (V_2). The difference ($V_2 - V_1$) in milliliters is equal to the volume of the solid. Density can now be calculated from Equation 5–4 in grams per milliliter.

The volumes of gases and liquids vary considerably with temperature; hence, when determining density, it is important to control and record the temperature at which the measurements are made. For example, 1 gram of water occupies a volume of 1 milliliter at 4°C and thus has a density of 1.0 g/mL. However, as the temperature of water increases, its volume expands. Therefore, at 20°C (room temperature), 1 gram of water occupies a volume of 1.002 mL and has a density of 0.998 g/mL.

The observation that a solid object either sinks, floats, or remains suspended when immersed in a liquid can be accounted for by the property of density. For instance, if the density of a solid is greater than that of the liquid in which it is immersed, the object sinks; if the solid's density is less than that of the liquid, it floats; and when the solid and liquid have equal densities, the solid remains suspended in the liquid medium. As we will see shortly, these observations provide a convenient technique for comparing the densities of solid objects.

Refractive Index

As we noted earlier, the bending of a light wave because of a change in velocity is called refraction. The phenomenon of refraction is apparent when we view an object that is immersed in a transparent medium such as water; because we are accustomed to thinking that light travels in a straight line, we often forget to account for refraction. For instance, suppose a ball is observed at the bottom of a swimming pool; the light rays reflected from the ball travel through the water and into the air to reach the eye. As the rays leave the water and enter the air, their velocity suddenly increases, causing them to be refracted. However, because of our assumption that light travels in a straight line, our eyes deceive us and make us think we see an object lying at a higher point than is actually the case. This phenomenon is illustrated in Figure 5–11.

refractive index
The ratio of the speed of light in a vacuum to its speed in a given medium

The ratio of the velocity of light in a vacuum to its speed in any medium determines the **refractive index** of that medium and is expressed as follows:

EQUATION 5–5

$$\text{Refractive Index} = \frac{\text{velocity of light in vacuum}}{\text{velocity of light in medium}}$$

For example, at 25°C the refractive index of water is 1.333. This means that light travels 1.333 times as fast in a vacuum as it does in water at this temperature. Like density, the refractive index is an intensive physical property of matter and characterizes a substance. However, any procedure used to determine a substance's refractive index must be performed under carefully controlled temperature and lighting conditions, because the refractive index of a substance varies with its temperature and the wavelength of light passing through it. Nearly all tabulated refractive indices are determined at a standard wavelength, usually 589.3 nanometers; this is the predominant wavelength emitted by sodium light and is commonly known as the sodium D light.

When a transparent solid is immersed in a liquid with a similar refractive index, light is not refracted as it passes from the liquid into the solid. For this

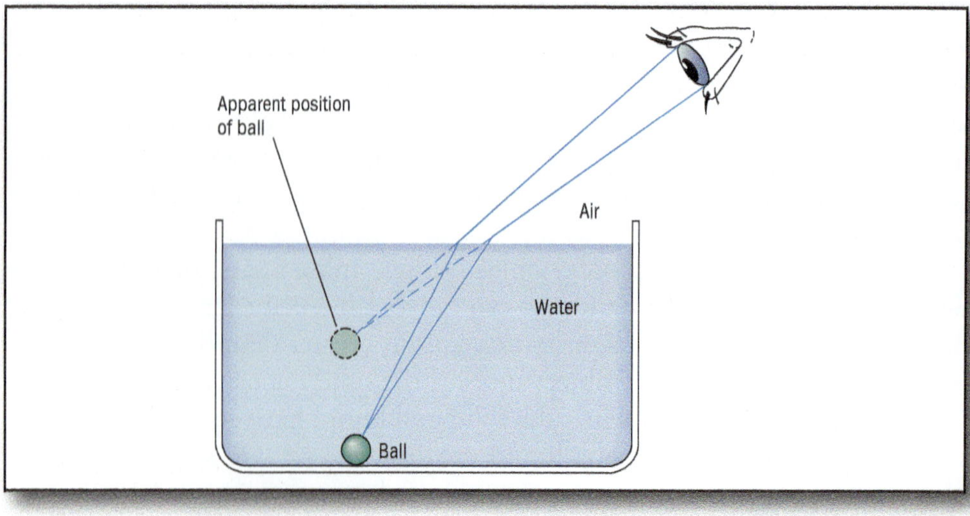

FIGURE 5–11
Light is refracted when it travels obliquely from one medium to another.

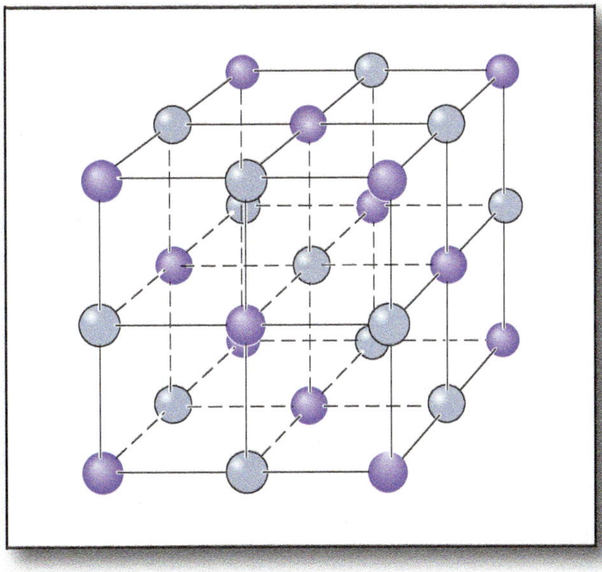

FIGURE 5–12 Diagram of a sodium chloride crystal. Sodium is represented by the dark spheres, chlorine by the light spheres.

reason, the eye cannot distinguish the liquid–solid boundary, and the solid seems to disappear from view. This observation, as we will see, offers the forensic scientist a rather simple method for comparing the refractive indices of transparent solids.

Normally, we expect a solid or a liquid to exhibit only one refractive index value for each wavelength of light; however, many crystalline solids have two refractive indices whose values depend in part on the direction from which the light enters the crystal with respect to the crystal axis. **Crystalline solids** have definite geometric forms because of the orderly arrangement of the fundamental particle of a solid, the atom.

crystalline solid
A solid in which the constituent atoms have a regular arrangement

In any type of crystal, the relative locations and distances between its atoms repeat throughout the solid. Figure 5–12 shows the crystalline structure of sodium chloride, or ordinary table salt. Sodium chloride is an example of a cubic crystal in which each sodium atom is surrounded by six chlorine atoms and each chlorine atom by six sodium atoms, except at the crystal surface. Not all solids are crystalline; some, such as glass, have their atoms arranged randomly throughout the solid; these materials are known as **amorphous solids.**

amorphous solid
A solid in which the constituent atoms or molecules are arranged in random or disordered positions

Most crystals, excluding those that have cubic configurations, refract a beam of light into two different light-ray components. This phenomenon, known as double refraction, can be observed by studying the behavior of the crystal calcite. When the calcite is laid on a printed page, the observer sees not one but two images of each word covered. The two light rays that create the double image are refracted at different angles, and each has a different refractive index value. The indices of refraction for calcite are 1.486 and 1.658, and subtracting the two values yields a difference of 0.172; this difference is known as **birefringence.** Thus, the optical properties of crystals provide points of identification that help characterize them.

birefringence
A difference in the two indices of refraction exhibited by most crystalline materials

Now that we have investigated various chemical and physical properties of objects, we are ready to apply some of these properties to the characterization of glass, an item of evidence frequently retrieved at crime scenes and often sent to the crime laboratory for examination.

Quick Review

- Temperature is a measure of heat intensity, or the amount of heat in a substance.
- In science, the most commonly used temperature scale is the Celsius scale, which is derived by assigning the freezing point of water a value of 0°C and its boiling point a value of 100°C.
- Weight is the force with which gravity attracts a body. Mass is a constant property of matter that reflects the amount of material present.
- The formula for calculating the density of an object is mass per unit volume.
- The ratio of the velocity of light in a vacuum to its speed in any medium determines the refractive index of that medium.
- Crystalline solids have definite geometric forms because of the orderly arrangement of their atoms.
- Crystalline solids refract a beam of light in two different light-ray components, resulting in double refraction. Birefringence is the numerical difference between these two refractive indices.

Forensic Analysis of Glass

Glass that is broken and shattered into fragments and minute particles during the commission of a crime can be used to place a suspect at the crime scene. For example, chips of broken glass from a window may lodge in a suspect's shoes or garments during a burglary; particles of headlight glass found at the scene of a hit-and-run accident may confirm the identity of a suspect vehicle. All of these possibilities require the comparison of glass fragments found on the suspect, whether a person or vehicle, with the shattered glass remaining at the crime scene.

Composition of Glass

Glass is a hard, brittle, amorphous substance composed of sand (silicon oxides) mixed with various metal oxides. When sand is mixed with other metal oxides, melted at high temperatures, and then cooled to a rigid condition without crystallization, the product is glass. Soda (sodium carbonate) is normally added to the sand to lower its melting point and make it easier to work with. Another necessary ingredient is lime (calcium oxide), which is added to prevent the "soda-lime" glass from dissolving in water. The forensic scientist is often asked to analyze soda-lime glass, which is used for manufacturing most window and bottle glass. Usually the molten glass is cooled on top of a bath of molten tin. This manufacturing process produces flat glass typically used for windows. This type of glass is called float glass. Leaded glass is made with lead oxide resulting in a heavy glass that refracts light. Leaded glass is most commonly found in prisms, lenses, and imitations jewelry.

The common metal oxides found in soda-lime glass are sodium, calcium, magnesium, and aluminum. In addition, a wide variety of special glasses can be made by substituting in whole or in part other metal oxides for the silica, sodium,

FIGURE 5–13 When tempered glass breaks, it usually holds together without splintering. *Stella Photography/Shutterstock*

and calcium oxides. For example, automobile headlights and heat-resistant glass, such as Pyrex, are manufactured by adding boron oxide to the oxide mix. These glasses are therefore known as borosilicates.

Another type of glass that the reader may be familiar with is **tempered glass**. This glass is made stronger than ordinary window glass by introducing stress through rapid heating and cooling of the glass surfaces. When tempered glass breaks, it does not shatter but rather fragments or "dices" into small squares with little splintering (see Figure 5–13). Because of this safety feature, tempered glass is used in the side and rear windows of automobiles sold in the United States. The windshields of all cars manufactured in the United States are constructed from **laminated glass**. This glass derives its strength by sandwiching one layer of plastic between two pieces of ordinary window glass.

tempered glass
Glass to which strength is added by introducing stress through rapid heating and cooling of the glass surfaces

laminated glass
Two sheets of ordinary glass bonded together with a plastic film

Comparing Glass Fragments

For the forensic scientist, comparing glass consists of finding and measuring the properties that will associate one glass fragment with another while minimizing or eliminating the possible existence of other sources. Considering the prevalence of glass in our society, it is easy to appreciate the magnitude of this analytical problem. Obviously, glass possesses its greatest evidential value when it can be individualized to one source. Such a determination, however, can be made only when the suspect and crime-scene fragments are assembled and physically fitted together. Comparisons of this type require piecing together irregular edges of broken glass as well as matching all irregularities and striations on the broken surfaces (see Figure 5–14). The possibility that two pieces of glass originating from different sources will fit together exactly is so unlikely as to exclude all other sources from practical consideration.

Unfortunately, most glass evidence is either too fragmentary or too minute to permit a comparison of this type. In such instances, the search for individual properties has proven fruitless. For example, the general chemical composition of

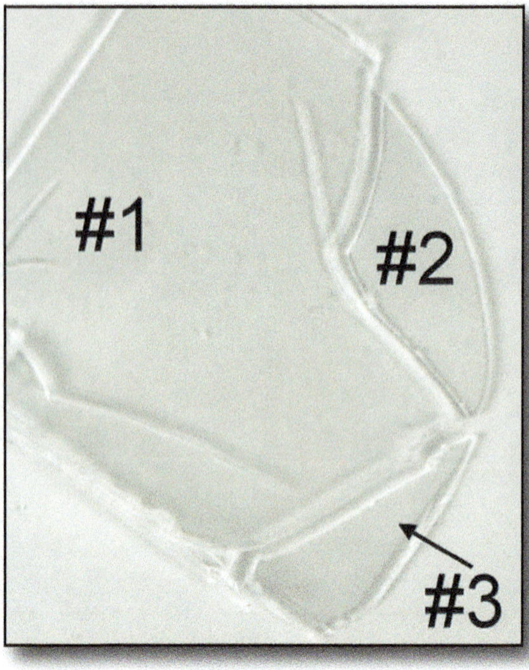

FIGURE 5–14 Match of broken glass. Note the physical fit of the edges. *Courtesy Sirchie Finger Print Laboratories, Inc., Youngsville, N.C., www.sirchie.com*

various window glasses within the capability of current analytical methods has so far been found to be relatively uniform among various manufacturers and thus offers no basis for individualization. However, as more sensitive analytical techniques are developed, trace elements present in glass may prove to be distinctive and measurable characteristics.

The physical properties of density and refractive indexes are used most successfully for characterizing glass particles. However, these properties are class characteristics, which cannot provide the sole criteria for individualizing glass to a common source. They do, however, give the analyst sufficient data to evaluate the significance of a glass comparison, and the absence of comparable density and refractive index values will certainly exclude glass fragments that originate from different sources.

Measuring and Comparing Density

Recall that a solid particle will either float, sink, or remain suspended in a liquid, depending on its density relative to the liquid medium. This knowledge gives the criminalist a rather precise and rapid method for comparing densities of glass.

In a method known as flotation, a standard/reference glass particle is immersed in a liquid; a mixture of bromoform and bromobenzene may be used. The composition of the liquid is carefully adjusted by adding small amounts of bromoform or bromobenzene until the glass chip remains suspended in the liquid medium. At this point, the standard/reference glass and liquid each have the same density. Glass chips of approximately the same size and shape as the standard/reference are now added to the liquid for comparison. If both the unknown and the standard/reference particles remain suspended in the liquid, their densities are equal to each other and to that of the liquid.[1] Particles of different densities either sink or float, depending on whether they are more or less dense than the liquid.

The density of a single sheet of window glass is not completely homogeneous throughout. It has a range of values that can differ by as much as 0.0003 g/mL. Therefore, in order to distinguish between the normal internal density variations of a single sheet of glass and those of glasses of different origins, it is advisable to let the comparative density approach but not exceed a sensitivity value of 0.0003 g/mL. The flotation method meets this requirement and can adequately distinguish glass particles that differ in density by 0.001 g/mL.

Determining and Comparing Refractive Index

Once glass has been distinguished by a density determination, different origins are immediately concluded. Comparable density results, however, require the added comparison of refractive indices. This determination is best accomplished by the immersion method. For this, glass particles are immersed in a liquid medium whose refractive index is adjusted until it equals that of the glass particles. At this point, known as the match point, the observer notes the disappearance of the **Becke line**, indicating minimum contrast between the glass and liquid medium. The Becke line is a bright halo observed near the border of a particle that is immersed in a liquid of a different refractive index. This halo disappears when the medium and fragment have similar refractive indices.

Beck line
A bright halo observed near the border of a particle immersed in a liquid of a different refractive index

The refractive index of an immersion fluid is best adjusted by changing the temperature of the liquid. Temperature control is, of course, critical to the success of the procedure. One approach is to heat the liquid in a special apparatus known as a hot stage (see Figure 5–15). The glass is immersed in a boiling liquid, usually

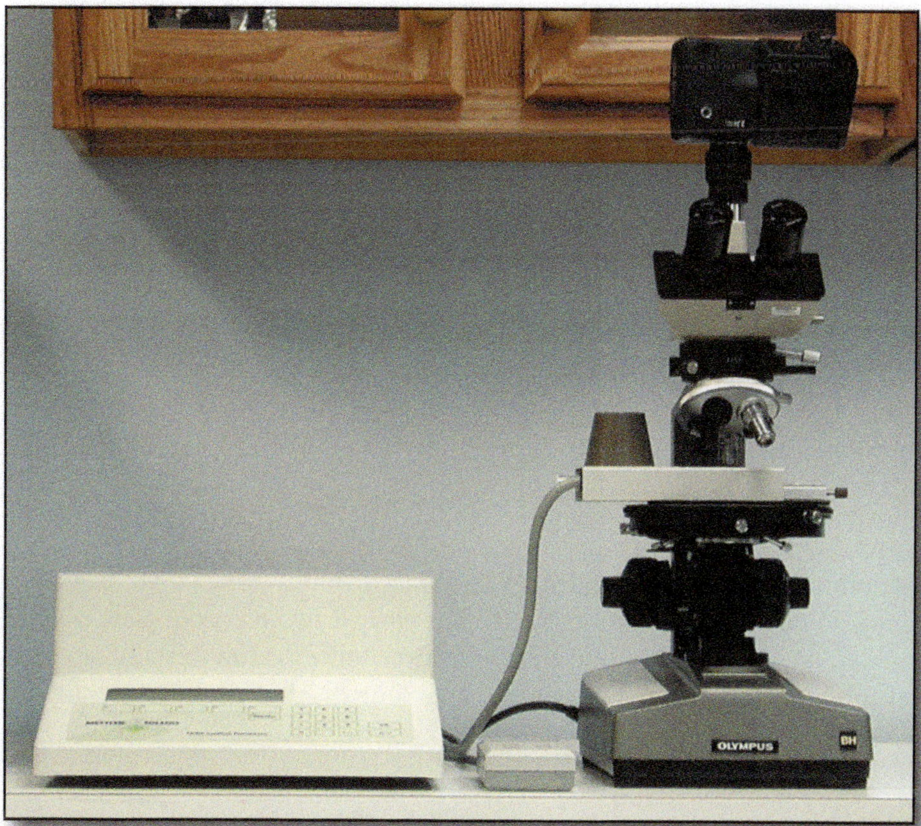

FIGURE 5–15 Hot-stage microscope. *Courtesy Chris Palenik*

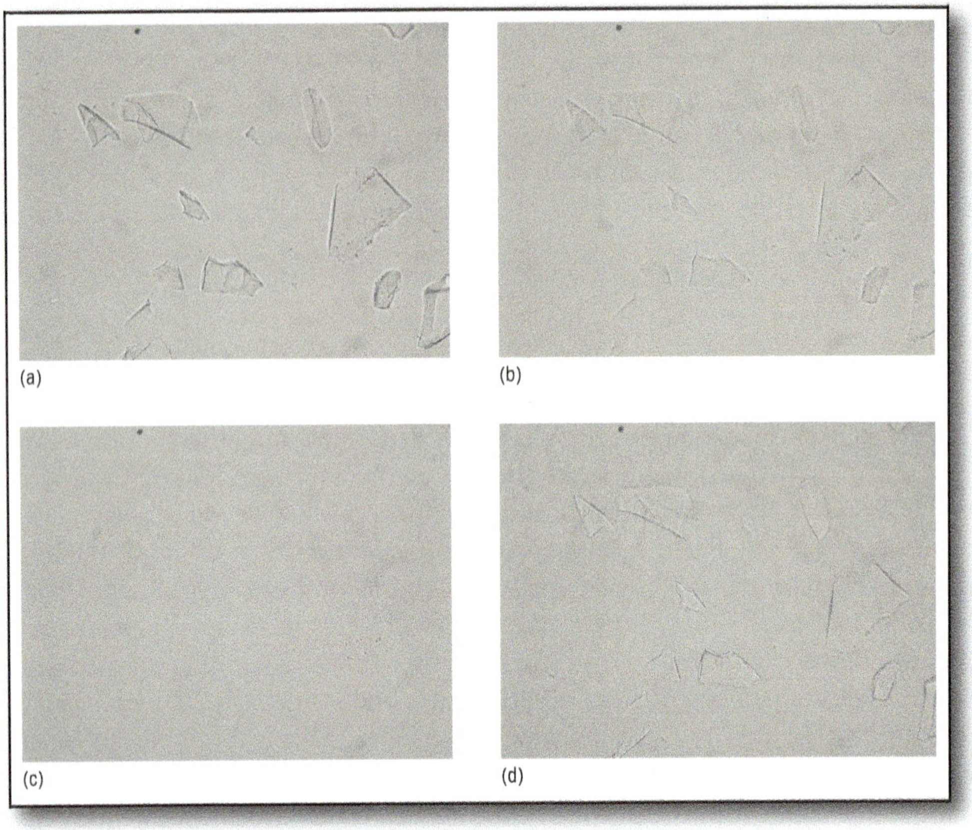

FIGURE 5–16 Determining the refractive index of glass. (a) Glass particles are immersed in a liquid of a much higher refractive index at a temperature of 20°C. (b) At 68°C the liquid still has a higher refractive index than the glass. (c) The refractive index of the liquid is closest to that of the glass at 100°C, as shown by the disappearance of the glass and the Becke lines. (d) At the higher temperature of 160°C, the liquid has a much lower index than the glass, resulting in significant edge contrast. The reference glass fragments shown here has refractive index of 1.529. *Courtesy Richard Saferstein, Ph.D.*

a silicone oil, and heated at the rate of 0.2°C per minute until the match point is reached. The examiner then observes the disappearance of the Becke line on minute glass particles that are illuminated with sodium D light or at other wavelengths of light. If all the glass fragments examined have similar match points, it can be concluded that they have comparable refractive indices (see Figure 5–16). Furthermore, the examiner can determine the refractive index value of the immersion fluid as it changes with temperature. With this information, the exact numerical value of the glass refractive index can be calculated at the match point temperature.[2]

As with density, glass fragments removed from a single sheet of plate glass may not have a uniform refractive index value; instead, their values may vary by as much as 0.0002. Hence, for comparison purposes, the difference in refractive index between a standard/reference and questioned glass must exceed this value. This allows the examiner to differentiate between the normal internal variations present in a sheet of glass and those present in glasses that originated from completely different sources.

Classification of Glass Samples

A significant difference in either density or refractive index proves that the glasses examined do not have a common origin. But what if two pieces of glass exhibit comparable densities and comparable refractive indices? How certain can one be that they did, indeed, come from the same source? After all, there are untold

Inside the Science

GRIM 3

An automated approach for measuring the refractive index of glass fragments by temperature control using the immersion method with a hot stage is with the instrument known as GRIM 3 (glass refractive index measurement)* (see Figure 1). The GRIM 3 is a personal computer/video system designed to automate the measurements of the match temperature and refractive index for glass fragments. This instrument uses a video camera to view the glass fragments as they are being heated. As the immersion oil is heated or cooled, the contrast of the video image is measured continually until a minimum, the match point, is detected (see Figure 2). The match point temperature is then converted to a refractive index using stored calibration data.

***Note:** There is a 5-degree error factor with this formula. This means that calculations are good to plus or minus 5 degrees of the actual value of the angle of impact. The measurements for length and width should be made with a ruler, micrometer, or photographic loupe.

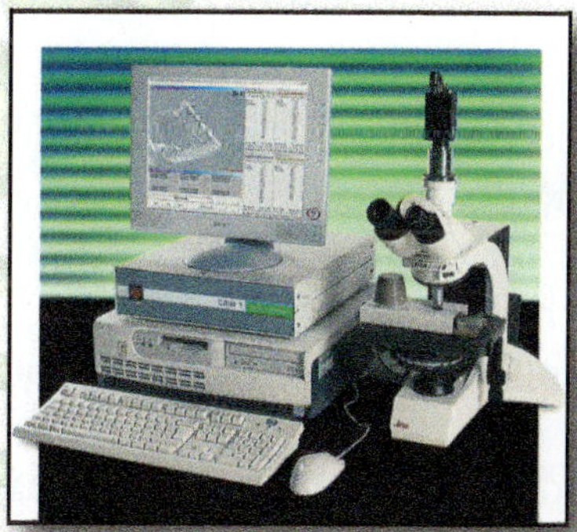

FIGURE 1 An automated system for glass fragment identification. *Courtesy Foster & Freeman, Worcestershire, U.K., www.fosterfreeman.co.uk*

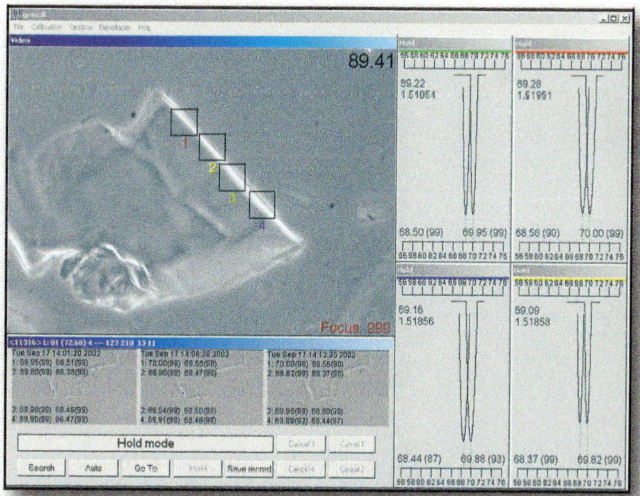

FIGURE 2 GRIM 3 identifies the refraction match point by monitoring a video image of glass fragment immersed in an oil. As the immersion oil is heated or cooled, the contrast of the image is measured continuously until a minimum, the match point, is detected. *Courtesy Foster & Freeman, Worcestershire, U.K., www.fosterfreeman.co.uk*

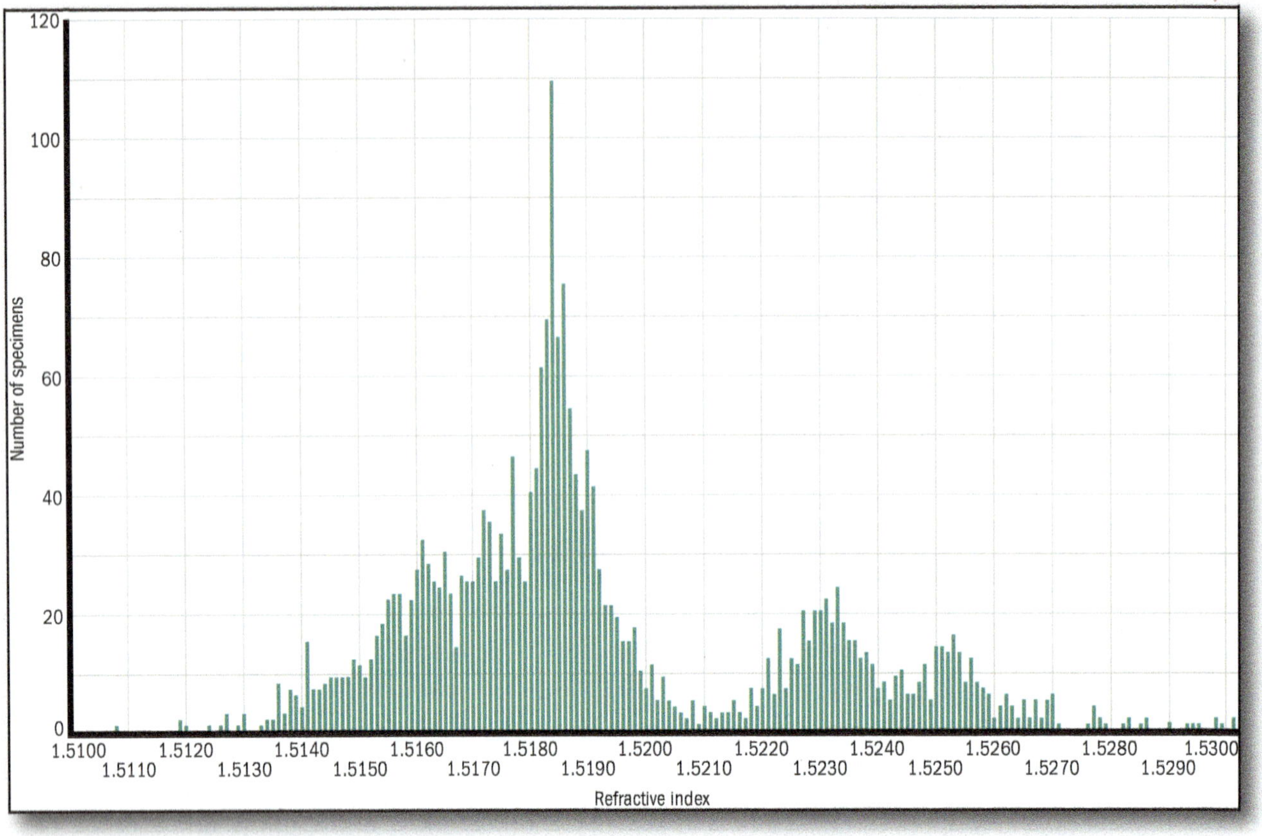

FIGURE 5-17 Frequency of occurrence of refractive index values (measured with sodium D light) for approximately two thousand flat glass specimens received by the FBI Laboratory. *Courtesy FBI Laboratory, Washington, D.C.*

millions of windows and other glass objects in this world. To provide a reasonable answer to this question, the FBI Laboratory has collected density and refractive index values from glass submitted to it for examination. What has emerged is a data bank correlating these values to their frequency of occurrence in the glass population of the United States. This collection is available to all forensic laboratories in the United States.[1]

Once a criminalist has completed a comparison of glass fragments, the criminalist can correlate their density and refractive index values to their frequency of occurrence and assess probability that the fragments came from the same source. Figure 5–17 shows the distribution of refractive index values (measured with sodium D light) for approximately 2,000 glasses analyzed by the FBI. The wide distribution of values clearly demonstrates that the refractive index is a highly distinctive property of glass and is thus useful for defining its frequency of occurrence and hence its evidential value. For example, a glass fragment with a refractive index value of 1.5290 is found in approximately only 1 out of 2,000 specimens, whereas glass with a value of 1.5180 occurs approximately in 22 glasses out of 2,000.

Although refractive index and density have been routinely used for the comparison of glass for some time, forensic scientists have long desired to extract additional information from glass fragments that would make their comparison more meaningful. The trace elemental composition of glass held a longtime attraction to forensic scientists for this purpose. However, until recently, the analytical instrumentation sensitive enough to develop a trace elemental profile from a glass fragment was too costly for most crime laboratories. This handicap has been overcome with the introduction of a technique that aims a high-energy laser pulse to vaporize a microscopic amount of glass, raising its temperature by thousands of degrees. As a result, the elements present in the glass are induced to emit light whose wavelengths correspond to the identity of the elements present.

The distinction between tempered and nontempered glass particles can be made by slowly heating and then cooling the glass (a process known as annealing). The change in the refractive index value for tempered glass upon annealing is significantly greater when compared to nontempered glass and thus serves as a point of distinction.[2]

Glass Fractures

Glass bends in response to any force that is exerted on any one of its surfaces; when the limit of its elasticity is reached, the glass fractures. Frequently, fractured window glass reveals information about the force and direction of an impact; such knowledge may be useful for reconstructing events at a crime-scene investigation and determining the direction of a projectile.

The penetration of ordinary window glass by a projectile, whether a bullet or a stone, produces a familiar fracture pattern in which cracks both radiate outward and encircle the hole, as shown in Figure 5-18. The radiating lines are appropriately known as **radial fractures**, and the circular lines are termed **concentric fractures**.

radial fracture
A crack in a glass that extends outward like the spoke of a wheel from the point at which the glass was struck

concentric fracture
A crack in a glass that forms a rough circle around the point of impact

FIGURE 5–18 Radial and concentric fracture lines in a sheet of glass. *Courtesy Sirchie Finger Print Laboratories, Inc., Youngsville, N.C., www.sirchie.com*

Often it is difficult to determine just from the size and shape of a hole in glass whether it was made by a bullet or by some other projectile. For instance, a small stone thrown at a comparatively high speed against a pane of glass often produces a hole very similar to that produced by a bullet. On the other hand, a large stone can completely shatter a pane of glass in a manner closely resembling the result of a close-range shot. However, in the latter instance, the presence of gunpowder deposits on the shattered glass fragments points to damage caused by a firearm.

When it penetrates glass, a high-velocity projectile such as a bullet often leaves a round, crater-shaped hole surrounded by a nearly symmetrical pattern of radial and concentric cracks. The hole is inevitably wider on the exit side (see Figure 5–19), and hence examining it is an important step in determining the direction of impact. However, as the velocity of the penetrating projectile decreases, the irregularity of the shape of the hole and of its surrounding cracks increases, so that at some point the hole shape will not help determine the direction of impact. At this point, examining the radial and concentric fracture lines may help determine the direction of impact.

When a force pushes on one side of a pane of glass, the elasticity of the glass permits it to bend in the direction of the force applied. Once the elastic limit is exceeded, the glass begins to crack. As shown in Figure 5–20, the first fractures form on the surface opposite that of the penetrating force and develop into radial lines. The continued motion of the force places tension on the front surface of the glass, resulting in the formation of concentric cracks. An examination of the edges of the radial and concentric cracks frequently reveals stress markings (Wallner lines) whose shape can be related to the side on which the window first cracked.

Stress marks, shown in Figure 5–21, are shaped like arches that are perpendicular to one glass surface and curved nearly parallel to the opposite surface. The importance of stress marks stems from the observation that the perpendicular edge always faces the surface on which the crack originated. Thus, in examining the stress marks on the edge of a radial crack near the point of impact,

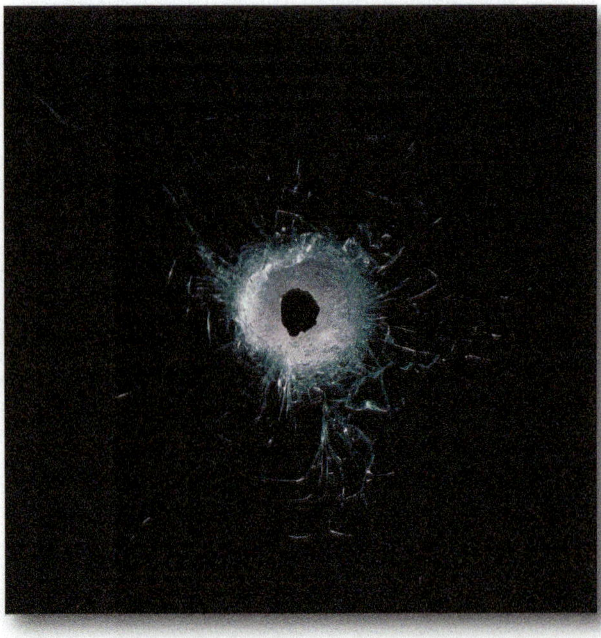

FIGURE 5–19 Crater-shaped hole made by a projectile passing through glass. The upper surface is the exit side of the projectile. *Don Farrall/Photodisc/Getty Images*

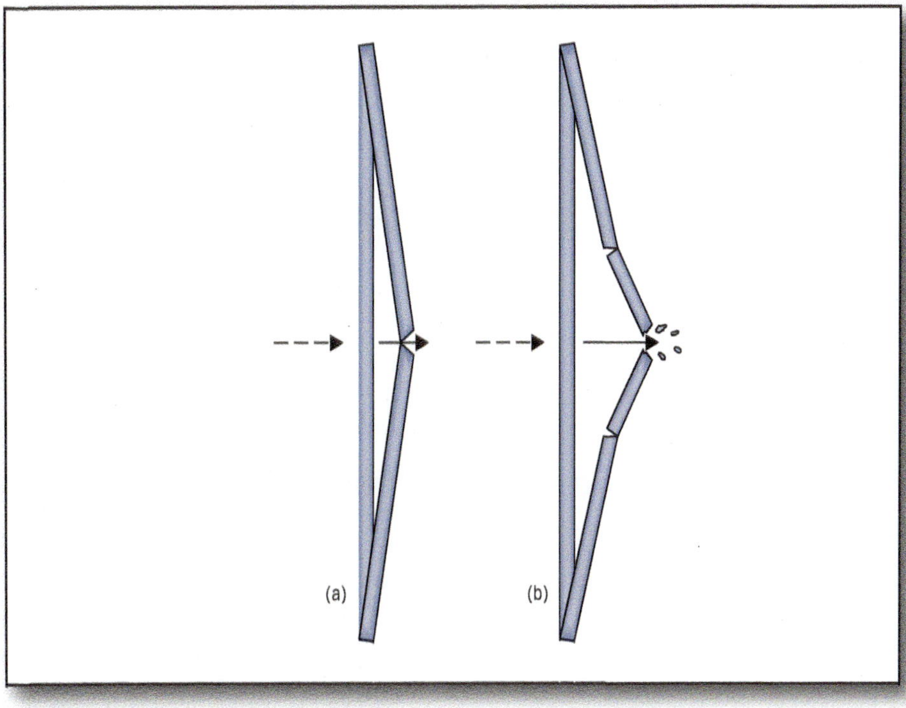

FIGURE 5–20 Production of radial and concentric fractures in glass. (a) Radial cracks are formed first, beginning on the side of the glass opposite to the destructive force. (b) Concentric cracks occur afterward, starting on the same side as the force.

FIGURE 5–21 Stress marks on the edge of a radial glass fracture. Arrow indicates direction of force. *Courtesy Richard Saferstein, Ph.D.*

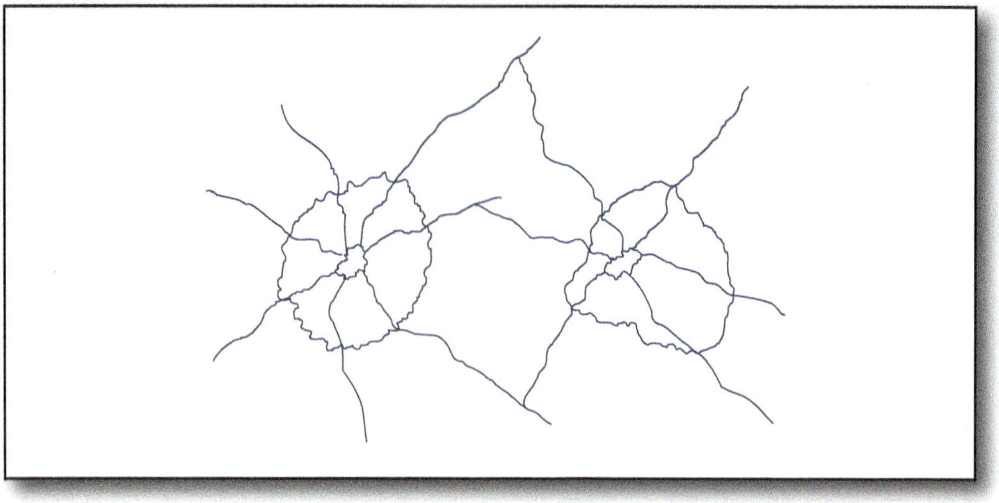

FIGURE 5–22
Two bullet holes in a piece of glass. The left hole preceded the right hole.

the perpendicular end is always found opposite the side from which the force of impact was applied. For a concentric fracture, the perpendicular end always faces the surface on which the force originated. A convenient way for remembering these observations is the 3R rule—Radial cracks form a Right angle on the Reverse side of the force. These facts enable the examiner to determine the side on which a window was broken. Unfortunately, the absence of radial or concentric fracture lines prevents these observations from being applied to broken tempered glass.

When there have been successive penetrations of glass, it is frequently possible to determine the sequence of impact by observing the existing fracture lines and their points of termination. A fracture always terminates at an existing line of fracture. In Figure 5–22, the fracture on the left preceded that on the right; we know this because the latter's radial fracture lines terminate at the cracks of the former.

Collection and Preservation of Glass Evidence

The gathering of glass evidence at the crime scene and from the suspect must be thorough if the examiner is to have any chance to individualize the fragments to a common source. If even the remotest possibility exists that fragments may be pieced together, every effort must be made to collect all the glass found. For example, evidence collection at hit-and-run scenes must include all the broken parts of the headlight and reflector lenses. This evidence may ultimately prove invaluable in placing a suspect vehicle at the accident scene by matching the fragments with glass remaining in the headlight or reflector shell of the suspect vehicle. In addition, examining the headlight's filaments may reveal whether an automobile's headlights were on or off before the impact (see Figure 5–23).

When an individual fit is improbable, the evidence collector must submit all glass evidence found in the possession of the suspect along with a sample of broken glass remaining at the crime scene. This standard/reference glass should

FIGURE 5–23 The presence of black tungsten oxide on the upper filament indicates that the filament was on when it was exposed to air. The lower filament was off, but its surface was coated with a yellow/white tungsten oxide, which was vaporized from the upper ("on") filament and condensed onto the lower filament. *Courtesy Richard Saferstein, Ph.D.*

always be taken from any remaining glass in the window or door frames, as close as possible to the point of breakage. About one square inch of sample is usually adequate for this purpose. The glass fragments should be packaged in solid containers to avoid further breakage. If the suspect's shoes and/or clothing are to be examined for the presence of glass fragments, they should be individually wrapped in paper and transmitted to the laboratory. The field investigator should avoid removing such evidence from garments unless absolutely necessary for its preservation.

When a determination of the direction of impact is desired, all broken glass must be recovered and submitted for analysis. Wherever possible, the exterior and interior surfaces of the glass must be indicated. When this is not immediately apparent, the presence of dirt, paint, grease, or putty may indicate the exterior surface of the glass.

Quick Review

- To compare glass fragments, a forensic scientist evaluates two important physical properties: density and refractive index.
- The immersion method is used to determine a glass fragment's refractive index. It involves immersing a glass particle in a liquid medium whose refractive index is adjusted by varying its temperature. At the refractive index match point the contrast between the glass and liquid is at a minimum.
- The flotation method is used to determine a glass fragment's density. It involves immersing a glass particle in a liquid whose density is carefully adjusted by adding small amounts of an appropriate liquid until the glass chip remains suspended in the liquid medium.
- By analyzing the radial and concentric fracture patterns in glass, the forensic scientist can determine the direction of impact by applying the 3R rule: Radial cracks form a Right angle on the Reverse side of the force.

Chapter Review

- Physical properties such as weight, volume, color, boiling point, and melting point describe a substance without reference to any other substance. A chemical property describes the behavior of a substance when it reacts or combines with another substance.
- Matter is anything that has mass and occupies space. An element is a fundamental particle of matter that cannot be broken down into simpler substances by chemical means.
- The smallest particle of an element that can exist and still retain its identity as that element is the atom.
- A compound is a pure substance composed of two or more elements. The smallest unit of a compound is a molecule.
- The three states of matter are solid, liquid, and gas (vapor).
- Sublimation is a change of state from a solid to a gas.
- A phase is a uniform body of matter distinguished from other matter by definite visible boundaries.
- Dispersion is the process of separating light into component colors. Because each color corresponds to a different range of frequencies or wavelengths of light, light rays of one color bend to a different degree than rays of all other colors.
- Refraction is the bending of light waves because of a change in velocity.
- Forensic scientists have learned to characterize different chemical substances by the type and quantity of light they absorb.
- Different types of electromagnetic radiation are distinguished from one another by their frequency or wavelength.
- As long as electromagnetic radiation is moving through space, its behavior can be described as that of a continuous wave. However, once radiation is absorbed by a substance, the model of light as a stream of discrete particles (photons) must be invoked to describe its behavior.
- Temperature is a measure of heat intensity, or the amount of heat in a substance.
- In science, the most commonly used temperature scale is the Celsius scale, which is derived by assigning the freezing point of water a value of 0°C and its boiling point a value of 100°C.
- Weight is the force with which gravity attracts a body. Mass is a constant property of matter that reflects the amount of material present.
- The formula for calculating the density of an object is mass per unit volume.
- The ratio of the velocity of light in a vacuum to its speed in any medium determines the refractive index of that medium.
- Crystalline solids have definite geometric forms because of the orderly arrangement of their atoms.

- Crystalline solids refract a beam of light in two different light-ray components, resulting in double refraction. Birefringence is the numerical difference between these two refractive indices.
- To compare glass fragments, a forensic scientist evaluates two important physical properties: density and refractive index.
- The immersion method is used to determine a glass fragment's refractive index. It involves immersing a glass particle in a liquid medium whose refractive index is adjusted by varying its temperature. At the refractive index match point the contrast between the glass and liquid is at a minimum.
- The flotation method is used to determine a glass fragment's density. It involves immersing a glass particle in a liquid whose density is carefully adjusted by adding small amounts of an appropriate liquid until the glass chip remains suspended in the liquid medium.
- By analyzing the radial and concentric fracture patterns in glass, the forensic scientist can determine the direction of impact by applying the 3R rule: Radial cracks form a Right angle on the Reverse side of the force.

Quick Lab: Glass and Density

Materials:

Meter sticks
Glass samples of various chemical compositions
Rulers/tape measures
Student worksheets

Procedure:

Density can be used to determine what type of glass evidence we are dealing with. Not all glass objects are made of the same type of glass, so it is important to determine what type of glass we have as evidence. Make sure when handling glass to use the forceps and not your fingers; be careful! First measure and record the mass of the piece of glass in grams using the scale. Fill the beaker with enough water that the glass will be covered when placed in it and record the amount in milliliters. Then place the piece of glass in the water and again record the new measurement from the beaker. Now subtract the first (just water) measurement from the second (water and glass) measurement; this is the volume of the glass. Determine the density of the glass by dividing the mass by the volume. Complete this for all the glass types.

Follow-Up Questions:

1. **What are some different objects that would be found as glass evidence?**
2. **Before your teacher reveals what density each type of glass was, try to guess which type of glass would be denser. Which type would be less dense? Why?**
3. **What is an intensive property?**

Review Questions

1. Which of the following is not a physical state of matter?
   a. solid
   b. plasma
   c. liquid
   d. vapor

2. If the density of a solid is greater than the liquid medium in which it is immersed, the object will
   a. sink.
   b. float.
   c. be suspended in the medium.
   d. lose density.

3. A bright halo that is observed near the border of a particle immersed in a liquid of a different refractive index is known as
   a. birefringence.
   b. the Becke line.
   c. a type line.
   d. the refraction point.

4. Wavelength is
   a. the distance between crests of adjacent waves.
   b. the speed of a wave.
   c. the distance from the crest of a wave to the trough.
   d. the number of waves that pass a given point in a second.

5. Refractive index measures the speed of light in a vacuum to its speed in
   a. air.
   b. water.
   c. glass.
   d. any given substance.

6. True or False: Visible light is only a small part of a large family of radiation that lies outside the range of the electromagnetic spectrum.

7. True or False: The physical properties of density and refractive index are used most successfully for characterizing glass particles.

8. True or False: A significant difference in either density or refractive index proves that the glasses examined have a common origin.

9. True or False: A high-velocity projectile, such as a bullet, will often leave a round, crater-shaped hole that is surrounded by a nearly symmetrical pattern of radial and concentric cracks.

10. True or False: The "3R rule" states that "radial cracks form a right angle on the reverse side of the force."

11. What is the difference between a physical property and a chemical property?

12. Define *matter* and *element*. How are matter and elements related?

13. What is the smallest particle of an element that can exist and still retain its identity?

14. What is a compound? Name three common compounds and give the chemical symbol for each.

15. What is the smallest unit of a compound?

16. Name the three states of matter. What chemical property of a substance determines its state?

17. What is sublimation?

18. What is a phase? Give an example of a two-phase system.

19. Define *dispersion* and explain how it is related to wavelength and frequency.

20. What is refraction?

21. Explain how color is related to the behavior of light. How do forensic scientists use this knowledge in their work?

22. What property distinguishes different types of electromagnetic radiation from one another?

23. Define *temperature* and name the two most common temperature scales.

24. What is weight? How does weight differ from mass?

25. How is the mass of an object determined? List three devices used for comparing the masses of different objects.

26. What is the formula for calculating the density of an object?

27. What is an intensive property? Name three intensive properties of matter.

28. What is the main ingredient in ordinary glass?

29. What kind of glass is used commonly in bottles and windows? How did this type of glass acquire its name?

30. What are tempered glass and laminated glass? On what part of an automobile is each type typically used?

31. What is the only way to individualize glass fragments found at a crime scene to a single source?

32. What physical properties are used most often to characterize glass particles? What is the main drawback of using these properties to characterize glass?

33. Define *flotation* and describe how it works.

34. Describe how a forensic scientist determines the refractive indices of suspect glass fragments.

35. How might a forensic scientist tell which of two fractures on a piece of glass was created earlier?

36. Define refractive index and explain how it is used in forensic glass analysis.

37. Describe and demonstrate the proper procedures to collect and preserve glass evidence.

Learn the Vocabulary

Use the information from this chapter along with your prior knowledge and experience to learn the key terms. Choose from the following activities:

1. Make flashcards with a term on the front and its definition on the back. Use the flashcards to quiz a partner on correctly pronouncing and defining the terms.
2. Make an audio or video recording of each term and its definition.
3. Work with a partner to write a sentence for each term. Read the sentences out loud to a partner or share them with the class.
4. Write a list of new terms and language you have encountered in this chapter. Reuse the terms to write sentences and then read the sentences out loud to a partner. Reuse the terms again by writing a short speech. The speech might explain the terms or teach a lesson using the terms. Give the speech to a partner or to the class.

Application and Critical Thinking

1. Use metric conversions to answer the following questions:
   a. Who is taller—someone who measures 6 feet, 3 inches, or someone who measures 1.9 meters?
   b. Which is heavier—a package that weighs 210 pounds or one that weighs 90 kilograms?
   c. Which holds more water—a 2-gallon bottle or an 8-liter bottle?
2. The following table shows the force of gravity on other planets relative to the force of gravity on the earth. For example, the force of gravity on the moon is only 0.166 (about one-sixth) that of the earth.

Table 1
The Force of Gravity on Other Planets

Planet	Mercury	Venus	Moon	Earth	Mars	Jupiter	Saturn	Uranus	Neptune	Pluto
Gravity Relative to Earth	0.378	0.907	0.166	1.000	0.377	2.533	1.064	0.889	1.125	0.067

   Using mathematical calculations, along with your knowledge of the difference between mass and weight, analyze the quantitative data and answer the following questions:
   a. Which weighs more—a 100-pound Great Dane on Pluto or a 25-pound terrier on Mars?
   b. Which weighs more—a quarter-pound hamburger on Jupiter or a 12-ounce steak on Venus?
   c. Which has a greater mass—a 300-pound football player on Saturn or an 800-pound gorilla on Mercury?
   d. If a car has a mass of 4,500 pounds on the earth, how much will it weigh on Neptune?

3. An accident investigator arrives at the scene of a hit-and-run collision. The driver who remained at the scene reports that the windshield or a side window of the car that struck him shattered on impact. The investigator searches the accident site and collects a large number of fragments of tempered glass. This is the only type of glass recovered from the scene. How can the glass evidence help the investigator locate the vehicle that fled the scene?

4. Indicate the order in which the bullet holes were made in the glass depicted in the figure below. Explain the reason for your answer.

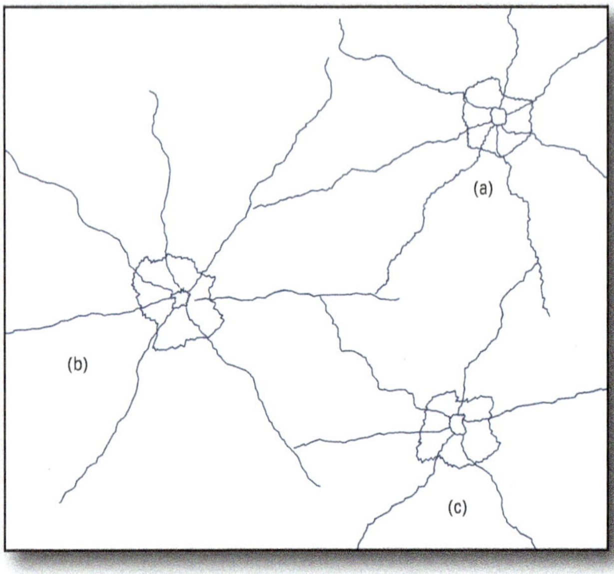

5. The right side of figure below depicts stress marks on the edge of a glass fracture caused by the application of force. Determine the direction of the projectile by examining these glass fractures. If this is a radial fracture, from which side of the glass (left or right) was the force applied? From which side was force applied if it is a concentric fracture? Explain the reason for your answers.

Courtesy Richard Saferstein, Ph.D.

6. In a brief essay, compare and contrast the composition of the following types of glass: soda lime, borosilicate, leaded, and tempered.

Laboratory Experiments

This activity requires the use of the following practices of science and engineering:

- Planning and carrying out investigations
- Analyzing and interpreting data
- Using appropriate safety equipment and practices during laboratory/classroom and field investigations as outlined in Texas Education Agency-approved safety standards

This activity consists of the following crosscutting concepts:

- **Patterns**—The density of glass allows them to be classified into types of glass.
- **Cause and effect**—The impact of the vehicle with the ATM caused glass to be left at the crime scene. The analysis of the glass can be used as evidence.
- **Structure and function**—The structure of the glass determines its density which is a characteristic property.

Before you begin, read the scenario and steps out loud with a partner. Ask for clarification as necessary to make sure you understand the experiment.

The Scenario

You are called to a hit and run car accident on November 16 at 8:35 AM. The vehicle collided with a bank ATM and the money was stolen. The cameras on the ATM were working, but video couldn't be used to determine the exact make and model of the vehicle, just that it was a sports car. The only evidence available is glass that appears to be from the headlight of the vehicle. A vehicle with a broken headlight was found in the apartment building's parking lot two blocks from the bank. Use your forensic science skills to demonstrate how to collect the glass evidence from the the crime scene, and preserve the glass evidence by storing it in a solid container. Then, you are to use density to determine if the glass found at the crime scene matches the glass from the suspected vehicle. Due to the small size of the glass pieces, the density must be determined using the flotation method.

SAFETY

Wear approriate PPE, such as eye protection, clothing protection, plastic gloves

MATERIALS

Glass samples (one from the car and one from the crime scene)
25-mL graduated cylinder
Beral Pipet
Deionized Water
Sugar

PROCEDURE

There are 2 types of glass samples available for testing:

a. Sample "a" is from a crime scene.

b. Sample "b" was taken from the suspect's vehicle.

1. Add 10 mL of distilled water to a test tube and clamp into a ring stand.
2. Add sample "a" to the water.
3. Using a spatula, add a pea-sized amount of sugar.
4. After each addition, swirl the tube gently to mix. Observe the glass chip.
5. Continue adding sugar in small amounts until the chip becomes suspended in the solution, somewhere in the middle.
6. Now add the glass fragment from sample "b." Allow it to equilibrate.

If glass chips stay suspended in the liquid for at least one minute, they compare by density. If only the crime scene glass sample remains suspended, while the questioned sample either sinks to the bottom or rises to the top, then the samples do not compare by density. Outline your findings and indicate if this is sufficient evidence to arrest the owner of the vehicle. Based on your laboratory findings, what questions would you ask of the vehicle owner?

Endnotes

1. As an added step, the analyst can determine the exact numerical density value of the particles of glass by transferring the liquid to a density meter, which will electrically measure and calculate the liquid's density. See A. P. Beveridge and C. Semen, "Glass Density Measurement Using a Calculating Digital Density Meter," *Canadian Society of Forensic Science Journal 12* (1979): 113.

2. A.R.Cassista and P.M.L.Sandercock, "Precision of Glass Refractive Index Measurements: Temperature Variation and Double Variation Methods, and the Value of Dispersion," *Canadian Society of Forensic Science Journal* 27 (1994): 203.

3. G. Edmondstone, "The Identification of Heat Strengthened Glass in Windshields," *Canadian Society of Forensic Science Journal* 30 (1997): 181.

Drugs

6

Uncredited/AP/Shutterstock

Key Terms

- anabolic steroids
- analgesic
- chromatography
- confirmation
- depressant
- hallucinogen
- infrared
- ion
- microcrystalline test
- monochromator
- narcotic
- physical dependence
- psychological dependence
- screening test
- spectrophotometry
- stimulant
- ultraviolet

El Chapo: Joaquin Guzman

In 2016, the U.S. government filed charges against Joaquin "El Chapo" Guzmán seeking $14 billion dollars in estimated wealth he earned from trafficking drugs for the Sinaloa Cartel. As the leader of the Cartel, Guzmán oversaw drug smuggling operations throughout the United States and Europe. Through the use of distribution cells in the United States and long-range tunnels near borders, Guzmán was able to export more drugs to the United States than any other trafficker in history. His leadership of the Cartel also brought immense wealth and power; Guzmán was ranked by Forbes as one of the most powerful people in the world between 2009 and 2013, while the Drug Enforcement Administration (DEA) estimated that he matched the influence and wealth of Pablo Escobar.

Guzmán was born in Sinaloa and raised in a poor farming family. He entered the drug trade, helping him grow marijuana for local dealers during his early adulthood. In the 1970s, Guzmán began working with Hector Luis Palma Salazar, one of the nation's rising drug lords, whom he helped map routes to move drugs through Sinaloa, and into the United States. He later supervised logistics for Miguel Angel Felix Gallardo, one of the Mexico's leading kingpins. During the 1980s, increased scrutiny on the South American Cartels allowed the Mexican Cartels to flourish. When Gallardo was finally arrested in 1989, Guzmán broke off on his own and started his own Cartel.

Guzmán was first captured in 1993 in Guatemala and was extradited to Mexico to face drug charges. There he was sentenced to 20 years in prison for murder and drug trafficking. He bribed prison guards and escaped from a federal maximum-security prison in 2001. The United States and Mexico offered a combined reward of $8.8 million for information leading to his capture, and he was re-arrested in Mexico in 2014. He escaped prior to formal sentencing in 2015, through a tunnel nearly a mile long under his jail cell. He was recaptured by Mexican authorities following a shoot-out in 2016, and was extradited to the United States a year later, where he was tried and convicted of 10 counts, including drug trafficking, weapons charges, and money laundering. He was sentenced in July 2019 to a life term of imprisonment plus 30 years to run consecutive to the life sentence for being a principal leader of a continuing criminal enterprise. He was also ordered to pay $12.6 billion in forfeiture.

Learning Objectives

After studying this chapter, you should be able to:

6.1 Explain psychological and physical dependency on drugs and its impact on society

6.2 Classify the commonly abused drugs

6.3 Explain the classification of drugs under the Controlled Substances Act

6.4 Describe the proper collection and preservation of drug evidence

6.5 Describe the laboratory tests normally used to perform a routine drug identification analysis

6.6 Discuss the process of chromatography as well as the difference between thin-layer chromatography and gas chromatography

6.7 Describe the utility of ultraviolet and infrared spectrophotometry for the identification of drugs

6.8 Describe the concept and utility of mass spectrometry for identification analysis

National Science Content Standards

Scientific Inquiry

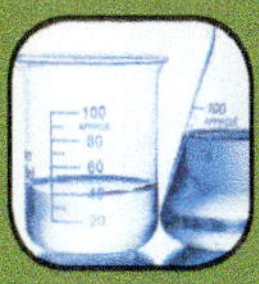
Physical Science

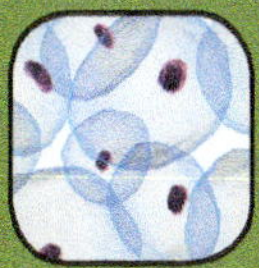
Life Science

Science and Technology

Science in Personal and Social Perspective

A *drug* can be defined as a natural or synthetic substance that is used to produce physiological or psychological effects in humans or other higher-order animals. However, this colorless clinical definition does not really tell us what drugs are; in their modern context, drugs mean something different to each person. To some, drugs are a necessity for sustaining and prolonging life; to others, drugs provide an escape from the pressures of life; to still others, they are a means of ending it.

Considering the wide application and acceptance of drugs in our society, it was perhaps inevitable that a segment of our population would abuse them. During the 1960s, successive waves of hallucinogens, amphetamines, and barbiturates found their way out of laboratories, pharmacies, and medicine chests and into the streets. During this decade, marijuana became the most widely used illicit drug in the United States, and alcohol consumption continued to rise—today 90 million Americans drink alcohol regularly, and 10 million of these are hopelessly addicted or have severe problems in coping with their drinking habits. In the 1970s, heroin addiction emerged as a national problem, and today the United States is in the midst of an epidemic of cocaine abuse.

FIGURE 6–1
Drug bust. *Juan Carlos Llorca/ AP/Shutterstock*

Drug abuse has grown from a problem generally associated with members of the lower end of the socioeconomic ladder to one that cuts across all social and ethnic classes of society. Today, approximately 23 million people in the United States use illicit drugs, including about a half million heroin addicts and nearly six million users of cocaine. In the United States, more than 75 percent of the evidence evaluated in crime laboratories is drug related. The deluge of drug specimens has forced the expansion of existing crime laboratories and the creation of new ones. For many concerned forensic scientists, the crime laboratory's preoccupation with drug evidence represents a serious distraction from time that could be devoted to evaluating evidence related to homicides and other types of serious crimes. However, the increasing caseloads associated with drug evidence have justified the expansion of forensic laboratory services. This expansion has increased the overall analytical capabilities of crime laboratories.

Drug Dependence

In assessing the potential danger of drugs, society has become particularly conscious of their effects on human behavior. In fact, the first drugs to be regulated by law in the early years of the 20th century were those deemed to have "habit-forming" properties. The early laws were aimed primarily at controlling opium and its derivatives, cocaine, and later marijuana. The ability of a drug to induce dependence after repeated use is submerged in a complex array of physiological and social factors.

Dependence on drugs exists in numerous patterns and in all degrees of intensity, depending on the nature of the drug, the route of administration, the dose, the frequency of administration, and the individual's rate of metabolism. Furthermore, nondrug factors play an equally crucial role in determining the behavioral patterns associated with drug use. The personal characteristics of the user, their expectations about the drug experience, society's attitudes and possible responses, and the setting in which the drug is used are all major determinants of drug dependence.

The question of how to define and measure a drug's influence on the individual and its danger to society is difficult to assess. The nature and significance of drug dependence must be considered from two overlapping points of view: the interaction of the drug with the individual, and the drug's impact on society. It will be useful to approach the problem from two distinctly different aspects of human behavior—**psychological dependence** and **physical dependence**.

psychological dependence
The conditioned use of a drug caused by underlying emotional needs

physical dependence
Physiological need for a drug brought about by its regular use and characterized by withdrawal sickness when administration of the drug abruptly stops

Psychological Dependence

The common denominator that characterizes all types of repeated drug use is the creation of a psychological dependence for continued use of the drug. It is important to discard the unrealistic image that all drug users are hopeless "addicts" who are social dropouts. Most users present quite a normal appearance and remain both socially and economically integrated in the life of the community.

The reasons why some people abstain from drugs while others become moderately or heavily involved are difficult if not impossible to delineate. Psychological needs arise from numerous personal and social factors that inevitably

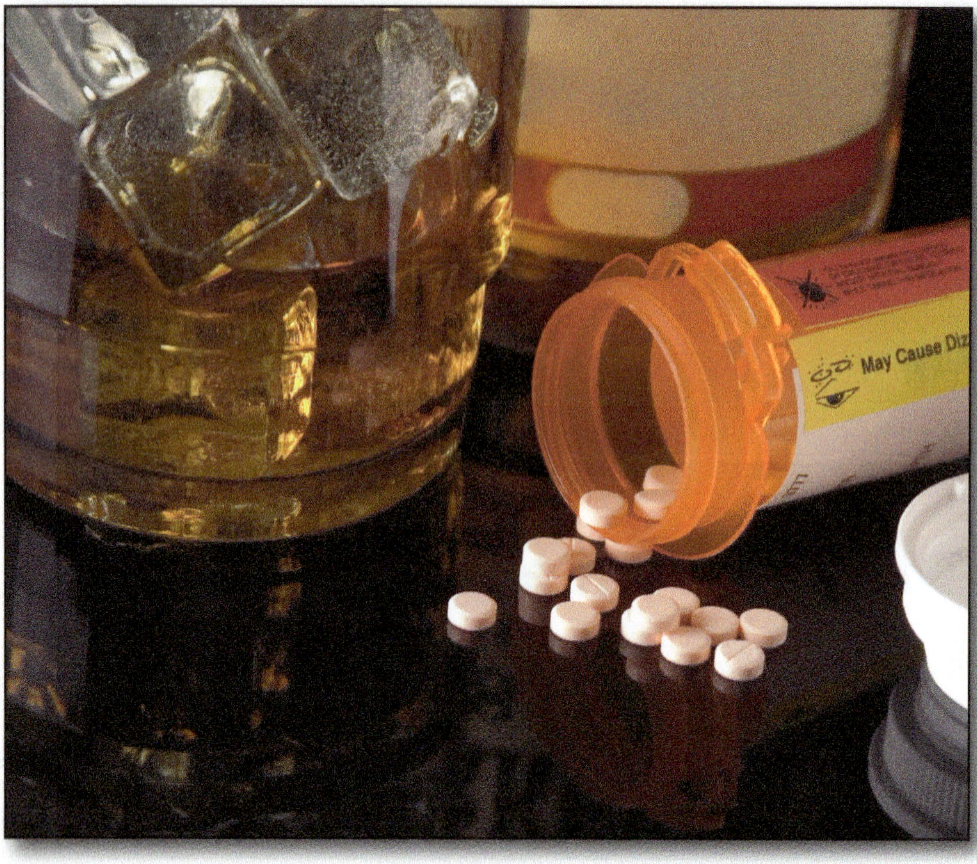

FIGURE 6–2 The drug of choice varies from person to person. *Courtesy Donna Cuic/ Shutterstock*

stem from the individual's desire to create a sense of well-being and to escape from reality. In some cases, the individual may seek relief from personal problems or stressful situations, or may be trying to sustain a physical and emotional state that permits an improved level of performance. Whatever the reasons, the underlying psychological needs and the desire to fulfill them create a conditioned pattern of drug abuse (see Figure 6–2).

The intensity of the psychological dependence associated with a drug's use is difficult to define and largely depends on the nature of the drug used. For drugs such as alcohol, heroin, amphetamines, barbiturates, and cocaine, continued use will likely result in a high degree of involvement. Other drugs, such as marijuana and codeine, appear to have a considerably lower potential for the development of psychological dependence. However, this does not imply that repeated abuse of drugs deemed to have a low potential for psychological dependence is safe or will always produce low psychological dependence. We have no precise way to measure or predict the impact of drug abuse on the individual. Even if a system could be devised for controlling the many possible variables affecting a user's response, the unpredictability of the human personality would still have to be considered.

Our general knowledge of alcohol consumption should warn us of the fallacy of generalizing when attempting to describe the danger of drug abuse. Obviously, not all alcohol drinkers are psychologically addicted to the drug; most are "social" drinkers who drink in reasonable amounts and on an irregular basis. Many people have progressed beyond this stage and consider alcohol a necessary crutch for dealing with life's stresses and anxieties. However, alcohol abusers exhibit a wide range of behavioral patterns, and to a large extent the determination of the degree of psychological dependence must be made individually. Likewise, it would be wrong to generalize that all users of marijuana can at worst develop a low degree of dependence on the drug. A wide range of factors also influence marijuana's effect, and heavy users of the drug expose themselves to the danger of developing a high degree of psychological dependence.

Physical Dependence

Although emotional well-being is the primary motive leading to repeated and intensive use of a drug, certain drugs, taken in sufficient dose and frequency, can produce physiological changes that encourage their continued use. Once the user abstains from such a drug, severe physical illness follows. The desire to avoid this withdrawal sickness, or abstinence syndrome, ultimately causes physical dependence, or addiction. Hence, for the addict who is accustomed to receiving large doses of heroin, the thought of abstaining and encountering body chills, vomiting, stomach cramps, convulsions, insomnia, pain, and hallucinations is a powerful inducement for continued drug use.

Interestingly, some of the more widely abused drugs have little or no potential for creating physical dependence. Drugs such as marijuana, LSD, and cocaine create strong anxieties when their repeated use is discontinued; however, no medical evidence attributes these discomforts to physiological reactions that accompany withdrawal sickness. On the other hand, use of alcohol, heroin, and barbiturates can result in development of physical dependence.

Physical dependence develops only when the drug user adheres to a regular schedule of drug intake; that is, the interval between doses must be short enough so that the effects of the drug never wear off completely. For example, the interval between injections of heroin for the drug addict probably does not exceed six to eight hours. Beyond this time the addict begins to experience the uncomfortable symptoms of withdrawal. Many heroin users avoid taking the drug regularly for fear of becoming physically addicted to its use. Similarly, the risk of developing physical dependence on alcohol becomes greatest when the consumption is characterized by a continuing pattern of daily use in large quantities.

Table 6–1 categorizes some of the more commonly abused drugs according to their effect on the body and summarizes their tendency to produce psychological dependence and to induce physical dependence with repeated use.

Societal Aspects of Drug Use

The social impact of drug dependence is directly related to the extent to which the user has become preoccupied with the drug. Here, the most important element is the extent to which drug use has become interwoven in the fabric of the user's life. The more frequently the drug satisfies the person's need, the greater the likelihood that the person will become preoccupied with its use, with a consequent neglect of individual and social responsibilities. Personal health, economic relationships, and family obligations may all suffer as the drug-seeking behavior increases in frequency and intensity and dominates the individual's life. Extreme drug dependence may lead to behavior that has serious implications for the public's safety, health, and welfare.

Table 6–1
The Potential of Some Commonly Abused Drugs to Produce Dependence with Regular Use

Drug	Psychological Dependence	Physical Dependence
Narcotics		
Morphine	High	Yes
Heroin	High	Yes
Methadone	High	Yes
Codeine	Low	Yes
Depressants		
Barbiturates (short-acting)	High	Yes
Barbiturates (long-acting)	Low	Yes
Alcohol	High	Yes
Methaqualone (Quaalude)	High	Yes
Meprobamate (Miltown, Equanil)	Moderate	Yes
Diazepam (Valium)	Moderate	Yes
Chlordiazepoxide (Librium)	Moderate	Yes
Stimulants		
Amphetamines	High	Unknown
Cocaine	High	No
Caffeine	Low	No
Nicotine	High	Yes
Hallucinogens		
Marijuana	Low	No
LSD	Low	No
Phencyclidine (PCP)	High	No

Drug dependence in its broadest sense involves much of the world's population. As a result, a complex array of individual, social, cultural, legal, and medical factors ultimately influence society's decision to prohibit or impose strict controls on a drug's distribution and use. Invariably, society must weigh the beneficial aspects of the drug against the ultimate harm its abuse will do to the ndividual and to society as a whole. Obviously, many forms of drug dependence do not carry sufficient adverse social consequences to warrant their prohibition, as illustrated by the widespread use of such drug-containing substances as tobacco and coffee. Although the heavy and prolonged use of these drugs may eventually damage body organs and injure an individual's health, there is no evidence that they result in antisocial behavior, even with prolonged or excessive use. Hence, society is willing to accept the widespread use of these substances.

We are certainly all aware of the disastrous failure in the United States to prohibit the use of alcohol during the 1920s and the current debate on whether marijuana should be legalized. Each of these issues emphasizes the delicate balance between individual desires and needs and society's concern with the consequences of drug abuse; moreover, this balance is continuously subject to change and re-evaluation. For example, the four most common drugs associated with crimes are cocaine, heroin, marijuana, and amphetamines. To reduce the occurrence of crimes connected with these drugs, it is important to develop an explanation of why there is a connection and propose a solution to reduce the connection. Possible solutions should be testable using scientific ideas, principles, and theories creating data that can be used to create models to combat the problem.

Quick Review

- A drug is a natural or synthetic substance that is used to produce physiological or psychological effects in humans or other animals.
- Nondrug factors that play a part in drug dependence include the personal characteristics of the user, the user's expectations about the drug experience, society's attitudes and possible responses, and the setting in which the drug is used.
- Physical dependence is defined as the physiological need for a drug that has been brought about by its regular use. Psychological dependence is the conditioned use of a drug caused by underlying emotional needs.

Types of Drugs

Narcotic Drugs

The term **narcotic** is derived from the Greek word *narkotikos*, meaning numbness or deadening. Although pharmacologists classify narcotic drugs as substances that relieve pain and produce sleep, the term narcotic has become popularly associated with any drug that is socially unacceptable. As a consequence of this incorrect usage, many drugs are improperly called narcotics.

narcotic
A drug that induces sleep and depresses vital body functions such as blood pressure, pulse rate, and breathing rate

This confusion has produced legal definitions that differ from the pharmacological actions of many drugs. For example, until the early 1970s, most drug laws in the United States incorrectly designated marijuana as a narcotic. Even today, federal law classifies cocaine as a narcotic drug, although pharmacologically, cocaine is actually a powerful central nervous system stimulant, possessing properties opposite those normally associated with the depressant effects of a narcotic.

Opiates Medical professionals apply the term opiate to most of the drugs properly classified as narcotics. Opiates behave pharmacologically like morphine, a painkiller derived from opium—a gummy, milky juice exuded through a cut made in the unripe pod of the Asian poppy (Papaver somniferium). Although morphine is readily extracted from opium, the most commonly used opiumbased drug is heroin, which is produced by reacting morphine with acetic anhydride or acetyl chloride (see Figure 6–3). Heroin's high solubility in water makes its street preparation for intravenous administration rather simple, because only by injection are heroin's effects felt almost instantaneously and with maximum sensitivity. The solution is drawn into a syringe or eyedropper for injection under the skin (see Figure 6–4).

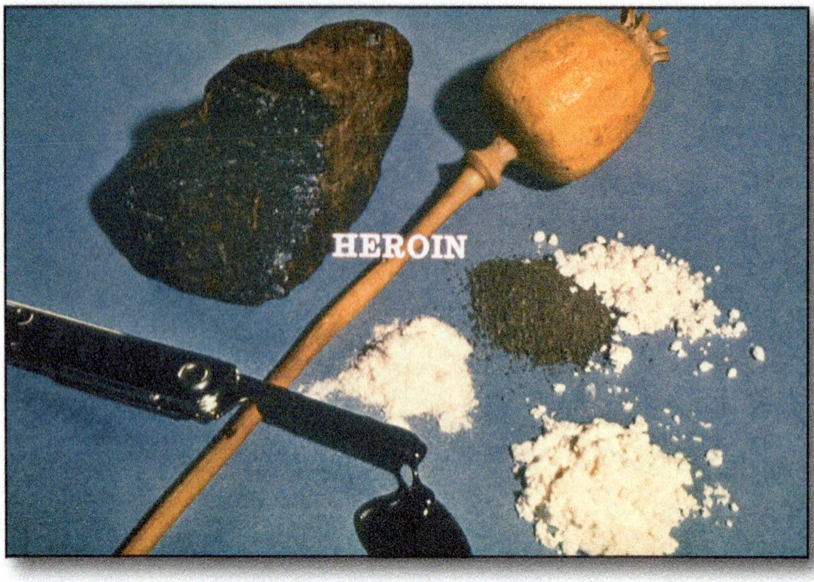

FIGURE 6–3
The opium poppy and its derivatives. Shown are the poppy plant, crude and smoking opium, codeine, heroin, and morphine.
Courtesy Pearson Education

FIGURE 6–4
Heroin paraphernalia.
Courtesy Drug Enforcement Administration, Washington, D.C.

analgesic
A substance that lessens or eliminates pain

Heroin and other narcotic drugs are **analgesics**—that is, they relieve pain by depressing the central nervous system. Besides being a powerful analgesic, heroin produces a "high" that is accompanied by drowsiness and a deep sense of wellbeing. The effect is short, generally lasting only three to four hours. Regular use of heroin—or any other narcotic drug—invariably leads to physical dependence, with all of its dire consequences.

Codeine is also present in opium, but it is usually prepared synthetically from morphine. It is commonly used as a cough suppressant in prescription cough syrup. Codeine, only one-sixth as strong as morphine, is not an attractive street drug for addicts.

Inside the Science

What's in That Bag?

The content of a typical heroin bag is an excellent example of the uncertainty attached to buying illicit drugs. For many years into the 1960s and early 1970s, the average bag contained 15 to 20 percent heroin. Currently, the average purity of heroin obtained in the illicit U.S. market is approximately 35 percent. The addict rarely knows or cares what comprises the other 65 percent or so of the material. Traditionally, quinine has been the most common diluent of heroin. Like heroin, it has a bitter taste and was probably originally used to obscure the actual potency of a heroin preparation from those who wished to taste-test the material before buying it. Other diluents commonly added to heroin are starch, lactose, procaine (Novocain), and mannitol.

Synthetic Opiates A number of narcotic drugs are not naturally derived from opium. However, because they have similar physiological effects on the body as the opium narcotics, they are also commonly referred to as opiates.

In 1995, the U.S. Food and Drug Administration (FDA) approved for use the pain-killing drug OxyContin. The active ingredient in OxyContin is oxycodone, a synthetic closely related to morphine and heroin in its chemical structure. OxyContin is an analgesic narcotic that has effects similar to those of heroin. It is prescribed to a million patients for treatment of chronic pain, with doctors writing close to seven million OxyContin prescriptions each year. The drug is compounded with a time-release formulation that the manufacturer initially believed would reduce the risk of abuse and addiction. This has not turned out to be the case. It is estimated that close to a quarter of a million individuals abuse the drug.

Because it is a legal drug that is diverted from legitimate sources, OxyContin is obtained differently from illegal drugs. Pharmacy robberies, forged prescriptions, and theft of the drug from patients with a legitimate prescription are ways in which abusers access OxyContin. Some abusers visit numerous doctors and receive prescriptions even though their medical condition may not warrant it.

Methadone is another well-known synthetic opiate. In the 1960s, scientists discovered that a person receiving methadone periodically in oral doses of 80–120 milligrams a day would not get high if they then took heroin or morphine. Clearly, although methadone is a narcotic pharmacologically related to heroin, its administration appears to eliminate the addict's desire for heroin while producing minimal side effects. Critics of the controversial methadone maintenance programs claim that methadone use is just substituting one narcotic drug for another, and supporters argue that this is the only known treatment for keeping the addict off heroin and offering some hope for eventual abstention from narcotics.

Physicians are increasingly prescribing methadone for pain relief. Unfortunately, in recent years, the wide availability of the drug for legitimate medical purposes has led to greater quantities of the drug being diverted into the illicit market. Methadone is being abused increasingly and is causing an alarming number of overdoses and deaths.

Hallucinogens

hallucinogen
A substance that induces changes in normal thought processes, perceptions, and moods

Hallucinogens are drugs that can cause marked alterations in normal thought processes, perceptions, and moods. Perhaps the most popular and controversial member of this class of drugs is marijuana.

Marijuana Marijuana easily qualifies as the most widely used illicit drug in the United States today. For instance, more than 43 million Americans have tried marijuana, according to the latest surveys, and almost half that number may be regular users. Marijuana is a preparation derived from the plant Cannabis. Most botanists believe there is only one species of the plant, Cannabis sativa L. The marijuana preparation normally consists of crushed leaves mixed in varying proportions with the plant's flower, stem, and seed. The plant secretes a sticky resin known as hashish. The resinous material can also be extracted from the plant by soaking in a solvent such as alcohol. On the illicit-drug market, hashish usually appears in the form of compressed vegetation containing a high percentage of resin. A potent form of marijuana is known as sinsemilla. This is made from the unfertilized flowering tops of the female Cannabis plants, attained by removing all male plants from the growing field at the first sign of their appearance. It follows that the production of sinsemilla requires a great deal of attention and care, and the plant is therefore cultivated on small plots.

The Cannabis plant contains a chemical known as tetrahydrocannabinol, or THC, which produces the psychoactive effects experienced by users. The THC content of Cannabis varies in different parts of the plant. The greatest concentration is usually found in the resin. Declining concentrations are typically found in the flowers and leaves, respectively. Little THC is found in the stem, roots, or seeds of the plant. The potency and resulting effect of the drug fluctuate, depending on the relative proportion of these plant parts in the marijuana mixture consumed by the user. The most common method of administration is by smoking either the dried flowers and leaves, or various preparations of hashish (see Figure 6–5). Marijuana is also occasionally taken orally, typically baked in sweets such as brownies or cookies.

Any study of marijuana's effect on humans must consider the potency of the marijuana preparation. An interesting insight into the relationship between dosage level and marijuana's pharmacological effect was presented in the first report of the National Commission on Marijuana and Drug Abuse:

> *At low, usual "social" doses. the user may experience an increased sense of well-being; initial restlessness and hilarity followed by a dreamy, carefree state of relaxation; alteration of sensory perceptions including expansion of space and time; a more vivid sense of touch, sight, smell, taste, and sound; a feeling of hunger, especially a craving for sweets; and subtle changes in thought formation and expression. To an unknowing observer, an individual in this state of consciousness would not appear noticeably different from his normal state.*
>
> *At higher, moderate doses these same reactions are intensified but the changes in the individual would still be scarcely noticeable to an observer. At very high doses, psychotomimetic phenomena may be experienced. These include distortion of body image, loss of personal identity, sensory and mental illusions, fantasies, and hallucinations.*[1]

In addition to its widespread illegal use, accumulating evidence suggests that marijuana has potential medical uses. Two promising areas of research are marijuana's reduction of excessive eye pressure in glaucoma and the lessening of nausea caused by powerful anticancer drugs. Marijuana may also be useful as a muscle relaxant.

No current evidence suggests that experimental or intermittent use causes physical or psychological harm. Marijuana does not cause physical dependence. However, the risk of harm lies instead in heavy, long-term use, particularly of the more potent preparations. Heavy users can develop a strong psychological dependence on the drug. Some effects of marijuana use include increased heart rate, dry mouth, reddened eyes, impaired motor skills and concentration, and, frequently, hunger and an increased desire for sweets.

FIGURE 6–5
Several rolled marijuana cigarettes lie on a pile of crushed dried marijuana leaves next to a tobacco cigarette. *Courtesy Drug Enforcement Administration, Washington, D.C.*

Inside the Science

Marijuana and Hashish

Marijuana is a weed that grows wild under most climatic conditions. The plant grows to a height of 5 to 15 feet and is characterized by an odd number of leaflets on each leaf. Normally, each leaf contains five to nine leaflets, all with serrated or saw-tooth edges (see Figure 1).

The potency of marijuana depends on its form. Marijuana in the form of loose vegetation has an average THC content of about 3–4.5 percent. The more potent sinsemilla form averages about 6–12 percent in THC content.

FIGURE 1 **The marijuana leaf.** *Courtesy Drug Enforcement Administration, Washington, D.C.*

Hashish preparations average about 2–8 percent THC. On the illicit-drug market, hashish usually appears in the form of compressed vegetation containing a high percentage of resin (see Figure 2). A particularly potent form of hashish is known as *liquid hashish* or *hashish oil.* Hashish in this form is normally a viscous substance, dark green with a tarry consistency.

FIGURE 2 **Blocks of hashish leaves and flowering tops of the marijuana plant.** *James King-Holmes/Science Source*

Liquid hashish is produced by efficiently extracting the THC-rich resin from the marijuana plant with an appropriate solvent, such as alcohol. Liquid hashish typically varies between 8 and 22 percent in THC content. Because of its extraordinary potency, one drop of the material can produce a "high."

Case Files

A Brief History of Marijuana

Marijuana and its related products have been in use legally and illegally for almost three thousand years. The first reference to medical use of marijuana is in a pharmacy book written about 2737 B.C. by the Chinese emperor Shen Nung, who recommended it for "female weakness, gout, rheumatism, malaria, beriberi, constipation, and absent-mindedness." In China, at that time, and even today, the marijuana or hemp plant was also a major source of fiber for rope production. Marijuana's mood-altering powers probably did not receive wide attention until about 1000 B.C., when it became an integral part of Hindu culture in India. After A.D. 500, marijuana began creeping westward, and references to it began to appear in Persian and Arabian literature.

The plant was probably brought to Europe by Napoleon's soldiers when they returned from Egypt in the early 19th century. In Europe, the drug excited the interest of many physicians who foresaw its application for treating a wide range of ailments. At this time, it also found some use as a painkiller and mild sedative. In later years, these applications were either forgotten or ignored.

Marijuana was first introduced into the United States around 1920. The weed was smuggled by Mexican laborers across the border into Texas. American soldiers also brought the plant in from the ports of Havana, Tampico, and Veracruz. Although its use was confined to a small segment of the population, its popularity quickly spread from the border and Gulf states into most major U.S. cities. By 1937, the federal government and 46 states had laws prohibiting the use or possession of marijuana. Under most of these laws, marijuana was subject to the same rigorous penalties applicable to morphine, heroin, and cocaine and was often erroneously designated a narcotic.

Abuse of the hallucinogen phencyclidine, commonly called PCP, has recently grown to alarming proportions. Because this drug can be synthesized by simple chemical processes, it is manufactured surreptitiously for the illicit market in socalled clandestine laboratories (see Figure 6–6). These laboratories range from large, sophisticated operations to small labs located in a garage or bathroom. Small-time operators normally have little or no training in chemistry and employ "cookbook" methods to synthesize the drug. Some of the more knowledgeable and experienced operators have been able to achieve clandestine production levels that approach a commercial level of operation.

FIGURE 6–6 Scene from a clandestine drug laboratory.
Courtesy Drug Enforcement Administration, Washington, D.C.

Other Hallucinogens A substantial number of other substances with widely varying chemical compositions are also used recreationally because of their hallucinogenic properties. These include both naturally occurring substances such as mescaline and psilocybin and synthetically created drugs including lysergic acid diethylamide (LSD) and phencyclidine (PCP).

LSD is synthesized from lysergic acid, a substance derived from ergot, which is a type of fungus that attacks certain grasses and grains. The drug appears in a variety of forms—as a pill, added to a cube of sugar, or absorbed onto a small piece of paper—and is taken orally. Its hallucinogenic effects were first described by the Swiss chemist Albert Hofmann after he accidentally ingested some of the material in his laboratory in 1943. LSD produces marked changes in mood, leading to laughing or crying at the slightest provocation. Feelings of anxiety and tension almost always accompany LSD use. LSD is very potent; as little as 25 micrograms is enough to start vivid visual hallucinations that can last for about 12 hours. Although physical dependence does not develop with continued use, the individual user may be prone to flashbacks and psychotic reactions even after use is discontinued.

Phencyclidine is often mixed with other drugs, such as LSD or amphetamines, and is sold as a powder ("angel dust"), capsule, or tablet, or as a liquid sprayed on plant leaves. The drug is smoked, ingested, or sniffed. Following oral intake of moderate doses (1–6 milligrams), the user first experiences feelings of strength and invulnerability, along with a dreamy sense of detachment. However, the user soon becomes unresponsive, confused, and agitated. Depression, irritability, feelings of isolation, audio and visual hallucinations, and sometimes paranoia accompany PCP use. Severe depression, tendencies toward violence, and suicide accompany long-term daily use of the drug. In some cases, the PCP user experiences sudden schizophrenic behavior days after the drug has been taken.

Depressants

depressant
A substance that slows down, or depresses, the functions of the central nervous system

Depressants are drugs that slow down, or depress, the central nervous system. Several types of drugs fall under this category, including the most widely used drug in the United States—alcohol.

Alcohol (Ethyl Alcohol) Many people overlook the fact that alcohol is a drug; its major behavioral effects derive from its depressant action on the central nervous system. In the United States, the alcohol industry annually produces more than one billion gallons of spirits, wine, and beer for which 90 million consumers pay nearly $40 billion. Unquestionably, these and other statistics support the fact that alcohol is the most widely used and abused drug.

The behavioral patterns of alcohol intoxication vary and depend in part on such factors as social setting, amount consumed, and the personal expectation of the individual with regard to alcohol. When alcohol enters the body's bloodstream, it quickly travels to the brain, where it suppresses the brain's control of thought processes and muscle coordination.

Low doses of alcohol tend to inhibit the mental processes of judgment, memory, and concentration. The drinker's personality becomes expansive, and the drinker exudes confidence. When taken in moderate doses, alcohol reduces coordination substantially,

Inside the Science

Synthetic Cannabis

Synthetic cannabinoids are chemicals designed to mimic the pharmacological effects of naturally occurring cannabinoids. These drugs are generally sold in retail establishments or over the Internet as herbal procedures, potpourri, or incense. Users generally spray the chemicals onto botanical materials and inhale the drug through burning or smoking. Synthetic cannabinoids derive their pharmacological activity from their affinity toward cannabinoid (CB) receptor sites in the brain. Early in their availability, these synthetics went by the common names "K2" and "spice"; however, currently their chemical composition and names have become quite varied as clandestine laboratories have become adept and innovative in modifying their chemical structures seeking to circumvent control by drug laws. Because the chemical structure of synthetic cannabinoids do not resemble marijuana constituents, they cannot be detected by routine drug screening tests. The symptomology associated with the use of synthetic cannabinoids can result in anxiety, agitation, and nausea. A federal law, the Food and Drug Administration Safety and Innovation Act, broadly covers any material that contains a synthetic cannabinoid.

inhibits orderly thought processes and speech patterns, and slows reaction times. Under these conditions, the ability to walk or drive becomes noticeably impaired. In the next chapter, we examine in greater detail the relationship between alcohol blood levels and driving ability. Higher doses of alcohol may cause the user to become highly irritable and emotional; displays of anger and crying are not uncommon. Extremely high doses may cause an individual to lapse into unconsciousness or even a comatose state that may precede a fatal depression of circulatory and respiratory functions.

Barbiturates Barbiturates are derivatives of barbituric acid, a substance first synthesized by a German chemist, Adolf von Bayer, more than a hundred years ago. They are commonly referred to as "downers" because they relax the user, create a feeling of well-being, and produce sleep. Like alcohol, barbiturates suppress the vital functions of the central nervous system. Twenty-five barbiturate derivatives are currently used in medical practice in the United States; however, five—amobarbital, secobarbital, phenobarbital, pentobarbital, and butabarbital—tend to be used for most medical applications.

Normally, barbiturate users take these drugs orally. The average sedative dose is about 10–70 milligrams. When taken in this fashion, the drug enters the blood through the walls of the small intestine. Some barbiturates, such as phenobarbital, are classified as long-acting barbiturates. They are absorbed into the bloodstream more slowly than others and therefore produce less pronounced effects than faster-acting barbiturates. The slow action of phenobarbital accounts for its low incidence of abuse. Apparently, barbiturate abusers prefer the faster-acting varieties—secobarbital, pentobarbital, and amobarbital.

Since the early 1970s, a nonbarbiturate depressant, methaqualone (Quaalude), has appeared on the illicit-drug scene. Methaqualone is a powerful sedative and muscle relaxant that possesses many of the depressant properties of barbiturates. When taken in prescribed amounts, barbiturates are relatively safe, but in instances of extensive and prolonged use, physical dependence can develop.

Antipsychotics and Antianxiety Drugs Although antipsychotics and antianxiety drugs can be considered depressants, they differ from barbiturates in the extent of their actions on the central nervous system. Generally, these drugs produce a relaxing tranquility without impairing high-thinking faculties or inducing sleep. Antipsychotics such as reserpine and chlorpromazine have been used to reduce the anxieties and tensions of mental patients. Antianxiety drugs are commonly prescribed to deal with the everyday tensions of many healthy people. These drugs include meprobamate (Miltown), chlordiazepoxide (Librium), and diazepam (Valium).

In the past 35 years, the use of these drugs—particularly antianxiety drugs—has grown dramatically. Medical evidence shows that these drugs produce psychological and physical dependence with repeated and high levels of usage. For this reason, the widespread prescribing of antianxiety drugs to overcome the pressures and tensions of life has worried many who fear the creation of a legalized drug culture.

"Huffing" Since the early 1960s, "huffing," the practice of sniffing materials containing volatile solvents (airplane glue or model cement, for example), has grown in popularity. Another dimension has recently been added to the problem with the increasing number of incidents involving the sniffing of aerosol gas propellants, such as freon. All materials abused by huffing contain volatile or gaseous substances that are primarily central nervous system depressants. Although toluene (a solvent used in airplane glue) seems to be the most popular solvent to sniff, others can produce comparable physiological effects. These chemicals include naphtha, methyl ethyl ketone (antifreeze), gasoline, and trichloroethylene (dry-cleaning solvent).

The usual immediate effects of huffing are a feeling of exhilaration and euphoria combined with slurred speech, impaired judgment, and double vision. Finally, the user may experience drowsiness and stupor, with these depressant effects slowly wearing off as the user returns to a normal state. Most experts believe that users become physiologically dependent on the effects achieved by huffing. However, little evidence suggests that solvent inhalation is addictive. But huffers expose themselves to the dangers of liver, heart, and brain damage from the chemicals they have inhaled. Even worse, sniffing of some solvents, particularly halogenated hydrocarbons such as freon and related gases, is accompanied by a significant risk of death.

Stimulants

The term **stimulant** refers to a range of drugs that stimulate, or speed up, the central nervous system.

stimulant
A substance that speeds up, or stimulates, the central nervous system

Amphetamines Amphetamines are a group of synthetic stimulants that share a similar chemical structure and are commonly referred to in the terminology of the drug culture as "uppers" or "speed." They are typically taken either orally or via intravenous injection and provide a feeling of well-being and increased alertness that is followed by a decrease in fatigue and a loss of appetite. However, these apparent benefits of the drug are accompanied by restlessness and instability or apprehension, and once the stimulant effect wears off, depression may set in.

In the United States, the most serious form of amphetamine abuse stems from intravenous injection of amphetamine or its chemical derivative, methamphetamine (see Figure 6–7). The desire for a more intense amphetamine experience is the primary motive for this route of administration. The initial sensation of a "flash" or "rush," followed by an intense feeling of pleasure, constitutes the principal appeal of the intravenous route for the user. During a "speed binge," the individual may inject 500–1,000 milligrams of amphetamines every two to three hours. Users have reported experiencing a euphoria that produces hyperactivity, with a feeling of clarity of vision as well as hallucinations. As the effect of the amphetamines wears off, the individual lapses into a period of exhaustion and may sleep continuously for one or two days. Following this, the user often experiences a prolonged period of severe depression, lasting from days to weeks.

A smokable form of methamphetamine known as "ice" is reportedly in heavy demand in some areas of the United States. Ice is prepared by slowly evaporating a methamphetamine solution to produce large, crystal-clear "rocks." Like crack cocaine (discussed next), ice is smoked and produces effects similar to those of

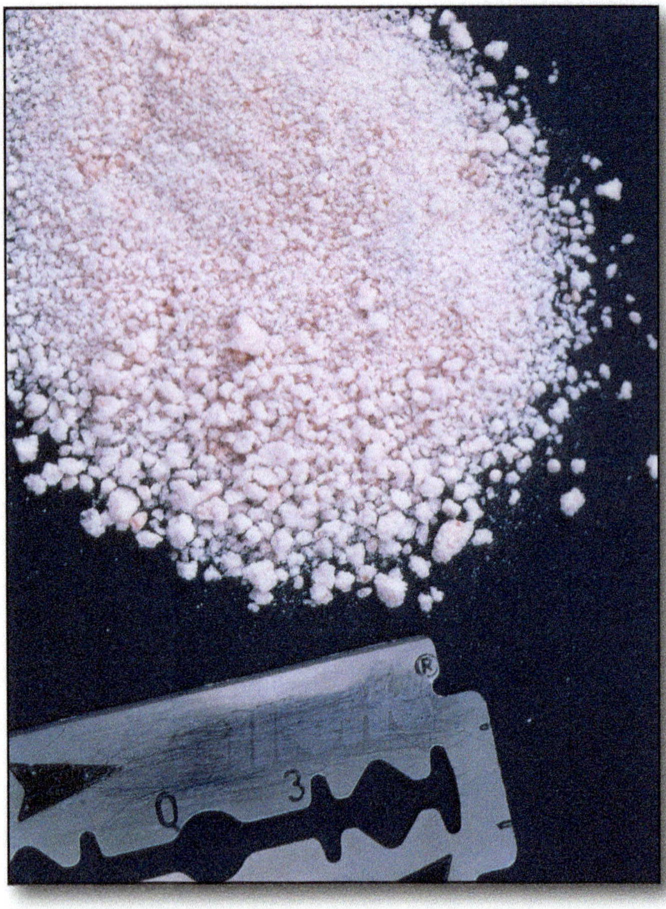

FIGURE 6–7
Granular amphetamine beside a razor blade.
Cordelia Molloy/Science Source

crack cocaine, but the effects last longer. Once the effects of ice wear off, users often become depressed and may sleep for days. Chronic users exhibit violent destructive behavior and acute psychosis similar to paranoid schizophrenia. Repeated use of amphetamines leads to a strong psychological dependence, which encourages their continued administration.

Cocaine Between 1884 and 1887, pioneering psychologist Sigmund Freud created something of a sensation in European medical circles by describing his experiments with a new drug. He reported a substance of seemingly limitless potential as a source of "exhilaration and lasting euphoria" that permitted "intensive mental or physical work [to be] performed without fatigue." He wrote, "It is as though the need for food and sleep was completely banished."

The object of Freud's enthusiasm was cocaine, a stimulant extracted from the leaves of *Erythroxylon coca*, a plant grown in the Andes mountains of South America as well as in tropical Asia (see Figure 6–8). Most commonly, cocaine is sniffed or "snorted" and absorbed into the body through the mucous membranes of the nose, but it is sometimes injected. Cocaine is a powerful stimulant to the central nervous system, and its effects resemble those caused by amphetamines—namely, increased alertness and vigor accompanied by suppression of hunger, fatigue, and boredom. Cocaine produces a feeling of euphoria by stimulating a pleasure center in the base of the brain, in an area connected to nerves that are responsible for emotions. It stimulates this pleasure

FIGURE 6–8
Coca leaves and illicit forms of cocaine. *Courtesy Drug Enforcement Administration, Washington, D.C.*

center to a far greater degree than it would ever normally be stimulated. Some regular users of cocaine report accompanying feelings of restlessness, irritability, and anxiety. Cocaine used chronically or at high doses can have toxic effects. Cocaine-related deaths result from cardiac arrest or seizures followed by respiratory arrest.

A particularly potent form of cocaine known as "crack" can be produced by mixing cocaine with baking soda and water and then heating the resulting solution. This material is then dried and broken into tiny chunks that dealers sell as crack "rocks" that are sufficiently volatile to be smoked. The faster the cocaine level rises in the brain, the greater the euphoria, and the fastest way to attain a rise in the brain's cocaine level is to smoke crack. Inhaling the cocaine vapor delivers the drug to the brain in less than 15 seconds—about as fast as injecting it and much faster than snorting it. The dark side of crack, however, is that the euphoria fades quickly as the cocaine levels rapidly drop, leaving the user feeling depressed, anxious, and pleasureless. The desire to return to the euphoric feeling is so intense that crack users quickly develop a habit for the drug that is almost impossible to overcome. Only a small percentage of crack abusers are ever cured of this drug habit. When a person uses large amounts of crack cocaine numerous times, the person usually develops a sense of paranoia. Paronoid delusions cause the person to lose their sense of reality, leaving them trapped in a world full of voices, whispers, and suspicions. Sufferers come to believe that they are being followed and that their drug use is being watched.

In the United States, cocaine abuse is on the rise. Many people are using cocaine apparently to improve their ability to work and to keep going when tired. Although there is no evidence of physical dependency accompanying cocaine's repeated use, abstention from cocaine after prolonged use brings on severe bouts of mental depression that produce a very strong compulsion to resume using the drug. In fact, laboratory experiments with animals have demonstrated that, of all the commonly abused drugs, cocaine produces the strongest psychological compulsions for continued use.

The United States spends millions of dollars annually in attempting to control cultivation of the coca leaf in various South American countries and to prevent the trafficking of cocaine into the United States. Three-quarters of the

Inside the Science

Bath Salts

It has become trendy in the drug culture to abuse a group of illicit substances known as "bath salts." These drugs are a mix of chemical derivatives derived from cathinone, a naturally occurring substance found in the khat plant. Cathinone is a stimulant having about half the potency of amphetamine. The allure of abusing cathinone and its chemical derivatives is to simulate the high associated with methamphetamine and cocaine abuse.

Synthetic derivatives of cathinone are commonly sold in powder, crystal, and liquid forms, but they are available also as tablets and capsules. They are sold in packages to be snorted, ingested, smoked, or injected. Like the side effects associated with methamphetamine or cocaine abuse, bath salts can induce agitation, violent behavior, and paranoia on the part of the user.

A federal law, the Food and Drug Administration Safety and Innovation Act, outlaws two synthetic cathinones, mephidrone and 3,4-methylenedioxypyrovalerone (MDPV), that are common constituents of bath salts.

cocaine smuggled into the United States was refined in clandestine laboratories in Colombia. The profits are astronomical. Peruvian farmers may be paid $200 for enough coca leaves to make one pound of cocaine. The refined cocaine is worth $1,000 when it leaves Colombia and sells at retail in the United States for up to $20,000.

Club Drugs

The term *club drugs* refers to synthetic drugs that are often used at nightclubs, bars, and raves (all-night dance parties). Substances that are used as club drugs include, but are not limited to, MDMA (Ecstasy), GHB (gamma hydroxybutyrate), Rohypnol "Roofies" (flunitrazepam), ketamine, and methamphetamine. These drugs have become popular at the dance scene as a way to stimulate the rave experience. A high incidence of use has been found among teens and young adults.

GHB and Rohypnol are central nervous system depressants that are often connected with drug-facilitated sexual assault, rape, and robbery. Effects accompanying the use of GHB include dizziness, sedation, headache, and nausea. Recreational users have reported euphoria, relaxation, disinhibition, and increased libido (sex drive). Rohypnol causes muscle relaxation, loss of consciousness, and an inability to remember what happened during the hours after ingesting the drug. This is particularly a concern in a sexual assault because victims are physically unable to resist the attack. Unsuspecting victims become drowsy or dizzy. experiences memory loss, blackouts, and disinhibition. Drugs such as Rohypnol and GHB are odorless, colorless, and tasteless, and thus remain undetected when slipped into a drink.

Methylenedioxymethamphetamine, also known as MDMA or Ecstasy, is a synthetic, mind-altering drug that exhibits many hallucinogenic and amphetamine-like effects. Ecstasy was originally patented as an appetite suppressant and was later discovered to induce feelings of happiness and relaxation. Recreational drug users find that Ecstasy enhances self-awareness and decreases inhibitions. However, seizures, muscle breakdown, stroke, kidney failure, and cardiovascular system failure often accompany chronic abuse of Ecstasy. In addition, chronic use of Ecstasy leads to serious damage to the areas of the brain responsible for thought and memory. Ecstasy increases heart rate and blood pressure; produces muscle tension, teeth grinding, and nausea; and causes psychological difficulties such as confusion, severe anxiety, and paranoia. The drug can cause significant increases in body temperature from the combination of the drug's stimulant effect with the often hot, crowded atmosphere of a rave club.

Ketamine is primarily used in veterinary medicine as an animal anesthetic. When used by humans, the drug can cause euphoria and feelings of unreality accompanied by visual hallucinations. Ketamine can also cause impaired motor function, high blood pressure, amnesia, and mild respiratory depression.

Anabolic Steroids

anabolic steroids
Synthetic compounds, chemically related to the male sex hormone testosterone, that are used to promote muscle growth

Anabolic steroids are synthetic compounds that are chemically related to the male sex hormone testosterone. Testosterone has two different effects on the body. It promotes the development of secondary male characteristics (androgenic effects), and it accelerates muscle growth (anabolic effects). Efforts to promote muscle growth and to minimize the hormone's androgenic effects have led to the synthesis of numerous anabolic steroids. However, a steroid free of the accompanying harmful side effects of an androgen drug has not yet been developed.

Incidence of steroid abuse first received widespread public attention when both amateur and professional athletes were discovered using these substances to enhance their performance. Interestingly, current research on male athletes given anabolic steroids has generally found little or, at best, marginal evidence of enhanced strength or performance. Although the full extent of anabolic steroid abuse by the general public is not fully known, the U.S. government is sufficiently concerned to regulate the availability of these drugs to the general population and to severely punish individuals for illegal possession and distribution of anabolic steroids. In 1991, anabolic steroids were classified as controlled dangerous substances, and the Drug Enforcement Administration was given enforcement power to prevent their illegal use and distribution (see Figure 6–9).

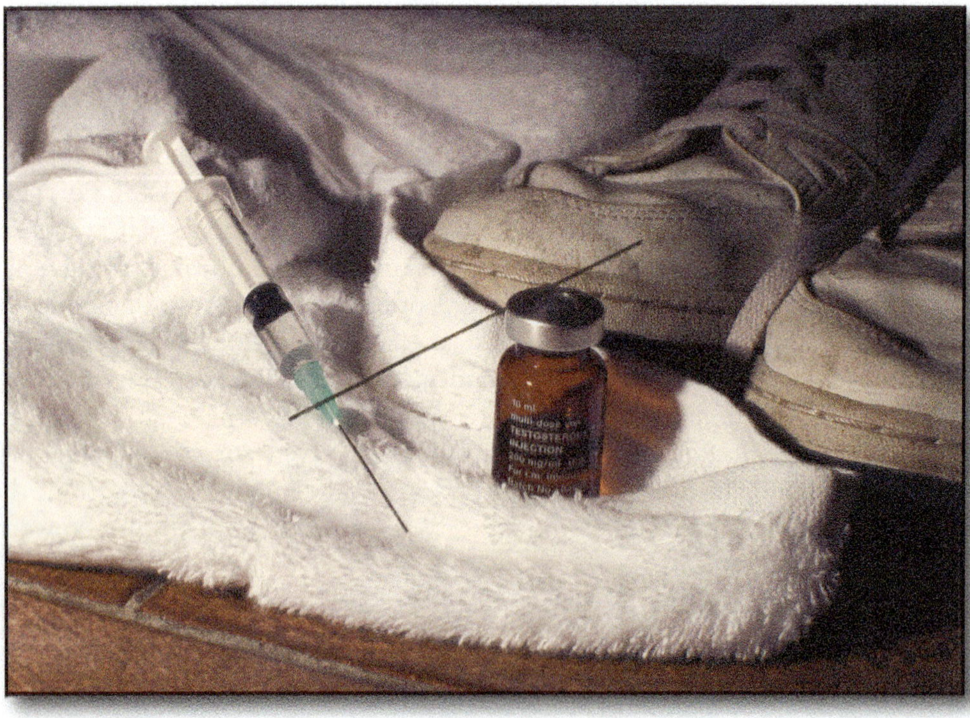

FIGURE 6–9
Anabolic steroids: a vial of testosterone and a syringe. Testosterone, the male sex hormone, is sometimes abused by athletes for its protein-building (anabolic) effect. *SPL/Science Source*

Anabolic steroids are usually taken by individuals who are unfamiliar with the harmful medical side effects. Liver cancer and other liver malfunctions have been linked to steroid use. These drugs also cause masculinizing effects in females, infertility, and diminished sex drive in males. For teenagers, anabolic steroids result in premature halting of bone growth. Anabolic steroids can also cause unpredictable effects on mood and personality, leading to unprovoked acts of anger and destructive behavior. Depression is also a frequent side effect of anabolic steroid abuse.

Quick Review

- Narcotic drugs are analgesics, meaning they relieve pain by depressing the central nervous system.
- The most common source for narcotic drugs is opium. Morphine is extracted from opium and used to synthesize heroin.
- Opiates are not derived from opium or morphine, but they have the same physiological effects on the body. Examples of opiates include methadone and OxyContin (oxycodone).
- Hallucinogens cause marked changes in normal thought processes, perceptions, and moods. Marijuana is the most well-known drug in this class. Other hallucinogens include LSD, mescaline, PCP, psilocybin, and MDMA (Ecstasy).
- Depressants decrease the activity of the central nervous system, calm irritability and excitability, and produce sleep. Depressants include alcohol (ethanol), barbiturates, tranquilizers, and various substances that can be sniffed, such as airplane glue or model cement.
- Stimulants increase the activity of the central nervous system and are taken to increase alertness and activity. Stimulants include amphetamines, sometimes known as "uppers" or "speed," and cocaine, which in its freebase form is known as *crack*.
- Club drugs are synthetic drugs that are used at nightclubs, bars, and raves (all-night dance parties). Some club drugs act as stimulants; others have depressant effects.
- Anabolic steroids are synthetic compounds that are chemically related to the male sex hormone testosterone. Anabolic steroids are often abused by individuals who are interested in accelerating muscle growth.

Drug-Control Laws

The provisions of drug laws are of particular interest to the criminalist, because they may impose specific analytical requirements on drug analysis. For example, the severity of a penalty associated with the manufacture, distribution, possession, and use of a drug may depend on the weight of the drug or its concentration in a mixture. In such cases, the chemist's report must contain all information that is needed to properly charge a suspect under the provisions of the existing law.

The provisions of any drug-control law are an outgrowth of national and local law enforcement requirements and customs, as well as the result of moral and political philosophies. These factors have produced a wide spectrum of national and local drug-control laws. Although their detailed discussion is beyond the intended scope of this book, a brief description of the U.S. federal law known as the Controlled Substances Act will illustrate a legal drug classification system that has been created to prevent and control drug abuse. Many states have modeled their own drug-control laws after this act, an important step in establishing uniform drug-control laws throughout the United States.

Controlled Substances Act

The federal law establishes five schedules of classification (as outlined next) for controlled dangerous substances on the basis of a drug's potential for abuse, potential for physical and psychological dependence, and medical value. This classification system is extremely flexible in that the U.S. attorney general has the authority to add, delete, or reschedule a drug as more information becomes available. In addition, controlled dangerous substances listed in schedules I and II are subject to manufacturing quotas set by the attorney general. For example, eight billion doses of amphetamines were manufactured in the United States in 1971. In 1972, production quotas reduced amphetamine production approximately 80 percent below 1971 levels.

Control Mechanisms and Penalties The criminal penalties for unauthorized manufacture, sale, or possession of controlled dangerous substances are related to the schedules as well. The most severe penalties are associated with drugs listed in schedules I and II. For example, for drugs included in schedules I and II, a first offense is punishable by up to 20 years in prison and/or a fine of up to $1 million for an individual or up to $5 million for other than individuals. Table 6–2 summarizes the control mechanisms and penalties for each schedule of the Controlled Substances Act.

Schedule I. Schedule I drugs have a high potential for abuse, have no currently accepted medical use in the United States, and/or lack accepted safety for use in treatment under medical supervision. Drugs controlled under this schedule include heroin, marijuana, methaqualone, and LSD.

Schedule II. Schedule II drugs have a high potential for abuse, a currently accepted medical use or a medical use with severe restrictions, and a potential for severe psychological or physical dependence. Schedule II drugs include opium and its derivatives not listed in schedule I, cocaine, methadone, phencyclidine (PCP), most amphetamine preparations, and most barbiturate preparations containing amobarbital, secobarbital, and pentobarbital. Dronabinol, the synthetic equivalent of the active ingredient in marijuana, has been placed in schedule II in recognition of its growing medical uses in treating glaucoma and chemotherapy patients.

Schedule III. Schedule III drugs have less potential for abuse than those in schedules I and II, a currently accepted medical use in the United States, and a potential for low or moderate physical dependence or high psychological dependence. Schedule III controls, among other substances, all barbiturate preparations (except phenobarbital) not covered under schedule II and certain codeine preparations. Anabolic steroids were added to this schedule in 1991.

Schedule IV. Schedule IV drugs have a low potential for abuse relative to schedule III drugs and have a current medical use in the United States; their abuse may lead to limited dependence relative to schedule III drugs. Drugs controlled in this schedule include propoxyphene (Darvon), phenobarbital, and tranquilizers such as meprobamate (Miltown), diazepam (Valium), and chlordiazepoxide (Librium).

Schedule V. Schedule V drugs must show low abuse potential, have medical use in the United States, and have less potential for producing dependence than schedule IV drugs. Schedule V controls certain opiate drug mixtures that contain nonnarcotic medicinal ingredients.

Controlled dangerous substances listed in schedules I and II are subject to manufacturing quotas set by the attorney general. For example, 8 billion doses of amphetamines were manufactured in the United States in 1971. In 1972, production quotas were established reducing amphetamine production approximately 80 percent below 1971 levels.

Other Provisions of the Act

The criminal penalties for unauthorized manufacture, sale, or possession of controlled dangerous substances are related to the schedules as well. The most severe penalties are associated with drugs listed in schedules I and II. For example, for drugs included in schedules I and II, a first offense is punishable by up to 20 years in prison and/or a fine of up to $1 million for an individual or up to $5 million for other than individuals. Table 6–2 summarizes the control mechanisms and penalties for each schedule of the Controlled Substances Act.

The Controlled Substances Act stipulates that an offense involving a controlled substance *analog*, a chemical substance substantially similar in chemical structure to a controlled substance, triggers penalties as if it were a controlled substance listed in schedule I. This section is designed to combat the proliferation of so-called *designer drugs*—substances that are chemically related to some controlled drugs and are pharmacologically very potent. These substances are manufactured by skilled individuals in clandestine laboratories, with the knowledge that their products will not be covered by the schedules of the Controlled Substances Act. For instance, fentanyl is a powerful narcotic that is commercially marketed for medical use and is also listed as a controlled dangerous substance. This drug is about one hundred times as potent as morphine. A number of substances chemically related to fentanyl have been synthesized by underground chemists and sold on the street. The first such substance encountered

Table 6–2

Control Mechanisms of the Controlled Substances Act

Schedule	Registration	Record Keeping	Manufacturing Quotas	Distribution Restrictions	Dispensing Limits	Import–Export		Security	Manufacturer/Distributor Enforcement Administration	Criminal Penalties for Individual Trafficking (First Offense)
						Narcotic	Nonnarcotic			
I	Required	Separate	Yes	Order forms	Research use only	Permit	Permit	Vault/safe	Yes	0–20 years/$1 million
II	Required	Separate	Yes	Order forms	Rx: written; no Refills	Permit	Permit	Vault/safe	Yes	0–20 years/$1 million
III	Required	Readily retrievable	No, but some drugs limited by schedule II quotas	Records required	Rx: written or oral; with medical authorization refills up to 5 times in 6 months	Permit	Declaration	Secure storage area	Yes, narcotic No, nonnarcotic	0–5 years/$250,000
IV	Required	Readily retrievable	No, but some drugs limited by schedule II quotas	Records required	Rx: written or oral; with medical authorization refills up to 5 times in 6 months	Permit	Declaration	Secure storage area	Manufacturer only, narcotic No, nonnarcotic	0–3 years/$250,000
V	Required	Readily retrievable	No, but some drugs limited by schedule II quotas	Records required	Over-the-counter (Rx drugs limited to MD's order) refills up to 5 times	Permit to import; declaration to export	Declaration	Secure storage area	Manufacturer only, narcotic No, nonnarcotic	0–1 year/$100,000

Source: Courtesy Drug Enforcement Administration, Washington, D.C.

was sold under the street name China White. These drugs have been responsible for more than a hundred overdose deaths in California and nearly 20 deaths in western Pennsylvania. As designer drugs, such as China White, are identified and linked to drug abuse, they are placed in appropriate schedules.

The Controlled Substances Act also reflects an effort to decrease the prevalence of clandestine drug laboratories designed to manufacture controlled substances. The act regulates the manufacture and distribution of *precursors*, the chemical compounds used by clandestine drug laboratories to synthesize abused drugs. Targeted precursor chemicals are listed in the definition section of the Controlled Substances Act. Severe penalties are provided for a person who possesses a listed precursor chemical with the intent to manufacture a controlled substance or who possesses or distributes a listed chemical knowing, or having reasonable cause to believe, that the listed chemical will be used to manufacture a controlled substance. In addition, precursors to PCP, amphetamines, and methamphetamines are enumerated specifically in schedule II, making them subject to regulation in the same manner as other schedule II substances.

Collection and Preservation of Drug Evidence

Preparation of drug evidence for submission to the crime laboratory is normally a relatively simple task, accomplished with minimal precautions in the field. The field investigator is responsible for ensuring that the evidence is properly packaged and labeled for delivery to the laboratory.

Considering the countless forms and varieties of drug evidence seized, it is not practical to prescribe any single packaging procedure for fulfilling these requirements. Generally, common sense is the best guide in such situations, keeping in mind that the package must prevent the loss and/or cross-contamination of the contents. Often, the original container in which the drug was seized will suffice to meet these requirements. Specimens suspected of containing volatile solvents, such as those involved in glue-sniffing cases, must be packaged in an airtight container to prevent evaporation of the solvent.

All packages must be marked with sufficient information to ensure identification by the officer in future legal proceedings and to establish the chain of custody. To aid the drug analyst, the investigator should supply any background information that may relate to a drug's identity. Analysis time can be markedly reduced when this information is at the disposal of the chemist. For the same reason, the results of drug-screening tests used in the field must also be transmitted to the laboratory. Although these tests may indicate the presence of a drug and may help the officer establish probable cause to search and arrest a suspect, they do *not* offer conclusive evidence of a drug's identity.

Quick Review

- Federal law establishes five schedules of classification for controlled dangerous substances on the basis of a drug's potential for abuse, potential for physical and psychological dependence, and medical value.

Forensic Drug Analysis

One only has to look into the evidence vaults of crime laboratories to appreciate the assortment of drug specimens that confront the criminalist. The presence of a huge array of powders, tablets, capsules, vegetable matter, liquids, pipes, cigarettes, cookers, and syringes is testimony to the vitality and sophistication of the illicit-drug market. If outward appearance is not evidence enough of the difficult analytical chore facing the forensic chemist, consider the complexity of the drug preparations themselves. Usually these contain active drug ingredients of unknown origin and identity, as well as additives—for example, sugar, starch, and quinine—that dilute their potency and stretch their value on the illicit-drug market. Do not forget that illicit-drug dealers are not hampered by government regulations that ensure the quality and consistency of a product.

When a forensic chemist picks up a drug specimen for analysis, the forensic chemist should be prepared for all contingencies. The analysis must leave no room for error, because its results will have a direct bearing on the process of determining the guilt or innocence of a defendant. There is no middle ground in drug identification—either the specimen is a specific drug or it is not—and once a positive conclusion is drawn, the chemist must be prepared to support and defend the validity of the results in a court of law.

Screening and Confirmation

The challenge or difficulty of forensic drug identification comes in selecting analytical procedures that will ensure specific identification of a drug. Presented with a substance of unknown origin and composition, the forensic chemist must develop a plan of action that will ultimately yield the drug's identity. This plan, or scheme of analysis, is divided into two phases.

screening test
A preliminary test used to reduce the number of possible identities of an unknown substance

First, faced with the prospect that the unknown substance may be any one of a thousand or more commonly encountered drugs, the analyst must use **screening tests**, also referred to as presumptive tests, to reduce these possibilities to a small and manageable number. The presumptive procedures to identify unknown substances is often accomplished by subjecting the unknown substances to a series of color tests that produce characteristic colors for the more commonly encountered illicit drugs. Even if these tests produce negative results, their value lies in having excluded certain drugs from further consideration.

Once the number of possibilities has been reduced substantially, the second phase of the analysis is devoted to pinpointing and confirming the drug's identity. In an era in which crime laboratories receive voluminous quantities of drug evidence, it is impractical to subject a drug to all the chemical and instrumental tests available. Indeed, it is more realistic to view these techniques as a large analytical arsenal. The chemist, aided by training and experience, must choose tests that will most conveniently identify a particular drug.

confirmation
A single test that specifically identifies a substance

Forensic chemists often use a specific test, known as a confirmatory procedure, to identify an unknown drug substance to the exclusion of all other known chemical substances. A confirmatory test is a single test that identifies a substance and is known as a **confirmation**. The analytical scheme sometimes consists of a series of nonspecific or presumptive tests. Each test in itself is insufficient to prove the drug's identity; however, the proper analytical scheme encompasses a combination of test results that characterize one and only one chemical substance—the drug under investigation. Furthermore, experimental evidence must confirm that the probability of any other substance responding in an identical manner to the scheme selected is so small as to be beyond any reasonable scientific certainty.

Another consideration in selecting an analytical technique is the need for either a *qualitative* or a *quantitative* determination. The former relates just to the identity of the material, whereas the latter refers to the percentage combination of the components of a mixture. Hence, a qualitative identification of a powder may reveal the presence of heroin and quinine, whereas a quantitative analysis may conclude the presence of 10 percent heroin and 90 percent quinine.

Obviously, a qualitative identification must precede any attempt at quantitation, for little value is served by attempting to quantitate a material without first determining its identity. Essentially, a qualitative analysis of a material requires the determination of numerous properties using a variety of analytical techniques. On the other hand, a quantitative measurement is usually accomplished by precise measurement of a single property of the material.

Forensic scientists normally rely on several presumptive and confirmatory laboratory procedures as they relate to toxicological analysis for a routine drug-identification scheme. These tests can include color tests, microcrystalline tests, chromatography, spectrophotometry, and mass spectrometry.

Color Tests

Many drugs yield characteristic colors when brought into contact with specific chemical reagents. Not only do these tests provide a useful indicator of a drug's presence, but they are also used by investigators in the field to examine materials suspected of containing a drug (see Figure 6–10).[2] However, presumptive tests are useful for screening purposes only and are never taken as conclusive identification of unknown drugs.

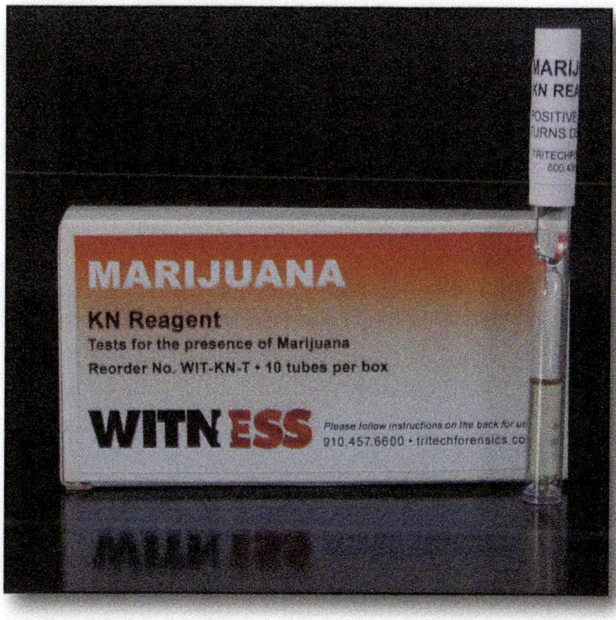

FIGURE 6–10
A field color test kit for marijuana. *Courtesy Tri-Tech Forensics, Inc., Southport, NC*

Five primary color test reagents are as follows:

1. **Marquis.** The reagent turns purple in the presence of heroin and morphine and most opium derivatives. Marquis also becomes orange-brown when mixed with amphetamines and methamphetamines.
2. **Dillie-Koppanyi.** This is a valuable screening test for barbiturates, in whose presence the reagent turns violet-blue in color.
3. **Duquenois-Levine.** This is a valuable color test for marijuana, performed by adding a series of chemical solutions, to the suspect vegetation. A positive result is shown by a purple color when chloroform is added.
4. **Van Urk.** The reagent turns blue-purple in the presence of LSD. However, owing to the extremely small quantities of LSD in illicit preparations, this test is difficult to conduct under field conditions.
5. **Scott Test.** This is a color test for cocaine. A powder containing cocaine turns a cobalt thiocyanate solution blue. Upon addition of hydrochloric acid, the blue color is transformed to a clear pink color. Upon addition of chloroform, if cocaine is present, the blue color reappears in the chloroform layer.

Quick Review

- Analysts use screening tests to determine the identity of drugs present in a sample. These tests reduce the number of possible drugs to a small and manageable number.
- A series of color tests produce characteristic colors for the more commonly encountered illicit drugs.
- After preliminary testing, forensic chemists use more specific tests to identify a drug substance to the exclusion of all other known chemical substances.

Microcrystalline Tests

microcrystalline test
A test that identifies a specific substance based on the color and shape of crystals formed when the substance is mixed with specific reagents

A technique considerably more specific than color tests is the **microcrystalline test**. A drop of a chemical reagent is added to a small quantity of the drug on a microscopic slide. After a short time, a chemical reaction ensues, producing a crystalline precipitate. The size and shape of the crystals, under microscope examination, are highly characteristic of the drug. Crystal tests for cocaine and methamphetamine are illustrated in Figure 6–11.

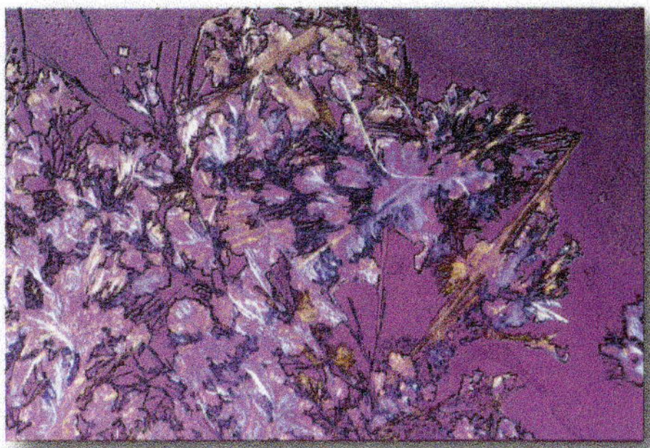
(a)

(b)

FIGURE 6–11 (a) A photograph of a cocaine crystal formed in platinum chloride (400x). (b) A photomicrograph of a methamphetamine crystal formed in gold chloride (400x). *Astrid & Hanns-Frieder Michler/Science Source; Pasieka/Science Source*

Over the years, analysts have developed hundreds of crystal tests to characterize the most commonly abused drugs. These tests are rapid and often do not require the isolation of a drug from its diluents; however, because diluents can sometimes alter or modify the shape of the crystal, the examiner must develop experience in interpreting the results of the test.

Most color and crystal tests are largely empirical—that is, scientists do not fully understand why they produce the results that they do. From the forensic chemist's point of view, this is not important. When the tests are properly chosen and used in proper combination, their results constitute an analytical scheme that is characteristic for one and only one drug.

Chromatography

Chromatography is a means of separating and tentatively identifying the components of a mixture. It is particularly useful for analyzing drug specimens, which may be diluted with practically any material in order to increase the quantity of the product available to prospective customers. The task of identifying an illicitdrug preparation would be arduous without the aid of chromatographic methods to first separate the mixture into its components.

chromatography
Any of several analytical techniques for separating organic mixtures into their components by attraction to a stationary phase while being propelled by a moving phase

Theory of Chromatography The theory of chromatography is based on the fact that chemical substances tend to partially escape into the surrounding environment when dissolved in a liquid or when absorbed on a solid surface. For example, if a beaker of water is covered with a bell jar, as shown in Figure 6–12, gas molecules (represented by green balls) escape from the water into the surrounding enclosed air. The molecules that remain are said to be in the liquid phase; the molecules that have escaped into the air are said to be in the gas phase.

As the gas molecules escape into the surrounding air, they accumulate above the water; random motion carries some of them back into the water. Eventually, a point is reached at which the number of molecules leaving the water equals the number returning. At this time, the liquid and gas phases are in equilibrium. If the temperature of the water is increased, the equilibrium state readjusts itself to a point at which more gas molecules move into the gas phase.

FIGURE 6–12
Evaporation of a liquid.

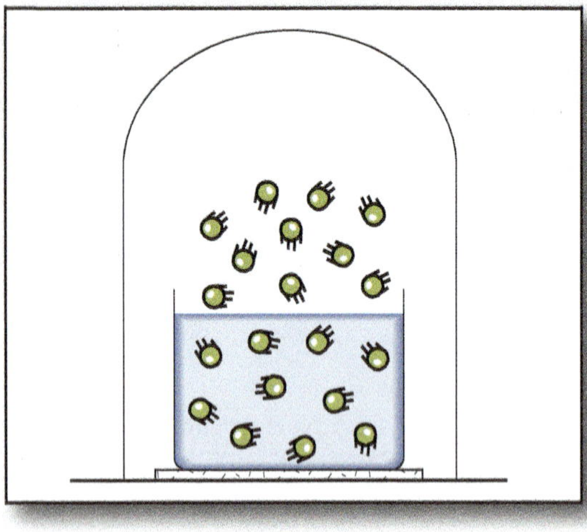

This behavior was first observed in 1803 by British chemist William Henry. His explanation of this phenomenon, known appropriately as Henry's law, may be stated as follows:

> *When a volatile chemical compound is dissolved in a liquid and is brought to equilibrium with air, there is a fixed ratio between the concentration of the volatile compound in air and its concentration in the liquid, and this ratio remains constant for a given temperature.*[3]

The distribution of a gas between the liquid and gas phases is determined by the solubility of the gas; that is, how easily the gas dissolves in the liquid. The higher its solubility, the greater the tendency of the gas molecules to remain in the liquid phase. If two different gases are simultaneously dissolved in the same liquid, each reaches a state of equilibrium with the surrounding air independently of the other. For example, as shown in Figure 6–13, gas A (green molecules) and gas B (blue molecules) are both dissolved in water. At equilibrium, gas A has a greater number of molecules dissolved in the water than does gas B. This is so because gas A is more soluble in water than gas B.

Thin-Layer Chromatography Thin-layer chromatography (TLC) uses a solid stationary phase and a moving liquid phase to separate the constituents of a mixture. Thin-layer chromatography is a powerful tool for solving many of

FIGURE 6–13
At equilibrium, there are more gas A molecules (green molecules) than gas B molecules (blue molecules) in the liquid phase.

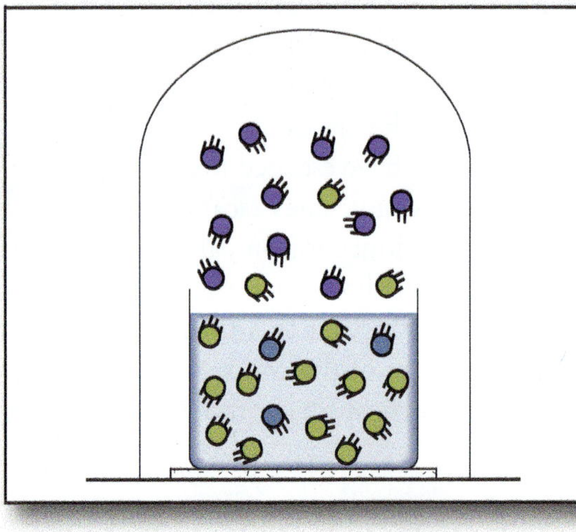

Inside the Science

The Chromatographic Process

In Figures 6–12 and 6–13, both phases—liquid and gas—were kept stationary; that is, they were not moving. During a chromatographic process, however, this is not the case. Instead, one phase is always made to move continuously in one direction over a stationary or fixed phase. For example, in Figure 6–13, chromatography will occur only when the air is forced to move continuously in one direction over the water. Because gas B (blue molecules) has a greater percentage of its molecules in the moving phase than does gas A (green molecules), the molecules of gas B will travel over the liquid at a faster pace than those of gas A. Eventually, when the moving phase has advanced a reasonable distance, the molecules of gas B will become entirely separated from those of gas A, and the chromatographic process will be complete. This process is illustrated in Figure 1.

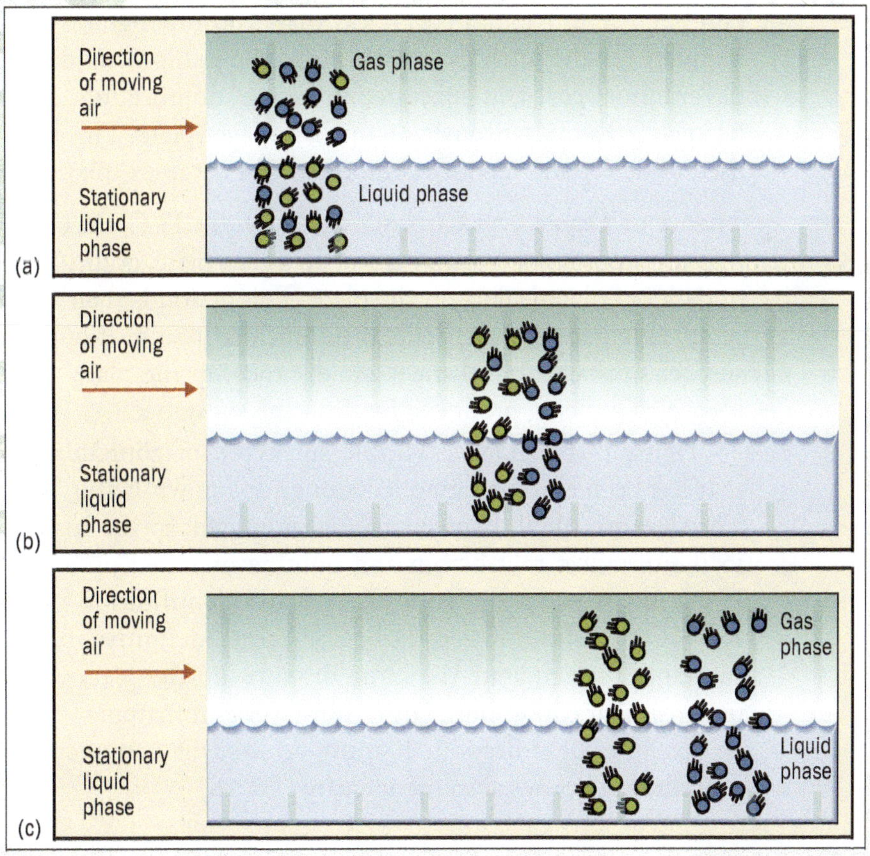

FIGURE 1 In this illustration of chromatography, the blue molecules have a greater affinity for the upper phase and hence will be pushed along at a faster rate by the moving air. Eventually, the two sets of molecules will separate from each other, completing the chromatographic process.

Inside the Science (CONTINUED)

Simply, we can think of chromatography as being analogous to a race between chemical compounds. At the starting line, all the participating substances are mixed together; however, as the race progresses, materials that prefer the moving phase slowly pull ahead of those substances that prefer to remain in the stationary phase. Finally, at the end of the race, all the participants are separated, each crossing the finish line at different times.

The different types of chromatographic systems are as varied as the number of stationary and moving-phase combinations that can be devised. However, two chromatographic processes—gas chromatography and thin-layer chromatography—are most applicable for solving many analytical problems in the crime laboratory.

the analytical problems presented to the forensic scientist. The method is both rapid and sensitive; moreover, less than 100 micrograms of suspect material are required for the analysis. In addition, the equipment necessary for TLC work has minimal cost and space requirements. Importantly, numerous samples can be analyzed simultaneously on one thin-layer plate. This technique is principally used to detect and identify components in complex mixtures.

Theory of Thin-Layer Chromatography. In TLC, the components of a suspect mixture are separated as they travel up a glass plate, eventually appearing as a series of dark or colored spots on the plate. This action is then compared to a standard sample separation of a specific drug, such as heroin. If both the standard and the suspect substance travel the same distance up the plate, they can tentatively be identified as being produced by the same substance.

Figure 6–14 shows a sample suspected of containing heroin and quinine that has been chromatographed alongside known heroin and quinine standards. The distance the unknown material migrated up the suspect plate is compared to the distances that heroin and quinine migrated up a standard sample plate. If the distances are the same, a tentative identification can be made. However, such an identification cannot be considered definitive, because numerous other substances can migrate the same distance up the plate when chromatographed under similar conditions. Thus, thin-layer chromatography alone cannot provide an absolute identification; it must be used in conjunction with other testing procedures to prove absolute identity.

TLC in Practice. A thin-layer plate is prepared by coating a glass plate or plastic backing with a thin film of a granular material, usually silica gel or aluminum oxide. This granular material serves as the solid stationary phase and is usually held in place on the plate with a binding agent such as plaster of Paris. If the sample to be analyzed is a solid, it must first be dissolved in a suitable solvent and a few microliters of the solution spotted with a capillary tube onto the granular surface near the lower edge

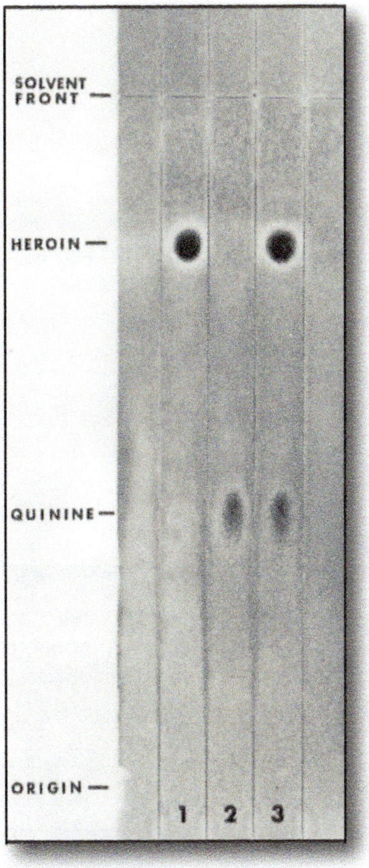

FIGURE 6–14
Chromatograms of known heroin (1) and quinine (2) standards alongside suspect sample (3).
Courtesy Richard Saferstein, Ph.D.

of the plate. A liquid sample may be applied directly to the plate in the same manner. The plate is then placed upright into a closed chamber that contains a selected liquid, with care that the liquid does not touch the sample spot.

The liquid slowly rises up the plate by capillary action. This rising liquid is the moving phase in thin-layer chromatography. As the liquid moves past the sample spot, the components of the sample become distributed between the stationary solid phase and the moving liquid phase. The components with the greatest affinity for the moving phase travel up the plate faster than those that have greater affinity for the stationary phase. When the liquid front has moved a sufficient distance (usually 10 cm), the development is complete, and the plate is removed from the chamber and dried (see Figure 6–15). An example of the chromatographic separation of ink is shown in Figure 6–16.

Because most compounds are colorless, no separation will be noticed after development unless the materials are visualized. To accomplish this, the plates are placed under ultraviolet light, revealing fluorescent materials (those that emit visible light when exposed to light of a shorter wavelength) as bright spots on a dark background. When a fluorescent dye has been incorporated into the solid phase, nonfluorescent substances appear as dark spots against a fluorescent background when exposed to the ultraviolet light.

In a second method of visualization, the plate is sprayed with a chemical reagent that reacts with the separated substances and causes them to form colored spots. Figure 6–17 shows the chromatogram of a marijuana extract that has been separated into its components by TLC and visualized by having been sprayed with a chemical reagent.

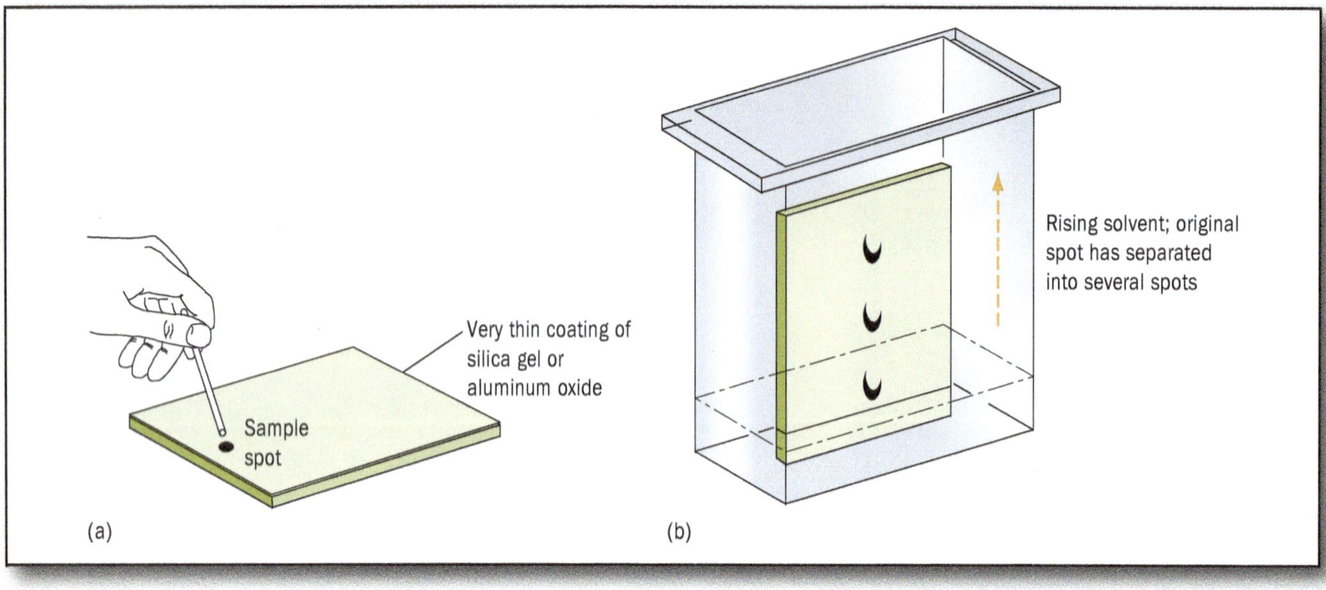

FIGURE 6–15 (a) In thin-layer chromatography, a liquid sample is spotted onto the granular surface of a gel-coated plate. (b) The plate is placed into a closed chamber that contains a liquid. As the liquid rises up the plate, the components of the sample distribute themselves between the coating and the moving liquid. The mixture is separated, with substances with a greater affinity for the moving liquid traveling up the plate at a faster speed.

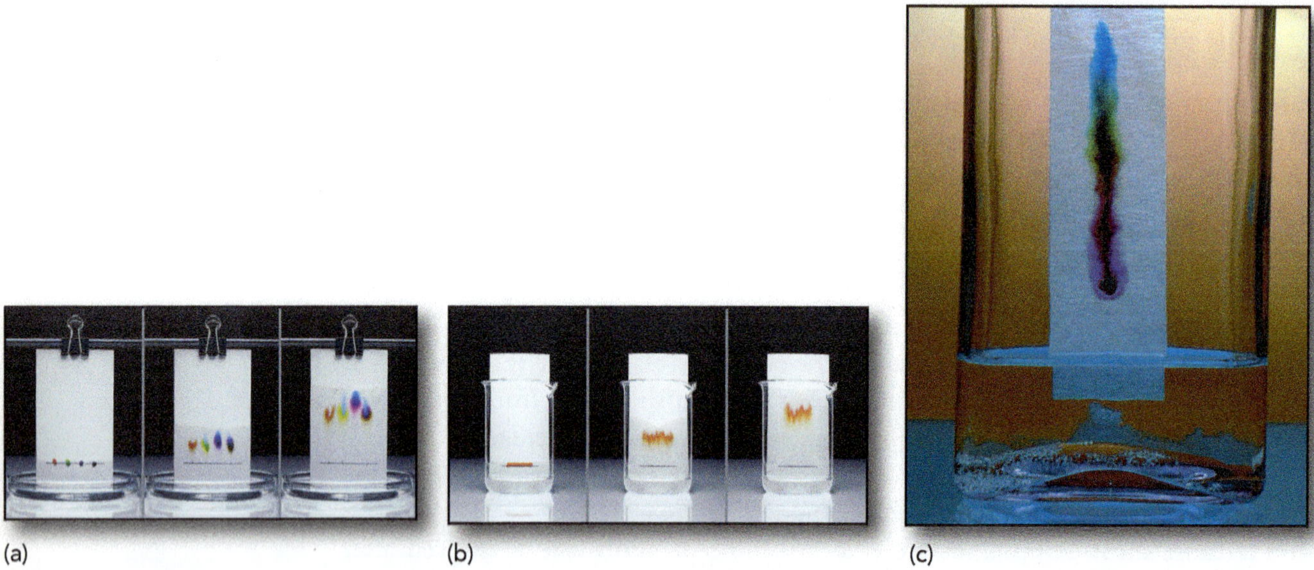

FIGURE 6–16 (a) The liquid phase begins to move up the stationary phase. (b) Liquid moves past the ink spot carrying the ink components up the stationary phase. (c) The moving liquid has separated the ink into its several components. *Courtesy Turtle Rock Scientific/Science Source Photos a and b; Richard Megna/Fundamental Photographs*

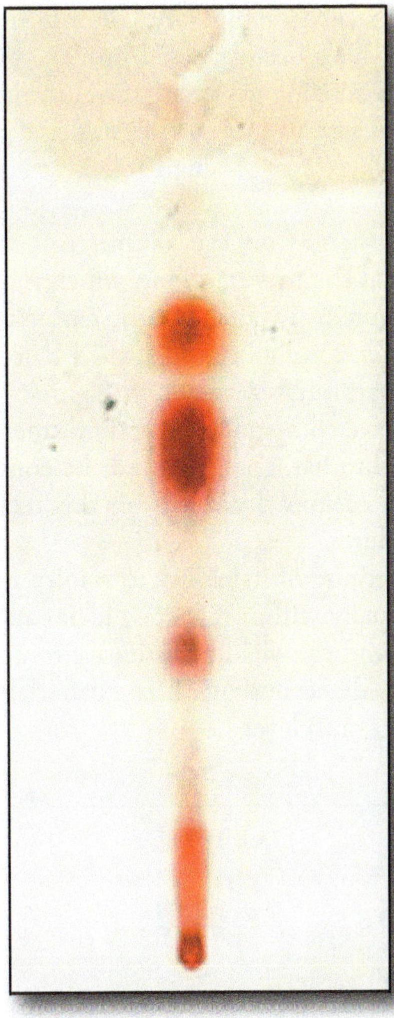

FIGURE 6–17
Thin-layer chromatogram of a marijuana extract. *Courtesy Sirchie Finger Print Laboratories, Inc., Youngsville, N.C.,* www.sirchie.com

The distance a spot has traveled up a thin-layer plate can be assigned a numerical value known as the R_f value. This value is defined as the distance traveled by the component divided by the distance traveled by the moving liquid phase. For example, in Figure 6–16 the moving phase traveled 10 centimeters up the plate before the plate was removed from the tank. After visualization, the heroin spot moved 8 centimeters, for an R_f value of 0.8; the quinine migrated 4 centimeters, for an R_f value of 0.4.

Gas Chromatography (GC) Gas chromatography (GC) separates mixtures based on their distribution between a stationary liquid phase and a moving gas phase. In GC, the moving phase is called the *carrier gas*, which flows through a column constructed of stainless steel or glass. The stationary phase is a thin film of liquid within the column.

Two types of columns are used: the *packed column* and the *capillary column*. With the packed column, the stationary phase is a thin film of liquid fixed onto small granular particles packed into the column. This column, usually constructed of stainless steel or glass, is 2 to 6 meters long and about 3 millimeters in diameter. Capillary columns are composed of glass and are much longer than packed columns—15 to 60 meters long. These types of columns are very narrow, ranging from 0.25 to 0.75 millimeter in diameter. Capillary columns can be made narrower than packed columns because their stationary liquid phase is actually coated as a very thin film directly onto the column's inner wall.

As the carrier gas flows through the packed or capillary column, it carries with it the components of a mixture that have been injected into the column. Components with a greater affinity for the moving gas phase travel through the column more quickly than those with a greater affinity for the stationary liquid phase. Eventually, after the mixture has traversed the length of the column, it emerges separated into its components.

The time required for a component to emerge from the column from the time of its injection into the column is known as the *retention time*, which is a useful identifying characteristic of a material. Figure 6–18(a) shows the chromatogram of two barbiturates; each barbiturate has tentatively been identified by comparing its retention time to those of known barbiturates, shown in Figure 6–18(b). However, because other substances may have comparable retention times under similar chromatographic conditions, gas chromatography cannot be considered an absolute means of identification. Conclusions derived from this technique must be confirmed by other testing procedures.

Gas chromatography is widely used because of its ability to resolve a highly complex mixture into its components, usually within minutes. It has an added advantage in that it is extremely sensitive and can yield quantitative results. Gas chromatography has sufficient sensitivity to detect and quantitate materials at the nanogram (0.000000001 gram or 1 × 109 gram) level.[4]

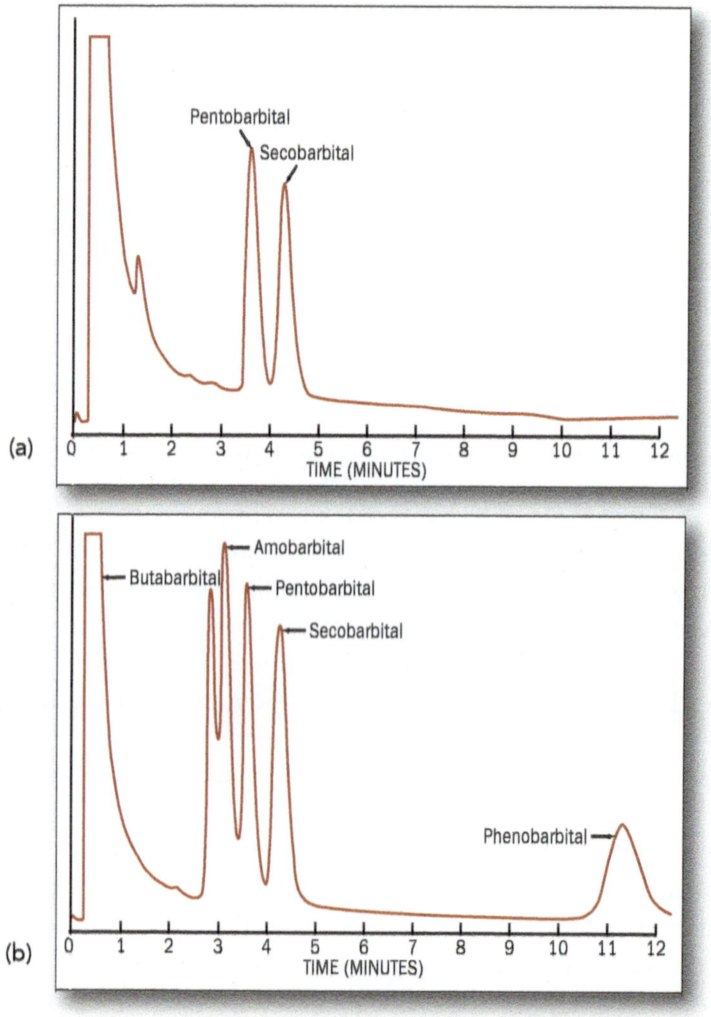

FIGURE 6–18
(a) An unknown mixture of barbiturates is identified by comparing its retention times to (b), a known mixture of barbiturates. *Courtesy Varian Inc., Palo Alto, Calif.*

Inside the Science

The Gas Chromatograph

A simplified scheme of the gas chromatograph is shown in Figure 1. The operation of the instrument can be summed up briefly as follows: The carrier gas is fed into the column at a constant rate. The carrier gas is chemically inert and is generally nitrogen or helium. The sample under investigation is injected as a liquid into a heated injection port with a syringe, where it is immediately vaporized and swept into the column by the carrier gas. The column itself is heated in an oven in order to keep the sample in a vapor state as it travels through the column. In the column, the components of the sample travel in the direction of the carrier gas flow at speeds that are determined by their distribution between the stationary and moving phases. If the analyst has selected the proper liquid phase and has made the column long enough, the components of the sample will be completely separated as they emerge from the column.

As each component emerges from the column, it enters a detector. One type of detector uses a flame to ionize the emerging chemical substance, thus generating an electrical signal. The signal is recorded on a strip-chart recorder as a function of time. This written record of the separation is called a chromatogram. A gas chromatogram is a plot of the recorder response (vertical axis) versus time (horizontal axis). A typical chromatogram shows a series of peaks, each peak corresponding to one component of the mixture.

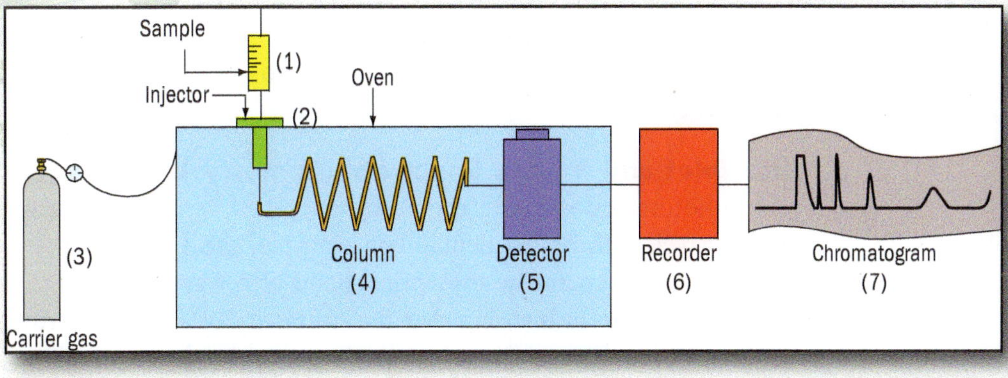

FIGURE 1 Basic gas chromatography. Gas chromatography permits rapid separation of complex mixtures into individual compounds and allows identification and quantitative determination of each compound. As shown, a sample is introduced by a syringe (1) into a heated injection chamber (2). A constant stream of nitrogen gas (3) flows through the injector, carrying the sample into the column (4), which contains a thin film of liquid. The sample is separated in the column, and the carrier gas and separated components emerge from the column and enter the detector (5). Signals developed by the detector activate the recorder (6), which makes a permanent record of the separation by tracing a series of peaks on the chromatograph (7). The time of elution identifies the component present, and the peak area identifies the concentration.

Spectrophotometry

The technique of chromatography is particularly suited for analyzing illicit drugs, because it can separate a drug from other substances that may be present in the drug preparation. However, chromatography has the drawback of not being able to specifically identify the material under investigation. For this reason, other analytical tools are frequently used to identify drugs. These include the technique of **spectrophotometry**, which can identify a substance by exposing it to a specific type of electromagnetic radiation.

spectrophotometry An analytical method for identifying a substance by its selective absorption of different wavelengths of light

Theory of Spectrophotometry We have already observed in the description of color that an object does not absorb all the visible light it is exposed to; instead, it selectively absorbs some frequencies and reflects or transmits others. Similarly, the absorption of other types of electromagnetic radiation by chemical substances is also selective. Selective absorption of a substance is measured by an instrument called a *spectrophotometer*, which produces a graph or *absorption spectrum* that depicts the absorption of light as a function of wavelength or frequency. The absorption of ultraviolet (UV), visible, and infrared (IR) radiation is particularly applicable for obtaining qualitative data pertaining to the identification of drugs.

Absorption at a single wavelength or frequency of light is not 100 percent complete—some radiation is transmitted or reflected by the material. Just how much radiation a substance absorbs is defined by a fundamental relationship known as Beer's law, shown in Equation (6–1):

EQUATION 6–1

$$A = kc$$

Here, A symbolizes the absorption or the quantity of light taken up at a single frequency, c is the concentration of the absorbing material, and k is a proportionality constant. This relationship shows that the quantity of light absorbed at any frequency is directly proportional to the concentration of the absorbing species; the more material you have, the more radiation it will absorb. By defining the relationship between absorbance and concentration, Beer's law permits spectrophotometry to be used as a technique for quantification.

ultraviolet Invisible long frequencies of light beyond violet in the visible spectrum

Ultraviolet and Visible Spectrophotometry **Ultraviolet (UV)** and visible spectrophotometry measure the absorbance of UV and visible light as a function of wavelength or frequency. For example, the UV absorption spectrum of heroin shows a maximum absorption band at a wavelength of 278 nanometers (see Figure 6–19). This shows that the simplicity of a UV spectrum facilitates its use as a tool for determining a material's probable identity. For instance, a white powder may have a UV spectrum comparable to heroin and therefore may be tentatively identified as such. (Fortunately, sugar and starch, common diluents of heroin, do not absorb UV light.)

This technique, however, does not provide a definitive result; other drugs or materials may have a UV absorption spectrum similar to that of heroin. Nevertheless, UV spectrophotometry is often useful in establishing the probable

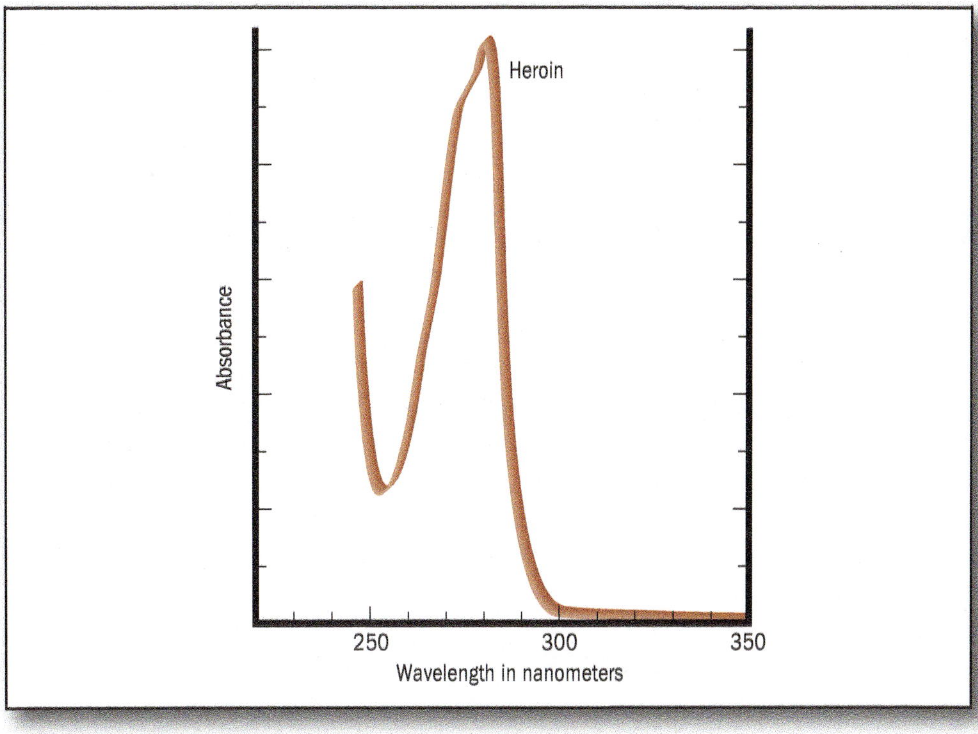

FIGURE 6–19
Ultraviolet spectrum of heroin.

identity of a drug. For example, if an unknown substance yields a UV spectrum that resembles that of amphetamine (see Figure 6–20), thousands of substances are immediately eliminated from consideration, and the analyst can begin to identify the material from a relatively small number of possibilities. A comprehensive collection of UV drug spectra provides an index that can rapidly be searched in order to tentatively identify a drug or, failing that, at least to exclude certain drugs from consideration.

Infrared Spectrophotometry In contrast to the simplicity of a UV spectrum, absorption in the **infrared** region provides a far more complex pattern. Figure 6–21 depicts the IR spectra of heroin and secobarbital. Here, the absorption bands are so numerous that each spectrum can provide enough characteristics to identify a substance specifically. Different materials always have distinctively different infrared spectra; each IR spectrum is therefore equivalent to a "fingerprint" of that substance and no other. This technique is one of the few tests available to the forensic scientist that can be considered specific in itself for identification. The IR spectra of thousands of organic compounds have been collected, indexed, and cataloged as invaluable references for identifying organic substances. The selective absorption of light by drugs in the UV and IR regions of the electromagnetic spectrum provides a valuable technique for characterizing drugs.

infrared
Invisible short frequencies of light before red in the visible spectrum

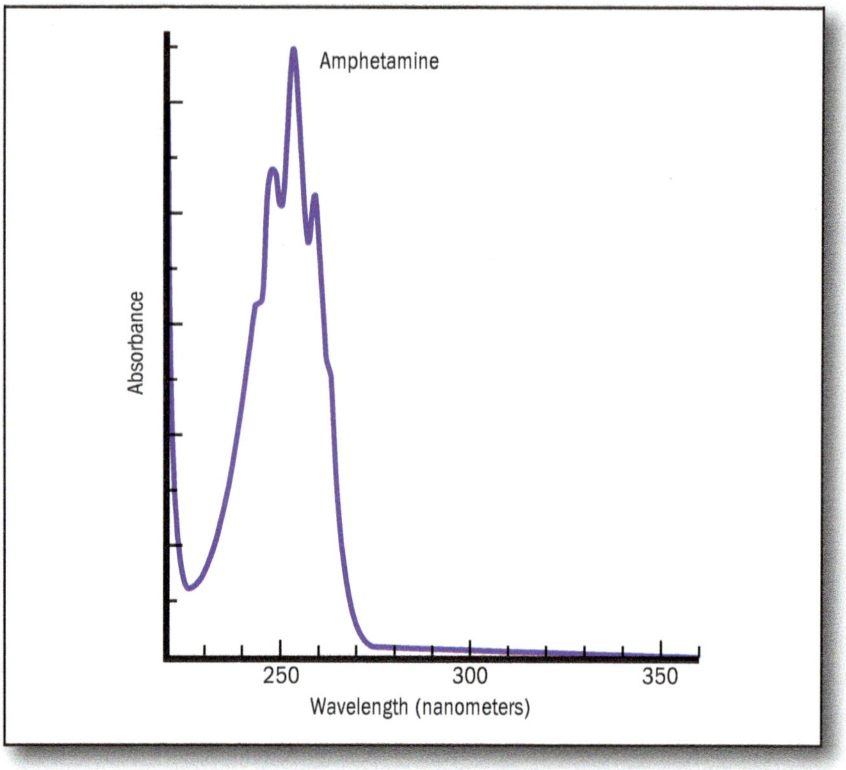

FIGURE 6–20
Ultraviolet spectrum of amphetamine.

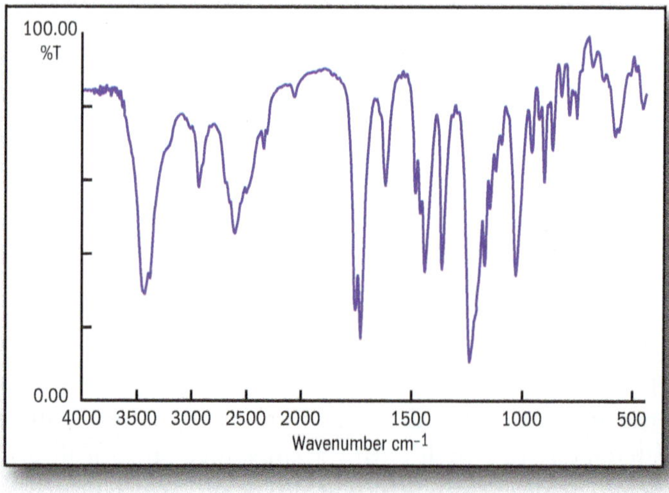

(a)

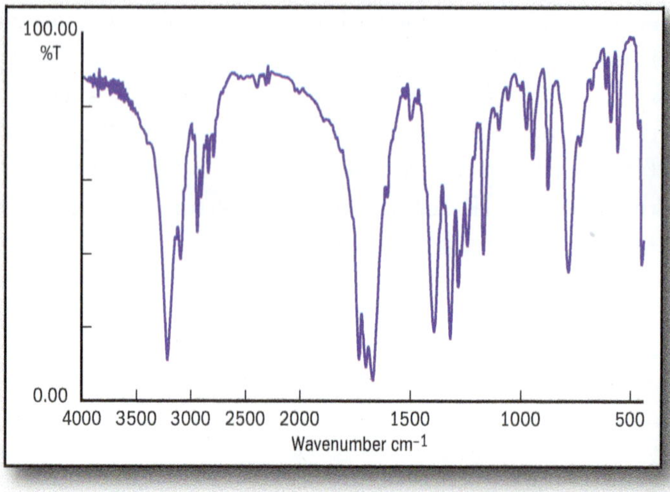

(b)

FIGURE 6–21
(a) Infrared spectrum of heroin.
(b) Infrared spectrum of secobarbital.

monochromator
A device for isolating individual wavelengths or frequencies of light

Inside the Science

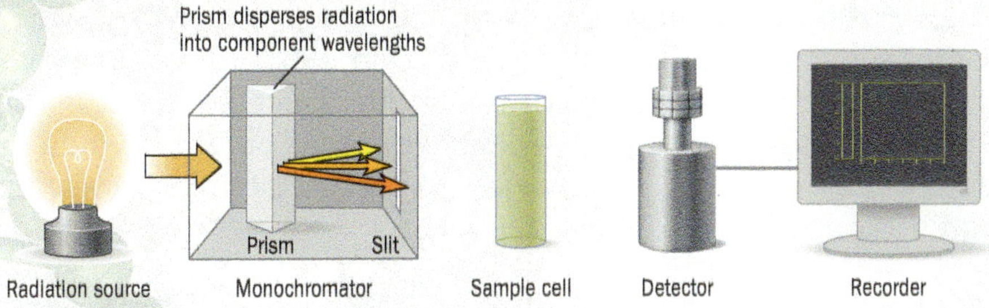

The spectrophotometer measures and records the absorption spectrum of a chemical. The basic components of a simple spectrophotometer are the same regardless of whether it is designed to measure the absorption of UV, visible, or IR radiation. These components are illustrated diagrammatically in the figure above. They include (1) a radiation source, (2) a monochromator or frequency selector, (3) a sample holder, (4) a detector to convert electromagnetic radiation into an electrical signal, and (5) a recorder to produce a record of the signal.

The choice of source varies with the type of radiation desired. For visible radiation, an ordinary tungsten bulb provides a convenient source of radiation. In the UV region, a hydrogen or deuterium discharge lamp is normally used, and a heated molded rod containing a mixture of rare-earth oxides is a good source of IR light.

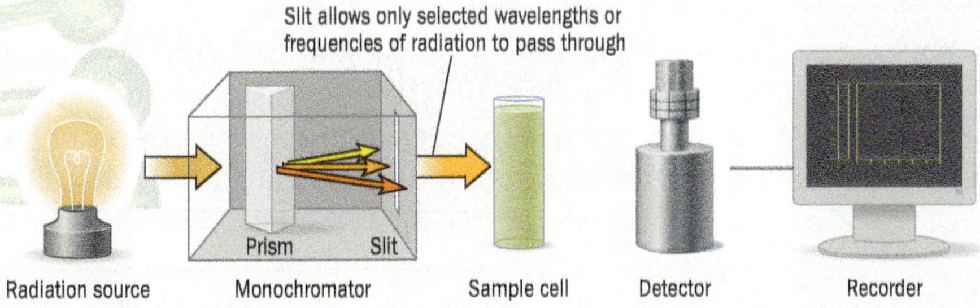

The function of the **monochromator** is to select a single wavelength or frequency of light from the source—monochromatic light. Some inexpensive spectrophotometers pass the light through colored glass filters to remove all radiation from the beam except for a desired range of wavelengths. More precise spectrophotometers may use a prism or diffraction grating to disperse radiation into its component wavelengths or frequencies.

Inside the Science (CONTINUED)

A diffraction grating is made by scratching thousands of parallel lines on a transparent surface such as glass. As light passes through the narrow spacings between the lines, it spreads out and produces a spectrum similar to that formed by a prism. The desired wavelength is obtained when the dispersed radiation is focused onto a narrow slit that permits only selected wavelengths to pass through.

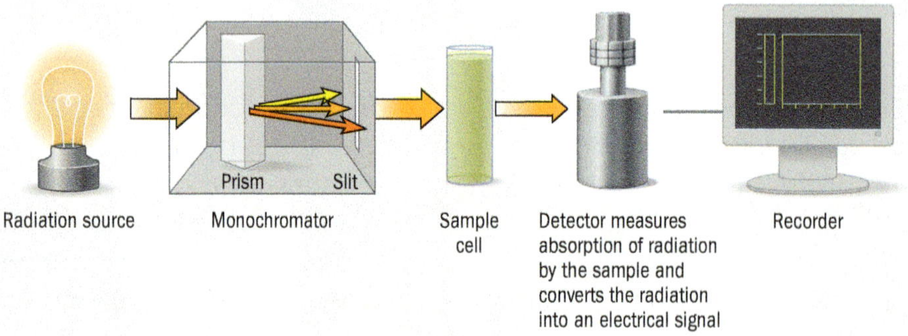

Most laboratory infrared spectrophotometers use Fourier transform analysis to measure the wavelengths of light at which a material absorbs in the infrared spectrum. This approach does not use any dispersive elements that select single wavelengths or frequencies of light emitted from a source; instead, the heart of a Fourier transform infrared (FT-IR) spectrometer is the Michelson interferometer. The interferometer uses a beam-splitting prism and two mirrors, one movable and one stationary, to direct light toward a sample. As the wavelengths pass through the sample and reach a detector, they are all measured simultaneously. A mathematical operation, the Fourier transform method, is used to decode the measured signals and record the wavelength data. These Fourier calculations are rapidly carried out by a computer. In a matter of seconds, a computer-operated FT-IR instrument can

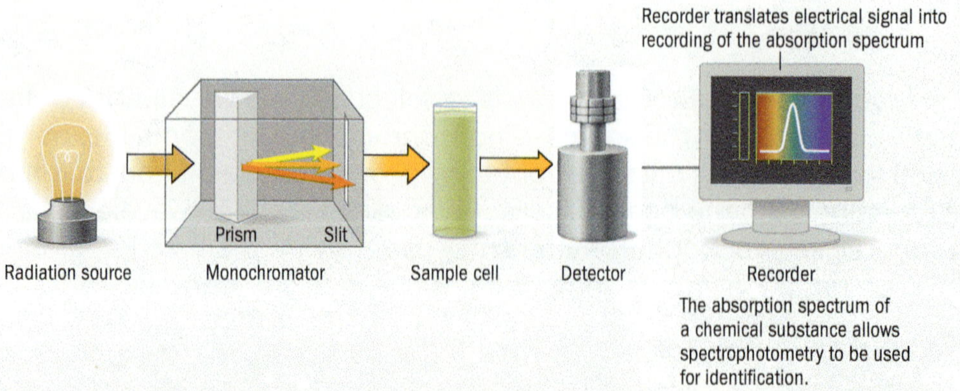

Inside the Science (CONTINUED)

produce an infrared absorption pattern compatible to one generated by a prism instrument.

Sample preparation varies with the type of radiation being studied. Absorption spectra in the UV and visible regions are usually obtained from samples that have been dissolved in an appropriate solvent. Because the cells holding the solution must be transparent to the light being measured, glass cells are used in the visible region and quartz cells in the ultraviolet region. Practically all substances absorb in some region of the IR spectrum, so sampling techniques must be modified to measure absorption in this spectral region; special cells made out of sodium chloride or potassium bromide are commonly used because they do not absorb light over a wide range of the IR portion of the electromagnetic spectrum.

The detector measures the quantity of radiation that passes through the sample by converting it to an electrical signal. UV and visible spectrophotometers use photoelectric tube detectors. A signal is generated when the photons strike the tube surface to produce a current that is directly proportional to the intensity of the light transmitted through the sample. When this signal is compared to the intensity of light that is transmitted to the detector in the absence of an absorbing material, the absorbance of a substance can be determined at each wavelength or frequency of light selected. The signal from the detection system is then fed into a recorder, which plots absorbance as a function of wavelength or frequency. Modern spectrophotometers are designed to trace an entire absorption spectrum automatically.

Mass Spectrometry

A previous section discussed the operation of the gas chromatograph. This instrument is one of the most important tools in a crime laboratory. Its ability to separate the components of a complex mixture is unsurpassed. However, gas chromatography has one important drawback—its inability to produce specific identification (confirmatory), providing only presumptive results. A forensic chemist cannot unequivocally state the identification of a substance based solely on a retention time as determined by the gas chromatograph. Fortunately, coupling the gas chromatograph to a mass spectrometer has largely overcome this problem.

A mixture's components are first separated on the gas chromatograph. A direct connection between the gas chromatograph column and the mass spectrometer then allows each component to flow into the spectrometer as it emerges from the gas chromatograph. In the mass spectrometer, the material enters a high-vacuum chamber where a beam of high-energy electrons is aimed at the sample molecules. The electrons collide with the molecules, causing them to lose electrons and to acquire a positive charge. These positively charged molecules, or **ions**, are unstable or are formed with excess energy and almost instantaneously decompose

ion
An atom or molecule bearing a positive or negative charge

into numerous smaller fragments. The fragments then pass through an electric or magnetic field, where they are separated according to their masses. The unique feature of mass spectrometry is that under carefully controlled conditions, no two substances produce the same fragmentation pattern. In essence, one can think of this pattern as a "fingerprint" of the substance being examined (see Figure 6–22).

The technique thus provides a specific means for identifying a chemical structure. It is also sensitive to minute concentrations. Mass spectrometry is most widely used to identify drugs; however, further research is expected to yield significant applications to identifying other types of physical evidence. Figure 6–23 illustrates the mass spectra of heroin and cocaine; here, each line represents a fragment of a different mass (actually the ratio of mass to charge), and the line height reflects the relative abundance of each fragment. Note how different the

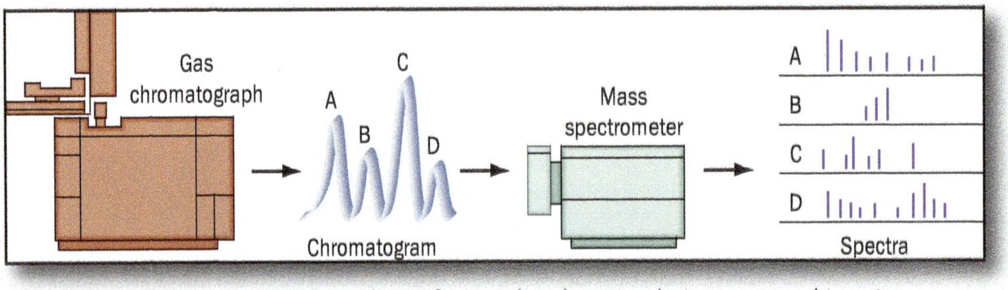

FIGURE 6–22 How GC/MS works. Left to right, the sample is separated into its components by the gas chromatograph, and then the components are ionized and identified by characteristic fragmentation patterns of the spectra produced by the mass spectrometer. *Courtesy Agilent Technologies, Inc., Palo Alto, Calif.*

FIGURE 6–23 (a) Mass spectrum of heroin. (b) Mass spectrum of cocaine.

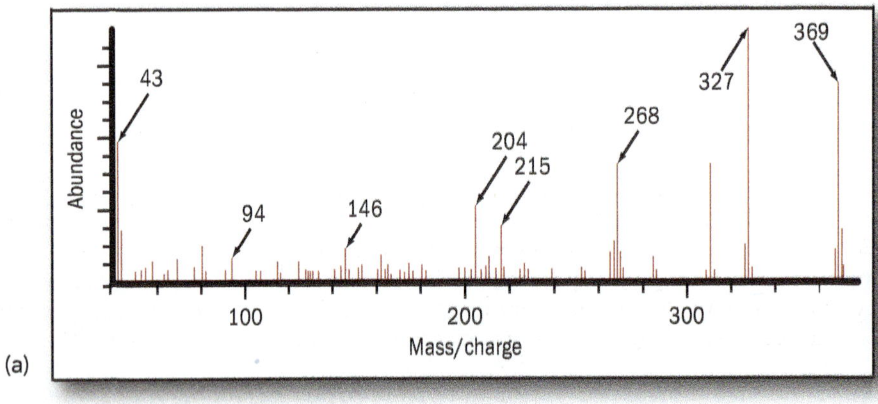

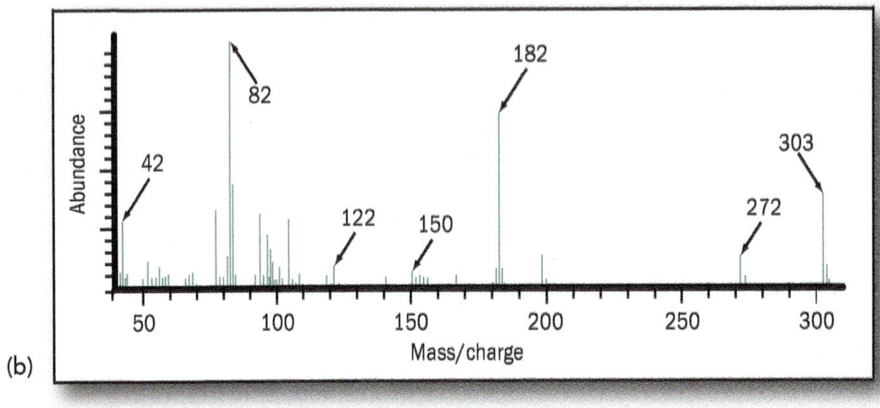

fragmentation patterns of heroin and cocaine are. Each mass spectrum is unique to each drug and therefore provides a specific test for identifying that substance.

The combination of the gas chromatograph and mass spectrometer (GC/MS) is further enhanced when a computer is added to the system. The integrated gas chromatograph/mass spectrometer/computer system provides the ultimate in speed, accuracy, and sensitivity. With the ability to record and store in its memory several hundred mass spectra, such a system can detect and identify substances present in only one-millionth-of-a-gram quantities. Furthermore, the computer can be programmed to compare an unknown spectrum against a comprehensive library of mass spectra stored in its memory. The advent of personal computers and microcircuitry has made it possible to design mass spectrometer systems that can fit on a small table. Such a unit is pictured in Figure 6–24. With data obtained from a GC/MS determination, a forensic analyst can, with one instrument, separate the components of a complex drug mixture, interpret the results from presumptive and confirmatory laboratory procedures, unequivocally identify each substance present in the mixture, and determine the implications of the results.

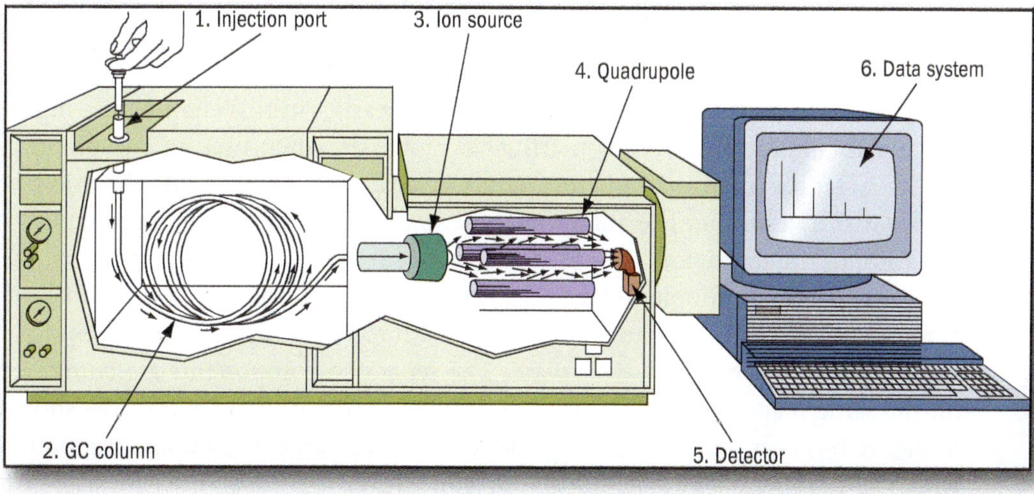

FIGURE 6–24 A tabletop mass spectrometer. (1) The sample is injected into a heated inlet port, and a carrier gas sweeps it into the column. (2) The GC column separates the mixture into its components. (3) In the ion source, a filament wire emits electrons that strike the sample molecules, causing them to fragment as they leave the GC column. (4) The quadrupole, consisting of four rods, separates the fragments according to their mass. (5) The detector counts the fragments passing through the quadrupole. The signal is small and must be amplified. (6) The data system is responsible for total control of the entire GC/MS system. It detects and measures the abundance of each fragment and displays the mass spectrum. *Courtesy Agilent Technologies, Inc., Palo Alto, Calif.*

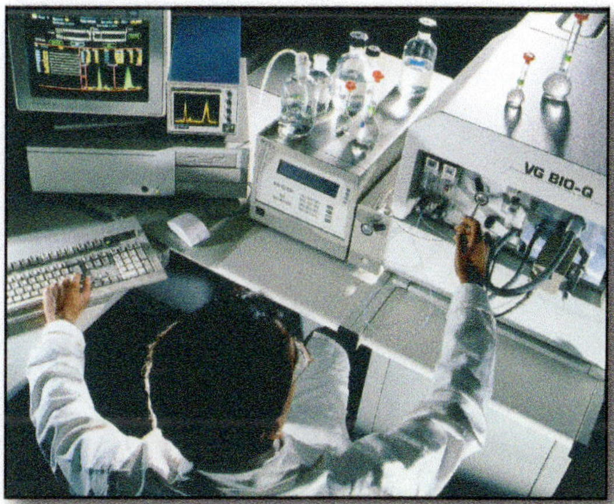

FIGURE 6–25
A gas chromatograph interfaced to a mass spectrometer in a forensic laboratory. A sample injection robot transfers samples from small vials to the gas chromatograph.
Geoff Tompkinson/Science Source

The Identification of Marijuana

Enforcement of laws prohibiting the sale and use of marijuana accounts for a high percentage of drug arrests in the United States. Any trial or hearing involving a seizure of marijuana requires identification of the material before the issue of guilt or innocence can be decided.

Unlike most other drugs received by the crime laboratory, marijuana (Cannabis sativa L.) possesses botanical features that impart identifiable characteristics. Because most marijuana specimens consist of small leaf fragments, their identification must be partially based on botanical features observed under the microscope by a trained expert. This approach is further augmented with a chemical test that will independently confirm the findings of the botanical examination.

The identification of marijuana by microscopic methods depends largely on observing short hairs shaped like "bear claws" on the upper side of the leaf (see the SEM photo in Figure 8–14). These hairs are known as cystolithic hairs. Further verification of the identity of marijuana is confirmed by the presence of longer, nonglandular hairs on the opposite side of the leaf.

The Duquenois-Levine color test, described earlier in this chapter, is a highly but not totally specific test for marijuana. However, when used in combination with a botanical examination, the results constitute a specific identification of marijuana. In addition, the analyst may be unable to obtain a microscopic identification of the marijuana leaf, as in the case of hashish or hashish oil. Here, the color test has to be supplemented by another examination, preferably thin-layer chromatography. This method involves separating chemical constituents found in the suspect resin on a thin-layer plate. The separated components are compared on the same plate to those obtained from a known marijuana extract, as shown in Figure 6–17. In this manner, a positive TLC comparison, used in conjunction with the Duquenois-Levine color test, constitutes a specific identification for marijuana.

Quick Review

- Chromatography is a means of separating and tentatively identifying the components of a mixture.
- TLC uses a solid stationary phase, usually coated onto a glass plate, and a mobile liquid phase to separate the components of the mixture.
- Gas chromatography (GC) separates mixtures on the basis of their distribution between a stationary liquid phase and a mobile gas phase.
- Spectrophotometry is the study of the absorption of light by chemical substances.
- Most forensic laboratories use ultraviolet (UV) and infrared (IR) spectrophotometers to characterize chemical compounds.
- IR spectrophotometry provides a far more complex pattern than UV spectrophotometry. Because different materials have distinctively different infrared spectra, each IR spectrum is equivalent to a "fingerprint" of that substance.
- Mass spectrometry characterizes organic molecules by observing their fragmentation pattern after their collision with a beam of high-energy electrons.
- Infrared spectrophotometry and mass spectrophotometry typically are used to specifically identify a drug substance.

Chapter Review

- A drug is a natural or synthetic substance that is used to produce physiological or psychological effects in humans or other animals.
- Nondrug factors that play a part in drug dependence include the personal characteristics of the user, the user's expectations about the drug experience, society's attitudes and possible responses, and the setting in which the drug is used.
- Physical dependence is defined as the physiological need for a drug that has been brought about by its regular use. Psychological dependence is the conditioned use of a drug caused by underlying emotional needs.
- Narcotic drugs are analgesics, meaning they relieve pain by depressing the central nervous system.
- The most common source for narcotic drugs is opium. Morphine is extracted from opium and used to synthesize heroin.
- Opiates are not derived from opium or morphine, but they have the same physiological effects on the body. Examples of opiates include methadone and OxyContin (oxycodone).
- Hallucinogens cause marked changes in normal thought processes, perceptions, and moods. Marijuana is the most well-known drug in this class. Other hallucinogens include LSD, mescaline, PCP, psilocybin, and MDMA (Ecstasy).
- Depressants decrease the activity of the central nervous system, calm irritability and excitability, and produce sleep. Depressants include alcohol (ethanol), barbiturates, tranquilizers, and various substances that can be sniffed, such as airplane glue or model cement.
- Stimulants increase the activity of the central nervous system and are taken to increase alertness and activity. Stimulants include amphetamines, sometimes known as "uppers" or "speed," and cocaine, which in its freebase form is known as crack.
- Club drugs are synthetic drugs that are used at nightclubs, bars, and raves (all-night dance parties). Some club drugs act as stimulants; others have depressant effects.
- Anabolic steroids are synthetic compounds that are chemically related to the male sex hormone testosterone. Anabolic steroids are often abused by individuals who are interested in accelerating muscle growth.
- Federal law establishes five schedules of classification for controlled dangerous substances on the basis of a drug's potential for abuse, potential for physical and psychological dependence, and medical value.
- Analysts use screening tests to determine the identity of drugs present in a sample. These tests reduce the number of possible drugs to a small and manageable number.

- A series of color tests produce characteristic colors for the more commonly encountered illicit drugs.
- After preliminary testing, forensic chemists use more specific tests to identify a drug substance to the exclusion of all other known chemical substances.
- Chromatography is a means of separating and tentatively identifying the components of a mixture.
- TLC uses a solid stationary phase, usually coated onto a glass plate, and a mobile liquid phase to separate the components of the mixture.
- Gas chromatography (GC) separates mixtures on the basis of their distribution between a stationary liquid phase and a mobile gas phase.
- Spectrophotometry is the study of the absorption of light by chemical substances.
- Most forensic laboratories use ultraviolet (UV) and infrared (IR) spectrophotometers to characterize chemical compounds.
- IR spectrophotometry provides a far more complex pattern than UV spectrophotometry. Because different materials have distinctively different infrared spectra, each IR spectrum is equivalent to a "fingerprint" of that substance.
- Mass spectrometry characterizes organic molecules by observing their fragmentation pattern after their collision with a beam of high-energy electrons.
- Infrared spectrophotometry and mass spectrophotometry typically are used to specifically identify a drug substance.

Quick Lab: Chromatography

Materials:

Roll of chromatography paper
5 different types of markers/pens
Thin piece of wire about 5 inches long
Beaker
Water

Procedure:

Chromatography is a process that can be used to identify different substances that make up a mixture. In forensics this process can be used on many types of evidence. In this activity you will perform paper chromatography to identify the components of different types of ink. Cut a piece of chromatography paper that is about 5 inches long. Select one of the markers/pens and place a dot at one end of the paper, leaving a small gap between the dot and end of paper. At the other end, push the wire through the paper so that the paper is hanging off of the wire with the dot end down. Fill the bottom of the beaker with water. Place the wire across the top of the beaker with the paper hanging in the beaker so that the dot end of the paper is in the water, but the dot is just above the water. Allow

5 minutes for the separation to occur. When the ink has stopped separating, take the paper out of the beaker and let it dry. Repeat the process for all the markers/pens. Your teacher could give you an unknown ink to test and compare to the samples you have just tested.

Follow-Up Questions:

1. **Did all of the inks separate during the process? If not, why do you think certain inks did not separate?**
2. **Besides ink, what other types of evidence could chromatography be used with?**
3. **Did you find a match to the unknown ink? Was it exactly the same as one of the other inks? If not, what was different?**

Quick Lab: Drug Screening Test

Materials:

Ward's Natural Science Crime Scene Drug Bust Kit

Procedure:

Often drugs are part of an investigation in forensics. Suspects may need to be tested to see if they have been using or may be on some type of drug. To do this, a screening test is completed. Follow the directions of the kit to see how some screening tests are completed.

Follow-up questions can be found in the kit handout section.

Quick Lab: What Is the White Powder?

Materials:

Table salt
Baking soda
Cornstarch
Sugar
Flour
Water
Vinegar
Iodine
Bunsen burner/candle
Aluminum foil
Tongs Slides
Toothpicks

Procedure:

Unknown substances may be found at crime scenes. When this happens, it is important to be able to identify what substance is actually present. In this activity you will test five white powders to look for differences among them in order to identify an unknown sample. First, mix each powder individually with water. Do this by placing a small amount of the powder on a slide and adding a drop of water to it; mix with a toothpick. Record what you observe. Do this for each powder. Repeat this process twice more, once using vinegar and once using iodine instead of water. Do not get iodine on anything but the slide; it stains! Next create a foil tray that will hold a small sample of powder. Carefully light the candle or the Bunsen burner; be sure to have eye protection on for this part of the activity. Place a small amount of the powder on the tray and use the tongs to hold it over the candle for a minute. Record your observations. Repeat this for each powder. When you are finished, your teacher may choose to give you an unknown powder sample for you to test and determine which powder it is.

Follow-Up Questions:

1. **Of the different tests performed, which one would be the best to use to identify an unknown powder?**
2. **In this activity you tested pure samples. Would this be the case if you were investigating a crime in the real world? Explain your answer.**
3. **Were you able to identify the unknown sample provided by your teacher?**
4. **How does this lab relate to drug tests in a forensic laboratory? Use the data obtained from this lab to develop an explanation of how drug tests are used to identify unknown substances collected at crime scenes.**

Review Questions

1. Chromatography, the distribution of a gas between the liquid and gas phases is determined by
   a. the solubility of the gas in the liquid.
   b. the volume of the gas in the container.
   c. the density of the gas relative to the liquid.
   d. the mass of the gas relative to the liquid.

2. Spectrophotometry uses which light source?
   a. Visible
   b. UV
   c. Infrared
   d. All of the above

3. The higher the solubility of a gas in a liquid, the greater the tendency of the gas molecules to
   a. move from a liquid phase to a gaseous or vapor phase.
   b. remain in the liquid phase.
   c. move from a liquid phase to a solid state.
   d. disperse.

4. All of the following are chromatographic processes found to be applicable for solving analytical problems encountered in the crime laboratory except
   a. thin-layer chromatography.
   b. gas chromatography.
   c. GC/MS.
   d. solid-state chromatography.

5. Which of the following tests can be considered specific in itself for identification purposes?
   a. Mass spectrometry
   b. Gas spectrometry
   c. Gas chromatography
   d. Electromagnetic radiation

6. True or False: Although cocaine is legally classified as a narcotic, pharmacologically it is actually a powerful central nervous system stimulant.

7. True or False: The quantity of a substance separated by gas chromatography can be determined by its R_f value.

8. True or False: The gas chromatography/mass spectrometry (GC/MS) combination produces a fragmentation pattern that serves as a virtual "fingerprint" of a chemical substance because, with few exceptions, no two substances fragment in the same fashion.

9. True or False: Chromatography is a means of separating and tentatively identifying the components of a mixture.

10. True or False: Infrared spectrophotometry allows for the identification of different materials because different organic substances always produce distinctive infrared spectra.

11. What is a drug? How has drug use affected the growth of crime laboratories in the United States?

12. Name three nondrug factors that play a part in drug dependence.

13. Define physical dependence and psychological dependence.

14. What is the pharmacological definition of a narcotic?

15. What is the source of most narcotic analgesics? Name two popular drugs prepared from this substance.

16. Name two synthetic opiates and describe the purpose for which each typically is used.

17. What is a hallucinogen? Name three commonly used hallucinogens.

18. What is the most widely used illicit drug in the United States? What is the active ingredient in this drug?

19. Arrange the following parts or products of the Cannabis plant in order of THC content, from highest to lowest concentration of THC: flowers, leaves, resin, seeds, stem.

20. List three potential medical uses of marijuana.

21. What is angel dust and what are the negative consequences of long-term use?

22. What is the most widely abused drug in the United States?

23. In what class of drugs do alcohol and barbiturates belong? What is the main physiological effect of such drugs?

24. What is a stimulant? Name two widely used stimulants.

25. Name two potent forms of methamphetamine. How is each of these drugs typically taken into the body?

26. What popular stimulant is derived from a plant that grows in the Andes mountains of South America?

27. What is crack, and how is it produced?

28. Name club drugs belonging to three different classes of drugs and indicate the class to which each belongs.

29. What are anabolic steroids, and why were they developed?

30. What is the difference between a screening test and a confirmation test?

31. Name two types of empirical tests used to identify drugs. Why are these tests referred to as empirical?

32. What is the difference between a qualitative evaluation and a quantitative evaluation?

33. Why is chromatography particularly well suited to the needs of a drug analyst?

34. Name three distinct advantages of gas chromatography in the identification of drugs.

35. What is the main drawback of gas chromatography in the identification of drugs?

36. What phenomenon forms the basis of spectrophotometry?

37. What is the main advantage of infrared spectrophotometry over ultraviolet or visible-light spectrophotometry?

38. With what analytical device is a gas chromatograph often connected to analyze drug mixtures, and why?

Learn the Vocabulary

Use the information from this chapter along with your prior knowledge and experience to learn the key terms. Choose from the following activities:

1. Make flashcards with a term on the front and its definition on the back. Use the flashcards to quiz a partner on correctly pronouncing and defining the terms.
2. Make an audio or video recording of each term and its definition.
3. Work with a partner to write a sentence for each term. Read the sentences out loud to a partner or share them with the class.
4. Write a list of new terms and language you have encountered in this chapter. Reuse the terms to write sentences and then read the sentences out loud to a partner. Reuse the terms again by writing a short speech. The speech might explain the terms or teach a lesson using the terms. Give the speech to a partner or to the class.

Application and Critical Thinking

1. An individual who has been using a drug for an extended period of time suddenly finds himself unable to secure more of the drug. He acts nervous and irritable and is hyperactive. He seems almost desperate to find more of the drug, but experiences no sickness, pain, or other outward physical discomfort. Based on his behavior, what drugs might he possibly have been using? Explain your answer.

2. Following are descriptions of behavior that are characteristic among users of certain classes of drugs. For each description, indicate the class of drug (narcotics, stimulants, and so on) for which the behavior is most characteristic. For each description, also name at least one drug that produces the described effects.
   a. slurred speech, slow reaction time, impaired judgment, reduced coordination
   b. intense emotional responses, anxiety, altered sensory perceptions
   c. alertness, feelings of strength and confidence, rapid speech and movement, decreased appetite
   d. drowsiness, intense feeling of well-being, relief from pain

3. Following are descriptions of four hypothetical drugs. According to the Controlled Substances Act, under which drug schedule would each substance be classified?
   a. This drug has a high potential for psychological dependence, it currently has accepted medical uses in the United States, and the distributor is not required to report to the U.S. Drug Enforcement Administration.
   b. This drug has medical use in the United States, is not limited by manufacturing quotas, and may be exported without a permit.
   c. This drug must be stored in a vault or safe, requires separate record keeping, and may be distributed with a prescription.
   d. This drug may not be imported or exported without a permit, is subject to manufacturing quotas, and currently has no medical use in the United States.

4. A police officer stops a motorist who is driving erratically and notices a bag of white powder on the front seat of the car that he suspects contains heroin. The officer brings the bag to you, a forensic scientist in the local crime lab. Name one screening test that you might perform to determine the presence of heroin. Assuming the powder tests positive for heroin, what should you do next?

5. The figure on the next page shows a chromatogram from the presumptive analysis of a known mixture of barbiturates. Based on this figure, interpret the results and answer the following questions:
   a. What barbiturate detected by the chromatogram had the longest retention time?
   b. Which barbiturate had the shortest retention time?
   c. What is the approximate retention time of amobarbital?
   d. What are the implications of the GC results? Why would it be necessary to follow up with MS analysis?
   e. Would it be necessary to perform a confirmatory test? What confirmatory test should be used? What would be the implications of any results obtained?

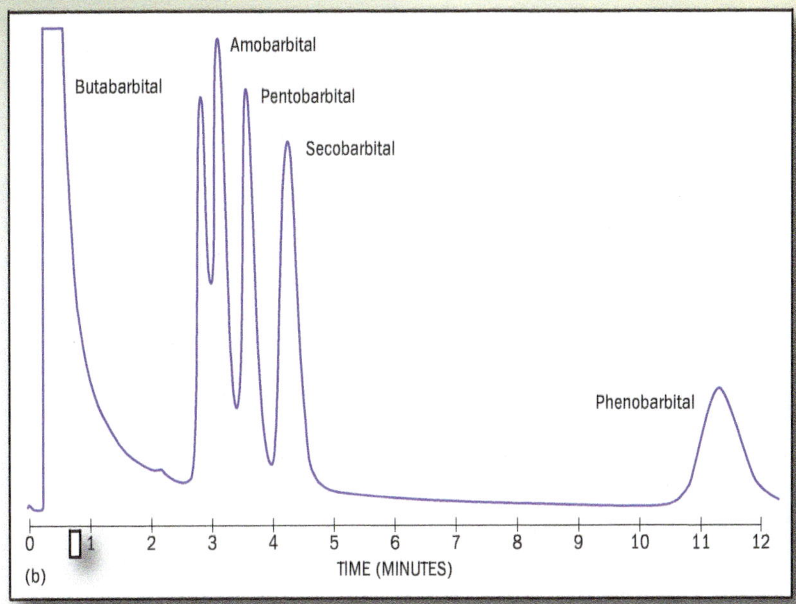

6. When investigating a potential warehouse for storing illegal drugs, the police collected a variety of drugs. The drugs were tested with presumptive color tests to determine their possible identity. The test tubes shown in the figure display the positive color tests. Match the drug on the right with the color test on the left. Name the test and describe the implications of the results for each presumptive test.

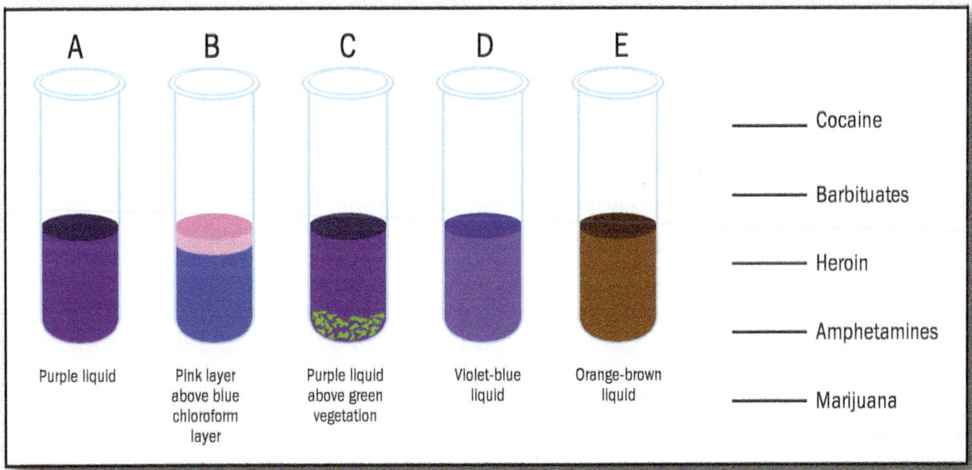

7. Create a poster or slide presentation that classifies controlled substances using the schedules under the Controlled Substances Act.

8. Research and explain in writing how you would identify unknown substances using presumptive and confirmatory procedures as they relate to toxicological analysis. Present your findings to the class.

9. U.S Customs and Border Protection tracks all drug seizures occurring along U.S. borders. By accessing the U.S. Customs and Border protection website you can see current and historical data of drug seizures by weight and type. Use the data and models provided to develop and support a logical explanation of why there are so many drug seizures. In writing, propose a logical solution for the number of drug seizures that is consistent with scientific ideas, principles, and theories.

Laboratory Experiments

This activity requires the use of the following practices of science and engineering:

- Analyzing and interpreting data
- Constructing explanations
- Using appropriate safety equipment and practices during laboratory/classroom and field investigations as outlined in Texas Education Agency-approved safety standards

This activity consists of the following crosscutting concepts:

- **Patterns**—The reactivity of the different drugs allows for their classification.
- **Structure and function**—The chemical structure of the powder determines their reactivity.

Before you begin, read the scenario and steps out loud with a partner. Ask for clarification as necessary to make sure you understand the experiment.

The Scenario

Two women had crossed the U.S. border into Nogales, Mexico, for a day of shopping. When returning to the United States the women were stopped by customs and randomly searched. A white powder was found in a small canister in one of the women's purses. The woman indignantly says it is an over-the-counter drug for her headaches and is powdered for quick dissolving. You have been asked to run a test to determine if she is telling the truth and whether to call the DEA or let the woman go. Based on the results, what do you advise them to do? Be prepared to provide evidence to support your decision.

Safety Precautions

- Goggles must be worn at all times in the laboratory.
- Do not ingest any of the over-the-counter drug samples during this laboratory. The samples are for laboratory use only, have been stored with other non-food-grade laboratory chemicals, and are not meant for human consumption.
- Hydrochloric acid solution, although it is used in dilute form in this experiment, is toxic by ingestion and inhalation and is corrosive to skin and eyes. In the event of skin contact, rinse well with water.
- Iron(III) nitrate solution may be a skin/tissue irritant. In the event of skin contact, rinse well with water.
- Universal indicator solution is a flammable, alcohol-based solution. Keep it away from flames.

MATERIALS

Spot plate 48 well
Thin stem beral pipets
Mortar and pestle
Tooth picks
Acetaminophen
Excedrin®
Aspirin
Ibuprofen
Bufferin®
Unknown powder
Distilled or deionized water
Ethanol, C_2H_5OH, 100%, can use specially denatured ethanol
Universal Indicator
Hydrochloric acid solution, HCl, 0.5 M
Iron(III) nitrate solution, $Fe(NO_3)_3$, 0.2 M
pH indicator papers, narrow range
Universal indicator solution, with color chart

PROCEDURE

1. In the first row of the spot plate, place a small amount of acetaminophen (no bigger than a grain of rice) into each of the first four depressions. Label the row.
2. Repeat step 1 for each of the other five powders. Label each row.
3. Examine each of the powders, noting color and texture. Make a data table and record your observations on it.
4. Examine each of the powders under the microscope. Record your observations.
5. Add five drops of distilled water to each powder in column #1 of the spot plate. Record your observations.
6. Add five drops of ethanol to each powder in column #2 of the spot plate. Record your observations.
7. Add one drop of universal indicator to each of the depressions in column #1. Use a different toothpick to stir each one. Record the color and pH of each powder. Note whether the substances are acidic or basic.
8. Determine an accurate pH of each drug by dipping one end of a strip of the appropriate narrow range pH paper into each of the corresponding wells in column #1. (If the approximate pH is 3, then use a narrow range pH paper such as pH 3–6.) Use the color chart on the pH paper container to determine the pH. Record your results.
9. Add two drops of HCl to each of the powders in column #3. Record your observations.
10. Add two drops of the ferric nitrate solution to each of the powders in column #4. Use a different toothpick to stir each one. Record your observations.
11. Carefully discard all solutions into the sink and rinse the plate with water.
12. Describe your results to a partner or to the class. Compare your results with a partners. If your results differ, ask questions to determine why your results differed.

Endnotes

1. Excerpt from "Marijuana Pharmacological Effect" from MARIHUANA: A SIGNAL OF MISUNDERSTANDING. Published by The New American Library, Inc., 1972.
2. Field-test color kits for drugs can be purchased from various commercial manufacturers.
3. 1803 by British chemist William Henry
4. Powers of 10 are quite useful and simple for handling large or small numbers. The exponent expresses the number of places the decimal point must be moved. If it is positive, the decimal point is moved to the right; if it is negative, the decimal point is moved to the left. Thus, to express 1×10^{-9} as a number, the decimal point is simply moved nine places to the left of 1.

Forensic Toxicology

7

JT Vintage/Glasshouse Images/Alamy Stock Photo

Key Terms

absorption
acid
alveoli
anticoagulant
artery
base
capillary
excretion
fuel cell detector
metabolism
oxidation
pH scale
preservative
toxicologist
vein

What Killed Napoleon?

Napoleon I, emperor of France, was sent into exile on the remote island of St. Helena by the British after his defeat at the Battle of Waterloo in 1815. St. Helena was hot, unsanitary, and rampant with disease. There, Napoleon was confined to a large reconstructed agricultural building known as Longwood House. Boredom and unhealthy living conditions gradually took their toll on Napoleon's mental and physical state. He began suffering from severe abdominal pains and experienced swelling of the ankles and general weakness of his limbs. From the fall of 1820, Napoleon's health began to deteriorate rapidly until death arrived on May 5, 1821. An autopsy concluded the cause of death to be stomach cancer.

It was inevitable that dying under British control, as Napoleon did, would bring with it numerous conspiratorial theories to account for his death. One of the more fascinating inquiries was conducted by a Swedish dentist, Sven Forshufvud, who systematically correlated the clinical symptoms of Napoleon's last days to those of arsenic poisoning. For Forshufvud, the key to unlocking the cause of Napoleon's death rested with Napoleon's hair. Forshufvud arranged to have Napoleon's hair measured for arsenic content by neutron activation analysis and found it consistent with arsenic poisoning over a lengthy period of time. Nevertheless, the cause of Napoleon's demise is still a matter for debate and speculation. Other Napoleon hairs collected in 1805 and 1814 have also shown high concentrations of arsenic, giving rise to the speculation that Napoleon was innocently exposed to arsenic. Even hair collected from Napoleon's three sisters, son, and first wife show significant levels of arsenic. Some question whether Napoleon even had clinical symptoms associated with arsenic poisoning. In truth, forensic science may never be able to answer the question "What killed Napoleon?"

Learning Objectives

After studying this chapter, you should be able to:

7.1 Explain how alcohol is absorbed into the bloodstream, transported throughout the body, and eliminated by oxidation and excretion

7.2 Describe the process by which alcohol is excreted in the breath via the lungs

7.3 Discuss the methods used to determine alcohol intoxication

7.4 Discuss the process involved in the analysis of blood for alcohol

7.5 Explain how the "implied consent" law recommended by the NHTSA addressed the constitutional issues raised against blood-alcohol laws

7.6 Describe the role of the forensic toxicologist and the techniques they use to identify substances with a forensic toxicology result

National Science Content Standards

- Scientific Inquiry
- Science and Technology
- Physical Science
- Science in Personal and Social Perspective
- Life Science
- History and Nature of Science

The Role of Forensic Toxicology

It is no secret that in spite of the concerted efforts of law enforcement agencies to prevent distribution and sale of illicit drugs, thousands die every year from intentional or unintentional administration of drugs, and many more innocent lives are lost as a result of the erratic and frequently uncontrollable behavior of individuals under the influence of drugs. But one should not automatically attribute these occurrences to the wide proliferation of illicit-drug markets. For example, in the United States alone, drug manufacturers produce enough sedatives and antidepressants each year to provide every man, woman, and child with about 40 pills. All of the statistical and medical evidence shows ethyl alcohol, a legal over-the-counter drug, to be the most heavily abused drug in Western countries.

toxicologist
An individual whose job is to detect and identify drugs and poisons in body fluids, tissues, and organs

Because the uncontrolled use of drugs has become a worldwide problem affecting all segments of society, the role of the **toxicologist** has taken on new and added significance. Toxicologists detect and identify drugs and poisons in body fluids, tissues, and organs. Their services are required not only in such legal institutions as crime laboratories and medical examiners' offices; they also reach into hospital laboratories—where identifying a drug overdose may represent the difference between life and death—and into various health facilities that monitor the intake of drugs and other toxic substances. Primary examples include performing blood tests on children exposed to leaded paints and analyzing the urine of addicts enrolled in methadone maintenance programs.

The role of the forensic toxicologist is limited to matters that pertain to violations of criminal law. However, responsibility for performing toxicological services in a criminal justice system varies considerably throughout the United States. In systems with a crime laboratory independent of the medical examiner, this responsibility may reside with one or the other or may be shared by both. Some systems, however, take advantage of the expertise of government health department laboratories and assign this role to them. Nevertheless, whatever facility handles this work, its caseload will reflect the prevailing popularity of the drugs that are abused in the community. In most cases, this means that the forensic toxicologist handles numerous requests to determine the presence of alcohol in the body.

Forty percent of all traffic deaths in the United States, nearly 17,500 fatalities per year, are alcohol related, along with more than 2 million injuries each year requiring hospital treatment. This highway death toll, as well as the untold damage to life, limb, and property, shows the dangerous consequences of alcohol abuse (see Figure 7–1). Because of the prevalence of alcohol in the toxicologist's work, we will begin by taking a closer look at how the body processes and responds to alcohol.

FIGURE 7–1 Alcohol consumption increases the risk of traffic accidents and fatalities. A crashed car is on display as part of a Don't Drink and Drive campaign. *Mosab Bilto/ Shutterstock*

Quick Review

- Forensic toxicologists detect and identify drugs and poisons in body fluids, tissues, and organs in matters that pertain to violations of criminal laws.
- Ethyl alcohol is the most heavily abused drug in Western countries.

Toxicology of Alcohol

The subject of the analysis of alcohol immediately confronts us with the primary objective of forensic toxicology—detecting and isolating drugs in the body to determine their influence on human behavior. Knowing how the body metabolizes alcohol provides the key to understanding its effects on human behavior. In the case of alcohol, however, the problem is further complicated by practical considerations. The predominant role of the automobile in our society has mandated the imposition of laws to protect the public from the drinking driver. This has meant that toxicologists have had to devise rapid and specific procedures for measuring the degree of alcohol intoxication. The methods used must be suitably designed to test hundreds of thousands of motorists annually without causing them undue physical harm or unreasonable inconvenience, while at the same time providing a reliable diagnosis that can be supported and defended within the framework of the legal system.

The Metabolism of Alcohol

All toxins that enter the body are eventually broken down by chemicals in the body and transformed into other chemicals that are easier to eliminate. This process of transformation of toxins, called **metabolism**, consists of three basic steps: absorption, distribution, and elimination.

metabolism
The transformation of a chemical in the body to other chemicals to facilitate its elimination from the body

absorption
Passage of alcohol across the wall of the stomach and small intestine into the bloodstream

Absorption and Distribution Alcohol, or ethyl alcohol, is a colorless liquid normally diluted with water and consumed as a beverage. Alcohol appears in the blood within minutes after it has been consumed and slowly increases in concentration while it is being absorbed from the stomach and the small intestine into the bloodstream. During the **absorption** phase, alcohol and other toxins slowly enter the body's bloodstream and are carried to all parts of the body. When the absorption period is completed, the alcohol, a toxin, becomes distributed uniformly throughout the watery portions of the body—that is, throughout about two-thirds of the body volume. Fat, bones, and hair are low in water content and therefore contain little alcohol, whereas alcohol concentration in the rest of the body is fairly uniform. After absorption is completed, a maximum alcohol level is reached in the blood, and the postabsorption period begins. Then, the alcohol concentration slowly decreases until a zero level is again reached.

Many factors determine the rate at which alcohol is absorbed into the bloodstream, including the total time taken to consume the drink, the alcohol content of the beverage, the amount consumed, and the quantity and type of food present in the stomach at the time of drinking. With so many variables, it is difficult to predict just how long the absorption process will require. For example, beer is absorbed more slowly than an equivalent concentration of alcohol in water, apparently because of the carbohydrates present in beer. Also, alcohol consumed on an empty stomach is absorbed faster than an equivalent amount of alcohol taken when there is food in the stomach. The longer the total time required for complete absorption to occur, the lower the peak alcohol concentration in the blood (see Figure 7–2). Depending on a combination of factors, maximum blood-alcohol concentration may not be reached until two or three hours have elapsed from the time of consumption. However, under normal social drinking conditions, it takes anywhere from 30 to 90 minutes from the time of the final drink until the absorption process is completed.

oxidation
The combination of oxygen with other substances to produce new products

excretion
Elimination of alcohol from the body in an unchanged state; alcohol is normally excreted in breath and urine

Elimination As the alcohol is circulated by the bloodstream, the body begins to eliminate it. The toxin, alcohol, is eliminated through two mechanisms—**oxidation** and **excretion**. Nearly all of the alcohol (95–98 percent) consumed is eventually oxidized to carbon dioxide and water. Oxidation takes place almost entirely in the liver. Here, in the presence of the enzyme alcohol dehydrogenase, the alcohol is converted into acetaldehyde and then to acetic acid. The acetic acid is subsequently oxidized in practically all parts of the body to carbon dioxide and water. The remaining alcohol is excreted unchanged in the breath, urine, and perspiration. Most significantly, the amount of alcohol exhaled in the breath is in direct proportion to the concentration of alcohol in the blood. This observation has had a tremendous impact on the technology and procedures used for blood-alcohol testing. The development of instruments to reliably measure breath for its alcohol content has made possible the testing of millions of people in a rapid, safe, and convenient manner.

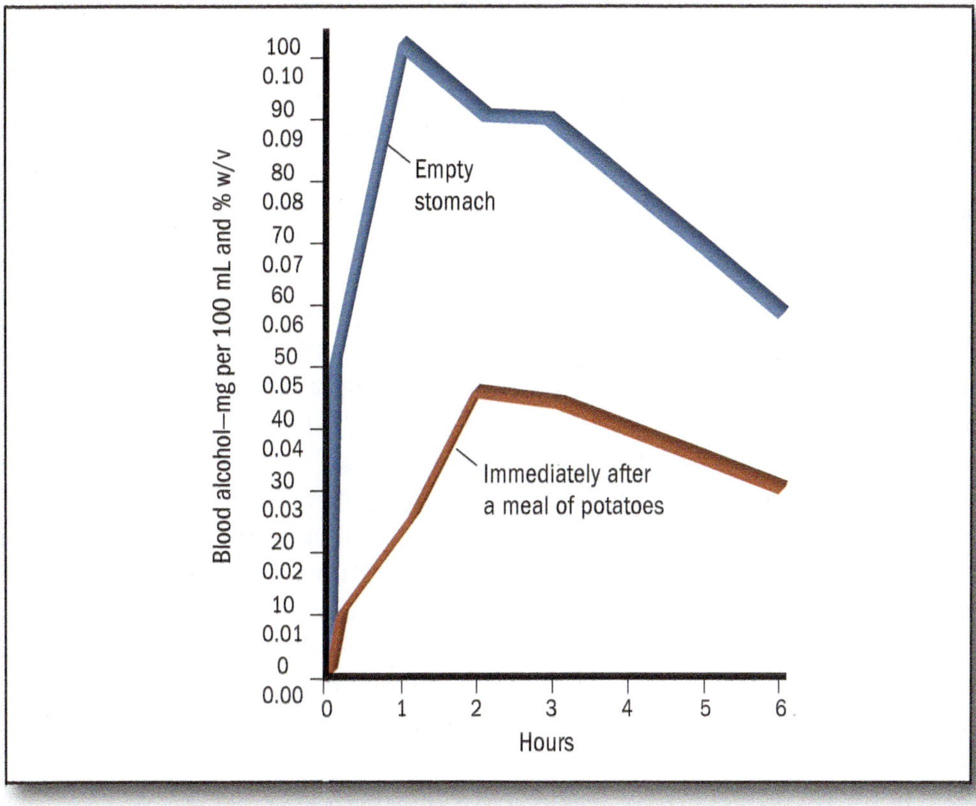

FIGURE 7–2
Blood-alcohol concentrations after ingestion of 2 ounces of pure alcohol mixed in 8 ounces of water (equivalent to about 5 ounces of 80-proof vodka). *Courtesy U.S. Department of Transportation, Washington, D.C.*

The fate of alcohol in the body is therefore relatively simple—namely, absorption into the bloodstream, distribution throughout the body's water, and finally, elimination by oxidation and excretion. The elimination or "burn-off" rate of alcohol varies in different individuals; 0.015 percent w/v (weight per volume) per hour seems to be an average value once the absorption process is complete.[1] However, this figure is an average that varies by as much as 30 percent among individuals.

Blood-Alcohol Concentration Logically, the most obvious measure of intoxication would be the amount of liquor a person has consumed. Unfortunately, most arrests are made after the fact, when such information is not available to legal authorities; furthermore, even if these data could be collected, numerous related factors, such as body weight and the rate of alcohol's absorption into the body, are so variable that it would be impossible to prescribe uniform standards that would yield reliable alcohol intoxication levels for all individuals.

Theoretically, for a true determination of the quantity of alcohol impairing an individual's normal body functions, it would be best to remove a portion of brain tissue and analyze it for alcohol content. For obvious reasons, this cannot be done on living subjects. Consequently, toxicologists concentrate on the blood, which provides the medium for circulating alcohol throughout the body, carrying it to all tissues, including the brain. Fortunately, experimental evidence supports this approach and shows blood-alcohol concentration to be directly proportional to the concentration of alcohol in the brain. From the medicolegal point of view, blood-alcohol levels have become the accepted standard for relating alcohol intake to its effect on the body.

As noted earlier, alcohol becomes concentrated evenly throughout the watery portions of the body. This knowledge can be useful for the toxicologist analyzing a body for the presence of alcohol. If blood is not available, as in some postmortem situations, a medical examiner can select a water-rich organ or fluid—for example, the brain, cerebrospinal fluid, or vitreous humor—to determine the body's alcohol content to a reasonable degree of accuracy.

artery
A blood vessel that carries blood away from the heart

vein
A blood vessel that carries blood toward the heart

capillary
A tiny blood vessel that receives blood from arteries and carries it to veins, and across whose walls exchange of materials between the blood and the tissues takes place

Alcohol in the Circulatory System

The extent to which an individual may be under the influence of alcohol is usually determined by measuring the quantity of alcohol present in the blood system. Normally, this is accomplished in one of two ways: (1) by analyzing the blood for its alcohol content or (2) by measuring the alcohol content of the breath. In either case, the significance and meaning of the results can better be understood when the movement of alcohol through the circulatory system is studied.

Humans, like all vertebrates, have a closed circulatory system, which consists basically of a heart and numerous arteries, capillaries, and veins. An **artery** is a blood vessel carrying blood away from the heart, and a **vein** is a vessel carrying blood back toward the heart. **Capillaries** are tiny blood vessels that connect the arteries with the veins. The exchange of materials between the blood and the other tissues takes place across the thin walls of the capillaries. A schematic diagram of the circulatory system is shown in Figure 7–3.

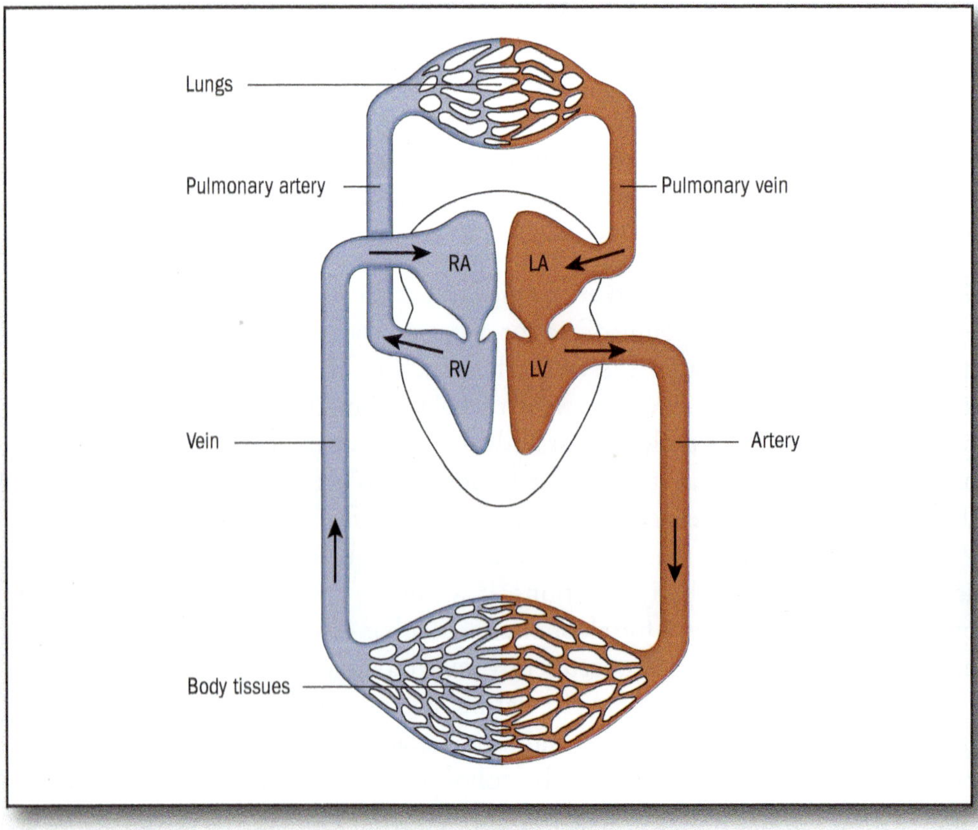

FIGURE 7–3
Simplified diagram of the human circulatory system. Dark vessels contain oxygenated blood; light vessels contain deoxygenated blood.

Ingestion and Distribution Let us now trace the movement of alcohol through the human circulatory system. After alcohol is ingested, it moves down the esophagus into the stomach. About 20 percent of the alcohol is absorbed through the stomach walls into the portal vein of the blood system. The remaining alcohol passes into the blood through the walls of the small intestine. Once in the blood, the alcohol is carried to the liver, where enzymes begin to break it down.

As the blood (still carrying the alcohol) leaves the liver, it moves up to the heart. The blood enters the upper right chamber of the heart, called the right atrium (or auricle), and is forced into the lower right chamber of the heart, known as the right ventricle. Having returned to the heart from its circulation through the tissues, the blood at this time contains very little oxygen and much carbon dioxide. Consequently, the blood must be pumped up to the lungs, through the pulmonary artery, to be replenished with oxygen.

Aeration In the lungs, the respiratory system bridges with the circulatory system so that oxygen can enter the blood and carbon dioxide can leave it. As shown in Figure 7–4, the pulmonary artery branches into capillaries lying close to tiny pear-shaped sacs called **alveoli**. The lungs contain about 250 million alveoli, all located at the ends of the bronchial tubes. The bronchial tubes connect to the windpipe (trachea), which leads up to the mouth and nose (see Figure 7–5). At the surface of the alveolar sacs, blood flowing through the capillaries comes in contact with fresh oxygenated air in the sacs.

alveoli
Small sacs in the lungs through whose walls air and other gases are exchanged between the breath and the blood

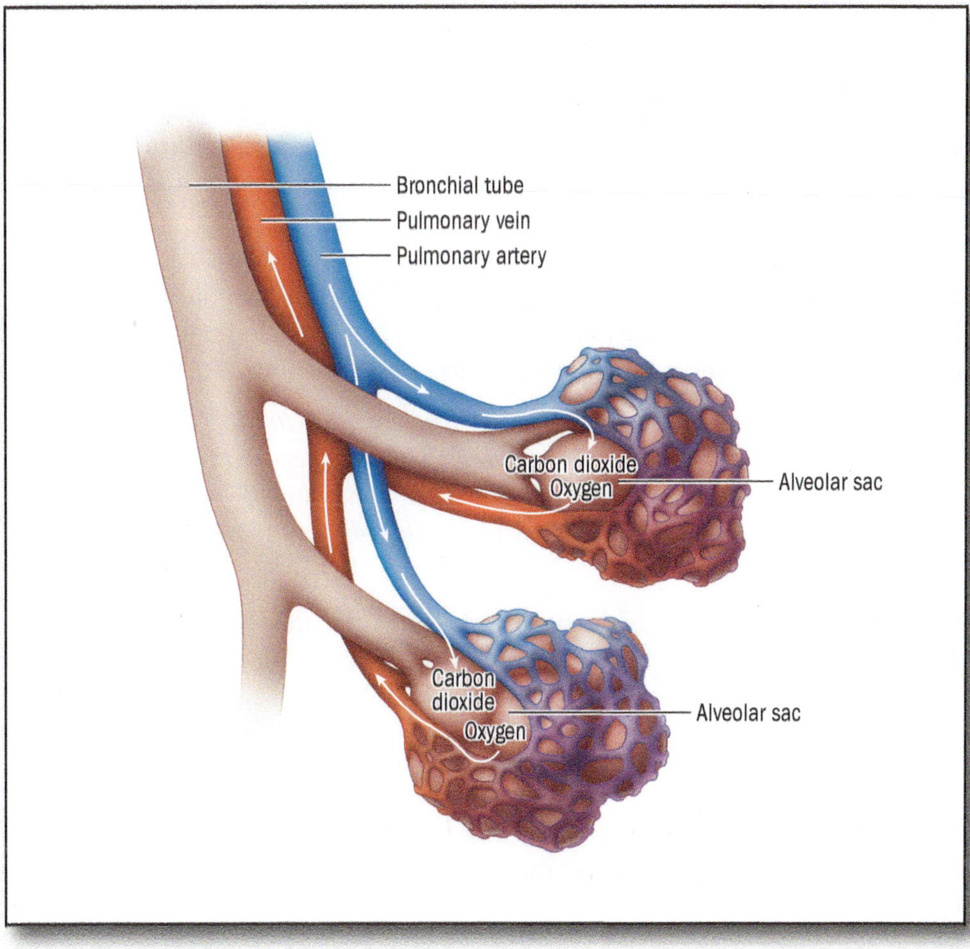

FIGURE 7–4
Gas exchange in the lungs. Blood flows from the pulmonary artery into vessels that lie close to the walls of the alveoli sacs. Here the blood gives up its carbon dioxide and absorbs oxygen. The oxygenated blood leaves the lungs via the pulmonary vein and returns to the heart.

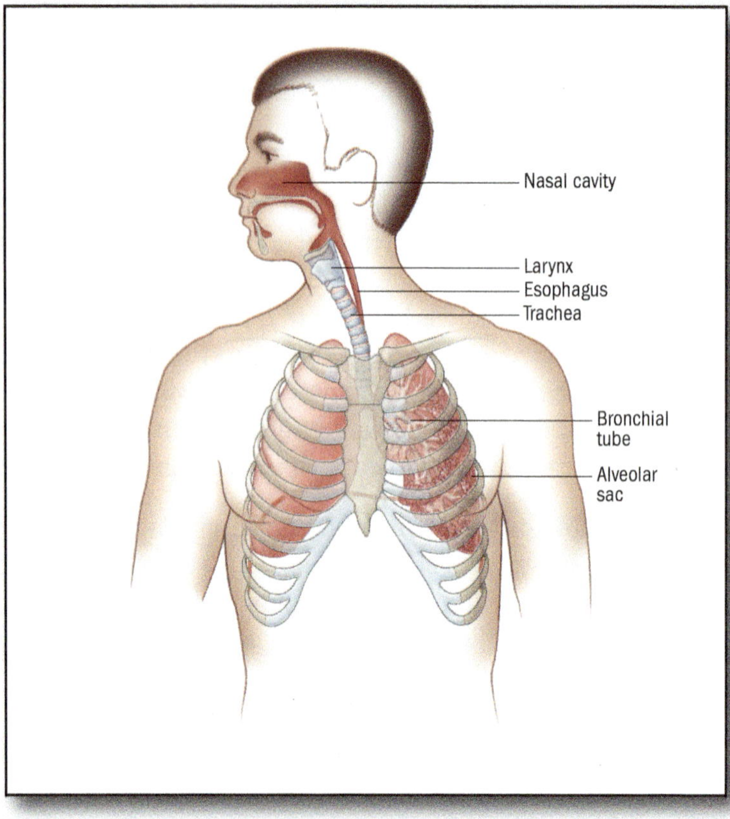

FIGURE 7–5
The respiratory system. The trachea connects the nose and mouth to the bronchial tubes. The bronchial tubes divide into numerous branches that terminate in the alveoli sacs in the lungs.

A rapid exchange now takes place between the fresh air in the sacs and the spent air in the blood. Oxygen passes through the walls of the alveoli into the blood while carbon dioxide is discharged from the blood into the air. If, during this exchange, alcohol or any other volatile substance is in the blood, it too will pass into the alveoli. During breathing, the carbon dioxide and alcohol are expelled through the nose and mouth, and the alveoli sacs are replenished with fresh oxygenated air breathed into the lungs, allowing the process to begin all over again.

The distribution of alcohol between the blood and alveolar air is similar to the example of a gas dissolved in an enclosed beaker of water, as described in Chapter 6. Here again, one can use Henry's law (see page 216) to explain how the alcohol will divide itself between the air and blood. Henry's Law may now be restated as follows:

> *When a volatile chemical (alcohol) is dissolved in a liquid (blood) and is brought to equilibrium with air (alveolar breath), there is a fixed ratio between the concentration of the volatile compound (alcohol) in air (alveolar breath) and its concentration in the liquid (blood), and this ratio is constant for a given temperature.*[2]

The temperature at which the breath leaves the mouth is normally 34°C. At this temperature, the ratio of alcohol in the blood to alcohol in alveolar air is approximately 2,100 to 1. In other words, 1 milliliter of blood contains nearly the same amount of alcohol as 2,100 milliliters of alveolar breath. Henry's law thus becomes a basis for relating breath to blood-alcohol concentration.

Recirculation and Absorption Now let's return to the circulating blood. After emerging from the lungs, the oxygenated blood is rushed back to the upper-left chamber of the heart (left atrium) by the pulmonary vein. When the left atrium contracts, it forces the blood through a valve into the left ventricle, which is the lower left chamber of the heart. The left ventricle then pumps the freshly oxygenated blood into the arteries, which carry the blood to all parts of the body. Each of these arteries, in turn, branches into smaller arteries, which eventually connect with the numerous tiny capillaries embedded in the tissues. Here the alcohol moves out of the blood and into the tissues. The blood then runs from the capillaries into tiny veins that fuse to form larger veins. These veins eventually lead back to the heart to complete the circuit.

During absorption, the concentration of alcohol in the arterial blood is considerably higher than the concentration of alcohol in the venous blood. One typical study revealed a subject's arterial blood-alcohol level to be 41 percent higher than the venous level 30 minutes after the last drink.[3] This difference is thought to exist because of the rapid diffusion of alcohol into the body tissues from venous blood during the early phases of absorption. Because the administration of a blood test requires drawing venous blood from the arm, this test is clearly to the advantage of a subject who may still be in the absorption stage. However, once absorption is complete, the alcohol becomes equally distributed throughout the blood system.

Quick Review

- Alcohol appears in the blood within minutes after it has been taken by mouth. It slowly increases in concentration while it is being absorbed from the stomach and the small intestine into the bloodstream.
- When all of the alcohol has been absorbed, a maximum alcohol level is reached in the blood, and the postabsorption period begins. During postabsorption, the alcohol concentration slowly decreases until a zero level is reached.
- Elimination of alcohol throughout the body is accomplished through oxidation and excretion. Oxidation takes place almost entirely in the liver, while alcohol is excreted unchanged in the breath, urine, and perspiration.
- Breath testing devices operate on the principle that the ratio between the concentration of alcohol in alveolar breath and its concentration in blood is fixed.

Testing for Intoxication

From a practical point of view, the idea of drawing blood from a vein to test motorists suspected of being under the influence of alcohol simply does not provide a convenient method for monitoring drivers. The need to transport the suspect to a location where a medically qualified person can draw blood would be costly and time consuming, considering the hundreds of tests that the average police department must conduct every year. The methods used must be designed to test hundreds of thousands of motorists annually without causing them undue physical harm or unreasonable inconvenience, while providing a reliable diagnosis that can be supported and defended within the framework of the legal system. This means that toxicologists have had to devise rapid and specific procedures for measuring a driver's degree of alcohol intoxication that can be easily administered in the field.

Breath Testing for Alcohol

The most widespread method for rapidly determining alcohol intoxication is breath testing. A breath tester is simply a device for collecting and measuring the alcohol content of alveolar breath. As we saw earlier, alcohol is expelled, unchanged, in the breath of a person who has been drinking. A breath test measures the alcohol concentration in the pulmonary artery by measuring its concentration in alveolar breath. Thus, breath analysis provides an easily obtainable specimen along with a rapid and accurate result.

Breath-test results obtained during the absorption phase may be higher than results obtained from a simultaneous analysis of venous blood. However, the former are more reflective of the concentration of alcohol reaching the brain and therefore more accurately reflect the effects of alcohol on the subject. Again, once absorption is complete, the difference between a blood test and a breath test should be minimal.

Breath-Test Instruments The first widely used instrument for measuring the alcohol content of alveolar breath was the "*Breathalyzer,*" developed in 1954 by R. F. Borkenstein, who was a captain in the Indiana State Police. Starting in the 1970s, the Breathalyzer was phased out and replaced by other instruments. Like the Breathalyzer, they assume that the ratio of alcohol in the blood to alcohol in alveolar breath is 2,100 to 1 at a mouth temperature of 34°C. Unlike the Breathalyzer, modern breath testers are free of chemicals. These devices include infrared light–absorption devices (described in the Inside the Science feature on pages 261–263) and **fuel cell detectors** (see Inside the Science on page 264).

fuel cell detector
A detector in which a chemical reaction involving alcohol produces electricity

Infrared and fuel-cell-based breath testers are microprocessor controlled, so all an operator has to do is to press a start button; the instrument automatically moves through a sequence of steps and produces a readout of the subject's test results. These instruments also perform self-diagnostic tests to ascertain whether they are in proper operating condition.

Considerations in Breath Testing An important feature of breath-testing instruments is that they can be connected to an external alcohol standard or simulator in the form of either a liquid or a gas. The liquid simulator comprises a known concentration of alcohol in water. It is heated to a controlled temperature, and the vapor formed above the liquid is pumped into the instrument. Drygas standards typically consist of a known concentration of alcohol mixed with an inert gas and compressed in cylinders. The external standard is automatically sampled by the breath-testing instrument before and/or after the subject's breath sample is taken and recorded. Thus, the operator can check the accuracy of the instrument against the known alcohol standard.

Inside the Science

Infrared Light Absorption

In principle, infrared instruments operate no differently from the spectrophotometers described in Chapter 6. An evidential testing instrument that incorporates the principle of infrared light absorption is shown in Figure 1. Any alcohol present in the subject's breath flows into the instrument's breath chamber. As shown in Figure 2, a beam of infrared light is aimed through the chamber. A filter is used to select a wavelength of infrared light that alcohol will absorb. As the infrared light passes through the

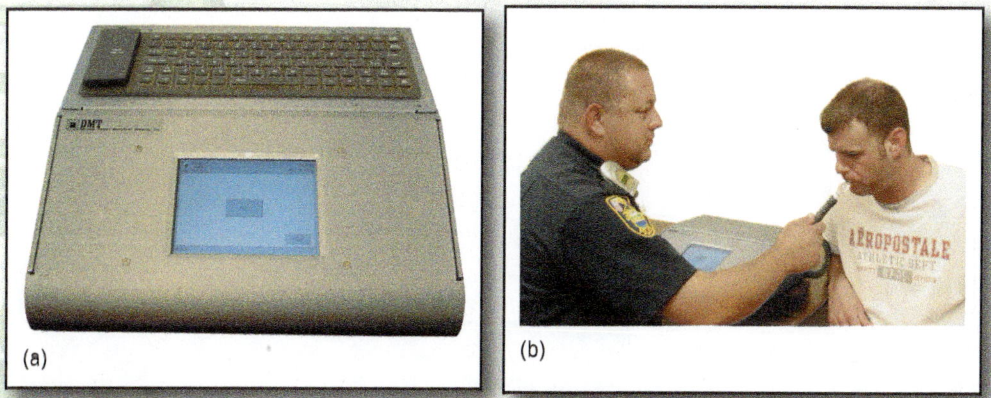

FIGURE 1 (a) An infrared breath testing instrument—the Data Master DMT. (b) A subject blowing into the DMT breath tester. *Intoximeters, Inc.*

chamber, it interacts with the alcohol, which causes the light to decrease in intensity. The decrease in light intensity is measured by a photoelectric detector that gives a signal proportional to the concentration of alcohol present in the breath sample. This information is processed by an electronic microprocessor, and the percent blood-alcohol concentration is displayed on a digital readout. Also, the blood-alcohol level is printed on a card to produce a permanent record of the test result. Most infrared breath testers aim a second infrared beam into the same chamber to check for acetone or other chemical interferences on the breath. If the instrument detects differences in the relative response of the two infrared beams that does not conform to ethyl alcohol, the operator is immediately informed of the presence of an "interferant."

Inside the Science (CONTINUED)

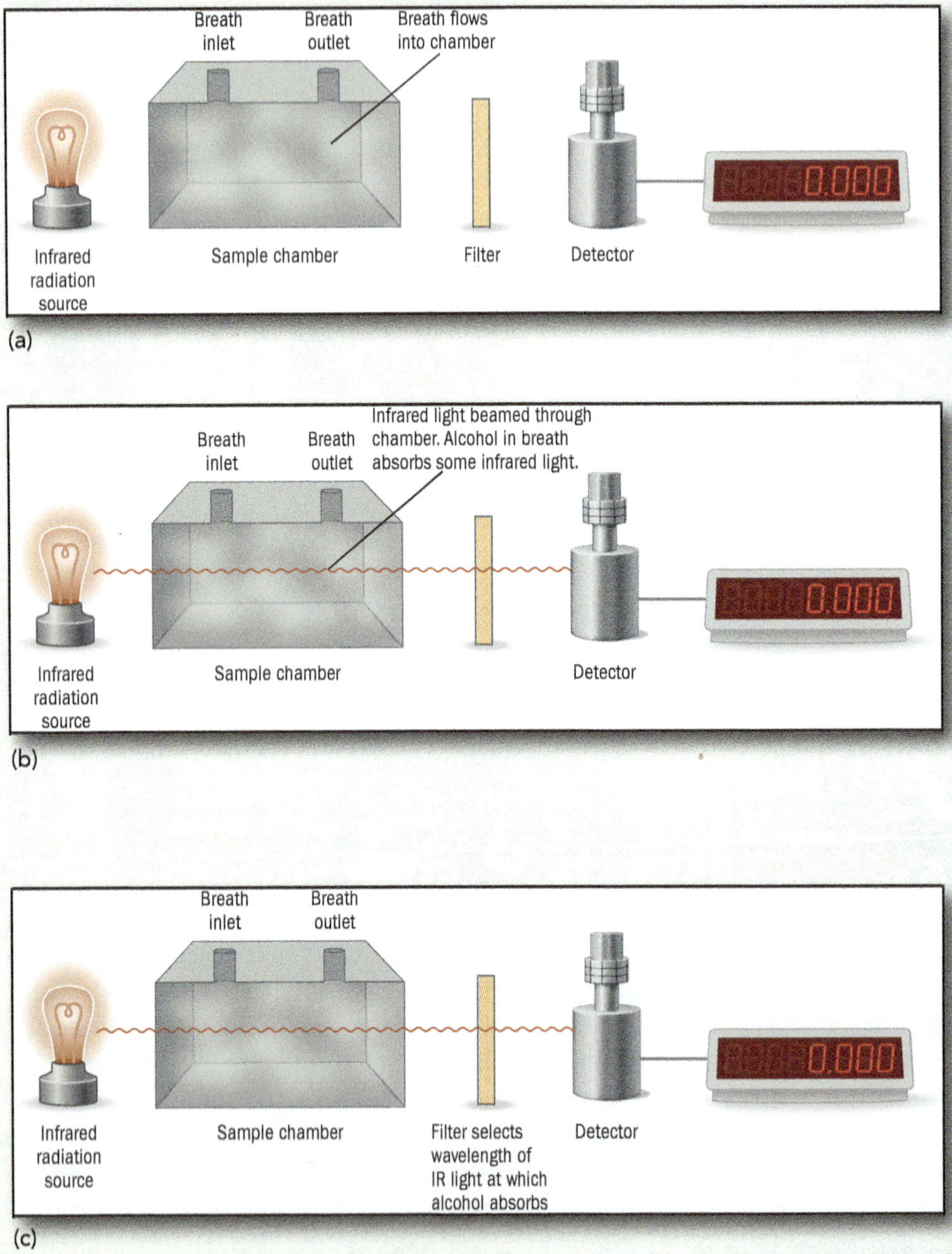

Inside the Science (CONTINUED)

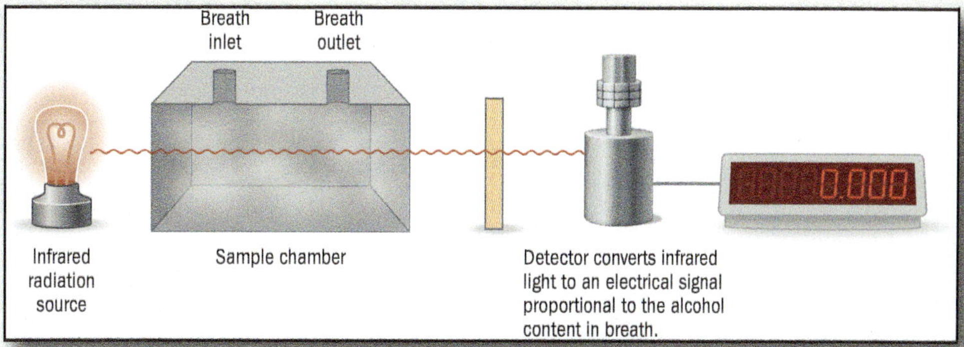

(d)

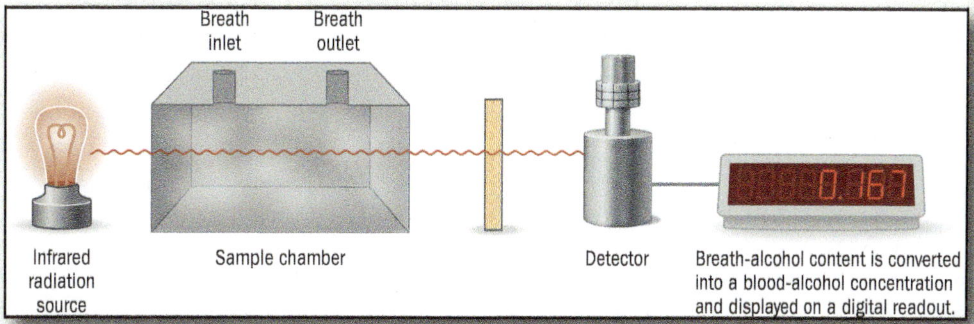

(e)

FIGURE 2

Inside the Science

The Fuel Cell

A fuel cell converts energy arising from a chemical reaction into electrochemical energy. A typical fuel cell consists of two platinum electrodes separated by an acid or base-containing porous membrane. A platinum wire connects the electrodes and allows a current to flow between them. In the alcohol fuel cell, one of the electrodes is positioned to come into contact with a subject's breath sample.

If alcohol is present in the breath, a reaction at the electrode's surface converts the alcohol to acetic acid. One by-product of this conversion is free electrons, which flow through the connecting wire to the opposite electrode, where they interact with atmospheric oxygen to form water (see the figure). The fuel cell also requires the migration of hydrogen ions across the acidic porous membrane to complete the circuit. The strength of the current flow between the two electrodes is proportional to the concentration of alcohol in the breath.

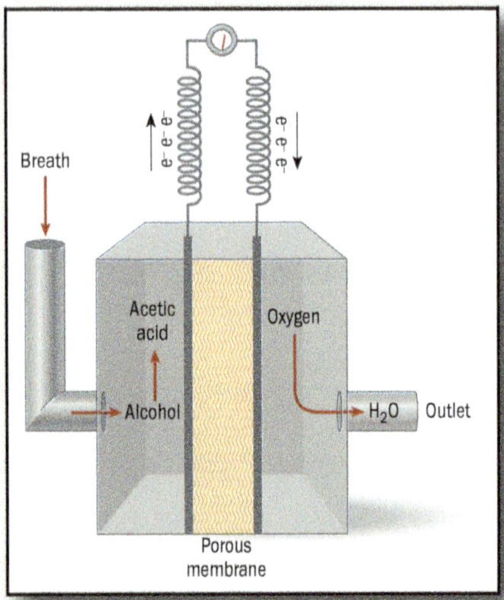

Schematic diagram of a fuel cell using alcohol as a fuel.

The key to the accuracy of a breath-testing device is to ensure that the unit captures the alcohol in the alveolar (deep-lung) breath of the subject. This is typically accomplished by programming the unit to accept no less than 1.1 to 1.5 liters of breath from the subject. Also, the subject must blow for a minimum time (such as 6 seconds) with a minimum breath flow rate (such as 3 liters per minute).

Another feature of these instruments is the *slope detector*. As the subject blows into the instrument, the breath-alcohol concentration initially will rise steadily as a function of time. The instrument accepts a breath sample only when consecutive breath measurements show little or no rate of change in breath-alcohol concentration. This approach ensures that the breath sample being measured is alveolar or deep-lung breath and thus most closely relates to the true blood-alcohol concentration of the subject being tested.

A breath-test operator must take other steps to ensure that the breath-test result truly reflects the actual blood-alcohol concentration of the subject. A major consideration is to avoid measuring "mouth alcohol" resulting from regurgitation, belching, or recent intake of an alcoholic beverage. Also, the recent gargling of an alcohol-containing mouthwash can lead to the presence of mouth alcohol. As a result, the alcohol concentration detected in the exhaled breath is higher than the concentration in the alveolar breath. To avoid this possibility, the operator must not allow the subject to take any foreign material into their mouth for at least 15 to 20 minutes before the breath test. Likewise, the subject should be observed not to have belched or regurgitated during this period. Mouth alcohol has been shown to dissipate after 15 to 20 minutes from its inception.

Independent measurement of duplicate breath samples taken within a few minutes of each other is another extremely important check of the integrity of the breath test. Acceptable agreement between the two tests taken minutes apart significantly reduces the possibility of errors arising from the operator, mouth alcohol, instrument component failures, and spurious electric signals.

Field Sobriety Testing

A police officer who suspects that an individual is under the influence of alcohol usually conducts a series of preliminary tests before ordering the suspect to submit to an evidential breath or blood test. These preliminary, or field sobriety, tests are normally performed to ascertain the degree of the suspect's physical impairment and whether an evidential test is justified.

Field sobriety tests usually consist of a series of psychophysical tests and a preliminary breath test (if such devices are authorized and available for use). A portable handheld, roadside breath tester is shown in Figure 7–6. This device, about the size of a pack of cigarettes, weighs 5 ounces and uses a fuel cell to measure the alcohol content of a breath sample. The fuel cell absorbs the alcohol from the breath sample, oxidizes it, and produces an electrical current proportional to the breath-alcohol content. This instrument can typically perform for three to five years before the fuel cell needs to be replaced. It's been approved for use as an evidential breath tester by the National Highway Traffic Safety Administration.

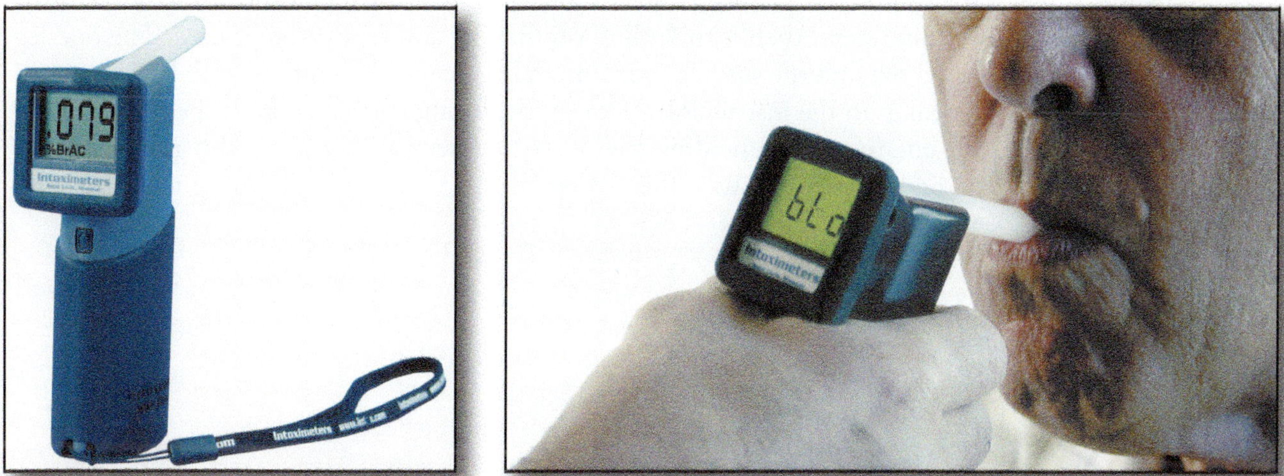

FIGURE 7–6 (a) The Alco-Sensor FST. (b) A subject blowing into the roadside tester device. *Courtesy Intoximeters, Inc.*

Horizontal-gaze nystagmus, walk and turn, and the one-leg stand constitute a series of reliable and effective psychophysical tests. Horizontal-gaze nystagmus is an involuntary jerking of the eye as it moves to the side. A person experiencing nystagmus is usually unaware that the jerking is happening and is unable to stop or control it. The subject being tested is asked to follow a penlight or some other object with their eye as far to the side as the eye can go. The more intoxicated the person is, the less the eye has to move toward the side before jerking or nystagmus begins. Usually, when a person's blood-alcohol concentration is in the range of 0.10 percent, the jerking begins before the eyeball has moved 45 degrees to the side (see Figure 7–7). Higher blood-alcohol concentration causes jerking at smaller angles. Also, if the suspect has taken a drug that also causes nystagmus (such as phencyclidine, barbiturates, and other depressants), the nystagmus onset angle may occur much earlier than would be expected from alcohol alone.

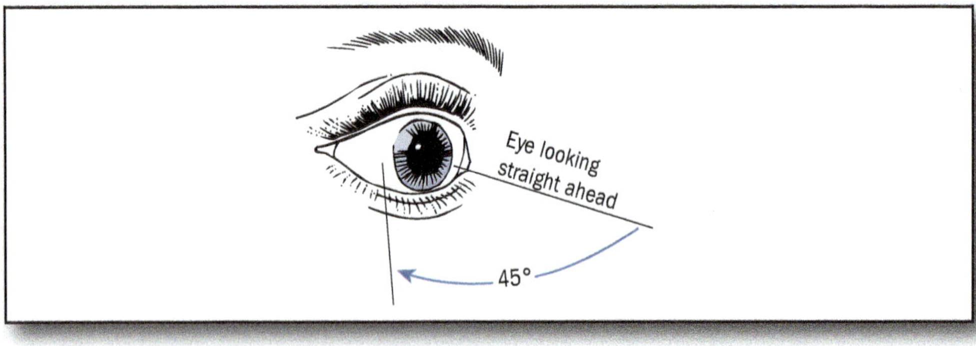

FIGURE 7–7 When a person's blood-alcohol level is in the range of 0.10 percent, jerking of the eye during the horizontal-gaze nystagmus test begins before the eyeball has moved 45 degrees to the side.

Walk and turn and the one-leg stand are divided-attention tasks, testing the subject's ability to comprehend and execute two or more simple instructions at one time. The ability to understand and simultaneously carry out more than two instructions is significantly affected by increasing blood-alcohol levels. Walk and turn requires the suspect to maintain balance while standing heel-to-toe and at the same time listening to and comprehending the test instructions. During the walking stage, the suspect must walk a straight line, touching heel-to-toe for nine steps, then turn around on the line and repeat the process. The one-leg stand requires the suspect to maintain balance while standing with heels together listening to the instructions. During the balancing stage, the suspect must stand on one foot while holding the other foot several inches off the ground for 30 seconds; simultaneously, the suspect must count out loud during the 30-second time period.

Quick Review

- Modern breath testers are free of chemicals. They include infrared light absorption devices and fuel cell detectors.
- The key to the accuracy of a breath testing device is to ensure that the unit captures the alcohol in the alveolar (deep-lung) breath of the subject.
- Many breath testers collect a set volume of breath and expose it to infrared light. The instrument measures alcohol concentration in breath by measuring the degree of the interaction of the light with alcohol in the collected breath sample.
- Law enforcement officers use field sobriety tests to estimate a motorist's degree of physical impairment by alcohol and to determine whether an evidential test for alcohol is justified.
- The horizontal-gaze nystagmus test, walk and turn, and the one-leg stand are all considered reliable and effective psychophysical tests for alcohol impairment.

The Analysis of Blood for Alcohol

Gas chromatography offers the toxicologist the most widely used approach for determining alcohol levels in blood. Under proper gas chromatographic conditions, alcohol can be separated from other volatiles in the blood. By comparing the resultant alcohol peak area to ones obtained with known blood-alcohol standards, the investigator can calculate the alcohol level with a high degree of accuracy (see Figure 7–8).

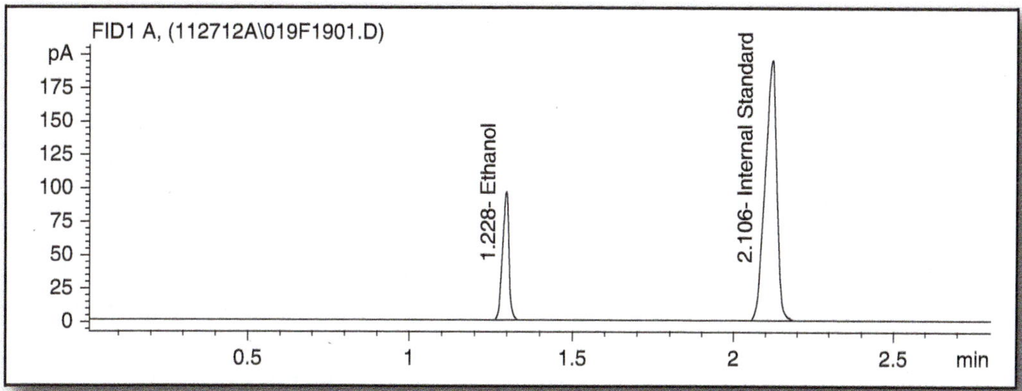

FIGURE 7–8 Gas chromatogram showing ethyl alcohol (ethanol) in whole blood.

Another procedure for alcohol analysis involves the oxidation of alcohol to acetaldehyde. This reaction is carried out in the presence of the enzyme alcohol dehydrogenase and the coenzyme nicotin-amide-adenine dinucleotide (NAD). As the oxidation proceeds, NAD is converted into another chemical species, NADH. The extent of this conversion is measured by a spectrophotometer and is related to alcohol concentration. This approach to blood-alcohol testing is normally associated with instruments used in a clinical or hospital setting. On the other hand, forensic laboratories normally use gas chromatography for determining blood-alcohol content.

Collection and Preservation of Blood

Blood must always be drawn under medically accepted conditions by a qualified individual. It is important to apply a nonalcoholic disinfectant before the suspect's skin is penetrated with a sterile needle or lancet. It is important to negate any argument that an alcoholic disinfectant may have inadvertently contributed to a falsely high blood-alcohol result. Nonalcoholic disinfectants such as aqueous benzalkonium chloride (Zepiran), aqueous mercuric chloride, or povidoneiodine (Betadine) are recommended for this purpose.

Once blood is removed from an individual, it is best preserved sealed in an airtight container after adding an anticoagulant and a preservative. The blood should be stored in a refrigerator until delivery to the toxicology laboratory. The addition of an **anticoagulant**, such as EDTA or potassium oxalate, prevents clotting; a **preservative**, such as sodium fluoride, inhibits the growth of microorganisms capable of destroying alcohol.

anticoagulant
A substance that prevents coagulation or clotting of blood

preservative
A substance that stops the growth of microorganisms in blood

One study performed to determine the stability of alcohol in blood removed from living individuals found that the most significant factors affecting alcohol's stability in blood are storage temperature, the presence of a preservative, and the time of storage.[4] Not a single blood specimen examined showed an increase in alcohol level with time. Failure to keep the blood refrigerated or to add sodium fluoride resulted in a substantial decline in alcohol concentration. Longer storage times also reduced blood-alcohol levels. Hence, failure to adhere to any of the proper preservation requirements for blood works to the benefit of the suspect and to the detriment of society.

The collection of postmortem blood samples for alcohol determination requires added precautions as compared to collection from living subjects. Ethyl alcohol may be generated in a deceased individual as a result of bacterial action. Therefore, it is best to collect a number of blood samples from different body sites. For example, blood may be removed from the heart and from the femoral (leg) and cubital (arm) veins. Each sample should be placed in a clean, airtight container containing an anticoagulant and sodium fluoride preservative and should be refrigerated. Blood-alcohol levels attributed solely to alcohol consumption should result in nearly similar results for all blood samples collected from the same person. Alternatively, collection of vitreous humor and urine is recommended. Vitreous humor and urine usually do not suffer from postmortem ethyl alcohol production to any significant extent.

Quick Review

- Gas chromatography is the most widely used approach for determining blood-alcohol levels in forensic laboratories.
- An anticoagulant should be added to a blood sample to prevent clotting; a preservative should be added to inhibit the growth of microorganisms capable of destroying alcohol.

Alcohol and the Law

Constitutionally, every state in the United States must establish and administer statutes regulating the operation of motor vehicles. Although such an arrangement might encourage diverse laws defining permissible blood-alcohol levels, this has not been the case. Since the 1930s, both the American Medical Association and the National Safety Council have exerted considerable influence in persuading the states to establish uniform and reasonable blood-alcohol standards.

Blood-Alcohol Laws

The American Medical Association and the National Safety Council initially recommended that a person with a blood-alcohol concentration in excess of 0.15 percent w/v was to be considered under the influence of alcohol.[5] However, continued experimental studies showed a clear correlation between drinking and driving impairment at blood-alcohol levels much below 0.15 percent w/v. These findings eventually led to a lowering of the blood-concentration standard for intoxication from 0.15 percent w/v to its current 0.08 percent w/v.

In 1992, the U.S. Department of Transportation (DOT) recommended that states adopt 0.08 percent blood-alcohol concentration as the legal measure of drunk driving. This recommendation was enacted into federal law in 2000. All 50 states have now established *per se* laws, meaning that any individual meeting or exceeding a defined blood-alcohol level (usually 0.08 percent) shall be deemed intoxicated. No other proof of alcohol impairment is necessary. The 0.08 percent level applies only to noncommercial drivers, as the federal government has set the maximum allowable blood-alcohol concentration for commercial truck and bus drivers at 0.04 percent.

Several Western countries have also set 0.08 percent w/v as the blood-alcohol level above which it is an offense to drive a motor vehicle, including Canada, Italy, Switzerland, and the United Kingdom. Finland, France, Germany, Ireland, Japan, the Netherlands, and Norway have a 0.05 percent limit. Australian states have adopted a 0.05 percent blood-alcohol concentration level. Sweden has lowered its blood-alcohol concentration limit to 0.02 percent.

As shown in Figure 7–9, a person is about four times as likely to become involved in an automobile accident at the 0.08 percent level as a sober individual. At the 0.15 percent level, the chances are 25 times as much for involvement in an automobile accident as compared to a sober driver. The reader can estimate the relationship of blood-alcohol levels to body weight and the quantity of 80-proof liquor consumed by referring to Figure 7–10.

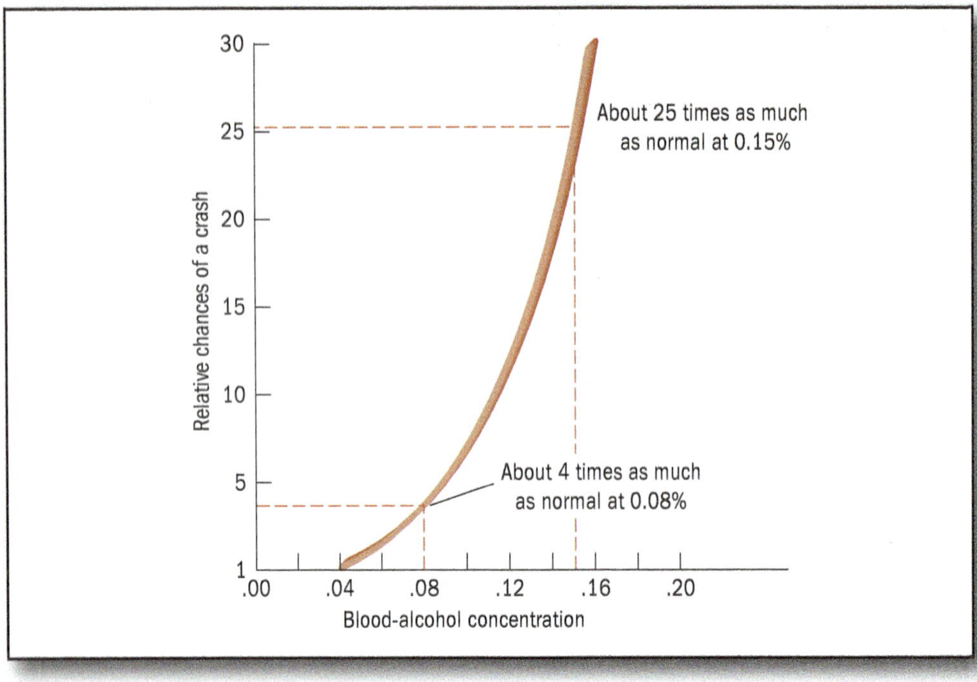

FIGURE 7–9
Diagram of increased driving risk in relation to blood-alcohol concentration. *Courtesy U.S. Department of Transportation, Washington, D.C.*

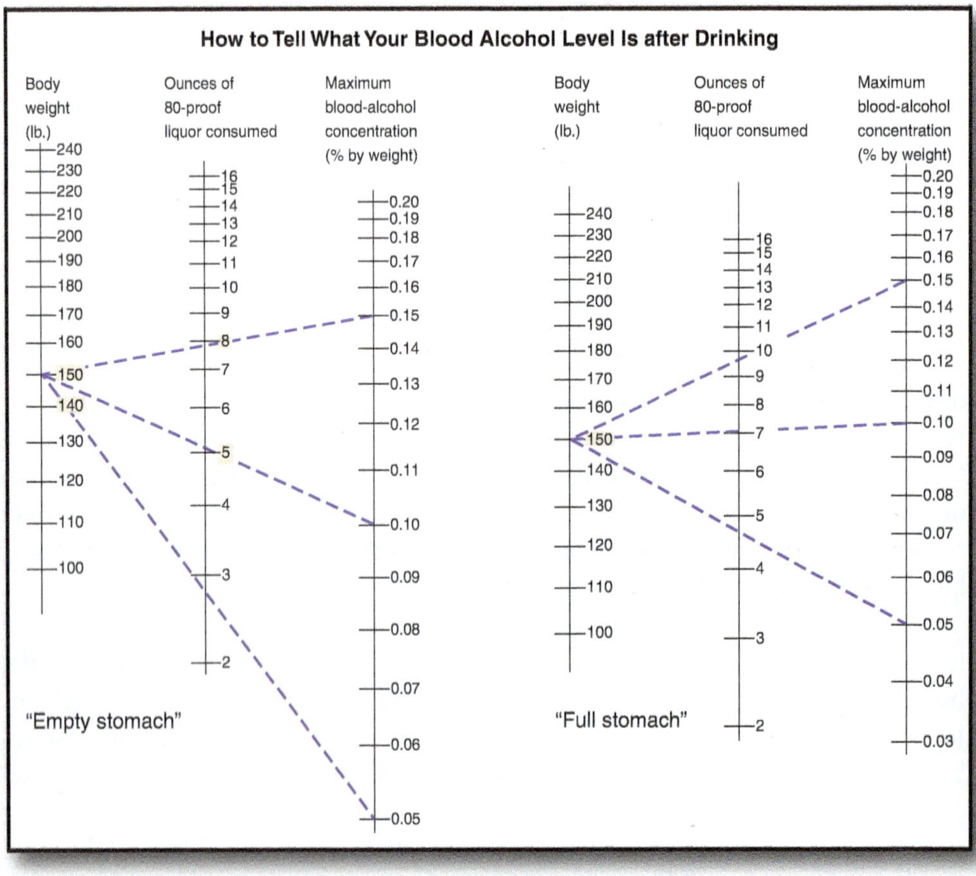

FIGURE 7–10
To use this diagram, lay a straightedge across the suspect's weight and the number of ounces they have consumed on an empty or full stomach. The point where the edge hits the right-hand column is one's maximum blood-alcohol level. The rate of elimination of alcohol from the bloodstream is approximately 0.015 percent per hour. Therefore, to calculate one's actual blood-alcohol level, subtract 0.015 from the number in the right-hand column for each hour from the start of drinking.

Constitutional Issues

The Fifth Amendment to the U.S. Constitution guarantees all citizens protection against *self-incrimination*—that is, against being forced to make an admission that would prove one's own guilt in a legal matter. Because consenting to a breath test for alcohol may be considered a form of self-incrimination, the National Highway Traffic Safety Administration (NHTSA) recommended an *implied consent* law to prevent a person from refusing to take a test on those constitutional grounds. This law states that the operator of a motor vehicle on a public highway must either consent to a test for alcohol intoxication, if requested, or lose their license for some designated period—usually six months to one year.

The leading case relating to the constitutionality of collecting a blood specimen for alcohol testing, as well as for obtaining other types of physical evidence from a suspect without consent, is *Schmerber v. California*.[6] While being treated at a Los Angeles hospital for injuries sustained in an automobile collision, Schmerber was arrested for driving under the influence of alcohol. Despite Schmerber's objections, a physician took a blood sample from him at the direction of the police. Schmerber was convicted of driving while intoxicated, and he subsequently appealed the decision. The case eventually reached the U.S. Supreme Court, where Schmerber argued that his privilege against self-incrimination had been violated by the introduction of the results of the blood test at his trial. The Court ruled against him, reasoning that the Fifth Amendment prohibits only compelling a suspect to give *testimonial* evidence that may prove to be self-incriminating; being compelled to furnish *physical evidence*, such as fingerprints, photographs, measurements, and blood samples, the Court ruled, was not protected by the Fifth Amendment.

The Court also addressed the question of whether Schmerber was subjected to an unreasonable search and seizure by the taking of a blood specimen without a search warrant. In the 1966 decision, the Court upheld the blood removal, reasoning that the natural body elimination of alcohol created an emergency situation allowing for a warrantless search. The Court revisited this issue once again forty-seven years after Schmerber in the case of *Missouri v. McNeely*.[7] Here, the Court addressed the issue as to whether the natural elimination of alcohol in blood categorically justifies a warrantless intrusion. The Court noted that advances in communication technology now allow police to obtain a warrant quickly by phone, e-mail, or teleconferencing:

> *In those drunk-driving investigations where police officers can reasonably obtain a warrant before a blood sample can be drawn without significantly undermining the efficacy of the search, the Fourth Amendment mandates that they do so... . In short, while the natural dissipation of alcohol in the blood may support a finding of exigency in a specific case, as it did in Schmerber, it does not do so categorically. Whether a warrantless blood test of a drunkdriving suspect is reasonable must be determined case by case based on the totality of the circumstances.*

Quick Review

- The current legal measure of drunk driving in the United States is a blood-alcohol concentration of 0.08 percent, or 0.08 grams of alcohol per 100 milliliters of blood.
- An implied-consent law states that the operator of a motor vehicle on a public highway must either consent to a test for alcohol intoxication, if requested, or lose their license for some designated period—usually six months to one year.

The Role of the Toxicologist

Once the forensic toxicologist ventures beyond the analysis of alcohol, the toxicologist encounters an encyclopedic maze of drugs and poisons. Even a cursory discussion of the problems and handicaps imposed on toxicologists is enough to develop a sense of appreciation for their accomplishments and ingenuity.

Forensic toxicologists also use toxicological analysis and controlled substance analysis to determine if these substances played a role in the death of the individual. Both toxicological analysis and controlled substance analysis involve testing bodily fluids. The samples are taken from the body postmortem in a lab. These samples include blood, hair, urine, and bile. The method of collection of these bodily samples includes placing them into individually labeled and sealed containers. The chain of custody must be maintained. Forensic toxicologists might use different types of spectrophotometry, chromatography, immunological testing, and other types of tests to analyze the drugs present in the body postmortem.

Challenges Facing the Toxicologist

The toxicologist is presented with body fluids and/or organs and asked to perform a controlled substance analysis or toxicological analysis to examine them for drugs and poisons. The toxicological analysis and controlled substance analysis will provide insight into the impact on the body the toxins have. For example, a toxicological analysis or controlled substance analysis will show the level of drugs/toxins found in the body. Scientists can then analyze these levels to determine their impact on the body (if the dosage was therapeutic or harmful, e.g.: overdose). A controlled substance analysis is a laboratory test done to specifically test a controlled substance and determine what substance it is. Controlled substances are opioids, anabolic steroids, stimulants, depressants, and hallucinogens. A toxicological analysis is an overall analysis to detect for toxins. Once the analysis is complete, the next step is to determine the impact on the body. If the toxicologist is fortunate, which is not often, some clue as to the type of toxic substance present may develop from the victim's symptoms, a postmortem pathological examination, an examination of the victim's personal effects, or the nearby presence of empty drug containers or household chemicals. Without such supportive information, the toxicologist must use general screening procedures with the hope of narrowing thousands of possibilities to one.

If this task does not seem monumental, consider that the toxicologist is not dealing with drugs at the concentration levels found in powders and pills. By the time a drug specimen reaches the toxicology laboratory, it has been dissipated

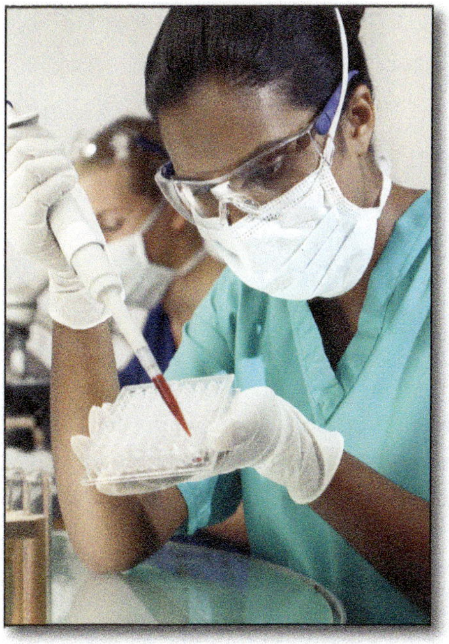

FIGURE 7–11
Scientist analyzing blood samples. *Darren Baker/ Shutterstock*

and distributed throughout the body. The drug analyst may have gram or milligram quantities of material to work with, but the toxicologist must be satisfied with nanogram or at best microgram amounts, acquired only after careful extraction from body fluids and organs.

Furthermore, the body is an active chemistry laboratory, and no one can appreciate this observation more than a toxicologist. Few substances enter and completely leave the body in the same chemical state. The drug that is injected is not always the substance extracted from the body tissues. Therefore, a thorough understanding of how the body alters or metabolizes the chemical structure of a drug is essential in detecting its presence.

It would, for example, be futile and frustrating to search exhaustively for heroin in the human body. This drug is almost immediately metabolized to morphine on entering the bloodstream. Even with this information, the search may still prove impossible unless the examiner also knows that only a small percentage of morphine is excreted unchanged in urine. For the most part, morphine becomes chemically bonded to body carbohydrates before elimination in urine. Thus, successful detection of morphine requires that its extraction be planned in accordance with a knowledge of its chemical fate in the body.

Another example of how one needs to know how a drug metabolizes itself in the body is exemplified by the investigation of the death of Anna Nicole Smith. In her case, the sedative chloral hydrate was a major contributor to her death and its presence was detected by its active metabolite, trichloroethanol (see the Case File on page 275).

Last, when and if the toxicologist has surmounted all of these obstacles and has finally detected, identified, and quantitated a drug or poison, the toxicologist must assess the substance's toxicity. Fortunately, there is published information relating to the toxic levels of most drugs. However, even when such data are available, their interpretation must assume that the victim's physiological behavior agrees with that of subjects of previous studies. Such an assumption may not be entirely valid without knowing the subject's case history. No experienced toxicologist would be surprised to find an individual tolerating a toxic level of a drug that would have killed most other people.

Case Files

Celebrity Toxicology Michael Jackson: The Demise of a Superstar

A call to 911 had the desperate tone of urgency. The voice of a young man implored an ambulance to hurry to the home of pop star Michael Jackson. The unconscious performer was in cardiac arrest and was not responding to CPR. The 50-year-old Jackson was pronounced dead upon arrival at a regional medical center. When the initial autopsy results revealed no signs of foul play, rumors immediately began to swirl around a drug-related death. News media coverage showed investigators carrying bags full of medical supplies out of the Jackson residence. Hence, it came as no surprise that the forensic toxicology report accompanying Jackson's autopsy showed that the entertainer had died of a drug overdose.

Apparently, Jackson had become accustomed to receiving sedatives to help him sleep. On the morning of his death, his physician gave Mr. Jackson a tab of valium. At 2 AM, he administered the sedative lorazepam, and at 3 AM the physician administered another sedative, midazolam. Those drugs were administered again at 5 AM and 7:30 AM, but Mr. Jackson still was unable to sleep. Finally, at about 10:40 AM, Jackson's doctor gave him 25 milligrams of propofol, at which point Mr. Jackson went to sleep. Propofol's is a powerful sedative whose principal use is the maintenance of surgical anesthesia. All of the drugs administered to Jackson were sedatives that act in concert to depress the activities of the central nervous system. Hence, it comes as no surprise that this drug cocktail resulted in cardiac arrest and death.

Justin Sullivan/Pool/AP Images

Case Files

Accidental Overdose: The Tragedy of Anna Nicole Smith

Rumors exploded in the media when former model, Playboy playmate, reality television star, and favorite tabloid subject Anna Nicole Smith was found unconscious in her hotel room at the Seminole Hard Rock Hotel & Casino in Hollywood, Florida. She was taken to Memorial Legal Hospital, where she was declared dead at age 39. Analysis of Smith's blood postmortem revealed an array of prescribed medications. Most pronounced was a toxic level of a metabolite of the sedative chloral hydrate. A part of the contents of the toxicology report from Smith's autopsy are shown here.

FINAL PATHOLOGICAL DIAGNOSES:

I. ACUTE COMBINED DRUG INTOXICATION

A. Toxic/legal drug:

Chloral Hydrate (Noctec)

1.	Trichloroethanol (TCE)	75 mg/L (active metabolite)
2.	Trichloroacetic acid (TCA) (inactive metabolite)	85 mg/L

B. Therapeutic drugs:

1.	Diphenhydramine (Benadryl)	0.11 mg/L
2.	Clonazepam (Klonopin)	0.04 mg/L
3.	Diazepam (Valium)	0.21 mg/L
4.	Nordiazepam (metabolite)	0.38 mg/L
5.	Temazepam (metabolite)	0.09 mg/L
6.	Oxazepam	0.09 mg/L
7.	Lorazepam	0.022 mg/L

C. Other non-contributory drugs present (atropine, topiramate, ciprofloxacin, acetaminophen)

A. Although many of the drugs present were detected at levels consistent with typical doses of the prescribed medications, it was their presence in combination with chloral hydrate that exacerbated the toxic level of chloral hydrate. The lethal combination of these prescription drugs caused failure of both her circulatory and respiratory systems and resulted in her death. The investigators determined that the overdose of chloral hydrate and other drugs was accidental and not a suicide. This was due to the nonexcessive levels of most of the prescription medications and the discovery of a significant amount of chloral hydrate still remaining in its original container; had she intended to kill herself she would have likely downed it all. Anna Nicole Smith was a victim of accidental overmedication.

Manuel Balce Ceneta/PA Images/Alamy Stock Photo

Case Files

Death by Tylenol

In 1982, two firefighters from a Chicago suburb were casually discussing four bizarre deaths that had recently taken place in a neighboring area. As they discussed the circumstances of the deaths, they realized that each of the victims had taken Tylenol. Their suspicions were immediately reported to police investigators. Tragically, before the general public could be alerted, three more victims died after taking poison-laced Tylenol capsules. Seven individuals, all in the Chicago area, were the first victims to die from what has become known as product tampering.

A forensic chemical analysis of Tylenol capsules recovered from the victims' residences showed that the capsules were filled with potassium cyanide in a quantity ten thousand times what was needed to kill an average person. It was quickly determined that the cyanide was not introduced into the bottles at the factory. Instead, the perpetrator methodically emptied each of 20 to 30 capsules and then refilled them with potassium cyanide. The tampered capsules were rebottled, carefully repackaged, and placed on the shelves of six different stores. The case of the Tylenol murders remains unsolved, and the $100,000 reward offered by Tylenol's manufacturer remains unclaimed.

Collection and Preservation of Toxicological Evidence

The toxicologist's capabilities depend directly on input from the attending physician, medical examiner, and police investigator. It is a tribute to forensic toxicologists, who must often labor under conditions that do not afford such cooperation, that they can achieve the high level of proficiency that they do.

Generally, with a deceased person, the medical examiner decides what biological specimens must be shipped to the toxicology laboratory for analysis. However, a living person suspected of being under the influence of a drug presents a completely different problem, and few options are available. When possible, both blood and urine are taken from any suspected drug user. The entire urine void (sample) is collected and submitted for toxicological analysis. Preferably, two consecutive voids should be collected in separate sterile specimen containers. Ideally, the urine sample should be analyzed within 1 hour. If this is not possible, the sample container should be sealed and placed in a sealed plastic bag and stored in a refrigerator at a temperature of 4°C. Urine samples should not be kept for more than 24 hours.

When a licensed physician or registered nurse is available, a sample of blood should also be collected. The amount of blood taken depends on the type of examination to be conducted. Comprehensive toxicological tests for drugs and poisons can conveniently be carried out on a minimum of 10 milliliters of blood. A determination solely for the presence of alcohol will require much less—approximately

5 milliliters of blood. However, many therapeutic drugs, such as tranquilizers and barbiturates, taken in combination with a small, nonintoxicating amount of alcohol, produce behavioral patterns resembling alcohol intoxication. For this reason, the toxicologist must be given enough blood to perform a comprehensive analysis for drugs in cases of low alcohol concentrations.

Techniques Used in Toxicology

For the toxicologist, the upsurge in drug use and abuse has meant that the overwhelming majority of fatal and nonfatal toxic agents are drugs. Not surprisingly, a relatively small number of drugs—namely, those discussed in Chapter 6—comprise nearly all the toxic agents encountered. Of these, alcohol, marijuana, and cocaine account for 90 percent or more of the drugs encountered in a typical toxicology laboratory.

Like the drug analyst, the toxicologist must devise an analytical scheme to detect, isolate, and identify a toxic substance. The first chore is to remove and isolate drugs and other toxic agents from the biological materials submitted as evidence. Because drugs constitute a large portion of the toxic materials found, a good deal of effort must be devoted to their extraction and detection. Many different procedures are used, and a useful description of them would be too detailed for this text. We can best understand the underlying principle of drug extraction by observing that many drugs fall into the categories of acids and bases.

Acids and Bases Although several definitions exist for these two classes, a simple one states that an **acid** is a compound capable of donating a hydrogen ion (or a hydrogen atom minus its electron) to another compound with reasonable ease. Conversely, a **base** is a molecule capable of accepting a hydrogen ion shed by an acid. The idea of acidity and basicity can be expressed in terms of a simple numerical value that relates to the concentration of the hydrogen ion (H^+) in a liquid medium such as water. Chemists use the **pH scale** to do this. This scale runs from 0 to 14:

pH =
0 1 2 3 4 5 6 7 8 9 10 11 12 13 14
Increasing acidity Neutral Increasing basicity

Normally, water is neither acid nor basic—in other words, it is neutral, with a pH of 7. However, when an acidic substance—for example, sulfuric acid or hydrochloric acid—is added to the water, it adds excess hydrogen ions, and the pH value becomes less than 7. The lower the number, the more acidic the water. Similarly, when a basic substance—for example, sodium hydroxide or ammonium hydroxide—is added to water, it removes hydrogen ions, thus making water basic. The more basic the water, the higher its pH value.

By controlling the pH of a water solution into which blood, urine, or tissues are dissolved, the toxicologist can control the type of drug that is recovered. For example, acid drugs are easily extracted from an acidified water solution (pH less than 7) with organic solvents such as chloroform. Similarly, basic drugs are readily removed from a basic water solution (pH greater than 7) with organic solvents. This simple approach gives the toxicologist a general technique for extracting and categorizing drugs. Some of the more commonly encountered drugs may be classified as follows:

Acid Drugs: Barbiturates, Acetylsalicylic acid (aspirin)
Basic Drugs: Phencyclidine, Methadone, Amphetamines, Cocaine

acid
A compound capable of donating a hydrogen ion (H^+) to another compound

base
A compound capable of accepting a hydrogen ion (H^+)

pH scale
A scale used to express the basicity or acidity of a substance. A pH of 7 is neutral; lower values are acidic and higher values basic

Screening and Confirmation Once the specimen has been extracted and divided into acidic and basic fractions, the toxicologist can identify the drugs present. The strategy for identifying abused drugs entails a two-step approach: screening and confirmation (see Figure 7–12). This approach results in presumptive and confirmatory laboratory procedures as they relate to toxicological analysis. A presumptive procedure, called a screening test, normally gives quick insight into the likelihood that a specimen contains an unknown substance like a drug. This test allows a toxicologist to examine a large number of specimens within a short period of time for a wide range of drugs. Any positive results from a screening test are tentative at best and must be verified with a confirmation test. A confirmatory procedure, such as the ones described below, help forensic scientists identify unknown substances that resulted in a positive screening test.

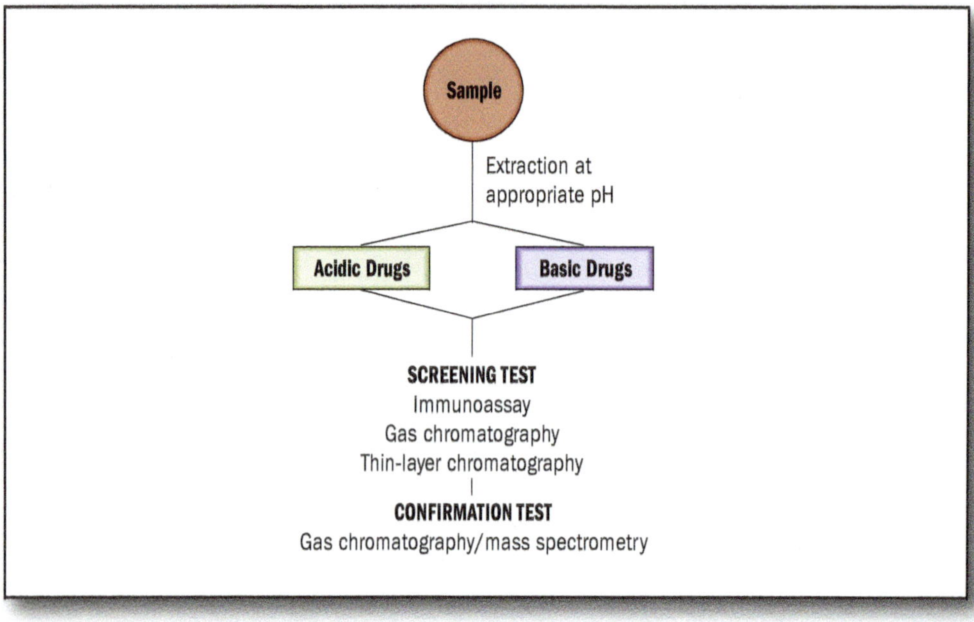

FIGURE 7–12
Biological fluids and tissues are extracted for acidic and basic drugs by controlling the pH of a water solution in which they are dissolved. Once this is accomplished, the toxicologist analyzes for drugs by using screening and confirmation test procedures.

Screening Tests. The three most widely used screening tests are thin-layer chromatography (TLC), gas chromatography (GC), and immunoassay. The techniques of GC and TLC are described in Chapter 6. The third technique, immunoassay, has proven to be a useful screening tool in toxicology laboratories. Its principles are very different from any of the analytical techniques we have discussed so far. Basically, immunoassay is based on specific drug antibody reactions. We will learn about this concept in Chapter 9. The primary advantage of immunoassay is its ability to detect small concentrations of drugs in body fluids and organs. In fact, this technique provides the best approach for detecting the low drug levels normally associated with smoking marijuana.

Confirmation Tests. A positive screening test may be due to a substance's close chemical structure to an abused drug. For this reason, the toxicologist must follow up any positive screening test with a confirmation test. Because of the potential impact of the results of a drug finding on an individual, only the most conclusive confirmation procedures should be used.

Gas chromatography/mass spectrometry is generally accepted as the confirmation test of choice. As we learned in Chapter 6, the combination of gas chromatography and mass spectrometry provides a one-step confirmation test of unequaled sensitivity and specificity. Figure 7–13 illustrates the process. After being introduced to the gas chromatograph, the sample is separated into its components. When the separated sample component leaves the column of the gas chromatograph, it enters the mass spectrometer, where it is bombarded with high-energy electrons. This bombardment causes the sample to break up into fragments, producing a fragmentation pattern or mass spectrum for each sample. For most compounds, the mass spectrum represents a unique "fingerprint" pattern that can be used for identification.

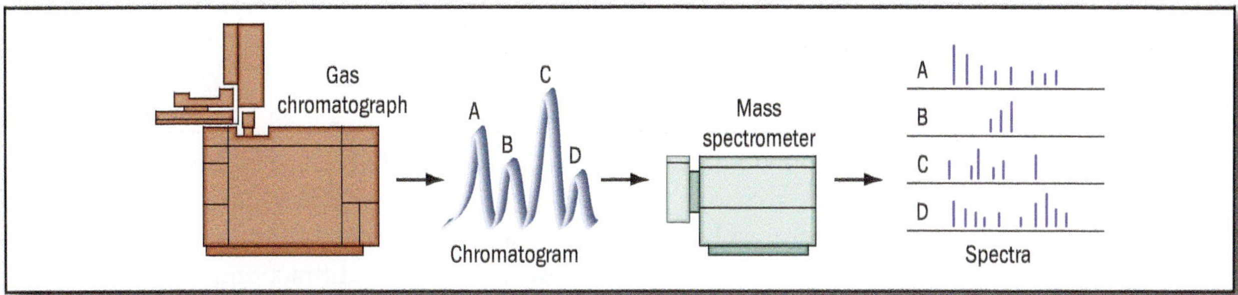

FIGURE 7–13 The combination of the gas chromatograph and the mass spectrometer enables forensic toxicologists to separate the components of a drug mixture and provides specific identification of a drug substance.

There is tremendous interest in drug-testing programs conducted not only in criminal matters but for industry and government as well. Urine testing for drugs is becoming common for job applicants and employees in the workplace. Likewise, the U.S. military has an extensive drug urine-testing program for its members. Many urine-testing programs rely on private laboratories to perform the required analyses. In any case, when the test results form the basis for taking action against an individual, both a screening and confirmation test must be incorporated into the testing protocol to ensure the integrity of the laboratory's conclusions.

Detecting Drugs in Hair When a forensic toxicological examination on a living person is required, practicality limits available specimens to blood and urine. Most drugs remain in the bloodstream for about 24 hours; in urine, they normally are present up to 72 hours. However, it may be necessary to go further back in time to ascertain whether a subject has been abusing a drug. If so, the only viable alternative to blood and urine is head hair.

Hair is nourished by blood flowing through capillaries located close to the hair root. Drugs present in blood diffuse through the capillary walls into the base of the hair and become permanently entrapped in the hair's hardening protein structure. As the hair continues to grow, the drug's location on the hair shaft becomes a historical marker for delineating drug intake. Given that the average human head hair grows at the rate of 1 centimeter per month, analyzing segments of hair for drug content may define the timeline for drug use, dating it back over a period of weeks, months, or even years, depending on the hair's length.

However, caution is required in interpreting the timeline. The chronology of drug intake may be distorted by drugs penetrating the hair's surface as a result of environmental exposure, or drugs may enter the hair's surface through sweat. Nevertheless, drug hair analysis is the only viable approach for measuring long-term abuse of a drug.

Detecting Nondrug Poisons Although forensic toxicologists devote most of their efforts to detecting drugs, they also test for a wide variety of other toxic substances. Some of these are rare elements, not widely or commercially available. Others are so common that virtually everyone is exposed to nontoxic amounts of them every day.

Heavy Metals. One group of poisons once commonly encountered in criminal cases of murder are known as heavy metals. They include arsenic, bismuth, antimony, mercury, and thallium. These days, however, the forensic toxicologist only occasionally encounters heavy metals because severe environmental protection regulations restrict their availability to the general public. Nevertheless, as the following case study makes clear, their use is by no means only a historical curiosity.

To screen for many of these metals, the investigator may dissolve the suspect body fluid or tissue in a hydrochloric acid solution and insert a copper strip into the solution. This process is known as the Reinsch test. The appearance of a silvery or dark coating on the copper indicates the presence of a heavy metal. Such a finding must be confirmed by analytical techniques suitable for inorganic analysis—namely, emission spectroscopy, or X-ray diffraction. These procedures are discussed in more detail in Chapters 12 and 14.

Carbon Monoxide. Unlike heavy metals, carbon monoxide still represents one of the most common poisons encountered in a forensic laboratory. When carbon monoxide enters the human body, it is primarily absorbed by the red blood cells, where it combines with hemoglobin to form carboxyhemoglobin. An average red blood cell contains about 280 million molecules of hemoglobin. Oxygen normally combines with hemoglobin, which transports the oxygen throughout the body. However, if a high percentage of the hemoglobin combines with carbon monoxide, not enough is left to carry sufficient oxygen to the tissues, and death by asphyxiation quickly follows.

There are two basic methods for measuring the concentration of carbon monoxide in the blood. Spectrophotometric methods examine the visible spectrum of blood to determine the amount of carboxyhemoglobin relative to oxyhemoglobin or total hemoglobin; also, a volume of blood can be treated with a reagent to liberate the carbon monoxide, which is then measured by gas chromatography.

The amount of carbon monoxide in blood is generally expressed as percent saturation. This represents the extent to which the available hemoglobin has been converted to carboxyhemoglobin. The transition from normal or occupational levels of carbon monoxide to toxic levels is not sharply defined. It depends, among other things, on the age, health, and general fitness of each individual. In a healthy middle-aged individual, a carbon monoxide blood saturation greater than 50–60 percent is considered fatal. However, in combination with alcohol or other depressants, fatal levels may be significantly lower. For instance, a carbon monoxide saturation of 35–40 percent may prove fatal in the presence of a blood-alcohol concentration of 0.20 percent w/v. Interestingly, chain smokers may have a constant carbon monoxide level of 8–10 percent from the carbon monoxide in cigarette smoke.

Inhaling automobile fumes is a relatively common way to commit suicide. A garden or vacuum cleaner hose is often used to connect the tail-pipe with the

Case Files

Joann Curley: Caught by a Hair

A vibrant young woman named Joann Curley rushed to the Wilkes-Barre (Pennsylvania) General Hospital—her husband, Bobby, was having an attack and required immediate medical attention. Bobby was experiencing a burning sensation in his feet, numbness in his hands, a flushed face, and intense sweating. After being discharged, Bobby experienced another bout of debilitating pain and numbness. He was admitted to another hospital. There doctors observed extreme alopecia, or hair loss.

Test results of Bobby's urine showed high levels of the heavy metal thallium in his body. Thallium, a rare and highly toxic metal that was used decades ago in substances such as rat poison and to treat ringworm and gout, was found in sufficient quantities to cause Bobby's sickness. The use of thallium was banned in the United States in 1984. Now, at least, Bobby could be treated. However, before Bobby's doctors could treat him for thallium poisoning, he experienced cardiac arrest and slipped into a coma. Joann Curley made the difficult decision to remove her husband of 13 months from life support equipment. He died shortly thereafter.

Investigators also learned that Bobby had changed his life insurance to list his wife, Joann, as the beneficiary of his $300,000 policy. Based on this information, police consulted a forensic toxicologist in an effort to glean as much from the physical evidence in Bobby Curley's body as possible. The toxicologist conducted segmental analysis of Bobby's hair, an analytical method based on the predictable rate of hair growth on the human scalp: an average of 1 centimeter per month. Bobby had approximately 5 inches (12.5 centimeters) of hair, which represents almost 12 months of hair growth. Each section tested represented a specific period of time in the final year of Bobby's life.

The hair analysis confirmed that Bobby Curley was poisoned with thallium. The first few doses were small, which probably barely made him sick at the time. Gradually, over a year or more, Bobby was receiving more doses of thallium until he finally succumbed to a massive dose three or four days before his death. After careful scrutiny of the timeline, investigators concluded that only Joann Curley had access to Bobby during each of these intervals. She also had motive, in the amount of $300,000.

Presented with the timeline and the solid toxicological evidence against her, Joann Curley pleaded guilty to murder. As part of her plea agreement, she provided a 40-page written confession of how she haphazardly dosed Bobby with some rat poison she found in her basement. She admitted that she murdered him for the money she would receive from Bobby's life insurance policy.

vehicle's interior, or the engine is allowed to run in a closed garage. A level of carbon monoxide sufficient to cause death accumulates in five to ten minutes in a closed single-car garage.

The level of carbon monoxide in the blood of a victim found dead at the scene of a fire is significant in ascertaining whether foul play has occurred. High levels of carbon monoxide in the blood prove that the victim breathed the combustion products of the fire and was alive when the fire began. Many attempts at covering up a murder by setting fire to a victim's house or car have been uncovered in this manner.

The Significance of Toxicological Findings

Once a drug is found and identified, the toxicologist assesses its influence on the behavior of the individual. Interpreting the results of a toxicology finding is one of the toxicologist's more difficult chores. Recall that many countries have designated a specific blood-alcohol level at which an individual is deemed under the influence of alcohol. These levels were established as a result of numerous studies conducted over several years to measure the effects of alcohol levels on driving performance. However, no such legal guidelines are available to the toxicologist who must judge how a drug other than alcohol affects an individual's performance or physical state.

For many drugs, blood concentration levels are readily determined and can be used to estimate the pharmacological effects of the drug on the individual. Often, when dealing with a living person, the toxicologist has the added benefit of knowing what a police officer may have observed about an individual's behavior and motor skills. For a deceased person, drug levels in various body organs

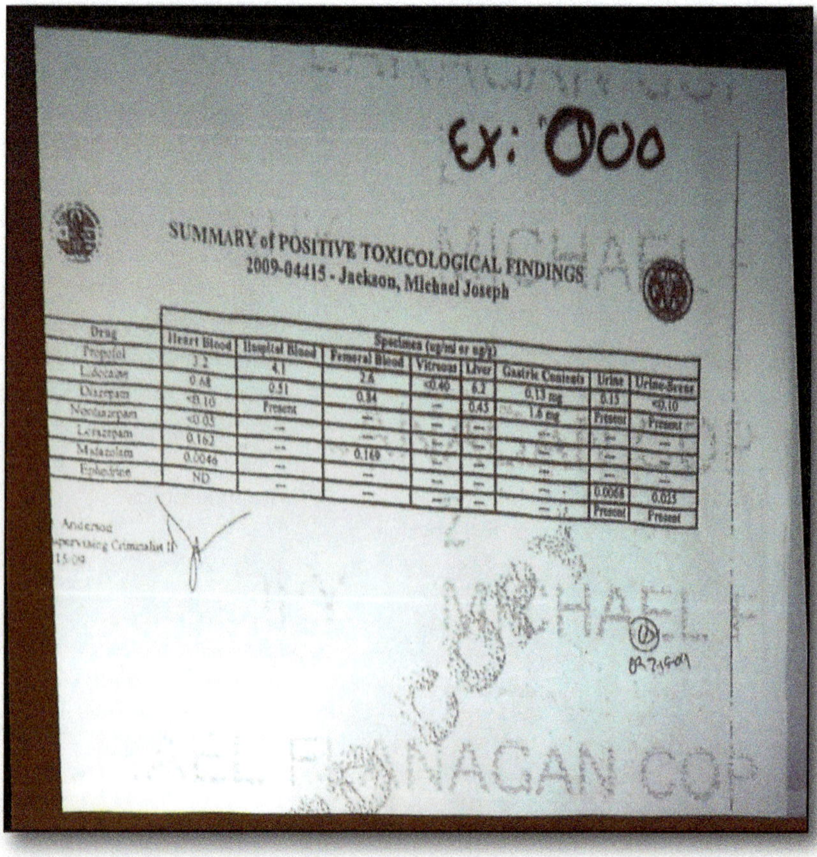

FIGURE 7–14 Toxicological report for Michael Jackson. *Courtesy Reuters/Alamy Stock Photos*

and tissues provide additional information about the individual's state at the time of death. However, before drawing conclusions about drug-induced behavior, the analyst must consider other factors, including the age, physical condition, and tolerance of the drug user.

With prolonged use of a drug, an individual may become less responsive to a drug's effects and tolerate blood-drug concentrations that would kill a casual drug user. Therefore, knowledge of an individual's history of drug use is important in evaluating drug concentrations. Another consideration is additive or synergistic effects of the interaction of two or more drugs, which may produce a highly intoxicated or comatose state even though none of the drugs alone is present at high or toxic levels. The combination of alcohol with barbiturates or narcotics is a common example of a potentially lethal drug combination.

The concentration of a drug present in urine is a poor indicator of how extensively an individual's behavior or state is influenced by the drug. Urine is formed outside the body's circulatory system, and consequently drug levels can build up in it over a relatively long period of time. Some drugs are found in the urine one to three days after they have been taken and long after their effects on the user have disappeared. Nevertheless, the value of this information should not be discounted. Urine drug levels, like blood levels, are best used by law enforcement authorities and the courts to corroborate other investigative and medical findings regarding an individual's condition. Hence, for an individual arrested for suspicion of being under the influence of a drug, a toxicologist's determinations supplement the observations of the arresting officer, including the results of a drug influence evaluation (discussed next).

For a deceased person, the medical examiner or coroner must establish a cause of death. However, before a conclusive determination is made, the examining physician depends on the forensic toxicologist to demonstrate the presence or absence of a drug or poison in the tissues or body fluids of the deceased. Only through the combined efforts of the toxicologist and the medical examiner (or coroner) can society be assured that death investigations achieve high professional and legal standards.

Quick Review

- The forensic toxicologist must devise an analytical scheme to detect, isolate, and identify toxic drug substances extracted from biological fluids, tissues, and organs.
- A screening test gives quick insight into the likelihood that a specimen contains a drug substance. Positive results arising from a screening test are tentative at best and must be verified with a confirmation test.
- The most widely used screening tests are thin-layer chromatography, gas chromatography, and immunoassay. Gas chromatography/mass spectrometry is generally accepted as the confirmation test of choice.
- Once a drug is extracted and identified, a toxicologist may be required to judge the drug's effect on an individual's natural performance or physical state.

Chapter Review

- Forensic toxicologists detect and identify drugs and poisons in body fluids, tissues, and organs in matters that pertain to violations of criminal laws.
- Ethyl alcohol is the most heavily abused drug in Western countries.
- Alcohol appears in the blood within minutes after it has been taken by mouth. It slowly increases in concentration while it is being absorbed from the stomach and the small intestine into the bloodstream.
- When all of the alcohol has been absorbed, a maximum alcohol level is reached in the blood, and the postabsorption period begins. During postabsorption, the alcohol concentration slowly decreases until a zero level is reached.
- Elimination of alcohol throughout the body is accomplished through oxidation and excretion. Oxidation takes place almost entirely in the liver, while alcohol is excreted unchanged in the breath, urine, and perspiration.
- Breath-testing devices operate on the principle that the ratio between the concentration of alcohol in alveolar breath and its concentration in blood is fixed.
- Modern breath testers are free of chemicals. They include infrared light absorption devices and fuel cell detectors.
- The key to the accuracy of a breath-testing device is to ensure that the unit captures the alcohol in the alveolar (deep-lung) breath of the subject.
- Many breath testers collect a set volume of breath and expose it to infrared light. The instrument measures alcohol concentration in breath by measuring the degree of the interaction of the light with alcohol in the collected breath sample.
- Law enforcement officers use field sobriety tests to estimate a motorist's degree of physical impairment by alcohol and to determine whether an evidential test for alcohol is justified.
- The horizontal-gaze nystagmus test, walk and turn, and the one-leg stand are all considered reliable and effective psychophysical tests for alcohol impairment.
- Gas chromatography is the most widely used approach for determining blood-alcohol levels in forensic laboratories.
- An anticoagulant should be added to a blood sample to prevent clotting; a preservative should be added to inhibit the growth of microorganisms capable of destroying alcohol.
- The current legal measure of drunk driving in the United States is a blood-alcohol concentration of 0.08 percent, or 0.08 grams of alcohol per 100 milliliters of blood.

- An implied-consent law states that the operator of a motor vehicle on a public highway must either consent to a test for alcohol intoxication, if requested, or lose their license for some designated period—usually six months to one year.
- The forensic toxicologist must devise an analytical scheme to detect, isolate, and identify toxic drug substances extracted from biological fluids, tissues, and organs.
- A screening test gives quick insight into the likelihood that a specimen contains a drug substance. Positive results arising from a screening test are tentative at best and must be verified with a confirmation test.
- The most widely used screening tests are thin-layer chromatography, gas chromatography, and immunoassay. Gas chromatography/mass spectrometry is generally accepted as the confirmation test of choice.
- Once a drug is extracted and identified, a toxicologist may be required to judge the drug's effect on an individual's natural performance or physical state.

Quick Lab: pH Test

Materials:

pH paper with color indicator chart
Liquid samples of acid and base substances (examples: tap water, bottled water, soda, ice tea, baking soda in water, salt water, lime juice, cleaning solution, liquid soap)

Procedure:

pH can help toxicologists determine what type of toxin may be present in a sample. This activity will allow you to identify the pH of some common substances. Before you begin, write a guess of what you think the pH of each sample will be. Now, take a piece of pH paper and dip into the first sample for a few seconds. Compare the pH paper to the color indicator chart and determine what the pH is of the sample. Record your observations. Repeat this for each sample.

Follow-Up Questions:

1. **How many samples did you guess correctly before testing them? Which ones were they?**
2. **Create a pH chart. Use the labels *Acid, Base,* and *Neutral.* Also label where each item you tested would be on your chart.**
3. **A toxicologist would test samples of what to determine the pH of the toxin in a body?**

Review Questions

1. About 95–98 percent of alcohol is oxidized to what two substances?
   a. carbon dioxide and dehydrogenase
   b. water and acetic acid
   c. acetaldehyde and acetic acid
   d. water and carbon dioxide

2. Carbon monoxide combines with what component of blood?
   a. carboxyhemoglobin
   b. hemoglobin
   c. oxyhemoglobin
   d. white blood cells

3. With a blood-alcohol level at 0.15 percent, the chance for involvement in an automobile accident is
   a. 10 times as great.
   b. 25 times as great.
   c. 50 times as great.
   d. 75 times as great.

4. Which of the following drugs would not have a pH of increasing basicity?
   a. phencyclidine
   b. amphetamines
   c. cocaine
   d. barbiturates

5. What percentage of alcohol is absorbed through the stomach walls into the portal vein of the blood system?
   a. 20 percent
   b. 70 percent
   c. 1 percent
   d. 10 percent

6. True or False: The role of the forensic toxicologist is limited to matters that pertain to violations of the law.

7. True or False: A breath test, used to measure alcohol, reflects the alcohol concentration in the pulmonary vein.

8. True or False: After a screening test has been used to determine the identity of an abused drug, the confirmation test of choice is thin-layer chromatography.

9. True or False: The concentration of a drug present in urine is an excellent indicator of how extensively an individual's behavior or state is influenced by the drug.

10. True or False: The technique used most widely by forensic toxicologists for detecting alcohol in blood is the gas chromatograph.

11. What is a toxicologist? Name three settings in which a toxicologist often works.

12. What is the most widely abused drug in Western countries?

13. Define *metabolism*.

14. List and describe the three stages of alcohol's fate in the human body.

15. Name at least three factors that influence the rate at which alcohol is absorbed into the bloodstream.

16. Alcohol that is not oxidized is expelled unchanged in what bodily excretions?

17. The amount of alcohol exhaled in the breath is in direct proportion to what?

18. For a longer total time required for complete absorption, will the peak blood-alcohol concentration be higher or lower?

19. List and describe the functions of the three types of blood vessels in the circulatory system.

20. Through the walls of which organ is most alcohol absorbed into the bloodstream? In what other organ does the remainder of absorption occur?

21. What are alveoli and what role do they play in circulation?

22. What scientific observation forms the theoretical basis for breath testing?

23. Briefly describe how a fuel cell detector measures blood-alcohol concentration.

24. What is mouth alcohol and how does it affect the accuracy of a breath test? Name three potential sources of mouth alcohol.

25. What is a field sobriety test? What two general types of test are included in a field sobriety test?

26. What type of disinfectant must be applied to a subject's skin before drawing blood? Why must such a disinfectant be applied?

27. What type of container best ensures the preservation of blood samples?

28. What two substances should be added to a blood sample after collection and why?

29. How do each of the following factors affect alcohol's stability in blood: storage temperature, the presence of a preservative, and the time of storage?

30. When collecting postmortem blood samples for alcohol determination, why is it best to collect a number of blood samples from different body sites?

31. What blood-alcohol concentration is the current legal measure of drunk driving in the United States, as established by the National Highway Traffic Safety Administration?

32. What is a *per se* law?

33. What is an implied-consent law? Why were such laws implemented?

34. Describe two challenges toxicologists face in detecting drugs and determining their toxicity.

35. What three drugs account for 90 percent or more of the drugs encountered in a typical toxicology laboratory?

36. What is the first task of a forensic toxicologist when establishing an analytical scheme to detect and identify drugs?

37. What are acids and bases? How are they used to extract and categorize drugs?

38. What are the three screening tests most widely used by forensic toxicologists? What is the confirmation test of choice?

39. Which of the following is not classified as a heavy metal?
    a. lead
    b. arsenic
    c. mercury
    d. thallium

40. Explain how inhaling carbon monoxide can cause death.

41. Why is the concentration of a drug present in urine a poor indicator of how extensively an individual's behavior or state is influenced by the drug?

42. Explain the necessary precautions for proper preservation of biological samples in the forensic laboratory.

43. Forensic toxicologists use different types of tests to determine the identity and amount of toxins in the body. Write a brief response to the following questions.
    a. What is the difference between the method of collection for a toxicological analysis and a controlled substance analysis?
    b. How do forensic toxicologists use these tests to determine the impact of different drugs/toxins on the body?
    c. What is the difference between the method of collection for a toxicological analysis and a controlled substance analysis?
    d. How do forensic toxicologists use these tests to determine the impact of different drugs/toxins on the body?

Learn the Vocabulary

Use the information from this chapter along with your prior knowledge and experience to learn the key terms. Choose from the following activities:

1. Make flashcards with a term on the front and its definition on the back. Use the flashcards to quiz a partner on correctly pronouncing and defining the terms.

2. Make an audio or video recording of each term and its definition.

3. Work with a partner to write a sentence for each term. Read the sentences out loud to a partner or share them with the class.

4. Write a list of new terms and language you have encountered in this chapter. Reuse the terms to write sentences and then read the sentences out loud to a partner. Reuse the terms again by writing a short speech. The speech might explain the terms or teach a lesson using the terms. Give the speech to a partner or to the class.

Application and Critical Thinking

1. Answer the following questions about driving risk associated with drinking and blood-alcohol concentration:
    a. Randy is just barely legally intoxicated. How much more likely is he to have an accident compared with someone who is sober?
    b. Marissa, who has been drinking, is 15 times as likely to have an accident as her sober friend, Christine. What is Marissa's approximate blood-alcohol concentration?

c. After several drinks, Charles is ten times as likely to have an accident as a sober person. Is he more or less intoxicated than James, whose blood-alcohol level is 0.10?

d. Under the original blood-alcohol standards recommended by the NHTSA, a person considered just barely legally intoxicated was how much more likely to have an accident than a sober individual?

2. Following is a description of four individuals who have been drinking. Use critical-thinking skills to analyze scientific explanations in the data collected. In a written response, use logical reasoning to identify any patterns and possible sources of error. Suggest any limitations to using this method. In a brief written summary, rank them from highest to lowest blood-alcohol concentration:

   a. John, who weighs 200 pounds and has consumed eight 8-ounce drinks on a full stomach
   b. Frank, who weighs 170 pounds and has consumed four 8-ounce drinks on an empty stomach
   c. Gary, who weighs 240 pounds and has consumed six 8-ounce drinks on an empty stomach
   d. Stephen, who weighs 180 pounds and has consumed six 8-ounce drinks on a full stomach

3. Following is a description of four individuals who have been drinking. In which (if any) of the following countries would each be considered legally drunk: the United States, Australia, and Sweden?

   a. Bill, who weighs 150 pounds and has consumed three 8-ounce drinks on an empty stomach
   b. Sally, who weighs 110 pounds and has consumed three 8-ounce drinks on a full stomach
   c. Rich, who weighs 200 pounds and has consumed six 8-ounce drinks on an empty stomach
   d. Carrie, who weighs 140 pounds and has consumed four 8-ounce drinks on a full stomach

4. You are a forensic scientist who has been asked to test two blood samples. You know that one sample is suspected of containing barbiturates and the other contains no drugs; however, you cannot tell the two samples apart. Describe how you would use the concept of pH to determine which sample contains barbiturates. Explain your reasoning.

5. You are investigating an arson scene and you find a corpse in the rubble, but you suspect that the victim did not die as a result of the fire. Instead, you suspect that the victim was murdered earlier, and that the blaze was started to cover up the murder. How would you go about determining whether the victim died before the fire?

6. Three individuals are stopped at a DUI checkpoint at ten p.m. Before performing the on-site breath test, the checkpoint officer asks each of them about their activities before they were stopped in order to estimate the expected blood-alcohol reading estimation (and to gauge their honesty). Use the testimony of each person that follows and the diagram in Figure 7–9 to estimate the expected BAC (blood-alcohol concentration). Indicate whether this is above or below the legal limit for operating a vehicle.

   a. Larry Questring had a large meal at an Italian restaurant with his wife. He ingested 8 ounces of 80-proof Frangelico between seven p.m. and nine p.m. Larry weighs approximately 180 pounds. Estimated BAC at ten p.m.
   b. Paul Raunetz is driving home from a bar. He drove there after finishing his workday without stopping to eat. At the bar, he had four shots of tequila (1 ounce each, 80 proof) and five shots of whiskey (1 oz each, 80 proof) between six p.m. and nine p.m. Paul weighs approximately 130 pounds. Estimated BAC at ten p.m.

c. Thomas Wold is returning home after an NFL game that started at six p.m. and ended at nine p.m. While watching the game, he ingested two hot dogs, a hot pretzel, peanuts, cheese fries, two diet sodas, ice cream, and six gin-and-tonics. Each gin-and-tonic contained 2 ounces of 80-proof gin. Thomas weighs approximately 240 pounds. Estimated BAC at ten p.m.

7. Choose one of the following toxins to research and, in a written summary, explain the absorption, distribution, metabolization, and elimination of the toxin through the human body: alcohol, prescription drugs, controlled substances, or carbon monoxide.

8. Describe one presumptive laboratory procedure related to toxicological analysis.

9. Describe one confirmatory laboratory procedure related to toxicological analysis.

Laboratory Experiments

This activity requires the use of the following practices of science and engineering:

- Asking questions and defining problems
- Developing and using models
- Planning and carrying out investigations
- Analyzing and interpreting data
- Using mathematics and computational thinking
- Constructing explanations
- Engaging in argument
- Obtaining, evaluating, and communicating information
- Using appropriate safety equipment and practices during laboratory/classroom investigations and field investigationsas outlined in Texas Education Agency-approved safety standards

This activity consists of the following crosscutting concepts:

- **Patterns**—The relationship between alcohol content and percent transmittance allows the student to analyze the data for the unknown samples.
- **Cause and effect**—This is demonstrated by analyzing the effect of time on the transmission of the unknown samples.

Before you begin, read the scenario and steps out loud with a partner. Ask for clarification as necessary to make sure you understand the experiment.

The Scenario

Police respond to a single vehicle accident at the corner of 29th and Main St. on Saturday May 25 at 11:02 PM. The vehicle had collided with a power pole with damage to the front passenger side with the airbags deployed. The driver of vehicle is a male 18 years old with minor injuries. The front passenger is a female 17 years old and was taken to a local hospital for head trauma and possible broken ribs. Both rear passengers, one male 17 years old and a female 16 years old, were uninjured. Due to nature of the accident, the driver is assumed to be under the influence. You have been called to the accident scene to administer a Breathalyzer test to the driver and the rear passengers. Using the procedure below, collect qualitative and quantitative data using the International System of Units (SI) and United States customary units with accuracy and precision to gather evidence to determine if the teenagers had been drinking during the evening of the accident.

EXPERIMENT

Safety

Compounds which contain chromium (VI) are toxic and suspected carcinogens. Wear gloves when performing this lab and wash your hands thoroughly with soap and water after the lab.
Dispose of any solution with your Breathalyzer solution (containing Chromium) in the designated waste containers. ***Do not*** pour any solution containing chromium down the sink.

MATERIALS

- Dichromate Reagent
- mouthwash containing alcohol
- small disposable drinking cups
- plastic drinking straw
- 50 mL plastic syringe with sparge tube
- several 100 mL volumetric flasks
- 10 mL graduated pipet
- fixed-wavelength single-beam spectrophotometer or colorimeter with two sample cuvets and caps

PROCEDURE

1. Turn on power to the spectrometer, set the wavelength to 440 nm, and allow the instrument to "warm up."

Spectrometer Preparation

2. If the instrument permits, set the signal to 0 %T.
3. Rinse and then fill a "blank" cuvet about three-fourths full with deionized water, stopper, wipe the exterior with a soft tissue, and place in the spectrometer's sample compartment, being sure to note precisely the cuvet's orientation (and to reproducibly orient the cuvet in all subsequent measurements). Gently close the lid, and set the signal to 100 %T. Don't empty this cuvet, as it will be needed for setting 100 %T in subsequent measurements.

Standard Calibration

4. Set the spectrometer's wavelength selector to the λ_{max} value identified above, and then set 0 %T and 100 %T as previously described.
5. Add to a "sample" cuvet precisely 3.00 mL of the dichromate reagent and measure its %T.
6. Add 0.0500 mL of ethanol standard to the dichromate reagent cuvet, cap, and gently invert several times to ensure thorough mixing. Allow the solution to stand for 2 min, and then measure its %T.
7. Repeat step 6 for three further 0.0500 mL additions of ethanol standard.
8. Make a calibration curve of the % T vs. the percentage of alcohol. (% by volume)

Breath Analysis

9. Select one member of your group to be the lead forensic investigator. Select three members of your group to represent the driver and the 2 passengers.
10. The investigator should instruct the 3 group members to exit the lab, noting the time. The investigator should tell them to rinse with the provided mouthwash labeled driver, passenger #1 male, and passenger #2 female.
11. Follow the investigator's instructions to prepare a sample cuvet with a fresh 3.00 mL portion of dichromate reagent and measure its % T.
12. Continue to follow the investigator's instructions to affix a clean drinking straw to the "needle-end" of a 50 mL plastic syringe (plunger removed), inhale deeply, and then exhale slowly by blowing through the straw until your lungs feel empty. Quickly insert the syringe plunger, remove the straw, and cap the open syringe needle port with the sparge tube. Depress the plunger to the 50 mL mark, and then immerse the sparge tube completely in the dichromate reagent cuvet. Sparge the reagent with the breath sample by slowly depressing the syringe plunger until all the sample is expelled, taking care to avoid loss of the reagent solution that may result from rapid sparging. Cap, mix, and let stand for 2 min prior to measuring the solution's %T.
13. Repeat steps 11 and 12 to examine the effect of time after rinsing on the amount of ethanol in breath.
14. Dispose of waste solutions in the provided waste container, return cleaned glassware (cuvets, flasks, etc.) to their original locations, and switch off power to the spectrometer.

DATA ANALYSIS

Using the breath analysis result and information from the calibration, compute the apparent blood alcohol concentration (BAC) of the accident victims providing the breath sample.

Conclusion

To what extent was the driver and the passenger intoxicated at the time of the accident? What effect would time since last drink have on the BAC? Explain your findings to a partner or to the class.

Endnotes

1. In the United States, laws that define blood-alcohol levels almost exclusively use the unit *percent weight per volume*—% w/v. Hence, 0.015 percent w/v is equivalent to 0.015 grams of alcohol per 100 milliliters of blood, or 15 milligrams of alcohol per 100 milliliters.
2. Henry's Law
3. R. B. Forney et al., "Alcohol Distribution in the Vascular System: Concentrations of Orally Administered Alcohol in Blood from Various Points in the Vascular System and in Rebreathed Air during Absorption," *Quarterly Journal of Studies on Alcohol* 25 (1964): 205.
4. G. A. Brown et al., "The Stability of Ethanol in Stored Blood," *Analytica Chemica* Acta 66 (1973): 271.
5. 0.15 percent w/v is equivalent to 0.15 grams of alcohol per 100 milliliters of blood, or 150 milligrams per 100 milliliters.
6. 384 U.S. 757 (1966).
7. 133 S. Ct. 932 (2013).

The Microscope

8

Information Wanted

The Suffolk County District Attorney, Massachusetts State Police, and Winthrop Police are seeking the public's help in identifying a female toddler whose body was found on the western shore of Deer Island on June 25, 2015. This computer-generated composite image depicts her as she may have appeared in life.

The child was approximately four years old at the time of her death, had brown eyes and brown hair, weighed about 30 pounds, and stood about 3½ feet tall.

Suffolk County
State Police Detective Unit

Mass State Police
Communication Section

Winthrop Police
Tip Line

Uncredited/AP/Shutterstock

Key Terms

- binocular
- condenser
- depth of focus
- eyepiece lens
- field of view
- microspectrophotometer
- monocular
- objective lens
- parfocal
- polarized light
- polarizer
- real image
- transmitted illumination
- vertical or reflected illumination
- virtual image

Baby Doe

On July 25, 2015, a woman was walking her dog along the beach on Deer Island outside Boston, Massachusetts. The dog stopped to sniff a contractor-style garbage bag and inside that bag the woman discovered the remains of a young girl, approximately 2–4 years old. The child, whom investigators believed to have been deceased for 1–2 days prior to discovery, was wrapped in two blankets and wearing a pair of white pants with black polka dots.

An investigation into the identity of the toddler ensued and authorities distributed a composite sketch of her face and a description the clothing she was wearing to the news media. The media dubbed the child, "Baby Doe," and the case garnered national attention.

Weeks went by, and though tips poured in, investigators were still unable to positively identify the girl. Detectives at the Massachusetts State Police began to pursue lesser known forensic methods that could indicate whether the child was local to the Boston area or whether her killers brought her body there in an effort to conceal the crime. They submitted, through the National Center for Missing and Exploited Children, two blankets used to wrap the body, the child's pants, and a sample of the child's hair, to the Laboratories and Scientific Services section of the U.S. Customs and Border Patrol in Houston for pollen analysis.

There, analysts determined that based on the recovered pollen and high quantities of soot, all of the analyzed samples suggest that the child was most likely from a suburban area in the Northeastern United States. They identified spores of oak and pine along with other species typically found in the deciduous forest of North America. Most telling, though, was the recovery of two different types of cedar pollen. The analyst determined that because cedar trees are not native to the Northeast, the only two places in that region where multiple species of cedar are known to grow together are the Arnold Arboretum of Harvard University in Boston and the Morris Arboretum of the University of Pennsylvania in Philadelphia. Although the analyst could not distinguish between the greater Philadelphia and Boston areas, it was determined that the Boston area may be the more likely candidate since the child was found there. This information helped investigators to focus their search in greater Boston and ultimately identify the child and the individuals responsible for her death. Read more about the case evidence in the Forensic Palynology section of this chapter.

Learning Objectives

After studying this chapter, you should be able to:

- 8.1 Outline the basics of a microscope
- 8.2 Identify the different parts of a compound microscope
- 8.3 Describe the features of a comparison microscope
- 8.4 Describe the features and utility of a stereoscopic microscope
- 8.5 Discuss how a polarizing microscope is designed to characterize polarized light
- 8.6 Discuss how a microspectrophotometer can be used to examine trace physical evidence
- 8.7 Describe a scanning electron microscope
- 8.8 Discuss how forensic palynology is used to link a crime scene and a person or object

National Science Content Standards

Scientific Inquiry

Life Science

Science and Technology

A microscope is an optical instrument that uses a lens or a combination of lenses to magnify and resolve the fine details of an object. The earliest methods for examining physical evidence in crime laboratories relied almost solely on the microscope to study the structure and composition of matter. Even the advent of modern analytical instrumentation and techniques has done little to diminish the usefulness of the microscope for forensic analysis. If anything, the development of the powerful scanning electron microscope promises to add a new dimension to forensic science heretofore unattainable within the limits of the ordinary light microscope.

Basics of the Microscope

virtual image
An image that cannot be seen directly, but can be seen only by a viewer looking through a lens

real image
An image that can be seen directly with the naked eye

The earliest and simplest microscope was the single lens commonly referred to as a magnifying glass. The handheld magnifying glass makes things appear larger than they are because of the way light rays are refracted, or bent, in passing from the air into the glass and back into the air. The magnified image is observed by looking through the lens, as shown in Figure 8–1. Such an image is known as a **virtual image**; it can be seen only by looking through a lens and cannot be viewed directly. This is distinguished from a **real image**, which can be seen directly, like the image that is projected onto a motion picture screen.

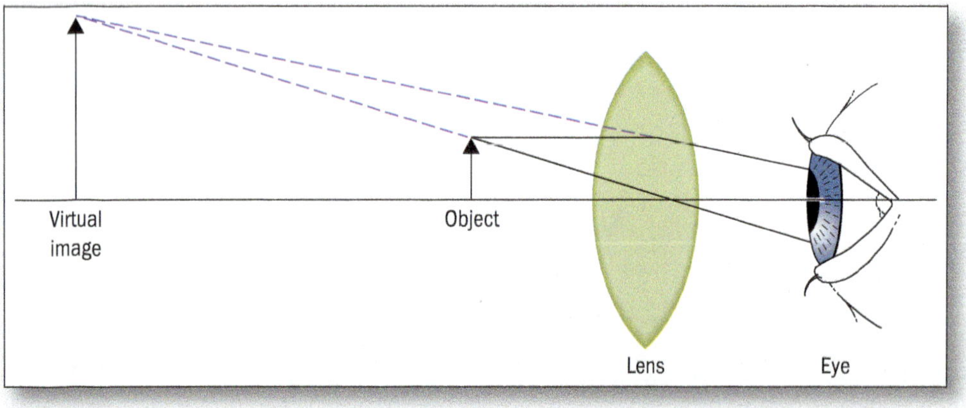

FIGURE 8–1 The passage of light through a lens, showing how magnification is obtained.

objective lens
The lower lens of a compound microscope; it is positioned directly over the specimen

eyepiece lens
The upper lens of a compound microscope, into which the viewer looks

The ordinary magnifying glass can achieve a magnification of about 5 to 10 times. Higher magnifying power is obtainable only with a compound microscope, constructed of two lenses mounted at each end of a hollow tube. The object to be magnified is placed under the lower lens, called the **objective lens**, and the magnified image is viewed through the upper lens, known as the **eyepiece lens**. As shown in Figure 8–2, the objective lens forms a real, inverted, magnified image of the object. The eyepiece, acting just like a simple magnifying glass, further magnifies this image into a virtual image, which is seen by the eye. The combined magnifying power of both lenses can produce an image magnified up to 1,500 times.

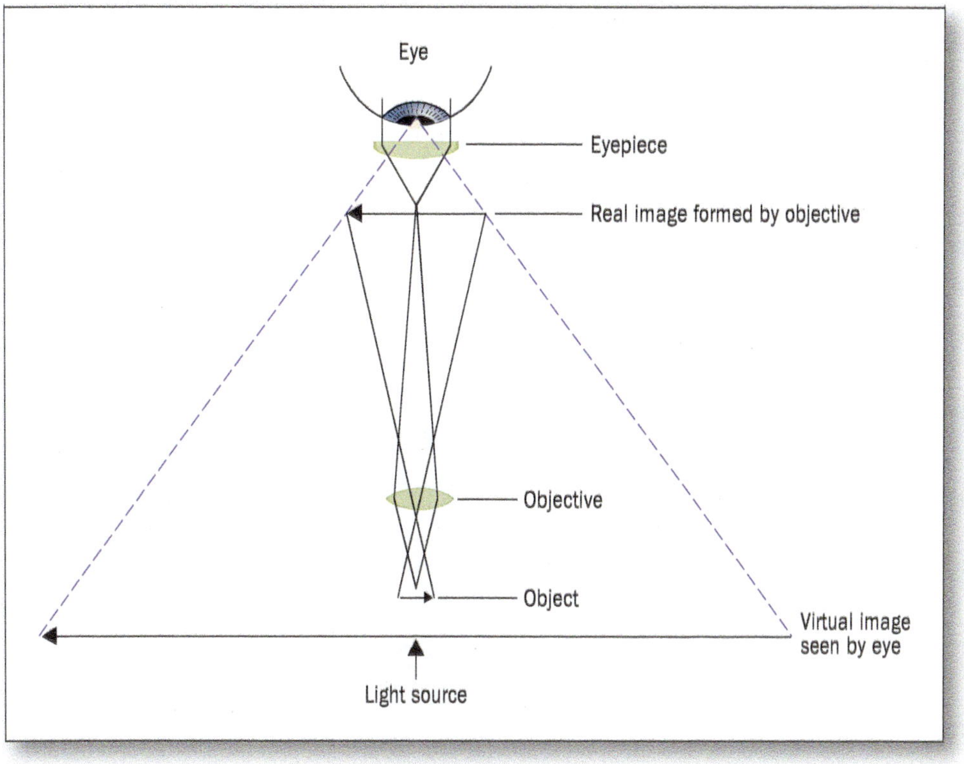

FIGURE 8–2 The principle of the compound microscope. The passage of light through two lenses forms the virtual image of the object seen by the eye.

The optical principles of the compound microscope are incorporated into the basic design of different types of light microscopes. The microscopes most applicable for examining forensic specimens are as follows:

- the compound microscope
- the comparison microscope
- the stereoscopic microscope
- the polarizing microscope
- the microspectrophotometer

After describing these five microscopes, we will talk about a completely different approach to microscopy, the scanning electron microscope (SEM). This instrument focuses a beam of electrons, instead of visible light, onto the specimen to produce a magnified image. The principle and design of this microscope permit magnifying powers as high as 100,000 times.

Quick Review

- A microscope is an optical instrument that uses a lens or a combination of lenses to magnify and resolve the fine details of an object.
- Two types of images are virtual images and real images. A real image can be seen directly with the naked eye; a virtual image can be seen only by a viewer looking through a lens.
- The eyepiece lens is the upper lens through which the viewer looks. The objective lens is the lower lens that is positioned directly over the specimen.

The Compound Microscope

The parts of the compound microscope are illustrated in Figure 8–3. Basically, this microscope consists of a mechanical system, which supports the microscope, and an optical system. The optical system illuminates the object under investigation and passes the light through a series of lenses to form an image of the specimen on the retina of the eye. The optical path of light through a compound microscope is shown in Figure 8–4.

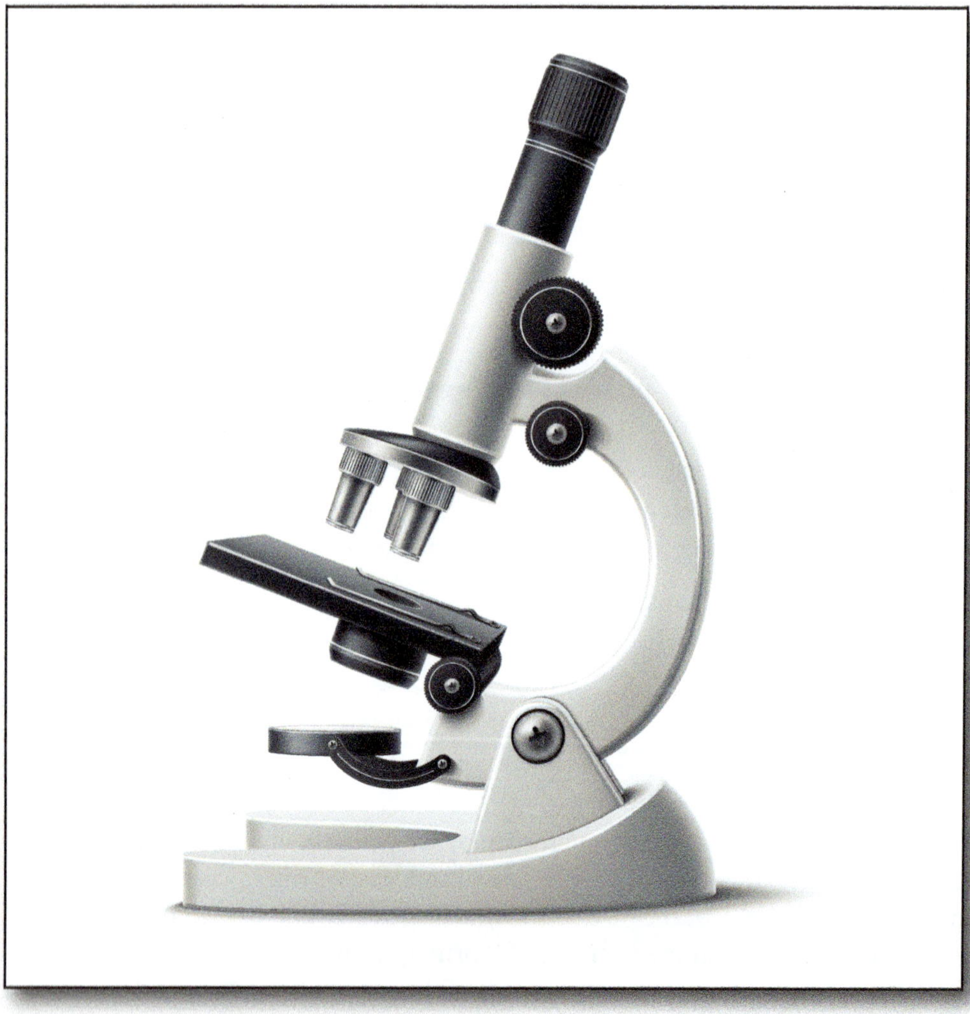

FIGURE 8–3 Parts of the compound microscope: (1) base, (2) arm, (3) stage, (4) body tube, (5) coarse adjustment, (6) fine adjustment, (7) illuminator, (8) condenser, (9) objective lens, and (10) eyepiece lens. *Courtesy Leica Microsystems, Buffalo, N.Y., www. leica-microsystems.com*

Construction of the Compound Microscope

The mechanical system of the compound microscope is composed of six parts:

Base The support on which the instrument rests.

Arm A C-shaped upright structure, hinged to the base, that supports the microscope and acts as a handle for carrying.

Stage The horizontal plate on which the specimens are placed for study. The specimens are normally mounted on glass slides that are held firmly in place on the stage by spring clips.

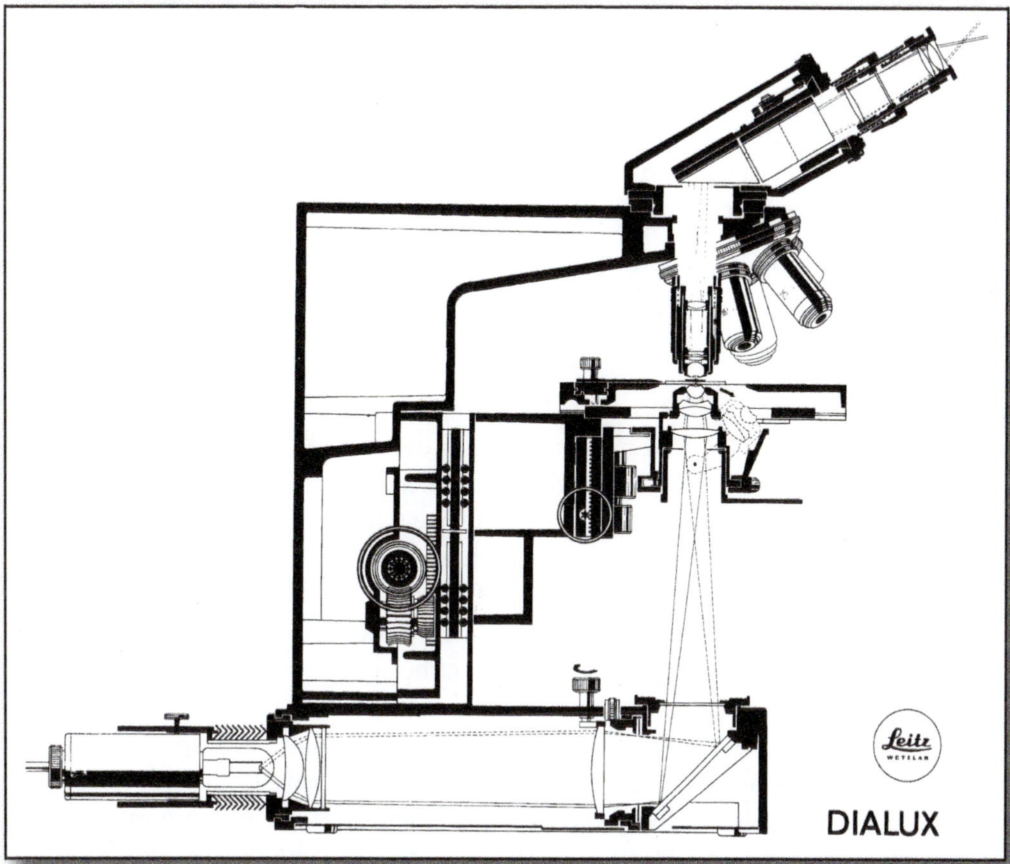

FIGURE 8–4
Optics of the compound microscope. *Courtesy Leica Microsystems, Buffalo, N.Y., www.leica-microsystems.com*

Body Tube A cylindrical hollow tube on which the objective and eyepiece lenses are mounted at opposite ends. This tube is merely a corridor through which light passes from one lens to another.

Coarse Adjustment This knob focuses the microscope lenses on the specimen by raising and lowering the body tube.

Fine Adjustment The movements effected by this knob are similar to those of the coarse adjustment but are of a much smaller magnitude.

The optical system is made up of four parts:

Illuminator Most modern microscopes use artificial light supplied by a lightbulb to illuminate the specimen being examined. If the specimen is transparent, the light is directed up toward and through the specimen stage from an illuminator built into the base of the microscope. This is known as **transmitted illumination.** When the object is opaque—that is, not transparent—the light source must be placed above the specimen so that it can reflect off the specimen's surface and into the lens system of the microscope. This type of illumination is known as **vertical or reflected illumination.**

Condenser The **condenser** is a lens system under the microscope stage that collects light rays from the base illuminator and focuses them on the specimen. The simplest condenser is known as the Abbé condenser. It consists of two lenses held together in a metal mount. The condenser also includes an iris diaphragm that can be opened or closed to control the amount of light passing into the condenser.

transmitted illumination
Light that passes up from the condenser and through the specimen; it is used to examine transparent specimens

vertical or reflected illumination
Illumination of a specimen from above; it is used to examine opaque specimens

condenser
The lens system under the microscope stage that collects light rays from the base illuminator and focuses them onto the specimen

parfocal
Describes a microscope such that when an image is focused with one objective in position, the other objective can be rotated into place and the field will remain in focus

monocular
Describes a microscope with one eyepiece

binocular
Describes a microscope with two eyepieces

Objective Lens This is the lens positioned closest to the specimen. To facilitate changing from one objective lens to another, several objectives are mounted on a revolving nosepiece or turret located above the specimen. Most microscopes are **parfocal**, meaning that when the microscope is focused with one objective in position, the other objective can be rotated into place by revolving the nosepiece while the specimen remains very nearly in correct focus.

Eyepiece or Ocular Lens This is the lens closest to the eye. A microscope with only one eyepiece is **monocular**; one constructed with two eyepieces (one for each eye) is **binocular**.

Properties of the Compound Microscope

Each microscope lens is inscribed with a number signifying its magnifying power. The image viewed by the microscopist will have a total magnification equal to the product of the magnifying power of the objective and eyepiece lenses. For example, an eyepiece lens with a magnification of 10 times (10×) used in combination with an objective lens of 10× has a total magnification power of 100×. Most forensic work requires a 10× eyepiece in combination with either a 4×, 10×, 20×, or 45× objective (see Figure 8–5). The respective magnifications will be 40×, 100×, 200×, and 450×.

In addition, each objective lens is inscribed with its numerical aperture (N.A.). The ability of an objective lens to resolve details into separate images instead of one blurred image is directly proportional to the numerical aperture value of the objective lens. For example, an objective lens of N.A. 1.30 can separate details at half the distance of a lens with an N.A. of 0.65. The maximum useful magnification of a compound microscope is approximately 1,000 times the N.A. of the objective being used. This magnification is sufficient to permit the eye to see all the detail that can be resolved. Any effort to increase the total magnification beyond this figure will yield no additional detail and is referred to as *empty magnification*.

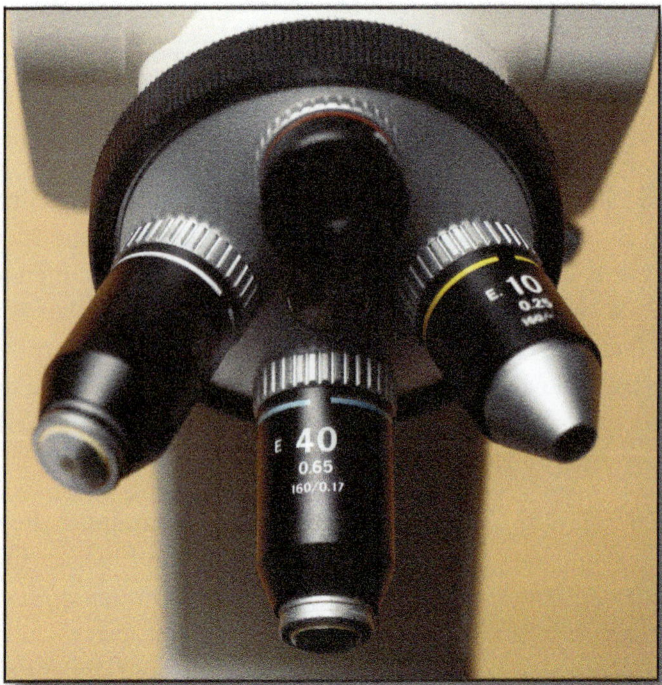

FIGURE 8–5
A microscope's revolving nosepiece shows the objective lenses available to view the specimen and their corresponding magnifications. *Ciaran Griffin/Stockbyte/Getty Images*

Although a new student of the microscope may be tempted to immediately choose the highest magnifying power available to view a specimen, the experienced microscopist weighs a number of important factors before choosing a magnifying power. A first consideration must be the size of the specimen area, or the **field of view**, that the examiner wishes to study. As magnifying power increases, the field of view decreases. Thus, it is best to first select a low magnification in which a good general overall view of the specimen is seen, and to switch later to a higher power in which a smaller portion of the specimen can be viewed in more detail.

field of view
The area of the specimen that can be seen after it is magnified

depth of focus
The thickness of a specimen that is entirely in focus under a microscope

The **depth of focus** is also a function of magnifying power. After a focus has been achieved on a specimen, the depth of focus defines the thickness of that specimen. Areas above and below this region will be blurred and can be viewed only when the focus is readjusted. Depth of focus decreases as magnifying power increases.

Quick Review

- Transparent illumination is light directed up toward and through a specimen from an illuminator built into the base of the microscope. It is used to view transparent specimens. Vertical or reflected illumination is light from a source above the specimen that reflects off the specimen's surface and into the lens system of the microscope. It is used to view opaque specimens.
- A condenser is a lens system under the microscope stage that collects light rays from the base illuminator and focuses them on the specimen.
- A monocular microscope has one eyepiece; a binocular microscope has two eyepieces.
- One determines the total magnification of a compound microscope by multiplying the magnifying power of the eyepiece lens by the magnifying power of the objective lens.
- When you increase a compound microscope's magnification, its field of view and depth of focus decrease.

The Comparison Microscope

Forensic microscopy often requires a side-by-side comparison of evidence, such as hair or fiber specimens. This kind of examination can best be performed with a comparison microscope, such as the one pictured in Figure 8–6.

Basically, the comparison microscope is two compound microscopes combined into one unit. The unique feature of its design is that it uses a bridge incorporating a series of mirrors and lenses to join two independent objective lenses into a single binocular unit. A viewer looking through the eyepiece lenses of the comparison microscope observes a circular field, equally divided into two parts by a fine line. The specimen mounted under the left-hand objective appears in the left half of the field, and the specimen under the right-hand objective appears in the right half of the field. It is important to closely match the optical characteristics of the objective lenses to ensure that both specimens are seen at equal

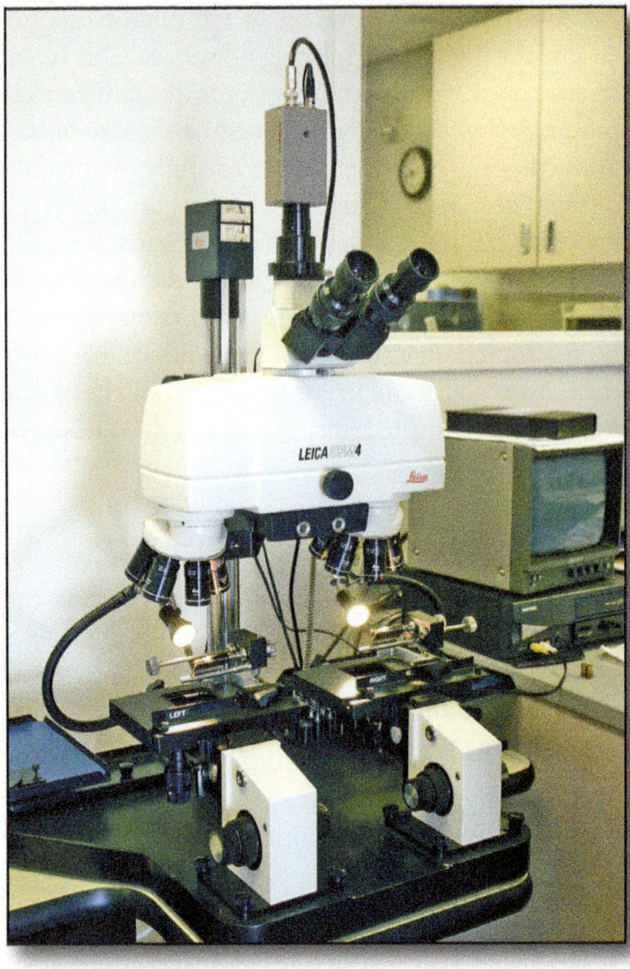

FIGURE 8–6
The comparison microscope—two independent objective lenses joined together by an optical bridge. *Leica Microsystems Inc*

magnification and with minimal but identical lens distortions. Comparison microscopes designed to compare opaque objects, such as bullets and cartridges, are equipped with vertical or reflected illumination. Comparison microscopes used to compare hairs or fibers use transmitted illumination.

Figure 8–7 shows the striation markings on two bullets that have been placed under the objective lenses of a comparison microscope. Modern firearms examination began with the introduction of the comparison microscope, with its ability to give the firearms examiner a side-by-side magnified view of bullets. Bullets that are fired through the same rifle barrel display comparable rifling markings on their surfaces. Matching the majority of striations present on each bullet justifies a conclusion that both bullets traveled through the same barrel.

Quick Review

- The comparison microscope consists of two independent objective lenses joined by an optical bridge to a common eyepiece lens. A viewer looking through the eyepiece lens observes the objects under investigation side-by-side in a circular field that is equally divided into two parts.
- Modern firearms examination began with the introduction of the comparison microscope, which gives the examiner a side-by-side magnified view of bullets.

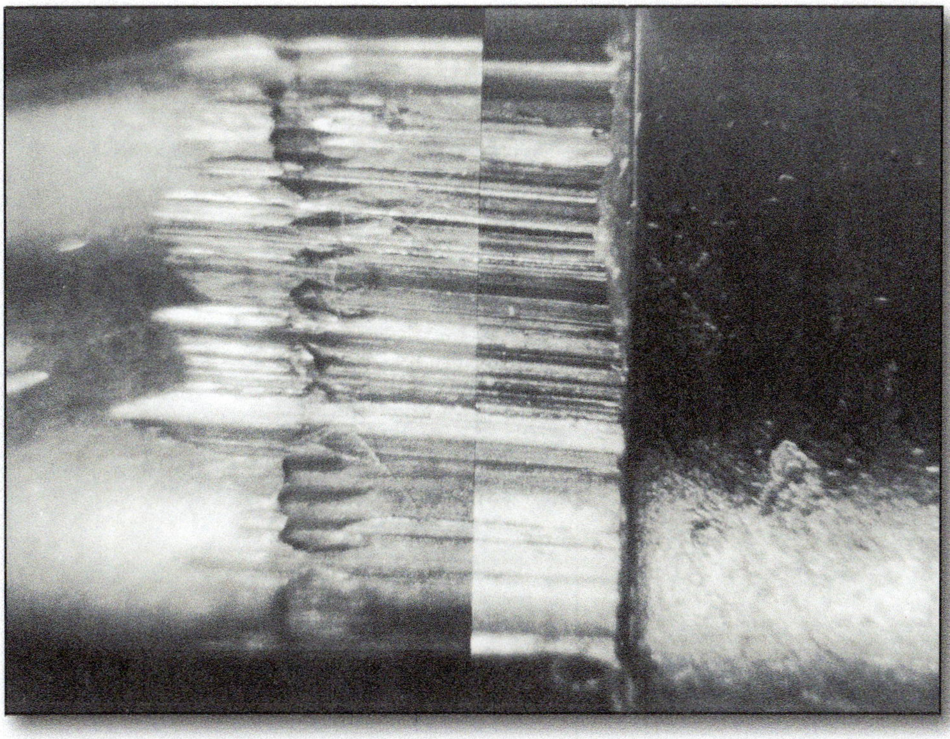

FIGURE 8–7
Photomicrograph taken through a comparison microscope. On the right are the striation markings on the test-fired bullet, fired through the suspect weapon. On the left are the markings of the crime-scene bullet. *Orlando/Three Lions/Hulton Archive/Getty Images*

The Stereoscopic Microscope

The details that characterize the structures of many types of physical evidence do not always require examination under very high magnifications. For such specimens, the stereoscopic microscope has proven quite adequate, providing magnifying powers from 10× to 125×. This microscope has the advantage of presenting a distinctive three-dimensional image of an object. Also, whereas the image formed by the compound microscope is inverted and reversed (upside-down and backward), the stereoscopic microscope is more convenient because prisms in its light path create a right-side-up image.

The stereoscopic microscope, shown in Figure 8–8, is actually two monocular compound microscopes properly spaced and aligned to present a three-dimensional image of a specimen to the viewer, who looks through both eyepiece lenses. The light path of a stereoscopic microscope is shown in Figure 8–9.

The stereoscopic microscope is undoubtedly the most frequently used and versatile microscope found in the crime laboratory. Its wide field of view and great depth of focus make it an ideal instrument for locating trace evidence in debris, garments, weapons, and tools. Furthermore, its potentially large working distance (the distance between the objective lens and the specimen) makes it ideal for microscopic examination of big, bulky items. When fitted with vertical illumination, the stereoscopic microscope becomes the primary tool for characterizing physical evidence as diverse as paint, soil, gunpowder residues, and marijuana.

FIGURE 8–8
A stereoscopic microscope.
Chuck Rausin/Shutterstock

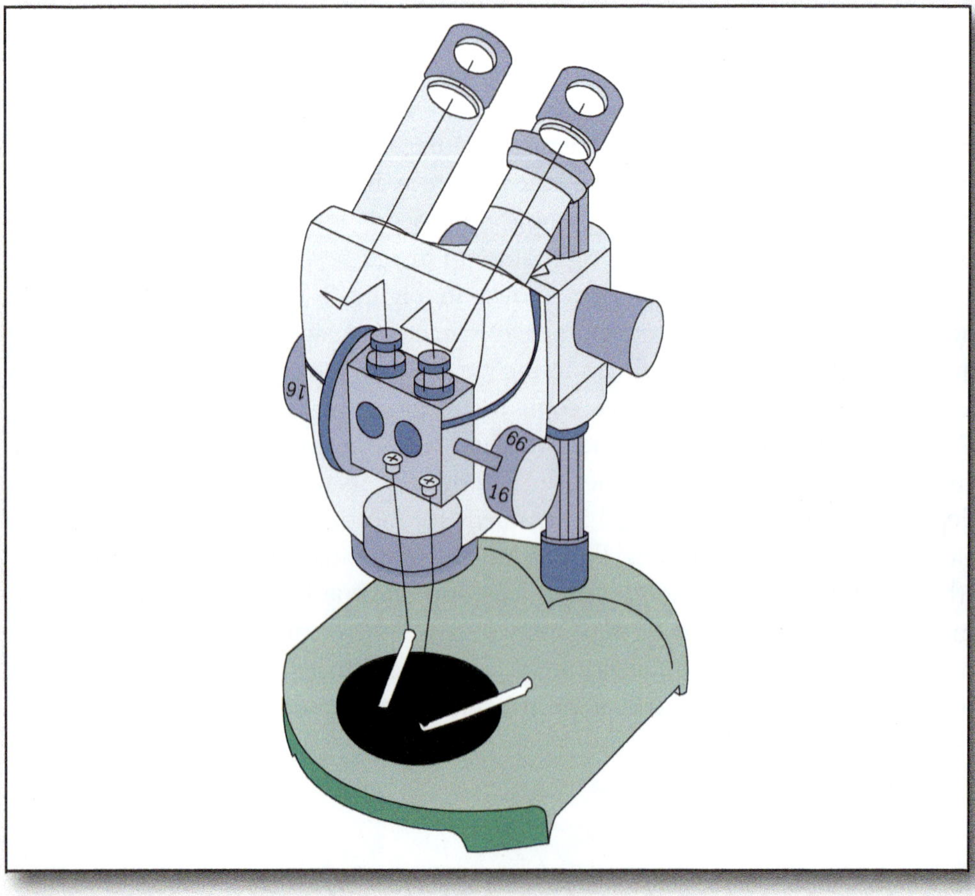

FIGURE 8–9
Schematic diagram of a stereoscopic microscope. This microscope is actually two separate monocular microscopes, each with its own set of lenses except for the lowest objective lens, which is common to both microscopes.

Quick Review

- The stereoscopic microscope consists of two monocular compound microscopes properly spaced and aligned to present a three-dimensional image of a specimen to the viewer, who looks through both eyepiece lenses.
- The large working distance of the stereoscopic microscope makes it ideal for microscopic examination of big, bulky items.

The Polarizing Microscope

Recall from Chapter 5 that light's wavelike motion in space can be invoked to explain many facets of its behavior. The waves that compose a beam of light can be pictured as vibrating in all directions perpendicular to the direction in which the light is traveling. However, when a beam of light passes through certain types of specially fabricated crystalline substances, it emerges vibrating in only one plane. Light that is confined to a single plane of vibration is said to be planepolarized. The device that creates **polarized light** in this manner is called a **polarizer**. A common example of this phenomenon is the passage of sunlight through polarized sunglasses. By transmitting light vibrating in the vertical plane only, these sunglasses eliminate or reduce light glare. Most glare consists of partially polarized light that has been reflected off horizontal surfaces and thus is vibrating in a horizontal plane.

polarized light
Light confined to a single plane of vibration

polarizer
A device that permits the passage of light waves vibrating in only one plane

Because polarized light appears no different to the eye from ordinary light, special means must be devised for detecting it. This is accomplished simply by placing a second polarizing crystal, called an analyzer, in the path of the polarized beam. As shown in Figure 8–10, if the polarizer and analyzer are aligned parallel to each other, the polarized light passes through and is seen by the eye. If, on the other hand, the polarizer and analyzer are set perpendicular to one another, or are "crossed," no light penetrates, and the result is total darkness or extinction.

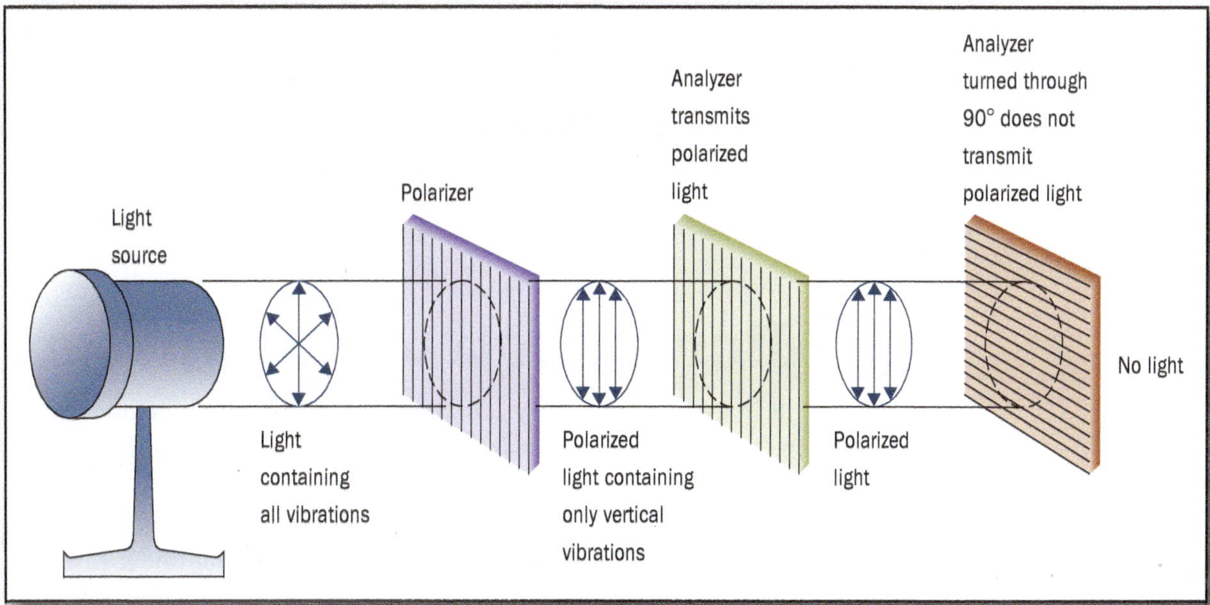

FIGURE 8–10 Polarization of light.

In this manner, a compound or stereoscopic microscope can be outfitted with a polarizer and analyzer to allow the viewer to detect polarized light. Such a microscope is known as a polarizing microscope. Essentially, the polarizer is placed between the light source and the sample stage to polarize the light before it passes through the specimen. The polarized light penetrating the specimen must then pass through an analyzer before it reaches the eyepiece and finally the eye. Normally the polarizer and analyzer are "crossed" so that when no specimen is in place, the field appears dark. However, introducing a specimen that polarizes light reorients the polarized light, allowing it to pass through the analyzer. This result produces vivid colors and intensity contrasts that make the specimen readily distinguishable.

The most obvious and important applications of this microscope relate to studying materials that polarize light. For example, as we learned in Chapter 5, many crystalline substances are birefringent; that is, they split a beam of light into two light-ray components of different refractive index values. What makes this observation particularly relevant to our discussion of the polarizing microscope is that the light beams are polarized at right angles to each other. Thus, polarizing microscopy has found wide application for the examination of birefringent minerals present in soil.

By using the immersion method and selecting the proper immersion liquids, a refractive index corresponding to each plane of polarized light can be determined. Thus, when a mineral is viewed under polarized light in a liquid that matches one of its refractive indices, the Becke line will no longer be visible. This information, plus observations on crystal color, form, and so on, makes it possible for the microscopist to identify the mineral. Similarly, criminalists use the fact that many synthetic fibers are birefringent to characterize them with a polarizing microscope.

Quick Review

- Light that is confined to a single plane of vibration is said to be planepolarized.
- The polarizing microscope made possible the examination of the interaction of plane-polarized light with matter.
- Polarizing microscopy is used to study birefringent materials, that is, materials that have a double refraction. Refractive index data provide information that helps identify minerals present in a soil sample or the identity of a synthetic fiber.

The Microspectrophotometer

Few instruments in a crime laboratory can match the versatility of the microscope. The microscope's magnifying power is indispensable for finding minute traces of physical evidence. Many items of physical evidence can be characterized by a microscopic examination of their morphological features. Likewise, the microscope can be used to study how light interacts with the material under investigation, or it can be used to observe the effects that other chemical substances have on such evidence. Each of these features allows an examiner to better characterize and identify physical evidence. Recently, linking the microscope to a computerized spectrophotometer has added a new dimension to its capability. This combination has given rise to a new instrument called the **microspectrophotometer**. In many respects, this is an ideal marriage from the forensic scientist's viewpoint. In Chapter 6, we saw how a chemist can use selective absorption of light by materials to characterize them. In particular, light in the ultraviolet, visible, and infrared regions of the electromagnetic spectrum is most helpful for this purpose. Unfortunately, in the past, forensic chemists were unable to take full advantage of the capabilities of spectrophotometry for examining trace evidence, because most spectrophotometers are not well suited for examining the very small particles frequently encountered as evidence.

microspectrophotometer An instrument that links a microscope to a spectrophotometer

With the development of the microspectrophotometer, a forensic analyst can view a particle under a microscope while a beam of light is directed at the particle to obtain its absorption spectrum. Depending on the type of light employed, an examiner can acquire either a visible or an IR spectral pattern of the substance being viewed under the microscope. The obvious advantage of this approach is to provide added information to characterize trace quantities of evidence. A microspectrophotometer designed to measure the uptake of visible light by materials is shown in Figure 8–11.

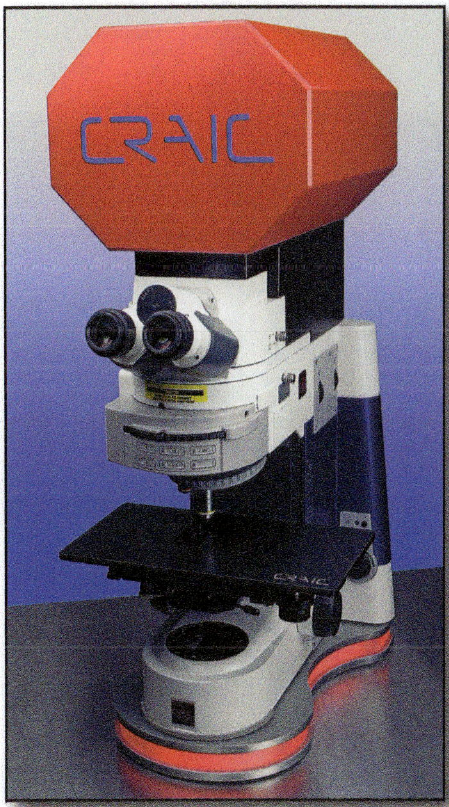

FIGURE 8–11
A visible-light microspectrophotometer.
Courtesy CRAIC Technologies Inc., Altadena, Calif.
www.microspectra.com

Visual comparison of color is usually one of the first steps in examining paint, fiber, and ink evidence. Such comparisons are easily obtained using a comparison microscope. A forensic scientist can use the microspectrophotometer to compare the color of materials visually while plotting an absorption spectrum for each item under examination. This displays the exact wavelengths at which each item absorbs in the visible-light spectrum. Occasionally colors that appear similar by visual examination show significant differences in their absorption spectra. An example of this approach is shown in Figure 8–12, in which the microspectrophotometer is used to distinguish counterfeit and authentic currency by comparing the spectral patterns of inked lines on the paper.

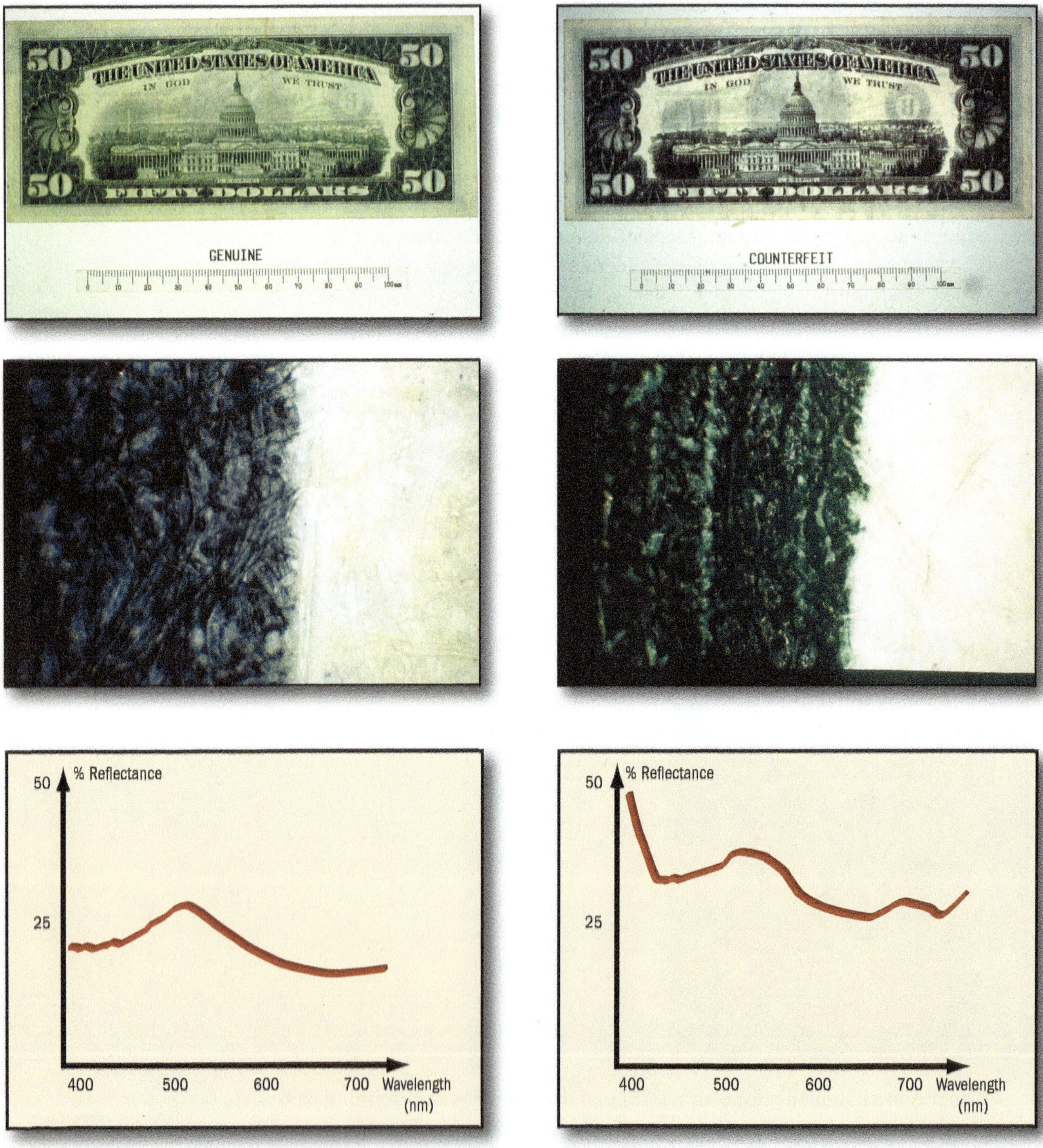

FIGURE 8–12 Two $50 bills are shown at top; one is genuine and the other is counterfeit. Below each bill is a microphotograph of an inked line present on each bill. Each line was examined under a visible-light microspectrophotometer. As shown, the visible absorption spectrum of each line is readily differentiated, thus allowing the examiner to distinguish a counterfeit bill from genuine currency. *Courtesy Peter W. Pfefferli, forensic scientist, Lausanne, Switzerland*

Another emerging technique in forensic science is the use of the infrared microspectrophotometer to examine fibers and paints. The "fingerprint" IR spectrum is unique for each chemical substance. Therefore, obtaining such a spectrum from either a fiber or a paint chip allows the analyst to better identify and compare the type of chemicals from which these materials are manufactured. With a microspectrophotometer, a forensic analyst can view a substance through the microscope and at the same time have the instrument plot the infrared absorption spectrum for that material.

Quick Review

- The microspectrophotometer is a spectrophotometer coupled with a light microscope. The device allows an examiner studying a specimen under the microscope to obtain the visible-light or IR spectrum of the material.

The Scanning Electron Microscope (SEM)

All the microscopes described thus far use light coming off the specimen to produce a magnified image. The scanning electron microscope (SEM) is, however, a special case in the family of microscopes (see Figure 8–13). The image is formed by aiming a beam of electrons onto the specimen and studying electron emissions on a closed TV circuit.

In the SEM, a hot tungsten filament emits a beam of electrons, which is focused by electromagnets onto the surface of the specimen. This primary electron beam causes the elements that make up the upper layers of the specimen to emit electrons known as secondary electrons. About 20–30 percent of the primary electrons rebound off the surface of the specimen. These electrons are known as *backscattered electrons.* The emitted electrons (both secondary and backscattered) are collected, and the amplified signal is displayed on a cathode-ray or TV tube. By scanning the primary electron beam across

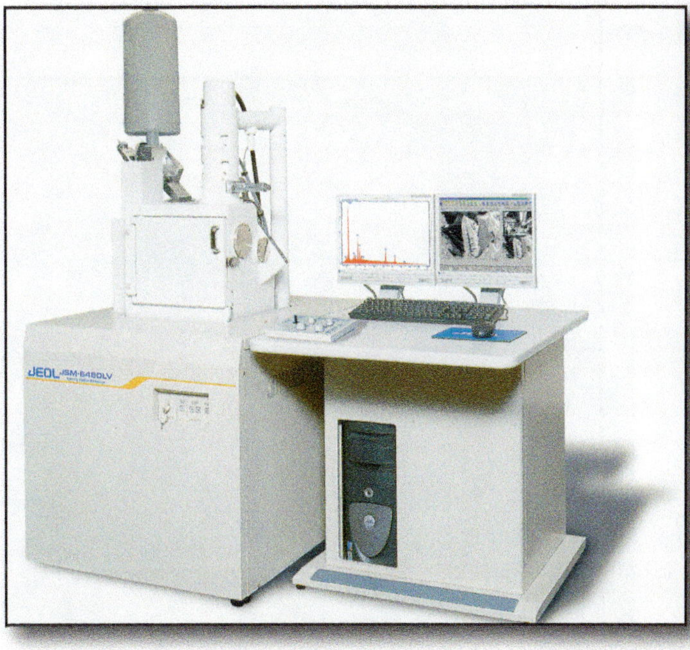

FIGURE 8–13
A scanning electron microscope. *Courtesy Jeol USA Inc., Peabody, Mass., www.jeolusa.com*

the specimen's surface in synchronization with the cathode-ray tube, the SEM converts the emitted electrons into an image of the specimen for display on the cathode-ray tube.

The major attractions of the SEM image are its high magnification, high resolution, and great depth of focus. In its usual mode, the SEM has a magnification that ranges from 10× to 100,000×. Its depth of focus is some 300 times better than optical systems at similar magnifications, and the resultant picture is almost stereoscopic in appearance. Its great depth of field and magnification are exemplified by the magnification of cystolithic hair on the marijuana leaf, as shown in Figure 8–14. An SEM image of a vehicle's headlight filaments may reveal whether the headlights were on or off at the time of a collision (see Figures 8–15 and 8–16).

Another facet of scanning electron microscopy has been the use of X-ray production to determine the elemental composition of a specimen. X-rays are generated when

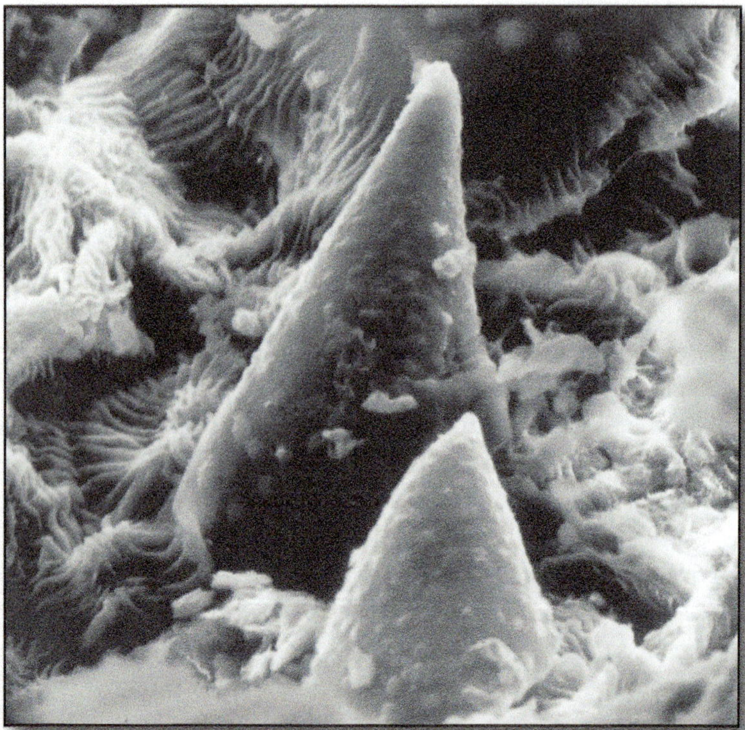

FIGURE 8–14
The cystolithic hairs of the marijuana leaf, as viewed with a screening electron microscope (800x).
Courtesy Richard Saferstein, Ph.D.

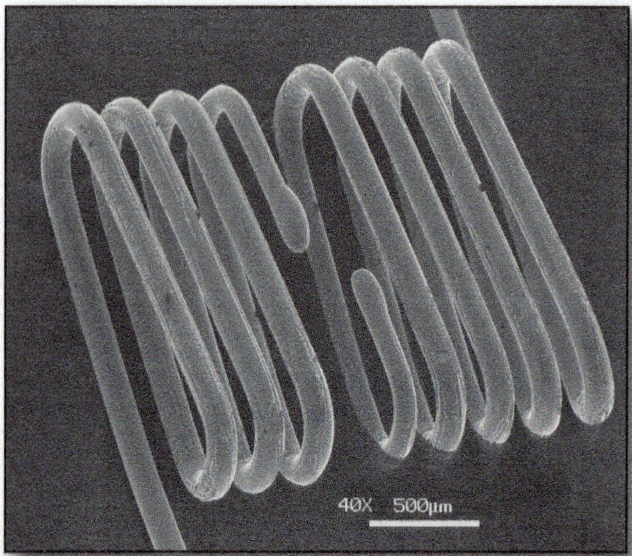

FIGURE 8–15
The melted ends of a hot filament break indicate that the headlights were on when an accident occurred.
Courtesy Jeol USA Inc., Peabody, Mass., www.jeolusa.com

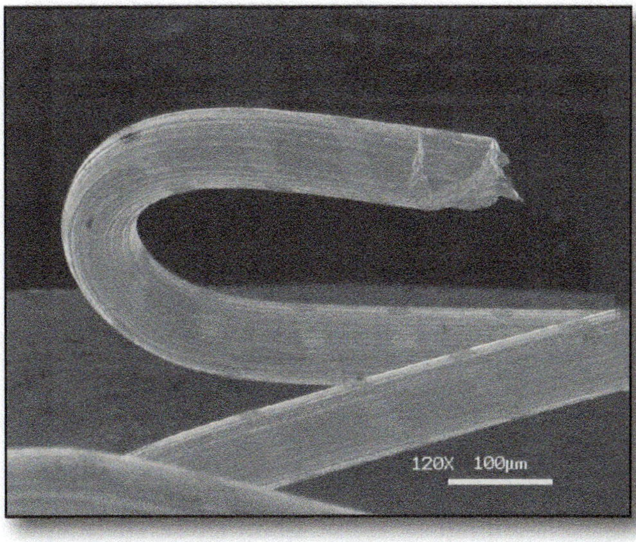

FIGURE 8–16
The sharp ends of a cold filament break indicate that the headlights were off when an accident occurred. *Courtesy Jeol USA, Inc., Peabody, Mass., www.jeolusa.com*

the electron beam of the scanning electron microscope strikes a target. When the SEM is coupled to an X-ray analyzer, the emitted X-rays can be sorted according to their energy values and used to build a picture of the elemental distribution in the specimen. Because each element emits X-rays of characteristic energy values, the X-ray analyzer can identify the elements present in a specimen. Furthermore, the element's concentration can be determined by measuring the intensity of the X-ray emission.

One application of scanning electron microscopy has been to determine whether a suspect has recently fired a gun. In this case, an attempt is made to remove any gunshot residue that remain on a shooter's hands by lifting them off with a piece of adhesive tape. The tape is then examined under the SEM for the presence of particles that may have originated from the bullet primer. These particles can be characterized by their size, shape, and elemental composition. As shown in Figure 8–17, when the sample of gunshot residue is exposed to a beam

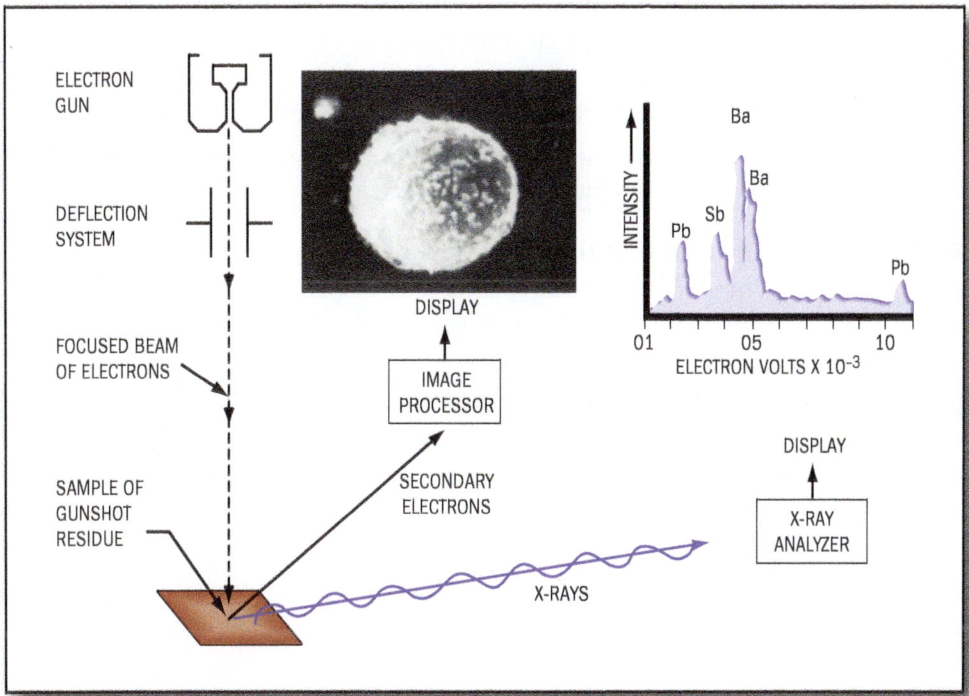

FIGURE 8–17 A schematic diagram of a scanning electron microscope displaying the image of a gunshot residue particle. Simultaneously, an X-ray analyzer detects and displays X-ray emissions from the elements lead (Pb), antimony (Sb), and barium (Ba) present in the particle. *Courtesy Richard Saferstein, Ph.D.*

of electrons from the scanning electron microscope, X-rays are emitted. These X-rays are passed into a detector, where they are converted into electrical signals. These signals are sorted and displayed according to the energies of the emitted X-rays. Through the use of this technique, the elements lead, antimony, and barium, frequently found in most primers, can be rapidly detected and identified.

Forensic Palynology: Pollen and Spores as Evidence

Of the many plant species on earth, more than half a million produce pollen or spores. The pollen or spores produced by each species has a unique type of ornamentation and morphology. This means that pollen or spores can be identified and used to provide links between a crime scene and a person or object if examined by a trained analyst. This technique is called forensic palynology and includes the collection and examination of pollen and spores connected with crime scenes, illegal activities, or terrorism. Microscopy is the principal tool used in the field of forensic palynology.

Characteristics of Spores and Pollen In nature, pollen grains are the singlecelled male gametophytes (reproductive cells) of seed-bearing plants. The pollen grain wall (*exine*) is durable because it protects and carries "the sperms" needed for plant reproduction. Spores consist of both the male and female gametes of plants such as algae, fungi, mosses, and ferns. Pollen-producing plants are either *anemophilous* (their pollen is dispersed by wind) or *entomophilous* (their pollen is carried and dispersed by insects or small animals). Fairly precise geographical locations can often be identified by the presence of different mixtures of airborne pollens produced by anemophilous plants. For example, it may be possible to identify a geographical origin by a profile of the pollen samples retrieved from a suspect's clothing simply by the type and percentages of airborne pollen grains. Entomophilous plants usually produce a small amount of pollen that is very sticky in nature. Therefore, this type of pollen is very rarely deposited on clothing or other objects except by direct contact with the plant. This information is useful when reconstructing the events of a crime because it indicates that the clothing, a vehicle, or other objects may have come into direct contact with plant types found at a crime scene.

Analysis of Spores and Pollen Both spores and pollen are microscopic in size and are produced by the adult plants and dispersed by the millions, and both can be analyzed using similar methods that use a variety of microscopic techniques. Using a compound light microscope with magnification capabilities up to 1,000×, analysts usually can identify pollen and spores as coming from a specific plant family or genus, and at times even a unique species. However, often the pollen or spores of related species may appear so similar that identification of the species is possible only with careful analysis using an SEM (see Figure 8–18).

Unique shapes, aperture type, and surface ornamentation are typically used to identify spore samples. Useful features for characterizing pollen grains include shape, apertures, and wall and surface sculpturing. Shapes of pollen grains include

FIGURE 8–18 Scanning electron micrograph of ragweed pollen magnified 950 times. *Mediscan/Alamy Stock Photo*

spheres, triangles, ellipses, hexagons, pentagons, and many other geometric variations. *Apertures* are the openings on pollen grains from which the pollen tube grows and carries the sperms to the egg to complete fertilization. *Sculpturing* of the pollen refers to the pattern of the pollen grain surface.

To avoid destruction or contamination, early collection of forensic pollen samples for analysis is important and should be completed as soon as possible at a crime scene by a trained palynologist. This kind of expert's first task is to calculate the estimated production and dispersal patterns of spores and pollen (called the *pollen rain*) for the crime scene or area of interest, and then using that information to produce a kind of "pollen fingerprint" of that location.

The information gained from the analysis of pollen and spore evidence has many possible uses. It can link a suspect or object to the crime scene or the victim, prove or disprove a suspect's alibi, include or exclude suspects, track the previous whereabouts of some item or suspect, or indicate the geographical origin of some item. In the past, pollen and spore evidence has also helped locate human remains and concealed burial sites, established the season or time of death of a victim, located the source areas of illegal drugs and fake pharmaceuticals, imprisoned terrorists, and proven the perpetration of illegal poaching or the adulteration of commercial foods.

Quick Review

- The scanning electron microscope (SEM) bombards a specimen with a beam of electrons instead of light to produce a highly magnified image. This produces X-ray emissions that can be used to characterize elements present in the material.
- Forensic palynology involves the collection and examination of pollen and spores connected with crime scenes, illegal activities, or terrorism. The scanning electron microscope is the principal tool used in the field of forensic palynology.

Case Files

Identifying Baby Doe

The body of a young girl was found on the shore of Deer Island outside Boston by a woman walking her dog (Figure 8–19). The child soon came to be known as "Baby Doe" by the local news media.

FIGURE 8–19 The beach on Deer Island where the child's body was discovered. *Craig F. Walker/The Boston Globe/Getty Image*

Two weeks later, police released a computer-generated composite image of the girl along with photos of the polkadot leggings and zebra-print blanket that were discovered with her body. In the days after the images were released, tips poured in as millions viewed the photos. Baby Doe's image was placed on billboards in 50 locations throughout the state. In the next three months, investigators followed up 210 leads but were unsuccessful in identifying her.

Investigators began considering the possibility that she washed up on the shore of Deer Island from a boat, or was transported to the area from somewhere else. In order to determine if they were focusing their search in the right place, they explored two forensic examination types which have the potential to indicate geographic origin: Palynology and Stable Isotope Analysis.

To perform the pollen analysis, investigators took vacuum samples from the blanket and pants, and forensic scientists were able to retrieve enough pollen to complete a standard pollen count (Figure 8–20). The pollen taxa encountered were expected to be found in the broad-leafed forests of the Northeastern United States.

The Massachusetts State Police also sent

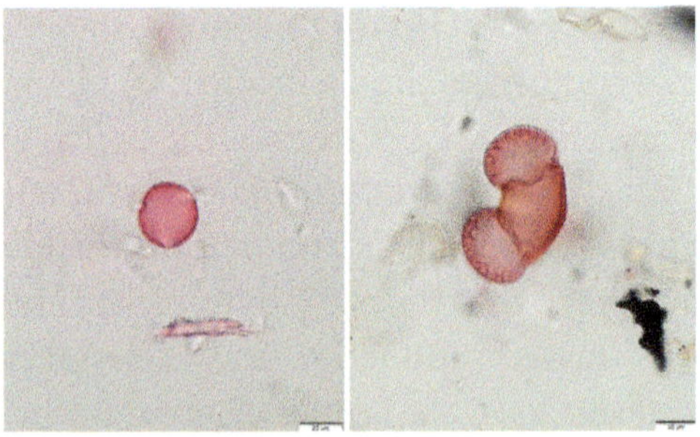

FIGURE 8–20 The pollen spores from the child's clothes were dominated by oak (Quercus), *left*, and pine (Pinus), *right*. *Courtesy of Massachusetts State Police*

samples to IsoForensics, a lab in Salt Lake City, Utah. Using stable isotopes in the girl's hair and teeth, analysts could examine the chemical characteristics she encountered throughout her life.

"What you're eating, what you're drinking, that's recorded in your tissues immediately," IsoForensics President Lesley Chesson said. "You're recording the isotopes of your drinking water into your hair, into your bones." Researchers use that information to paint a general picture of the person's life; where they lived, where they have been recently. An isotope analysis showed the girl spent time in New England before her death.

Efforts of law enforcement paid off when a tip led to the identity of Baby Doe in September 2015. The girl was positively identified as Bella Bond, 2 ½ years old, from Dorchester, MA. As a result of the tip, police subsequently charged Michael P. McCarthy, the child's mother's boyfriend, with her murder. He was convicted and sentenced to life in prison.

Case Files

Clues from the Cornfield

A case exemplifying the application of forensic palynology to a criminal investigation occurred when a victim was kidnapped, robbed, and then murdered in the eastern part of the American Midwest. The victim's car was stolen, but later abandoned when it got stuck in mud near a busy highway. The next night a drifter was arrested in a nearby town for breaking into a closed store. While in jail awaiting trial, the drifter told a fellow inmate about his car being stuck in the mud and that he would not be in jail but for that mishap. The other prisoner, hoping to work a deal for a lighter sentence, told this story to the sheriff.

During the investigation of the crime scene, one of the law enforcement agents noticed that there was a large field of mature corn (maize, Zea mays) growing between the dirt road, where the stolen car had been abandoned in the mud, and the nearby highway leading to the next town. The investigator wondered if traces of torn maize leaves on the suspect's clothing might link him to the crime scene. Fortunately, the drifter's shirt and pants had been removed and stored in sterile paper bags when he was arrested. As with all prisoners in that region, he had been given a pair of orange overalls to wear while in jail.

The shirt and pants were sent to a botanist who was asked to search for traces of maize leaves on the clothing. The botanist was also a palynologist, and thus also collected samples and searched for traces of pollen. The pollen samples provided the best results. The samples collected from the suspect's shirt revealed that the neck and shoulder region of the shirt had high concentrations of fresh maize pollen. The forensic sample collected from the pants also contained maize pollen, but in a lower percentage. The forensic pollen data indicated that the drifter had recently walked through a maize field, similar to the one between the abandoned car and the highway. As he walked through the field, he had brushed against blooming male tassels on the corn plants that were about head high. This accounted for the high amount of maize pollen found on the shoulder and neck area of the shirt. Lesser amounts of maize pollen also fell on his pants as he walked through the field. While the suspect awaited trial, additional evidence and several fingerprints from the victim's farm also linked him to the murder.

Source: V. M. Bryant and G. D. Jones, "Forensic Palynology: Current Status of a Rarely Used Technique in the United States of America," *Forensic Science International*, vol. 163 (2006), pp.183–197.

Chapter Review

- A microscope is an optical instrument that uses a lens or a combination of lenses to magnify and resolve the fine details of an object.
- Two types of images are virtual images and real images. A real image can be seen directly with the naked eye; a virtual image can be seen only by a viewer looking through a lens.
- The eyepiece lens is the upper lens through which the viewer looks. The objective lens is the lower lens that is positioned directly over the specimen.
- Transparent illumination is light directed up toward and through a specimen from an illuminator built into the base of the microscope. It is used to view transparent specimens. Vertical or reflected illumination is light from a source above the specimen that reflects off the specimen's surface and into the lens system of the microscope. It is used to view opaque specimens.
- A condenser is a lens system under the microscope stage that collects light rays from the base illuminator and focuses them on the specimen.
- A monocular microscope has one eyepiece; a binocular microscope has two eyepieces.
- One determines the total magnification of a compound microscope by multiplying the magnifying power of the eyepiece lens by the magnifying power of the objective lens.
- When you increase a compound microscope's magnification, its field of view and depth of focus decrease.
- The comparison microscope consists of two independent objective lenses joined by an optical bridge to a common eyepiece lens. A viewer looking through the eyepiece lens observes the objects under investigation side-by-side in a circular field that is equally divided into two parts.
- Modern firearms examination began with the introduction of the comparison microscope, which gives the examiner a side-by-side magnified view of bullets.
- The stereoscopic microscope consists of two monocular compound microscopes properly spaced and aligned to present a three-dimensional image of a specimen to the viewer, who looks through both eyepiece lenses.
- The large working distance of the stereoscopic microscope makes it ideal for microscopic examination of big, bulky items.
- Light that is confined to a single plane of vibration is said to be plane-polarized.
- The polarizing microscope made possible the examination of the interaction of plane-polarized light with matter.

- Polarizing microscopy is used to study birefringent materials, that is, materials that have a double refraction. Refractive index data provide information that helps identify minerals present in a soil sample or the identity of a synthetic fiber.
- The microspectrophotometer is a spectrophotometer coupled with a light microscope. The device allows an examiner studying a specimen under the microscope to obtain the visible-light or IR spectrum of the material.
- The scanning electron microscope (SEM) bombards a specimen with a beam of electrons instead of light to produce a highly magnified image. This produces X-ray emissions that can be used to characterize elements present in the material.
- Forensic palynology involves the collection and examination of pollen and spores connected with crime scenes, illegal activities, or terrorism. The scanning electron microscope is the principal tool used in the field of forensic palynology.

Quick Lab: Focusing the Microscope

Materials:

Compound light microscope
Prepared evidence slides

Procedure:

The microscope is one of the most important tools in the laboratory. The first step in using the microscope is being able to focus on what you want to see. Take a prepared slide and place it on the stage, securing it in place with the stage clips. Always start with the low-power objective lens in place. Use the coarse adjustment knob (the larger knob) to bring the object into view. To bring the object into focus, always focus upward to avoid crushing the slide with the lens. Once you can see the object, use the fine adjustment knob (the smaller knob) to focus the item. If the slide needs to be adjusted, many microscopes have stage adjustment knobs that will move the stage or slide. If your microscope does not, you will have to manually move the slide. Once you have the object focused in the center of your view, try to view the object under high power. To do this, move the high-power lens into place above the slide. Use the fine adjustment knob to focus the object; you should not have to use the coarse adjustment knob (doing so may break the slide).

Follow-Up Questions:

1. To determine total magnification, multiply the objective lens power by the eyepiece lens power. What was the total magnification of your microscope when you were viewing the object under low power? What about under high power?
2. Besides magnification, what did you observe about the real image (what your naked eye saw on the slide) compared to the virtual image (what you saw looking through the scope)?
3. Did you observe anything when you moved the slide on the stage while looking through the scope?

Quick Lab: Creating Wet-Mount Slides

Materials:

Compound light microscope
Slide with cover slip
Water with dropper
Sample object (examples: hair, soil, thread, paper)

Procedure:

To view some objects, it is better to create a wet-mount slide. This helps create a clearer view than a regular slide would allow. To create a wet-mount slide, place the sample you want to view on the slide. Now, place a drop of water on the sample. Take the cover slip and slowly cover the sample so that there are no air bubbles in the water; these would block your view of the sample. The slide is now ready to be viewed on the scope. Remember the steps to focus your slide.

Follow-Up Questions:

1. After looking at the sample as a wet-mount slide, create a regular slide with the same type of sample. Do you see a difference between the two types of slides and the images they produce?
2. Other than the sample you viewed, can you think of any other types of evidence that could be viewed using a wet-mount slide?

Review Questions

1. Which of the following is not a part of the mechanical system of a compound microscope?
   a. coarse adjustment
   b. objective lens
   c. body tube
   d. stage

2. Which microscope is the most frequently used and versatile microscope found in the crime laboratory?
   a. the comparison microscope
   b. the stereoscopic microscope
   c. the compound microscope
   d. the polarizing microscope

3. The major attractions of the scanning electron microscope (SEM) include all of the following except
   a. its high magnification.
   b. its ability to polarize light.
   c. its high resolution.
   d. its great depth of focus.

4. Which microscope can be used to determine whether a suspect has recently fired a gun?
   a. a scanning electron microscope
   b. a polarizing microscope
   c. a compound microscope
   d. a comparison microscope

5. The "fingerprint" IR spectrum can be seen using a
   a. scanning electron microscope.
   b. microspectrophotometer.
   c. polarizing microscope.
   d. stereoscopic microscope.

6. True or False: The mechanical system of a microscope is composed of six parts.

7. True or False: The magnification of a microscope with a 10× eyepiece and a 10× objective will be 1,000×.

8. True or False: The polarizing microscope is best used for forensic microscopy requiring a side-by-side comparison.

9. True or False: The stereoscopic microscope is the most frequently used and versatile microscope found in the crime laboratory.

10. True or False: With the development of the scanning electron microscope, a forensic analyst can now view a particle under a microscope while, at the same time, a beam of light is directed at the particle in order to obtain its absorption spectrum.

11. Name two types of images and explain the difference between them.

12. List and define the two types of lenses in an ordinary microscope.

13. Name two types of illumination and explain the difference between them. What types of specimens are viewed under each type of illumination?

14. What is a condenser? Name and describe the simplest type of condenser.

15. Define parfocal.

16. What is the difference between a monocular and a binocular microscope?

17. How does one determine the total magnification of a compound microscope?

18. What does numerical aperture describe?

19. What is the maximum useful magnification for a compound microscope? Why?

20. What happens to a compound microscope's field of view when you increase its magnification? How does this affect the way a microscopist views specimens under a compound microscope?

21. What is a comparison microscope? How is the image viewed through a comparison microscope different from one viewed through an ordinary or compound microscope?

22. When using a comparison microscope, why is it important to closely match the optical characteristics of the objective lenses?

23. What is the most frequently used and versatile microscope found in the crime laboratory? List three reasons why this type of microscope is so widely used.

24. What is plane-polarized light?

25. Name two types of physical evidence that are often characterized using a polarizing microscope.

26. What is a microspectrophotometer? What advantage does it enjoy over the spectrophotometer for analyzing physical evidence?

27. Briefly explain how a scanning electron microscope (SEM) produces an image.

28. What device can be coupled to an SEM to produce a picture of the elemental distribution of a specimen? Briefly describe how this combination of devices can be used to determine whether a suspect has recently fired a gun.

Learn the Vocabulary

Use the information from this chapter along with your prior knowledge and experience to learn the key terms. Choose from the following activities:

1. Make flashcards with a term on the front and its definition on the back. Use the flashcards to quiz a partner on correctly pronouncing and defining the terms.

2. Make an audio or video recording of each term and its definition.

3. Work with a partner to write a sentence for each term. Read the sentences out loud to a partner or share them with the class.

4. Write a list of new terms and language you have encountered in this chapter. Reuse the terms to write sentences and then read the sentences out loud to a partner. Reuse the terms again by writing a short speech. The speech might explain the terms or teach a lesson using the terms. Give the speech to a partner or to the class.

Application and Critical Thinking

1. A compound microscope has eyepiece lenses of 10×, 15×, and 20× magnification and objective lenses with 4×, 10×, 25×, 50×, and 100× magnifications. Determine the total magnification power of every possible combination of lenses. Which combinations of lenses would a forensic scientist be unlikely to use and why?

2. What type of microscope would a forensic scientist most likely use in each of the following situations? Explain your answers.
   a. examining two bullets to see if they were fired by the same gun
   b. determining the chemical composition of paint chips
   c. identifying birefringent minerals present in soil
   d. examining a large specimen, such as a tool or piece of clothing
   e. examining extremely tiny specimens that require very high magnification

Laboratory Experiments

This activity requires the use of the following practices of science and engineering:

- Analyzing and interpreting data
- Constructing explanations

This activity consists of the following crosscutting concepts:

- **Patterns**—The observed pattern of the fibers helps to distinguish their identity.
- **Cause and effect**—The Locard Principle is based on the idea that every contact leaves a trace. The presence or absence of hair evidence can be used to determine if a person was present in a given location.

Before you begin, read the scenario and steps out loud with a partner. Ask for clarification as necessary to make sure you understand the experiment.

The Scenario

On Sunday morning, a man's body was found in the laundry room of an apartment building on the eastside of town. The man, who was a resident of the building, had been badly beaten. Crime scene investigators recovered several different fibers from the victim's body. All but one of the fibers match fabrics in the victim's home. One fiber, a blue one, does not match anything belonging to the victim. Neighbors tell investigators that the victim was not a fan of blue, and probably did not own anything that color.

Residents, questioned by police, reveal that they saw a tall man of average build in a blue coat or jacket arguing with the man earlier in the evening. The night watchman reported that only residents entered or exited the building the night of the murder. Therefore, the police believe that the murderer lives in the apartment building.

Police begin to question all men in the building that match the description to determine if they own a blue coat. They also checked the labels to find out what kind of fibers make up the coat material. The following men are considered suspects:

- Michael has a blue sport coat made of wool
- Daniel is the owner of an expensive silk blue jacket

- Roman owns an old blue hoodie made of cotton
- Carter owns a blue rayon coat
- Ryan wears his polyester blue coat almost everyday

You have been called into the lab to analyze the fibers as empirical evidence and determine if there is a match.

PROCEDURE

1. Prepare a wet mount slide of the fiber removed from the victim by placing it on the slide, adding a drop of water and covering the fiber and water with a cover slip.
2. Examine the fiber under low, medium and high magnification of your microscope. Sketch what you see in the table below.
3. Prepare slides with the known samples of wool, rayon, silk, polyester, and cotton. Sketch each of these samples at the magnification that gives you the best view, and record these sketches in the data table.

   Note: All sketches should include the following:

   - Details of parts
   - Color—include shades and depth
   - Easy to view—not too small or large and centered
   - Labels on all parts that are sketched
   - Measurement in metric
   - Notes and descriptions of unusual things you observed
4. Write a description of the crime scene fabric, and using the data determine the type of fabric. Which suspect do you believe committed the crime based on your analysis of the evidence?

Crime Scene Fiber	*Low*	*Medium*	*High*

Known Fiber	*Low*	*Medium*	*High*
Wool			
Rayon			
Silk			
Polyester			
Cotton			

It appears that the fire was started in several locations: the kitchen, living room, and den.

Forensic Serology

9

Myung J. Chun UPI Photo Service/Newscom

Key Terms

acid phosphatase
agglutination
allele
antibody
antigen
antiserum
aspermia
chromosome
deoxyribonucleic acid (DNA)
egg
enzyme
erythrocyte
gene
genotype
hemoglobin
heterozygous
homozygous
hybridoma cells
locus
luminol
monoclonal antibodies
oligospermia
phenotype
plasma
polyclonal antibodies
precipitin
serology
serum
sperm
X chromosome
Y chromosome
zygote

O. J. Simpson—A Mountain of Evidence

On June 12, 1994, police arrived at the home of Nicole Simpson only to view a horrific scene.

The bodies of O. J. Simpson's estranged wife and her friend Ron Goldman were found on the path leading to the front door of Nicole's home. Both bodies were covered in blood and had suffered deep knife wounds. Nicole's head was nearly severed from her body. This was not a well-planned murder. A trail of blood led away from the murder scene. Blood was found in O. J. Simpson's Bronco. Blood drops were on O. J.'s driveway and in the foyer of his home. A blood-soaked sock was located in O. J. Simpson's bedroom, and a bloodstained glove rested outside his residence.

As DNA was extracted and profiled from each bloodstained article, a picture emerged that seemed to irrefutably link Simpson to the murders. A trail of DNA leaving the crime scene was consistent with O. J.'s profile, as was the DNA found entering Simpson's home. Simpson's DNA profile was found in the Bronco along with that of both victims. The glove contained the DNA profiles of Nicole and Ron, and the sock had Nicole's DNA profile. At trial, the defense team valiantly fought back. Miscues in evidence collection were craftily exploited. The defense strategy was to paint a picture of not only an incompetent investigation, but one that was tinged with dishonest police planting evidence. The strategy worked. O. J. Simpson was acquitted of murder.

Learning Objectives

After studying this chapter, you should be able to:

- 9.1 Explain the nature of blood and the concept of antigen–antibody interactions
- 9.2 Explain the application of serology in typing whole blood
- 9.3 Describe forensic tests used to characterize a stain as blood
- 9.4 Discuss the principles of heredity
- 9.5 Summarize the laboratory tests necessary to characterize seminal stains
- 9.6 Describe the collection and analysis of physical evidence in a rape investigation

National Science Content Standards

Scientific Inquiry

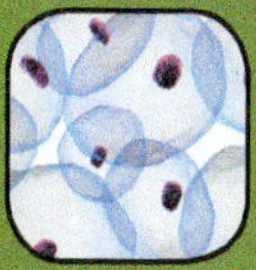
Life Science

Science and Technology

History and Nature of Science

In 1901, Karl Landsteiner announced one of the most significant discoveries of the 20th century—the typing of blood—a finding that 29 years later earned him a Nobel Prize (see Figure 9–1). For years physicians had attempted to transfuse blood from one individual to another. Their efforts often ended in failure because the transfused blood tended to coagulate, or clot, in the body of the recipient, causing instantaneous death. Landsteiner was the first to recognize that all human blood was not the same; instead, he found that blood is distinguishable by its group or type.

FIGURE 9–1 Portrait of Karl Landsteiner (1868–1943), Austrian-US pathologist and Nobel Laureate. *NLM/Science Source*

Out of Landsteiner's work came the classification system that we call the A-B-O system. Now physicians had the key for properly matching the blood of a donor to that of a recipient. One blood type cannot be mixed with a different blood type without disastrous consequences. This discovery, of course, had important implications for blood transfusion and the millions of lives it has since saved.

Meanwhile, Landsteiner's findings opened a new field of research in the biological sciences. Others began to pursue the identification of additional characteristics that could further differentiate blood. By 1937, the Rh factor in blood had been demonstrated and, shortly thereafter, numerous blood factors or groups were discovered. More than 100 different blood factors have been identified. However, the ones in the A-B-O system are still the most important for properly matching a donor and recipient for a transfusion.

Until the early 1990s, forensic scientists focused on blood factors, such as A-B-O, as offering the best means for linking blood to an individual. What made these factors so attractive was that in theory no two individuals, except for identical twins, could be expected to have the same combination of blood factors. In other words, blood factors are controlled genetically and have the potential of being a highly distinctive feature for personal identification. What makes this observation so relevant is the great frequency of bloodstains at crime scenes, especially crimes of the most serious nature— homicides, assaults, and rapes. Consider, for example, a transfer of blood between the victim and assailant during a struggle; that is, the victim's blood is transferred to the suspect's garment or vice versa. If the criminalist could individualize human blood by identifying all of its known factors, the result would be strong evidence for linking the suspect to the crime scene.

deoxyribonucleic acid (DNA)
The molecules that carry the body's genetic information

The advent of DNA technology has dramatically altered the approach of forensic scientists toward individualization of bloodstains and other biological evidence. The search for genetically controlled blood factors in bloodstains has been abandoned in favor of characterizing biological evidence by select regions of our **deoxyribonucleic acid (DNA)**, which carries the body's genetic information. As a result, the individualization of dried blood and other biological evidence has become a reality and has significantly altered the role that crime laboratories play in criminal investigations. In fact, as we will learn in the next chapter, the high sensitivity of DNA analysis has even altered the types of materials collected from crime scenes in the search for DNA.

The next chapter is devoted to discussing recent breakthroughs in associating blood and semen stains with a single individual through characterization of DNA. This chapter focuses on underlying biological concepts that forensic scientists historically relied on as they sought to characterize and individualize biological evidence before the dawning of the age of DNA.

The Nature of Blood

The word *blood* refers to a complex mixture of cells, enzymes, proteins, and inorganic substances. The fluid portion of blood is called **plasma**; it is composed principally of water and accounts for 55 percent of blood content. Suspended in the plasma are solid materials consisting chiefly of several types of cells—red blood cells (**erythrocytes**), white blood cells (leukocytes), and platelets. The solid portion of blood accounts for 45 percent of its content. Blood clots when a protein in the plasma known as fibrin traps and enmeshes the red blood cells. If the clotted material were removed, a pale yellowish liquid known as **serum** would be left.

plasma
The fluid portion of unclotted blood

erythrocyte
A red blood cell

serum
The liquid that separates from the blood when a clot is formed

Obviously, considering the complexity of blood, any discussion of its function and chemistry would have to be extensive, extending beyond the scope of this text. It is certainly far more relevant at this point to concentrate our discussion on the blood components that are directly pertinent to the forensic aspects of blood identification—the red blood cells and the blood serum.

Antigens and Antibodies

Red blood cells transport oxygen from the lungs to the body tissues and remove carbon dioxide from tissues by transporting it back to the lungs, where it is exhaled. However, for reasons unrelated to the red blood cell's transporting mission, on the surface of each cell are millions of characteristic chemical structures called **antigens**. Antigens impart specific characteristics to the red blood cells. Blood antigens are grouped into systems depending on their relationship to one another. More than 15 blood antigen systems have been identified to date; of these, the A-B-O and Rh systems are the most important.

antigen
A substance, usually a protein, that stimulates the body to produce antibodies against it

If an individual is type A, this simply indicates that each red blood cell has A antigens on its surface; similarly, all type B individuals have B antigens, and the red blood cells of type AB individuals contain both A and B antigens. Type O individuals have neither A nor B antigens on their cells. Hence, the presence or absence of A and B antigens on the red blood cells determines a person's blood type in the A-B-O system.

Another important blood antigen has been designated as the *Rh factor*, or D antigen. People with the D antigen are said to be *Rh positive*; those without this antigen are *Rh negative*. In routine blood banking, the presence or absence of the three antigens—A, B, and D—must be determined in testing compatibility of the donor and recipient.

antibody
A protein in the blood serum that destroys or inactivates a specific antigen

antiserum
Blood serum that contains specific antibodies

agglutination
The clumping together of red blood cells by the action of an antibody

Serum is important because it contains proteins known as **antibodies**. The fundamental principle of blood typing is that for every antigen, there exists a specific antibody. Each antibody symbol contains the prefix *anti-*, followed by the name of the antigen for which it is specific. Hence, anti-A is specific only for A antigen, anti-B for B antigen, and anti-D for D antigen. The serum-containing antibody is referred to as the **antiserum**, meaning a serum that reacts against something (antigens).

An antibody reacts only with its specific antigen and no other. Thus, if serum containing anti-B is added to red blood cells carrying the B antigen, the two will combine, causing the antibody to attach itself to the cell. Antibodies are normally *bivalent*—that is, they have two reactive sites. This means that each antibody can simultaneously be attached to antigens located on two different red blood cells. This creates a vast network of cross-linked cells usually seen as clumping or **agglutination** (see Figure 9–2).

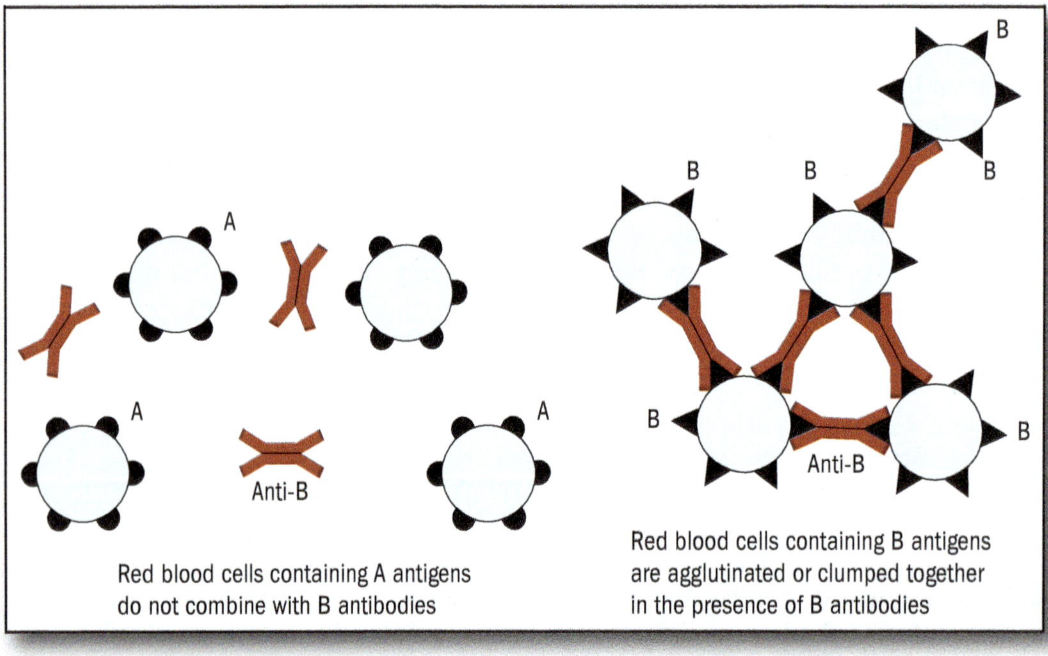

FIGURE 9–2 Agglutination.

Let's look a little more closely at this phenomenon. In normal blood, shown in Figure 9–3(a), antigens on red blood cells and antibodies coexist without destroying each other because the antibodies present are not specific toward any of the antigens. However, suppose a foreign serum added to the blood introduces a new antibody. This results in a specific antigen–antibody reaction that immediately causes the red blood cells to link together, or agglutinate, as shown in Figure 9–3(b).

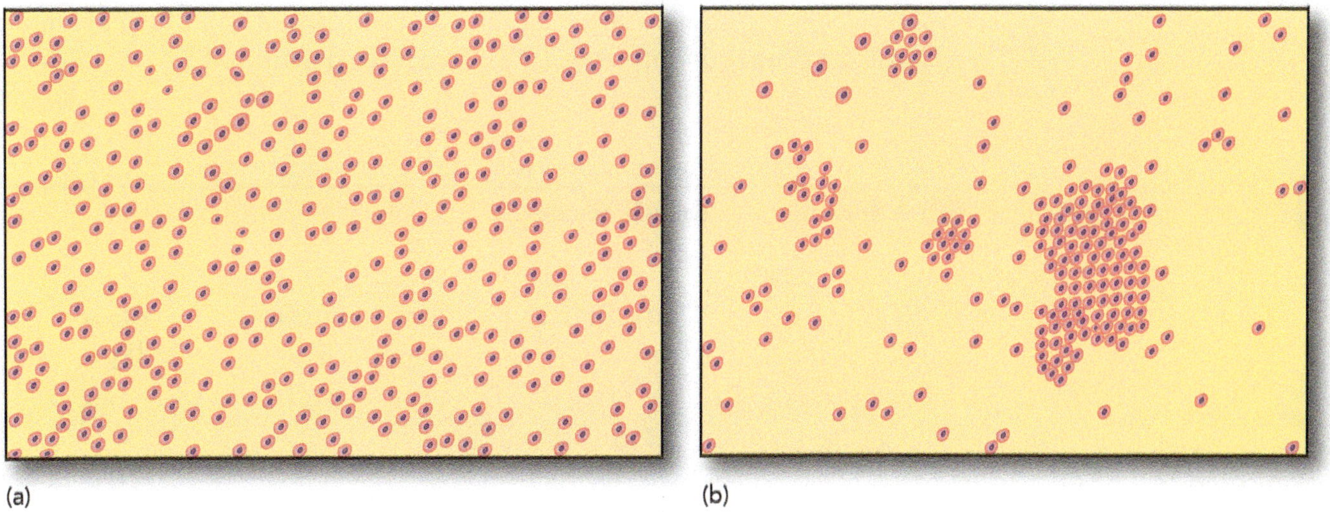

FIGURE 9–3 (a) Microscopic view of normal and red blood cells (500×). (b) Microscopic view of agglutinated red blood cells (500×).

Evidently, nature has taken this situation into account, because when we examine the serum of type A blood, we find anti-B and no anti-A. Similarly, type B blood contains only anti-A, type O blood has both anti-A and anti-B, and type AB blood contains neither anti-A nor anti-B. The antigen and antibody components of normal blood are summarized in the following table:

Blood Type	Antigens on Red Blood Cells	Antibodies in Serum
A	A	Anti-B
B	B	Anti-A
AB	AB	Neither anti-A nor anti-B
O	Neither A nor B	Both anti-A and anti-B

The reasons for the fatal consequences of mixing incompatible blood during a transfusion should now be quite obvious. For example, the transfusion of type A blood into a type B patient will cause the natural anti-A in the blood of the type B patient to react promptly with the incoming A antigens, resulting in agglutination. In addition, the incoming anti-B of the donor will react with the B antigens of the patient.

Blood Typing

The term **serology** describes a broad scope of laboratory tests that use specific antigen and serum antibody reactions. The most widespread application of serology is the typing of whole blood for its A-B-O identity. In determining the A-B-O blood type, only two antiserums are needed—anti-A and anti-B. For routine blood typing, both of these antiserums are commercially available.

serology
The study of antigen–antibody reactions

Table 9–1 summarizes how the identity of each of the four blood groups is established when the blood is tested with anti-A and anti-B serum. Type A blood is agglutinated by anti-A serum; type B blood is agglutinated by anti-B serum;

type AB blood is agglutinated by both anti-A and anti-B; and type O blood is not agglutinated by either the anti-A or anti-B serum (see Figure 9–4).

Table 9–1
Identification of Blood with Known Antiserum

Anti-A Serum + Whole Blood	Anti-B Serum + Whole Blood	Antigen Present	Blood Type
+	–	A	A
–	+	B	B
+	+	A and B	AB
–	–	Neither A nor B	O

Note: + shows agglutination; – shows absence of agglutination.

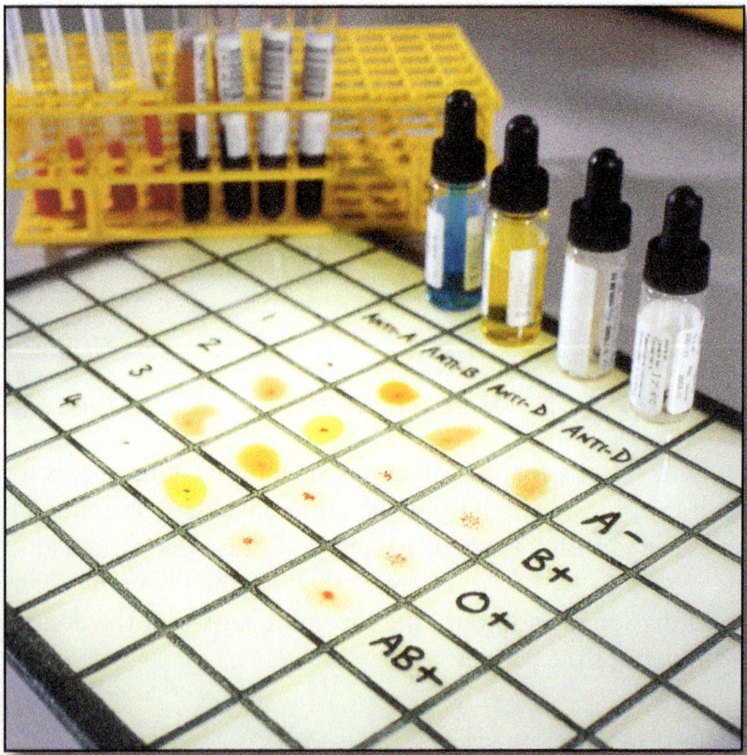

FIGURE 9–4 A blood type test for types A, B, AB, and O. Commercial antisera are systematically applied to a questioned blood in order to determine blood type. *Tek Image/ Science Source.*

The identification of natural antibodies present in blood offers another way to determine blood type. Testing blood for the presence of anti-A and anti-B requires using red blood cells that have known antigens. Again, these cells are commercially available. Hence, when A cells are added to a blood specimen, agglutination occurs only in the presence of anti-A. Similarly, B cells agglutinate only in the presence of anti-B. All four A-B-O types can be identified in this manner by testing blood with known A and B cells, as summarized in Table 9–2.

Table 9–2
Identification of Blood with Known Cells

A Cells + Blood	B Cells + Blood	Antibody Present	Blood Type
+	–	Anti-A	B
–	+	Anti-B	A
+	+	Both anti-A and anti-B	O
–	–	Neither anti-A nor anti-B	AB

Note: + shows agglutination; – shows absence of agglutination.

The population distribution of blood types varies with location and ancestry throughout the world. In the United States, a typical distribution is as follows:

O	**A**	**B**	**AB**
43%	42%	12%	3%

Quick Review

- Serology involves a broad scope of laboratory tests that use specific antigen and serum antibody reactions.
- An antibody reacts or agglutinates only with its specific antigen. The concept of specific antigen–antibody reactions has been applied to techniques for detecting abused drugs in blood and urine.
- Every red blood cell contains either an A antigen, a B antigen, or no antigen (this is called type O). The type of antigen on one's red blood cells determines one's A-B-O blood type. People with type A blood have A antigens on their red blood cells, those with type B blood have B antigens, and those with type O blood have no antigens on their red blood cells.

Immunoassay Techniques

The concept of a specific antigen–antibody reaction is finding application in other areas unrelated to blood typing. Most significantly, this approach has been extended to the detection of drugs in blood and urine. Antibodies that react with drugs do not exist naturally; however, they can be produced in animals such as rabbits by first combining the drug with a protein and injecting this combination into the animal. This drug–protein complex acts as an antigen stimulating the animal to produce antibodies (see Figure 9–5). The recovered blood serum of the animal now contains antibodies that are specific or nearly specific to the drug.

Currently, thousands of individuals regularly submit to urinalysis tests for the presence of drugs of abuse. These individuals include military personnel, transportation industry employees, police and corrections personnel, and subjects requiring

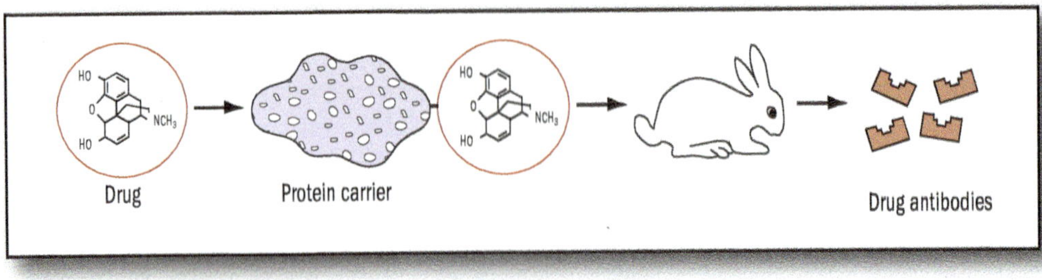

FIGURE 9–5 Stimulating production of drug antibodies.

preemployment drug screening. Immunoassay testing for drugs has proven quite suitable for handling the large volume of specimens that must be rapidly analyzed for drug content on a daily basis. Testing laboratories have access to many commercially prepared sera arising from animals being injected with any one of a variety of drugs. A particular serum that has been added to a urine specimen is designed to interact with opiates, cannabinoids, cocaine, amphetamines, phencyclidine, barbiturates, methadone, or other drugs. A word of caution: immunoassay is only presumptive in nature, and its result must be confirmed by additional testing. Specifically, the confirmation test of choice is gas chromatography-mass spectrometry, which is described in more detail in Chapter 6.

Inside the Science

Enzyme-Multiplied Immunoassay Technique (EMIT)

Several immunological assay techniques are commercially available for detecting drugs through an antigen–antibody reaction. One such technique, the *enzyme-multiplied immunoassay technique (EMIT)*, has gained widespread popularity among toxicologists because of its speed and high sensitivity for detecting drugs in urine.

A typical EMIT analysis begins by adding to a subject's urine antibodies that bind to a particular type or class of drug being looked for. This is followed by adding to the urine a chemically labeled version of the drug. As shown in the figure, a competition will ensue between the labeled and unlabeled drug (if it's present in the subject's urine) to bind with the antibody. If this competition does occur in a person's urine, it signifies that the urine screen test was positive for the drug being tested. For example, to check someone's urine for methadone, the analyst would add methadone antibodies and chemically labeled methadone to the urine. Any methadone present in the urine immediately competes with the labeled methadone to bind with the methadone antibodies. The quantity of chemically labeled

Inside the Science (CONTINUED)

methadone left uncombined is then measured, and this value is related to the concentration of methadone originally present in the urine.

One of the most frequent uses of EMIT in forensic laboratories has been for screening the urine of suspected marijuana users. The primary pharmacologically active agent in marijuana is tetrahydrocannabinol, or THC. To facilitate the elimination of THC, the body converts it to a series of substances called *metabolites* that are more readily excreted. The major THC metabolite found in urine is a substance called *THC-9-carboxylic acid*. Antibodies against this metabolite are prepared for EMIT testing. Normally the urine of marijuana users contains a small quantity of THC-9-carboxylic acid (less than one-millionth of a gram); however, this level is readily detected by EMIT.

The greatest problem with detecting marijuana in urine is interpretation of the test results. Although marijuana use results in the detection of THC metabolites, it is difficult to determine when the individual actually used marijuana. In individuals who use marijuana frequently, detection is possible within two to five days after the last use of the drug. However, some individuals may yield positive results up to 30 days after the last use of marijuana.

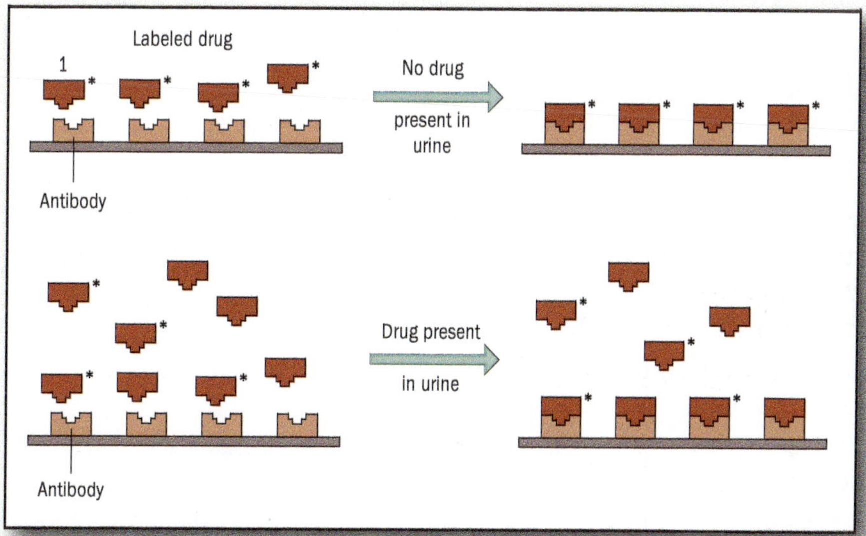

In the EMIT assay, a drug that may be present in a urine specimen will compete with added labeled drug for a limited number of antibody binding sites. The labeled drugs are indicated by an asterisk. Once the competition for antibody sites is completed, the number of remaining unbound labeled drug is proportional to the drug's concentration in urine.

Inside the Science

Polyclonal and Monoclonal Antibodies

As we have seen in the previous section, when an animal such as a rabbit or mouse is injected with an antigen, the animal responds by producing antibodies designed to bind to the invading antigen. However, the process of producing antibodies designed to respond to foreign antigens is complex. For instance, an antigen typically has structurally different sites to which an antibody may bind. Thus, in the presence of a specific antigen, an animal produces many different antibodies, all of which are designed to attack some particular site on the antigen of interest. These antibodies are known as **polyclonal antibodies**. However, the disadvantage of polyclonal antibodies is that an animal can produce antibodies that vary in composition over time. As a result, different batches of polyclonals may vary in their specificity and their ability to bind to a particular antigen site.

Modern forensic technologies occasionally require antibodies that are more uniform in their composition and attack power than the traditional polyclonals. Forensic scientists thus need a way to produce antibodies designed to attack one and only one site on an antigen. Such antibodies are known as **monoclonal antibodies**. How can such monoclonals be produced?

The process begins by injecting a mouse with the antigen of interest. In response, the mouse's spleen cells produce antibodies to fight off the invading antigen. The spleen cells are removed from the animal and are fused to fast-growing blood cancer cells to produce **hybridoma cells**. The hybridoma

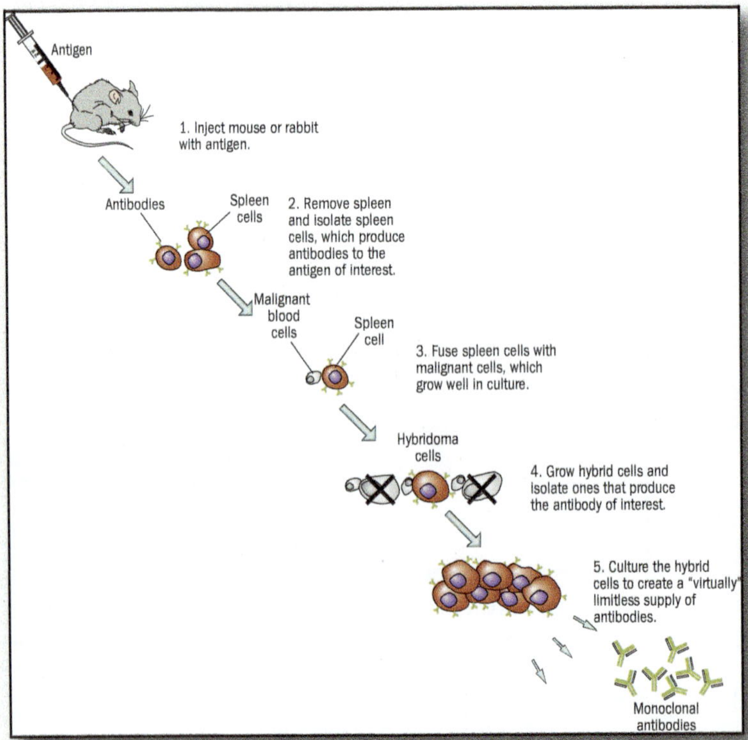

Steps required to produce monoclonal antibodies.

Inside the Science (CONTINUED)

cells are then allowed to multiply and are screened for their specific antibody activity. Hybridoma cells bearing the antibody activity of interest are then selected and cultured. The rapidly multiplying cancer cells linked to the selected antibody cells produce identical monoclonal antibodies in a limitless supply (see the figure).

Monoclonal antibodies are being incorporated into commercial forensic test kits with increasing frequency. Many immunoassay test kits for abused drugs are being formulated with monoclonal antibodies. Also, a recently introduced test for semen that incorporates a monoclonal antibody has found wide popularity in crime laboratories.

As a side note, in 1999 the U.S. Food and Drug Administration approved a monoclonal drug treatment for cancer. Rituxin is a nontoxic monoclonal antibody designed to attack and destroy cancerous white blood cells containing an antigen designated as CD20. Other monoclonal drug treatments are in the pipeline. Monoclonals are finally beginning to fulfill their long-held expectation as medicine's version of the "magic bullet."

Quick Review

- To produce antibodies capable of reacting with drugs, the analyst combines a specific drug with a protein and injects this combination into an animal such as a rabbit. This drug–protein complex acts as an antigen, stimulating the animal to produce antibodies. The recovered blood serum of the animal now contains antibodies that are specific or nearly specific to the drug.
- When an animal is injected with an antigen, its body produces a series of different antibodies, all of which are designed to attack some particular site on the antigen of interest. These antibodies are known as polyclonal antibodies.
- A more uniform and specific collection of antibodies designed to combine with a single antigen site can be manufactured. Such antibodies are known as monoclonal antibodies.

polyclonal antibodies
Antibodies produced by injecting animals with a specific antigen; a series of antibodies are produced responding to a variety of different sites on the antigen

monoclonal antibodies
A collection of identical antibodies that interact with a single antigen site

hybridoma cells
Fused spleen and tumor cells; used to produce identical monoclonal antibodies in a limitless supply

Forensic Characterization of Bloodstains

The criminalist must answer the following questions when examining dried blood: (1) Is it blood? (2) From what species did the blood originate? (3) If the blood is human, how closely can it be associated with a particular individual?

Color Tests

reagent
A mixture or substance used to test for a substance based on its biological or chemical activity

hemoglobin
A red blood cell protein that transports oxygen in the bloodstream; it is responsible for the red color of blood

The determination of blood is best made by means of a preliminary color test. This type of presumptive test for various biologicals gives scientists clarity if the substance at the scene is blood. Various **reagents**, which are mixtures or substances used that produce a reaction in the presence of a particular substance. For many years, the most common test was the benzidine color test. However, because benzidine has been identified as a known carcinogen, its use has generally been discontinued, and the chemical phenolphthalein is usually substituted in its place (this test is also known as the Kastle-Meyer color test).[1]

Both the benzidine and Kastle-Meyer color tests are based on the observation that blood **hemoglobin** possesses peroxidase-like activity. Peroxidases are enzymes that accelerate the oxidation of several classes of organic compounds when combined with peroxides. For example, when a bloodstain, phenolphthalein reagent, and hydrogen peroxide are mixed together, oxidation of the hemoglobin in the blood produces a deep pink color.

The Kastle-Meyer test is not a specific test for blood; some vegetable materials, for instance, may turn Kastle-Meyer pink. These substances include potatoes and horseradish. However, such materials will probably not be encountered in criminal situations, and thus from a practical point of view, a positive Kastle-Meyer test is highly indicative of blood. Field investigators have found Hemastix strips a useful presumptive field test for blood. Designed as a urine dipstick test for blood, the strip can be moistened with distilled water and placed in contact with a suspect bloodstain. The appearance of a green color indicates blood.

Tetramethylbenzidine (TMB) is a non-carcinogenic derivative of benzidine and can be used as a screening test for blood. A blue-green color is produced when TMB reacts with the hemoglobin in blood. TMB and phenolphthalein (Kastle-Meyer) are blood presumptive tests for various biologicals.

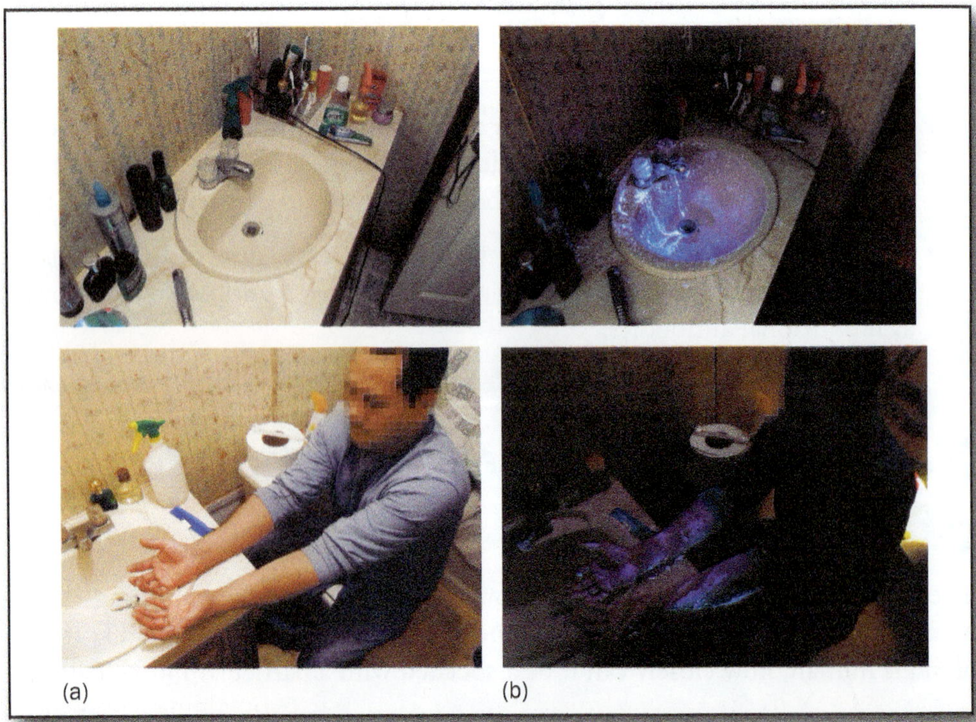

FIGURE 9–6
(a) Sink and arms before applications of Bluestar reagent.
(b) Bluestar applications reveals bloodstain patterns. *Courtesy Johnnie Sutphin with BLUESTAR*

Luminol, Bluestar®, and Amido Black

In some cases, blood may not be readily visible due to attempts to remove the presence of blood. In these situations, latent blood patterns may be visualized by using chemical methods to enhance the presence of blood. One example of a presumptive identification test for blood is the **luminol** test.[2] Unlike the benzidine and Kastle-Meyer tests, the reaction of luminol with blood produces light rather than color. By spraying luminol reagent onto a suspect item, investigators can quickly screen large areas for bloodstains. The sprayed objects must be located in a darkened area while being viewed for the emission of light (luminescence); any bloodstains produce a faint blue glow (see Figure 9–6).

luminol
The most sensitive chemical test that is capable of presumptively detecting bloodstains diluted to as little as 1 in 100,000; its reaction with blood emits light and thus requires the result to be observed in a darkened area

A relatively new reagent, Bluestar, is now available to be used in place of luminol (www.bluestar-forensic.com). Bluestar is easy to mix in the field. Its reaction with blood can be observed readily without having to create complete darkness. The luminol and Bluestar tests are extremely sensitive—capable of detecting bloodstains diluted to as little as 1 in 100,000. For this reason, spraying large areas such as carpets, walls, flooring, or the interior of a vehicle may reveal blood traces or patterns that would have gone unnoticed under normal lighting conditions (see Figure 9–7). It is important to note that luminol and Bluestar do not interfere with any subsequent DNA testing.[3]

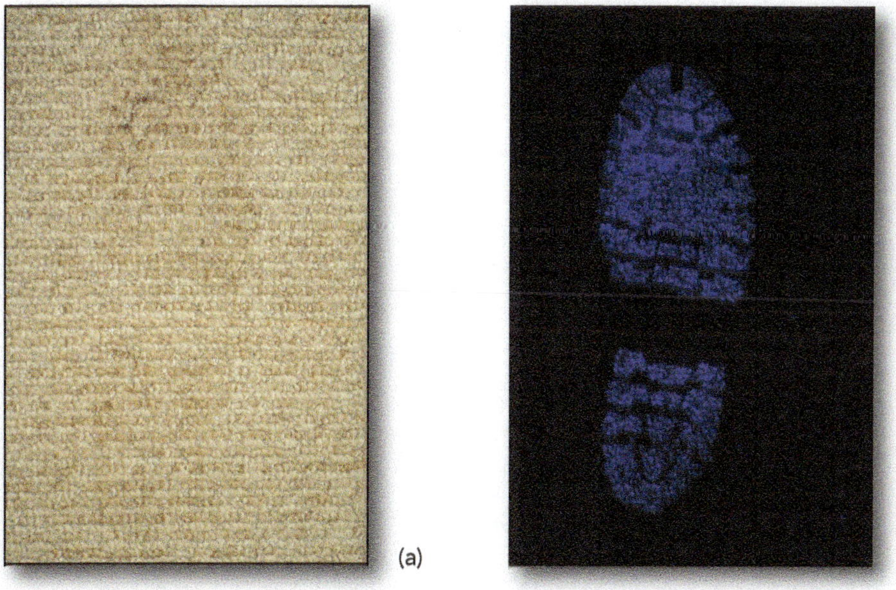

(a) (b)

FIGURE 9–7 (a) A section of a carpet under normal light showing a faint footprint in blood. (b) Same section of the carpet after spraying with luminol. *Courtesy Sirchie Fingerprint Laboratories, Youngsville, NC, www.sirchie.com*

Amido Black is another reagent that can be used to presumptively identify blood. This dark blue stain can be applied by spraying it on the suspected blood, dipping an item into the stain, or by a wash technique. Once the stain is applied, the area must be rinsed to remove any background coloration leaving the protein in the protein in the blood with dark blue ridges. Amido black is note specific for blood so confirmatory tests are necessary.

Microcrystalline Tests

The identification of blood can be made more specific if microcrystalline tests are performed on the material. Several tests are available; the two most popular ones are the Takayama and Teichmann tests. Both depend on the addition of specific chemicals to the blood to form characteristic crystals containing hemoglobin derivatives. Crystal tests are far less sensitive than color tests for blood identification and are more susceptible to interference from contaminants that may be present in the stain.

Precipitin Test

precipitin
An antibody that reacts with its corresponding antigen to form a precipitate

Once the stain has been characterized as blood, the serologist determines whether the blood is of human or animal origin. The standard test is the **precipitin** test. Precipitin tests are based on the fact that when animals (usually rabbits) are injected with human blood, antibodies form that react with the invading human blood to neutralize its presence. The investigator can recover these antibodies by bleeding the animal and isolating the blood serum, which contains antibodies that specifically react with human antigens. For this reason, the serum is known as human antiserum. In the same manner, by injecting rabbits with the blood of other known animals, virtually any kind of animal antiserum can be produced. Antiserums are commercially available for humans and for a variety of commonly encountered animals—for example, dogs, cats, and deer.

Several techniques have been devised for performing precipitin tests on bloodstains. The classic method is to layer an extract of the bloodstain on top of the human antiserum in a capillary tube. Human blood, or, for that matter, any protein of human origin in the extract, reacts specifically with antibodies present in the antiserum, as indicated by the formation of a cloudy ring or band at the interface of the two liquids (see Figure 9–8).

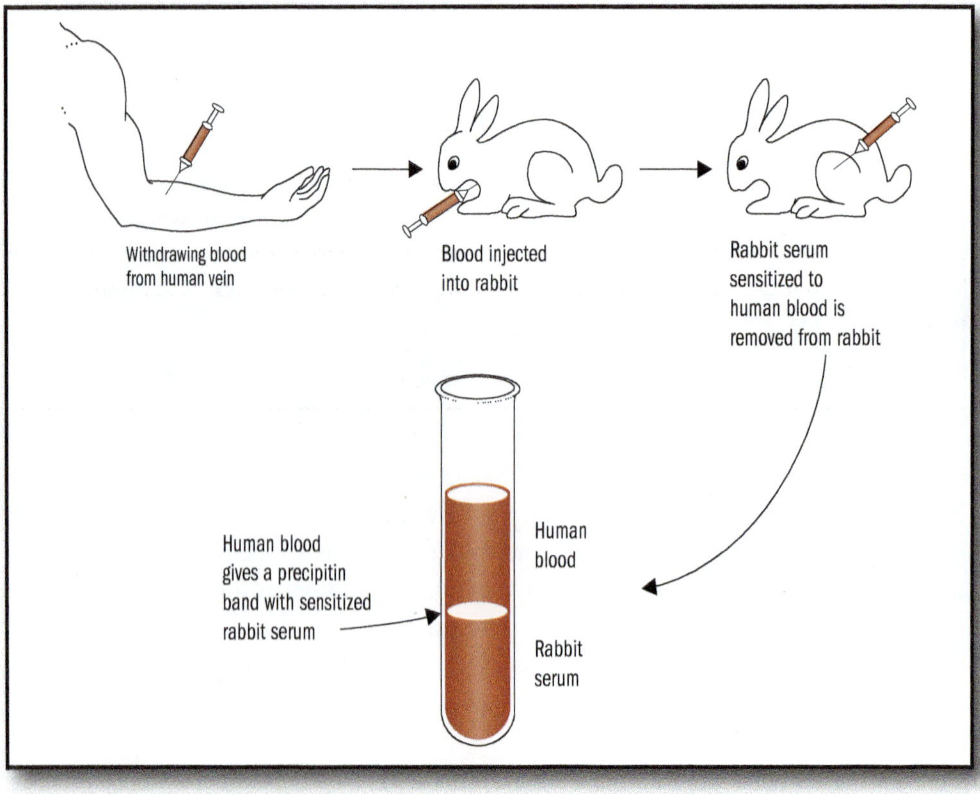

FIGURE 9–8
The precipitin test.

Gel Diffusion

Another method, called *gel diffusion*, takes advantage of the fact that antibodies and antigens diffuse or move toward one another on a plate coated with a gel medium made from a natural polymer called agar. The extracted bloodstain and the human antiserum are placed in separate holes opposite each other on the gel. If the blood is human, a line of precipitation forms where the antigens and antibodies meet.

Similarly, the antigens and antibodies can be induced to move toward one another under the influence of an electrical field. In the *electrophoretic method* (examined in detail in Chapter 10), an electrical potential is applied to the gel medium; a specific antigen–antibody reaction is denoted by a line of precipitation formed between the hole containing the blood extract and the hole containing the human antiserum (see Figure 9–9).

The precipitin test is very sensitive and requires only a small amount of blood for testing. Human bloodstains dried for 10 to 15 years and longer may still give a positive precipitin reaction. Even extracts of tissue from mummies four to five thousand years old have given positive reactions with this test. Furthermore, human bloodstains diluted by washing in water and left with only a faint color may still yield a positive precipitin reaction (see Figure 9–10).

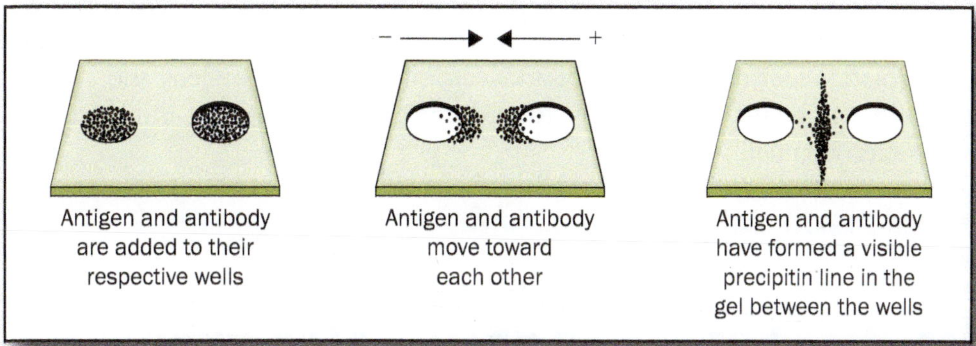

FIGURE 9–9 Antigens and antibodies moving towards one another under the influence of an electrical potential.

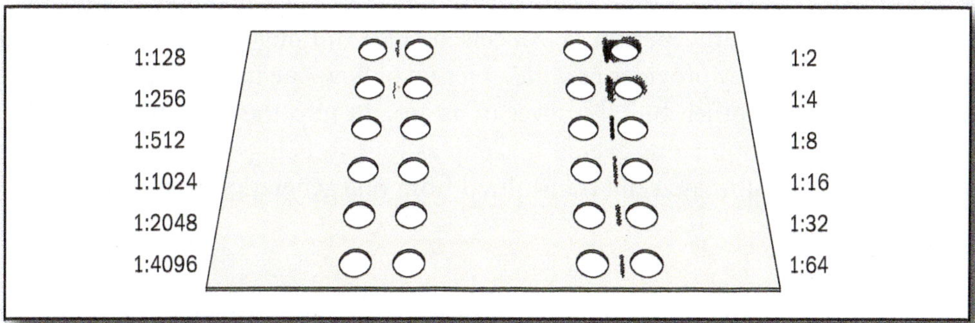

FIGURE 9–10 Results of the precipitin test of dilutions of human serum up to 1 in 4,096 against a human antiserum. A reaction is visible for blood dilutions up to 1 in 256.

Once it has been determined that the bloodstain is human, an effort must be made to associate or disassociate the stain with a particular individual. Until the mid-1990s, routine characterization of bloodstains included the determination of A-B-O types; however, the widespread use of DNA profiling or typing has relegated this subject to one of historical interest only.

Quick Review

- The criminalist must answer the following questions when examining dried blood: (1) Is it blood? (2) From what species did the blood originate? (3) If the blood is human, how closely can it be associated with a particular individual?
- The determination of blood is best made with a preliminary color test. A positive result from the Kastle-Meyer color test is highly indicative of blood.
- The luminol test is used to find trace amounts of blood at crime scenes.
- The precipitin test uses antisera normally derived from rabbits that have been injected with the blood of a known animal to determine the species origin of a questioned bloodstain.
- Materials undergoing electrophoresis are forced to move across a gel-coated plate under the influence of an electrical potential. Antigens and antibodies can be induced to move toward one another under the influence of an electrical field.

Principles of Heredity

All of the antigens and polymorphic enzymes and proteins that have been described in previous sections are genetically controlled traits. That is, they are inherited from parents and become a permanent feature of a person's biological makeup from the moment the person is conceived. Determining the identity of these traits, then, not only provides us with a picture of how one individual compares to or differs from another, but also gives us an insight into the basic biological substances that determine our overall makeup as human beings and the mechanism by which those substances are transmitted from one generation to the next.

Genes and Chromosomes

gene
The basic unit of heredity, consisting of a DNA segment located on a chromosome

Hereditary material is transmitted via microscopic units called **genes**. The gene is the basic unit of heredity. Each gene by itself or in concert with other genes controls the development of a specific characteristic in the new individual; the genes determine the nature and growth of virtually every body structure.

The genes are positioned on **chromosomes**, threadlike bodies that appear in the nucleus of every body cell (see Figure 9–11). Almost all human cells contain 46 chromosomes, mated in 23 pairs (see Figure 9–12). The only exceptions are the human reproductive cells, the **egg** and **sperm**, which contain 23 unmated chromosomes. During fertilization, a sperm and egg combine so that each contributes 23 chromosomes to form the new cell (**zygote**). Hence, the new individual begins life properly with 23 mated chromosome pairs. Because the genes are positioned on the chromosomes, the new individual inherits genetic material from each parent.

chromosome
A threadlike structure in the cell nucleus, along which the genes are located

egg
The female reproductive cell

sperm
The male reproductive cell

zygote
The cell arising from the union of an egg and a sperm cell

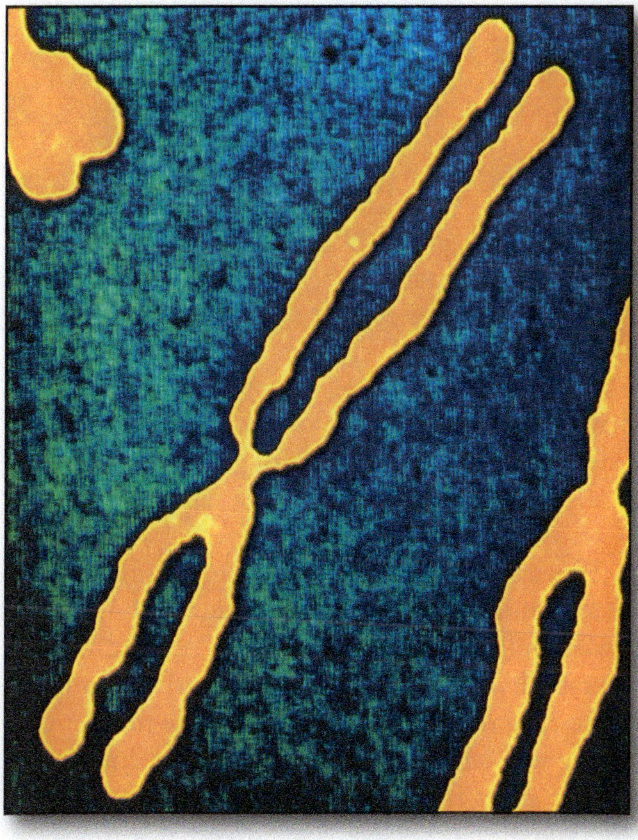

FIGURE 9–11
Computer-enhanced photomicrograph image of human chromosomes. *Alfred Pasieka/ Science Source*

Actually, two dissimilar chromosomes are involved in the determination of sex. The egg cell always contains a long chromosome known as the **X chromosome**; the sperm cell may contain either a long X chromosome or a short **Y chromosome**. When an X-carrying sperm fertilizes an egg, the new cell is XX and develops into a female. A Y-carrying sperm produces an XY fertilized egg and develops into a male. Because the sperm cell determines the nature of the chromosome pair, we can say that the father biologically determines the sex of the child.

X chromosome
The female sex chromosome

Y chromosome
The male sex chromosome

Alleles Just as chromosomes come together in pairs, so do the genes they bear. The position a gene occupies on a chromosome is its **locus**. Genes that govern a given characteristic are similarly positioned on the chromosomes inherited from the mother and father. Thus, a gene for eye color on the mother's

locus
The physical location of a gene on a chromosome

FIGURE 9–12 Fluorescent light micrograph of the 46 chromosomes from a normal human female. *L. Willatt/ Science Source*

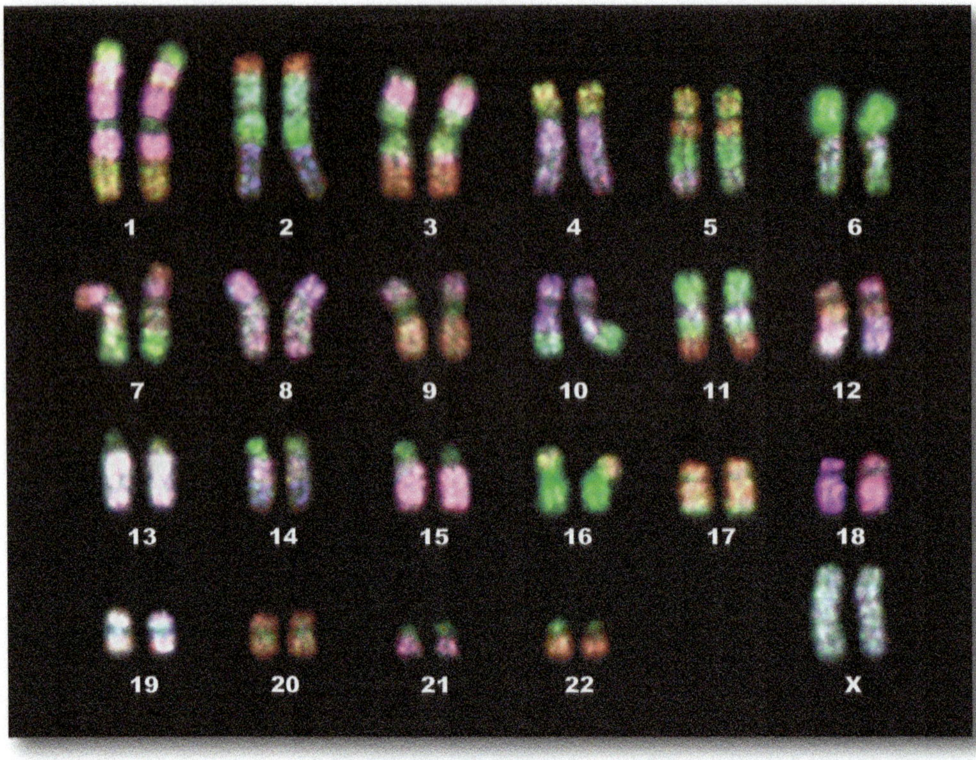

allele
Any of several alternative forms of a gene located at the same point on a particular pair of chromosomes

homozygous
Having two identical allelic genes on two corresponding positions of a pair of chromosomes

heterozygous
Having two different alleles on two corresponding positions of a pair of chromosomes

chromosome will be aligned with a gene for eye color on the corresponding chromosome inherited from the father. Alternative forms of genes that influence a given characteristic and are aligned with one another on a chromosome pair are known as **alleles**.

Inheritance of blood type offers a simple example of allele genes in humans. An individual's blood type is determined by three genes, designated A, B, and O. A gene pair made up of two similar alleles—for example, AA and BB—is said to be **homozygous**. For example, if the chromosome inherited from the father carries the A gene and the chromosome inherited from the mother carries the same gene, the offspring will have an AA combination. Thus, when an individual inherits two similar genes from their parents, there is no problem in determining the blood type of that person. An individual with an AA combination will always be type A, a BB will be type B, and an OO will be type O.

A gene pair made up of two different alleles—AO, for example—is said to be **heterozygous**. For example, if the chromosome from one parent carries the A gene and the chromosome from the other parent carries the O gene, the genetic makeup of the offspring will be AO. When two different genes are inherited, one gene will be dominant—that is, the characteristic coded for by that gene is expressed. The other gene will be recessive—that is, its characteristics remain hidden. In the case of blood types, A and B genes are dominant and the O gene is recessive. Thus, with an AO combination, A is always dominant over O, and the individual is typed as A. Similarly, a BO combination is typed as B. In the case of AB, the genes are *codominant*, and the individual's blood type will be AB. The recessive characteristics of

O appear only when both recessive genes are present in combination OO, which is typed simply as O.

Genotypes and Phenotypes A pair of allele genes together constitutes the **genotype** of the individual. However, no known laboratory test can determine an individual's A-B-O genotype. For example, a person's outward characteristic, or **phenotype**, may be type A, but this does not tell us whether the genotype is AA or AO. The genotype can be determined only by studying the family history of the individual. If the genotypes of both parents are known, that of their possible offspring can be forecast.

genotype
The particular combination of genes present in the cells of an individual

phenotype
The physical manifestation of a genetic trait such as shape, color, or blood type

An easy way to determine an individual's genotype is to construct a Punnett square. To do this, write along a horizontal line the two genes of the male parent, and in the vertical column write the two kinds of female genes present, as shown. In our example, we assume the male parent is type O and therefore has to be an OO genotype; the female parent is type AB and can be only an AB genotype:

Mother's genotype \ Father's genotype	O	O
A		
B		

Next, write in each box the corresponding gene contributed from the female and then from the male. The squares will contain all the possible genotype combinations that the parents can produce in their offspring:

	O	O
A	AO	AO
B	BO	BO

In this case, 50 percent of the offspring are likely to be AO and the other 50 percent BO. These are the only genotypes possible from this combination. Because O is recessive, 50 percent of the offspring will probably be type A and 50 percent type B. From this example, we can see that no blood group gene can appear in a child unless it is present in at least one of the parents.

Paternity Testing

Although the genotyping of blood factors has useful applications for studying the transmission of blood characteristics from one generation to the next, it has no direct relevance to criminal investigations. It does, however, have important implications in disputed-paternity cases, which are normally encountered in civil, not criminal, courts.

Many cases of disputed paternity can be resolved by comparing the blood group genotypes of the suspected parents and offspring. For instance, suppose the male in the preceding Punnett square example is suspected of fathering a child by the female. If the child has type AB blood, the suspected father will be cleared because a type O father and a type AB mother cannot have a type AB child. On the other hand, if the child has type A or type B, the most that can be said is that the suspect may be the father. This does not mean that he is the father, just that he is not excluded based on blood typing. Obviously, many other males also have type O blood. Of course, the more blood group systems that are tested, the better the chances of excluding an innocent male from involvement. Conversely, if no discrepancies are found between the offspring and the suspect father, the more certain one can be that the suspect is indeed the father.

In fact, routine paternity testing involves characterizing blood factors other than A-B-O. For example, the HLA (human leukocyte antigen) test relies on identifying a complex system of antigens on white blood cells. If this test cannot exclude a suspect as the father of a child, the chances are better than 90 percent that he is the father. Paternity-testing laboratories have implemented DNA test procedures that can raise the odds of establishing paternity beyond 99 percent.

Quick Review

- The gene is the basic unit of heredity. A chromosome is a threadlike structure in the cell nucleus, along which the genes are located.
- Most human cells contain 46 chromosomes, arranged in 23 mated pairs. The only exceptions are the human reproductive cells, the egg and sperm, which contain 23 unmated chromosomes.
- During fertilization, a sperm and egg combine so that each contributes 23 chromosomes to form the new cell, or zygote, that develops into the offspring.
- An allele is any of several alternative forms of genes that influence a given characteristic and that are aligned with one another on a chromosome pair.
- A heterozygous gene pair is made up of two different alleles; a homozygous gene pair is made up of two similar alleles.
- When two different genes are inherited, the characteristic coded for by a dominant gene is expressed. The characteristic coded for by a recessive gene remains hidden.
- A genotype is the particular combination of genes present in the cells of an individual. A phenotype is the physical manifestation of a genetic trait.

Forensic Characterization of Semen

Many cases received in a forensic laboratory involve sexual offenses, making it necessary to examine exhibits for the presence of seminal stains. The normal male releases 2.5 to 6 milliliters of seminal fluid during an ejaculation. Each milliliter contains 100 million or more spermatozoa, the male reproductive cells.

The forensic examination of articles for seminal stains can actually be considered a two-step process. First, before any tests can be conducted, the stain must be located (see Figure 9–13). Considering the number and soiled condition of outer garments, undergarments, and possible bedclothing submitted for examination, this may prove to be an arduous task. Once located, the stain must be subjected to tests that will prove its identity. This will most likely include testing for the DNA profile of the individual from whom it originated.

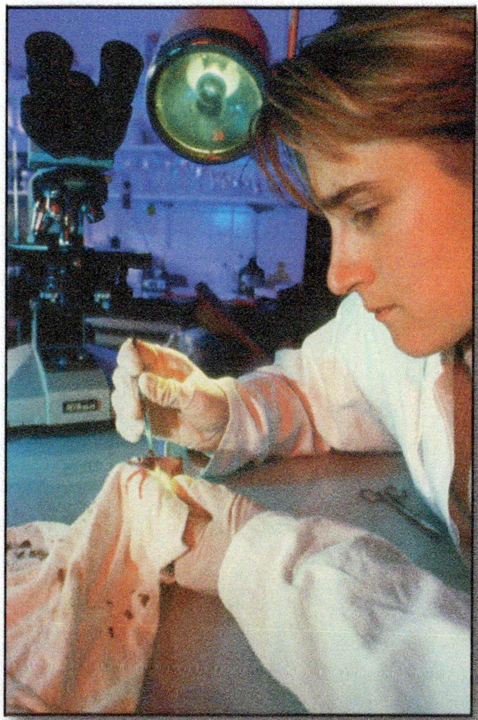

FIGURE 9–13
A forensic analyst examines a garment for blood and semen deposits. *Dr. Jurgen Scriba/Science Source*

Testing for Seminal Stains

Often seminal stains are visible on a fabric because they exhibit a stiff, crusty appearance. However, reliance on such appearance for locating the stain is unreliable and is useful only when the stain is in an obvious area. If the fabric has been washed or contains only minute quantities of semen, visual examination offers little chance of detecting the stain. The best way to locate and at the same time characterize a seminal stain is to perform the *acid phosphatase color test.*

Acid Phosphatase Test **Acid phosphatase** is an **enzyme** that is secreted by the prostate gland into seminal fluid. Its concentrations in seminal fluid are up to 400 times those found in any other body fluid. Its presence can easily be detected when it comes in contact with an acidic solution of sodium alpha naphthylphosphate and Fast Blue B dye. Also, 4-methyl umbelliferyl phosphate (MUP) fluoresces (emits light) under UV light when it comes in contact with acid phosphatase.

acid phosphatase
An enzyme found in high concentration in semen

enzyme
A protein that acts as a catalyst for certain specific reactions

The utility of the acid phosphatase test is apparent when it becomes necessary to search many garments or large fabric areas for seminal stains. Simply moistening a filter paper with water and rubbing it lightly over the suspect area transfers any acid phosphatase present to the filter paper. Placing a drop or two of the sodium alpha naphthylphosphate and Fast Blue B solution on the paper produces a purple color that indicates the acid phosphatase enzyme. In this manner, any fabric or surface can be systematically searched for seminal stains.

If it is necessary to search extremely large areas—for example, a bedsheet or carpet—the article can be tested in sections, narrowing the location of the stain with each successive test. Alternatively, the garment can be pressed against a suitably sized piece of moistened filter paper. The paper is then sprayed with MUP solution. Semen stains appear as strongly fluorescent areas under UV light. A negative reaction can be interpreted as absence of semen. Although some vegetable and fruit juices (such as cauliflower and watermelon), fungi, contraceptive creams, and vaginal secretions give a positive response to the acid phosphatase test, none of these substances normally reacts with the speed of seminal fluid. A reaction time of less than 30 seconds is considered a strong indication of semen.

Microscopic Examination of Semen Semen can be unequivocally identified by the presence of spermatozoa. When spermatozoa are located through a microscope examination, the stain is definitely identified as having been derived from semen. *Spermatozoa* are slender, elongated structures 50–70 microns long, each with a head and a thin flagellate tail (see Figure 9–14). The criminalist can normally locate them by immersing the stained material in a small volume of water. Rapidly stirring the liquid transfers a small percentage of the spermatozoa present into the water. A drop of the water is dried onto a microscope slide, then stained and examined under a compound microscope at a magnification of approximately 400×.[4]

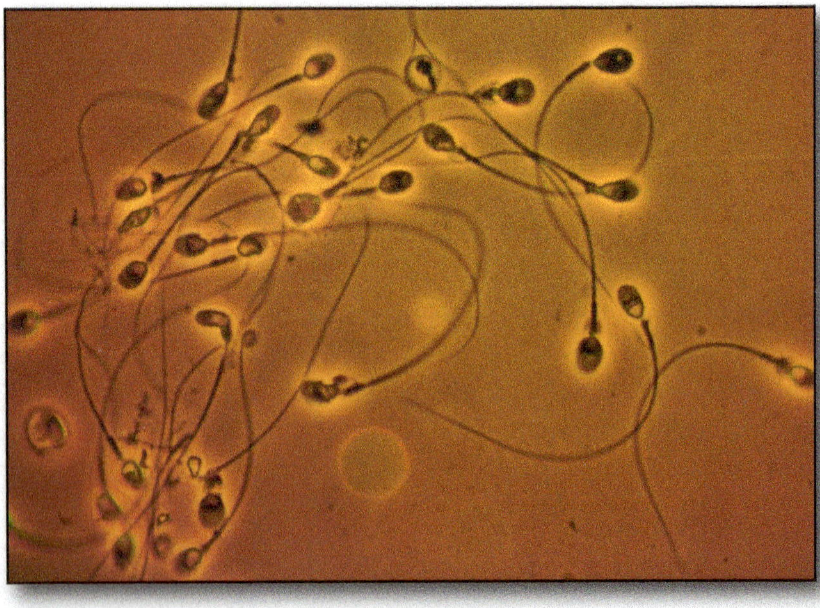

FIGURE 9–14
Photomicrograph of human spermatozoa (300x).
John Walsh/Science Source

Considering the extremely large number of spermatozoa found in seminal fluid (the normal male releases 250–600 million spermatozoa during ejaculation), the chance of locating one should be very good; however, this is not always true. One reason is that spermatozoa bind tightly to cloth materials.[5] Also, spermatozoa are extremely brittle when dry and easily disintegrate if the stain is washed or when the stain is rubbed against another object, as happens frequently in the handling and packaging of this type of evidence. Furthermore, sexual crimes may involve males who have an abnormally low sperm count, a condition known as **oligospermia**, or who have no spermatozoa at all in their seminal fluid (**aspermia**). Significantly, aspermatic individuals are increasing in numbers because of the growing popularity of vasectomies.

oligospermia
An abnormally low sperm count

aspermia
The absence of sperm; sterility in males

Prostate-Specific Antigen (PSA) Forensic analysts often must examine stains or swabs that they suspect contain semen (because of the presence of acid phosphatase) but that yield no detectable spermatozoa. How, then, can one unequivocally prove the presence of semen? The solution to this problem came with the discovery in the 1970s of a protein called *p30* or *prostate specific antigen (PSA)*. At first, this protein was thought to be prostate specific and hence a unique identifier of semen. However, additional research has shown that low levels of p30 may be detectable in other human tissues. A more reasonable approach to the unequivocal identification of semen is to use a positive PSA (p30) in combination with an acid phosphatase color test with a reaction time of less than 30 seconds.[6]

When p30 is isolated and injected into a rabbit, it stimulates the production of polyclonal antibodies (anti-p30). The serum collected from these immunized rabbits can then be used to test suspected semen stains. As shown in Figure 9–15, the stain extract is placed in one well of an electrophoretic plate and the anti-p30 in an opposite well. When an electric potential is applied, the antigens and antibodies move toward each other. The formation of a visible line midway between the two wells shows the presence of p30 in the stain and indicates that the stain was seminal.

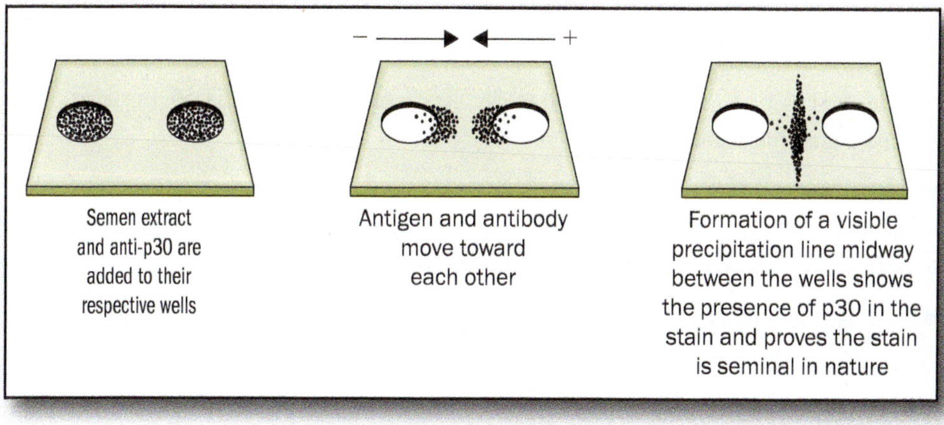

FIGURE 9–15
PSA testing by electrophoresis.

A more elegant approach to identifying PSA (p30) is shown in Figure 9–17. First, a monoclonal PSA antibody is attached to a dye and placed on a porous membrane. Next, an extract from a sample suspected of containing PSA is placed on the membrane. If PSA is present in the extract, it combines with the monoclonal PSA antibody to form a PSA antigen–monoclonal PSA antibody complex. This complex migrates along the membrane, where it interacts with a polyclonal PSA antibody embedded in the membrane. The antibody–antigen–antibody "sandwich" that forms is apparent by the presence of a colored line (see Figure 9–16). This monoclonal antibody technique is about 100 times as sensitive as the electrophoretic method for detecting PSA.

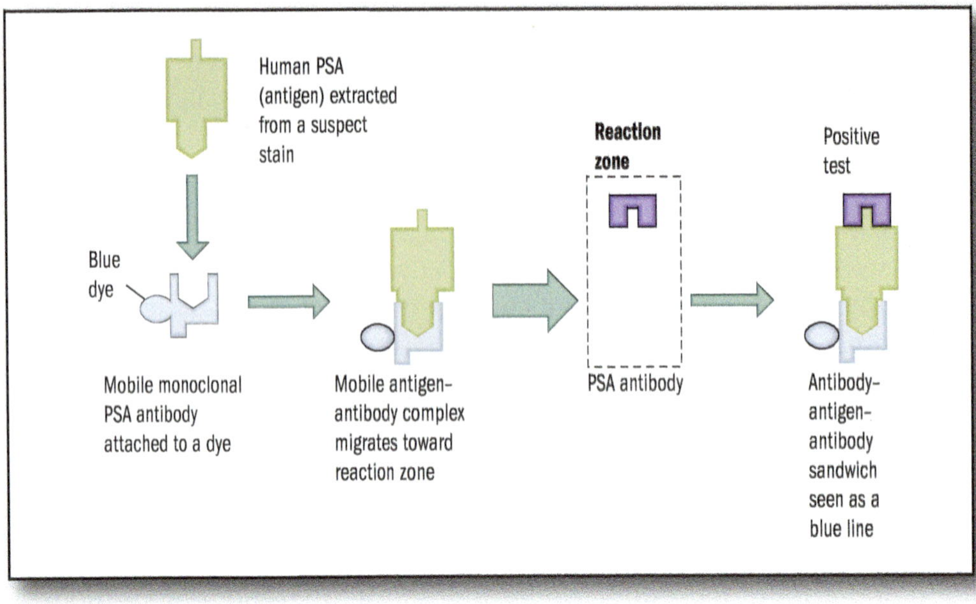

FIGURE 9–16 An antibody-antigen-antibody sandwich or complex is seen as a colored band. This signifies the presence of PSA in the extract of a stain. and positively identifies human semen.

Once the material is proven to be semen, the next task is to associate the semen as closely as possible with an individual. As we will learn in Chapter 10, forensic scientists can link seminal material to one individual with DNA technology. Just as important is the knowledge that this technology can exonerate many of those wrongfully accused of sexual assault.

Quick Review

- The best way to locate and characterize a seminal stain is to perform the acid phosphatase color test.
- The presence of spermatozoa proves that a sample stain contains semen. The presence of the protein called prostate specific antigen (PSA), also known as p30, indicates that a stain contains semen.
- Forensic scientists can link seminal material to an individual by DNA typing.

Collection and Preservation of Rape Evidence Seminal constituents on a rape victim are important evidence that sexual intercourse has taken place, but their absence does not necessarily mean that a rape did not occur. Physical injuries such as bruises and bleeding tend to confirm that a violent assault occurred. Furthermore, the forceful physical contact between victim and assailant may result in a transfer of physical evidence—blood, semen, hairs, and fibers. The presence of such evidence helps forge a vital link in the chain of circumstances surrounding a sexual crime.

To protect this kind of evidence, all the outer garments and undergarments from the people involved should be carefully removed and packaged separately in paper (not plastic) bags. Place a clean bedsheet on the floor and lay a clean paper sheet over it. The victim must remove her shoes before standing on the paper. Have the person disrobe while standing on the paper in order to collect any loose foreign material falling from the clothing. Collect each piece of clothing as it is removed and place in separate paper bags to avoid cross-contamination. Carefully fold the paper sheet so that all foreign materials are contained inside. If appropriate, bedding or the object on which the assault took place should be submitted to the laboratory for processing.

Items suspected of containing seminal stains must be handled carefully. Folding an article through the stain may cause it to flake off, as will rubbing the stained area against the surface of the packaging material. If, under unusual circumstances, it is not possible to transport the stained article to the laboratory, the stained area should be cut out and submitted with an unstained piece as a substrate control.

In the laboratory, analysts try to link seminal material to a donor(s) using DNA typing. Because an individual may transfer their DNA types to a stain through perspiration, investigators must handle stained articles with care, minimizing direct personal contact. The evidence collector must wear disposable latex gloves when such evidence must be touched.

The rape victim must undergo a medical examination as soon as possible after the assault. At this time, the appropriate items of physical evidence are collected by trained personnel. Evidence collectors should have an evidence-collection kit from the local crime laboratory (see Figures 9–17A, 9–17B, and 9–17C).

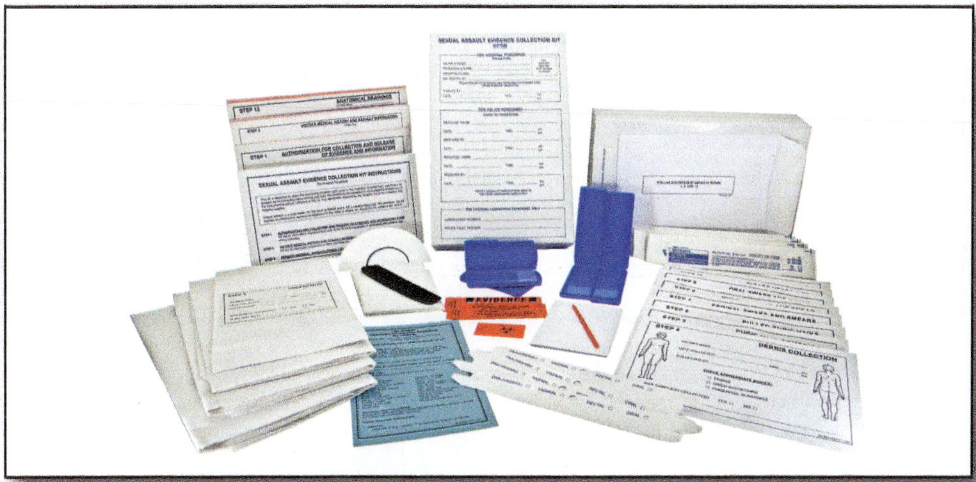

FIGURE 9–17A Victim rape collection kit with instructions, and forms for medical history and assault information. Envelope for foreign materials. Collection bags for outer clothing and underpants. Envelopes for debris, pubic hair combings. Envelope for pulled pubic hair. Envelopes for vaginal swabs and rectal swabs with swap boxes and microscope slides. Envelopes for oral swabs and smear with microscope and a swab box, Envelope for known saliva sample. Known blood sample envelope. Anatomical drawings. *Courtesy Tri-Tech, Inc., Southport, N.C., www.tritechusa.com*

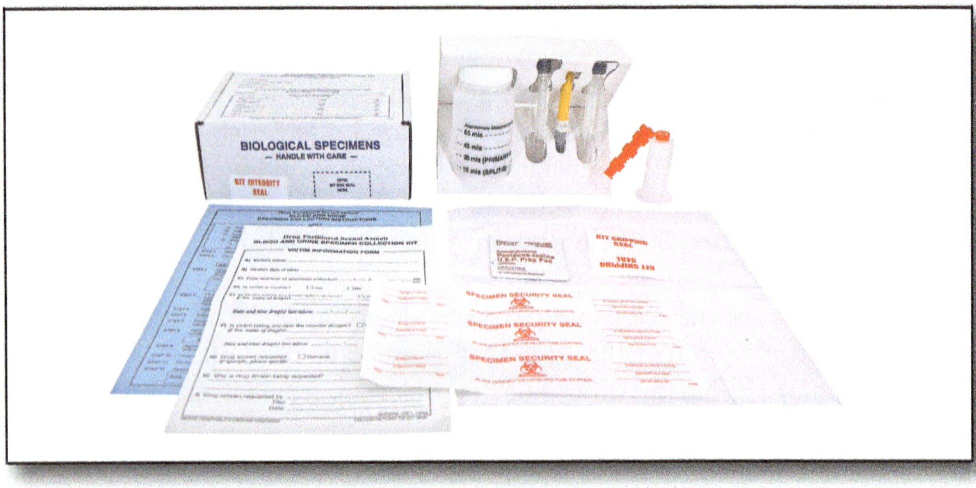

FIGURE 9–17B Drug Facilitated Sexual Assault Evidence Toxicology Kit containing a blood tube and urine specimen bottle holder. *Courtesy Tri-Tech, Inc., Southport, N.C., www.tritechusa.com*

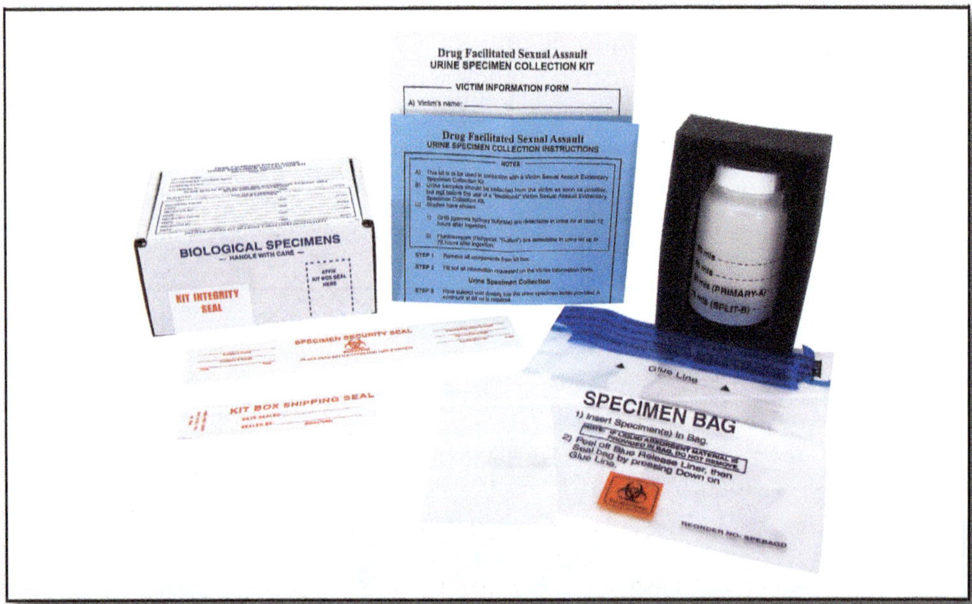

FIGURE 9–17C Drug Facilitated Sexual Assault Evidence Toxicology Kit containing a urine specimen bottle. *Courtesy Tri-Tech, Inc., Southport, N.C., www.tritechusa.com*

The following items of physical evidence are to be collected:

1. **Pubic combings.** Place a paper towel under the buttocks and comb the pubic area for loose or foreign hairs.
2. **Pubic hair standard/reference samples.** Cut 15 to 20 full-length hairs from the pubic area at the skin line.
3. **External genital dry-skin areas.** Swab with at least one dry swab and one moistened swab.

4. **Vaginal swabs and smear.** Using two swabs simultaneously, carefully swab the vaginal area and let the swabs air-dry before packaging. Using two additional swabs, repeat the swabbing procedure and smear the swabs onto separate microscope slides, allowing them to air-dry before packaging.
5. **Cervix swabs.** Using two swabs simultaneously, carefully swab the cervix area and let the swabs air-dry before packaging.
6. **Rectal swabs and smear.** To be taken when warranted by case history. Using two swabs simultaneously, swab the rectal canal, smearing one of the swabs onto a microscope slide. Allow both samples to air-dry before packaging.
7. **Oral swabs and smear.** To be taken if oral–genital contact occurred. Use two swabs simultaneously to swab the buccal area and gum line. Using both swabs, prepare one smear slide. Allow both swabs and the smear to air-dry before packaging.
8. **Swabs of body areas, such as breasts,** suspected of being in contact with DNA arising from touching or saliva.
9. **Head hairs.** Cut at skin line a minimum of five full-length hairs from each of the following scalp locations: center, front, back, left side, and right side. It is recommended that a total of at least 25 hairs be cut and submitted to the laboratory.
10. **Blood sample.** Collect at least 20 milliliters in a vacuum tube containing the preservative EDTA. The blood sample can be used for toxicological analysis if required for a drug-facilitated sexual investigation (see pages 205–206).
11. **Fingernail scrapings.** Scrape the undersurface of the nails with a dull object over a piece of clean paper to collect debris. Use separate paper, one for each hand.
12. **All clothing.** Package as described earlier.
13. **Urine specimen.** Collect 30 milliliters or more of urine from the victim for the purpose of conducting a drug toxicological analysis for Rohypnol, GHB, and other substances associated with drug-facilitated sexual assaults (see pages 205–206).

Often during the investigation of a sexual assault, the victim reports that a perpetrator engaged in biting, sucking, or licking of areas of the victim's body. As we will learn in the next chapter, the tremendous sensitivity associated with DNA technology offers investigators the opportunity to identify a perpetuator's DNA types from saliva residues collected off the skin. The most efficient way to recover saliva residues from the skin is to first swab the suspect area with a rotating motion using a cotton swab moistened with distilled water. A second, dry swab is then rotated over the skin to recover the moist remains on the skin's surface from the wet swab. The swabs are air-dried and packaged together as a single sample.

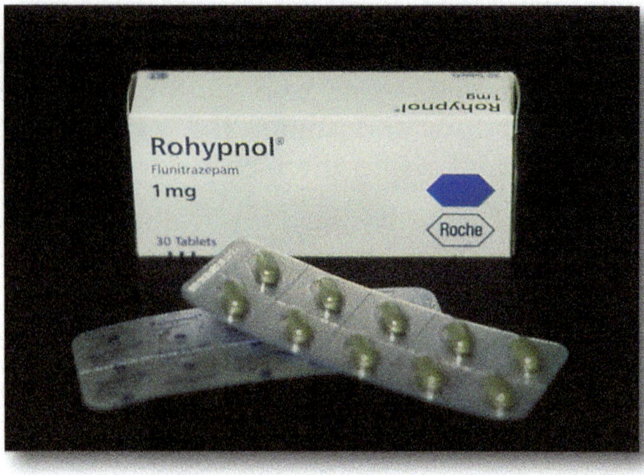

FIGURE 9–18 Box and blister packs of Rohypnol tranquilizer pills. This has gained notoriety as a "date rape" drug. Analysts may test for its presence in a victim's blood or urine. *Josh Sher/Science Source*

If a suspect is apprehended, the following items are routinely collected:

1. **All clothing and any other items believed to have been worn at the time of assault.**
2. **Pubic hair combings.**
3. **Pulled head and pubic hair standard/reference samples.**
4. **A penile swab taken within 24 hours of the assault, when appropriate to the case history.**
5. **A blood sample or buccal swab for DNA-typing purposes.**

The advent of DNA profiling has forced investigators to rethink what items are evidential in a sexual assault. As we will learn in Chapter 10, DNA levels in the range of one-billionth of a gram are now routinely characterized in crime laboratories. In the past, scant attention was paid to the underwear recovered from a male who was suspected of being involved in a sexual assault; seminal constituents on a man's underwear had little or no investigative value. Today, the sensitivity of DNA analysis has created new areas of investigation. It is possible to link a victim and an assailant by analyzing biological material recovered from the interior front surface of a male suspect's underwear. This is especially important when investigations have failed to yield the presence of suspect's DNA on exhibits recovered from the victim.

The persistence of seminal constituents in the vagina may become a factor when trying to ascertain the time of an alleged sexual attack. Although the presence of spermatozoa in the vaginal cavity provides evidence of intercourse, important information regarding the time of sexual activity can be obtained from the knowledge that nonmotile sperm may be found in a living female for up to three days after intercourse and occasionally up to six days later. However, intact sperm (sperm with tails) are not normally found 16 hours after intercourse but have been found as late as 72 hours after intercourse. The likelihood of finding

Case Files

DNA Transfer

A common mode of DNA Transfer occurs when skin cells from the walls of the victim's vagina are transferred onto the suspect during intercourse. Subsequent penile contact with the inner surface of the suspect's underwear often leads to the recovery of the female victim's DNA from the underwear's inner surface. The power of DNA is illustrated in a case in which the female victim of a rape had consensual sexual intercourse with a male partner before being assaulted by a different male. DNA extracted from the inside front area of the suspect's underwear revealed a female DNA profile matching that of the victim. The added bonus in this case was finding male DNA on the same underwear that matched that of the consensual partner.

Source: Gary G. Verret, "Sexual Assault Cases with No Primary Transfer of Biological Material from Suspect to Victim: Evidence of Secondary and Tertiary Transfer of Biological Material from Victim to Suspect's Undergarments," Proceedings of the Canadian Society of Forensic Science, Toronto, Ontario, November 2001.

seminal acid phosphatase in the vaginal cavity markedly decreases with time following intercourse, with little chance of identifying this substance 48 hours after intercourse.6 Hence, with the possibility of the prolonged persistence of both spermatozoa and acid phosphatase in the vaginal cavity after intercourse, investigators should determine when and if voluntary sexual activity last occurred before the sexual assault. This information will be useful for evaluating the significance of finding these seminal constituents in the female victim. Blood or buccal swabs for DNA analysis are to be taken from any consensual partner having sex with the victim within 72 hours before the assault.

Another significant indicator of recent sexual activity is PSA. This semen marker normally is not detected in the vaginal cavity beyond 72 hours following intercourse.

Quick Review

- A rape victim must undergo a medical examination as soon as possible after the assault. At that time clothing, hairs, and vaginal and rectal swabs can be collected for subsequent laboratory examination.

Chapter Review

- Serology involves a broad scope of laboratory tests that use specific antigen and serum antibody reactions.
- An antibody reacts or agglutinates only with its specific antigen. The concept of specific antigen–antibody reactions has been applied to techniques for detecting abused drugs in blood and urine.
- Every red blood cell contains either an A antigen, a B antigen, or no antigen (this is called type O). The type of antigen on one's red blood cells determines one's A-B-O blood type. People with type A blood have A antigens on their red blood cells, those with type B blood have B antigens, and those with type O blood have no antigens on their red blood cells.
- To produce antibodies capable of reacting with drugs, the analyst combines a specific drug with a protein and injects this combination into an animal such as a rabbit. This drug–protein complex acts as an antigen, stimulating the animal to produce antibodies. The recovered blood serum of the animal now contains antibodies that are specific or nearly specific to the drug.
- When an animal is injected with an antigen, its body produces a series of different antibodies, all of which are designed to attack some particular site on the antigen of interest. These antibodies are known as polyclonal antibodies.
- A more uniform and specific collection of antibodies designed to combine with a single antigen site can be manufactured. Such antibodies are known as monoclonal antibodies.
- The criminalist must answer the following questions when examining dried blood: (1) Is it blood? (2) From what species did the blood originate? (3) If the blood is human, how closely can it be associated with a particular individual?
- The determination of blood is best made with a preliminary color test. A positive result from the Kastle-Meyer color test is highly indicative of blood.
- The luminol test is used to find trace amounts of blood at crime scenes.
- The precipitin test uses antisera normally derived from rabbits that have been injected with the blood of a known animal to determine the species origin of a questioned bloodstain.
- Materials undergoing electrophoresis are forced to move across a gel-coated plate under the influence of an electrical potential. Antigens and antibodies can be induced to move toward one another under the influence of an electrical field.
- The gene is the basic unit of heredity. A chromosome is a threadlike structure in the cell nucleus, along which the genes are located.
- Most human cells contain 46 chromosomes, arranged in 23 mated pairs. The only exceptions are the human reproductive cells, the egg and sperm, which contain 23 unmated chromosomes.
- During fertilization, a sperm and egg combine so that each contributes 23 chromosomes to form the new cell, or zygote, that develops into the offspring.

- An allele is any of several alternative forms of genes that influence a given characteristic and that are aligned with one another on a chromosome pair.
- A heterozygous gene pair is made up of two different alleles; a homozygous gene pair is made up of two similar alleles.
- When two different genes are inherited, the characteristic coded for by a dominant gene is expressed. The characteristic coded for by a recessive gene remains hidden.
- A genotype is the particular combination of genes present in the cells of an individual. A phenotype is the physical manifestation of a genetic trait.
- The best way to locate and characterize a seminal stain is to perform the acid phosphatase color test.
- The presence of spermatozoa proves that a sample stain contains semen. The presence of the protein called prostate specific antigen (PSA), also known as p30, indicates that a stain contains semen.
- Forensic scientists can link seminal material to an individual by DNA typing.
- A rape victim must undergo a medical examination as soon as possible after the assault. At that time clothing, hairs, and vaginal and rectal swabs can be collected for subsequent laboratory examination.

Quick Lab: Luminol Test

Materials:

Luminol (powder needs to be mixed with water)
Spray bottle
Simulated blood
Piece of wood or flooring
UV light source
Alternate Light Sources
ALS protective goggles

Procedure:

Apply some blood to the wood/flooring. Then try to completely clean it, as if you were trying to cover up a crime. If the teacher does not have the luminol mixed for you, follow instructions on how to mix it. If you do not understand the instructions, ask for clarification. Using the spray bottle, apply some luminol to the wood/flooring that you cleaned. Keep the room dark for this step. You may shine the UV light on the area where you sprayed the luminol; this may help if you do not see a reaction right away.

Follow-Up Questions:

1. **Did you observe any reaction when the room was dark? When you shined the UV light on the wood/flooring? If so, what did you observe?**
2. **How does luminol detect bloodstains?**
3. **What is luminescence?**

Quick Lab: Blood Presumptive Tests

Materials:

Kastle-Meyer color test
Tetramethylbenzidine (TMB)
Simulated bodily fluids (blood, urine)
Pieces of cloth
Pieces of cloth with stains
Spray bottles
Protective goggles

Procedure:

To conduct the blood presumptive tests for various biologicals, obtain two clean pieces of cloth. Using appropriate tools and equipment, prepare reference sample of known positive and negative reactions. Obtain two samples of cloth containing simulated bodily fluid stains. Spray one sample using the Kastle-Meyer color test (phenolphthalein) and the other using the TMB. Compare results to the reference samples. Interpret the results.

Follow-Up Questions:

1. **Interpret the results of the Kastle-Meyer color test and the TMB test.**
2. **What conclusions can be made based on the results of each test?**
3. **What should be the next step in the analysis process?**

Review Questions

1. Type AB blood contains
   a. anti-A antibodies and B antigens.
   b. anti-A antigens and anti-B antibodies.
   c. both A and B antigens.
   d. both anti-A and anti-B antibodies.

2. All nucleated human cells contain
   a. 64 chromosomes.
   b. 46 chromosomes.
   c. 32 chromosomes.
   d. 23 chromosomes.

3. Which of the following questions must the criminalist be prepared to answer when examining dried blood?
   a. Is it blood?
   b. From what species did the blood originate?
   c. How closely can human blood be associated with a single individual?
   d. all of the above

4. The determination of whether a substance is blood is best made by means of a preliminary color test such as the Kastle-Meyer color test, which uses the chemical

a. benzidine.

b. p30.

c. phenolphthalein.

d. precipitin.

5. In which phenotype pairings can the genotypes of the individuals be directly known?

a. type AB and type O

b. type A and type B

c. type B and type O

d. type A and type AB

6. True or False: The advent of DNA technology has dramatically altered the approach forensic scientists have taken toward the individualization of bloodstains and other biological evidence.

7. True or False: The fundamental principle of blood typing is that for every antigen, there exists a specific antibody.

8. True or False: The standard test used to determine whether a blood stain is of human or animal origin is the precipitin test.

9. True or False: Today it is possible for forensic scientists to successfully link seminal material to one individual with DNA technology.

10. True or False: The presence or absence of four antigens determines an individual's blood type in the A-B-O system.

11. What technique supplanted blood typing for associating bloodstain evidence with a particular individual?

12. What is plasma? What percentage of blood content does plasma account for?

13. Which of the following types of cells are not contained in plasma?

a. phagocytes

b. leukocytes

c. erythrocytes

d. platelets

14. What are antigens and antibodies? What part of the blood contains antibodies?

15. Describe how antibodies and antigens determine one's A-B-O blood type. How can you identify how the red blood cell antigens and antibodies relate to human blood types?

16. What is the fourth important antigen other than A, B, and O?

17. What happens when serum containing B antibodies is added to red blood cells carrying the B antigen? Will the same thing happen if serum containing B antibodies is added to red blood cells carrying the A antigen? Explain your answer.

18. Briefly describe how antibodies capable of reacting with drugs are produced in animals.

19. What is the greatest problem associated with detecting marijuana in urine?

20. What is the difference between monoclonal and polyclonal antibodies?

21. Name the most common color test for blood and briefly describe how it identifies bloodstains.

22. Briefly describe how luminol is used to detect bloodstains.

23. What is the standard test used to determine whether blood is of human or animal origin? Briefly state the principle underlying the test.

24. Which technique takes advantage of the fact that antibodies and antigens move toward one another on a plate coated with medium made from a natural polymer called agar?

25. In what technique can antigens and antibodies be induced to move toward one another under the influence of an electrical field?

26. Define gene and chromosome.

27. How many chromosomes do most human cells contain, and how are chromosomes arranged in the cell? What cells are the exception to this rule? How are these cells different from all other cells?

28. Describe how genetic material is transferred from parents to offspring.

29. What is an allele?

30. What is the difference between a heterozygous gene pair and a homozygous gene pair?

31. What is the difference between a dominant and a recessive gene?

32. Define genotype and phenotype. What is the only way to determine an individual's genotype?

33. The best way to locate and characterize a seminal stain is to perform what test? In what situation is this test particularly useful?

34. Define oligospermia and aspermia.

35. The presence of what protein indicates that a sample stain contains semen? What two techniques are used to detect this protein?

36. Besides swabbing for semen constituents, what other bodily fluids should be collected from a rape victim during a medical examination?

37. What items should be collected from the suspected perpetrator of a sexual assault?

38. Explain how reagents are used to chemically enhance latent blood patterns.

Learn the Vocabulary

Use the information from this chapter along with your prior knowledge and experience to learn the key terms. Choose from the following activities:

1. Make flashcards with a term on the front and its definition on the back. Use the flashcards to quiz a partner on correctly pronouncing and defining the terms.

2. Make an audio or video recording of each term and its definition.

3. Work with a partner to write a sentence for each term. Read the sentences out loud to a partner or share them with the class.

4. Write a list of new terms and language you have encountered in this chapter. Reuse the terms to write sentences and then read the sentences out loud to a partner. Reuse the terms again by writing a short speech. The speech might explain the terms or teach a lesson using the terms. Give the speech to a partner or to the class.

Application and Critical Thinking

Mother's genotype \ Father's genotype	A	O
A		
B		

Mother's genotype \ Father's genotype	A	B
A		
A		

Mother's genotype \ Father's genotype	B	O
B		
O		

1. Complete the following Punnett squares and answer the questions that follow.
   a. Among this group of people, which blood type(s) is most likely to be expressed? Which blood type is least likely to be expressed? Explain your answers.
   b. What blood type is Greg most likely to have? Explain your answer.
   c. Which person is least likely to have type B blood? Explain your answer.
   d. In which, if any, of these people may a recessive gene express itself? Explain your answer.

2. Suppose that Greg and Maura have a child. Construct all of the possible Punnett squares for that child and answer the following questions:
   a. What blood type(s) is the child most likely to have?
   b. What genotype(s) is the child most likely to have?
   c. If Maura and Greg have the same genotype, what blood type is the child most likely to have? What are the chances that the child will have type O blood? Explain your answers.

3. Use critical-thinking skills to analyze the following scenario. Police investigating the scene of a sexual assault recover a large blanket that they believe may contain useful physical evidence. They take it to the laboratory of forensic serologist Scott Alden, asking him to test it for the presence of semen. Using his senses and observations, Scott spots empirical evidence. Noticing faint pink stains on the blanket, Scott asks the investigating detective if he is aware of anything that might recently have been spilled on the blanket. The detective reports that an overturned bowl of grapes and watermelon was found at the scene, as well as a broken glass that had contained wine. After the detective departs, Scott chooses and administers what he considers the best test for analyzing the piece of evidence in his possession. Three minutes after completion of the test, Scott uses scientific explanations to determine the source of the stain. The blanket shows a positive reaction. What blood presumptive tests for various biologicals did Scott choose and what was his conclusion? Explain your answer in writing.

4. Criminalist Cathy Richards is collecting evidence from the victim of a sexual assault. She places a sheet on the floor, asks the victim to disrobe, and places the clothing in a paper bag. After collecting pubic combings and pubic hair samples, she takes two vaginal swabs, which she allows to air-dry before packaging. Finally, Cathy collects blood, urine, and scalp hair samples from the victim. What mistakes, if any, did she make in collecting this evidence?

5. Police have been called to the scene of a hit-and-run accident on a busy rural highway. This is a major highway, but the surrounding area is not densely populated. There is a lot of wildlife in the area, including deer, raccoons, and skunks. The victim has been transferred to a local hospital, and police have interviewed several witnesses. All witnesses recall the same make and model of car, a silver compact. Police recover possible bloodstains from the bumper of three vehicles (Vehicles A, B, and C). The Kastle-Meyer presumptive color test for blood was positive for all stains.

When questioned about the blood, suspects A, B, and C all claim the blood originated from wildlife they accidentally hit. Your job as the forensic scientist assigned to the case is to first determine whether the blood samples are animal or human blood. Also, determine whether the blood on any of the suspected vehicles matches the victim's blood type of AB, Rh negative.

a. Use the results of each test to determine whether the blood on each suspect's bumper is of human origin.

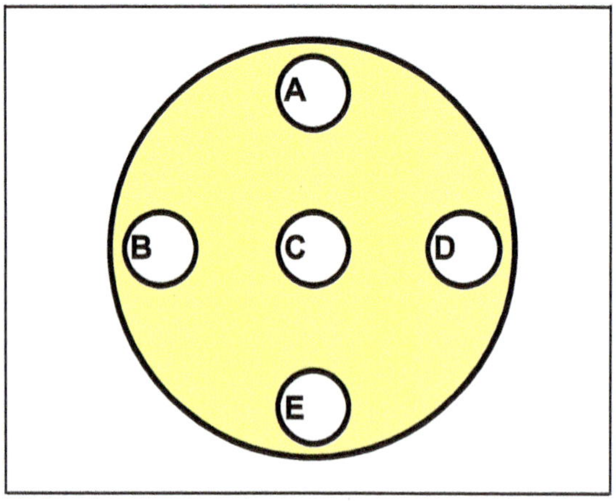

A = human antiserum,
B = racoon antiserum,
D = deer antiserum,
E = skunk antiserum

Blood sample extract from Vehicle A is placed in the center (C):

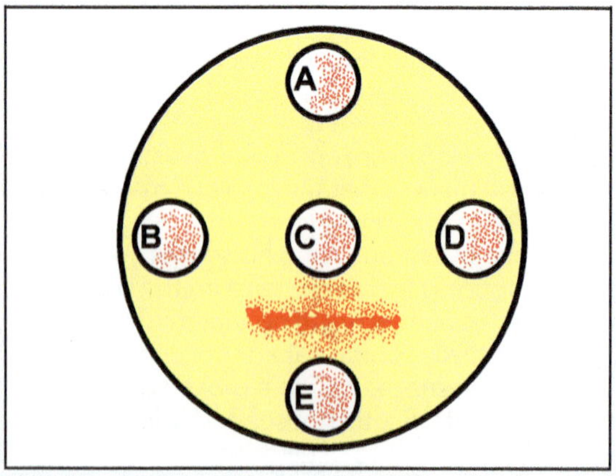

Vehicle A Blood Origin

Blood sample extract from Vehicle B is placed in the center (C):

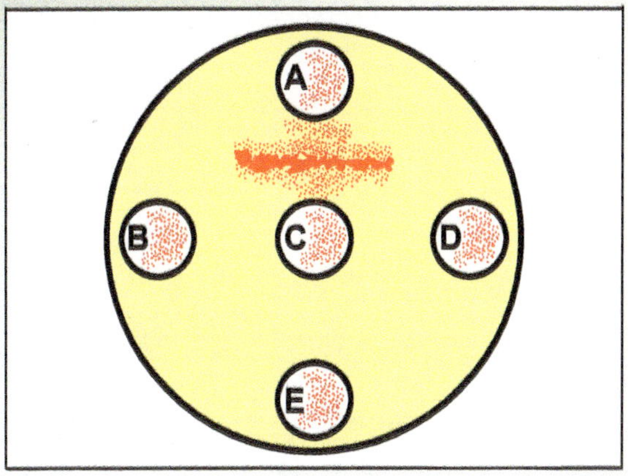

Vehicle B Blood Origin

Blood sample extract from Vehicle C is placed in the center (C):

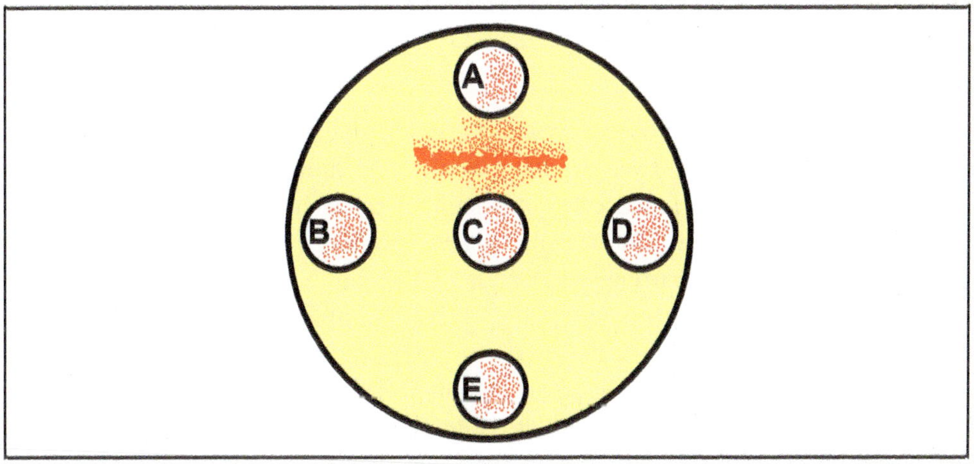

Vehicle C Blood Origin

b. Using the following blood-typing results, indicate the apparent blood type recovered from Vehicles B and C. Determine whether the blood from the car of each suspect is consistent with the victim's blood type of AB negative.

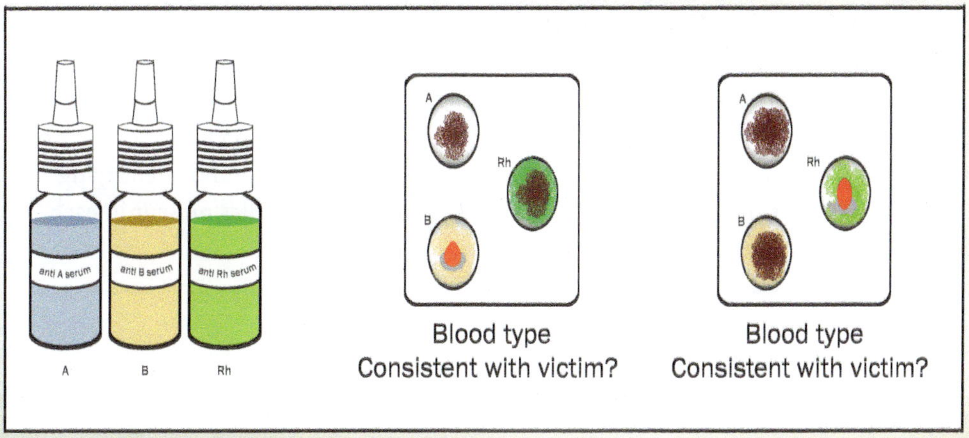

Laboratory Experiments

This activity requires the use of the following practices of science and engineering:

- Asking questions and defining problems
- Planning and carrying out investigations
- Analyzing and interpreting data
- Constructing explanations
- Engaging in argument
- Obtaining, evaluating, and communication information
- Using appropriate safety equipment and practices as outlined in Texas Education Agency-approved safety standards

This activity consists of the following crosscutting concepts:

- **Patterns**—The observed patterns in blood type allow for organization and classification.

Before you begin, read the scenario and steps out loud with a partner. Ask for clarification as necessary to make sure you understand the experiment.

The Scenario

Jennifer is jogging along a residential street one evening near midnight. Even though she has been warned this practice is not safe, Jennifer enjoys being outside at night and chooses to ignore the warning. About 30 minutes after leaving her house, a tall man grabs Jennifer from behind, and tries to force her into a car parked on the street. Jennifer fights desperately with the assailant because she fears for her life. In her struggles, she scratches him several times on the top of the head. Finally, he releases her and turns away. Bravely, she follows after him a short distance and watches him run into a fourplex at 2121 Mocassin Street. However, she never gets a good look at his face. Frightened, Jennifer turns around and runs home.

When she gets home, Jennifer calls the police and tells them what has happened.

Following their instructions, Jennifer meets the police officers at Parkview Hospital Emergency Room, where she is carefully examined. The ER doctor finds Jennifer to be in pretty good shape. He takes samples of dried blood from under her fingernails.

Testing of the blood reveals that it is type O–. Armed with this information, the police obtain a proper warrant and visit the residents of 2121 Mocassin Street. They find the fourplex is occupied by four men: Suspect #1 Mr. Bell, Suspect #2 Mr. Barry, Suspect #3 Mr. Alvarez, and Suspect # 4 Mr. Wu. The four individuals are brought to police headquarters for questioning and to obtain blood samples.

Although samples of the blood are being sent to an outside lab for DNA analysis, you have been asked to determine the ABO and Rh blood type of the four samples by analyzing and identifying the antigens and antibodies present in each.

THE EXPERIMENT

Blood Type Analysis

Suspect	*Anti-A*	*Anti-B*	*Anti-Rh*	*Blood Type*
1. Mr. Bell				
2. Mr. Barry				
3. Mr. Alvarez				
4. Mr. Wun				

1. Can any of the suspects be eliminated?
2. Police also found a hair on Jennifer's jogging shorts that did not belong to her. What should the police do?
3. What other evidence could be obtained to narrow down the suspects and possibly identify the assailant?
4. Is blood type class or individual evidence?
5. What type of antigens and antibodies would a person with type AB- blood type have?

Endnotes

1. S. Tobe et al., "Evaluation of Six Presumptive Tests for Blood, Their Specificity, Sensitivity, and Effect on High Molecular-Weight DNA," *Journal of Forensic Sciences* 52 (2007): 102.
2. The luminol reagent is prepared by mixing 0.1 gram 3-amino-phthalhydrazide and 5.0 grams sodium carbonate in 100 milliliters distilled water. Before use, 0.7 gram sodium perborate is added to the solution.
3. A. M. Gross et al., "The Effect of Luminol on Presumptive Tests and DNA Analysis Using the Polymerase Chain Reaction," *Journal of Forensic Sciences* 44 (1999): 837.
4. J. P. Allery et al., "Cytological Detection of Spermatozoa: Comparison of Three Staining Methods," *Journal of Forensic Sciences* 46 (2001): 349.
5. In one study, only a maximum of 4 sperm cells out of 1,000 could be extracted from a cotton patch and observed under the microscope. Edwin Jones (Ventura County Sheriff's Department, Ventura, Calif.), personal communication.
6. R. Dziak et al. "Providing Evidence-Based Opinions on Time Since Intercourse (TSI) Based on Body Fluid Testing Results of Internal Samples," *Canadian Society of Forensic Science Journal* 44 (2011): 59.

DNA: The Indispensable Forensic Science Tool

10

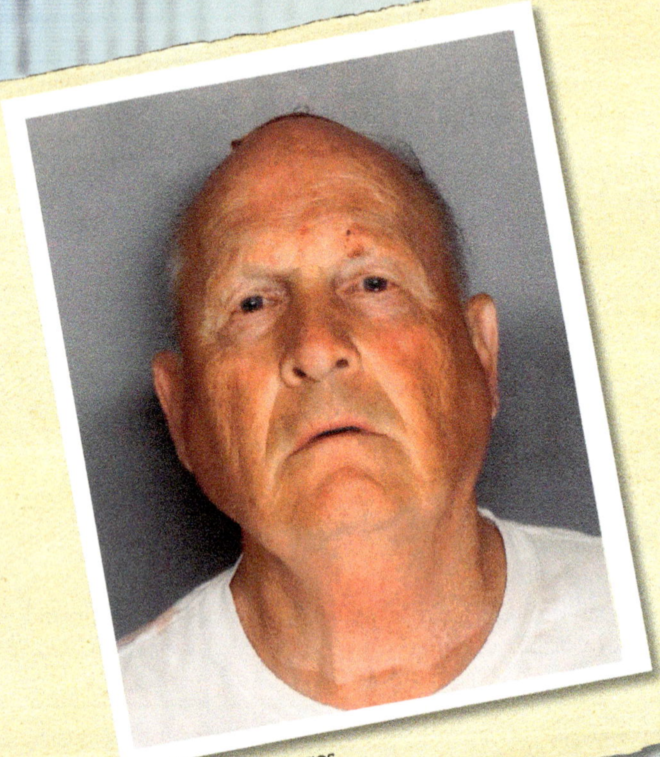

Handout/Getty images

Key Terms

amelogenin gene
amino acids
buccal cells
chromosome
complementary base pairing
deoxyribonucleic acid (DNA)
electrophoresis
epithelial cells
human genome
hybridization
low copy number
mitochondria
multiplexing
nucleotide
picogram
polymer
polymerase chain reaction (PCR)
primer
proteins
replication
restriction enzymes
restriction fragment length polymorphisms (RFLPs)
sequencing
short tandem repeat (STR)
substrate control
tandem repeat
touch DNA
Y-STRs

The Golden State Killer

The crime spree began in 1974 in the East Area of Sacramento, California. A serial rapist was at work and over the next 12 years he would be linked to additional attacks in Contra Costa County, Stockton, and Modesto. His modus operandi was unique. He was a particularly organized criminal, carefully stalking his victims and becoming familiar with their homes and schedules before executing his attacks. His crimes were so prolific that he garnered several names in the press. Known as the "East Area Rapist" in Northern California and "The Original Night Stalker" in the South, DNA testing revealed three decades later that the crimes were all the work of one man.

Over the course of his criminal career, he is suspected of committing at least 13 murders, more than 50 rapes, and over 100 burglaries. Advancements in DNA testing are credited with linking several rapes and murders in 2001. At that time, authorities had the DNA profile of the killer, but they still didn't know his identity. His DNA wasn't present in the Combined DNA Indexing System (CODIS) database and neither was the DNA of any close relatives. Investigators painstakingly made their way through a list of possible suspects, collecting samples and eventually ruling out each one. In 2016, the Federal Bureau of Investigation and local law enforcement agencies held a news conference to announce a renewed nationwide effort, offering a $50,000 reward for his capture.

In 2017, investigators heard about a new and promising method for identifying individuals based on their DNA. It was being used to help reunite families who had been separated, as in cases of adoption. It was called genetic genealogy, and it involved examining similarities shared by related people for clues about the donor's lineage, as discussed in Chapter 3. By tracing the DNA back to a family tree using parentage and relatedness databases like GED match, investigators could narrow the pool of potential suspects considerably. Investigators, with the help of a genetic genealogist, traced the DNA to a common ancestor and constructed family trees using birth records, newspaper clippings, and social media profiles.

It was through this method authorities identified 72-year-old United States Navy veteran and former police officer Joseph James DeAngelo as the Golden State Killer. He was charged with eight counts of first-degree murder. On August 21, 2020, DeAngelo was sentenced to life imprisonment without the possibility of parole.

Learning Objectives

After studying this chapter, you should be able to:

- **10.1** Explain DNA and its structure
- **10.2** Explain how the amino acid sequence in a protein chain is determined by the structure of DNA
- **10.3** Explain the phenomenon of DNA replication and the impact of the PCR technique on duplicating DNA strands
- **10.4** Discuss STR analysis and the concept of electrophoresis
- **10.5** Describe the difference between nuclear and mitochondrial DNA
- **10.6** List the necessary procedures for the proper preservation of bloodstained evidence for laboratory DNA analysis

National Science Content Standards

Scientific Inquiry

Science and Technology

Physical Science

Science in Personal and Social Perspective

Life Science

History and Nature of Science

Earth/ Space Science

Understanding DNA

deoxyribonucleic acid (DNA)
The molecules carrying the body's genetic information; DNA is double stranded in the shape of a double helix

The discovery of **deoxyribonucleic acid (DNA)**, the deciphering of its structure, and the decoding of its genetic information were turning points in our understanding of the underlying concepts of inheritance. Now, with incredible speed, as molecular biologists unravel the basic structure of genes, we can create new products through genetic engineering and develop diagnostic tools and treatments for genetic disorders.

For a number of years, these developments were of seemingly peripheral interest to forensic scientists. All that changed when, in 1985, what started out as a more or less routine investigation into the structure of a human gene led to the discovery that portions of the DNA structure of certain genes are as unique to each individual as fingerprints. Alec Jeffreys and his colleagues at Leicester University, England, who were responsible for these revelations, named the process for isolating and reading these DNA markers *DNA fingerprinting*. As researchers uncovered new approaches and variations to the original Jeffreys technique, the terms *DNA profiling* and *DNA typing* came to be applied to describe this relatively new technology.

This discovery caught the imagination of the forensic science community, because forensic scientists have long searched for ways to definitively link biological evidence such as blood, semen, hair, and tissue to a single individual. Although conventional testing procedures had gone a long way toward narrowing the source of biological materials, individualization remained an elusive goal. DNA typing has allowed forensic scientists to accomplish this goal. Although the technique is still relatively new, DNA typing has become routine in public crime laboratories. It also has been made available to interested parties through the services of a number of skilled private laboratories. In the United States, courts have overwhelmingly admitted DNA evidence and accepted the reliability of its scientific underpinnings.

What Is DNA?

chromosomes
A threadlike structure in the cell nucleus composed of DNA, along which the genes are located

Inside each of 60 trillion cells in the human body are strands of genetic material called **chromosomes**. Arranged along the chromosomes, like beads on a thread, are nearly 25,000 genes. The gene is the fundamental unit of heredity. It instructs body cells to make proteins that determine everything including hair color and susceptibility to diseases. Each gene is composed of DNA. The function of a DNA molecule is to carry the genetic information for an organism. DNA contains instructions for organism growth and development. It also has instructions for making proteins.

Although DNA was first discovered in 1868, scientists were slow to understand and appreciate its fundamental role in inheritance. Painstakingly, researchers developed evidence that DNA was probably the substance by which genetic instructions are passed from one generation to the next. However, the first major breakthrough in comprehending how DNA works did not occur until the early 1950s, when two researchers, James Watson and Francis Crick, deduced the structure of DNA (see Figure 10–1). It turns out that DNA is an extraordinary molecule skillfully designed to control the genetic traits of all living cells, plant and animal.

FIGURE 10–1
The discoverers of the structure of DNA. James Watson at left and Francis Crick, seen with their model of part of a DNA molecule.
Courtesy Photo Researchers Inc.

The Structure of DNA Before examining the implications of Watson and Crick's discovery, let's see how DNA is constructed. DNA is a **polymer**. A polymer is a very large molecule made by linking a series of repeating units, or monomers. In this case, the units are known as **nucleotides**.

polymer
A substance composed of a large number of atoms; these atoms are usually arranged in repeating units, or monomers

Nucleotides. A nucleotide is composed of a sugar molecule, a phosphorus atom surrounded by four oxygen atoms, and a nitrogen-containing molecule called a base. Figure 10–2 shows how nucleotides can be strung together to form a DNA strand. In this figure, S designates the sugar component, which is joined with a phosphate group to form the backbone of the DNA strand. Projecting from the backbone are the bases.

nucleotide
A repeating unit of DNA consisting of one of four bases— adenine, guanine, cytosine, or thymine— attached to a phosphate–sugar group

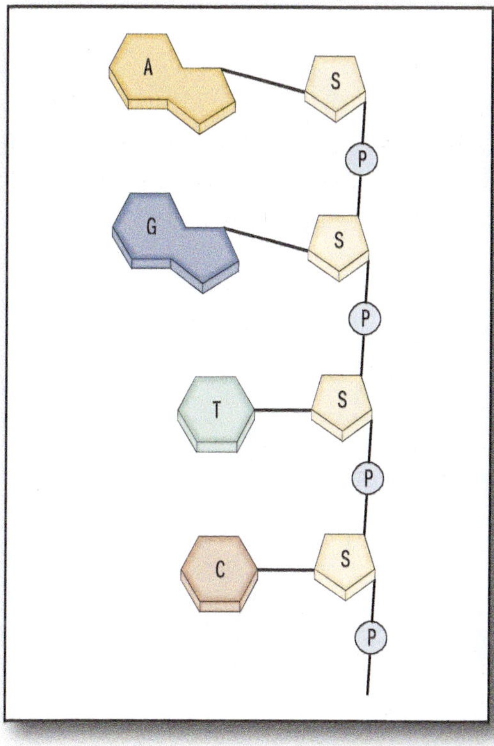

FIGURE 10–2
How nucleotides can be linked to form a DNA strand. S designates the sugar component, which is joined with phosphate groups (P) to form the backbone of DNA. Projecting from the backbone are four bases: A, adenine; G, guanine; T, thymine; and C, cytosine.

The key to understanding how DNA works is to appreciate the fact that only four types of bases are associated with DNA: adenine, cytosine, guanine, and thymine. To simplify our discussion of DNA, we will designate each of these bases by the first letter of their names. Hence, A will stand for adenine, C for cytosine, G for guanine, and T for thymine.

Again, notice in Figure 10–2 how the bases project from the backbone of DNA. Also, although this figure shows a DNA strand of four bases, keep in mind that in theory there is no limit to the length of the DNA strand; a DNA strand can be composed of a long chain with millions of bases. This information was well known to Watson and Crick by the time they set about to detail the structure of DNA. Their efforts led them to discover that the DNA molecule is composed of two DNA strands coiled into a double helix. This can be thought of as resembling two wires twisted around each other.

As Watson and Crick manipulated scale models of DNA strands, they realized that the only way the bases on each strand could be properly aligned with each other in a double-helix configuration was to place base A opposite T and G opposite C. Watson and Crick had solved the puzzle of the double helix and presented the world with a simple but elegant picture of DNA (see Figure 10–3).

complementary base pairing
The specific pairing of base A with T and base C with G in double-stranded DNA

Complementary Base Pairing. The only arrangement possible in the double-helix configuration is the pairing of bases A to T and G to C, a concept that has become known as **complementary base pairing**. Although A–T and G–C pairs are always required, there are no restrictions on how the bases are sequenced on a DNA strand. Thus, one can observe the sequences T–A–T–T or G–T–A–A or G–T–C–A. When these sequences are joined with their complements in a double-helix configuration, they pair as follows:

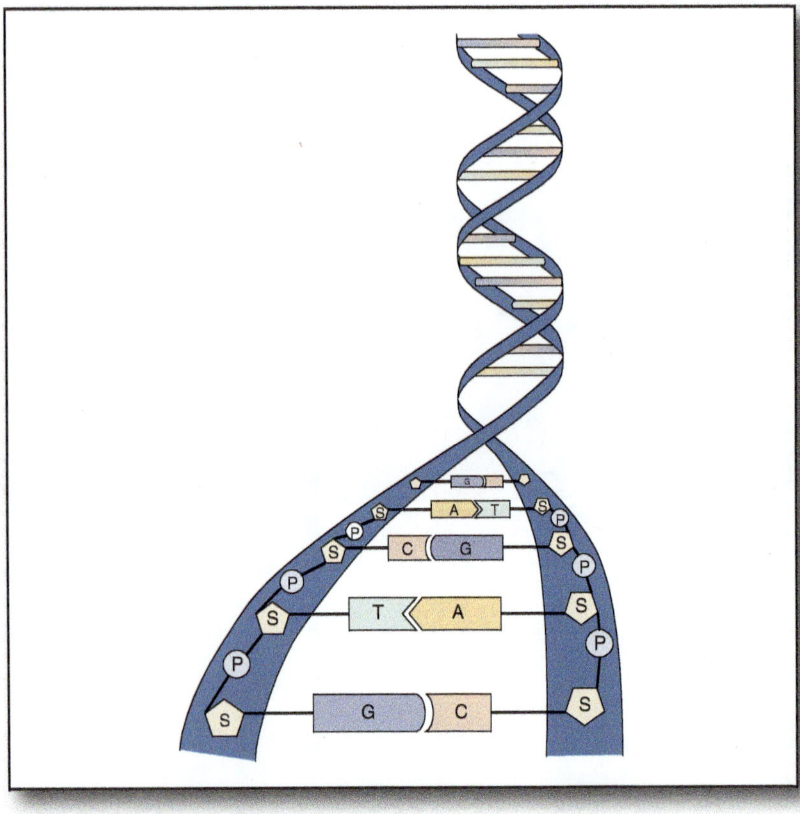

FIGURE 10–3
A representation of a DNA double helix. Notice how bases G and C pair with each other, as do bases A and T. This is the only arrangement in which two DNA strands can align with each other in a double-helix configuration. *Courtesy Photo Researchers, Inc.*

Any base can follow another on a DNA strand, which means that the possible number of different sequence combinations is staggering. Consider that the average human chromosome has DNA containing 100 million base pairs. All of the human chromosomes taken together contain about 3 billion base pairs. From these numbers, we can begin to appreciate the diversity of DNA and hence the diversity of living organisms. DNA is like a book of instructions. The alphabet used to create the book is simple enough: A, T, G, and C. The order in which these letters are arranged defines the role and function of a DNA molecule.

DNA at Work The inheritable traits that are controlled by DNA arise out of its ability to direct the production of complex molecules called **proteins**. Proteins are made by linking a combination of **amino acids**. Although thousands of proteins exist, they can all be derived from a combination of up to 20 known amino acids. The sequence of amino acids in a protein chain determines the shape and function of the protein.

proteins
Polymers of amino acids that play basic roles in the structure and function of living things

amino acids
The building blocks of proteins

Let's look at one example: The protein hemoglobin is found in our red blood cells. It carries oxygen to our body cells and removes carbon dioxide from these cells. One of the four amino acid chains of "normal" hemoglobin is shown in Figure 10–4(a). Studies of individuals who have sickle-cell anemia show that this inheritable disorder arises from the presence of "abnormal" hemoglobin in their red blood cells. An amino acid chain for "abnormal" hemoglobin is shown in Figure 10–4(b). Note that the sole difference between "normal" and "abnormal," or sickle-cell hemoglobin, arises from the substitution of one amino acid for another in the protein chain (see Figure 10–4).

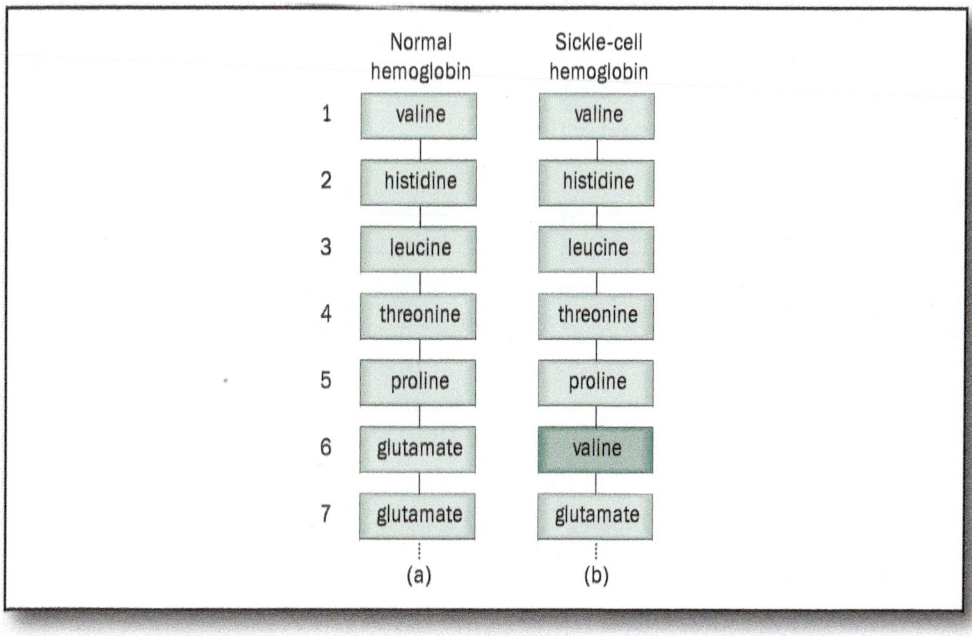

FIGURE 10–4 (a) A string of amino acids composes one of the protein chains of hemoglobin. (b) Substitution of just one amino acid for another in the protein chain results in sickle-cell hemoglobin.

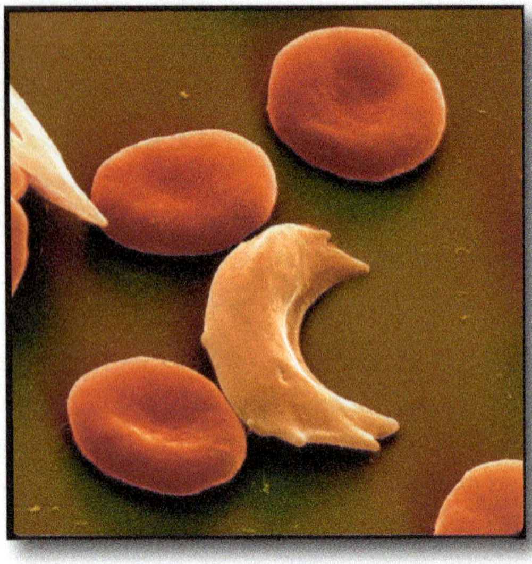

FIGURE 10–5
Sickle cell anemia is an inherited anomaly in which the red blood cells contain an abnormal form of hemoglobin. Decreased oxygen supply causes normal red blood cells to change their form into sickles.
Courtesy Photo Researchers, Inc.

The genetic information that determines the amino acid sequence for every protein manufactured in the human body is stored in DNA in a genetic code that relies on the sequence of bases along the DNA strand. The alphabet of DNA is simple—*A, T, G,* and *C*—but the key to deciphering the genetic code is to know that each amino acid is coded by a sequence of three bases. Thus, the amino acid alanine is coded by the combination *C–G–T*; the amino acid aspartate is coded by the combination *C–T–A*; and the amino acid phenylalanine is coded by the combination *A–A–A*. With this code in hand, we can now see how the amino acid sequence in a protein chain is determined by the structure of DNA. Consider the DNA segment

–C–G–T–C–T–A–A–A–A–C–G–T–

The triplet code in this segment translates into

[C–G–T] – [C–T–A] – [A–A–A] – [C–G–T]

or the protein chain

alanine – aspartate – phenylalanine – alanine

Interestingly, this code is not restricted to humans. Almost all living cells studied to date use the same genetic code as the language of protein synthesis.[1]

If we look at the difference between "normal" and sickle-cell hemoglobin (see Figure 10–4), we see that the latter is formed by substituting one amino acid (valine) for another (glutamate). Within the DNA segment that codes for the production of normal hemoglobin, the letter sequence is

– [C–C–T] – [G–A–G] – [G–A–G] –
proline glutamate glutamate

Individuals with sickle-cell disease carry the sequence

– [C–C–T] – [G–T–G] – [G–A–G] –
proline valine glutamate

Thus, we see that a single base or letter change (T has been substituted for A in valine) is the underlying cause of sickle-cell anemia, demonstrating the delicate chemical balance between health and disease in the human body.

As scientists unravel the base sequences of DNA, they obtain a greater appreciation for the roles of proteins in the chemistry of life. Already the genes responsible for hemophilia, Duchenne muscular dystrophy, and Huntington's disease have been located. Once scientists have isolated a disease-causing gene, they can determine the protein that the gene has directed the cell to manufacture. By studying these proteins—or the absence of them—scientists will be able to devise a treatment for genetic disorders.

A 13-year project to determine the order of bases on all 23 pairs of human chromosomes (also called the **human genome**) is now complete. Knowing the location on a specific chromosome at which DNA codes for production of a particular protein is useful for diagnosing and treating genetic diseases. This information is also crucial for understanding the underlying causes of cancer. Also, comparing the human genome with that of other organisms will help us understand the role and implications of evolution.

human genome The order of bases on all 23 pairs of human chromosomes

Quick Review

- The gene is the fundamental unit of heredity. Each gene is composed of DNA specifically designed to control the genetic traits of our cells.
- DNA is constructed as a very large molecule made by linking a series of repeating units called nucleotides.
- Four types of bases are associated with the DNA structure: adenine (*A*), guanine (*G*), cytosine (*C*), and thymine (*T*).
- The bases on each strand of DNA are aligned in a double-helix configuration so that adenine pairs with thymine and guanine pairs with cytosine. This concept is known as complementary base pairing.
- The order in which the base pairs are arranged defines the role and function of a DNA molecule.

The Combined DNA Indexing System (CODIS)

Perhaps the most significant investigative tool to arise from a DNA-typing program is CODIS (Combined DNA Index System), a computer software program developed by the FBI that maintains local, state, and national databases of DNA profiles from convicted offenders, unsolved crime-scene evidence, and profiles of missing people. CODIS allows crime laboratories to compare DNA types recovered from crime-scene evidence to those of convicted sex offenders and other convicted criminals.

Thousands of CODIS matches have linked serial crimes to each other and have solved crimes by allowing investigators to match crime-scene evidence to known convicted offenders. This capability is of tremendous value to investigators in cases in which the police have not been able to identify a suspect. The CODIS concept has already had a significant impact on police investigations in various states, as shown in the Case Files feature on page 99.

Ancestry-based Databases

Genetic genealogy databases have started being used in the identification of missing persons and identifying suspects in previously unsolved cases referred to as cold cases. Genealogists work in conjunction with law enforcement officials to analyze the ancestry-based databases of individuals who have sent their DNA to be analyzed to trace their family trees. When an unknown DNA sample is compared to the ancestry-based database, a connection can be made to potential relatives that have submitted their DNA to the database.

Quick Review

- CODIS is a computer software program developed by the FBI that maintains local, state, and national databases of DNA profiles from convicted offenders, unsolved crime-scene evidence, and profiles of missing people.
- Ancestry-based databases are used in the identification of missing persons and identifying suspects in cold cases.

Replication of DNA

Once the double-helix structure of DNA was discovered, how DNA duplicated itself before cell division became apparent. The concept of base pairing in DNA suggests the analogy of positive and negative photographic film. Each strand of DNA in the double helix has the same information; one can make a positive print from a negative or a negative from a positive.

The Process of Replication

replication
The synthesis of new DNA from existing DNA

DNA **replication**—the synthesis of new DNA from existing DNA—begins with the unwinding of the DNA strands in the double helix. Each strand is then exposed to a collection of free nucleotides. Letter by letter, the double helix is re-created as the nucleotides are assembled in the proper order, as dictated by the principle of base pairing (*A* with *T* and *G* with *C*). The result is the emergence of two identical copies of DNA where before there was only one (see Figure 10–6). A cell can now pass on its genetic identity when it divides.

Many enzymes and proteins are involved in unwinding the DNA strands, keeping the two DNA strands apart, and assembling the new DNA strands. For example, DNA polymerases are enzymes that assemble a new DNA strand in the proper base sequence determined by the original or parent DNA strand. DNA polymerases also "proofread" the growing DNA double helices for mismatched base pairs, which are replaced with correct bases.

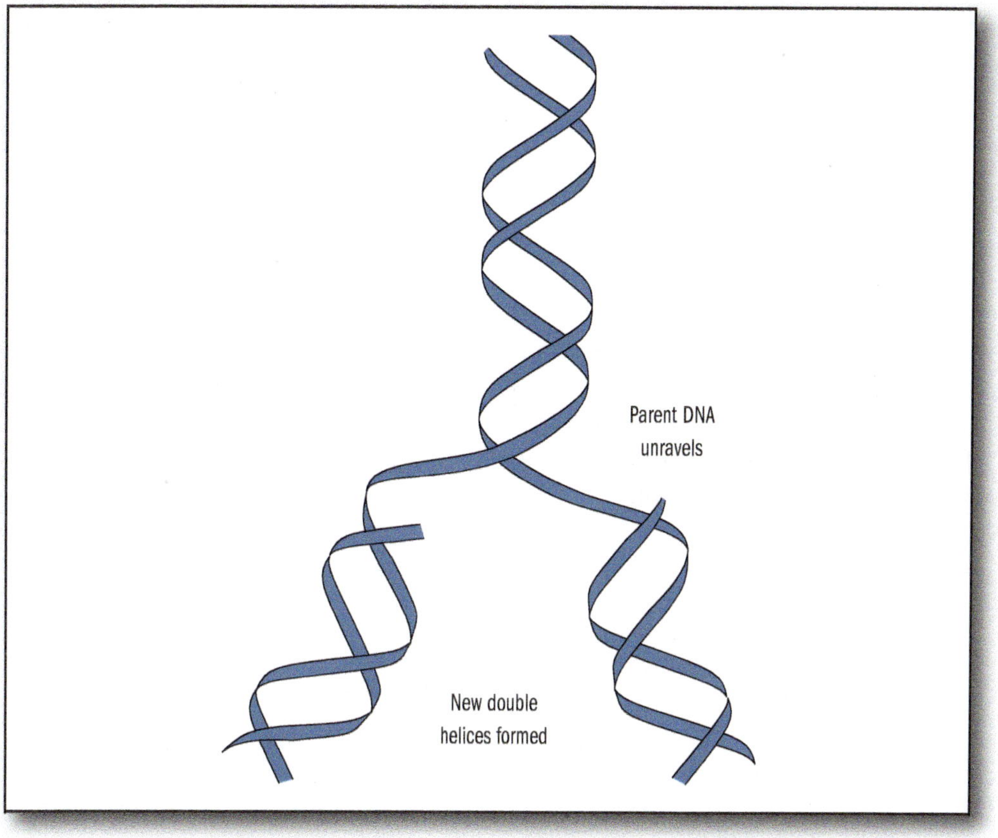

FIGURE 10–6
Replication of DNA. The strands of the original DNA molecule are separated, and two new strands are assembled.

Quick Review

- DNA replication begins with the unwinding of the DNA strands in the double helix. The double helix is re-created as the nucleotides are assembled in the proper order (A with T and G with C). Two identical copies of DNA emerge from the process.
- PCR (polymerase chain reaction) is a technique for replicating or copying a portion of a DNA strand outside a living cell.
- Recombinant DNA is a process by which the DNA of an organism is altered by inserting into it fragments of DNA from another organism (see Inside the Science on the following pages).

Inside the Science

Recombinant DNA

The relationship between the base letters on a DNA strand and the type of protein specified for manufacture by the sequence of these letters is called the genetic code. Once a particular DNA site has been identified as controlling the production of a certain protein, molecular biologists can take advantage of the natural chemical-producing abilities of the DNA site. This undertaking has given rise to the technology known as recombinant DNA.

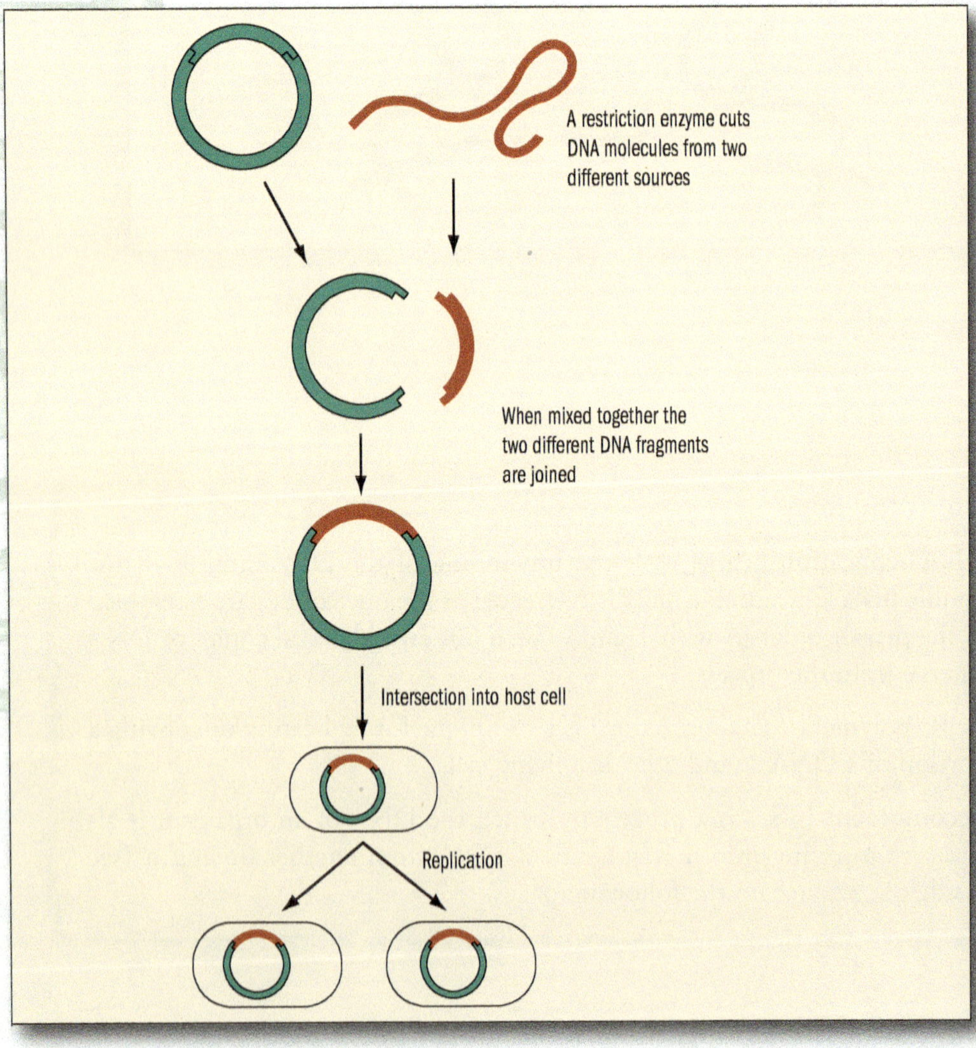

Inside the Science (CONTINUED)

Recombinant DNA relies on the ability of certain chemicals, known as restriction enzymes, to cut DNA into fragments that can later be incorporated into another DNA strand. Restriction enzymes can be thought of as highly specialized scissors that cut a DNA molecule when they recognize a specific sequence of bases. At present, more than 150 restriction enzymes are commercially available. Thus, molecular biologists have a great deal of flexibility in choosing the portion of a DNA strand they wish to cut out.

Once a portion of the DNA strand has been cut out with the aid of a restriction enzyme, the next step in the recombinant DNA process is to insert the isolated DNA segment into a foreign DNA strand (normally, bacterium DNA is selected). Many types of bacteria contain DNA shaped in a circle. A restriction enzyme is used to cut open the circular DNA; then the foreign DNA is spliced in to re-form the circle (see Figure on the previous page). The newly fashioned DNA is reintroduced into the bacterial cells. As the bacteria multiply rapidly in their usual fashion, copies of altered DNA are passed on to all descendants.

The commercial implications of recombinant DNA technology are enormous. For example, the gene that produces human growth hormone has been introduced into goldfish and carp, and the gene that produces growth hormone in rainbow trout has been introduced into carp. In each case, the gene-altered fish have grown significantly faster and larger than their natural relatives. If altered bacteria are infused with the DNA segment that makes human insulin, for example, the bacteria make human insulin. Because bacteria multiply so rapidly, it is not long before significant amounts of insulin can be recovered and used to treat diabetes. In this manner, other naturally occurring substances can be produced in commercial quantities for the treatment of human ailments. Likewise, plant genetic engineering holds promise for increasing global food production.

DNA Typing with Tandem Repeats

tandem repeat
Region of a chromosome that contains multiple copies of a core DNA sequence arranged in a repeating fashion

Geneticists have discovered that portions of the DNA molecule contain sequences of letters that are repeated numerous times. In fact, more than 30 percent of the human genome is composed of repeating segments of DNA. These repeating sequences, or **tandem repeats**, seem to act as filler or spacers between the coding regions of DNA. Although these repeating segments do not seem to affect our outward appearance or control any other basic genetic function, they are nevertheless part of our genetic makeup, inherited from our parents in the manner illustrated by the Punnett square (page 343). The origin and significance of these tandem repeats is a mystery, but to forensic scientists they offer a means of distinguishing one individual from another through DNA typing.

Restriction Fragment Length Polymorphisms

Forensic scientists first began applying DNA technology to human identity in 1985. From the beginning, attention has focused on the tandem repeats of the genome. These repeats can be visualized as a string of connected boxes with each box having the same core sequence of DNA bases (see Figure 10–7). All humans have the same type of repeats, but there is tremendous variation in the number of repeats that each of us has.

restriction fragment length polymorphisms (RFLPs)
Different fragment lengths of base pairs that result from cutting a DNA molecule with restriction enzymes

Up until the mid-1990s, the forensic community aimed its efforts at characterizing repeat segments known as **restriction fragment length polymorphisms (RFLPs)**. A number of different RFLPs were selected by the forensic science community for performing DNA typing. Typically a core sequence is 15 to 35 bases long and repeats itself up to one thousand times. These repeats are cut out of the DNA double helix by a restriction enzyme that acts like a pair of scissors. Once the DNA molecules have been cut up by the restriction enzyme, the resulting fragments were sorted out by separating the fragments by a technique known as **electrophoresis**. Figure 10–8 illustrates a portion of a pair of chromosomes.

RFLP DNA typing has the distinction of being the first scientifically accepted protocol in the United States used for the forensic characterization of DNA. However, its utility has been short lived. New technology incorporating PCR has supplanted RFLP. In its short history, perhaps RFLP's most startling impact related to the impeachment trial of President Bill Clinton. The whole complexion of the investigation regarding the relationship of the president with a White House intern, Monica Lewinsky, changed when it was revealed that Ms. Lewinsky possessed a dress that she claimed was stained with the president's semen. The FBI Laboratory was asked to compare the DNA extracted from the

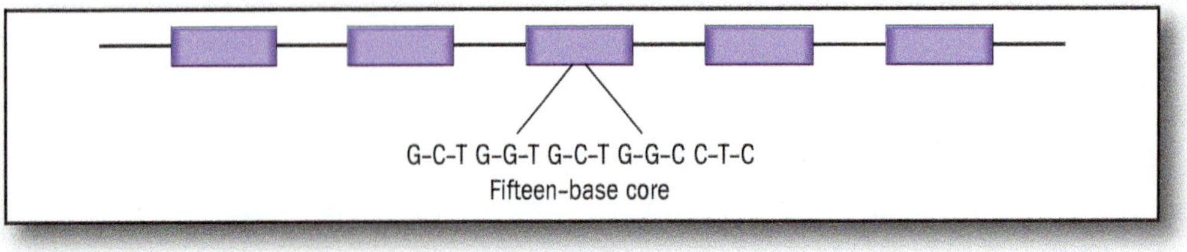

FIGURE 10–7
A DNA segment consisting of a series of repeating DNA units. In this illustration, the fifteen-base core can repeat itself hundreds of times. The entire DNA segment is typically hundreds to thousands of bases long.

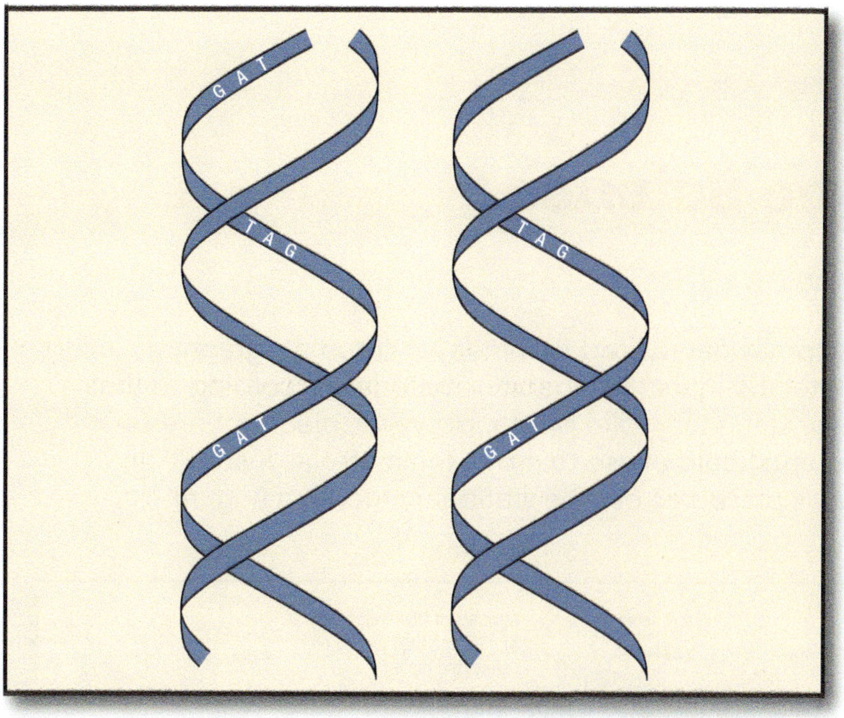

FIGURE 10-8
Intertwined strands of DNA representing segments of two chromosomes. Note that the chromosome segment on the left contains three repeating sequences of T-A-G, while the chromosome segment on the right has two repeating sequences of T-A-G.

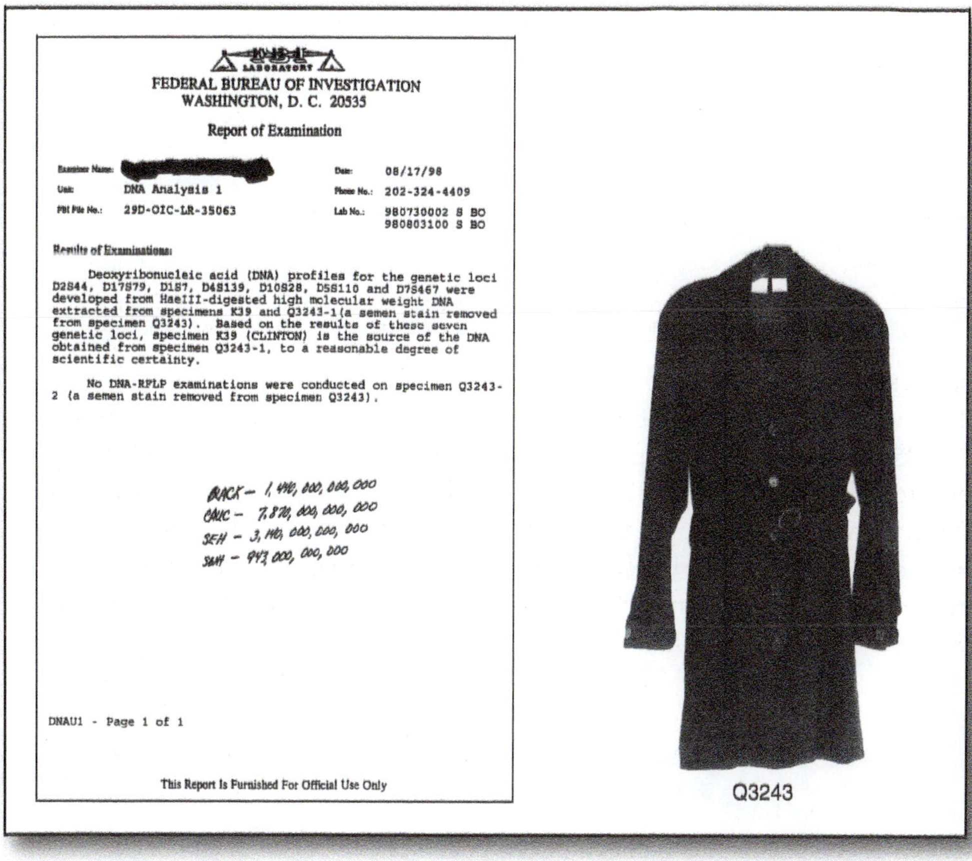

LABORATORY
FEDERAL BUREAU OF INVESTIGATION
WASHINGTON, D. C. 20535

Report of Examination

Examiner Name: [redacted]
Date: 08/17/98
Unit: DNA Analysis 1
Phone No.: 202-324-4409
FBI File No.: 29D-OIC-LR-35063
Lab No.: 980730002 S BO
980803100 S BO

Results of Examinations:

Deoxyribonucleic acid (DNA) profiles for the genetic loci D2S44, D17S79, D1S7, D4S139, D10S28, D5S110 and D7S467 were developed from HaeIII-digested high molecular weight DNA extracted from specimens K39 and Q3243-1(a semen stain removed from specimen Q3243). Based on the results of these seven genetic loci, specimen K39 (CLINTON) is the source of the DNA obtained from specimen Q3243-1, to a reasonable degree of scientific certainty.

No DNA-RFLP examinations were conducted on specimen Q3243-2 (a semen stain removed from specimen Q3243).

BLACK – 1,440,000,000,000
CAUC – 7,870,000,000,000
SEH – 3,140,000,000,000
SWH – 943,000,000,000

DNAU1 - Page 1 of 1

This Report Is Furnished For Official Use Only

FIGURE 10–9
The dress and the FBI Report of Examination for a semen stain located on the dress.
Federal Bureau of Investigation

dress stain with that of the president. An RFLP match was obtained between the president's DNA and the stain. The combined frequency of occurrence for the seven DNA types found was nearly one in eight trillion, an undeniable link. The dress and a copy of the FBI DNA report are shown in Figure 10–9.

Inside the Science

Electrophoresis

Electrophoresis is somewhat related to thin-layer chromatography (discussed in Chapter 12) in that it separates materials according to their migration rates on a stationary solid phase. However, electrophoresis does not use a moving liquid phase to move the material; instead, an electrical potential is placed across the stationary medium (Figure 1).

FIGURE 1

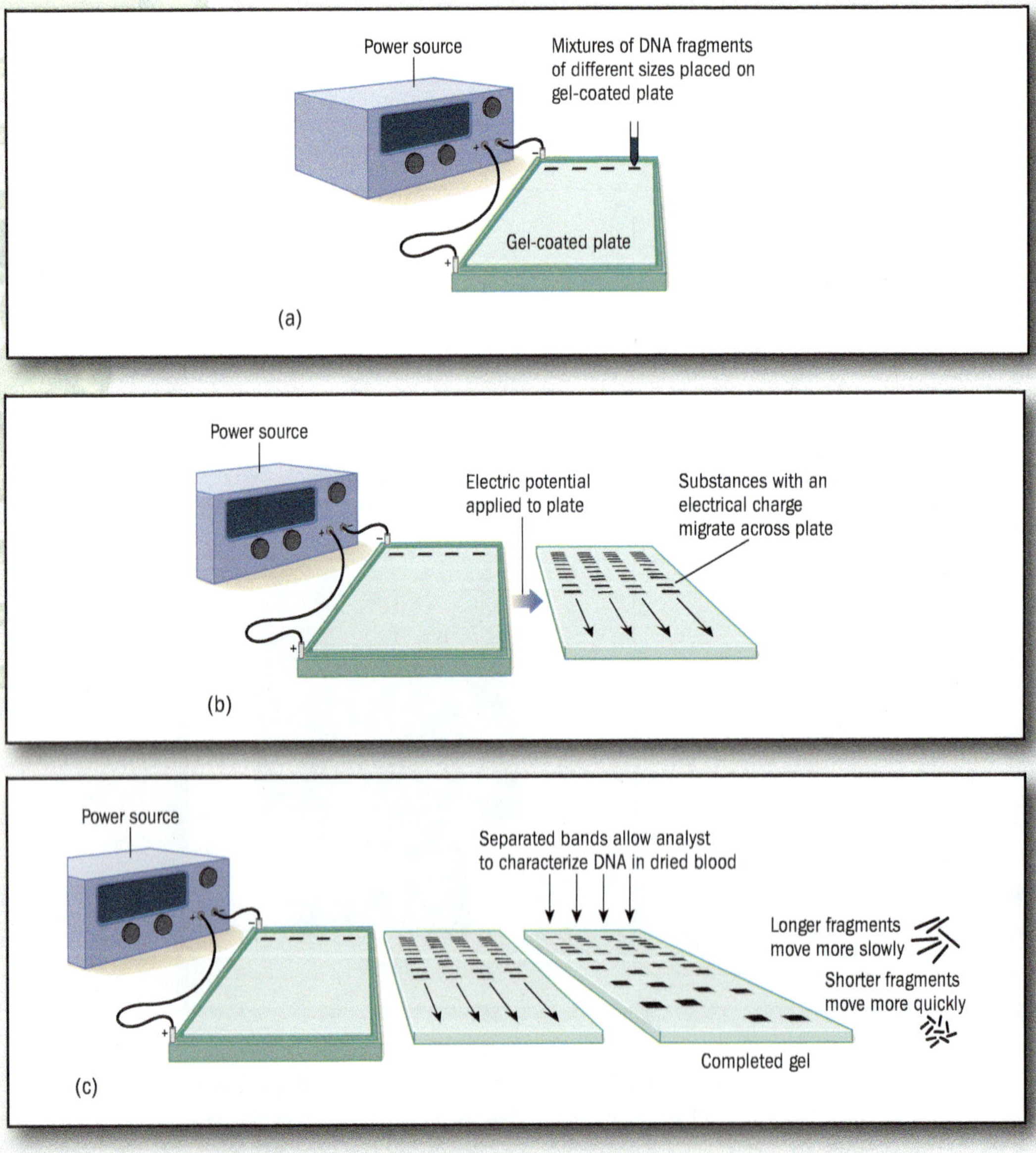

Inside the Science (CONTINUED)

The technique is particularly useful for separating and identifying complex biochemical mixtures. In forensic science, electrophoresis is most useful for characterizing proteins and DNA in dried blood.

Forensic serologists have developed several electrophoretic procedures for characterizing DNA in dried blood. Mixtures of DNA fragments can be separated by gel electrophoresis by taking advantage of the fact that the rate of movement of DNA Inside the Science across a gel-coated plate depends on the molecule's size. Smaller DNA fragments move faster along the plate than larger DNA fragments. After completing the electrophoresis run, the separated DNA is stained with a suitable developing agent for visual observation (see Figure 2).

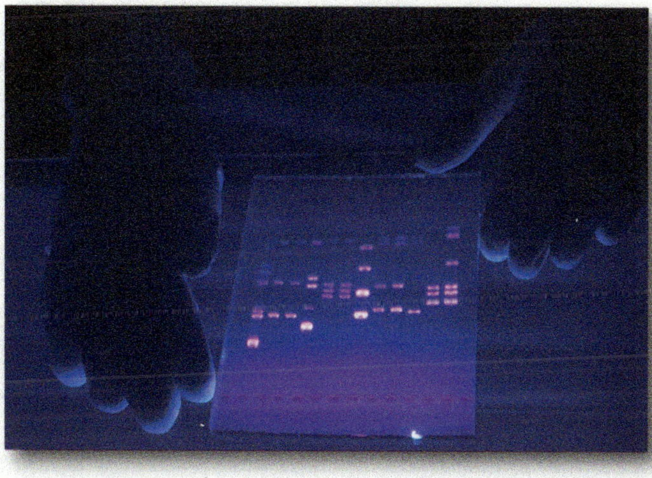

FIGURE 2 DNA fragments separated by gel electrophoresis are visualized under a UV light. *Courtesy Explorer/Science Source*

The separation of STRs can typically be carried out on a flat gel-coated electrophoretic plate, as described earlier. However, the need to reduce analysis time and to automate sampling and data collection has led to the emergence of *capillary electrophoresis* as the preferred technology for characterization of STRs. Capillary electrophoresis is carried out in a thin glass column rather than on the surface of a coated-glass plate.

Capillary electrophoresis technology has evolved from the traditional flat gel electrophoresis approach. The separation of DNA segments is carried out on the interior wall of a glass capillary tube that is kept at a constant voltage. The size of the DNA fragments determines the speed at which they move through the column.

Quick Review

- Portions of the DNA molecule contain sequences of bases that are repeated numerous times. These tandem repeats offer a means of distinguishing one individual from another through DNA typing.
- Length differences associated with relatively long repeating DNA strands—called restriction fragment length polymorphisms (RFLPs)—formed the basis for one of the first DNA-typing procedures.
- In the laboratory, DNA molecules are cut up by a restriction enzyme, and the resulting fragments are sorted out by electrophoresis.
- Materials undergoing electrophoresis are forced to move across a gel-coated plate under the influence of an electrical potential. Substances such as DNA can be separated and characterized using electrophoresis.
- A typical DNA fragment pattern shows two bands (one RFLP from each chromosome).

Polymerase Chain Reaction (PCR)

polymerase chain reaction (PCR)
A technique for replicating or copying a portion of a DNA strand outside a living cell

For nearly a decade, RFLP was the dominant DNA-typing procedure in the United States. However, its utility quickly ended by the mid-1990s. What caused this change? The answer is quite simple: the emergence of a revolutionary and elegant technique known as **polymerase chain reaction**, or PCR. Put simply, PCR is a technique designed to copy or multiply DNA strands outside a living cell. For the forensic scientist, who is often presented with minute quantities of materials, the opportunity to multiply the quantity of sample available for analysis was too good to pass up.

In PCR, small quantities of DNA or broken pieces of DNA found in crimescene evidence can be copied with the aid of a DNA polymerase. The copying process can be accomplished in an automated fashion using a DNA Thermal Cycler (see Figure 10–10). Each cycle of the PCR technique results in a doubling of the DNA. Within a few hours, 30 cycles can multiply DNA a billionfold. Once DNA copies are in hand, they can be analyzed by any of the methods of modern molecular biology. The ability to multiply small bits of DNA opens new and exciting avenues for forensic scientists to explore. It means that sample size is no longer a limitation in characterizing DNA recovered from crime-scene evidence.

The PCR Process

PCR is the outgrowth of knowledge gained from an understanding of how DNA strands naturally replicate within a cell. The most important feature of PCR is the knowledge that an enzyme called DNA polymerase can be directed to synthesize a specific region of DNA. In a relatively straightforward manner, PCR can be used to repeatedly duplicate or amplify a strand of DNA millions

FIGURE 10–10
The DNA Thermal Cycler, an instrument that automates the rapid and precise temperature changes required to copy a DNA strand. Within a matter of hours, DNA can be multiplied a billionfold.
Applied Biosystems

of times. As an example, let's consider a segment of DNA that we want to duplicate by PCR:

*–G–T–C–T–C–A–G–C–T–T–**C–C–A–G**–*
*–**C–A–G–A**–G–T–C–G–A–A–G–G–T–C–*

To perform PCR on this DNA segment, short sequences of DNA on each side of the region of interest must be identified. In the example shown here, the short sequences are designated by boldface letters in the DNA segment. These short DNA segments must be available in a pure form known as a **primer** if the PCR technique is going to work.

primer
A short strand of DNA used to target a region of DNA for replication by PCR

The first step in PCR is to heat the DNA strands to about 94°C. At this temperature, the double-stranded DNA molecules separate completely:

–G–T–C–T–C–A–G–C–T–T–C–C–A–G–

–C–A–G–A–G–T–C–G–A–A–G–G–T–C–

The second step is to add the primers to the separated strands and allow the primers to combine, or hybridize, with the strands by lowering the test-tube temperature to about 60°C.

–G–T–C–T–C–A–G–C–T–T–C–C–A–G–
C–A–G–A

C–C–A–G
–C–A–G–A–G–T–C–G–A–A–G–G–T–C–

The third step is to add the DNA polymerase and a mixture of free nucleotides (A, C, G, T) to the separated strands (see Figure 10–11). When the test tube is heated to 72°C, the polymerase enzyme directs the rebuilding of a doublestranded DNA molecule, extending the primers by adding the appropriate bases, one at a time, resulting in the production of two complete pairs of double-stranded DNA segments:

–G–T–C–T–C–A–G–C–T–T–C–C–A–G–
C–A–G–A–G–T–C–G–A–A–G–G–T–C–
–G–T–C–T–C–A–G–C–T–T–C–C–A–G
–C–A–G–A–G–T–C–G–A–A–G–G–T–C–

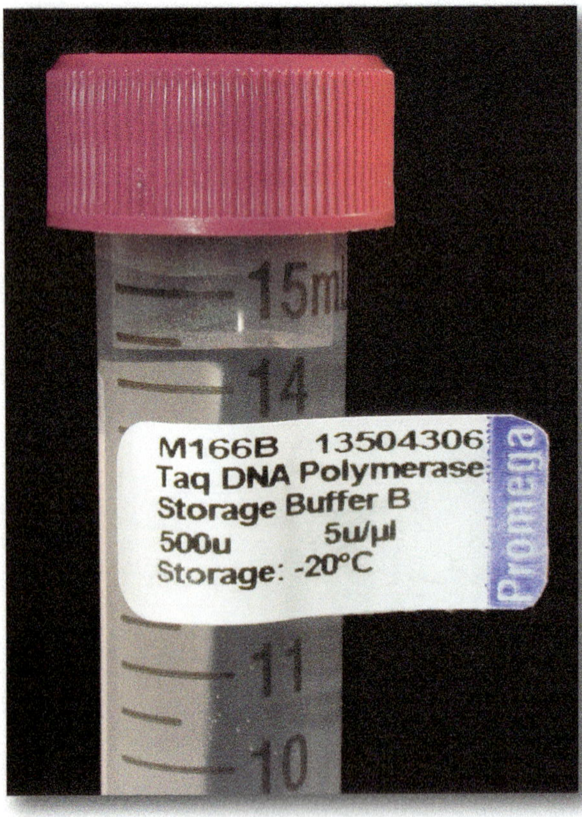

FIGURE 10–11 Vial of DNA Polymerase used in PCR for replication of DNA. *Courtesy Beth Plowes-Proteapix*

This completes the first cycle of the PCR technique, which results in a doubling of the number of DNA strands from one to two. The cycle of heating, cooling, and strand rebuilding is then repeated, resulting in a further doubling of the DNA strands. On completion of the second cycle, four double-stranded DNA molecules have been created from the original double-stranded DNA sample. Typically, 28 to 32 cycles are carried out to yield more than one billion copies of the original DNA molecule. Each cycle takes less than two minutes.

Advantages of PCR

Why couldn't the PCR technology be applied to RFLP DNA typing? Simply put, the RFLP strands are too long, often containing thousands of bases. PCR is best used with DNA strands that are no longer than a couple of hundred bases. The obvious solution to this problem is to characterize DNA strands that are much shorter than RFLPs.

Another advantage in moving to shorter DNA strands is that they would be expected to be more stable and less subject to degradation brought about by adverse environmental conditions. The long RFLP strands tend to break apart under adverse conditions not uncommon at crime scenes.

From the forensic scientist's viewpoint, PCR offers a third distinct advantage in that it can amplify minute quantities of DNA, thus overcoming the limited-sample-size problem often associated with crime-scene evidence. With PCR, less than one-billionth of a gram of DNA is required for analysis. Consequently, PCR can characterize DNA extracted from small quantities of blood, semen, and saliva. The extraordinary sensitivity of PCR allows forensic analysts to characterize small quantities of DNA that could never be detected by RFLP. For instance, PCR has been applied to the identification of saliva residues found on envelopes, stamps, soda cans, and cigarette butts.

Quick Review

- Polymerase chain reaction (PCR) can amplify minute quantities of DNA. The technique evolved from an understanding of how DNA strands naturally replicate within a cell.
- PCR technology cannot be applied to RFLP DNA typing because RFLP strands are too long, often numbering in the thousands of bases. PCR is best used with DNA strands that are no longer than a couple of hundred bases.
- Long RFLP strands tend to break apart under the adverse conditions at many crime scenes. The shorter DNA strands used in PCR are more stable and less subject to degradation caused by adverse environmental conditions.

Short Tandem Repeats (STRs)

The latest method of DNA typing, **short tandem repeat (STR)** analysis, has emerged as the most successful and widely used DNA-profiling procedure. STRs are locations (loci) on the chromosome that contain short sequence elements that repeat themselves within the DNA molecule. They serve as helpful markers for identification because they are found in great abundance throughout the human genome.

short tandem repeat (STR)
A region of a DNA molecule that contains short segments of three to seven repeating base pairs

STRs normally consist of repeating sequences of three to seven bases; the entire strand of an STR is also very short, less than 450 bases long. These strands are significantly shorter than those encountered in the RFLP procedure. This means that STRs are much less susceptible to degradation and are often recovered from bodies or stains that have been subject to extreme decomposition. Also, because of their shortness, STRs are an ideal candidate for multiplication by PCR, thus overcoming the limited-sample-size problem often associated with crime-scene evidence. Only one-billionth of a gram or less of DNA is required—1/50 to 1/100 the amount normally required for RFLP analysis.

To understand the utility of STRs in forensic science, let's look at one commonly used STR known as TH01. This DNA segment contains the repeating sequence *A–A–T–G*. Seven TH01 variants have been identified in the human genome. These variants contain 5 to 11 repeats of *A–A–T–G*. Figure 10–12 illustrates two such TH01 variants, one containing 6 repeats and the other containing 8 repeats of *A–A–T–G*.

During a forensic examination, TH01 is extracted from biological materials and amplified by PCR as described earlier. The ability to copy an STR means that extremely small amounts of the molecule can be detected and analyzed. Once the

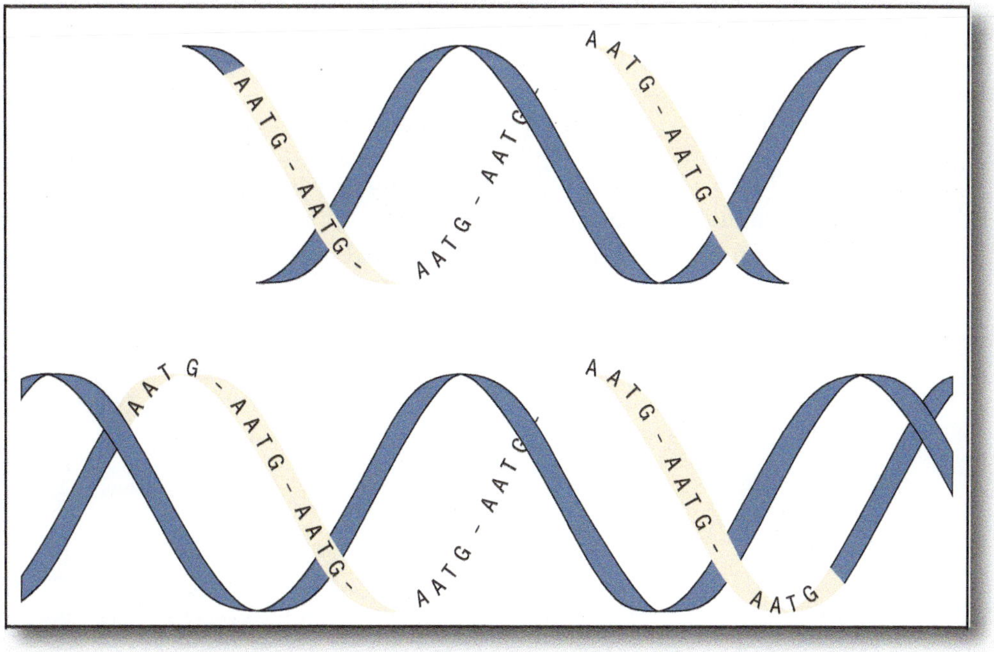

FIGURE 10–12 Variants of the short tandem repeat TH01. The upper DNA strand contains six repeats of the sequence *A-A-T-G*; the lower DNA strand contains eight repeats of the sequence *A-A-T-G*. This DNA type is known as TH01 6, 8. *Courtesy Dorling Kindersley Limited*

STRs have been copied or amplified, they are separated by electrophoresis. By examining the distance the STR has migrated on the electrophoretic plate, one can determine the number of *A–A–T–G* repeats in the STR. Every person has two STR types for TH01, one inherited from each parent. Thus, for example, one may find in a semen stain TH01 with six repeats and eight repeats. This combination of TH01 is found in approximately 3.5 percent of the population.

Multiplexing

What makes STRs so attractive to forensic scientists is that hundreds of different types of STRs are found in human genes. The more STRs one can characterize, the smaller the percentage of the population from which these STRs can emanate. This gives rise to the concept of **multiplexing**. Using PCR technology, one can simultaneously extract and amplify a combination of different STRs.

multiplexing
A technique that simultaneously detects more than one DNA marker in a single analysis

One STR system on the commercial market is the STR Blue Kit. This kit provides the necessary materials for amplifying and detecting three STRs (a process called *triplexing*)—D3S1358, vWA, and FGA. The design of the system ensures that the size of the STRs does not overlap, thereby allowing each marker to be viewed clearly on an electrophoretic gel. In the United States, the forensic science community has standardized on 13 STRs for entry into a national database known as the Combined DNA Index System (CODIS).

When an STR is selected for analysis, not only must the identity and number of core repeats be defined, but the sequence of bases flanking the repeats must also be known. This knowledge allows commercial manufacturers of STR-typing kits to prepare the correct primers to delineate the STR segment to be amplified by PCR. Figure 10–13 illustrates how appropriate primers are used to define the region of DNA to be amplified. Also, a mix of different primers aimed at different STRs will be used to simultaneously amplify a multitude of STRs (multiplexing). In fact, one STR kit on the commercial market can simultaneously make copies of 15 different STRs (see Figure 10–14).

DNA Typing with STRs

The 20 CODIS STRs are listed in Table 10–1 along with their probabilities of identity. The probability of identity is a measure of the likelihood that two individuals selected at random will have an identical STR type. The smaller the value of this probability, the more discriminating the STR. A high degree of discrimination and even individualization can be attained by analyzing a combination of

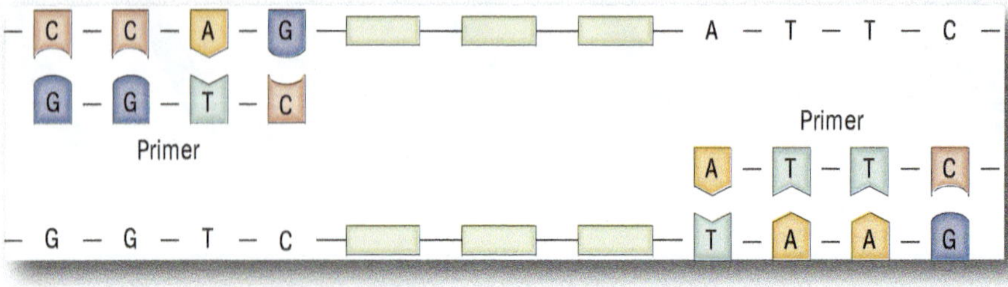

FIGURE 10–13 Appropriate primers flanking the repeat units of a DNA segment must be selected and put in place to initiate the PCR process.

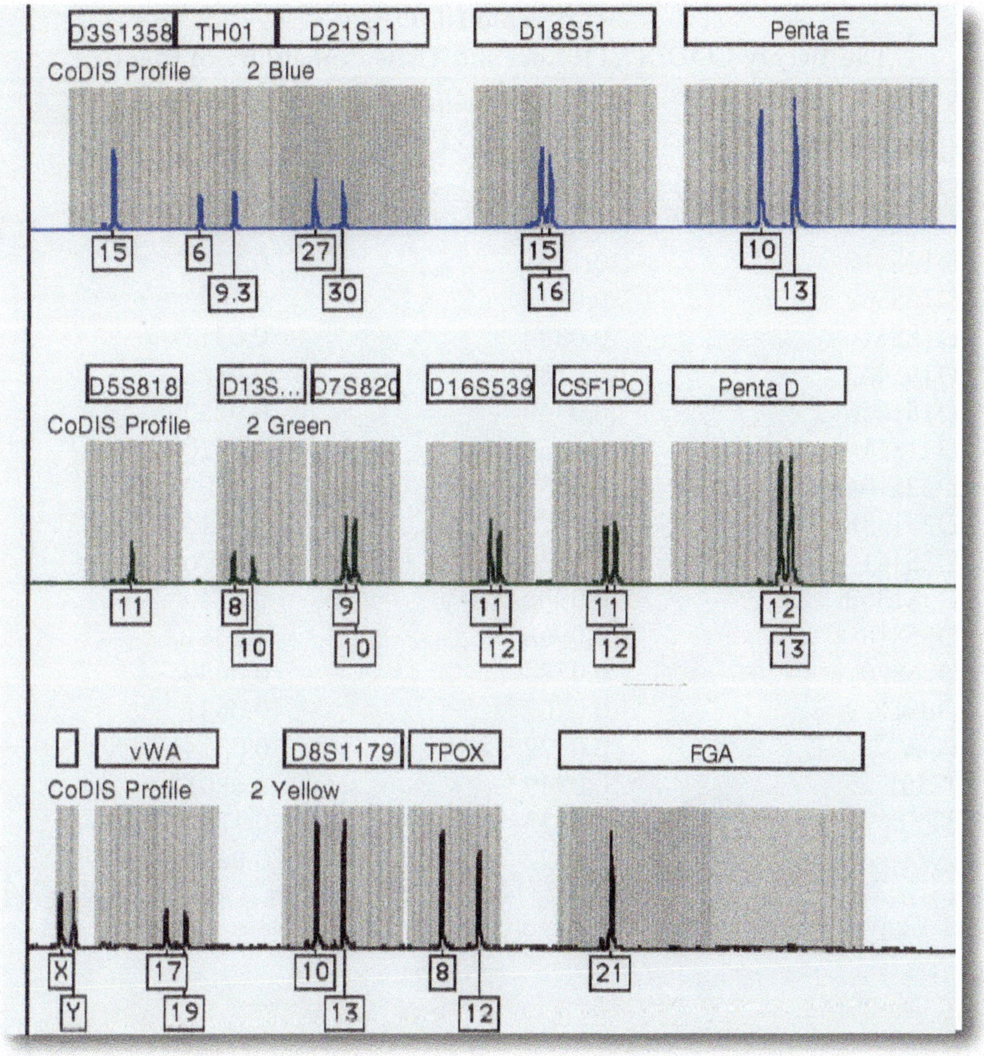

FIGURE 10–14 STR profile for 15 loci. *Promega corporation*

STRs (multiplexing). Because STRs occur independently of each other, the probability of biological evidence having a particular combination of STR types is determined by the product of their frequency of occurrence in a population. This combination is referred to as the *product rule.* Hence, the greater the number of STRs characterized, the smaller the frequency of occurrence of the analyzed sample in the general population.

The combination of the first three STRs shown in Table 10–1 typically produces a frequency of occurrence of about 1 in 5,000. A combination of the first six STRs typically yields a frequency of occurrence in the range of 1 in two million for the Caucasian population, and if the top nine STRs are determined in combination, this frequency declines to about 1 in one billion. The combination of the first 13 core loci shown in Table 10–1 typically produces frequencies of occurrence that measure in the range of 1 in 575 trillion for European Americans and 1 in 900 trillion for African Americans. For the combined 20 core loci, probabilities of occurrence often exceed 1 in every quadrillion or even quintillion. Importantly, several commercially available kits allow forensic scientists to profile STRs in the kinds of combinations cited here.

Table 10–1
The Twenty CODIS STR Loci and Their Probability of Identities

STR Loci	African American	U.S. Caucasian
CSF1PO	0.0850	0.1333
D10S1248	0.0693	0.0943
D12S391	0.0377	0.0231
D13S317	0.1451	0.0761
D16S539	0.0727	0.1043
D18S51	0.0322	0.0311
D19S433	0.0388	0.0862
D1S1656	0.0340	0.0223
D21S11	0.0453	0.0520
D22S1045	0.0559	0.1309
D2S1338	0.0225	0.0316
D2S441	0.1030	0.0976
D3S1358	0.0984	0.0749
D5S818	0.0968	0.1341
D7S820	0.0784	0.0680
D8S1179	0.0762	0.0631
FGA	0.0322	0.0384
TH01	0.0949	0.0801
TPOX	0.0833	0.1757
vWA	0.0622	0.0650

Source: The Future of Forensic DNA Testing: Predictions of the Research and Development Working Group. Washington, D.C.: National Institute of Justice, Department of Justice, 2000, p. 41.

Analytical Procedure for Generating a DNA Profile

In order to analyze a biological sample to obtain a DNA profile, the DNA must first be extracted from the nucleus of the cell and separated from the other cellular components. This is accomplished through the extraction process. The extraction process occurs in three basic steps: break down the cell membrane components, using enzymes to break down the proteins that help to condense the DNA, and separating the DNA from the rest of the cellular components.

Once DNA has been isolated from the other cellular debris, quantification of the sample occurs. Quantification is the process of determining the average concentration of DNA in a sample. Amplification is the process of creating more copies of the DNA sample to be used for further analysis. Amplification of DNA is most commonly accomplished by PCR. The final step in the process used to create a DNA profile is to separate the DNA by using capillary electrophoresis.

Capillary Electrophoresis

The separation of STRs can typically be carried out on a flat gel-coated electrophoretic plate, as described earlier. However, the need to reduce analysis time and to automate sampling and data collection has led to the emergence of *capillary electrophoresis* as the preferred technology for characterization of STRs. Capillary

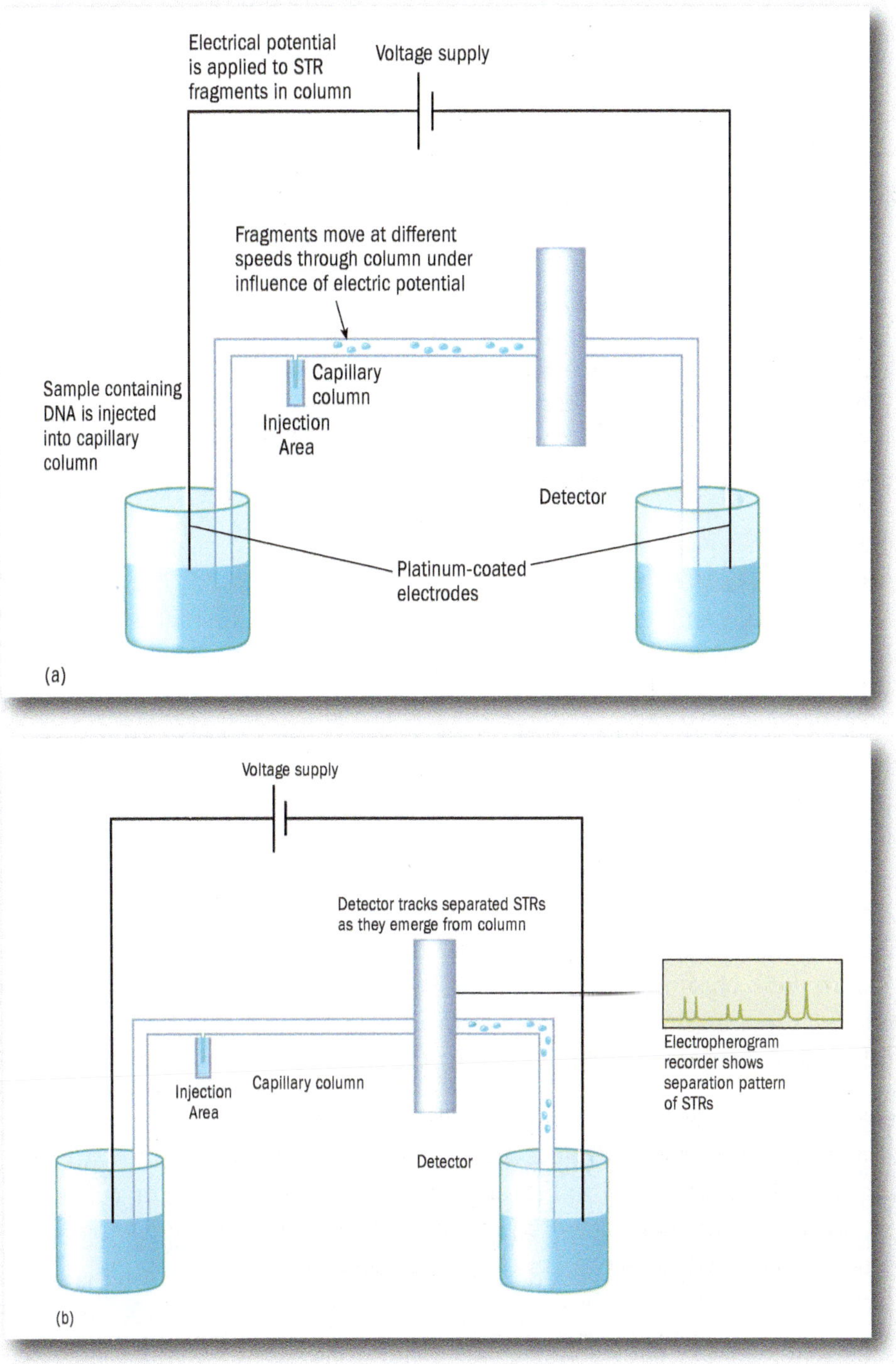

FIGURE 10–15 Capillary electrophoresis technology has evolved from the traditional flat gel electrophoresis approach. The separation of DNA segments is carried out on the interior wall of a glass capillary tube coated with a gel polymer and kept at a constant voltage. The size of the DNA fragments determines the speed at which they move through the column. This figure illustrates the separation of three sets of STRs (triplexing).

electrophoresis is carried out in a thin glass column rather than on the surface of a coated-glass plate.

As illustrated in Figure 10–15, each end of the column is immersed in a reservoir of buffer liquid that also holds electrodes (coated with platinum) to supply high-voltage energy. The column is coated with a gel polymer, and the DNA-containing sample solution is injected into one end of the column with a syringe. The STR fragments then move through the column under the influence of an electrical potential at a speed that is related to the length of the STR fragments. The other end of the column is connected to a detector that tracks the separated STRs as they emerge from the column. As the DNA peaks pass through the detector, they are recorded on a display known as an electropherogram (see image on page 409). The components of the electropherogram show the separation of the alleles present for each STR loci analyzed using electrophoresis.

Single Nucleotide Polymorphisms (SNPs)

Single Nucleotide Polymorphisms (SNPs) provide one of the most common types of genetic variations in the human population. To be classified as SNPs (pronouced "snips"), a variation must occur in at minimum 1 percent of the population. Scientists have identified more than 600 million SNPs. These variations occur when a single nucleotide, such as adenine (A), is replaced with the nucleotide guanine (G) in a particular section of DNA. SNPs are unique in that they typically occur in the sequences of DNA in between the genes.

SNPs are useful in situations where DNA samples may be too degraded or fragmented to use STR typing, such as in some missing persons cases or mass disasters.

Sex Identification Using STR

amelogenin gene
A genetic locus useful for determining sex

Manufacturers of commercial STR kits typically used by crime laboratories provide one additional piece of useful information along with STR types: the sex of the DNA contributor. The focus of attention here is the **amelogenin gene** located on both the X and Y chromosomes. This gene, which is actually the gene for tooth pulp, has an interesting characteristic in that it is shorter by six bases in the X chromosome than in the Y chromosome. Hence, when the amelogenin gene is amplified by PCR and separated by electrophoresis, males, who have an X and a Y chromosome, show two bands; females, who have two X chromosomes, have just one band. Typically, these results are obtained in conjunction with STR types.

Y-STRs
Short tandem repeats located on the human Y chromosome

Another tool in the arsenal of the DNA analyst is the ability to type STRs located on the Y chromosome. The Y chromosome is male specific and is always paired with the X chromosome. More than 20 different **Y-STR** markers have been identified, and a commercial kit allows for the characterization of 17 Y chromosome STRs. When can it be advantageous to seek out Y-STR types? Generally, Y-STRs are useful for analyzing blood, saliva, or a vaginal swab that is a mix originating from more than one male. For example, Y-STRs prove useful when multiple males are involved in a sexual assault. Further simplifying the analysis is that any DNA in the mixture that originates from a female will not show.

Keep in mind that STR types derived from the Y chromosome originate only from this single male chromosome. A female subject, or one with an XX chromosome pattern, does not contribute any DNA information. Also, unlike a conventional STR analysis that is derived from two chromosomes and typically shows two bands or peaks, a Y-STR has only one band or peak for each STR type.

For example, the traditional STR DNA pattern may prove to be overly complex in the case of a vaginal swab containing the semen of two males. Each STR type would be expected to show four bands, two from each male. Also complicating the appearance of the DNA profile may be the presence of DNA from skin cells emanating from the walls of the vagina. In this circumstance, homing in on the Y chromosome greatly simplifies the appearance and interpretation of the DNA profile. Thus, when presented with a DNA mixture of two males and one female, each STR type would be expected to show six bands. However, the same mixture subjected to Y-STR analysis would show only two bands (one band for each male) for each Y-STR type.

Case Files

Reopening the Boston Strangler Case

Albert DeSalvo's rape and murder of eleven women shocked the country in the early 1960's. Better known as the Boston Strangler, DeSalvo would gain entry to primarily single females' apartments using disguises and tricks. Boston Police finally apprehended DeSalvo through a witness sketch, and he confessed to the crimes through detailed descriptions. Despite his confession, DeSalvo later recanted and forensic experts have contested his guilt. DeSalvo was charged with life in prison for unrelated sexual assaults, but never was charged with the Boston Strangler killings. Albert DeSalvo died in jail in 1973, further complicating the situation.

Through funding by the National Institute of Justice, the Boston Strangler case was reopened and examined with new DNA testing techniques. The Boston Police Department's cold case squad used DNA found on one of the Strangler's victims, Mary Sullivan, to link the crimes back to DeSalvo. Forensic scientists focused their research on the (Y) chromosome, as it relates back to every male in a paternal lineage. DeSalvo's nephew provided the key DNA used to tie Albert to the killings, and ultimately a positive STR match was made. Detectives confirmed these results after exhuming DeSalvo's body; they concluded that the odds of someone else committing these crimes were 1 in 220 billion. The Boston Strangler Case has finally been resolved.

When gauging the significance of a Y-STR match between questioned and known specimens, one should take into consideration that all male paternal relatives (e.g., brothers, father, male offspring, and uncles) would be expected to have the same Y-STR profile. Another advantage of employing STR technology is to extend the success of detecting evidential DNA from vaginal swabs collected from rape victims. Casework experience has demonstrated significant difficulties in obtaining traditional STR DNA profiles for the male donor from vaginal swabs collected after three to four days after intercourse. However, the application of Y-STR technology often extends the routine postcoital detection time to five days for the male donor.

Significance of DNA Typing

STR DNA typing has become an essential and basic investigative tool in the law enforcement community. The technology has progressed at a rapid rate and in only a few years has surmounted numerous legal challenges so that DNA typing is now vital evidence for resolving violent crimes and sex offenses. DNA evidence is impartial, implicating the guilty and exonerating the innocent.

In a number of well-publicized cases, DNA evidence has exonerated individuals who have been wrongly convicted and imprisoned. To exonerate an individual, DNA samples from the original case must be analyzed using modern-day observational and experimental testing techniques that may not have been available at the time of the original analysis of the evidence. Results must be collected and evaluated. Scientific explanations and solutions must be critiqued using empirical evidence, logical reasoning, and critical thinking. When DNA from a suspect initially convicted of a crime is determined to not match the crime scene, they must be exonerated. Critical thinking on the part of the analyst is a key to exonerating the wrongly convicted. The importance of DNA analyses in criminal investigations has also placed added burdens on crime laboratories to improve their quality-assurance procedures and to ensure the correctness of their results. In fact, in several well-publicized instances, the accuracy of DNA tests conducted by government-funded laboratories has been called into question.

Quick Review

- STRs are locations on the chromosome that contain short sequences that repeat themselves within the DNA molecule. They serve as useful markers for identification because they are found in great abundance throughout the human genome.
- The entire strand of an STR is very short, less than 450 bases long. This makes STRs much less susceptible to degradation, and they are often recovered from bodies or stains that have been subjected to extreme decomposition.
- The more STRs one can characterize, the smaller the percentage of the population from which a particular combination of STRs can emanate. This gives rise to the concept of multiplexing, in which the forensic scientist can simultaneously extract and amplify a combination of different STRs.
- With STR, as little as 125 picograms of DNA is required for analysis—1/100 the amount normally required for RFLP analysis.
- To generate a DNA profile, DNA must be extracted from the cell's nucleus; separated from the other cellular components; quantified; amplified; and separated.
- Single Nucleotide Polymorphisms (SNPs) are one of the most common types of genetic variations in humans. SNPs are usefully when analyzing DNA samples that are too degraded to analyze using STR analysis techniques.

Mitochondrial DNA

Typically, when one describes DNA in the context of a criminal investigation, the subject is assumed to be the DNA in the nucleus of a cell. Actually, a human cell contains two types of DNA—nuclear and mitochondrial. The first constitutes the 23 pairs of chromosomes in the nuclei of our cells. Each parent contributes to the genetic makeup of these chromosomes. Mitochondrial DNA (mtDNA), on the other hand, is found outside the nucleus of the cell and is inherited solely from the mother.

mitochondria
Small structures outside the nucleus that supply energy to the cell

Mitochondria are cell structures found in all human cells (see Figure 10–16). They are the power plants of the body, providing about 90 percent of the energy that the body needs to function. A single mitochondrion contains several loops of DNA, all of which are involved in energy generation. Further, because each cell

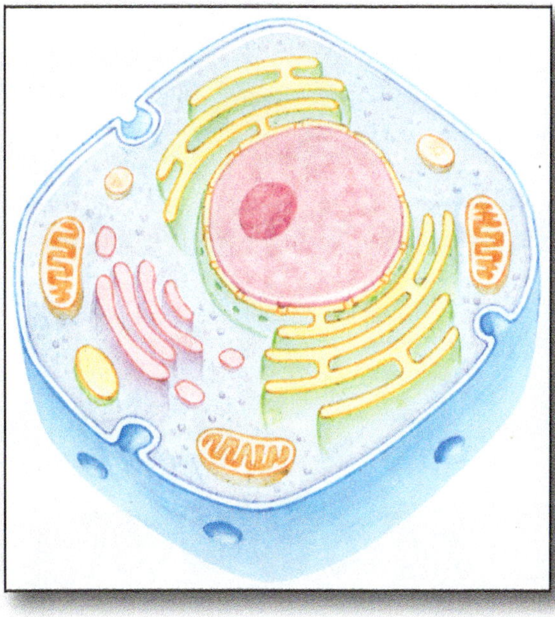

FIGURE 10–16 Illustration of a human cell containing a nucleus and cell structures involved in cell functions within the cell's cytoplasm. Mitochondria units in the shape of a loop make the energy for the cell. *Dorling Kindersley/Getty Images*

in our bodies contains hundreds to thousands of mitochondria, there are hundreds to thousands of mtDNA copies in a human cell. This compares to just one set of nuclear DNA located in that same cell.

Forensic scientists rely on mtDNA to identify a subject when nuclear DNA is significantly degraded, such as in charred remains, or when nuclear DNA may be present in only very small quantities (such as in a hair shaft). Interestingly, when authorities cannot obtain a reference sample from an individual who may be long deceased or missing, an mtDNA reference sample can be obtained from any maternally related relative. However, all individuals of the same maternal lineage will be indistinguishable by mtDNA analysis.

Although mtDNA analysis is significantly more sensitive than nuclear DNA profiling, forensic analysis of mtDNA is more rigorous, time consuming, and costly than nuclear DNA profiling. For this reason, only a handful of public and private forensic laboratories receive evidence for mtDNA determination. The FBI Laboratory strictly limits the types of cases in which it will apply mtDNA technology.

As was previously discussed, nuclear DNA is composed of a continuous linear strand of nucleotides (*A, C, G,* and *T*). By contrast, mtDNA is constructed in a circular or loop configuration. Each loop contains enough *A, C, G,* and *T* (approximately 16,569 total nucleotides) to make up 37 genes involved in mitochondrial energy generation.

Two regions of mtDNA have been found to be highly variable in the human population. These two regions have been designated hypervariable region I (HV1) and hypervariable region II (HV2), as shown in Figure 10–17. As indicated previously, the process for analyzing HV1 and HV2 is tedious. It involves generating many copies of these DNA hypervariable regions by PCR and then determining the order of the *A–T–C–G* bases constituting the hypervariable regions. This process is known as **sequencing**. The FBI Laboratory, the Armed Forces DNA Identification Laboratory, and other laboratories have collaborated to compile an mtDNA population database containing the base sequences from HV1 and HV2.

sequencing
A procedure used to determine the order of the base pairs that constitute DNA

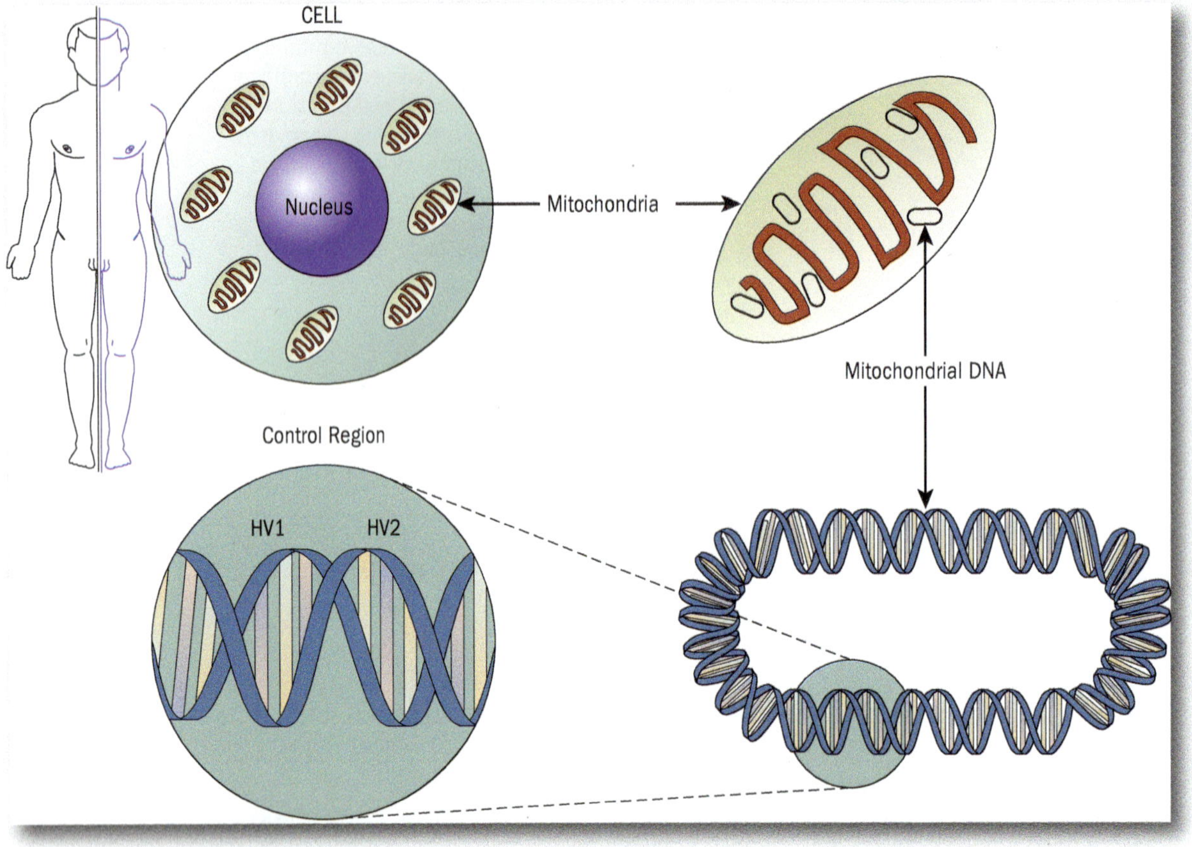

FIGURE 10–17 Every cell in the body contains hundreds of mitochondria, which provide energy to the cell. Each mitochondrion contains numerous copies of DNA shaped in the form of a loop. Distinctive differences between individuals in their mitochondrial DNA makeup are found in two specific seqments of the control region on the DNA loop known as HV1 and HV2.

Once the sequences of the hypervariable regions from a case sample are obtained, most laboratories simply report the number of times these sequences appear in the mtDNA database maintained by the FBI. The mtDNA database contains about five thousand sequences. This approach permits an assessment of how common or rare an observed mtDNA sequence is in the database.

Interestingly, many of the sequences that have been determined in casework are unique to the existing database, and many types are present at frequencies of no greater than 1 percent in the database. Thus it is often possible to demonstrate how uncommon a particular mtDNA sequence is. However, even under the best circumstances, mtDNA typing does not approach STR analysis in its discrimination power. Thus, mtDNA analysis is best reserved for samples for which nuclear DNA typing is simply not possible.

The first time mtDNA was admitted as evidence in a U.S. court was in 1996 in the case of *State of Tennessee v. Paul Ware*. Here, mtDNA was used to link two hairs recovered from the crime scene to the defendant. Interestingly, in this case, blood and semen evidence was absent. Mitochondrial DNA analysis also plays a key role in the identification of human remains. An abundant amount of mtDNA is generally found in skeletal remains. Importantly, mtDNA reference samples are available from family members sharing the same mother, grandmother, great-grandmother, and so on.

One of the most publicized cases performed on human remains was the identification of the individual buried in the tomb of the Vietnam War's unknown soldier. The remains lying in the tomb were believed to belong to First Lt. Michael J. Blassie, whose A-37 warplane was shot down near An Loc, South Vietnam, in 1972. In 1984, the U.S. Army Central Identification Laboratory failed to identify the remains by physical characteristics, personal artifacts, or blood-typing results from hairs. The remains were subsequently placed in the tomb. In 1998, at the insistence of the Blassie family, the remains were disinterred for mtDNA analysis, and the results were compared to references from seven families thought to be associated with the case. The remains in the tomb were subsequently analyzed and confirmed to be consistent with DNA from Lt. Blassie's family.

Quick Review

- Mitochondrial DNA is located outside the cell's nucleus and is inherited from the mother.
- Two regions of mitochondrial DNA, HV1 and HV2, are sequenced for forensic typing purposes.
- Mitochondrial DNA typing does not approach STR analysis in its discrimination power and thus is best reserved for samples, such as hair, for which STR analysis may not be possible.

Collection and Preservation of Biological Evidence for DNA Analysis

Since the early 1990s, the advent of DNA profiling has vaulted biological crimescene evidence to a stature of importance that is eclipsed only by the fingerprint. In fact, the high sensitivity of DNA determinations has even changed the way police investigators define biological evidence.

Just how sensitive is STR profiling? Forensic analysts using currently accepted protocols can reach sensitivity levels as low as 125 **picograms**. Interestingly, a human cell has an estimated 7 picograms of DNA, which means that only eighteen DNA-bearing cells are needed to obtain an STR profile. However, modifications in the technology can readily extend the level of detection down to nine cells. A quantity of DNA that is below the normal level of detection is defined as a **low copy number**. (However, analysts must take extraordinary care in analyzing low-copy-number DNA and often may find that courts will not allow this data to be admissible in a criminal trial.) With this technology in hand, the horizon of the criminal investigator extends beyond the traditional dried blood or semen stain to include stamps and envelopes licked with saliva, a cup or can that has touched a person's lips, chewing gum, the sweat band of a hat, or a bedsheet containing dead skin cells.

picogram
One-trillionth of a gram, or 0.000000000001 gram

low copy number
Fewer than 18 DNA-bearing cells

Case Files

Using Mitochondrial DNA

In the fall of 1979, a 61-year-old patient wandered away from a U.S. Department of Veterans Affairs medical facility. Despite an extensive search, authorities never located the missing man. More than ten years later, a dog discovered a human skull in a wooded area near the facility. DNA Analysis Unit II of the FBI Laboratory received the case in the winter of 1999. The laboratory determined that the mitochondrial DNA profile from the missing patient's brother matched the mitochondrial DNA profile from the recovered skull and provided the information to the local medical examiner. Subsequently, the remains were declared to be those of the missing patient and returned to the family for burial.

Source: FBI Law Enforcement Bulletin 78 (2002): 21.

epithelial cells
The outer layer of skin cells

touch DNA
DNA from skin cells transferred onto the surface of an object by simple contact

Likewise, skin or **epithelial cells** transferred onto the surface of a weapon, the interior of a glove, or a pen have yielded DNA results.[2] The phenomenon of transferring DNA via skin cells onto the surface of an object has come to be called **touch DNA**. Again, keep in mind that, in theory, only eighteen skin cells deposited on an object are required to obtain a DNA profile.

The ultimate sensitivity goal in forensic DNA analysis is profiling DNA extracted from one human cell. Such an accomplishment seems close to fruition. Researchers have reported obtaining STR profiles from one or two cells and have profiled DNA from single dermal ridge fingerprints.[4] Although it's premature to imply that this technology, or a comparable one, is eligible for admission in criminal trials, one cannot exclude its use in criminal and forensic intelligence investigations (see Figure 10–18). Table 10–2 illustrates the power of DNA as a creator of physical evidence.

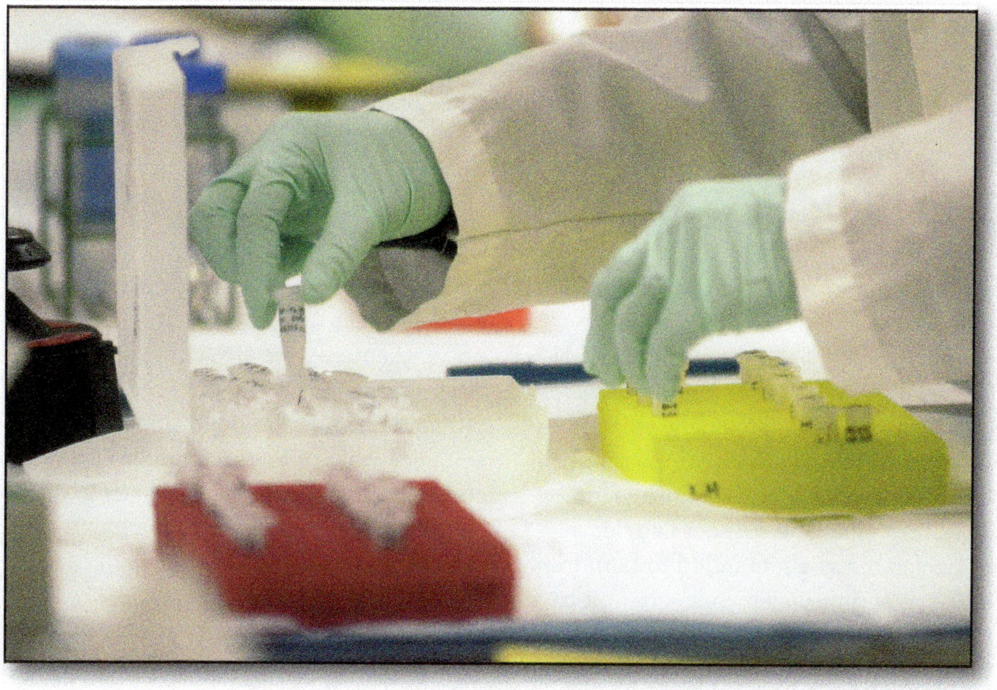

FIGURE 10–18 A criminalist inspects and extracts DNA from crime scene evidence at a DNA crime lab. *Courtesy Corbis*

Table 10–2
Location and Sources of DNA at Crime Scenes

Evidence	Possible Location of DNA on the Evidence	Source of DNA
Baseball bat or similar weapon	Handle, end	Sweat, skin, blood, tissue
Hat, bandanna, or mask	Inside	Sweat, hair, dandruff
Eyeglasses	Nose or ear pieces, lens	Sweat, skin
Facial tissue, cotton swab	Surface area	Mucus, blood, sweat, semen, ear wax
Dirty laundry	Surface area	Blood, sweat, semen Saliva
Toothpick	Tips	Saliva
Used cigarette	Cigarette butt	Saliva
Stamp or envelope	Licked area	Saliva
Tape or ligature	Inside/outside surface	Skin, sweat
Bottle, can, or glass	Sides, mouthpiece	Saliva, sweat
Used condom	Inside/outside surface	Semen, vaginal or rectal cells
Blanket, pillow, sheet	Surface area	Sweat, hair, semen, urine, saliva
"Through and through" bullet	Outside surface	Blood, tissue
Bite mark	Person's skin or clothing	Saliva
Fingernail, partial fingernail	Scrapings	Blood, sweat, tissue

Source: National Institute of Justice, U.S. Department of Justice.

Case Files

Cold-Hit Identification

In 1990, a series of attacks on elderly victims were committed in Goldsboro, North Carolina, by an unknown individual dubbed the Night Stalker. During one such attack in March, an elderly woman was brutally raped and almost murdered. Her daughter's early arrival home saved the woman's life. The suspect fled, leaving behind materials intended to burn the residence and the victim in an attempt to conceal the crime.

In July 1990, another elderly woman was raped and murdered in her home. Three months later, a third elderly woman was raped and stabbed to death. Her husband was also murdered. Although their house was set alight in an attempt to cover up the crime, fire and rescue personnel pulled the bodies from the house before it was engulfed in flames. DNA analysis of biological evidence collected from vaginal swabs from the three rape victims enabled authorities to conclude that the same perpetrator had committed all three crimes. However, there was no suspect.

More than ten years after these crimes were committed, law enforcement authorities retested the biological evidence from all three cases using newer DNA technology and entered the DNA profiles into North Carolina's DNA database. The DNA profile developed from the crime-scene evidence was compared to thousands of convicted-offender profiles already in the database.

In April 2001, a "cold hit" was made with an individual in the convicted-offender DNA database. The perpetrator had been convicted of shooting into an occupied dwelling, an offense that requires inclusion in the North Carolina DNA database. The suspect was brought into custody for questioning and was served with a search warrant to obtain a sample of his blood. That sample was analyzed and compared to the crime-scene evidence, confirming the DNA database match. When confronted with the DNA evidence, the suspect confessed to all three crimes.

Source: National Institute of Justice, "Using DNA to Solve Cold Cases," NIJ special report, 2002, http://www.ojp.usdoj.gov/nij/ pubs_sum/194197.htm.

Collection of Biological Evidence

However, before investigators become enamored with the wonders of DNA, they should first realize that the crime scene must be treated in the traditional manner. Before the collection of evidence begins, biological evidence should be photographed close up and its location relative to the entire crime scene recorded through notes, sketches, and photographs. If the shape and position of bloodstains may provide information about the circumstances of the crime, an expert must immediately conduct an on-the-spot evaluation of the blood evidence. The significance of the position and shape of bloodstains can best be ascertained when the expert has an on-site overview of the entire crime scene and can better reconstruct the movement of the individuals involved. No attempt should be made to disturb the blood pattern before this phase of the investigation is completed.

The evidence collector must handle all body fluids and biologically stained materials with a minimum amount of personal contact. All body fluids must be assumed to be infectious; hence, wearing disposable latex gloves while handling the evidence is required. Latex gloves also significantly reduce the possibility that the evidence collector will contaminate the evidence. These gloves should be changed frequently during the evidence-collection phase of the investigation. Safety considerations and avoidance of contamination also call for the wearing of face masks, a lab coat, eye protection, shoe covers, and possibly coveralls.

The deposition of DNA onto crime-scene objects via saliva, sweat, skin, blood, and semen has created a vast array of forensic evidence that is quite different from the traditional evidence collected at crime scenes prior to the DNA era (see Table 10–2).

Packaging of Biological Evidence

substrate control
An unstained object adjacent to an area on which biological material has been deposited

Biological evidence should not be packaged in plastic or airtight containers because accumulation of residual moisture could contribute to the growth of DNA-destroying bacteria and fungi. Each stained article should be packaged separately in a paper bag or a well-ventilated box. A red biohazard label must be attached to each container. If feasible, the entire stained article should be packaged and submitted for examination. If this is not possible, dried blood is best removed from a surface with a sterile cotton-tipped swab lightly moistened with distilled water from a dropper bottle. A portion of the unstained surface material near the recovered stain must likewise be removed or swabbed and placed in a separate package. This is known as a **substrate control**. The forensic examiner may use the substrate swab to confirm that the results of the tests performed were brought about by the stain and not by the material on which it was deposited. One point is critical, and that is that the collected swabs must not be packaged in a wet state. After the collection is made, the swab must be air-dried for approximately 5 to 10 minutes. Then, it is best to place it in a swab box (see Figure 10–19), which has a circular hole to allow air circulation. The swab box can then be placed in a paper or manila envelope.

All packages containing biological evidence should be refrigerated or stored in a cool location out of direct sunlight until delivery to the laboratory. However, one common exception is blood mixed with soil. Microbes present in soil rapidly degrade DNA. Therefore, blood in soil must be stored in a clean glass or plastic container and immediately frozen.

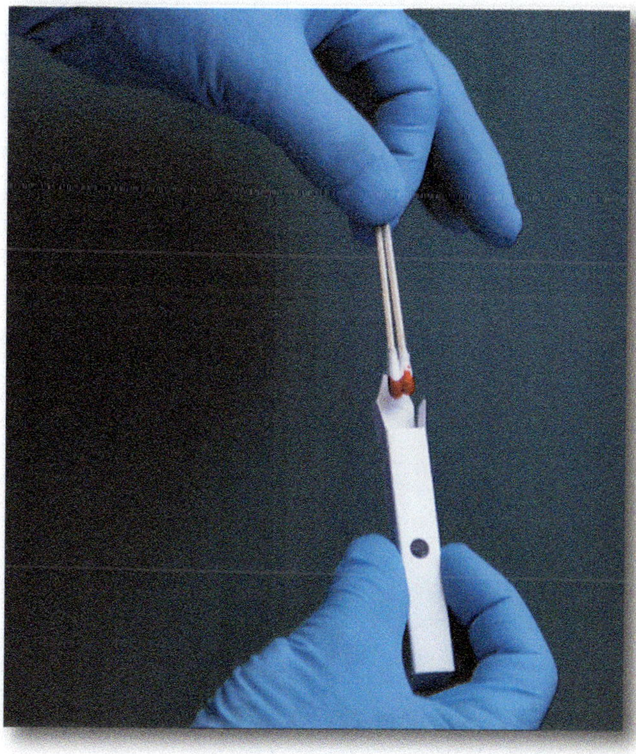

FIGURE 10–19 Air-dried swabs are placed in a swab box for delivery to the forensic laboratory. *Courtesy of Tri-Tech Forensics, Inc., Southport, NC*

Obtaining DNA Reference Specimens

buccal cells
Cells from the inner cheek lining

Biological evidence attains its full forensic value only when an analyst can compare each of its DNA types to known DNA samples collected from victims and suspects. The least intrusive method for obtaining a DNA standard/reference, one that nonmedical personnel can readily use, is the *buccal swab*. Cotton swabs are placed in the subject's mouth and the inside of the cheek is vigorously swabbed, resulting in the transfer of **buccal cells** onto the swab (see Figure 10–20).

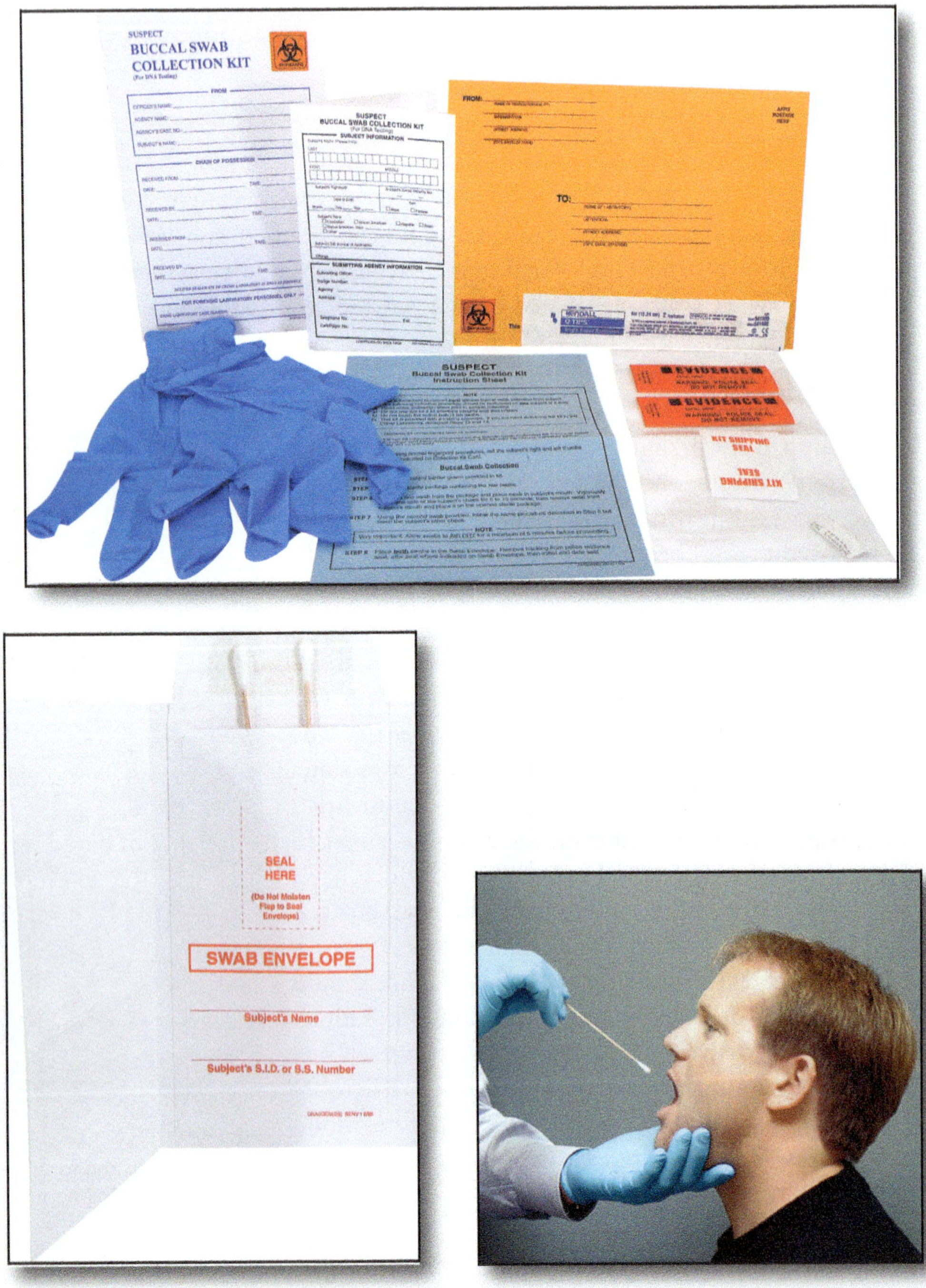

FIGURE 10–20 A buccal swab collection kit is designed for use by nonmedical personnel. The cotton-tipped swabs are placed in the subject's mouth and the inside of the cheek is vigorously swabbed, resulting in the transfer of buccal cells onto the cotton bulb of the swab. The kit is then delivered to the forensic laboratory. *Courtesy Tri-Tech Forensics, Inc., Southport, NC*

If an individual is not available to give a DNA standard/reference sample, some interesting alternatives are available, including a toothbrush, combs and hairbrushes, a razor, soiled laundry, used cigarette butts, and earplugs. Any of these items may contain a sufficient quantity of DNA for typing. Interestingly, as investigators worked to identify the remains of victims of the World Trade Center attack on September 11, 2001, the families of the missing were asked to supply the New York City DNA Laboratory with these types of items in an effort to match recovered DNA with human remains.

Contamination of DNA Evidence

One key concern during the collection of a DNA-containing specimen is contamination. Contamination can occur by introducing foreign DNA through coughing or sneezing onto a stain during the collection process, or there can be a transfer of DNA when items of evidence are incorrectly placed in contact with each other during packaging. Fortunately, an examination of DNA band patterns in the laboratory readily reveals the presence of contamination. For example, with an STR, one will expect to see a two-band pattern. More than two bands suggests a mixture of DNA from more than one source.

Crime-scene investigators can take some relatively simple steps to minimize contamination of biological evidence:

1. **Use disposable gloves.**
2. **Wear a face mask while collecting evidence, a lab coat, eye protection, as well as shoe covers.**
3. **Change gloves before handling each new piece of evidence.**
4. **Collect a substrate control for possible subsequent laboratory examination.**
5. **Pick up small items of evidence such as cigarette butts and stamps with clean forceps. Disposable forceps are to be used so that they can be discarded after a single evidence collection.**
6. **Always package each item of evidence in its own well-ventilated container.**

A common occurrence at crime scenes is to suspect the presence of blood but not be able to observe any with the naked eye. In these situations, the common test of choice is luminol or Bluestar (see pages 336–337). Interestingly, neither luminol nor Bluestar is expected to inhibit the ability to detect and characterize STRs.3 Therefore, luminol and Bluestar can be used to locate traces of blood and areas that have been washed nearly free of blood without compromising the potential for DNA typing.

Case Files

Contact Lens Evidence

A woman alleged that she had been held against her will and sexually assaulted by a male friend in an apartment. During the course of the assault, a contact lens was knocked from the victim's eye. After the assault, she escaped, but because she was afraid of the threats made by her attacker, she did not report the assault to the police for three days. When the police examined the apartment, they noted that it had been thoroughly cleaned. A vacuum cleaner bag was seized for examination, and several pieces of material resembling fragments of a contact lens were discovered within the bag.

In the laboratory, approximately 20 nanograms of human DNA was recovered from the contact lens fragments. Cells from both the eyeball and the interior of the eyelids are naturally replaced every 6 to 24 hours. Therefore, both are potential sources for the DNA found. The DNA profile originating from the fragments matched the victim, thus corroborating the victim's account of the crime. The estimated population frequency of occurrence for the nine matching STRs are approximately 1 in 850 million. The suspect subsequently pleaded guilty to the offense.

STR Locus	Victim's DNA Type	Contact Lens
D3S1358	15,18	15,18
FGA	24,25	24,25
vWA	17,17	17,17
TH01	6,7	6,7
F13A1	5,6	5,6
fes/fps	11,12	11,12
D5S818	11,12	11,12
D13S317	11,12	11,12
D7S820	10,12	10,12

Source: R. A. Wickenheiser and R. M. Jobin, "Comparison of DNA Recovered from a Contact Lens Using PCR DNA Typing," Canadian Society of Forensic Science Journal 32 (1999): 67.

Quick Review

- Biological evidence should not be packaged in plastic or airtight containers because the accumulation of residual moisture could contribute to the growth of blood-destroying bacteria and fungi. Each article should be packaged separately in a paper bag or in a well-ventilated box.
- The least intrusive method for obtaining a DNA standard/reference is the buccal swab. In this procedure, cotton swabs are placed in the subject's mouth and the inside of the cheek is vigorously swabbed, resulting in the transfer of cells from the inner cheek lining onto the swab.

Chapter Review

- The gene is the fundamental unit of heredity. Each gene is composed of DNA specifically designed to control the genetic traits of our cells.
- DNA is constructed as a very large molecule made by linking a series of repeating units called nucleotides.
- Four types of bases are associated with the DNA structure: adenine (*A*), guanine (*G*), cytosine (*C*), and thymine (*T*).
- The bases on each strand of DNA are aligned in a double-helix configuration so that adenine pairs with thymine and guanine pairs with cytosine. This concept is known as complementary base pairing.
- The order in which the base pairs are arranged defines the role and function of a DNA molecule.
- DNA replication begins with the unwinding of the DNA strands in the double helix. The double helix is re-created as the nucleotides are assembled in the proper order (*A* with *T* and *G* with *C*). Two identical copies of DNA emerge from the process.
- PCR (polymerase chain reaction) is a technique for replicating or copying a portion of a DNA strand outside a living cell.
- Recombinant DNA is a process by which the DNA of an organism is altered by inserting into it fragments of DNA from another organism.
- Portions of the DNA molecule contain sequences of bases that are repeated numerous times. These tandem repeats offer a means of distinguishing one individual from another through DNA typing.
- Length differences associated with relatively long repeating DNA strands—called restriction fragment length polymorphisms (RFLPs)—formed the basis for one of the first DNA-typing procedures.
- In the laboratory, DNA molecules are cut up by a restriction enzyme and the resulting fragments are sorted out by electrophoresis.
- Materials undergoing electrophoresis are forced to move across a gel-coated plate under the influence of an electrical potential. Substances such as DNA can be separated and characterized using electrophoresis.
- A typical DNA fragment pattern shows two bands (one RFLP from each chromosome).
- Polymerase chain reaction (PCR) can amplify minute quantities of DNA. The technique evolved from an understanding of how DNA strands naturally replicate within a cell.
- PCR technology cannot be applied to RFLP DNA typing because RFLP strands are too long, often numbering in the thousands of bases. PCR is best used with DNA strands that are no longer than a couple of hundred bases.

- Long RFLP strands tend to break apart under the adverse conditions at many crime scenes. The shorter DNA strands used in PCR are more stable and less subject to degradation caused by adverse environmental conditions.
- STRs are locations on the chromosome that contain short sequences that repeat themselves within the DNA molecule. They serve as useful markers for identification because they are found in great abundance throughout the human genome.
- The entire strand of an STR is very short, less than 450 bases long. This makes STRs much less susceptible to degradation, and they are often recovered from bodies or stains that have been subjected to extreme decomposition.
- The more STRs one can characterize, the smaller the percentage of the population from which a particular combination of STRs can emanate. This gives rise to the concept of multiplexing, in which the forensic scientist can simultaneously extract and amplify a combination of different STRs.
- With STR, as little as 125 picograms of DNA is required for analysis—1/100 the amount normally required for RFLP analysis.
- Mitochondrial DNA is located outside the cell's nucleus and is inherited from the mother.
- Two regions of mitochondrial DNA, HV1 and HV2, are sequenced for forensic typing purposes.
- Mitochondrial DNA typing does not approach STR analysis in its discrimination power and thus is best reserved for samples, such as hair, for which STR analysis may not be possible.
- CODIS is a computer software program developed by the FBI that maintains local, state, and national databases of DNA profiles from convicted offenders, unsolved crime-scene evidence, and profiles of missing people.
- Biological evidence should not be packaged in plastic or airtight containers because the accumulation of residual moisture could contribute to the growth of blood-destroying bacteria and fungi. Each article should be packaged separately in a paper bag or in a well-ventilated box.
- The least intrusive method for obtaining a DNA standard/reference is the buccal swab. In this procedure, cotton swabs are placed in the subject's mouth and the inside of the cheek is vigorously swabbed, resulting in the transfer of cells from the inner cheek lining onto the swab.

Review Questions

1. New technology incorporating has supplanted RFLP.
   a. PCR
   b. STR
   c. mtDNA
   d. CODIS

2. A/An is composed of a sugar molecule, a phosphorus-containing group, and a base
   a. amino acid
   b. protein
   c. chromosomes
   d. nucleotide

3. DNA is the synthesis of new DNA from existing DNA and begins with the unwinding of the DNA strands in the double helix.
   a. transcription
   b. translation
   c. replication
   d. cloning

4. The pairing of nucleotides *A* to *G* and *C* to *T* is known as
   a. mitochondrial DNA.
   b. complementary base pairing.
   c. polymerase chain reaction.
   d. tandem repeats.

5. The inheritable traits that are controlled by DNA arise out of its ability to direct the production of complex molecules called from smaller units called.
   a. chromosomes; genes
   b. lipids; triglycerides
   c. proteins; nucleic acids
   d. proteins; amino acids

6. True or False: As the fundamental unit of heredity, genes instruct the body cells to make proteins that determine everything from hair color to susceptibility of disease.

7. True or False: DNA replication is accomplished using a technique known as polymerase chain reaction.

8. True or False: DNA technology is useful in identification because no two humans, except for identical twins, have the same type of tandem repeats in a strand of DNA.

9. True or False: The latest, most successful and widely used DNA profiling procedure is the short tandem repeats.

10. True or False: Mitochondrial DNA is found outside the nucleus of the cell and is inherited solely from the mother.

11. What is DNA and why is it important to forensic scientists?

12. What are genes and what is their function? Of what are genes composed and where are they located?

13. With what discovery are James Watson and Francis Crick credited?
14. Describe the basic structure and function of the DNA molecule. What is the name given to this type of structure?
15. Name the four bases associated with DNA. How are these bases paired on the DNA molecule?
16. How are proteins made? What determines the shape and function of a protein molecule?
17. What is the human genome? Name two medical applications of information about the human genome.
18. What is PCR? Why is it useful to forensic scientists?
19. What is recombinant DNA? How is recombinant DNA technology used to treat diabetes?
20. What are tandem repeats? How are they useful to forensic scientists?
21. What is a short tandem repeat (STR)? Why are STRs so attractive to forensic scientists?
22. Name two processes by which a forensic scientist can separate STRs for characterization. Which process is preferred and why?
23. What gene is often used to determine the sex of a DNA contributor? What characteristic of the gene allows forensic scientists to make this determination?
24. Name one advantage and three disadvantages of mtDNA analysis compared to nuclear DNA profiling.
25. What is CODIS? How is CODIS useful to forensic scientists?
26. What type of gloves should an evidence collector wear when handling biological evidence? Name two reasons for wearing this type of glove.
27. What type of packaging should not be used for biological evidence? Why? What type of packaging should be used instead for articles containing biological evidence?
28. How should packages containing biological evidence be stored until they are delivered to a laboratory? Name one common exception and describe how it should be stored.
29. List four ways to minimize contamination of biological evidence.
30. What are some examples of biological samples that could be identified at a crime scene? How would you properly collect and preserve each item? Write a short written response.

Learn the Vocabulary

Use the information from this chapter along with your prior knowledge and experience to learn the key terms. Choose from the following activities:

1. Make flashcards with a term on the front and its definition on the back. Use the flashcards to quiz a partner on correctly pronouncing and defining the terms.
2. Make an audio or video recording of each term and its definition.

3. Work with a partner to write a sentence for each term. Read the sentences out loud to a partner or share them with the class.

4. Write a list of new terms and language you have encountered in this chapter. Reuse the terms to write sentences and then read the sentences out loud to a partner. Re-use the terms again by writing a short speech. The speech might explain the terms or teach a lesson using the terms. Give the speech to a partner or to the class.

Quick Lab: Buccal Swab

Materials:

Swab/Q-tip
Slide
Light microscope
Methylene blue stain

Procedure:

An important part of obtaining standard/reference samples is determining the DNA profile of any individuals involved in the investigation. One way to do this is by a buccal swab. Take the swab provided by your teacher and rub the inside of your check with it. Next, rub the swab on the slide and add a drop of methylene blue stain; be careful not to get the stain on you or anything but the slide. Place a cover slip on the slide and view it under the microscope. Using critical-thinking skills, analyze the structure of cheek cells through this experimental testing and complete the questions below. Be sure to practice proper collection and preservation techniques.

Follow-Up Questions:

1. **How many cells were you able to see on your slide? Were you surprised by the number you saw?**

2. **What on the slide was stained by the methylene blue?**

3. **Where in the cell is the DNA located?**

4. **How did you follow proper guidelines for the collection and preservation of your sample?**

Application and Critical Thinking

1. The following sequence of bases is located on one strand of a DNA molecule:

   C–G–A–A–T–C–G–C–A–A–T–C–G–A–C–C–T–G

   List the sequence of bases that will form complementary pairs on the other strand of the DNA molecule.

2. Police discover a badly decomposed body buried in an area where a man disappeared some years before. The case was never solved, nor was the victim's body ever recovered. As the lead investigator, you suspect that the newly discovered body is that of the victim. What is your main challenge in using DNA typing to determine whether your suspicion is correct? How would you go about using DNA technology to test your theory?

3. You are a forensic scientist performing DNA typing, a type of experimental testing, on a blood sample sent to your laboratory. DNA typing provides scientific explanations about the identity of an individual. While performing an STR analysis on the sample, you notice a four-band pattern. Using your critical-thinking skills determine what conclusion you should draw. Why?

4. Use critical-thinking skills to analyze the following empirical evidence to determine information about the attacker. A woman reports being mugged by a masked assailant, whom she scratched on the arm during a brief struggle. The victim is not sure whether the attacker was male or female. DNA analysts extract and amplify the amelogenin gene from the epithelial cells under the victim's fingernails (allegedly belonging to the attacker) and from a buccal swab of the victim. The sample is separated by gel electrophoresis with the result shown here. The victim's amelogenin DNA is in lane 2, and the amelogenin DNA from the fingernail scraping is in lane 4. What conclusion can you draw about the attacker from this result? How did you reach this conclusion?

5. At a crime scene you encounter each of the following items. For each item, indicate the potential sources of DNA. The five possible choices are saliva, skin cells, sweat, blood, and semen.

Donatas1205/ Shutterstock

Elena Elisseeva/ Shutterstock

Somcnaij/ Shutterstock

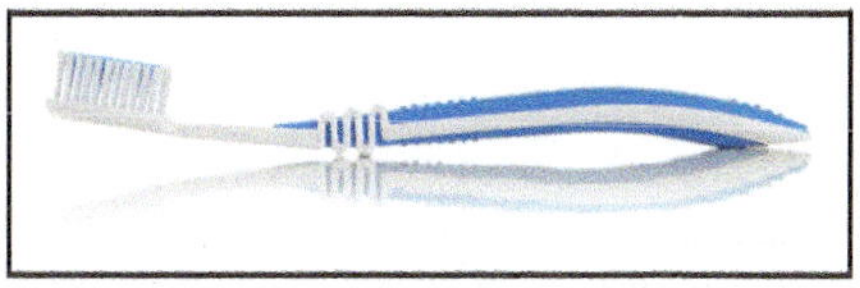
R-O-M-A/Shutterstock

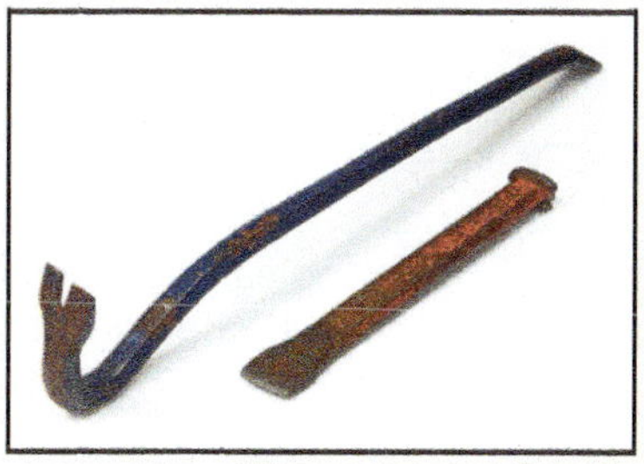
Birgit Reitz-Hofmann/ Shutterstock

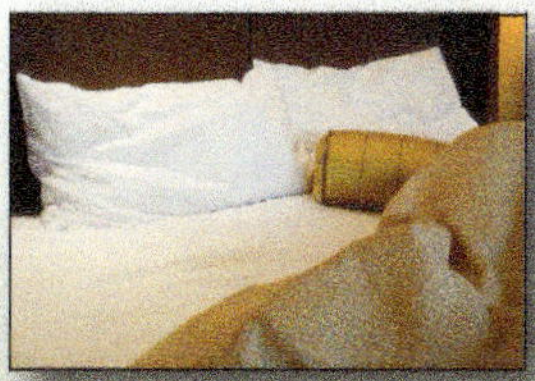
Lim Yong Hian/Shutterstock

Tarasyuk Igor/Shutterstock

Africa Studio/Shutterstock

6. The 15-STR locus DNA profile of a missing person, James Dittman, is given in the following table.

STR Loci	Allele
D3S1358	15
THO1	6, 9.3
D21S11	27
D18S51	15,16
PENTA E	10
D5S818	11
D13S807	10,13
D7S820	9,10
D16S539	11,12
CSF1PO	13
PENTA D	12,13
AMELOGENIN	XY
VWA	17, 19
D8S1179	10, 13
TPOX	8, 12
FGA	21

Decomposing remains were found deep in the woods near the missing person's house. DNA from these remains was extracted, amplified, and analyzed at 15 STR loci. Using critical-thinking skills, interpret the components of the electropherogram shown on the following page.

A. Analyze the evidence provided in the electropherogram. Evaluate the evidence using logical reasoning and critical thinking to propose a scientific explanation of whether or not the remains could belong to James Dittman.
B. Using experimental and observational testing, support your conclusion using scientific explanations and logical reasoning.
C. Critique your conclusion by verifying the empirical evidence that was used to support the conclusion of the remains belonging to James Dittman, or identifying at which STR loci the profiles differ if you concluded the remains do not belong to James Dittman.

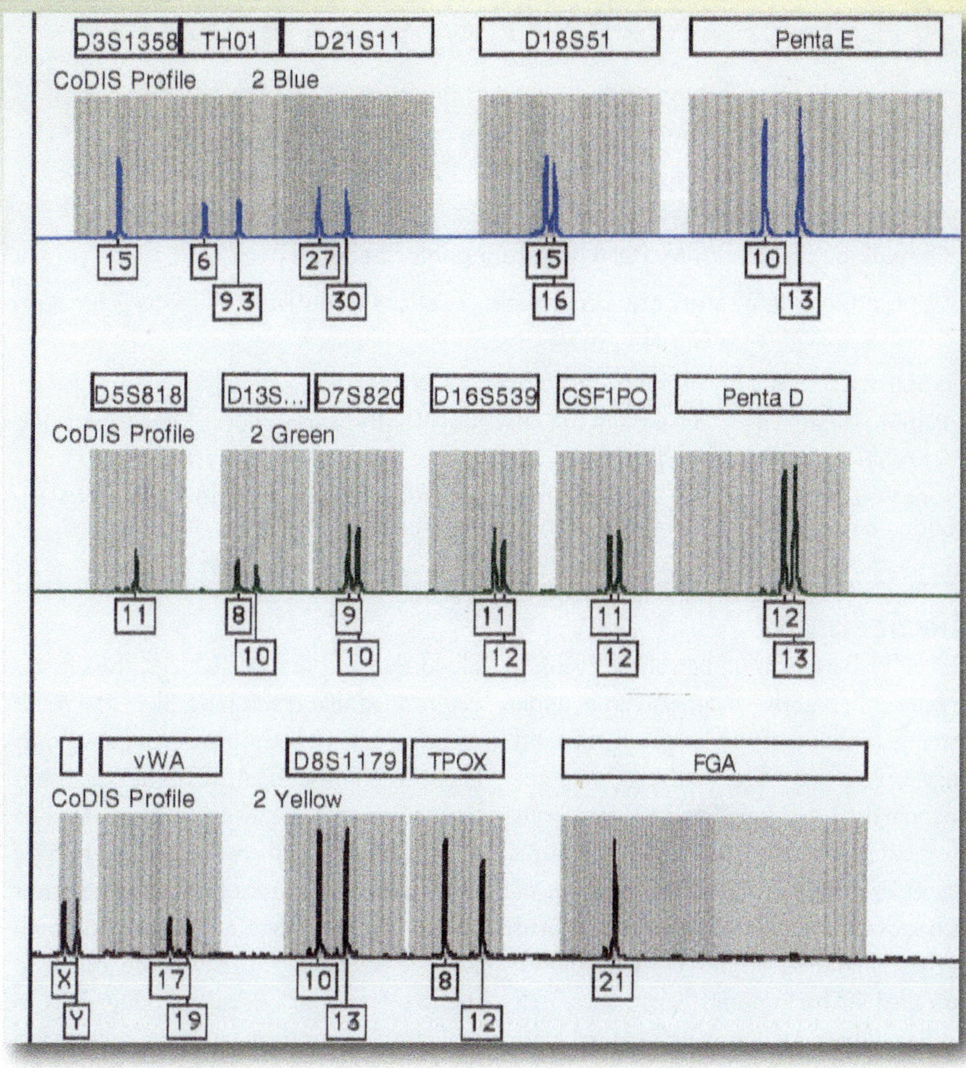

Richard Saferstein, Ph.D.

7. Explain the analytical procedure for generating a DNA profile. Be sure to include the process of extraction, quantification, amplification, and capillary electrophoresis in your explanation.

8. To evaluate different methodologies and scientific explanations, scientists often use logical reasoning. Choose one of the following methodologies and use your critical-thinking skills to explain in writing how it is used in the forensic analysis of DNA: short tandem repeats (STRs), Y-STRs, mitochondrial DNA, or single nucleotide polymorphisms (SNPs).

9. Research and explore the database systems associated with DNA and explain how they are used. Create a short presentation to share with your class.

Laboratory Experiments

This activity requires the use of the following practices of science and engineering:

- Asking questions and defining problems
- Developing and using models
- Analyzing and interpreting data
- Obtaining, evaluating, and communicating information
- Using appropriate safety equipment and practices as outlined in Texas Education Agency-approved safety standards

This activity consists of the following crosscutting concept:

- **Patterns**—Patterns in base pairs of DNA allow for the separation of fragments and comparison of samples.

Before you begin, read the scenario and steps out loud with a partner. Ask for clarification as necessary to make sure you understand the experiment.

The Scenario

Officer TJ Barry was on patrol the evening of November 28th when he came upon a strange pattern of markings on the Old Nogales Highway. The markings were approximately 2 feet long, stopped for about 5 feet, and the pattern repeated for about a quarter of a mile. Officer Barry got out of his patrol car to investigate and determined that the markings appeared to be blood. He called in backup and began to search the area. The markings were getting closer to the ditch so he began looking along the side of the road. He came upon the body of a man with no legs. It appeared as though the legs had been cut off. Further searching resulted in no other evidence. Based on the blood trail and the severed legs, the officers determined that it was most likely the result of being hit by a semitruck. Several hours later, in response to an APB, the weigh station entering New Mexico from Arizona reported that a semi-truck appeared to have human tissue on some of its tires. The driver denied hitting anything. As a top-notch forensic scientist, you have been called to the lab. Use your critical-thinking skills to evaluate the empirical evidence collected at the crime scene. Since DNA tissue provides scientific explanations about an individual's identity, determine if the DNA from the tissue on the truck matches the DNA of the victim.

DNA EVIDENCE EVALUATION

1. To make the simulation manageable, each simulation base represents approximately one thousand bases (one kilobase) that would be found in an actual DNA profile. To simulate the restriction digest, you must first cut out the strips of DNA sequences, then tape together the strips representing one sample of DNA. Next, mark the sample strip at the recognition sites for the restriction enzyme Hae III (GGCC). Then cut the strip all the way across between the center G and C of each restriction site.
2. Count the number of base pairs in each piece of DNA that you created. Record the base pair number on the back side of the DNA fragment.
3. On a large piece of paper label the y-axis with the number of base pairs. Use a ruler to ensure that the distances are uniform. Label the x-axis with the identification number.

4. Tape your DNA fragments to a chart, using the base pair numbers as a guideline for fragment placement.
5. Using your chart, compare the crime scene DNA to the unknown DNA samples. Two random samples have been added to prevent biases. If you think you have found a match between two samples, raise your hand and share your findings. If needed, use the chart as a reference for your explanation. Your instructor will inform you as to which sample was from the semi-truck after you report your findings. Do not hesitate to ask for clarification, as necessary.

Victim

GTCGACCGGTGACCGTGCGTACACAGTGCTCCGGATAGCT
GATAGCTCCGGTGCAGCTGGCCACTGGCACGCATGTGTCA
CGAGGCCTATCGACTATCGAGGCCAC

Unknown Sample 1

GTCCCAGCCGGACCGTACCGGTAGATCAGCCGGTAGATTA
TAGCGTGATGTGCAGGGTCGGCCTGGCATGGCCATCTAGT
CGGCCATCTAACTATCGCACTACAC

Unknown Sample 2

GTCGACCGGTGACCGTGCGTACACAGTGCTCCGGATAGCT
GATAGCTCCGGTGCAGCTGGCCACTGGCACGCATGTGTCA
CGAGGCCTATCGACTATCGAGGCCAC

Unknown Sample 3

GTCTACGTAATCGTAGCCATCCGGACAGTGTGCACGATCGT
ACATGCTACGTGCAGATGCATTAGCATCGGTAGGCCTGTCA
CACGTGCTAGCATGTACGATGCAC

Endnotes

1. Instructions for assembling proteins are carried from DNA to another region of the cell by ribonucleic acid (RNA). RNA is directly involved in the assembly of the protein, using the genetic code it received from DNA.
2. R. A. Wickenheiser, "Trace DNA: A Review, Discussion of Theory, and Application of the Transfer of Trace Quantities through Skin Contact," *Journal of Forensic Sciences* 47 (2002): 442.
3. A. M. Gross et al., "The Effect of Luminol on Presumptive Tests and DNA Analysis Using the Polymerase Chain Reaction," *Journal of Forensic Sciences* 44 (1999): 837.

Crime-Scene Reconstruction: Bloodstain Pattern Analysis

11

UPPA/Photoshot

Key Terms

angle of impact
area of convergence
area of origin
arterial spray
back spatter
cast-off
crime-scene reconstruction
drip trail pattern
expirated blood pattern
flow patterns
forward spatter
impact spatter
satellite spatter
satellite stain
skeletonization
transfer pattern
void

The Sam Sheppard Case: A Trail of Blood

Convicted in 1954 of bludgeoning his wife to death, Dr. Sam Sheppard achieved celebrity status when the storyline of TV's *The Fugitive* was apparently modeled on his efforts to seek vindication for the crime he professed not to have committed. Dr. Sheppard, a physician, claimed he was dozing on his living room couch when his pregnant wife, Marilyn, was attacked. Sheppard's story was that he quickly ran upstairs to stop the carnage, but was briefly knocked unconscious by the intruder. The suspicion that fell on Dr. Sheppard was fueled by the revelation that he was having an adulterous affair. At trial, the local coroner testified that a pool of blood on Marilyn's pillow contained the impression of a "surgical instrument." After Sheppard had been imprisoned for ten years, the U.S. Supreme Court set aside his conviction because of the "massive, pervasive, and prejudicial publicity" that had attended his trial.

In 1966, the second Sheppard trial commenced. This time, the same coroner was forced to back off from his insistence that the bloody outline of a surgical instrument was present on Marilyn's pillow. However, a medical technician from the coroner's office now testified that blood on Dr. Sheppard's watch was from blood spatter, indicating that Dr. Sheppard was wearing the watch in the presence of the battering of his wife. The defense countered with the expert testimony of eminent criminalist Dr. Paul Kirk. Dr. Kirk concluded that blood spatter marks in the bedroom showed the killer to be left-handed. Dr. Sheppard was right-handed.

Dr. Kirk further testified that Sheppard stained his watch while attempting to obtain a pulse reading. After less than 12 hours of deliberation, the jury failed to convict Sheppard. But the ordeal had taken its toll. Four years later Sheppard died, a victim of drug and alcohol abuse.

Learning Objectives

After studying this chapter, you should be able to:

11.1 Summarize the principles of crime-scene reconstruction and the personnel involved in reconstruction

11.2 Describe the general features of bloodstain formation

11.3 Discuss the methods to determine the area of convergence and area of origin for impact spatter patterns

11.4 Describe how various blood pattern types are created and which features of each pattern can be used to aid in reconstructing events at a crime scene

11.5 Describe the methods for documenting bloodstain patterns at a crime scene

National Science Content Standards

Scientific Inquiry

Physical Science

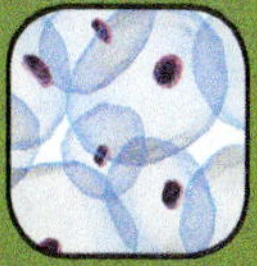
Life Science

Crime-Scene Reconstruction

Previous discussions dealing with the processes of identification and comparison have stressed laboratory work routinely performed by forensic scientists. However, there is another dimension to the role that forensic scientists play during the course of a criminal investigation: participating in a team effort to reconstruct events that occurred before, during, and after the commission of a crime.

Law enforcement personnel must take proper action to enhance all aspects of the crime scene search so as to optimize the crime-scene reconstruction. First, and most important, is securing and protecting the crime scene. Protecting the scene is a continuous endeavor from the beginning to the end of the search. Evidence that can be invaluable to reconstructing the crime can be unknowingly altered or destroyed by people trampling through the scene, rendering the evidence useless. The issue of possible contamination of evidence will certainly be attacked during the litigation process and could make the difference between a guilty and not-guilty verdict.

Before processing the crime scene for physical evidence, the investigator should make a preliminary examination of the scene as it was left by the perpetrator. Each crime scene presents its own set of circumstances. The investigator's experience and the presence or absence of physical evidence become critical factors in reconstructing a crime. The investigator captures the nature of the scene as a whole by performing an initial walk-through of the crime scene. During the walk-through, the investigator's task is to document observations and formulate how the scene should ultimately be processed. As the collection of physical evidence begins, any and all observations should be recorded through photographs, sketches, and notes. By carefully collecting physical evidence and thoroughly documenting the crime scene, the investigator can begin to unravel the sequence of events that took place during the commission of the crime.

Personnel Involved in Reconstruction

Because investigators consider many types of evidence when reconstructing a crime scene, reconstruction is a team effort that involves various professionals putting together many pieces of a puzzle. The team as a whole works to answer the typical "who, what, when, where, why, and how" of a crime scene. Often reconstruction requires the involvement of law enforcement personnel, a medical examiner, and/or a forensic scientist. All of these professionals contribute unique perspectives to develop the crime-scene reconstruction. Was more than one person involved? How was the victim killed? Were actions taken to cover up what took place? The positioning of the victim in a crime scene can often reveal pertinent information for the investigation. Trained medical examiners can examine the victim at a crime scene and determine whether the body has been moved after death by evaluating the livor distribution within the body (see page 120). For example, if livor has developed in areas other than those closest to the ground, the medical examiner can reason that the victim was probably moved after death. Likewise, the examiner can determine whether the victim was clothed after death because livor will not develop in areas of the body that are restricted by clothing.

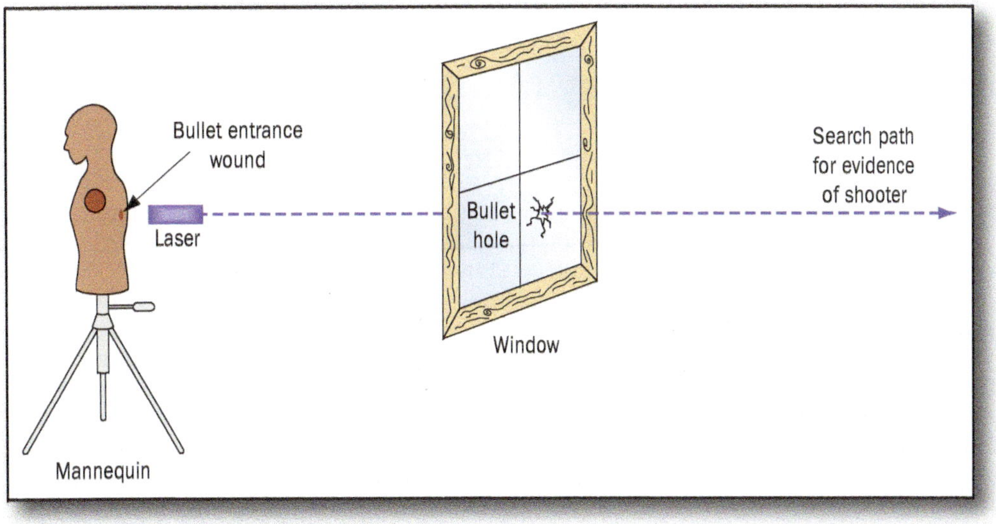

FIGURE 11–1 A laser beam is used to determine the search area for the position of a shooter who has fired a bullet through a window and wounded a victim. The bullet path is determined by lining up the victim's bullet wound with the bullet hole present in the glass pane.

A forensic scientist or trained crime-scene investigator can also bring special skills to the reconstruction of events that occurred during the commission of a crime. For example, a forensic scientist using a laser beam to plot the approximate bullet path in trajectory analysis can help determine the probable position of the shooter relative to that of the victim (see Figure 11–1). Other skills that a forensic scientist may employ during a crime-scene reconstruction analysis include determining the direction of impact of projectiles penetrating glass objects (see page 175); locating gunshot residues deposited on the victim's clothing for the purpose of estimating the distance of a shooter from a target (see pages 641–642); searching for primer residues deposited on the hands of a suspect shooter (see pages 642–643); and, as we will see from the discussion that follows, analyzing blood spatter patterns.

Crime-scene reconstruction is the method used to support a likely sequence of events at a crime scene by observing and evaluating physical evidence and statements made by individuals involved with the incident. The evidence may also include information obtained from re-enactments. Therefore, reconstructions have the best chance of being accurate when investigators use proper documentation and collection methods for all types of evidence.

Principles of Crime-Scene Reconstruction

Physical evidence left behind at a crime scene plays a crucial role in reconstructing the sequence of events surrounding the crime. Although the evidence alone may not describe everything that happened, it can support or contradict accounts given by witnesses and/or suspects. Information obtained from physical evidence can also generate leads and confirm the reconstruction of a crime to a jury. The collection, documentation, and interpretation of physical evidence is the foundation of a reconstruction. Reconstruction develops a likely sequence of events by the observation and evaluation of physical evidence as well as statements made by witnesses and input from those involved with the investigation of the incident. Analysis of all available data will help to create a workable model for reconstruction.

Case Files

Pistorius: Valentine's Day Murder

Oscar Pistorius overcame leaps and bounds while many others in his position have failed to achieve. Pistorius was a double amputee from the time he was a child; both of his legs had to be removed due to a medical condition he had from birth. Despite his physical disability, Pistorius chased the dream of becoming an Olympic runner. This dream became a reality in 2012, when Pistorius faced able-bodied opponents in the Summer Olympics. Oscar Pistorius has received endless notoriety for his athletic achievement and has been dubbed by the media as the "blade runner."

Pistorius and his super model girlfriend Reeva Steenkamp had been dating for over three months, and according to security logs from Pistorius's gated South African neighborhood, Steenkamp had been staying with him for multiple days. On Valentine's day, 2013, Steenkamp was found lying motionless on the ground, suffering multiple 9 mm pistol wounds to her body and head. Reeva Steenkamp had been shot by Pistorius multiple times through a locked wooden bathroom door. Pistorius claimed that he had fired the shots because he thought there was an intruder in the house. The Olympic runner appeared to be distraught and grieving heavily over his deceased girlfriend. Local South African prosecutors ultimately charged Pistorius with murder due to the circumstances of the event.

During the trial, the prosecution painted a picture of Pistorius as a violent and angry lover. Prosecutors claimed that Steenkamp ran into the bathroom and locked the door to seek protection from a spiteful Pistorius during a heated argument. Detectives placed Steenkamp in a defensive position, crouched behind the bathroom door with her arms and hands crossed in front of her face, as to protect herself from Pistorius. The prosecution also questioned why Pistorius would engage the intruder, rather than grabbing his girlfriend and leaving the residence. The defense called upon the support of Wollie Wolmarans, a ballistics expert, to support Pistorius's account of the incident. According to Wolmarans's reconstruction of the scene, wood splinter marks around Steenkamp's arm proved that she was reaching out to open the door when the incident occurred contrary to the prosecution's account of Steenkamp being crouched and protecting herself in the bathroom.

Oscar Pistorius was found guilty on a charge of culpable homicide, equivalent to manslaughter, after being acquitted of murder in the killing of his girlfriend. The presiding judge ruled that there was not enough evidence to support the contention that Pistorius knew that Steenkamp was behind the toilet door. In 2015, an appeals court overturned the verdict and convicted him of murder. Facing a 15-year jail sentence, Pistorius was sentenced to a jail term of six years.

Quick Review

- Crime-scene reconstruction relies on the combined efforts of medical examiners, criminalists, and law enforcement personnel to recover physical evidence and to sort out the events surrounding the occurrence of a crime.
- Examples of crime-scene reconstruction include determining whether a body was moved after death, determining whether a victim was clothed after death, analyzing bullet trajectory, analyzing blood spatter, determining the direction from which projectiles penetrated glass objects, estimating the distance of a shooter from a target, and locating primer residue on suspects.

General Features of Bloodstain Formation

Crimes involving violent contact between individuals are frequently accompanied by bleeding and resultant bloodstain patterns. Crime-scene analysts have come to appreciate that bloodstain patterns deposited on floors, walls, ceilings, bedding, and other relevant objects can provide valuable insights into events that occurred during the commission of a violent crime. The information one is likely to uncover as a result of bloodstain pattern interpretation includes the following:

- The direction from which blood originated
- The angle at which a blood droplet struck a surface
- The location or position of a victim at the time a bloody wound was inflicted
- The movement of a bleeding individual at the crime scene
- The minimum number of blows that struck a bleeding victim
- The approximate location of an individual delivering blows that produced a bloodstain pattern

The crime-scene investigator must not overlook the fact that the location, distribution, and appearance of bloodstains and spatters may be useful for interpreting and reconstructing the events that accompanied the bleeding. A thorough analysis of the significance of the position and shape of blood patterns with respect to their origin and trajectory is exceedingly complex and requires the services of an examiner who is experienced in such determinations. Most important, the interpretation of bloodstain patterns necessitates a carefully planned control experiment using surface materials comparable to those found at the crime scene. This chapter presents the basic principles and common deductions behind bloodstain pattern analysis to give the reader general knowledge to use at the crime scene.

Surface Texture

Surface texture is of paramount importance in the interpretation of bloodstain patterns; comparisons between standards and unknowns are valid only when identical surfaces are used. Forensic blood spatter analysts must analyze blood stain patterns based on surface type in order to determine the mechanism by which the patterns are created. In general, harder and nonporous surfaces (such as glass or smooth tile) result in less spatter. Blood spatter on these surfaces is usually round in shape due to the surface tension in the blood. Rough surfaces, such as a concrete floor or wood, **satellite stain** (see Figure 11–2). Blood spatter on these surfaces is more spread out and not round in shape because the surface tension in the blood is broken when it hits a rough surface.

satellite stain
A smaller bloodstain that originated during the formation of the parent stain as a result of blood impacting a surface

Direction and Angle of Impact

An investigator may discern the direction of travel of blood that struck an object by studying the stain's shape. As the stain becomes more elliptical in shape, its direction becomes more discernible because the pointed end of a

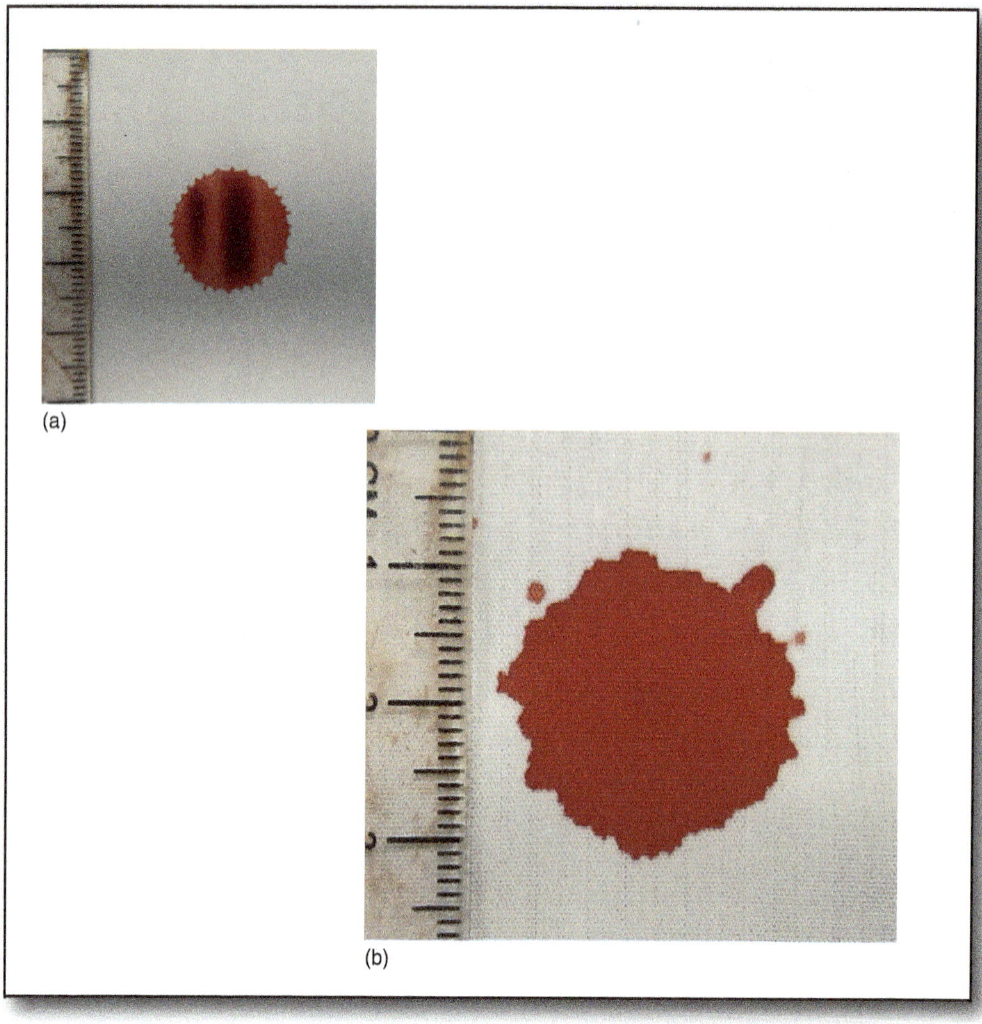

FIGURE 11–2 (a) A bloodstain from a single drop of blood that struck a paper surface after falling.
(b) A bloodstain from a single drop of blood that struck a cotton muslin sheet after falling.
Courtesy of Jeremiah Morris, Johnson County Sheriff's Office Criminalistics Laboratory

bloodstain faces its direction of travel. The distorted or disrupted edge of an elongated stain indicates the direction of travel of the blood drop. Satellite spatter around parent stains will have the pointed end facing against the direction of travel. In Figure 11–3, the bloodstain pattern was produced by several droplets of blood that were traveling from left to right before striking a flat, level surface.

It is possible to determine the impact angle of blood on a flat surface by measuring the degree of circular distortion of the stain. A drop deposited at an **angle of impact** of about 90 degrees (directly vertical to the surface) will be approximately circular in shape with no tail or buildup of blood. However, as the angle of impact deviates from 90 degrees, the stain becomes elongated in shape. Buildup of blood will occur when the angles are larger, whereas longer and longer tails will appear as the angle of impact becomes smaller (see Figure 11–4).

angle of impact
The angle formed between the path of a blood drop and the surface that it contacts

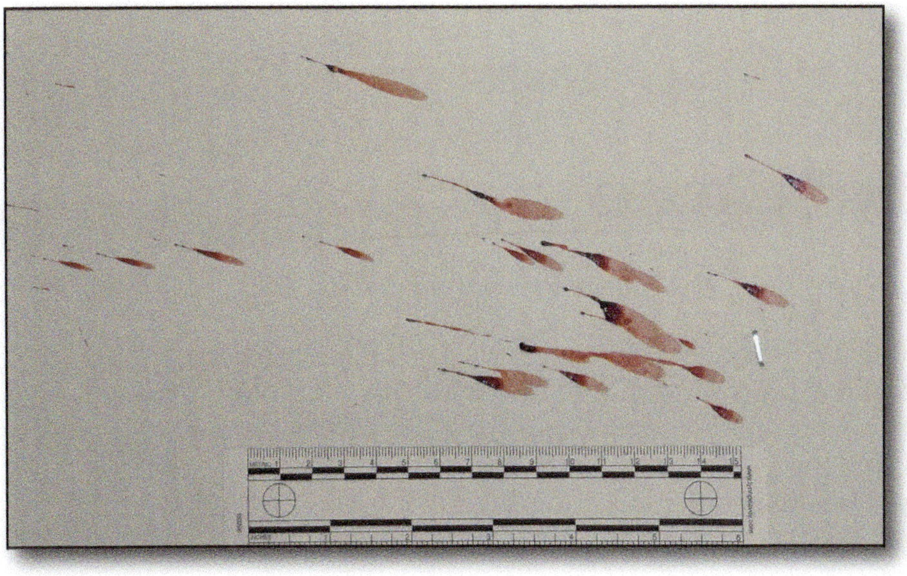

FIGURE 11–3
A blood-stain pattern produced by droplets of blood that were traveling from left to right. *Courtesy of Jeremiah Morris, Johnson County Sheriff's Office Criminalistics Laboratory*

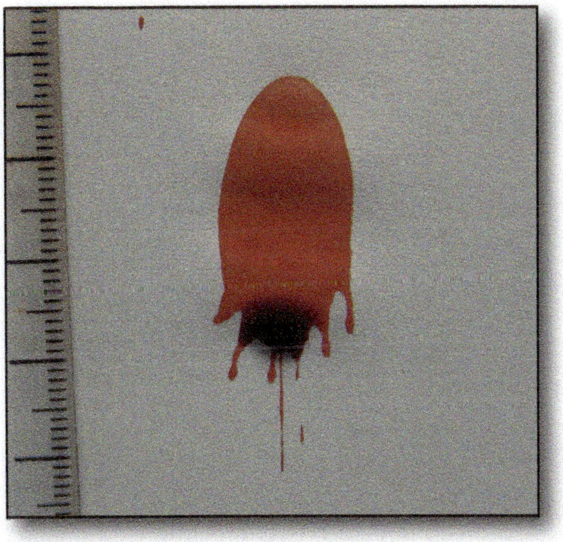

FIGURE 11–4 The higher pattern is of a single drop of human blood that fell 24 inches and struck hard, smooth cardboard at 50 degrees. On this drop the collection of blood shows the direction. The lower pattern is of a single drop of human blood that fell 24 inches and struck hard, smooth cardboard at 15 degrees. On this drop the tail shows the direction. *Courtesy of Jeremiah Morris, Johnson County Sheriff's Office Criminalistics Laboratory*

Inside the Science

Determining the Angle of Impact of Bloodstains

The distorted or disrupted edge of an elongated stain indicates the direction of travel of the blood drop. One may establish the location or origin of bloodshed by determining the directionality of the stain and the angle at which blood came into contact with the landing surface. To determine the angle of impact, calculate the stain's length-to-width ratio and apply the formula:

$$\sin A = \frac{\textit{width of blood stain}}{\textit{length of blood stain}}$$

where A = the angle of impact.

For example, suppose the width of a stain is 11 mm and the length is 22 mm. Then,

$$\sin A = \frac{11\text{ mm}}{22\text{ mm}} = (11\text{ mm} \div 22\text{ mm}) = 0.50$$

A scientific calculator having the trigonometric function will calculate that the inverse sine of 0.50 is equal to a 30-degree angle.

Note: The measurements for length and width should be made with a ruler, micrometer, or photographic loupe.

Quick Review

- Individual bloodstains can convey to the bloodstain analyst the directionality and angle of impact of the blood when it impacted a surface. Bloodstain patterns may convey to the analyst the location of victims or suspects, the movement of bleeding individuals, and the number of blows delivered.
- Surface texture is of paramount importance in the interpretation of bloodstain patterns; rounder drops generally are produced from smooth, nonporous surfaces, whereas rough surfaces create irregular-edged drops. However, correlations between standards and unknowns are valid only if identical surfaces are used.
- The direction of travel of blood striking an object may be discerned by the stain's shape. The pointed end of a bloodstain always faces in its direction of travel.
- The angle of impact of an individual bloodstain can be approximated by the degree of distortion or lengthening of the bloodstain, or it can be more effectively estimated using the ratio of width/length of the stain.

Bloodstain Spatter Patterns

A **spatter pattern** is a bloodstain resulting from an airborne blood drop created when external force is applied to liquid blood. The most common type of bloodstain pattern found at a crime scene is **impact spatter**. This pattern occurs when an object impacts the source of the blood.

Spatter projected outward and away from the source, such as an exit wound, is called **forward spatter**. **Backspatter**, sometimes called blow-back spatter, is blood projected backward from a source, such as an entrance wound, and potentially deposited on the object or person who created the impact. Impact spatter patterns consist of many drops radiating in direct lines from the origin of blood to the stained surface (see Figure 11–5).

spatter pattern
A bloodstain resulting from an airborne blood drop created when external force is applied to liquid blood

impact spatter
A bloodstain pattern resulting from an object striking liquid blood

forward spatter
A bloodstain pattern resulting from blood drops that can be produced when a projectile creates an exit wound

backspatter
A bloodstain pattern resulting from blood drops that can be produced when a projectile creates an entrance wound

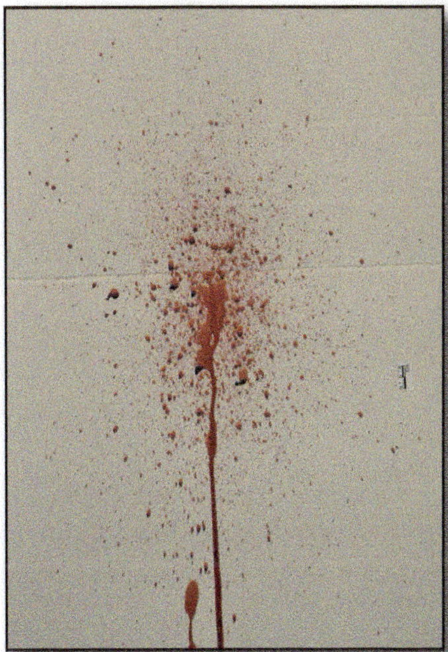

FIGURE 11–5
Spatter produced by an automatic weapon.
Courtesy of Jeremiah Morris, Johnson County Sheriff's Office Criminalistics Laboratory

In general, as the velocity of the force of the impact on the source of blood increases, so does the velocity of the blood drops emanating from the source. It is also generally true that as both the force and velocity of impact increase, the diameter of the resulting blood drops decreases.

Classifying Impact Pattern

In the past, bloodstain analysts used to interpret the droplet size of a bloodstain pattern to classify impact patterns by velocity. Because general principles of fluid dynamics apply to bloodstains, analysts knew generally what type of force was required to create droplets of varying size. However, the classifications of low, medium, and high velocity cannot illuminate the specific events that produced the stain pattern. For example, beatings can produce either high-velocity spatter or stain sizes that look more like low-velocity spatter. The existence of the low, medium, and high classifications only proved problematic for modern bloodstain pattern interpretation because these designations lend themselves, intentionally or unintentionally, to inferences about the type of velocity or mechanism used to create the stain. Forensic scientists have no way of determining the velocity of the object that created the stains or the velocity of the droplets themselves prior to contacting the target surface. As such, these references are no longer recommended by guidance bodies in the field. Proper interpretation for an impact pattern should encompass observations of stain size, shape, location, and distribution without any references to the force or mechanism that might have been used to create it.

Blood spatter patterns can arise from a number of distinctly different sources, which will be discussed in this chapter. Illustrations of patterns emanating from impact, cast-off, and projected pattern are shown in Figure 11–6.

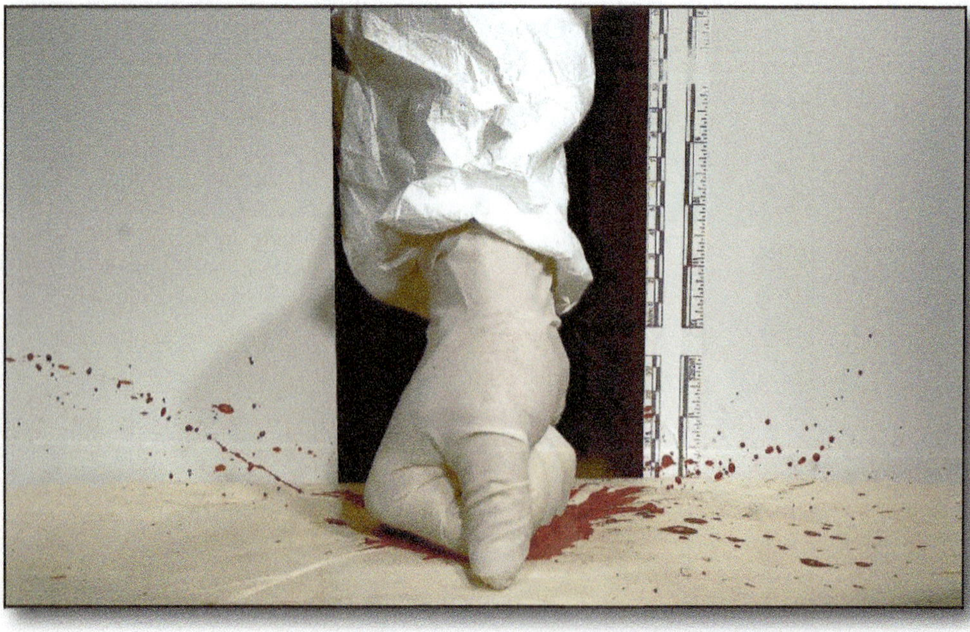

FIGURE 11–6
The action associated with producing an impact pattern.
Courtesy A.Y. Wonder

Origin-of-Impact Patterns

Impact spatter patterns can offer investigators clues that help determine the origin of the blood source and the position of the victim at the time of the impact.

Area of Convergence The **area of convergence** is the area on a two-dimensional plane from which the drops originated. This can be established by drawing straight lines through the long axis of several individual bloodstains, following the line of their tails. The intersection of these lines is the area of convergence, and the approximate area of origin will be on a line straight out from this area. Figure 11–7 illustrates how to draw lines to find an area of convergence.

An object hitting a source of blood numerous times will never produce exactly the same pattern each time. One may therefore determine the number of impacts by drawing the area of convergence for groups of stains from separate impacts.

area of convergence
The area on a two-dimensional plane where lines traced through the long axis of several individual bloodstains meet; this approximates the two-dimensional place from which the bloodstains were projected

Area of Origin It may also be important to determine the **area of origin** of a bloodstain pattern, the area in a three-dimensional space from which the blood was projected. This will show the position of the victim or suspect in space when the stain-producing event took place. The distribution of the drops in an impact pattern gives a general idea of the distance from the blood source to the blood-stained surface. Impact patterns produced at a distance close to the surface will appear as clustered stains. As the distance from the surface increases, so do the distribution and distance between drops.

area of origin
The location in three-dimensional space from which blood that produced a bloodstain originated. The location of the area of convergence and the angle of impact for each bloodstain is used to approximate this area

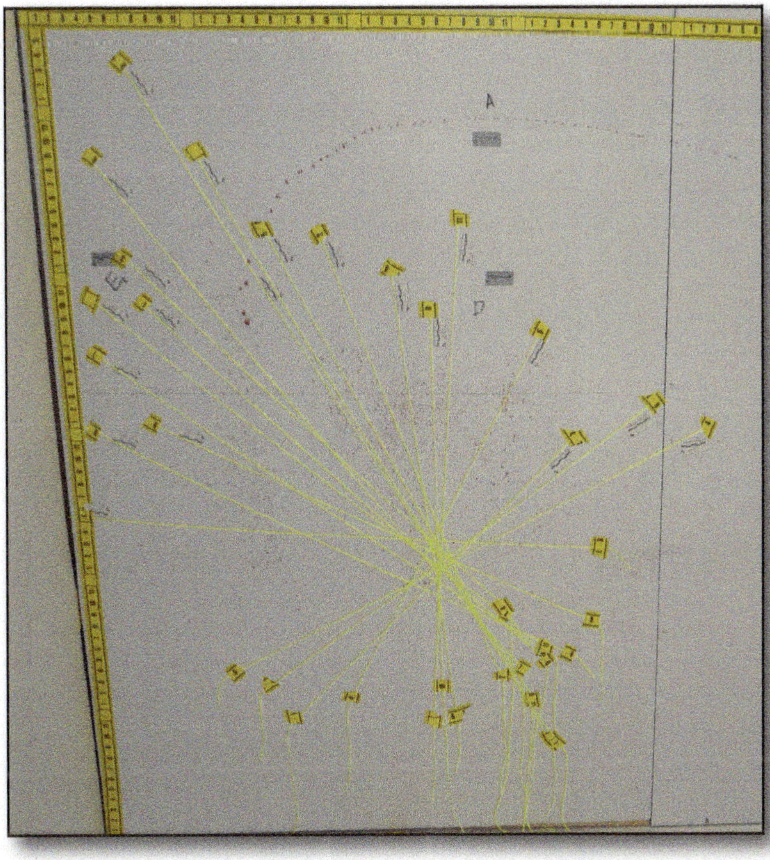

FIGURE 11–7 An illustration of stain convergence on a two-dimensional plane. Convergence represents the area from which the stains emanated. *Courtesy The Institute of Applied Forensic Technology, Ocoee, Florida*

A common method for determining the area of origin at the crime scene is called the string method. Figure 11-7 illustrates how to string lines to find an area of convergence.

1. **Find the area of convergence for the stain pattern.**
2. **Place a pole or stand as an axis coming from the area of convergence.**
3. **Attach one end of a string next to each droplet. Place a protractor next to each droplet and lift the string until it lines up with the determined angle of impact of the drop. Keeping the string in line with the angle, attach the other end of the string to the axis pole.**
4. **View the area of origin of the drops where the strings appear to meet. Secure the strings at this area.**

Quick Review

- An impact spatter pattern occurs when an object impacts a source of blood. This produces forward spatter projected forward from the source and back spatter projected backward from the source.
- The area of convergence is the point on a two-dimensional plane from which the drops of an impact spatter pattern originated. This area can be estimated by drawing straight lines through the long axis of several individual bloodstains, following the line of their tails.
- The area of origin of a bloodstain pattern is the area in three-dimensional space where blood was projected from, showing the position of the victim or suspect when the stain-producing event took place. The string method is commonly used at a crime scene to approximate the position of the area of origin.

More Bloodstain Spatter Patterns

Although forensic scientists focus mostly on the size, shape, location, and distribution of blood spatter, there are general guidelines that show how blood spatter looks when caused by different mechanisms. Forensic scientists analyze blood stain patterns based on the surface type in order to determine the mechanism by which the patterns are created. The following examples show more bloodstain spatter patterns.

FIGURE 11–8 A forward spatter pattern. *Courtesy of Jeremiah Morris, Johnson County Sheriff's Office Criminalistics Laboratory*

Gunshot Spatter

A shooting may leave a distinct gunshot spatter pattern. This may be characterized by both forward spatter from an exit wound and back spatter from an entrance wound. The presence of back spatter on a firearm or a shooter is dependent on the distance between the firearm and victim. Forward spatter generally leaves a pattern of very fine droplets (see Figure 11–8). Medium- and large-sized drops may also be observed within the spatter pattern.

The location of the injury, the size of the wound created, and the distance between the victim and the muzzle of the weapon all affect the amount of back spatter that occurs. Finding spatter containing the victim's blood on a suspect can help investigators place the suspect in the vicinity when the gun was discharged. Backspatter created by a firearm discharge generally contains fewer and smaller atomized stains than does forward spatter. A muzzle blast striking an entrance wound is expected to cause the formation of atomized blood.

Depending on the distance from the victim at which the gun was discharged, some backspatter may strike the shooter and enter the gun muzzle. This is called the drawback effect. Blood within the muzzle of a gun can place the weapon in the vicinity of the gunshot wound. The presence of blow-back spatter on a weapon's muzzle is consistent with the weapon being close to the victim at the time of firing.

Cast-Off Spatter

A **cast-off** pattern is created when a blood-covered object flings blood in an arc onto a nearby surface. This kind of pattern commonly occurs when a person pulls a bloody fist or weapon back between delivering blows to a victim (see Figure 11–9). The bloodstain tails will point in the direction in which the object was moving.

cast-off
A bloodstain pattern that is created when blood is flung from a blood-bearing object in motion onto a surface

The width of the cast-off pattern created by a bloody object may help suggest the kind of object that produced the pattern. The sizes of the drops are directly related to the size of the point from which they were propelled. Drops propelled from a small or pointed surface will be smaller and the pattern more linear; drops propelled from a large or blunt surface will be larger and the pattern wider. The volume of blood deposited on an object from the source also affects the size and number of drops in the cast-off pattern. The less blood on the object, the smaller the stains produced. The pattern may also suggest whether the blow that caused the pattern was directed from right to left or left to right. The pattern will point in the direction of the backward thrust, which will be opposite the direction of the blow. This could suggest which hand the assailant used to deliver the blows.

Cast-off patterns may also show the minimum number of blows delivered to a victim. Each blow should be marked by an upward-and-downward or forward-and-backward arc pattern (see Figure 11–9). By counting and pairing the patterns, one can estimate the minimum number of blows. An investigator should take into consideration that the first blow would only cause blood to pool to the area; it would not produce a cast-off pattern. Also, some blows may not come into contact with blood and so will not produce a pattern. The medical examiner is in the best position to estimate the number of blows a victim received.

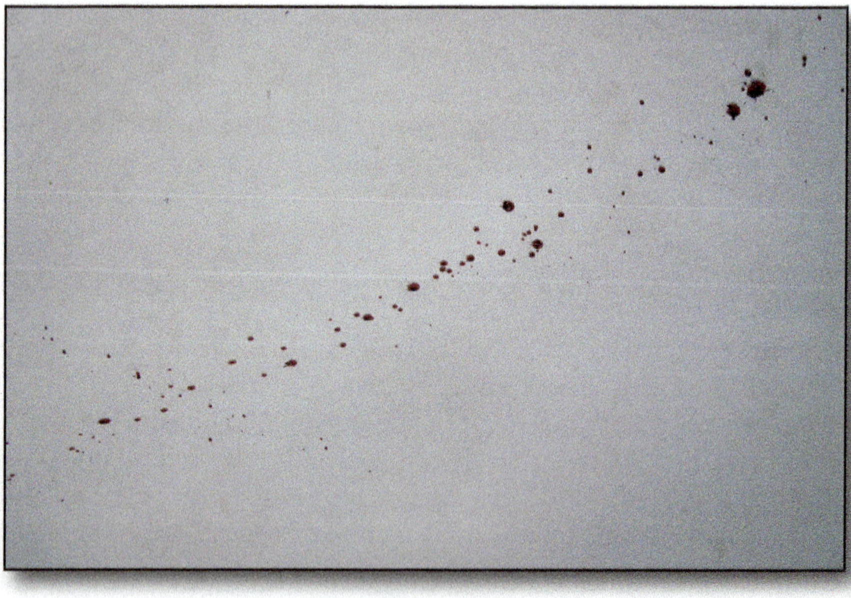

FIGURE 11–9 The cast-off pattern created from one backward and one forward motion of an overhand swing. The larger drops are away from the victim because they're made when the weapon holds the greatest amount of blood. The smaller spatters are directed toward the victim. *A.Y. Wonder*

Case Files

Blood Spatter Evidence

Stephen Scher banged on the door of a cabin in the woods outside Montrose, Pennsylvania. According to Scher, his friend, Marty Dillon, had just shot himself while chasing after a porcupine. The two had been skeet shooting at Scher's cabin, enjoying a friendly sporting weekend, when Dillon spotted a porcupine and took off out of sight. Scher heard a single shot and waited to hear his friend's voice. After a few moments, he chased after Dillon and found him lying on the ground near a tree stump, bleeding from a wound in his chest. Scher administered CPR after locating his dying friend, but he was unable to save Dillon, who later died from his injuries. Police found that Dillon's untied boot had been the cause of his shotgun wound. They determined that he had tripped while running with his loaded gun and shot himself. The griefstricken Scher aroused no suspicion, so the shooting was ruled an accident

Shortly thereafter, Scher moved from the area, divorced his wife, and married Dillon's widow. This was too suspicious to be ignored; police reopened the case and decided to reconstruct the crime scene. The reconstruction provided investigators with several pieces of blood evidence that pointed to Scher as Dillon's murderer.

Police noticed that Scher's boots bore the unmistakable spray of impact blood spatter, evidence that he was standing within an arm's length of Dillon when Dillon was shot. This pattern of bloodstains cannot be created while administering CPR, as Scher claimed had happened. The spatter pattern also clearly refuted Scher's claim that he did not witness the incident. In addition, the tree stump near Dillon's body bore the same type of blood spatter, in a pattern that indicated Dillon was seated on the stump and not running when he was shot. Finally, Dillon's ears were free of the blood spatter that covered his face, but blood was on his hearing protectors found nearby. This is a clear indication that he was wearing his hearing protectors when he was shot, and they were removed before investigators arrived. This and other evidence resulted in Scher's conviction for the murder of his longtime friend Marty Dillon.

FIGURE 11–10 A projected pattern found at a crime scene where a victim suffered injury to an artery. *BR Photo Addicted/Shutterstock*

Projected Pattern

projected pattern
A bloodstain pattern resulting from the ejection of blood under hydraulic pressure, typically from a breach in the circulatory system

Projected pattern is created when a victim suffers an injury to a main artery or the heart. The pressure of the continuing pumping of blood causes blood to spurt out of the injured area (see Figure 11–10). Commonly, the pattern shows large spurted stains for each time the heart pumps. Some radial spikes, satellite spatter, or flow patterns may be evident because of the large volume of blood being expelled with each spurt. Drops may also be seen on the surface in fairly uniform size and shape and in parallel arrangement (see Figure 11–10).

The lineup of the stains shows the victim's movement. Any vertical arcs or waves in the line show fluctuations in blood pressure. The site of the initial injury to the artery can be found where the pattern begins with the biggest spurt. Arterial patterns can also be differentiated because the oxygenated blood spurting from the artery tends to be a brighter red color than blood expelled from impact wounds. This may not be seen at crime scenes with dry blood present.

Expiration Patterns

expiration pattern
A bloodstain pattern resulting from blood forced by airflow out of the nose, mouth, or a wound

A pattern created by blood that is expelled from the mouth or nose from an internal injury is called an **expiration pattern**. If the blood that creates such a pattern is under great pressure, it produces very fine spatter. Expirated blood produces a stain cluster with irregular edges. The presence of bubbles in the drying drops can differentiate a pattern created by expirated blood from other types of bloodstains. Expirated blood also may be lighter in color than impact spatter as a result of being diluted by saliva. The presence of expirated blood gives an important clue to the injuries suffered and the events that took place at a crime scene.

Void Patterns

A **void** is created when an object blocks the deposition of blood spatter onto a surface or object (see Figure 11–11). It can also be created by limiting angles. The spatter is deposited onto the object or person instead. The blank space on the surface or object may give a clue to the size and shape of the missing object or person. Once the object or person is found, the missing piece of the pattern should fit in, much like a puzzle piece, with the rest of the pattern. Voids may help establish the body position of the victim or assailant at the time of the incident.

void
An absence of blood in an otherwise continuous bloodstain or bloodstain pattern

FIGURE 11–11 A void pattern is found behind the door where the surface of the door blocked the deposition of spatter on that area. This void, and the presence of spatter on the door, shows that the door was open when the spatter was deposited. *Courtesy of Jeremiah Morris, Johnson County Sheriff's Office Criminalistics Laboratory*

Quick Review

- Gunshot spatter can consist of both forward spatter from an exit wound and back spatter from an entrance wound; however, only back spatter will be produced if the bullet does not exit the body.
- A cast-off pattern is created when a blood-covered object flings blood in an arc onto a nearby surface. This kind of pattern commonly occurs when a person pulls a bloody fist or weapon back between delivering blows to a victim.
- The characteristic arterial spray spatter is created when a victim suffers an injury to a main artery or the heart and the pressure of the continuing pumping of blood projects blood out of the injured area in spurts, which are apparent in the pattern.
- Expirated blood is expelled from the mouth or nose and may appear as very fine spatter or large bloodstain clusters. This kind of pattern may contain bubbles of oxygen or be mixed with saliva.
- A void pattern features an area free of spatter where an object (or person) blocks the deposition of blood spatter onto a target surface or object. Because the spatter is deposited onto the object or person instead, the shape of the void may give a clue as to the size and shape of the missing object or person.

Other Bloodstain Patterns

Not all bloodstains at a crime scene appear as spatter patterns. The circumstances of the crime often create other types of stains that can be useful to investigators.

Contact/Transfer Stains

transfer stain
A bloodstain resulting from contact between a blood-bearing surface and another surface

When an object with blood on it touches another object that did not have blood on it, this produces a contact or **transfer stain**. Examples of transfers with features include fingerprints (see Figure 11–12), handprints, footprints, footwear prints, tool prints, and fabric prints in blood. These may provide further leads by offering individual characteristics.

The size and general shape of a tool may be seen in a simple transfer. This can lead to narrowing the possible tools by class characteristics. A transfer that shows a very individualistic feature may help point to the tool that made the pattern.

Simple transfer stains are produced when the bloody object makes contact with a surface and the object is removed without any further movement. Other transfers known as *wipe patterns* may be caused by movement of the bloody object across a surface. The pattern may lighten and "feather" as the pattern moves away from the initial contact point (see Figure 11–13). However, because "feathering" is also a function of the amount of pressure being applied to the surface, the analyst must interpret directionality with care. The direction of separate bloody transfers, such as footwear prints in blood, may show the movement of the suspect, victim, or others through the crime scene after the blood was present.

FIGURE 11–12 A transfer stain consisting of bloody fingerprints with apparent ridge detail. *Boris Zhitkov/ Alamy Stock Photo*

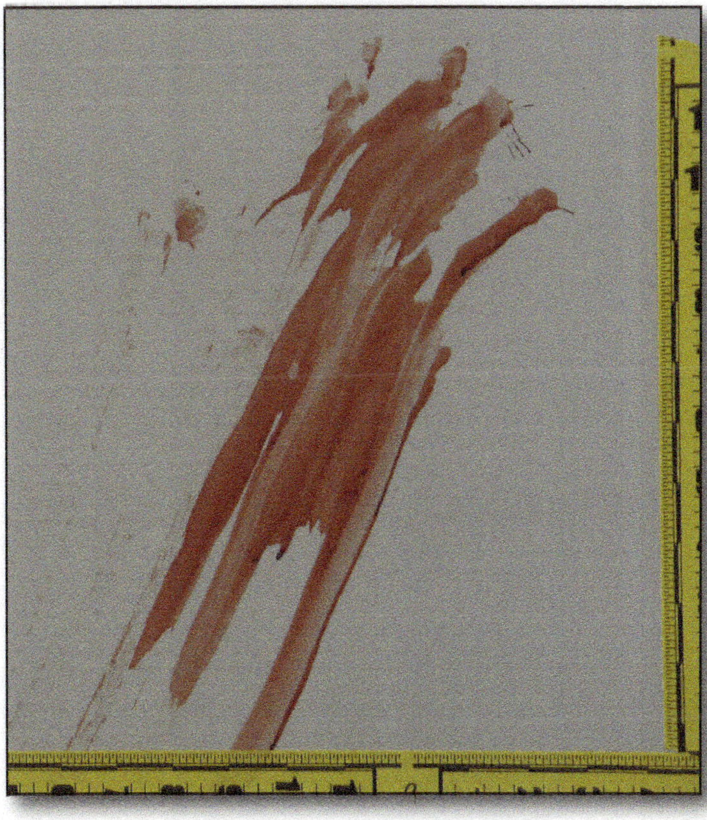

FIGURE 11–13 A wipe pattern. *Courtesy of Jeremiah Morris, Johnson County Sheriff's Office Criminalistics Laboratory*

The first transfer stain in a series will be dark and heavy with blood, whereas subsequent transfers will be increasingly lighter in color. The transfers get lighter as less and less blood is deposited from the transferring object's surface. Bloody shoe imprints may also suggest whether the wearer was running or walking. Running typically produces imprints with more space between them.

Flows

flow
A bloodstain resulting from the movement of a volume of blood on a surface due to gravity or movement of the target

Patterns made by drops or large amounts of blood flowing by the pull of gravity are called **flow**. Flows may be formed by single drops or large volumes of blood coming from an actively bleeding wound or blood deposited on a surface from an arterial spurt. Clotting of the blood's solid parts may occur when a flow extends onto an absorbent surface.

The flow direction may show movements of objects or bodies while the flow was still in progress or after the blood had dried. Figure 11–14 illustrates a situation in which movement of the surface while the flow was still in progress led to a specific pattern.

Interruption of a flow pattern may be helpful in assessing the sequence and passage of time between the flow and its interruption. If a flow found on an object or body does not appear consistent with the direction of gravity, one may surmise that the object or body was moved after the blood had dried.

Pools

A pool of blood occurs when blood collects in a level (not sloped) and undisturbed place. Blood that pools on an absorbent surface may be absorbed throughout the surface and diffuse, creating a pattern larger than the original pool. This often occurs with pools on beds or sofas.

FIGURE 11–14
The image below shows a flow pattern. *Courtesy of Jeremiah Morris, Johnson County Sheriff's Office Criminalistics Laboratory*

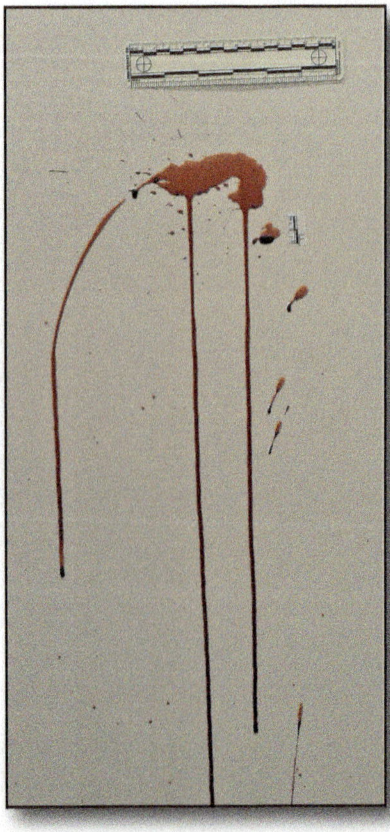

The approximate drying time of a pool of blood is related to the environmental condition of the scene. By experimentation, an analyst may be able to estimate the drying times of stains of different sizes. Small and large pools of blood can aid in reconstruction by providing an estimate of the time that elapsed since the blood was deposited. Considering the drying time of a blood pool can yield information about the timing of events that accompanied the incident.

The edges of a stain will dry to the surface, producing a **perimeter stain** (see Figure 11–15). This usually occurs within 50 seconds of deposition of drops, and longer for larger volumes of blood. If the central area of the pooled bloodstain is altered by wiping, the skeletonized perimeter will be left intact. This can be used to interpret whether movement or activity occurred shortly after the pool was deposited, or whether the perimeter had time to skeletonize before the movement occurred. This may be important for classifying the source of the original stain.

perimeter stain
An altered stain consisting of its edge characteristics, the central area having been partially or entirely removed

Drip Trail Patterns

A **drip trail pattern** is a series of drops that are separate from other patterns, formed by blood dripping off an object or injury. The stains form a kind of line or path usually made by the suspect after injuring or killing the victim, or they can show the movement of a wounded victim. It may simply show movement, lead to a discarded weapon, or provide identification of the suspect by the suspect's own blood. Investigators often see this type of pattern in stabbings during which the suspect cuts themself as a result of the force necessary to stab the victim. Figure 11–16 shows a drip trail pattern away from the center of action at a crime scene.

drip trail pattern
A bloodstain pattern resulting from the movement of a source of drip stains between two points

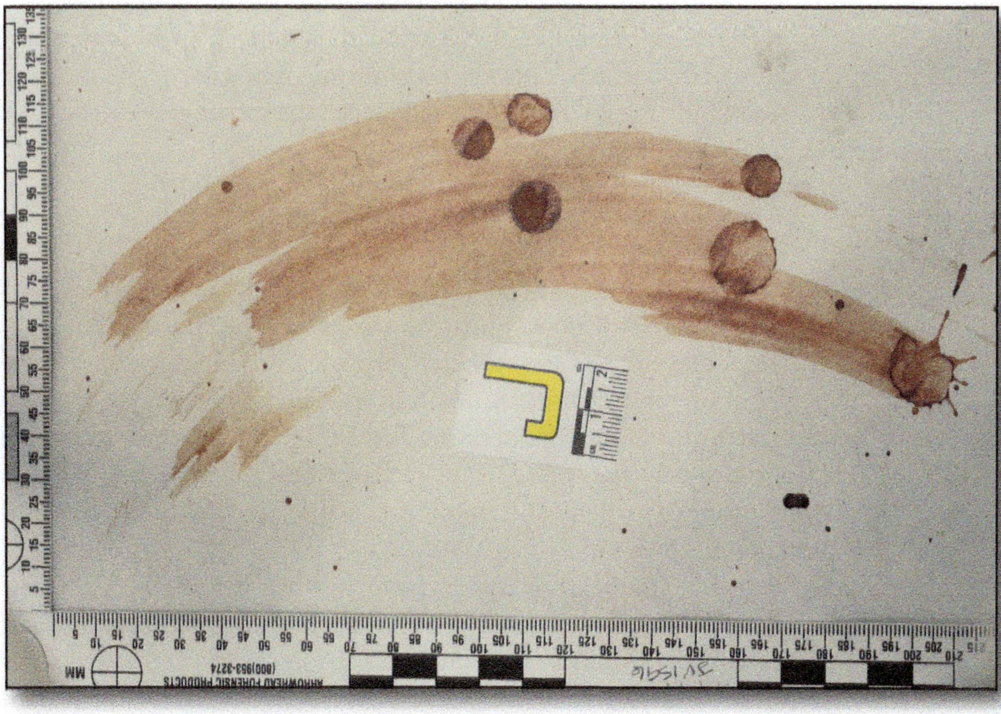

FIGURE 11–15
A perimeter stain is shown in a bloodstain that was disturbed after the edges had time to dry.

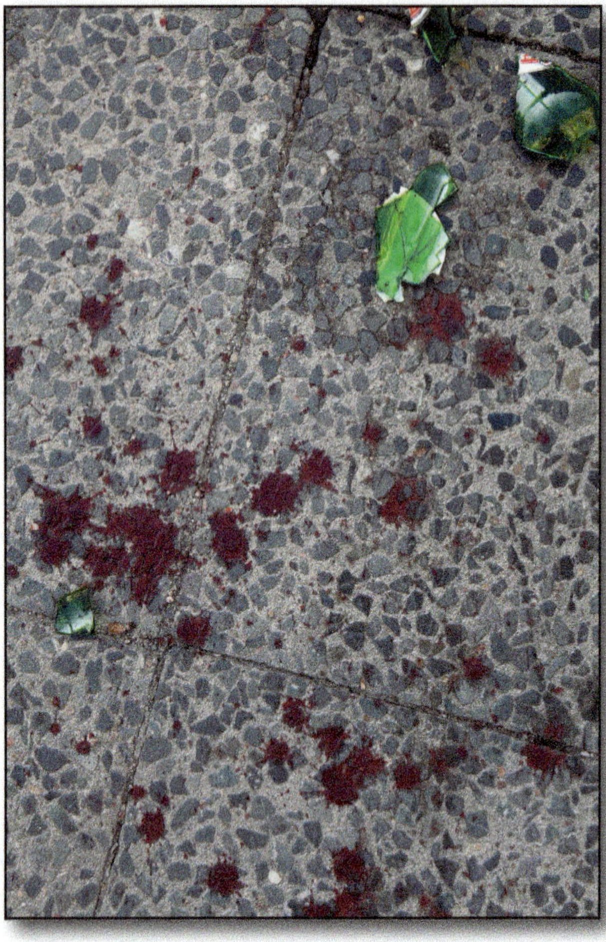

FIGURE 11–16
This is an example of a drip trail pattern from a crime scene.
Courtesy Marc Volk/ Getty Images

The shape of the stains in a drip trail pattern can help investigators determine the direction and speed at which a person was moving. The tails of the drops in a drip trail pattern point in the direction the person was moving. More circular stains are found where the person was moving slowly enough to not form tails. This information may be helpful in reconstruction.

Documenting Bloodstain Pattern Evidence

Blood spatter patterns of any kind can provide a great deal of information about the events that took place at a crime scene. For this reason, investigators should note, study, and photograph each pattern and drop. This must be done to accurately record the location of specific patterns and to distinguish the stains from which laboratory samples were taken. The photographs and sketches can also point out specific stains used in determining the direction of force, angle of impact, and area of origin.

Just as in general crime-scene photography, the investigator should create photographs and sketches of the overall pattern to show the orientation of the pattern to the scene. The medium-range documentation should include pictures and sketches of the whole pattern and the relationships between individual stains within the pattern. The close-up photographs and sketches should show the dimensions of each individual stain. Close-up photographs should be taken with a scale of some kind apparent in the photograph.

Bloodstain Reconstruction

An elderly male was found lying dead on his living room floor. He had been beaten about the face and head, then stabbed in the chest and robbed. The reconstruction of bloodstains found on the interior front door and the adjacent wall documented that the victim was beaten about the face with a fist and struck on the back of the head with his cane. A three-dimensional diagram and photograph illustrating the evidential bloodstain patterns is shown in Figures 1(a) and (b).

(a)

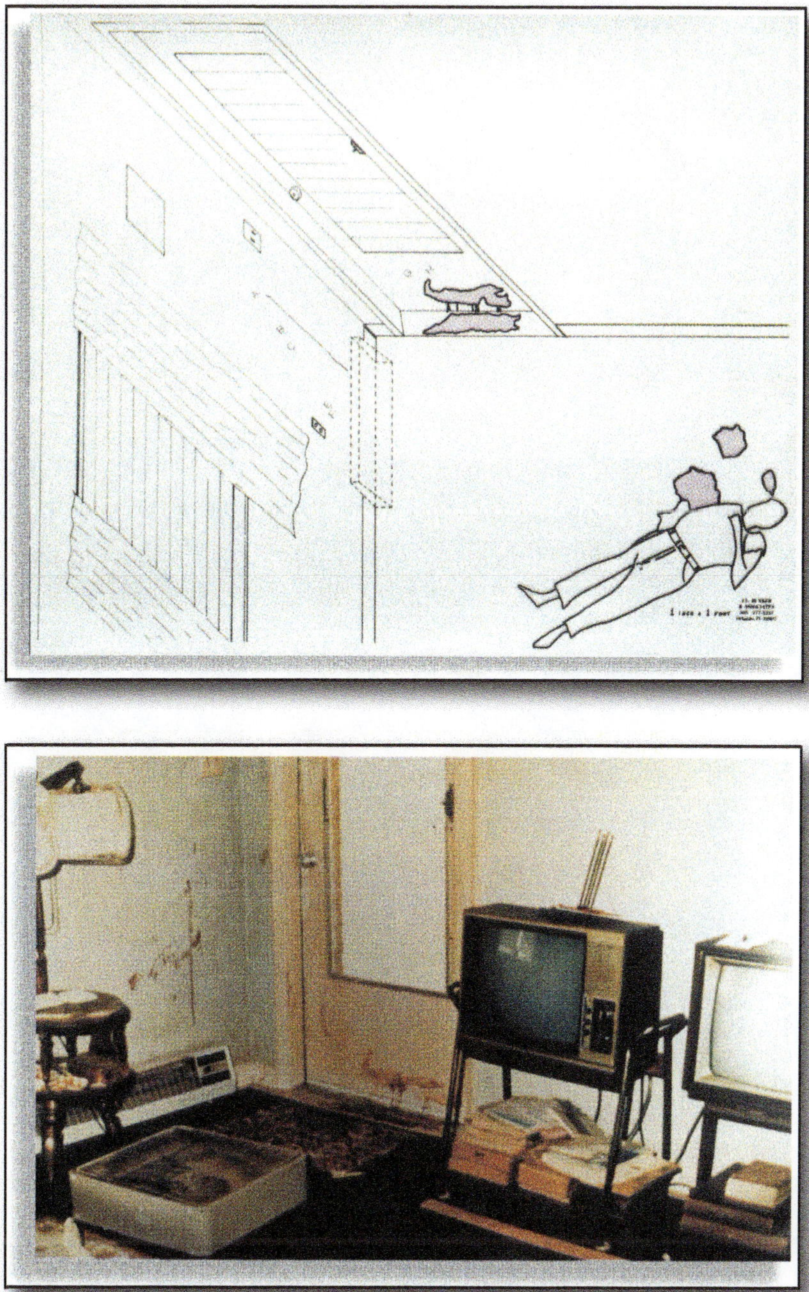

(b)

FIGURE 1 (a) A three-dimensional diagram illustrating bloodstain patterns that were located, documented, and reconstructed. (b) A crime-scene photograph of bloodstained areas. *Institute of Applied Forensic Technology*

(continued)

(continued)

A detail photograph of bloodstains next to the interior door is shown in Figure 2. Arrow 1 in Figure 2 points to the cast-off pattern directed left to right as blood was flung from the perpetrator's fist while inflicting blows. Arrow 2 in Figure 2 points to three transfer impression patterns directed left to right as the perpetrator's bloodstained hand contacted the wall as the fist blows were being inflicted on the victim. Arrow 3 in Figure 2 points to blood flow from the victim's wounds as he slumped against the wall.

Figure 3 contains a series of laboratory test patterns created to evaluate the patterns contained within Figure 2.

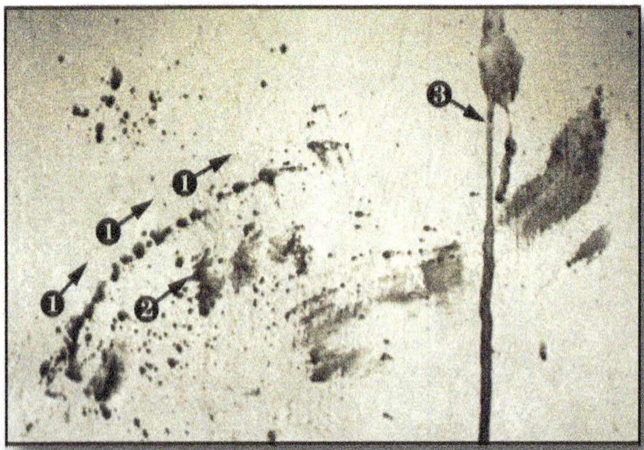

FIGURE 2 **Positions of impact spatter from blows that were inflicted on the victim's face.** *Institute of Applied Forensic Technology*

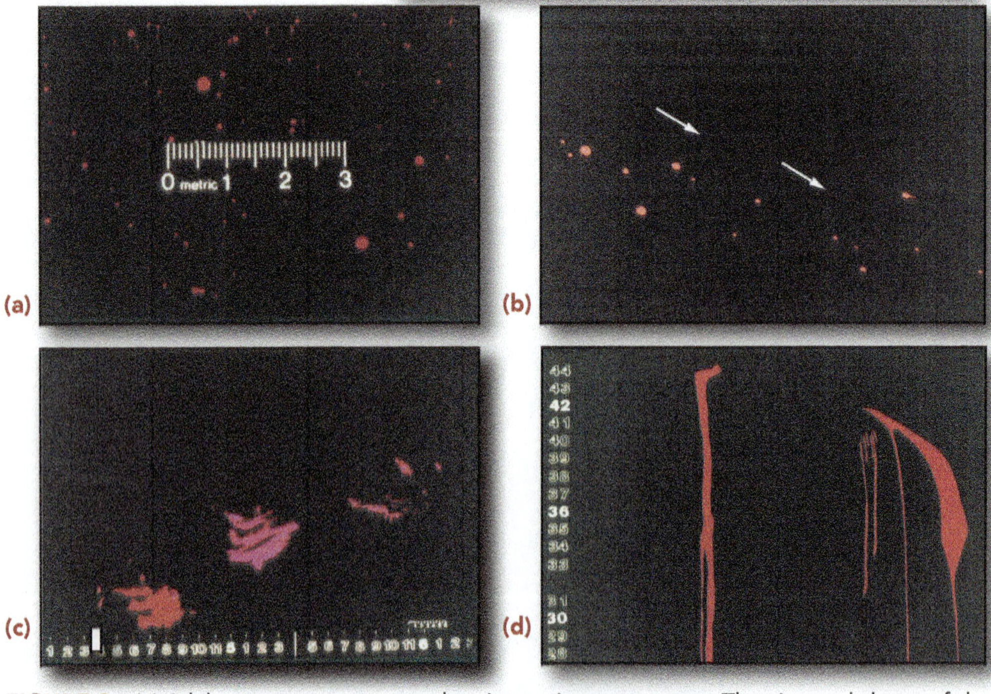

FIGURE 3 **(a) A laboratory test pattern showing an impact spatter. The size and shape of the stains demonstrate a forceful impact 90 degrees to the target. (b) A laboratory test pattern illustrating a cast-off pattern directed left to right from an overhead swing. (c) A laboratory test pattern showing a repetitive transfer impression pattern produced by a bloodstained hand moving left to right across the target. (d) A laboratory test pattern illustrating vertical flow patterns. The left pattern represents a stationary source; the right pattern was produced by left-to-right motion.** *Institute of Applied Forensic Technology*

(continued)

Figure 4 shows how the origin of individual impact spatter patterns located on the wall and door and emanating from the bleeding victim can be documented by the determination of separate areas of convergence.

A suspect was apprehended three days later, and he was found to have an acute fracture of the right hand. When he was confronted with the bloodstain evidence, the suspect admitted striking the victim, first with his fist, then with a cane, and finally stabbing him with a kitchen knife.

The suspect pleaded guilty to three first-degree felonies.

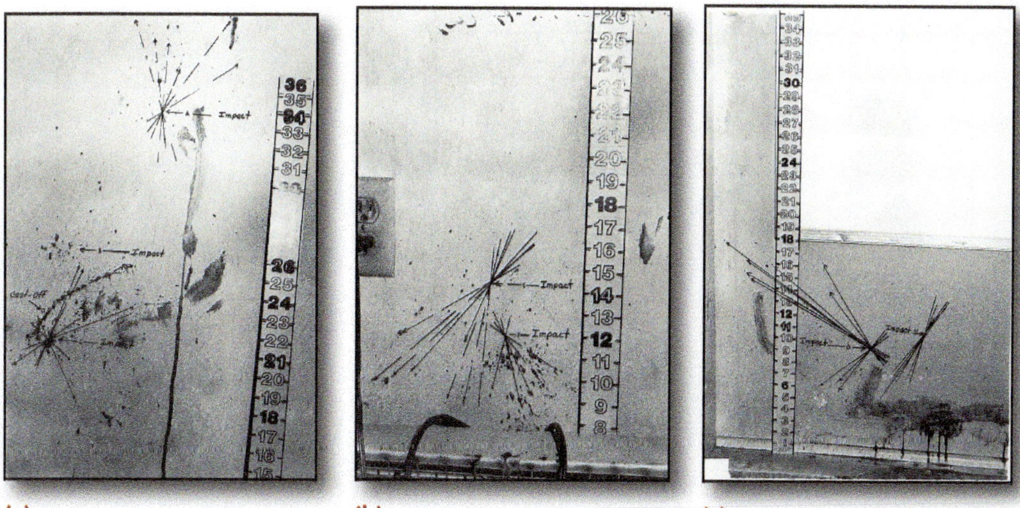

(a) (b) (c)

FIGURE 4 (a) A convergence of impact spatter patterns associated with beating with a fist. (b) The convergence of impact spatter associated with the victim falling to the floor while bleeding from the nose. (c) The convergence of impact spatter associated with the victim while face down at the door, being struck with a cane. *Institute of Applied Forensic Technology*

Two common methods of documenting bloodstain patterns place attention on the scale of the patterns. The *grid method* involves setting up a grid of squares of known dimensions over the entire pattern using string and stakes (see Figure 11–17). All overall, medium-range, and close-up photographs are taken with and without the grid. The second method, called the *perimeter ruler method*, involves setting up a rectangular border of rulers around the pattern and then placing a small ruler next to each stain. In this method, the large rulers show scale in the overall and medium-range photos, whereas the small rulers show scale in the close-up photographs (see Figure 11–18). Some investigation teams use tags in close-up photographs to show evidence numbers or other details.

An area-of-origin determination may be calculated at the discretion of the bloodstain analyst when the circumstances of the case warrant such a determination. All measurements of stains and calculations of angle of impact and point of origin should be recorded in crime-scene notes. Especially important stains can be roughly sketched within the notes.

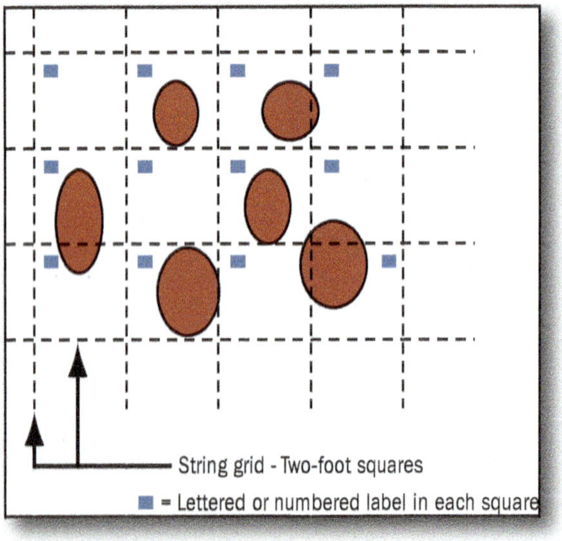

FIGURE 11–17
The grid method may be used for photographing bloodstain pattern evidence. *Courtesy R.R. Ogle, Jr., Crime Scene Investigation & Reconstruction, 2nd ed., Prentice Hall, Upper Saddle River, NJ, 2007*

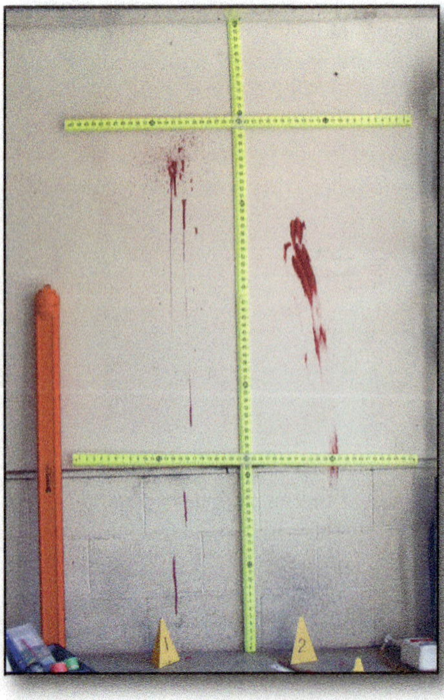

FIGURE 11–18
The perimeter ruler method may be used for photographing bloodstain pattern evidence. *Courtesy Evident, Union Hall, Va 24176-4025, www. evident crimescene.com*

Only some jurisdictions have a specialist on staff to decipher patterns either at the scene or from photographs at the lab. Therefore, it is important that all personnel be familiar with patterns to properly record and document them for use in reconstruction.

Bloodstain Pattern Analysis: Proceed with Caution

The field of bloodstain pattern analysis has proliferated within the forensic science community, as well as with practitioners who pride themselves as being called crime scene investigators. Within the law enforcement community, there are hundreds of investigators (not necessarily scientists) who are proud to be labeled blood spatter specialists. At this point in time, there is no measurement to judge the uniformity of training in this field within the United States. The closest training criteria stems from the Bloodstain Pattern Analysis Cortication program offered under the auspices of the International Association of Bloodstain Pattern Analysis. The certificate requires a minimum of 40 hours of education in an approved workshop. A minimum of three years of practice within the discipline of bloodstain pattern identification augments the classroom training. Under the best of circumstances, following the successful completion of the certificate requirements, many law enforcement agencies find it appropriate to insert these individuals into the crime-scene investigation process.

The complexity of blood spatter interpretation runs the gamut from the simple to the complex. What should be worrisome to the forensic science community and the judicial system is the apparent absence of uniformity in imposing quality assurance standards on interpreting bloodstain patterns. The forensic science community is already reeling from adverse and embarrassing publicity arising from the misidentification of fingerprints, bullet lead, bite mark impressions, and hair evidence. Warnings are already being clearly and loudly sounded about the error rates associated with bloodstain pattern analysis. In a series of studies sponsored by the National Institute of Justice (NIJ), very high error rates for blunt instrument spatter patterns in the range of 37 percent have been reported:[1] Where a bloodstain pattern classification was required, 13.1 percent of these classifications did not include the correct pattern for rigid surfaces and 23.4 percent for fabric surfaces.[2] Another NIJ study has demonstrated that fabrics may interact to, distort, and alter a bloodstain pattern in many different and complex ways compared to bloodstains on hard surfaces.[3] Given the highly subjective nature of bloodstain pattern interpretation and its potential significant error rate, a process of independent assessment of any bloodstain pattern interpretation by two or more examiners at the crime scene and in the laboratory is desirable.[4,5] The "Analysis, Comparison and Evaluation-Verification" (ACE-V) process utilized by fingerprint examiners (p. 136) may serve as a useful model for general advisory guidance in bloodstain pattern evidence verification.

Quick Review

- Transfer patterns are created when an object with blood on it makes simple contact with a surface or moves along a surface. The direction of movement may be shown by feathering of the pattern.
- Flows may originate from a single drop or a large amount of blood. The direction of the flow is caused by gravity, so the direction of a pattern may suggest the original position of the surface when the flow was formed.
- A pool is formed where large amounts of blood collect. The pool may be absorbed into the surface of deposition over time.
- The presence of skeletonization of the perimeter of a bloodstain suggests that the stain was disturbed after the edges had sufficient time to skeletonize.
- A drip trail pattern is separate from other patterns, and it is formed by single blood drops dripping off an object or injury.
- Photographs and sketches should first be created of the overall pattern to show orientation of the pattern to the scene.
- Medium-range and close-up photographs may use the grid method or perimeter ruler method to show the orientation and relative size of the pattern and individual stains.

Chapter Review

- Crime-scene reconstruction relies on the combined efforts of medical examiners, criminalists, and law enforcement personnel to recover physical evidence and to sort out the events surrounding the occurrence of a crime.
- Examples of crime-scene reconstruction include determining whether a body was moved after death, determining whether a victim was clothed after death, analyzing bullet trajectory, analyzing blood spatter, determining the direction from which projectiles penetrated glass objects, estimating the distance of a shooter from a target, and locating primer residue on suspects.
- Individual bloodstains can convey to the bloodstain analyst the directionality and angle of impact of the blood when it impacted a surface. Bloodstain patterns may convey to the analyst the location of victims or suspects, the movement of bleeding individuals, and the number of blows delivered.
- Surface texture is of paramount importance in the interpretation of bloodstain patterns; rounder drops generally are produced from smooth, nonporous surfaces, whereas rough surfaces create irregular-edged drops. However, correlations between standards and unknowns are valid only if identical surfaces are used.
- The direction of travel of blood striking an object may be discerned by the stain's shape. The pointed end of a bloodstain always faces in its direction of travel.
- The angle of impact of an individual bloodstain can be approximated by the degree of distortion or lengthening of the bloodstain, or it can be more effectively estimated using the ratio of width/length of the stain.
- An impact spatter pattern occurs when an object impacts a source of blood. This produces forward spatter projected forward from the source and back spatter projected backward from the source.
- The area of convergence is the point on a two-dimensional plane from which the drops of an impact spatter pattern originated. This area can be estimated by drawing straight lines through the long axis of several individual bloodstains, following the line of their tails.
- The area of origin of a bloodstain pattern is the area in three-dimensional space where blood was projected from, showing the position of the victim or suspect when the stain-producing event took place. The string method is commonly used at a crime scene to approximate the position of the area of origin.

- Gunshot spatter can consist of both forward spatter from an exit wound and back spatter from an entrance wound; however, only back spatter will be produced if the bullet does not exit the body.
- A cast-off pattern is created when a blood-covered object flings blood in an arc onto a nearby surface. This kind of pattern commonly occurs when a person pulls a bloody fist or weapon back between delivering blows to a victim.
- The characteristic arterial spray spatter is created when a victim suffers an injury to a main artery or the heart and the pressure of the continuing pumping of blood projects blood out of the injured area in spurts, which are apparent in the pattern.
- Expirated blood is expelled from the mouth or nose and may appear as very fine spatter or large bloodstain clusters. This kind of pattern may contain bubbles of oxygen or be mixed with saliva.
- A void pattern features an area free of spatter where an object (or person) blocks the deposition of blood spatter onto a target surface or object. Because the spatter is deposited onto the object or person instead, the shape of the void may give a clue as to the size and shape of the missing object or person.
- Transfer patterns are created when an object with blood on it makes simple contact with a surface or moves along a surface. The direction of movement may be shown by feathering of the pattern.
- Flows may originate from a single drop or a large amount of blood. The direction of the flow is caused by gravity, so the direction of a pattern may suggest the original position of the surface when the flow was formed.
- A pool is formed where large amounts of blood collect. The pool may be absorbed into the surface of deposition over time.
- The presence of skeletonization of the perimeter of a bloodstain suggests that the stain was disturbed after the edges had sufficient time to skeletonize.
- A drip trail pattern is separate from other patterns, and it is formed by single blood drops dripping off an object or injury.
- Photographs and sketches should first be created of the overall pattern to show orientation of the pattern to the scene.
- Medium-range and close-up photographs may use the grid method or perimeter ruler method to show the orientation and relative size of the pattern and individual stains.

Quick Lab: Blood Drop Analysis

Materials:

Simulated blood*
Dropper
Construction paper
Metric ruler or caliper
Scientific calculator

Procedure:

Forensic scientists analyze scientific explanations about drops of blood at a crime scene using observational testing and critical thinking to determine information to help reconstruct what occurred. One piece of data we can gather is the angle of impact for a blood drop. For this activity, start by creating some blood drops on the construction paper. Use the dropper to place drops of blood on the paper and be sure to squirt the blood at different angles onto the paper. Remember, you want drops, not a puddle. Once you have created about ten drops at different angles, let the blood dry. Once dry, use the ruler to measure the width and length of each drop in millimeters. Record your data. To determine the angle at which each blood drop impacted, use the calculator to solve the equation width/length = sin of the angle of impact. The sin of the angle of impact can be converted to the angle of impact by hitting the arcsin button ($\sin^{-1}$) on the calculator. This is the angle at which the blood drop fell onto the surface.

Follow-Up Questions:

1. **How would knowing the angle of impact of a blood drop be helpful in an investigation?**
2. **Did any of your drops fall at a 90-degree angle? If so, what shape were they?**
3. **Create a graph of showing angle of impact vs. blood drop length using the data from your ten blood drops. Explain whether there is a correlation between these two pieces of data.**

*Available from Forensics Source, www.forensicssource.com

Quick Lab: Blood Spatter Analysis

Materials:

Sheet of blank paper
Water-soluble red paint
Paintbrush and pointed stirring stick
Various surfaces (smooth, rough, porous, nonporous)

Procedure:

Cast-off blood spatter is often of great help to investigators in reconstructing events at the scene of a violent crime. In particular, the shape and appearance of cast-off bloodstains can indicate the relative positions of the victim and assailant during the crime and the type of weapon used. For this exercise, students should be matched up in teams of two. Each member of the pair, working separately, will make a series of spatters on various textured and non-textured surfaces with the red paint. Using the paintbrush, make three stains by flicking a small amount of red paint onto the paper. Hold the brush at differing angles to the paper when making each stain. One should be made at a shallow angle, the second at a steeper angle, and the third at an even steeper angle. Now make similar marks using the stirring stick instead of the paintbrush. Let the paint dry, then mark the stains A, B, C, D, E, and F. Be sure to record which instrument made each mark and the angle (shallow, steep, very steep) at which it was made. (It is best to alternate making marks with one instrument, then the other, rather than making all three marks with the paintbrush, then all with the stirring stick.) Now pair up with your partner and exchange sheets. Answer the following questions.

Follow-Up Questions:

1. **Which marks were made by the paintbrush and which were made by the stick? How do you know this?**
2. **For each instrument, list the stains from shallowest to steepest. How did you determine the relative angles of each stain?**
3. **What would the presence of shallow-angle bloodstains suggest about the relative position of victim and assailant? What would a steeper angle suggest?**
4. **How did the surface type affect the appearance of the blood stains?**
5. **How did you analyze the blood stain pattern to determine the mechanism by which the patterns were created?**

Review Questions

1. Violent contact between individuals at a crime scene frequently produces bleeding and results in the formation of

a. footwear impressions.

b. bloodstain patterns.

c. blood typing.

d. rigor mortis.

2. A drop of blood that strikes a surface at an angle of impact of approximately 90° will be close to ________ in shape.

a. elongated

b. elliptical

c. teardropped

d. circular

3. The ________ method is used at the crime scene to determine the area of origin.

a. perimeter ruler

b. grid

c. string

d. cast-off

4. The edges of a bloodstain will generally skeletonize within ________ of deposition and be left intact even if the central area of a bloodstain is altered by a wiping motion.

a. 50 seconds

b. 10 seconds

c. 50 minutes

d. 3 hours

5. True or False: Harder and less porous surfaces result in less spatter, whereas rough surfaces produce stains with more spatter and serrated edges.

6. True or False: Movement of a bloody object across a surface causes the pattern to darken as the object moves away from point of contact.

7. True or False: Footwear transfer patterns created by an individual who was running typically show imprints with more space between them as compared to those of an individual who was walking.

8. True or False: The direction of a flow pattern may show movements of objects or bodies while the flow was still in progress or after the blood has dried.

9. True or False: Each bloodstain pattern found at a crime scene does not have to be noted, studied, and photographed.

10. What is crime-scene reconstruction?

11. In which of the following ways can physical evidence aid in crime-scene reconstruction? (You may choose more than one response.)
    a. by supporting or contradicting statements given by witnesses and/or suspects
    b. by describing exactly what happened at the crime scene
    c. by confirming the accuracy of a crime-scene reconstruction to a jury
    d. by distinguishing between individual and class characteristics
    e. by generating leads that may help investigators solve the case

12. Give at least three examples of a specialized task a criminalist might perform in crime-scene reconstruction.

13. Which of the following is of paramount importance in the interpretation of bloodstain patterns?
    a. the direction of impact
    b. the surface texture
    c. the angle of impact
    d. the amount of blood

14. How can an investigator tell the direction of travel of blood from the shape of a bloodstain?

15. What is the difference between the shape of a bloodstain that impacts a surface at a low angle and one that impacts at a higher angle?

16. What is the difference between forward spatter and back spatter? Which is more likely to be deposited on the object or person creating the impact?

17. Define the terms *area of convergence* and *area of origin* and explain what each reveals to an investigator.

18. Name three factors that affect the amount of backward spatter produced by a gunshot wound.

19. What is the drawback effect? How is it helpful to an investigator?

20. What determines the size of blood drops in a cast-off pattern? Explain.

21. What is expirated blood? Name two ways to distinguish expirated blood from other types of bloodstains.

22. What is a void pattern? How might a void pattern be useful to investigators?

23. What is a transfer pattern? How is a simple transfer pattern created?

24. How does the first transfer pattern in a series differ from subsequent ones?

25. What is a flow pattern? What should one surmise if a flow found on an object or body does not appear consistent with the direction of gravity?

26. How can pools of blood aid in reconstructing a crime scene?

27. Explain how the shape of stains in a drip trail pattern can help investigators determine the direction and speed at which a person was moving.

28. Name and describe two methods for documenting bloodstain patterns.

Learn the Vocabulary

Use the information from this chapter along with your prior knowledge and experience to learn the key terms. Choose from the following activities:

1. Make flashcards with a term on the front and its definition on the back. Use the flashcards to quiz a partner on correctly pronouncing and defining the terms.
2. Make an audio or video recording of each term and its definition.
3. Work with a partner to write a sentence for each term. Read the sentences out loud to a partner or share them with the class.
4. Write a list of new terms and language you have encountered in this chapter. Reuse the terms to write sentences and then read the sentences out loud to a partner. Reuse the terms again by writing a short speech. The speech might explain the terms or teach a lesson using the terms. Give the speech to a partner or to the class.

Application and Critical Thinking

1. After looking at the bloodstains in the figure, answer the following questions:
   a. Which three drops struck the surface closest to a 90-degree angle? Explain your answer.
   b. Which three drops struck the surface farthest from a 90-degree angle? Explain your answer.
   c. In what direction were drops 2 and 7 traveling when they struck the surface? Explain your answer

2. Investigator Priscilla Wright arrives at a murder scene and finds the body of a victim who suffered a gunshot wound but sees no blood spatter on the wall or floor behind it. What should she conclude from this observation?

3. Investigator Terry Martin arrives at an assault scene and finds a cast-off pattern consisting of tiny drops of blood in a very linear arc pattern on a wall near the victim. What does this tell him about the weapon used in the crime?

Laboratory Experiments

This activity requires the use of the following practices of science and engineering:

- Analyzing and interpreting data
- Constructing explanations
- Engaging in argument from evidence
- Using mathematics and computational thinking
- Obtaining, evaluating, and communicating information
- Using appropriate safety equipment and practices as outlined in Texas Education Agency-approved safety standards

This activity consists of the following crosscutting concepts:

- **Patterns**—The observed patterns in blood type and in blood spatter analysis allows for organization and classification.
- **Cause and effect**—The crime scene can be reconstructed based on the blood spatter that resulted from the attack.

Before you begin, read the scenario and steps out loud with a partner. Ask for clarification as necessary to make sure you understand the experiment.

The Scenario

Police responded to a welfare check at 325 Talavera Way in Marana, Arizona. The neighbors reported to the police because the resident, Troy Drachman, hadn't been seen for several days, which they found unusual since they typically saw him daily walking his dog. Having knocked on the door with no response, the police looked into the window to try and see the resident. They discovered what appeared to be a bloody crime scene. The police forcibly entered the residence, where they found evidence of blood spatter on the living room walls and indication that a body had been dragged from the living room to the back door, where Mr. Drachman was found.

As a blood pattern expert, you have been called in to analyze the blood spatter evidence and write a report to summarize your findings. You will analyze the blood stain characteristics based on surface type, pattern type, identification of directionality, and mechanism, which includes area of convergence and angle of impact. There are 5 areas of interest: three areas that contain blood spatter of similar size (1–3), a series of small drops found near the door (4), and several drops of blood on the ceiling (5).

THE EVIDENCE

After running blood type analysis, you find that most of the blood at the crime is A+, the victim's blood type. One other blood type was found at the crime scene. Use the information below to deduce the blood type of the other blood found at the crime scene and the blood type of each of the 5 possible suspects.

Suspect	*Anti A*	*Anti B*	*Anti Rh*	*Blood Type*
1	–	+	+	
2	+	+	+	
3	–	–	+	
4	+	–	–	
5	+	+	+	
Crime Scene	+	+	+	

CRIME SCENE DIAGRAM: LIVING ROOM, DRACHMAN RESIDENCE

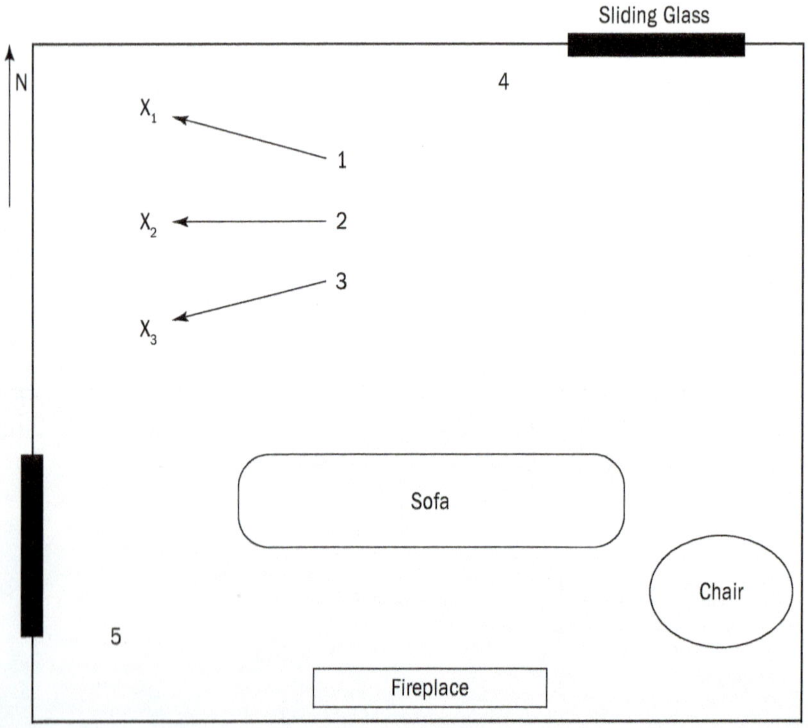

Blood drops 1–3 in the crime scene are impact blood droplets. The area of convergence for blood drops 1–3 are areas X_1, X_2, and X_3, respectively. Below are the actual blood drops. You will need to determine the angle at which the blood fell.

Area	*Representative blood drop*	*Angle from which blood fell*
X_1		
X_2		
X_3	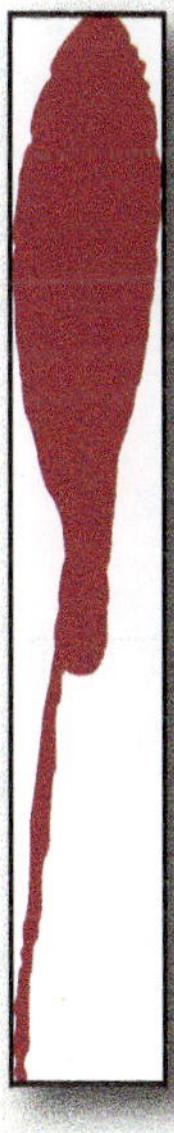	

Blood stain 4 and 5 in the crime scene diagram are given below. Figure 4 represents the blood stain found on the sliding glass door. Figure 5 represents the blood stain on the ceiling near the fireplace and the sofa. Explain what the blood stains are and how they are formed.

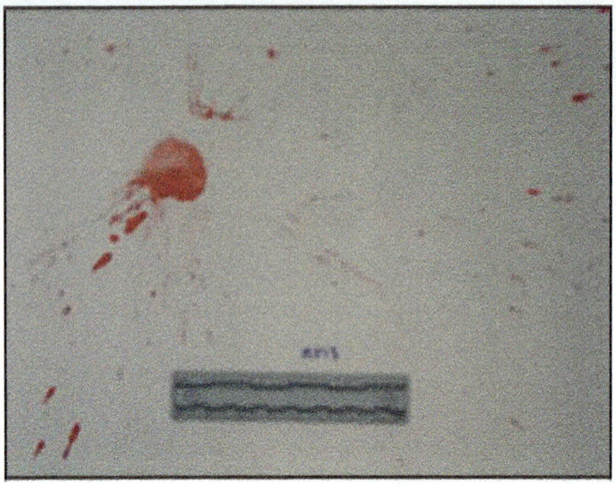

Courtesy Richard Saferstein, Ph.D.

BLOOD STAIN 4

Courtesy Richard Saferstein, Ph.D.

BLOOD STAIN 5

Endnotes

1. William Ristenpart et al., Quantitative Analysis of High Velocity Bloodstain Patters (2013), available at http://www.ncjrs.gov/pdffiles1/nij/grants/241744.pdf
2. Terry Laber et al., National Institute of Justice, Final Report, Reliability Assessment of Current Methods in Bloodstain Pattern Analysis (2014), available at http://www.ncjrs.gov/pdffiles1/nij/grants/247180.pdf
3. Stephen Michielsen et al., Bloodstain Patterns on Textile Surfaces: A Fundamental Analysis (2015), available at http://www.ncjrs.gov/pdffiles1/nij/grants/248671.pdf
4. The Forensic Science Regulator, Codes of Practice and Conduct: Bloodstain Pattern Analysis, FSR-C-102. Birmingham: Office (2015), available at https://www.gov.uk/government/uploads/system/uploads/attachment_data/file/484905/C102_Bloodstain_Pattern_Analysis_2015.pdf.
5. Rachel Zajac et al., "Contextual Bias: What Bloodstain Pattern Analysts Need to Know," *Journal of Bloodstain Pattern Analysis* 31 (2015): 7.

Trace Evidence I: Hairs and Fibers

12

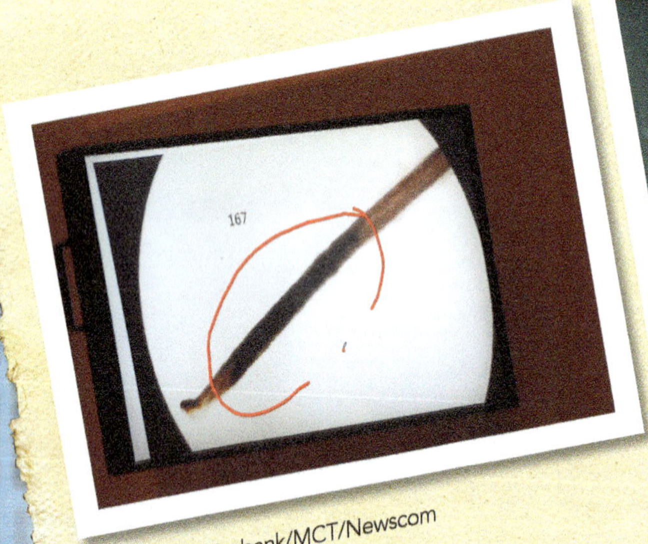

Courtesy Joe Burbank/MCT/Newscom

Courtesy Joe Burbank/Reuters

Key Terms

- anagen phase
- catagen phase
- cortex
- cuticle
- follicular tag
- macromolecule
- manufactured fibers
- medulla
- mitochondrial DNA
- molecule
- monomer
- natural fibers
- nuclear DNA
- polymer
- telogen phase

Casey Anthony: A Single Hair

On July 15, 2008, Caylee Anthony was reported missing by her maternal grandmother, Cindy Anthony, who said she had not seen Caylee for 31 days. After receiving varied explanations as to Caylee's whereabouts, Cindy began searching for the child herself. After retrieving her daughter Casey's car from the impound, she reported to police that it smelled like a dead body had been inside it. Once police were involved, it became clear that Casey Anthony was not being honest about what had happened to the young girl. After a lie about the child being taken by a nanny was quickly unraveled by police, Casey was arrested and charged with giving false statements to law enforcement, child neglect, and obstruction of a criminal investigation.

Caylee Anthony's body was recovered from a wooded area not far from the Anthony family home on December 11, 2008. The investigation into the disappearance of Caylee Anthony quickly turned into a murder investigation. Police examined Casey's car to ascertain whether or not the child had been placed in the trunk before her body had been dumped in the wooded area. They recovered one telling piece of evidence, a single hair was microscopically similar to that of Casey's daughter, Caylee. The hair was microscopically distinguishable from Casey's head hair. Mitochondrial DNA sequence analysis revealed similarity between the trunk hair, Caylee's hair, and Casey's hair, as one would expect from maternally inherited mitochondrial DNA.

An FBI analyst testified that the hair from the truck exhibited *root banding*, a phenomenon consistent with hair from a deceased individual's head. There was only one person who would have the Anthony mitochondrial DNA sequence who was recently deceased and suspected of being present in the trunk: Caylee. This hair was strong evidence that the child was in the trunk of her mother's car for a time before she was left in the wooded area where she was discovered five months later. It was proof that Casey Anthony knew more about the circumstances that led to Caylee's death than she previously led them to believe.

Learning Objectives

After studying this chapter, you should be able to:

12.1 Discuss the morphology of hair, including the three phases of hair growth

12.2 Discuss the considerations and questions concerning forensic hair examinations

12.3 Describe the proper collection and preservation of forensic hair evidence

12.4 Identify the different types of fibers

12.5 List the properties of fibers that are most useful for forensic comparisons

12.6 Describe the proper collection and preservation of forensic fiber evidence

National Science Content Standards

Scientific Inquiry

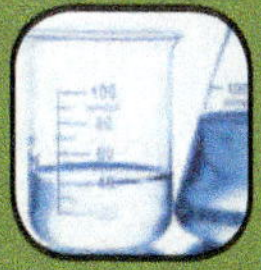

Physical Science

The trace evidence transferred between individuals and objects during the commission of a crime, if recovered, often corroborates other evidence developed during the course of an investigation. Although in most cases physical evidence cannot by itself positively identify a suspect, laboratory examination may narrow the origin of such evidence to a group that includes the suspect. Using many of the instruments and techniques we have already examined, the crime laboratory has developed a variety of procedures for comparing and tracing the origins of physical evidence. This chapter and those that follow discuss how to apply these techniques to the analysis of the types of physical evidence most often encountered at crime scenes. We begin with a discussion of hairs and fibers.

Forensic Examination of Hair

Hair is encountered as physical evidence in a wide variety of crimes. However, any review of the forensic aspects of hair examination must start with the observation that it is not yet possible to individualize a human hair to any single head or body through its morphology, or structural characteristics. Over the years, criminalists have tried to isolate the physical and chemical properties of hair that could serve as individual characteristics of identity. Partial success has finally been achieved by isolating and characterizing the DNA present in hair.

The importance of hair as physical evidence cannot be overemphasized. Its removal from the body often denotes physical contact between a victim and perpetrator and hence a crime of a serious or violent nature. When hair is properly collected at the crime scene and submitted to the laboratory along with enough standard/reference samples, it can provide strong corroborative evidence for placing an individual at a crime site. The first step in the forensic examination of hair logically starts with its color and structure, or morphology, and, if warranted, progresses to the more detailed DNA extraction, isolation, and characterization.

Morphology of Hair

Hair is an appendage of the skin that grows out of an organ known as the hair follicle. The length of a hair extends from its root or bulb embedded in the follicle, continues into the shaft, and terminates at the tip end. The shaft, which is composed of three layers—the **cuticle**, **cortex**, and **medulla**—is most intensely examined by the forensic scientist (see Figure 12–1).

cuticle
The scale structure covering the exterior of the hair

cortex
The main body of the hair shaft

medulla
A cellular column running through the center of the hair

The Cuticle Two features that make hair a good subject for establishing individual identity are its resistance to chemical decomposition and its ability to retain structural features over a long period of time. Much of this resistance and stability is attributed to the cuticle, a scale structure covering the exterior of the hair. The cuticle is formed by overlapping scales that always point toward the tip end of each hair. The scales form from specialized cells that have hardened (keratinized) and flattened in progressing from the follicle. There are three basic patterns that describe the appearance of the cuticle: coronal, spinous, and imbricate (see Figure 12–2).

Although the scale pattern is not a useful characteristic for individualizing human hair, the variety of patterns formed by non-human (animal) hair makes it an important feature for species identification. Figure 12–3 shows the scale

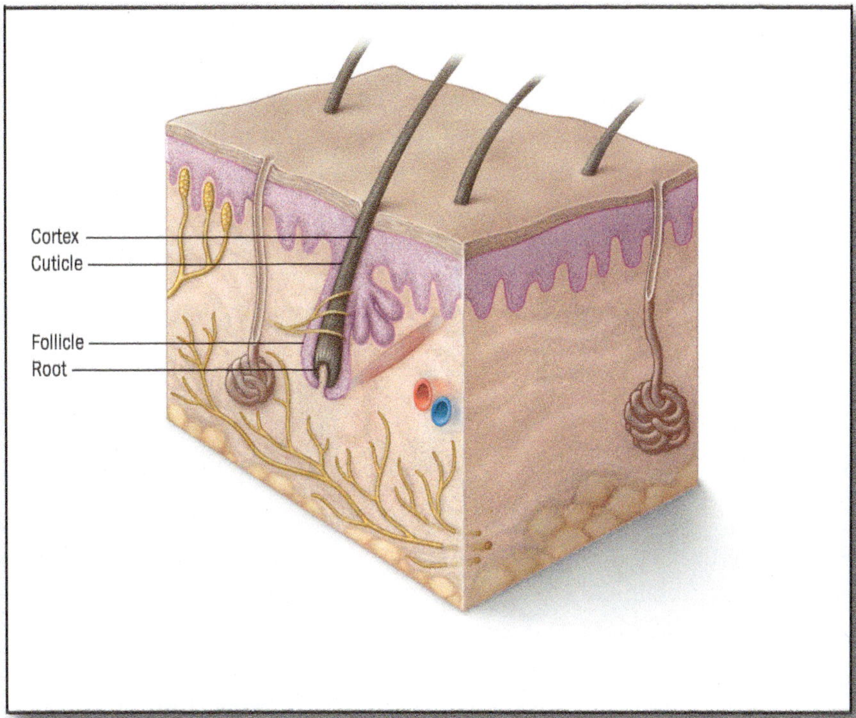

FIGURE 12–1 Cross section of skin showing hair growing out of a tubelike structure called the follicles.

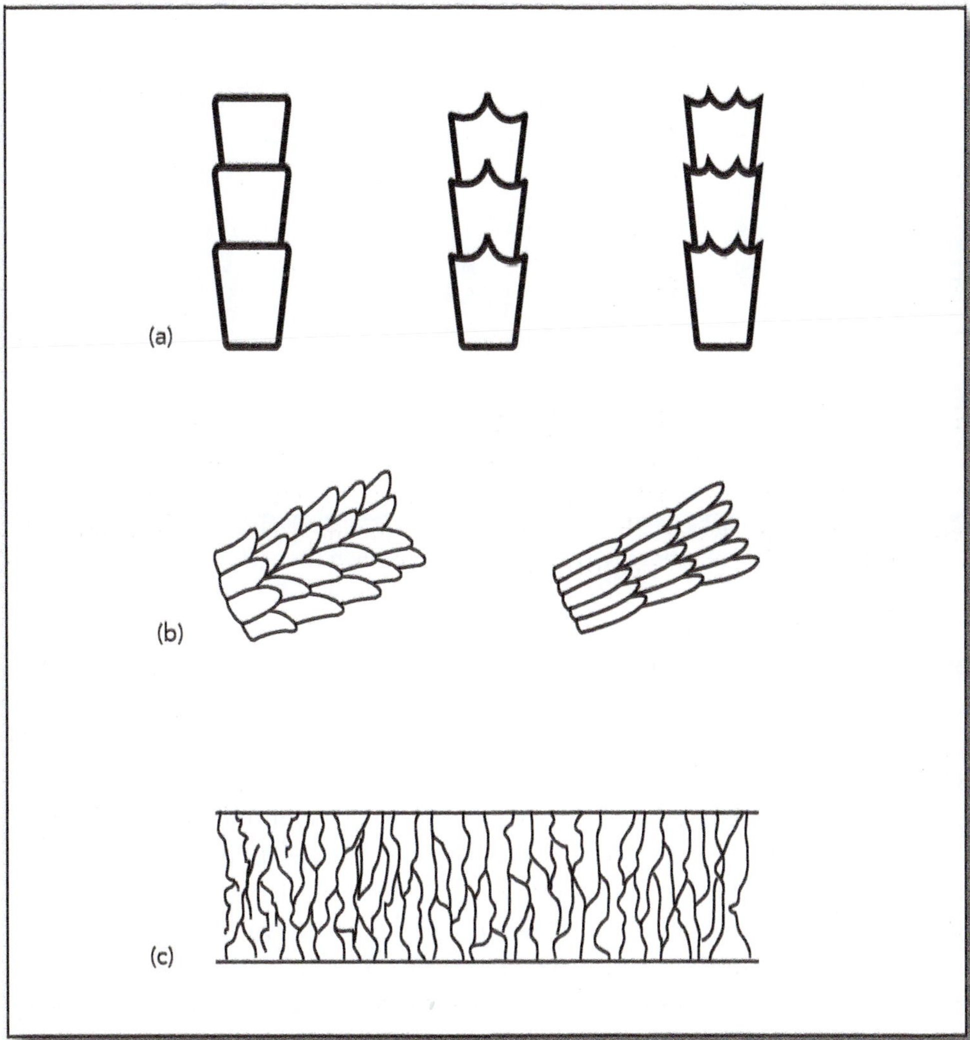

FIGURE 12–2
(a) The coronal, or crownlike scale pattern resembles a stack of paper cups.
(b) Spinous or petal-like scales are triangular in shape and protrude from the hair shaft.
(c)The imbricate or flattened-scale type consists of overlapping scales with narrow margins.
Courtesy Richard Saferstein Ph.D.

patterns of some animal hairs and of a human hair as viewed by the scanning electron microscope. Another method of studying the scale pattern of hair is to make a cast of its surface. This is done by embedding the hair in a soft medium, such as clear nail polish or softened vinyl. When the medium has hardened, the hair is removed, leaving a clear, distinct impression of the hair's cuticle, ideal for examination with a compound microscope.

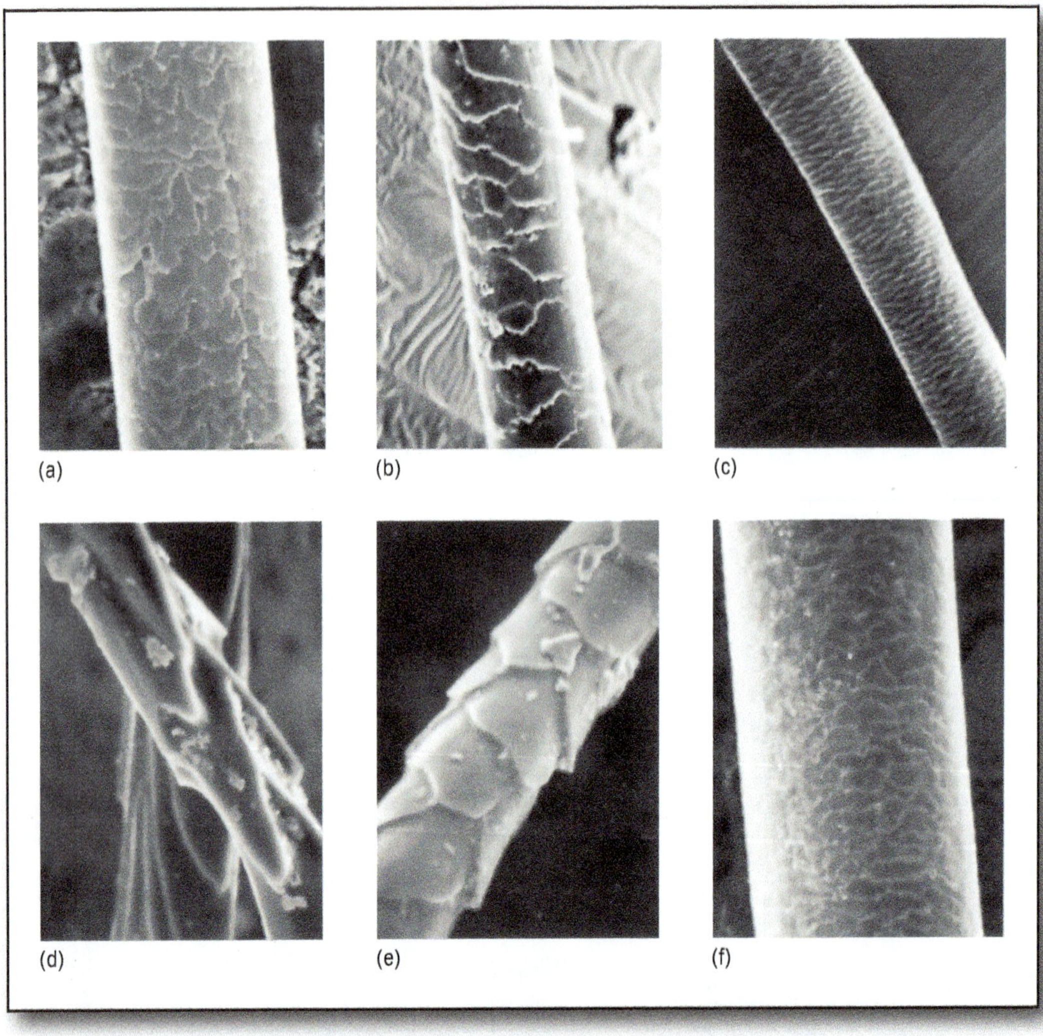

FIGURE 12–3 Scale patterns of various types of hair. (a) Human head hair (600x), (b) dog (1250x), (c) deer (120x), (d) rabbit (300x), (e) cat (2000x), and (f) horse (450x). *Courtesy Richard Saferstein, Ph.D.*

The Cortex Contained within the protective layer of the cuticle is the cortex, the main body of the hair shaft. The cortex is made up of spindle-shaped cortical cells aligned in a regular array, parallel to the length of the hair. The cortex derives its major forensic importance from the fact that it is embedded with the pigment granules that give hair its color. The color, shape, and distribution of these granules provide important points of comparison among the hairs of different individuals.

The structural features of the cortex are examined microscopically after the hair has been mounted in a liquid medium with a refractive index close to that of the hair. Under these conditions, the amount of light reflected off the hair's surface is minimized, and the amount of light penetrating the hair is optimized.

The Medulla The medulla is a collection of cells that looks like a central canal running through a hair. In many animals, this canal is a predominant feature, occupying more than half of the hair's diameter. The *medullary index* measures the diameter of the medulla relative to the diameter of the hair shaft and is normally expressed as a fraction. For humans, the index is generally less than one-third; for most other animals, the index is one-half or greater.

The presence and appearance of the medulla vary from individual to individual and even among the hairs of a given individual. Not all hairs have medullae, and when they do exist, the degree of medullation can vary. In this respect, medullae may be classified as being either continuous, interrupted, fragmented, or absent (see Figure 12–4). Human head hairs generally exhibit no medullae or have fragmented ones; they rarely show continuous medullation. One noted exception is people with Asian ancestry, who usually have head hairs with continuous medullae. Also, most animals have medullae that are either continuous or interrupted.

FIGURE 12–4 **Medulla patterns.** *Courtesy Richard Saferstein, Ph.D.*

Another interesting feature of the medulla is its shape. Humans, as well as many animals, have medullae that give a nearly cylindrical appearance. Other animals exhibit medullae that have a patterned shape. For example, the medulla of a cat can best be described as resembling a string of pearls, whereas members of the deer family show a medullary structure consisting of spherical cells occupying the entire hair shaft. Figure 12–5 illustrates medullary sizes and forms for a number of common animal hairs and a human head hair.

A searchable database on CD-ROM of the 35 most common animal hairs encountered in forensic casework is commercially available.[1] This database allows an examiner to rapidly search for animal hairs based on scale patterns and/or medulla type using a computer. A typical screen presentation arising from such a data search is shown in Figure 12–6.

The Root The root and other surrounding cells within the hair follicle provide the tools necessary to produce hair and continue its growth. Human head hair grows in three developmental stages, and the shape and size of the hair root is determined by the hair's current growth phase. The three phases of hair growth are the **anagen**, **catagen**, and **telogen phases**.

In the anagen phase (the initial growth phase), which may last up to six years, the root is attached to the follicle for continued growth, giving the root bulb a flame-shaped appearance (Figure 12–7[a]). When pulled from the root, some hairs in the anagen phase have a **follicular tag**. With the advent of DNA analysis, this follicular tag is important for individualizing hair.

Hair continues to grow, but at a decreasing rate, during the catagen phase, which can last anywhere from two to three weeks. In the catagen phase, roots typically take on an elongated appearance (Figure 12–7[b]) as the root bulb shrinks and is pushed out of the hair follicle. Once hair growth ends, the telogen phase begins and the root takes on a club-shaped appearance (Figure 12–7[c]). Over two to six months, the hair is pushed out of the follicle, causing the hair to be naturally shed.

anagen phase
The initial growth phase during which the hair follicle actively produces hair

catagen phase
A transition stage between the anagen and telogen phases of hair growth

telogen phase
The final growth phase in which hair naturally falls out of the skin

follicular tag
A translucent piece of tissue surrounding the hair's shaft near the root that contains the richest source of DNA associated with hair

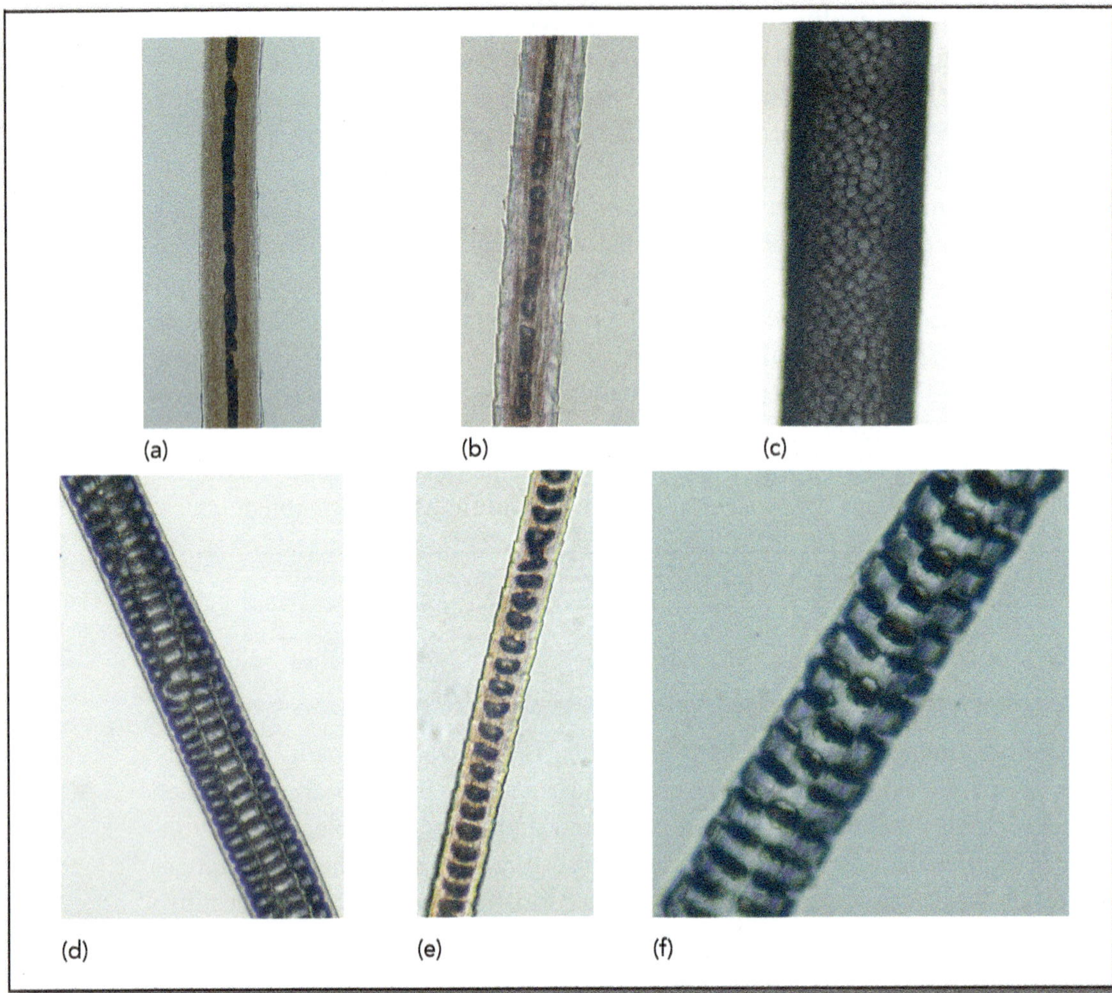

FIGURE 12–5 Medulla patterns for various types of hair. (a) Human head hair (400x), (b) dog (400x), (c) deer (500x), (d) rabbit (450x), (e) cat (400x), and (f) mouse (500x). *Courtesy Richard Saferstein, Ph.D.*

Identification and Comparison of Hair

Most often the prime purpose for examining hair evidence in a crime laboratory is to establish whether the hair is human or animal in origin, or to determine whether human hair retrieved at a crime scene compares with hair from a particular individual.

Stable Isotope Analysis When water is consumed, it leaves a chemical fingerprint in hair. Because people tend to drink and cook with their local water, which can vary by region, the signature left on the hair will be geographically unique. This can be helpful to investigators when they are faced with a murder victim that has not yet been identified. After exhausting all other avenues to identify the victim, investigators can use stable isotope information to indicate whether a person is local to the area or whether the body was transported there after death.

Scientists can analyze stable isotopes—different forms of the same chemical element—present in the hair. Because hair retains isotopic information, and grows about one centimeter each month, it can provide a personal chronology of where a person has been. If a person moves across the country over the course of a year, that movement will be reflected in the last 12 centimeters of hair growth.

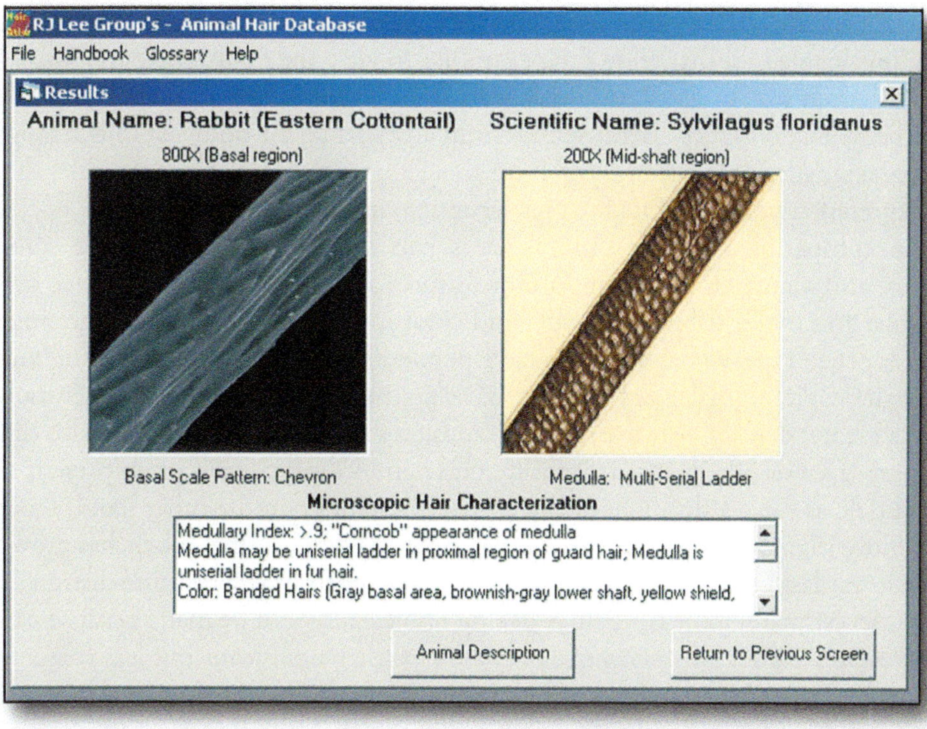

FIGURE 12–6 Information on rabbit hair contained within the Forensic Animal Hair Atlas. *Courtesy RJ Lee Group, Inc. Monroeville, Pa.*

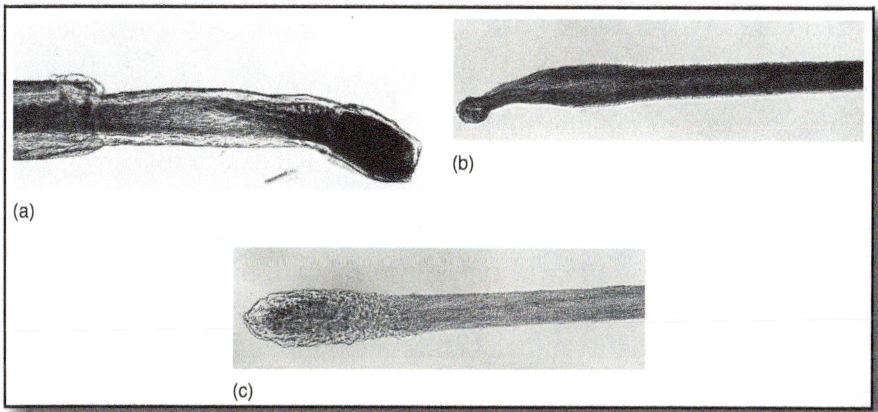

FIGURE 12–7 Hair roots in the: (a) anagen phase, (b) catagen phase, and (c) telogen phase (100x). *Virginia Division of Forensic Science*

This technology played a role in the identification of Bella Bond, discussed in more detail in Chapter 8. Investigators sent samples of her hair for stable isotope testing to help them determine if the child was local to the Boston area. After tests indicated that she was local, the Massachusetts State Police ramped up efforts to identify her by putting her image on billboards in the Boston metro area.

A careful microscopic examination of hair reveals morphological features that can distinguish human hair from animal hair. The hair of various animals also differs enough in structure that the examiner can often identify the species. Before reaching such a conclusion, however, the examiner must have access to a comprehensive collection of reference standards and the accumulated experience of hundreds of prior hair examinations. Scale structure, medullary index, and medullary shape are particularly important in hair identification.

The most common request when hair is used as forensic evidence is to determine whether hair recovered at the crime scene compares to hair removed from a suspect. In most cases, such a comparison relates to hair obtained from the scalp or pubic area. Ultimately, the evidential value of the comparison depends on the degree of probability with which the examiner can associate the hair in question with a particular individual.

Hair Characteristics In making a hair comparison, a comparison microscope is an invaluable tool that allows the examiner to view the questioned and known hair together, side by side. Any variations in the microscopic characteristics will thus be readily observed. Because hair from any part of the body exhibits a range of characteristics, it is necessary to have an adequate number of known hairs that are representative of all of its features when making a comparison.

In comparing hair, the criminalist is particularly interested in the color, length, and diameter. Other important features are the presence or absence of a medulla and the distribution, shape, and color intensity of the pigment granules in the cortex. Human hairs tend to have pigments distributed evenly throughout the hair, while animal hairs tend to have pigments distributed more centrally with a greater density near the medulla. A microscopic examination may also distinguish dyed or bleached hair from natural hair. A dyed color is often present in the cuticle as well as throughout the cortex. Bleaching, on the other hand, tends to remove pigment from the hair and to give it a yellowish tint. If hair has grown since it was last bleached or dyed, the natural-end portion will be quite distinct in color. An estimate of the time since dyeing or bleaching can be made because *hair grows approximately one centimeter per month*. Other significant but less frequent features may be observed in hair. For example, morphological abnormalities may be present because of certain diseases or deficiencies. Also, the presence of fungal and nit infections can further link a hair specimen to a particular individual.

Microscopic Examination of Hair A comparison microscope is an invaluable tool that allows the examiner to view the questioned and known hair together, side by side. Any variations in the microscopic characteristics will thus be readily observed. Because hair from any part of the body exhibits a range of characteristics, it is necessary to have an adequate number of known hairs that are representative of all of its features when making a comparison.

Although microscopic comparison of hairs has long been accepted as an appropriate approach for including and excluding questioned hairs against standard/reference hairs, many forensic scientists have long recognized that this approach is subjective and is highly dependent on the skills and integrity of the analyst, as well as the hair morphology being examined. However, until the advent of DNA analysis, the forensic science community had no choice but to rely on the microscope to carry out hair comparisons.

Any lingering doubts about the necessity of augmenting microscopic hair examinations with DNA analysis evaporated with the publication of an FBI study describing significant error rates associated with microscopic comparison of hairs.[2] Hair evidence submitted to the FBI for DNA analysis between 1996 and 2000 was examined both microscopically and by DNA analysis. Approximately 11 percent of the hairs (9 out of 80) in which FBI hair examiners found a positive microscopic match between questioned and standard/reference hairs were found to be nonmatches when they were later subjected to DNA analysis. The course of events is clear; microscopic hair comparisons must be regarded by police and courts as presumptive in nature, and all positive microscopic hair comparisons must be confirmed by DNA determinations.

Questions about Hair Examination A number of questions may be asked to further ascertain the present status of forensic hair examinations. The answers to these questions can be of great significance to the investigator working with hair evidence.

Can the Body Area from Which a Hair Originated Be Determined? Normally it is easy to determine the body area from which a hair came. For example, scalp hairs generally show little diameter variation and have a more uniform distribution of pigment when compared to other body hairs. Pubic hairs are short and curly, with wide variations in shaft diameter, and usually have continuous medullae. Beard hairs are coarse, are normally triangular in cross-section, and have blunt tips acquired from cutting or shaving.

hair analysis
There are morphological characteristics of hair that have been linked to individuals with African, Asian, and European ancestry. Microscopic hair analysis is based on the evaluation of these class and individual characteristics. While not dispositive of race or ancestry of the donor, evaluations of the morphology are the basis for any forensic hair comparison.

Can the Racial Origin of Hair Be Determined? In many instances, the examiner can distinguish hair originating from members of different ancestries; this is especially true of European and African head hair. African hairs are normally kinky, containing dense, unevenly distributed pigments. European hairs are usually straight or wavy, with very fine to coarse pigments that are more evenly distributed when compared to African hair. Sometimes a cross-sectional examination of hair may help identify ancestry. Cross-sections of hair from Europeans are oval to round in shape, whereas cross-sections of African hair are flat to oval in shape. However, all of these observations are general, with many possible exceptions. The criminalist must approach the determination of ancestry from hair with caution and a good deal of experience.

Can the Age and Sex of an Individual Be Determined from a Hair Sample? The age of an individual cannot be learned from a hair examination with any degree of certainty except with infant hair. Infant hairs are fine and short and have fine pigmentation. Although the presence of dye or bleach on the hair may offer some clue to sex, present hairstyles make these characteristics less valuable than they were in the past. The recovery of nuclear DNA either from tissue adhering to hair or from the root structure of the hair will allow a determination of whether the hair originated from a male or female.

Is It Possible to Determine Whether Hair Was Forcibly Removed from the Body? A microscopic examination of the hair root may establish whether the hair fell out or was pulled out of the skin. A hair root with follicular tissue (root sheath cells) adhering to it, as shown in Figure 12–8, indicates a hair that has been pulled out either by a person or by brushing or combing. Hair naturally falling off the body has a bulbous-shaped root free of any adhering tissue.

The absence of sheath cells cannot always be relied on for correctly judging whether hair has been forcibly pulled from the body. In some cases the root of a hair is devoid of any adhering tissue even when it has been pulled from the body. Apparently, an important consideration is how quickly the hair is pulled out of the head. Hairs pulled quickly from the head are much more likely to have sheath cells compared to hairs that have been removed slowly from the scalp.[3]

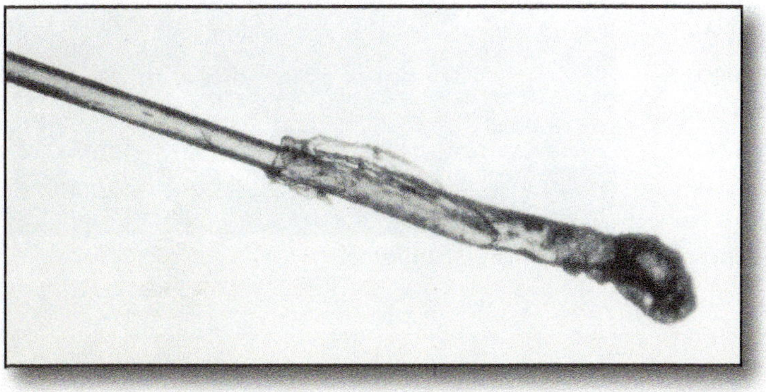

FIGURE 12–8 Forcibly removed head hair, with follicular tissue attached. *Courtesy Richard Saferstein, Ph.D.*

Case Files

The Central Park Jogger Case Revisited

On April 19, 1989, a young woman left her apartment around nine p.m. to jog in New York's Central Park. Nearly five hours later, she was found comatose lying in a puddle of mud in the park. She had been raped, her skull was fractured, and she had lost 75 percent of her blood. When the woman recovered, she had no memory of what happened to her. The brutality of the crime sent shock waves through the city and seemed to fuel a national perception that crime was running rampant and unchecked through the streets of New York.

Already in custody at the station house of the Central Park Precinct was a group of 14-and 15-year-old boys who had been rounded up leaving the park earlier in the night by police who suspected that they had been involved in a series of random attacks. Over the next two days, four of the teenagers gave videotaped statements, which they later recanted, admitting to participating in the attack. Ultimately, five of the teenagers were charged with the crime.

Interestingly, none of the semen collected from the victim could be linked to any of the defendants. However, according to the testimony of a forensic analyst, two head hairs collected from the clothing of one of the defendants microscopically compared to those of the victim, and a third hair collected from the same defendant's T-shirt microscopically compared to the victim's pubic hair. Besides these three hairs, a fourth hair was found microscopically similar to the victim's. This hair was recovered from the clothing of Steven Lopez, who was originally charged with rape but not prosecuted for the crime.

Hairs were the only pieces of physical evidence offered by the district attorney to directly link any of the teenagers to the crime. The five defendants were convicted and ultimately served from 9 to 13 years.

In August 1989, more than three months after the jogger attack, New York Police arrested a man named Matias Reyes, who pleaded guilty to murdering a pregnant woman, raping three others, and committing a robbery. In January 2002, Reyes also confessed to the Central Park attack. Follow-up tests revealed that Reyes's DNA compared to semen recovered from the jogger's body and her sock. Other DNA tests showed that the hairs offered into evidence at the original trial did not come from the victim, and so could not be used to link the teenagers to the crime as the district attorney had argued. After an 11-month reinvestigation of the original charges, a New York State Supreme Court judge dismissed all the convictions against the five teenage suspects in the Central Park jogger case.

Paolo Omero/Shutterstock

Is It Possible to Determine Whether Hair Came from a Deceased Individual? During the murder trial of Casey Anthony (see page 455), witnesses testified that a human head hair was recovered from the trunk of Casey's automobile. Examination showed that the hair was microscopically similar to that of Casey's daughter, Caylee. The hair was microscopically distinguishable from Casey's head hair. Mitochondrial DNA sequence analysis revealed similarity between the trunk hair, Caylee's hair, and Casey's hair, as one would expect from maternally inherited mitochondrial DNA. An FBI analyst testified that the hair from the truck exhibited "root banding," a phenomenon consistent with hair from a deceased individual's head.

As exemplified by the Casey Anthony case, a forensic examiner may on occasion be confronted with situations where it's important to know whether hair was deposited after the donor was deceased. Studies have noted that postmortem decomposition may be accompanied by a darkening or banding around the root area of the hair (see Figure 12–9). Eventually, the hair will break off at the area of the dark hair, leaving a discolored hair end with a point or brushlike appearance. Recently, it has been confirmed that the onset of postmortem changes to the root portion of hair was observed only in anagenic and catagenic hairs, and specifically within a short area of the root where the hair would have been beneath the scalp. Root banding was slow to occur in cold weather and progressed at a faster rate in warmer temperatures. Significantly, hairs in the telogen stage showed no evidence of postmortem root banding.[4]

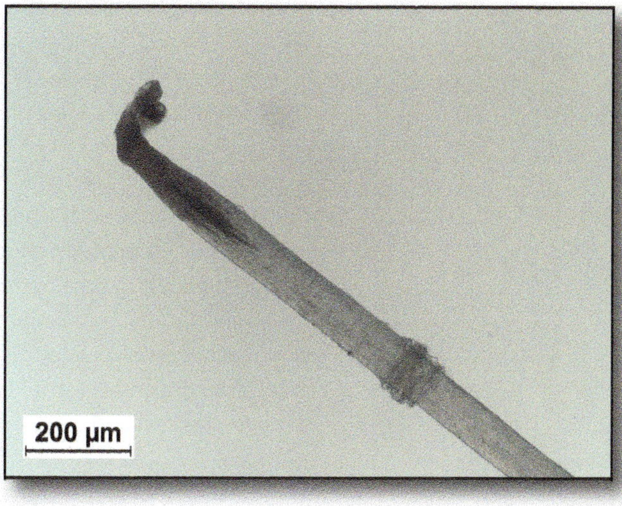

FIGURE 12–9 Hair with a banded root. Courtesy Chris Palenik of Microtrace LLC, Elgin, IL.

Are Efforts Being Made to Individualize Human Hair? As we saw in Chapter 10, forensic scientists routinely isolate and characterize individual variations in DNA. Forensic hair examiners can link human hair to a particular individual by characterizing the **nuclear DNA** in the hair root or in follicular tissue adhering to the root (see Figure 12–8). Recall that the follicular tag is the richest source of DNA associated with hair. In the absence of follicular tissue, an examiner must extract DNA from the hair root.

nuclear DNA
DNA that is present in the nucleus of a cell and that is inherited from both parents

The growth phase of hair is a useful predictor of the likelihood of successfully typing DNA in human hair.[5] Examiners have a higher success rate in extracting DNA from hair roots in the anagen phase or from anagen-phase hairs entering the catagen phase of growth. Telogen-phase hairs have an inadequate amount of DNA for typing. Because most hairs are naturally shed and are expected to be in the telogen stage, these observations do not portend well for hairs collected at crime scenes. However, some crime scenes are populated with forcibly removed hairs that are expected to be rich sources for nuclear DNA.

When a questioned hair does not have adhering tissue or a root structure amenable to isolation of nuclear DNA, there is an alternative—**mitochondrial DNA**. Unlike the nuclear DNA described earlier, which is located in the nuclei of practically every cell in our bodies, mitochondrial DNA is found in cellular material outside the nucleus. Interestingly, unlike nuclear DNA, which is passed down to us from both parents, mitochondrial DNA is transmitted only from mother to child. Importantly, many more copies of mitochondrial DNA are located in our cells as compared to nuclear DNA. For this reason,

mitochondrial DNA
DNA present in small structures (mitochondria) outside the nucleus of a cell; mitochondria supply energy to the cell; this form of DNA is inherited maternally (from the mother)

Case Files

Hair Evidence

The murder of Ennis Cosby, son of entertainer Bill Cosby, at first appeared unsolvable. It was a random act. When his car tire went flat, Ennis pulled off the road and called a friend on his cell phone to ask for assistance. Shortly thereafter, an assailant demanded money and, when Cosby didn't respond quickly enough, shot him once in the temple. Acting on a tip from a friend of the assailant, police investigators later found a .38-caliber revolver wrapped in a blue cap miles from the crime scene. Mikail Markhasev was arrested and charged with murder.

At trial, the district attorney introduced firearms evidence to show that the recovered gun had fired the bullet aimed at Cosby. However, a single hair also recovered from the cap dramatically linked Markhasev to the crime. Los Angeles Police Department forensic analyst Harry Klann identified six DNA markers from the follicular tissue adhering to the hair root that matched Markhasev's DNA. This particular DNA profile is found in one out of 15,500 members of the general population. Upon hearing all of the evidence, the jury deliberated and convicted Markhasev of murder.

Bill Cosby and his son Ennis Cosby. *Courtesy Andrea Mohin/Redux Pictures*

the success rate of finding and typing mitochondrial DNA is much greater from samples, such as hair, that have limited quantities of nuclear DNA. Hairs 1–2 centimeters long can be subjected to mitochondrial analysis with extremely high odds of success. This subject is discussed in greater detail in Chapter 10.

Can DNA Individualize a Human Hair? In some cases, the answer is yes. As we learned in Chapter 10, nuclear DNA produces frequencies of occurrence as low as one in billions or trillions. On the other hand, mitochondrial DNA cannot individualize human hair, but its diversity within the human population often permits exclusion of a significant portion of a population as potential contributors of a hair sample. Ideally, the combination of a positive microscopic comparison and an association through nuclear or mitochondrial DNA analysis strongly links a questioned hair and standard/reference hairs. However, a word of caution: mitochondrial DNA cannot distinguish microscopically similar hairs from different individuals who are maternally related.

Collection and Preservation of Hair Evidence

When questioned hairs are submitted to a forensic laboratory for examination, they must always be accompanied by an adequate number of standard/reference samples from the victim of the crime and from individuals suspected of having deposited hair at the crime scene. We have learned that hair from different parts of the body varies significantly in its physical characteristics. Likewise, hair from any one area of the body can also have a wide range of characteristics. For this reason, the questioned and standard/reference hairs must come from the same area of the body; one cannot, for instance, compare head hair to pubic hair. It is also important that the collection of standard/reference hair be carried out in a way to ensure a representative sampling of hair from any one area of the body.

Forensic hair comparisons generally involve either head hair or pubic hair. Collecting 25 full-length hairs from all areas of the scalp normally ensures a representative sampling of head hair. Likewise, a minimum collection of 25 full-length pubic hairs should cover the range of characteristics present in this type of hair. In rape cases, care must first be taken to comb the pubic area with a clean comb to remove all loose foreign hair present before the victim is sampled for standard/reference hair. The comb should then be packaged in a separate envelope.

Because a hair may vary in color and other morphological features over its entire length, the entire hair is collected. This requirement is best accomplished by either pulling the hair out of the skin or clipping it at the skin line. During an autopsy, hair samples are routinely collected from a victim of suspicious death. Because the autopsy may occur early in an investigation, the need for hair standard/reference samples may not always be apparent. However, one should never rule out the possible involvement of hair evidence in subsequent investigative findings. Failure to make this simple collection may result in complicated legal problems later.

Quick Review

- The hair shaft is composed of three layers called the cuticle, cortex, and medulla and is most intensely examined by the forensic scientist.
- When comparing strands of hair, the criminalist is particularly interested in matching the color, length, and diameter. Other important features for comparing hair are the presence or absence of a medulla and the distribution, shape, and color intensity of pigment granules in the cortex.
- The probability of detecting DNA in hair roots is more likely for hair being examined in its anagen or early growth phase as opposed to its catagen or telogen phases.
- The follicular tag, a translucent piece of tissue surrounding the hair's shaft near the root, is a rich source of DNA associated with hair. Mitochondrial DNA can also be extracted from the hair shaft.
- All positive microscopic hair comparisons must be confirmed by DNA analysis.

Forensic Examination of Fibers

Just as hair left at a crime scene can serve as identification, the same logic can reasonably be extended to the fibers that compose our fabrics and garments. Fibers may become important evidence in incidents that involve personal contact—such as homicide, assault, and sexual offenses—in which cross-transfers may occur between the clothing of suspect and victim. Similarly, the force of impact between a hit-and-run victim and a vehicle often leaves fibers, threads, or even whole pieces of clothing adhering to parts of the vehicle. Fibers may also become fixed in screens or glass broken in the course of a breaking-and-entering attempt.

Regardless of where and under what conditions fibers are recovered, their ultimate value as forensic evidence depends on the criminalist's ability to narrow their origin to a limited number of sources or even to a single source. Unfortunately, mass production of garments and fabrics has limited the value of fiber evidence in this respect, and only rarely do fibers recovered at a crime scene provide individual identification with a high degree of certainty.

Types of Fibers

For centuries, humans depended on fibers derived from natural sources such as plants and animals. However, early in the 20th century, the first manufactured fiber—rayon—became a practical reality, followed in the 1920s by the introduction of cellulose acetate. Since the late 1930s, scientists have produced dozens of new fibers. In fact, the development of fibers, fabrics, finishes, and other textileprocessing techniques has made greater advances since 1900 than in the preceding five thousand years of recorded history. Today, such varied items as clothing, carpeting, drapes, wigs, and even artificial turf attest to the predominant role that manufactured fibers have come to play in our culture and environment. When discussing forensic examination of fibers, it is convenient to classify them into two broad groups: *natural* and *manufactured*.

natural fibers
Fibers derived entirely from animal or plant sources

manufactured fibers
Fibers derived from either natural or synthetic polymers graph of cotton fiber (450x).

Natural Fibers **Natural fibers** are wholly derived from animal or plant sources. Animal fibers constitute most natural fibers encountered in crime laboratory examinations. These include hair coverings from such animals as sheep (wool), goats (mohair, cashmere), camels, llamas, alpacas, and vicuñas. Fur fibers include those obtained from animals such as mink, rabbit, beaver, and muskrat.

Forensic examination of animal fibers uses the same procedures discussed in the previous section for the forensic examination of animal hairs. Identification and comparison of such fibers relies solely on a microscopic examination of color and morphological characteristics. Again, a sufficient number of standard/reference specimens must be examined to establish the range of fiber characteristics of the suspect fabric.

By far the most prevalent plant fiber is cotton. The wide use of undyed white cotton fibers in clothing and other fabrics has made its evidential value almost meaningless, although the presence of dyed cotton in a combination of colors has, in some cases, enhanced its evidential significance. The microscopic view of cotton fiber shown in Figure 12–10 reveals its most distinguishing feature—a ribbonlike shape with twists at irregular intervals.

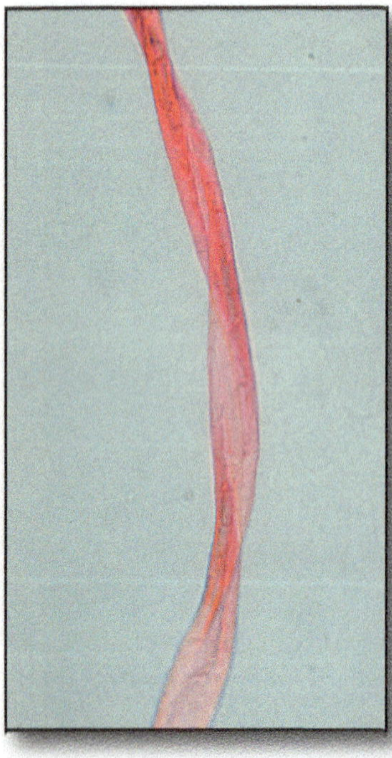

FIGURE 12–10 Photomicrograph of cotton fiber (450x). *Courtesy Richard Saferstein Ph.D.*

Manufactured (Synthetic) Fibers Beginning with the introduction of rayon in 1911 and the development of nylon in 1939, **manufactured fibers** have increasingly replaced natural fibers in garments and fabrics. Such fibers are marketed under hundreds of different trade names. To reduce consumer confusion, the U.S. Federal Trade Commission has approved "generic" or family names for the grouping of all manufactured fibers. Many of these generic classes are produced by several manufacturers and are sold under a confusing variety of trade names. For example, in the United States, polyesters are marketed under names that include Dacron, Fortrel, and Kodel. In England, polyesters are called Terylene. Table 12–1 lists major generic fibers, along with common trade names and their characteristics and applications.

The first machine-made fibers were manufactured from raw materials derived from cotton or wood pulp. These materials are processed, and pure cellulose is extracted from them. Depending on the type of fiber desired, the cellulose may be chemically treated and dissolved in an appropriate solvent before it is forced through the small holes of a spinning jet, or spinneret, to produce the fiber. Fibers manufactured from natural raw materials in this manner are classified as *regenerated fibers* and commonly include rayon, acetate, and triacetate, all of which are produced from regenerated cellulose.

Table 12–1
Major Generic Fibers

Major Generic Fiber	Description	Common Uses
Acetate Rayon	Acetate Cellulose fiber where 74%–92% of hydroxyl groups are acetylated	*Apparel*: Blouses, dresses, foundation garments, lingerie, linings, shirts, slacks, sportswear *Fabrics*: Brocade, crepe, double knits, faille, knitted jerseys, lace, satin, taffeta, tricot *Home Furnishings*: Draperies, upholstery *Other*: Cigarette filters, fiberfill for pillows, quilted products
Acrylic	Fiber comprised of linear macromolecules composed of at least 85% acrylonitrile monomer	*Apparel*: Dresses, infant wear, knitted garments, ski-wear, socks, sportswear, sweaters *Fabrics*: Fleece and pile fabrics, facing fabrics in bonded fabrics, simulated furs, jerseys *Home Furnishings:* Blankets, carpets, draperies, upholstery *Other*: Auto tops, awnings, hand-knitting and craft yarns, industrial and geotextile fabrics

Table 12–1
Major Generic Fibers (*continued*)

Major Generic Fiber	Description	Common Uses
Aramid	Aramid fibers are composed of a series of synthetic polymers in which repeating units containing large phenyl rings are linked together by amide groups.	Hot-gas filtration fabrics, protective clothing, military helmets, protective vests, structural composites for aircraft and boats, sailcloth, tires, ropes and cables, mechanical rubber goods, marine and sporting goods
Bicomponent	A bicomponent fiber is made of two materials, utilizing desired properties of each material. Such fibers can be created by extrusion spinning. One or both materials may remain in the finished product, or one material may be dissolved, leaving only one material remaining.	Uniform distribution of adhesive; fiber remains a part of structure and adds integrity; customized sheath materials to bond various materials; wide range of bonding temperatures; cleaner, environmentally friendly (no effluent); recyclable; lamination/molding/densification of composites
Lyocell	Lyocell is a form of rayon that consists of cellulose fiber made from dissolving bleached wood pulp using dry jet-wet spinning.	Dresses, slacks, and coats
Rayon	Rayon is made from purified cellulose from wood pulp which is made into a soluble compound. It is then dissolved and forced through a spinneret to produce filaments which are chemically solidified, resulting in fibers of nearly pure cellulose.	*Apparel:* Blouses, coats, dresses, jackets, lingerie, linings, millinery, rainwear, slacks, sports shirts, sportswear, suits, ties, work clothes *Home Furnishings:* Bedspreads, blankets, carpets, curtains, draperies, sheets, slipcovers, tablecloths, upholstery *Other:* Industrial products, medical-surgical products, nonwoven products, tire cord
Spandex	Spandex is a lightweight, synthetic fiber that is used to make stretchable clothing such as sportswear. It is made up of a long chain polymer called polyurethane, which is produced by reacting a polyester with a diisocyanate. The polymer is converted into a fiber using a dry spinning technique.	*Apparel* (in which stretch is desired): Athletic apparel, bathing suits, delicate laces, foundation garments, golf jackets, ski pants, slacks, support and surgical hose

Major Generic Fiber	Description	Common Uses
Melamine	Melamine fiber is a manufactured fiber in which the fiber-forming substance is a synthetic polymer composed of at least 50% by weight of a cross-linked melamine polymer.	*Fire-Blocking* Fabrics: Aircraft seating, fire blockers for upholstered furniture in high-risk occupancies (e.g., to meet California TB 133 requirements) *Protective Clothing:* Firefighters' turnout gear, insulating thermal liners, knit hoods, molten metal splash apparel, heat-resistant gloves *Filter Media:* High-capacity, high-efficiency, high-temperature baghouse air filters
Modacrylic	Modacrylic fibers are manufactured fibers in which the fiber-forming substance is any longchain synthetic polymer composed of less than 85%, but at least 35% weight acrylonitrile units except when the polymer qualifies as rubber.	*Apparel:* Deep-pile coats, trims, linings, simulated fur, wigs and hairpieces *Fabrics:* Fleece fabrics, industrial fabrics, knit-pile fabric backings, nonwoven fabrics *Home Furnishings:* Awnings, blankets, carpets, flame-resistant draperies and curtains, scatter rugs *Other:* Filters, paint rollers, stuffed toys
Nylon	Nylon is a generic name for a family of synthetic polymers, more specifically, aliphatic or semiaromatic polyamides in which at least 85% by weight of the amide-linkages (—CO—NH—) are attached directly to two aliphatic groups.	*Apparel:* Blouses, dresses, foundation garments, hosiery, lingerie and underwear, raincoats, ski and snow apparel, suits, windbreakers *Home Furnishings:* Bedspreads, carpets, draperies, curtains, upholstery *Other:* Air hoses, conveyor and seat belts, parachutes, racket strings, ropes and nets, sleeping bags, tarpaulins, tents, thread, tire cord, geotextiles
Olefin	Olefin fiber is a manufactured fiber in which the fiber-forming substance is any long-chain synthetic polymer composed of at least 85% by weight of ethylene, propylene, or other olefin units.	*Apparel:* Pantyhose, underwear, knitted sports shirts, men's half-hose, men's knitted sportswear, sweaters *Home Furnishings:* Carpet and carpet backing, slipcovers, upholstery, wall paper *Other:* Dye nets, filter fabrics, laundry and sandbags, geotextiles, automotive interiors, cordage, doll hair, industrial sewing thread

Table 12–1
Major Generic Fibers (*continued*)

Major Generic Fiber	Description	Common Uses
Polyester	Polyester is a synthetic polymer made of purified terephthalic acid (PTA) or its dimethyl ester dimethyl terephthalate (DMT) and monoethylene glycol (MEG). A manufactured fiber in which the fiber forming substance is any long-chain synthetic polymer composed of at least 85% by weight of an ester of a substituted aromatic carboxylic acid.	*Apparel:* Blouses, shirts, career apparel, children's wear, dresses, half-hose, insulated garments, ties, lingerie and underwear, permanent-press garments, slacks, suits *Home Furnishings:* Carpets, curtains, draperies, sheets, pillowcases *Other:* Fiberfill for various products, fire hose, power belting, ropes and nets, tire cord, sail, V-belts
PBI	Polybenzimidazole fibers (PBI) are fibers in which the fiber-forming substance is a long-chain aromatic polymer having recurring imidazole groups as one of the main structural repeat units in the polymer backbone. PBI is prepared from an aromatic tetraamine and an aromatic dicarboxylic acid or a derivative of it. The resin is then spun into fibers via a dry spinning process using dimethyl acetamide as a solvent.	Suitable for high-performance protective apparel such as firefighters' turnout coats, astronaut space suits, and applications in which fire resistance is important

polymer
A substance composed of a large number of atoms that are usually arranged in repeating units

Most of the fibers currently manufactured are produced solely from synthetic chemicals and are therefore classified as synthetic fibers. These include nylons, polyesters, and acrylics. The creation of synthetic fibers became a reality only when scientists developed a method of synthesizing long-chained molecules called **polymers**.

In 1930, chemists discovered an unusual characteristic of one of the polymers under investigation. When a glass rod in contact with viscous material in a beaker was slowly pulled away, the substance adhered to the rod and formed a fine filament that hardened as soon as it entered the cool air. Furthermore, the cold filaments could be stretched several times their extended length to produce a flexible, strong, and attractive fiber. This first synthetic fiber was improved and then marketed as nylon. Since then, fiber chemists have successfully synthesized new polymers and have developed more efficient methods for manufacturing them. These efforts have produced a multitude of synthetic fibers.

Polymers The polymer is the basic chemical substance of all synthetic fibers. Indeed, an almost unbelievable array of household, industrial, and recreational products is manufactured from polymers; these include plastics, paints, adhesives, and synthetic rubber. Polymers exist in countless forms and varieties and with the proper treatment can be made to assume different chemical and physical properties.

As we have already observed, chemical substances are composed of basic structural units called **molecules**. The molecules of most materials are composed of just a few atoms; for example, water, H_2O, has 2 atoms of hydrogen and 1 atom of oxygen. The heroin molecule, $C_{21}H_{23}O_5N$, contains 21 atoms of carbon, 23 atoms of hydrogen, 5 atoms of oxygen, and 1 atom of nitrogen. Polymers, on the other hand, are formed by linking a large number of molecules, so that a polymer often contains thousands or even millions of atoms. This is why polymers are often referred to as **macromolecules**, or "big" molecules.

molecule
Two or more atoms held together by chemical bonds

macromolecule
A molecule with a high molecular mass

Simply, a polymer can be pictured as resembling a long, repeating chain, with each link representing the basic structure of the polymer (see Figure 12–11). The repeating molecular units in the polymer, called **monomers**, are joined end to end, so that thousands link to form a long chain. What makes polymer chemistry so fascinating is the countless possibilities for linking different molecules. By simply varying the chemical structure of the monomers, and by devising numerous ways to weave them together, chemists have created polymers that exhibit different properties. This versatility enables polymer chemists to synthesize glues, plastics, paints, and fibers.

monomer
The basic unit of structure from which a polymer is constructed

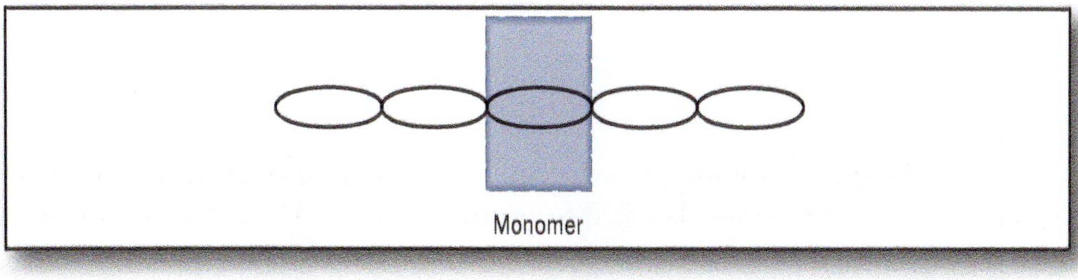

FIGURE 12–11 The chain-link model of a segment of a polymer molecule. The molecule may contain as many as several million monomer units or links.

Not all polymers are synthesized in the chemical laboratory; nature has produced polymers that humans have not yet been able to copy. For example, the proteins that form the basic structure of animal hairs, as well as of all living matter, are polymers, composed of thousands of amino acids linked in a highly organized arrangement and sequence. Similarly, cellulose (the basic ingredient of wood and cotton) and starch are both natural polymers built by the combination of several thousand carbohydrate monomers, as shown in Figure 12–12. Hence, the synthesis of manufactured fibers merely represents an extension of chemical principles that nature has used to produce hair and vegetable fibers.

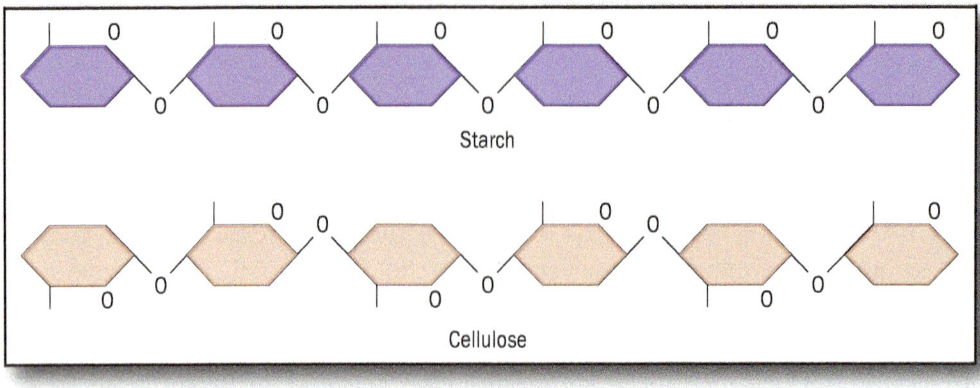

FIGURE 12–12 Starch and cellulose are natural carbohydrate polymers consisting of a large number of repeating units or monomers.

Identification and Comparison of Manufactured (Synthetic) Fibers

The evidential value of fibers lies in the criminalist's ability to trace their origin. Obviously, if the examiner is presented with fabrics that can be exactly fitted together at their torn edges, the fabrics must be of common origin.

More often, however, the criminalist obtains a limited number of fibers for identification and comparison. Generally, in these situations obtaining a physical match is unlikely, and the examiner must resort to a side-by-side comparison of the standard/reference and crime-scene fibers.

Microscopic Examination of Fibers The first and most important step in the examination is a microscopic comparison for color and diameter using a comparison microscope. Unless these two characteristics agree, there is little reason to suspect a match. Other morphological features that may aid in the comparison are lengthwise striations (lined markings) on the surface of some fibers and the pitting of the fiber's surface with delustering particles (usually titanium dioxide) added in the manufacturing process to reduce shine (see Figure 12–13).

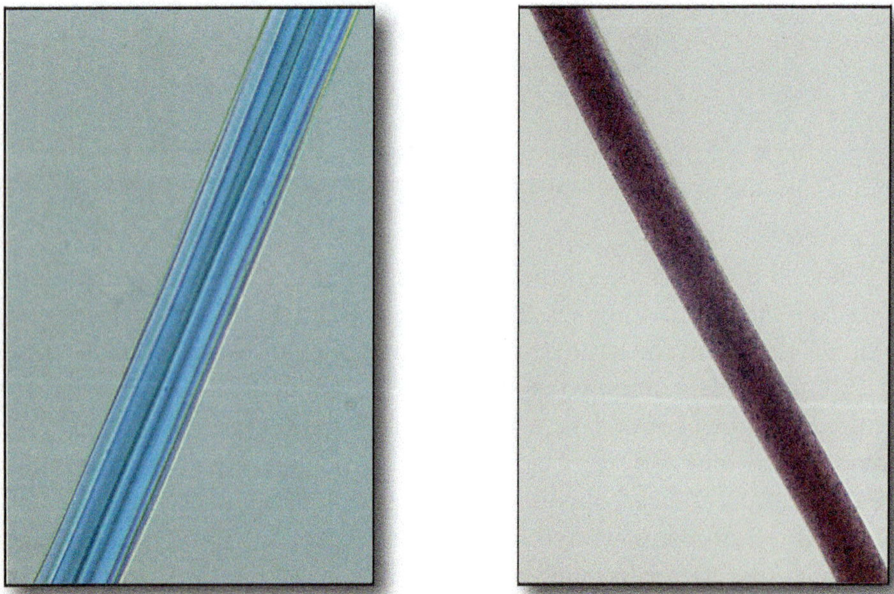

(a) (b)

FIGURE 12–13 Photomicrographs of synthetic fibers: (a) cellulose triacetate (450x) and (b) olefin fiber embedded with titanium dioxide particles (450x).
Courtesy Richard Saferstein, Ph.D.

The cross-sectional shape of a fiber may also help characterize the fiber (see Figure 12–14).[6] In the early 1980s, Wayne Williams was charged and tried for the murder of two individuals in the Atlanta, Georgia region. During the eight week trail, evidence linking Williams to those murders and to the murder of ten other individuals was introduced. An essential part of the government's case was the numerous fibers linking Williams to these murders. For example, unusually shaped yellow-green fibers discovered on a number of the murder victims were linked to a carpet in the Williams home. This fiber was a key element in proving Williams's guilt. A photomicrograph of this unusually shaped fiber is shown in Figure 12–15.

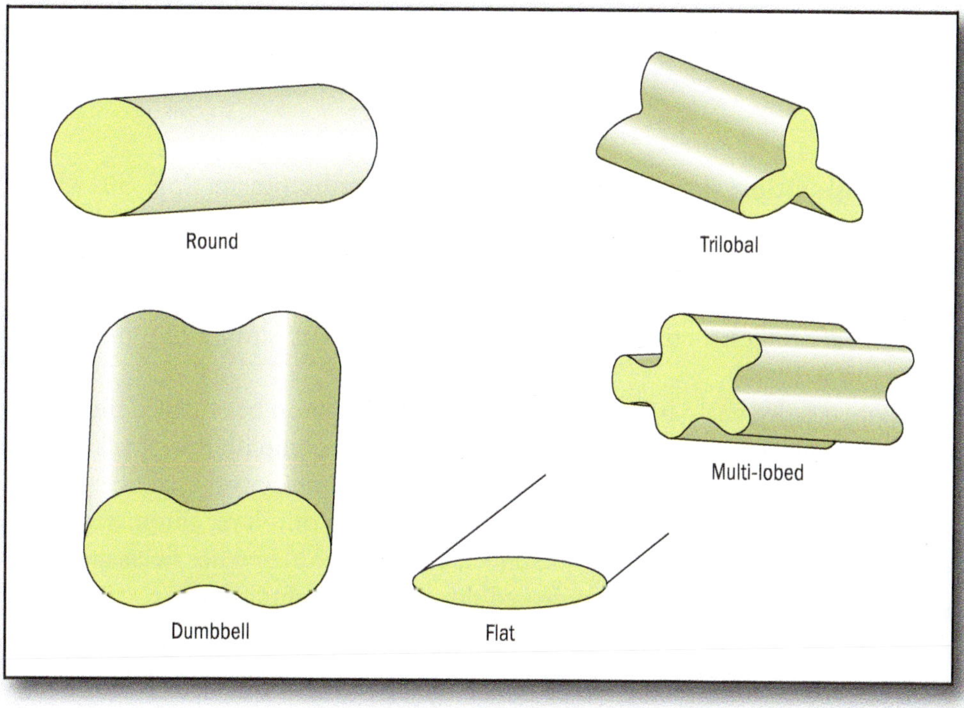

FIGURE 12–14
Cross-sectional shapes of fibers.

Although two fibers may seem to have the same color when viewed under the microscope, compositional differences may exist in the dyes that were applied to them during their manufacture. In fact, most textile fibers are impregnated with a mixture of dyes selected to obtain a desired shade or color. The significance of a fiber comparison is enhanced when the forensic examiner can show that the questioned and standard/reference fibers have the same dye composition.

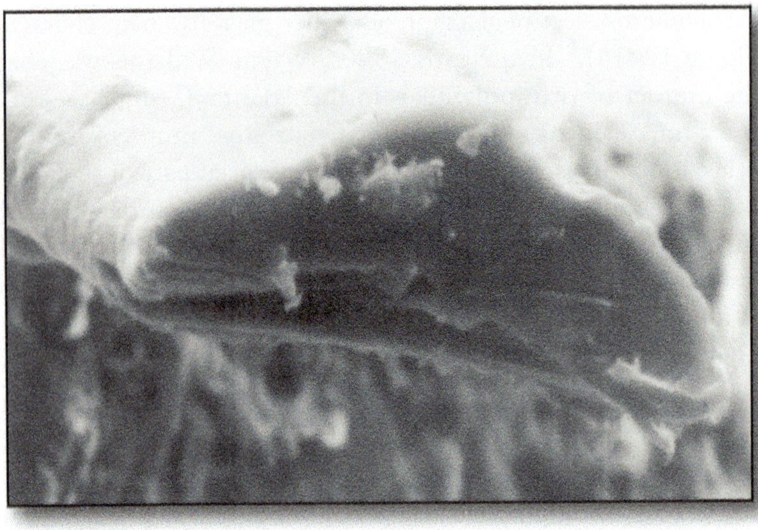

FIGURE 12–15 A scanning electron photomicrograph of the cross-section of a nylon fiber removed from a sheet used to transport the body of a murder victim. The fiber, associated with a carpet in Wayne Williams's home, was manufactured in 1971 in relatively small quantities. *Courtesy Federal Bureau of Investigation, Washington, D.C.*

Dye Composition The visible-light microspectrophotometer (pp. 304–306) is a convenient way for analysts to compare the colors of fibers through spectral patterns. This technique is not limited by sample size—a fiber as small as 1 millimeter long or less can be examined by this type of microscope. The examination is nondestructive and is carried out on fibers simply mounted on a microscope slide.

A more detailed analysis of the fiber's dye composition can be obtained through a chromatographic separation of the dye constituents. To accomplish this, small strands of fibers are compared for dye content by first extracting the dye off each fiber with a suitable solvent and then spotting the dye solution onto a thin-layer chromatography plate. The dye components of the questioned and standard/reference fibers are separated on the thin-layer plate and compared side by side for similarity.[7]

Chemical composition. Once this phase of the analysis is complete, and before any conclusion can be reached that two or more fibers compare, they must be shown to have the same chemical composition. In this respect, tests are performed to confirm that all of the fibers involved belong to the same broad generic class. Additionally, the comparison will be substantially enhanced if it can be demonstrated that all of the fibers belong to the same subclassification within their generic class. For example, at least four different types of nylon are available in commercial and consumer markets, including nylon 6, nylon 6–10, nylon 11, and nylon 6–6. Although all types of nylon have many properties in common, each may differ in physical shape, appearance, and dyeability because of modifications in basic chemical structure. Similarly, a study of more than two hundred different samples of acrylic fibers revealed that they could be divided into 24 distinguishable groups on the basis of their polymeric structure and microscopic characteristics.[8]

Textile chemists have devised numerous tests for determining the class of a fiber. However, unlike the textile chemist, the criminalist frequently does not have the luxury of having a substantial quantity of fabric to work with and must therefore select tests that will yield the most information with the least amount of material. Only a single fiber may be available for analysis, and often this may amount to no more than a minute strand recovered from a fingernail scraping of a homicide or rape victim.

Birefringence. A useful physical property of fibers is that many manufactured fibers exhibit double refraction or birefringence (discussed in Chapter 5). Synthetic fibers are manufactured by melting a polymeric substance or dissolving it in a solvent and then forcing it through the fine holes of a spinneret. The polymer emerges as a fine filament, with its molecules aligned parallel to the length of the filament (see Figure 12–16). Just as the regular arrangement of atoms produces a crystal, so will the regular arrangement of the fiber's polymers cause crystallinity in the finished fiber. This crystallinity makes a fiber stiff and strong and gives it the optical property of double refraction.

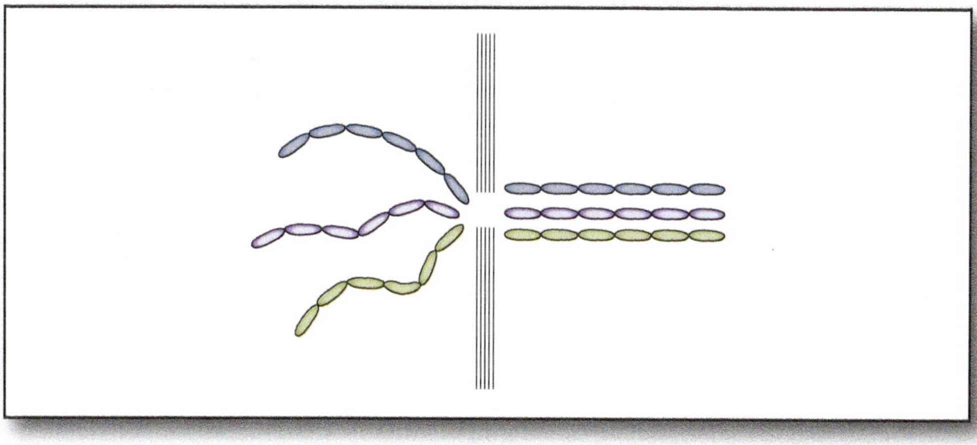

FIGURE 12–16 In the production of manufactured fibers, the bulk polymer is forced through small holes to form a filament in which all the polymers are aligned in the same direction.

Polarized white light passing through a synthetic fiber is split into two rays that are perpendicular to each other, causing the fiber to display polarization or interference colors when viewed under a polarizing microscope (see Figure 12–17). Depending on the class of fiber, each polarized plane of light has a characteristic index of refraction. This value can be determined by immersing the fiber in a fluid with a comparable refractive index and observing the disappearance of the Becke line under a polarizing microscope. Table 12–2 lists the two refractive indices of some common classes of fibers, along with their birefringence. The virtue of this technique is that a single microscopic fiber can be analyzed in a nondestructive manner.

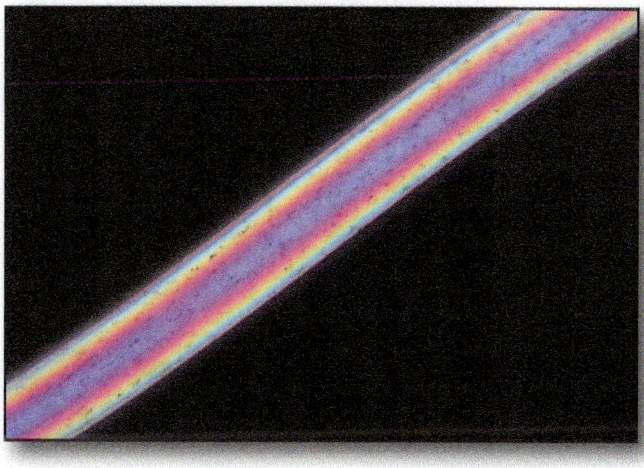

FIGURE 12–17 A photomicrograph of a nylon fiber displaying interference colors when observed between the crossed polars of a polarizing microscope (100x). *Courtesy Chris Palenik, Microtrace LLC, Elgin, IL.*

Infrared absorption. The polymers that compose a manufactured fiber, just as in any other organic substance, selectively absorb infrared light in a characteristic pattern. Infrared spectrophotometry thus provides a rapid and reliable method for identifying the generic class, and in some cases the subclasses, of fibers. The infrared microspectrophotometer combines a microscope with an infrared spectrophotometer. Such a combination makes possible the infrared analysis of a small single-strand fiber while it is being viewed under a microscope.[9]

Table 12–2
Refractive Indices of Common Textile Fibers

	Refractive Index		
Fiber	**Parallel**	**Perpendicular**	**Birefringence**
Acetate	1.478	1.477	0.001
Triacetate	1.472	1.471	0.001
Acrylic	1.524	1.520	0.004
Nylon			
Nylon 6	1.568	1.515	0.053
Nylon 6–6	1.582	1.519	0.063
Polyester			
Dacron	1.710	1.535	0.175
Kodel	1.642	1.540	0.102
Modacrylic	1.536	1.531	0.005
Rayon			
Cuprammonium rayon	1.552	1.520	0.032
Viscose rayon	1.544	1.520	0.024

Note: The listed values are for specific fibers, which explains the precise values given. In identification work, such precision is not practical; values within 0.02 or 0.03 of those listed will suffice.

Fabric Burn Test. Also referred to as a flame test, this method is used to identify the type of fibers in fabric. The burn, or flame test, is performed by burning a small section of the fabric and observing the flame as it burns, detecting any odors produced while burning, noting the color and composition of any smoke produced, then inspecting the residue once the flame has gone out. Take note of how the individual threads behave as they near the flame. Cotton fibers will ignite as they get close to the flame, while synthetic fibers tend to curl away from the heat source and melt.

Case Files

Jeffrey MacDonald: Fatal Vision

The grisly murder scene that confronted police on February 17, 1970, is one that cannot be wiped from memory. Summoned to the Fort Bragg residence of Captain Jeffrey MacDonald, a physician, police found the bludgeoned body of MacDonald's wife. She had been repeatedly knifed, and her face was smashed to a pulp. MacDonald's two children, ages 2 and 5, had been brutally and repeatedly knifed and battered to death. Suspicion quickly fell on MacDonald. To the eyes of investigators, the murder scene had a staged appearance. MacDonald described a frantic effort to subdue four intruders who had slashed at him with an ice pick. However, the confrontation left MacDonald with minor wounds and no apparent defense wounds on his arms. MacDonald then described how he had covered his slashed wife with his blue pajama top. Interestingly, when the body was removed, blue threads were observed under the body. In fact, blue threads matching the pajama top turned up throughout the house—19 in one

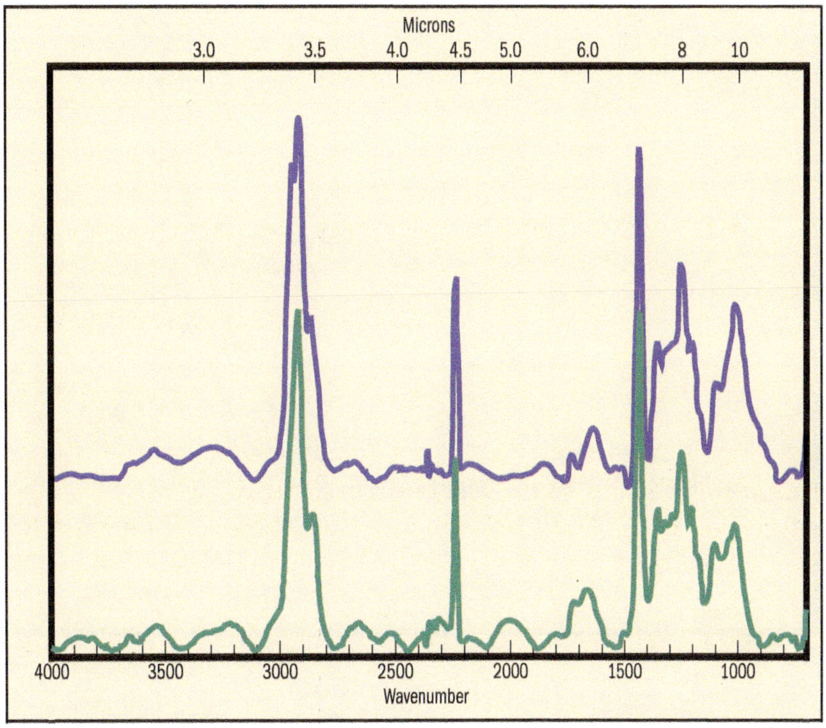

FIGURE 1 A fiber comparison made with an infrared spectrophotometer. The infrared spectrum of a fiber from Mrs. MacDonald's fall compares to a fiber recovered from a hairbrush in the MacDonald home. These fibers were identified as modacrylics, the most common type of synthetic fiber used in the manufacture of human hair goods. *Courtesy SA Michael Malone, FBI Laboratory, Washington, D.C. Courtesy SA Michael Malone, FBI Laboratory, Washington, D.C.*

Case Files

child's bedroom, including one beneath her fingernail, and two in the other child's bedroom. Eighty-one blue fibers were recovered from the master bedroom, and two were located on a bloodstained piece of wood outside the house. Later forensic examination showed that the 48 ice pick holes in the pajama top were smooth and cylindrical, a sign that the top was stationary when it was slashed. Also, folding the pajama top demonstrated that the 48 holes actually could have been made by 21 thrusts of an ice pick. This coincided with the number of wounds that MacDonald's wife sustained. As described in the book Fatal Vision, which chronicled the murder investigation, when MacDonald was confronted with adulterous conduct, he replied, "You guys are more thorough than I thought." MacDonald is currently serving three consecutive life sentences.

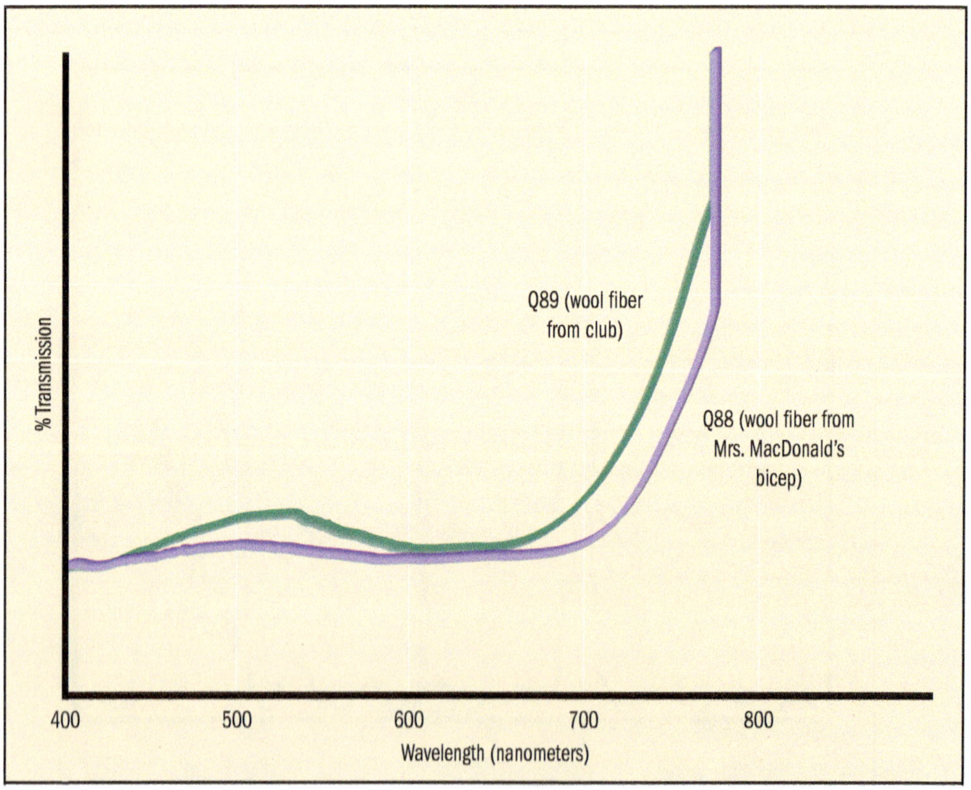

FIGURE 2 The visible-light spectrum for the woolen fiber recovered from Mrs. MacDonald's body is clearly different from that of the fiber recovered from the club used to assault her. *Courtesy SA Michael Malone, FBI Laboratory, Washington, D.C.*

Case Files

Eleven years after this conviction, MacDonald's attorneys filed a petition for a new trial, claiming the existence of "critical new" evidence. The defense asserted that wig fibers found on a hairbrush in the MacDonald residence were evidence that an intruder dressed in a wig entered the MacDonald home on the day of the murder. Subsequent examination of this claim by the FBI Laboratory focused on a blond fall frequently worn by MacDonald's wife. Fibers removed from the fall were shown to be consistent with fibers on the hairbrush. The examination included the use of infrared microspectrophotometry to demonstrate that the suspect's wig fibers were chemically identical to fibers found in the composition of the MacDonald fall (see the figure). Hence, although wig fibers were found at the crime scene, the source of these fibers could be accounted for—they came from Mrs. MacDonald's fall.

A fiber comparison made with an infrared spectrophotometer. The infrared spectrum of a fiber from Mrs. MacDonald's fall compares to a fiber recovered from a hairbrush in the MacDonald home. These fibers were identified as modacrylics, the most common type of synthetic fiber used in the manufacture of human hair goods.

The visible-light spectrum for the woolen fiber recovered from Mrs. MacDonald's body is clearly different from that of the fiber recovered from the club used to assault her.

Another piece of evidence cited by MacDonald's lawyers was a bluish-black woolen fiber found on the body of Mrs. MacDonald. They claimed that this fiber compared to a bluish-black woolen fiber recovered from the club used to assault her. These wool fibers were central to MacDonald's defense that the "intruders" wore dark-colored clothing. Initial examination showed that the fibers were microscopically indistinguishable. However, the FBI also compared the two wool fibers by visible-light microspectrophotometry. Comparison of their spectra clearly showed that their dye compositions differed, providing no evidence of outside intruders (see the figure). Ultimately, the U.S. Supreme Court denied the merits of MacDonald's petition for a new trial.

Fatal Vision Revisited

Eleven years after this conviction, MacDonald's attorneys filed a petition for a new trial, claiming the existence of "critical new" evidence.

The defense asserted that wig fibers found on a hairbrush in the MacDonald residence were evidence that an intruder dressed in a wig entered the MacDonald home on the day of the murder. Subsequent examination of this claim by the FBI Laboratory focused on a blond fall (a type of artificial hair extension) frequently

Jeffrey MacDonald in 1995 at Sheridan, Oregon, Federal Correctional Institution. *Shane Young/AP/Shutterstock*

Case Files

worn by MacDonald's wife. Fibers removed from the fall were shown to clearly match fibers on the hairbrush. The examination included the use of infrared microspectrophotometry to demonstrate that the suspect wig fibers were chemically identical to fibers found in the composition of the MacDonald fall (see Figure 1). Hence, although wig fibers were found at the crime scene, the source of these fibers could be accounted for—they came from Mrs. MacDonald's fall.

Another piece of evidence cited by MacDonald's lawyers was a bluish-black woolen fiber found on the body of Mrs. MacDonald. They claimed that this fiber compared to a bluish-black woolen fiber recovered from the club used to assault her. These wool fibers were central to MacDonald's defense that the "intruders" wore dark-colored clothing. Initial examination showed that the fibers were microscopically indistinguishable. However, the FBI also compared the two wool fibers by visible-light microspectrophotometry. Comparison of their spectra clearly showed that their dye compositions differed, providing no evidence of outside intruders (see Figure 2). Ultimately, the U.S. Supreme Court denied the merits of MacDonald's petition for a new trial.

Source: Data from B. M. Murtagh and M. P. Malone, "Fatal Vision Revisited," The Police Chief (June 1993): 15.

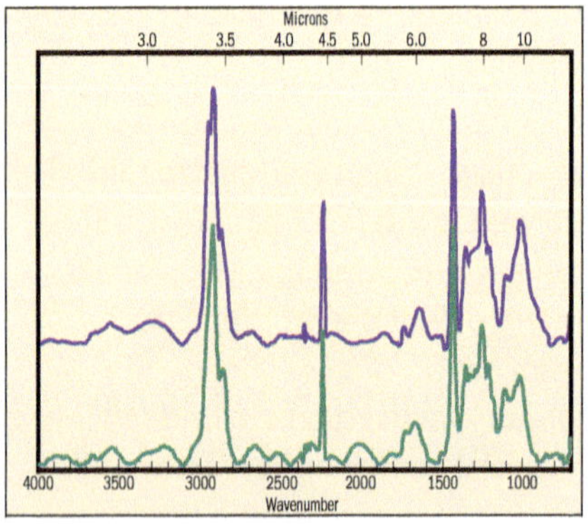

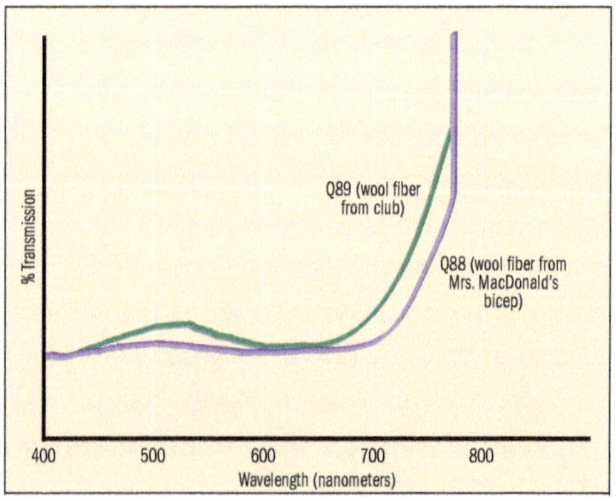

Significance of Fiber Evidence

Once a fiber match has been determined, the question of the significance of such a finding is bound to be raised. No analytical technique permits the criminalist to associate a fiber strand definitively to any single garment. Furthermore, except in the most unusual circumstances, no statistical databases are available for determining the probability of a fiber's origin. Considering the mass distribution of synthetic fibers and the constantly changing fashion tastes of our society, it is highly unlikely that such data will be available in the foreseeable future.

Despite these limitations, one should not discount or minimize the significance of a fiber association. An enormous variety of fibers exists in our society. By simply looking at the random individuals we meet every day, we can see how unlikely it is to find two different people wearing identically colored fabrics (with the exception of blue denims or white cottons). There are thousands of different-colored fibers in our environment. Combine this with the fact that forensic scientists compare not only the color of fibers but also their size, shape, microscopic appearance, chemical composition, and dye content, and one can now begin to appreciate how unlikely it is to find two indistinguishable colored fibers emanating from randomly selected sources.

Furthermore, the significance of a fiber association increases dramatically if the analyst can link two or more distinctly different fibers to the same object. Likewise, the associative value of fiber evidence is dramatically enhanced if it is accompanied by other types of physical evidence linking a person or object to a crime. As with most class evidence, the significance of a fiber comparison is dictated by the circumstances of the case; by the location, number, and nature of the fibers examined; and, most important, by the judgment of an experienced examiner.

Collection and Preservation of Fiber Evidence

As criminal investigators have become more aware of the potential contribution of trace physical evidence to the success of their investigations, they have placed greater emphasis on conducting thorough crime-scene searches for evidence of forensic value. Their skill and determination at carrying out these tasks is tested in the collection of fiber-related evidence. Fiber evidence can be associated with virtually any type of crime. It usually cannot be seen with the naked eye and thus easily can be overlooked by someone not specifically searching for it.

An investigator committed to optimizing the laboratory's chances for locating minute strands of fibers identifies and preserves potential "carriers" of fiber evidence. Relevant articles of clothing should be packaged carefully in paper bags. Each article must be placed in a separate bag to avoid cross-contamination of evidence. Scrupulous care must be taken to prevent articles of clothing from different people or from different locations from coming into contact. Such articles must not even be placed on the same surface before packaging. Likewise, carpets, rugs, and bedding are to be folded carefully to protect areas suspected of containing fibers. Car seats should be carefully covered with polyethylene sheets to protect fiber evidence, and knife blades should be covered to protect adhering fibers. If a body is thought to have been wrapped at one time in a blanket or carpet, adhesive tape lifts of exposed body areas may reveal fiber strands.

Occasionally the field investigator may need to remove a fiber from an object, particularly if loosely adhering fibrous material may be lost in transit to the laboratory. These fibers must be removed with a clean forceps and placed in a small sheet of paper, which, after folding and labeling, can be placed inside another container. Again, scrupulous care must be taken to prevent contact between fibers collected from different objects or from different locations.

In the laboratory, the search for fiber evidence on clothing and other relevant objects, as well as in debris, is time consuming and tedious and will test the skill and patience of the examiner. The crime-scene investigator can manage this task by collecting only relevant items for examination. The crime-scene investigator must pinpoint areas where a likely transfer of fiber evidence occurred and then ensure proper collection and preservation of these materials.

Quick Review

- Fibers may be classified into two broad groups: *natural* and *manufactured*.
- Most fibers currently manufactured are produced solely from synthetic chemicals and are therefore classified as *synthetic fibers*. They include nylons, polyesters, and acrylics.
- Microscopic comparisons between questioned and standard/reference fibers are initially undertaken for color and diameter characteristics. Other features that could be important in comparing fibers are striations on the surface of the fiber, the presence of delustering particles, and the cross-sectional shape of the fiber.
- The visible-light microspectrophotometer is a convenient way for analysts to compare the colors of fibers through spectral patterns.
- Infrared spectrophotometry and the polarizing microscope are reliable methods for identifying the chemical composition of fibers.
- Fiber evidence collected at each location should be placed in separate containers to avoid cross-contamination. Care must be taken to prevent articles of clothing from different people or from different locations from coming into contact.

Chapter Review

- The hair shaft is composed of three layers called the cuticle, cortex, and medulla and is most intensely examined by the forensic scientist.
- When comparing strands of hair, the criminalist is particularly interested in matching the color, length, and diameter. Other important features for comparing hair are the presence or absence of a medulla and the distribution, shape, and color intensity of pigment granules in the cortex.
- The probability of detecting DNA in hair roots is more likely for hair being examined in its anagen or early growth phase as opposed to its catagen or telogen phases.
- The follicular tag, a translucent piece of tissue surrounding the hair's shaft near the root, is a rich source of DNA associated with hair. Mitochondrial DNA can also be extracted from the hair shaft.
- All positive microscopic hair comparisons must be confirmed by DNA analysis.
- Fibers may be classified into two broad groups: natural and manufactured (synthetic).
- Most fibers currently manufactured are produced solely from *synthetic chemicals* and are therefore classified as synthetic fibers. They include nylons, polyesters, and acrylics.
- Microscopic comparisons between questioned and standard/reference fibers are initially undertaken for color and diameter characteristics. Other features that could be important in comparing fibers are striations on the surface of the fiber, the presence of delustering particles, and the cross-sectional shape of the fiber.
- The visible-light microspectrophotometer is a convenient way for analysts to compare the colors of fibers through spectral patterns.
- Infrared spectrophotometry and the polarizing microscope are reliable methods for identifying the chemical composition of fibers.
- Fiber evidence collected at each location should be placed in separate containers to avoid cross-contamination. Care must be taken to prevent articles of clothing from different people or from different locations from coming into contact.

Review Questions

1. The ______________ is important for the individualization of hair.
   a. cuticle
   b. cortex
   c. follicular tag
   d. medulla

2. The final growth phase in which hair naturally falls out of the skin is called the
   a. telogen phase.
   b. anagen phase.
   c. catagen phase.
   d. follicular phase.

3. The most prevalent plant fiber is
   a. hemp.
   b. cotton.
   c. wool.
   d. mohair.

4. Which of the following is not a manufactured fiber grouping?
   a. polyester
   b. rayon
   c. mohair
   d. spandex

5. In the examination of fibers, the first and most important step in the examination will be
   a. a microscopic comparison for color and diameter using a comparison microscope.
   b. a determination of whether the fiber is natural or manufactured.
   c. synthesizing long-chained molecules into a polymer.
   d. analyzing the individual characteristics of the material.

6. True or False: Because of advances in forensic technology and the equipment available, it is now possible to individualize human hair through its morphology.

7. True or False: Two of the features that make hair a good subject for establishing individual identity are its resistance to chemical decomposition and its ability to retain structural features over a long period of time.

8. True or False: Most often, when hair evidence is present in a criminal case, the primary purpose is to establish the identity of the individual when no other means is available.

9. True or False: The ultimate value of fibers as forensic evidence will depend on the criminalist's ability to narrow their origin to a limited number of sources or even to a single source.

10. True or False: Properties frequently used to identify fibers are refractive index and an IR spectrum.

11. What is hair and what organ produces it?

12. Name and briefly define the three layers of the hair shaft.

13. What two features make hair a good subject for establishing individual identity? To which layer of the hair shaft are much of these features attributed?

14. Why is the scale pattern of the cuticle an important feature for characterizing hair?

15. What is the main forensic importance of the cortex?

16. What is the difference between the medullae of human and animal hairs? Name one exception to this among humans.

17. Name the three phases of hair growth. A criminalist is more likely to collect DNA from hairs in which stage of growth? Why?

18. Why must microscopic human hair comparisons be undertaken with extreme caution?

19. In comparing hairs, what aspects of the hair is the criminalist particularly interested in matching? Name at least one other important feature that the criminalist may compare.

20. Which of the following questions cannot be answered with a microscopic examination of hair?
    a. whether a hair came from a 25-year-old or an infant
    b. whether a hair is from a man or a woman
    c. whether a hair is from a scalp or a beard
    d. whether the hair is consistent with European or African hair

21. What types of hairs found at a crime scene are most likely to provide useful DNA evidence? Why?

22. What part of a hair is most likely to yield useful DNA evidence?

23. Why must questioned hairs and standard/reference hairs being compared come from the same area of the body?

24. Why should the entire hair be collected when performing a hair comparison?

25. Define polymer and monomer. Why are polymers sometimes known as macromolecules?

26. What two morphological characteristics does a criminalist first compare when examining fibers with a microscope? What other features may be important in such a comparison?

27. What analytical technique does a criminalist use to analyze the composition of the dye in a fiber?

28. Describe three analytical techniques for comparing the color of two fibers.

29. What microscopic characteristics can be used to determine the origin of a human hair sample? Describe and illustrate an example of each.

30. How can you compare the microscopic characteristics of human and nonhuman hair, looking at medulla, pigment distribution, and scale patterns? Summarize your findings in writing.

31. How can you differentiate between natural and manufactured (synthetic) fibers?

Learn the Vocabulary

Use the information from this chapter along with your prior knowledge and experience to learn the key terms. Choose from the following activities:

1. Make flashcards with a term on the front and its definition on the back. Use the flashcards to quiz a partner on correctly pronouncing and defining the terms.
2. Make an audio or video recording of each term and its definition.
3. Work with a partner to write a sentence for each term. Read the sentences out loud to a partner or share them with the class.
4. Write a list of new terms and language you have encountered in this chapter. Reuse the terms to write sentences and then read the sentences out loud to a partner. Reuse the terms again by writing a short speech. The speech might explain the terms or teach a lesson using the terms. Give the speech to a partner or to the class.

Application and Critical Thinking

1. Indicate the phase of growth of each of the following hairs:
   a. the root is club-shaped
   b. the hair has a follicular tag
   c. the root bulb is flame-shaped
   d. the root is elongated
2. A criminalist studying a dyed sample hair notices that the dyed color ends about 1.5 centimeters from the tip of the hair. Approximately how many weeks before the examination was the hair dyed? Explain your answer.
3. Following are descriptions of several hairs; based on these descriptions, indicate the likely ancestry of the person from whom the hair originated.
   a. evenly distributed, fine pigmentation
   b. continuous medullation
   c. dense, uneven pigmentation
   d. wavy with a round cross-section
4. Criminalist Pete Evett is collecting fiber evidence from a murder scene. He notices fibers on the victim's shirt and trousers, so he places both of these items of clothing in a plastic bag. He also sees fibers on a sheet near the victim, so he balls up the sheet and places it in separate plastic bag. Noticing fibers adhering to the windowsill from which the attacker gained entrance, Pete carefully removes them with his fingers and places them in a regular envelope. What mistakes, if any, did Pete make while collecting this evidence? Describe the proper procedures to collect hair evidence at this crime scene.
5. For each of the following human hair samples, indicate the medulla pattern present.

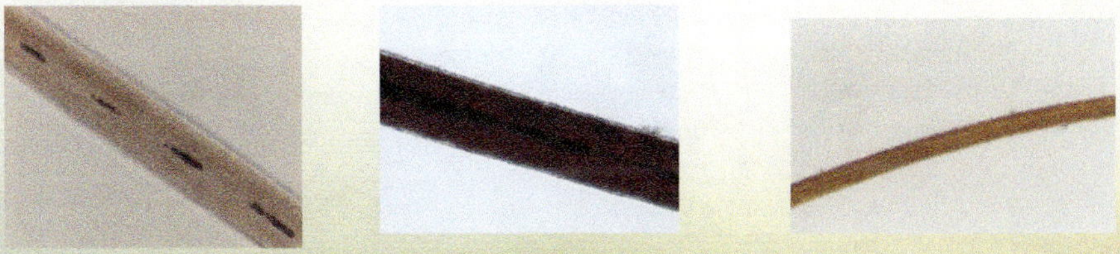

A.________ B.________ C.________

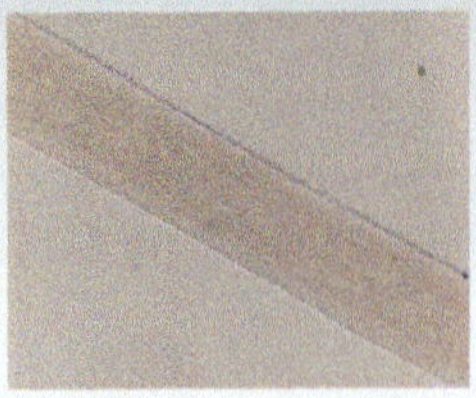
D.________

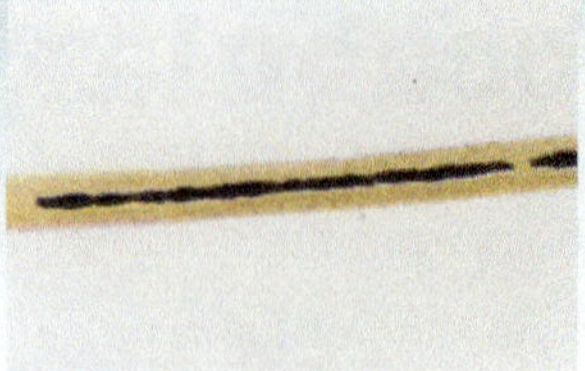
E.________

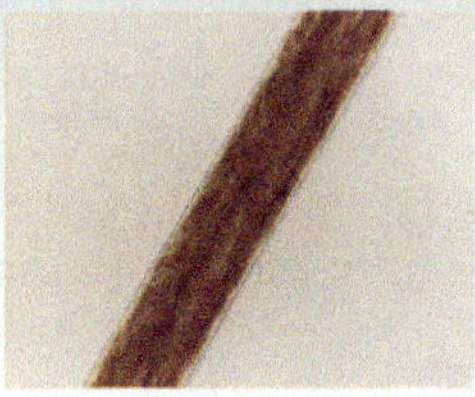
F.________

G.________

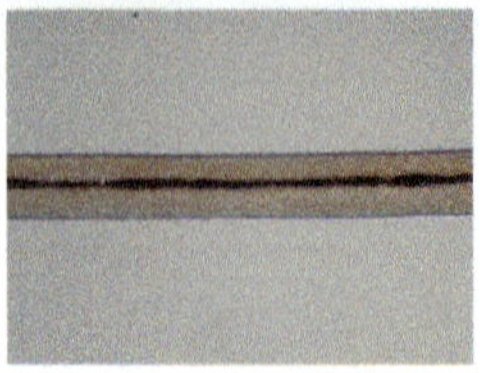
H.________

I.________

Courtesy Richard Saferstein, Ph.D.

6. The most common scale patterns found on hairs are generally classified as coronal, spinous, and imbricate. Examine the scale casts of animal hairs shown here and indicate the scale pattern of each.

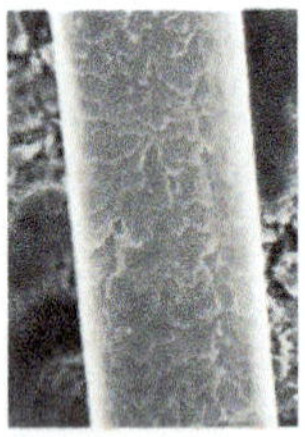
A.________

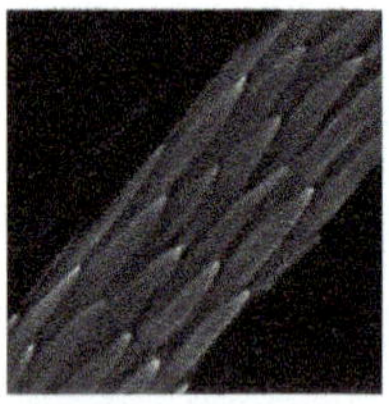
B.________

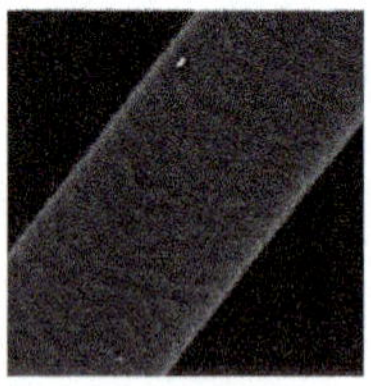
C.________

D.________

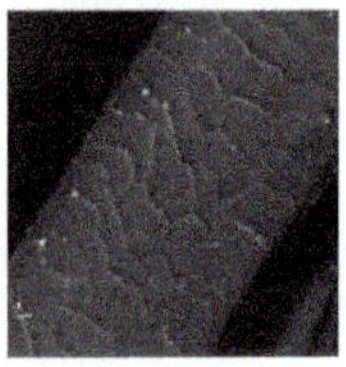
E.________

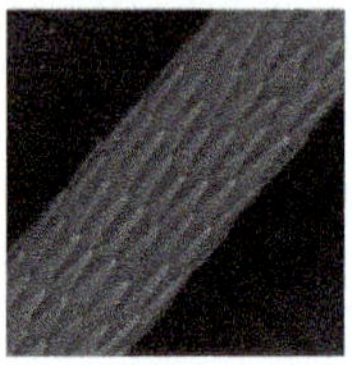
F.________

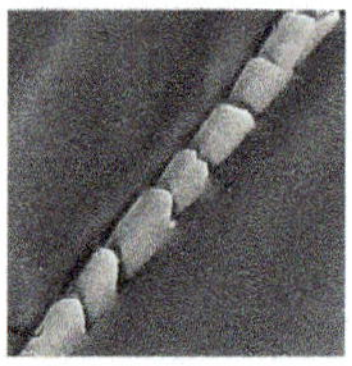
G.________

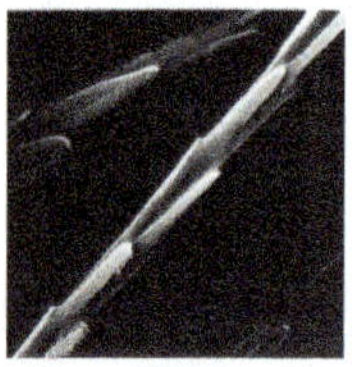
H.________

Courtesy Richard Saferstein, Ph.D.

7. A young child is kidnapped from her school playground. Shown on the left is a reference sample of the kidnapped child's hair. The only cars that left the parking lot before the child was discovered to be missing were those of four cafeteria workers. The car of each worker was searched and hairs collected. These recovered hairs are shown on the right. Which recovered hair, if any, is consistent with that of the victim and warrants further investigation?

Reference Hair from Victim

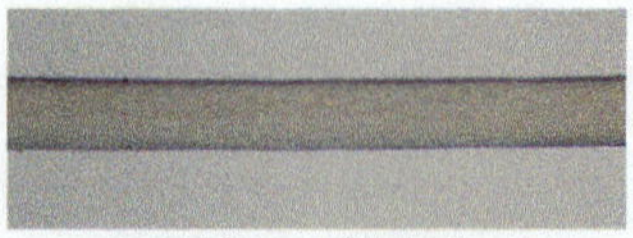
Hair from car of Worker A

Hair from car of Worker B

Hair from car of Worker C

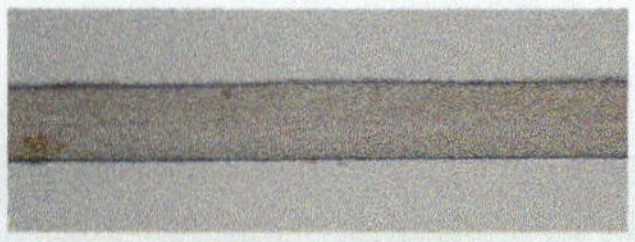
Hair from car of Worker D

Courtesy Richard Saferstein, Ph.D.

Laboratory Experiments

Teacher should supply students with sample hairs and fibers to be analyzed in this experiment. Students will demonstrate how to analyze hairs using a compound light microscope and analyze the fibers using flame testing and other chemical test to identify the fibers using forensic science methods. This activity requires the use of the following practices of science and engineering:

- Asking questions and defining problems
- Planning and carrying out investigations
- Analyzing and interpreting data
- Engaging in argument
- Obtaining, evaluating, and communication information
- Using appropriate safety equipment and practices as outlined in Texas Education Agency-approved safety standards

This activity consists of the following crosscutting concepts:

- **Patterns**—Patterns in the medulla allow for the classification of hair.
- **Cause and effect**—If hair is found to be present at a crime scene that does not belong to the victim the hair is most likely due to physical contact, possible with an assailant. (Locard's principle)

Before you begin, read the scenario and steps out loud with a partner. Ask for clarification as necessary to make sure you understand the experiment.

The Scenario

Crime scene investigators are called to the scene of a brutal attack. A young man with blond hair wearing a red polyester jacket has been killed by what appears to be blunt force trauma to the head. A number of eyewitnesses have identified an individual of Asian descent wearing a blue cotton hooded sweatshirt observed near the area where the body was found at about the time of the slaying. The figures below show photomicrographs of hairs found on the body that do not belong to the victim. You have been called in to conduct an expert analysis of this evidence and to advise the district attorney and police if there is enough physical evidence to consider the Asian individual a suspect.

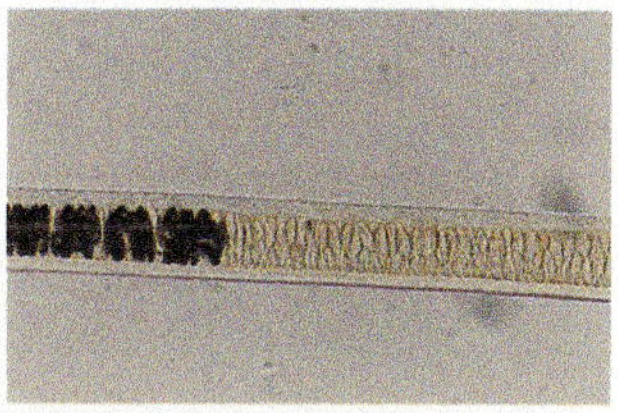
FIGURE 1

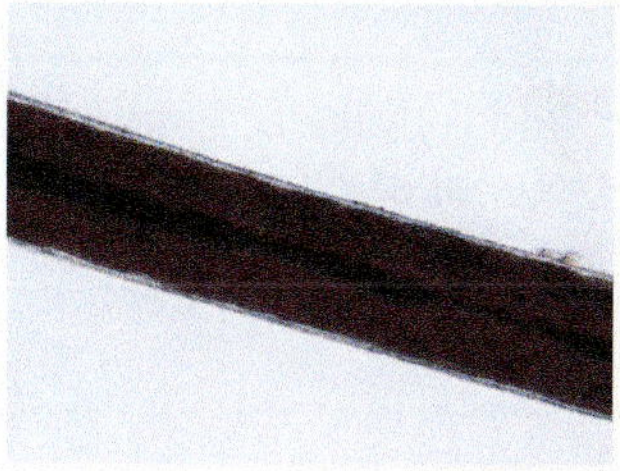
FIGURE 2

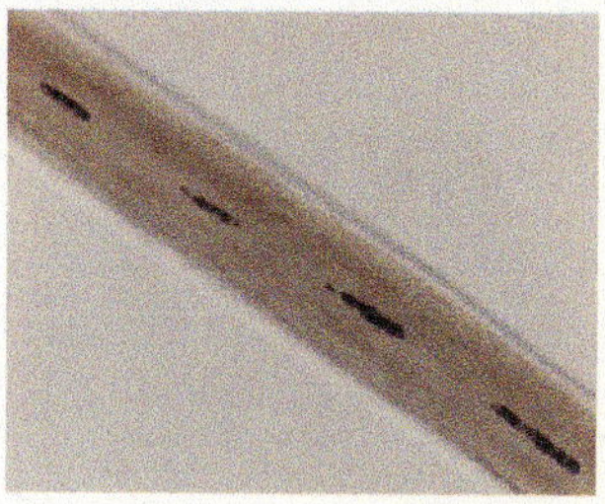
FIGURE 3

FIGURE 4

Courtesy Deedrick, D. W., and S. L. Koch, "Microscopy of Hair Part I: A Practical Guide and Manual for Human Hairs," Forensic Science Communications 6, no. 1 (2004), http://www.fbi.gov/about-us/lab/forensic-science-communications/fsc/jan2004

1. Demonstrate how to collect hair and fiber evidence at this crime scene.
2. Analyze the hair and fiber evidence using microscopy and flame testing.

HAIR EXAMINATION

Complete the following table.

Hair Sample	*Animal or Human*	*Medulla—continuous, interrupted, fragmented, absent*	*Pigment*	*Possible Ancestry*

FIBER ANALYSIS

Complete the following table.

Fiber Sample	*Natural or Synthetic*	*Flame test: approach/smoke/residue*

Endnotes

1. J. D. Baker and D. L. Exline, Forensic Animal Hair Atlas: A Searchable Database on CD-ROM. RJ Lee Group, Inc., 350 Hochberg Rd., Monroeville, PA 15146.
2. M. M. Houk and B. Budowle, "Correlation of Microscopic and Mitochondrial DNA Hair Comparisons," *Journal of Forensic Sciences* 47 (2002): 964.
3. L. A. King, R. Wigmore, and J. M. Twibell, "The Morphology and Occurrence of Human Hair Sheath Cells," *Journal of the Forensic Science Society* 22 (1982): 267.
4. C. A. Linch et al., "Evaluation of the Human Hair Root for DNA Typing Subsequent to Microscopic Comparison," *Journal of Forensic Sciences* 43 (1998): 305.
5. S. L. Koch, A. L. Michaud, and C. Mikell,, "Taphonomy of Hair-A Study of Postmortem Root Banding," *Journal of Forensic Sciences* 58 (2013): S52.
6. S. Palenik and C. Fitzsimons, "Fiber Cross-Sections: Part I," *Microscope* 38 (1990): 187.
7. D. K. Laing et al., "The Standardisation of Thin-Layer Chromatographic Systems for Comparisons of Fibre Dyes," *Journal of the Forensic Science Society* 30 (1990): 299.
8. M. C. Grieve, "Another Look at the Classification of Acrylic Fibres, Using FTIR Microscopy," *Science & Justice* 35 (1995): 179.
9. M. W. Tungol et al., "Analysis of Single Polymer Fibers by Fourier Transform Infrared Microscopy: The Results of Case Studies," *Journal of Forensic Sciences* 36 (1992): 1027.

Trace Evidence II: Metals, Paint, and Soil

13

Pierre Barlier/Abaca Press/Alamy Stock Photo

Key Terms

alpha particle	electron orbital	mineral
atomic mass	emission spectrum	neutron
atomic number	excited state	nucleus
beta particle	gamma ray	proton
continuous spectrum	isotope	pyrolysis
electron	line spectrum	radioactivity

The Green River Killer

This case takes its name from the Green River, which flows through Washington state and empties into Puget Sound in Seattle. In 1982, the bodies of five females were discovered in or near the river within six months. Most of the victims were known prostitutes who were strangled and apparently raped. As police focused their attention on an area known as Sea-Tac Strip, a haven for prostitutes, girls mysteriously disappeared with increasing frequency. By the end of 1986, the body count in the Seattle region rose to 40, all of whom were believed to have been murdered by the Green River Killer. As the investigation pressed on into 1987, the police renewed their interest in one suspect, Gary Ridgway, a local truck painter. Interestingly, in 1984 Ridgway had passed a lie detector test. Now with a search warrant in hand, police searched the Ridgway residence and also obtained hair and saliva samples from Ridgway. Again, insufficient evidence caused Ridgway to be released from custody. However, as the investigation proceeded, a DNA link between Ridgway and his victims eluded investigators. Ultimately, a careful microscopic search of Ridgway's clothing revealed the presence of paint spheres of various colors, which compared to spheres on the clothing of six of the victims. The paint was microscopically and chemically identified as Imron, a high-end specialty paint that was manufactured before 1984. This product had been used at the truck plant where Ridgway worked and was identified as dried paint spheres emanating from a spray paint. Two of the victims were further linked to Ridgway through DNA, further solidifying the case against Ridgway. Ridgway avoided the death penalty by confessing to the murders of 48 women.

Learning Objectives

After studying this chapter, you should be able to:

13.1 Describe the utility of trace elements for forensic comparison of various types of physical evidence

13.2 Describe the Emission Spectrum of Elements as it relates to the analysis of trace evidence

13.3 Explain how paint is examined, collected, and preserved in the laboratory

13.4 Discuss the processes in the forensic analysis of soil

13.5 Describe the proper collection of soil evidence

National Science Content Standards

Scientific Inquiry

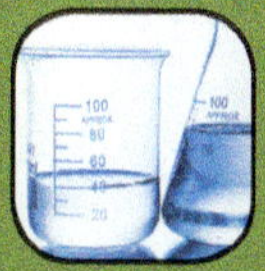
Science and Technology

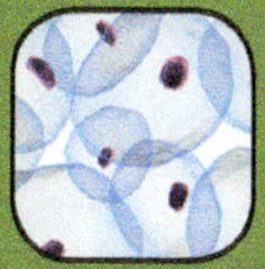
Physical Science

Science in Personal and Social Perspective

Life Science

Forensic Analysis of Trace Elements

Considering that most of our raw materials originate from the earth's crust, it is not surprising that they are rarely obtained in pure form; instead, they include numerous elemental impurities that usually have to be eliminated through industrial processing. However, in most cases it is not economically feasible to completely exclude all such minor impurities, especially when their presence will have no effect on the appearance or performance of the final product. For this reason, many manufactured products, and even most natural materials, contain small quantities of elements present in concentrations of less than 1 percent.

For the criminalist, the presence of *trace elements* is particularly useful because they provide "invisible" markers that may establish the source of a material or at least provide additional points for comparison. Glass fragments represent a valuable type of trace evidence; however, generally because of their minute size they present the criminalist with two distinct issues. First is classifying the type of glass being examined. Three types of glass are normally encountered as forensic evidence; float glass (windows, windshields), container glass (bottles, jars), and borosilicates (kitchenware). Figure 13–1 depicts an elemental analysis comparing three different glasses and clearly shows how they are distinguished by the intensity of the peaks associated with boron.

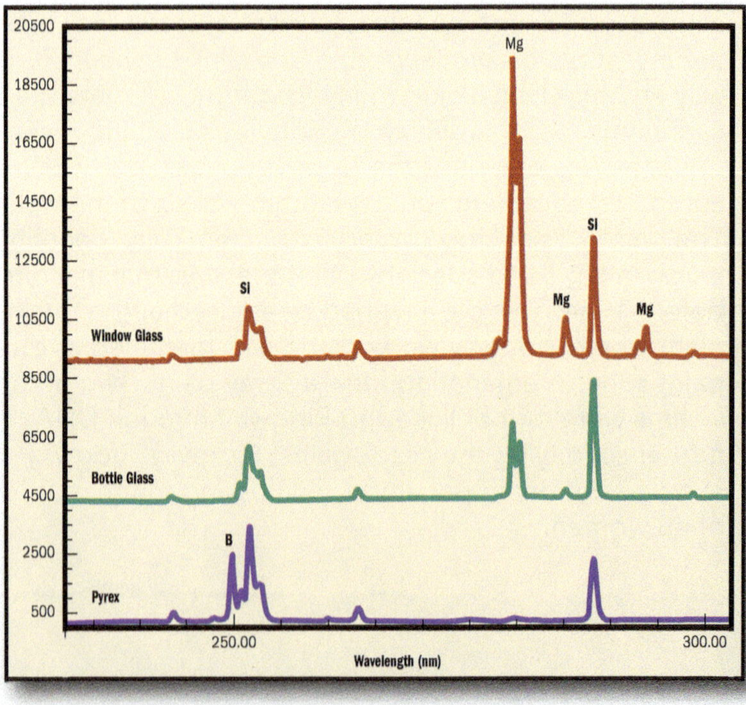

FIGURE 13–1 The presence of trace elements in glass as shown above can be used to identify glass types. *Courtesy Foster & Freeman*

Similarly, the comparison of trace elements present in glass may provide particularly meaningful data with respect to source or origin. Technological advances in the manufacture of glass has led to more uniformity in the final product. Unfortunately, this has the consequence of diminishing the value of the most important comparative physical property—refractive index. Fortunately, minor variations in the chemical composition of glass remain between and within batches because of the presence of natural contaminants in raw materials. Up to 25 different elements have been identified in glass. The forensic discrimination associated with glass comparisons can now be enhanced by combining elemental analysis with refractive index (see Figure 13–2). Forensic investigators have also examined the evidential value of trace elements present in soil, fibers, and paint, as well as in all types of metallic objects. One example of this application occurred with the examination of the bullet and bullet fragments recovered after the assassination of President Kennedy.

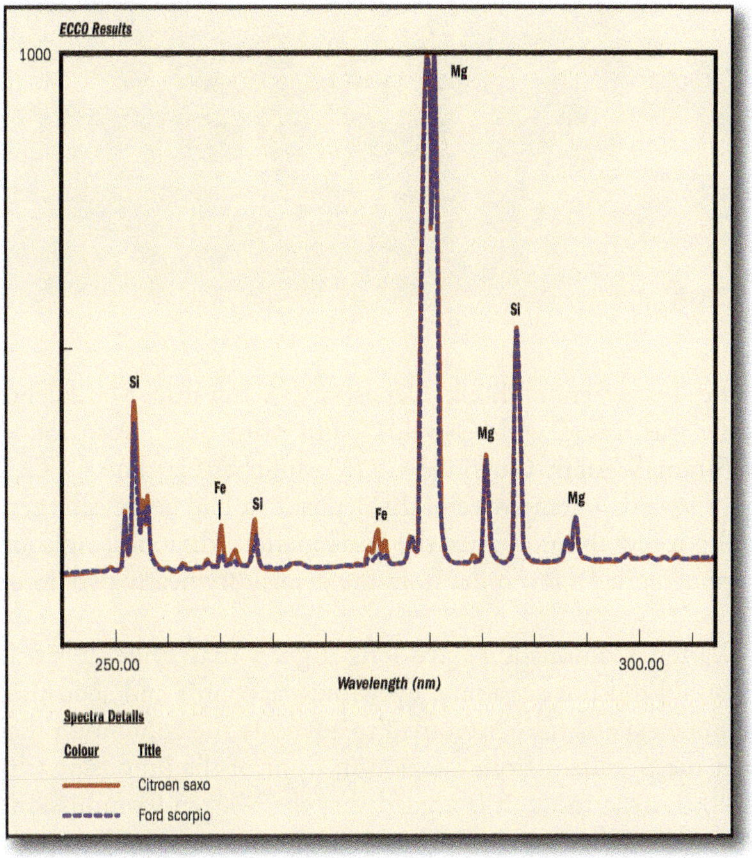

FIGURE 13–2 The presence of trace elements in glass as shown here can be used to discriminate between glass particles that are indistinguishable by other test protocols, such as refractive index. *Courtesy Foster & Freeman*

Evidence in the Assassination of President Kennedy

Ever since President Kennedy was killed in 1963, questions have lingered about whether Lee Harvey Oswald was part of a conspiracy to assassinate the president or, as the Warren Commission concluded, a lone assassin. In arriving at its conclusions, the Warren Commission reconstructed the crime as follows: Oswald fired three shots from behind the president while positioned in the Texas School Book Depository building. The president was struck by two bullets, with one bullet totally missing the president's limousine. One bullet hit the president in the back, exited his throat, and then went on to strike Governor Connally, who was sitting in a jump seat in front of the president. The bullet hit Connally first in his back, then exited his chest, struck his right wrist, and temporarily lodged in his left thigh. This bullet was later found on the governor's stretcher at the hospital. A second bullet in the skull fatally wounded the president (see Figure 13–3).

FIGURE 13–3
President John F. Kennedy, Governor John Connally of Texas, and Mrs. Jacqueline Kennedy ride through Dallas moments before the assassination.
Courtesy Pictorial Press Ltd/Alamy Stock Photo

In a room at the Texas School Book Depository, a 6.5-mm Mannlicher Carcano military rifle was found with Oswald's palm print on it. Also found were three spent 6.5-mm Western Cartridge Co./Mannlicher-Carcano (WCC/MC) cartridge cases. Oswald, an employee of the depository, had been seen there that morning and also a few minutes after the assassination, disappearing soon thereafter. He was apprehended a few miles from the depository nearly two hours after the shooting.

Critics of the Warren Commission have long argued that evidence exists that would prove Oswald did not act alone. Eyewitness accounts and acoustical data interpreted by some experts have been used to advocate the contention that someone else fired at the president from a region in front of the limousine (the so-called grassy knoll). Furthermore, it is argued that the Warren Commission's reconstruction of the crime relied on the assumption that only one bullet caused both the president's throat wound and Connally's back wound. Critics contend that such damage would have deformed and mutilated a bullet. Instead, the recovered bullet showed some flattening, no deformity, and only about 1 percent weight loss.

In 1977, at the request of the U.S. House of Representatives Select Committee on Assassinations, the bullet taken from Connally's stretcher along with bullet fragments recovered from the car and various wound areas were examined for trace element levels.

Lead alloys used for the manufacture of bullets contain an assortment of trace elements. For example, antimony is often added to lead as a hardening agent; copper, bismuth, and silver are other trace elements commonly found in bullet lead. In this case, the bullet and bullet fragments were compared for their antimony and silver content. Previous studies had amply demonstrated that the levels of these two elements are particularly important for characterizing WCC/MC bullets. Bullet lead from this type of ammunition ranges in antimony concentration from 20 to 1,200 parts per million (ppm) and 5 to 15 ppm in silver content.

As can be seen in Table 13–1, the samples designated Q1 and Q9 (the Connally stretcher bullet and fragments from Connally's wrist, respectively) are indistinguishable from one another in antimony and silver content. The samples Q2, Q4, Q5, and Q14 (Q4 and Q5 being fragments from Kennedy's brain, and Q2 and Q14 being fragments recovered from two different areas in the car) also are indistinguishable in antimony and silver content but are different from Q1 and Q9.

Table 13–1
Bullet and Bullet Fragments Examined in the Kennedy Assassination Investigation

Sample	Description
Q1	Connally stretcher bullet
Q9	Fragments from Connally's wrist
Q2	Large fragment from car
Q4, Q5	Fragments from Kennedy's brain
Q14	Small fragment found in car

Elemental analysis classified the bullets and fragments into two distinctly different groups. Q1 and Q9 were of similar composition containing 815 ppm[1] antimony and 9.3 ppm of silver, respectively. Q2, Q4, Q5, and Q14 fell into a second group comprised of 622 ppm of antimony and 8.1 ppm of silver, respectively. All the samples examined were consistent with WCC/MC bullet lead, although other sources could not be entirely ruled out.

[1]One part per million equals 0.0001 percent.

The conclusions derived from studying these results are as follows:

1. **There is evidence of only two bullets—one composed of 815 ppm antimony and 9.3 ppm silver, the other composed of 622 ppm antimony and 8.1 ppm silver.**
2. **Both bullets have a composition highly consistent with WCC/MC bullet lead, although other sources cannot entirely be ruled out.**
3. **The bullet found on the Connally stretcher also damaged Connally's wrist. The absence of bullet fragments from the back wounds of Kennedy and Connally prevented any effort at linking these wounds to the stretcher bullet.**

None of these conclusions can totally verify the Warren Commission's reconstruction of the assassination, but the results are at least consistent with the commission's findings. Further, in 2003, an ABC television broadcast showed the results of a 10-year 3-D computer animation study of the events of November 22, 1963. The animation graphically showed that the bullet wounds were completely consistent with Kennedy's and Connally's positions at the time of shooting, and that by following the bullet's trajectory backward they could be found to have originated from a narrow cone including only a few windows of the sixth floor of the Texas School Book Depository.

Principles and Technology of Trace Analysis

The Emission Spectrum of Elements

We have already observed that molecules can readily be characterized by their selective absorption of ultraviolet, visible, or infrared radiation. Equally significant to the analytical chemist is the knowledge that elements also selectively absorb and emit light. These observations form the basis of an important analytical technique designed to determine the elemental composition of materials—*emission spectroscopy*.

Types of Spectra The statement that elements emit light should not come as a total surprise, for one need only observe the common tungsten incandescent lightbulb or the glow of a neon light to confirm this observation. When the light emitted from a bulb or from any other light source is passed through a prism, it is separated into its component colors or frequencies. The resulting display of colors is called an **emission spectrum**. When sunlight or the light from an incandescent bulb is passed through a prism, a range of rainbow colors is produced. This emission spectrum is called a **continuous spectrum** because all the colors merge or blend into one another to form a continuous band. However, not all light sources produce such a spectrum.

emission spectrum Light emitted from a source and separated into its component colors or frequencies

continuous spectrum A type of emission spectrum showing a continuous band of colors all blending into one another

For example, if the light from a sodium lamp, a mercury arc lamp, or a neon light were passed through a prism, the resultant spectrum would consist not of a continuous band but of several individual colored lines separated by dark spaces. Here, each line represents a definite wavelength or frequency of light that is separate and distinct from all others present in the spectrum. This type of spectrum is called a line spectrum. Figure 13–4 shows the line spectra of three elements.

If a solid or liquid is vaporized and "excited" by exposure to a high temperature, each element present emits light composed of select frequencies that are characteristic of the element.

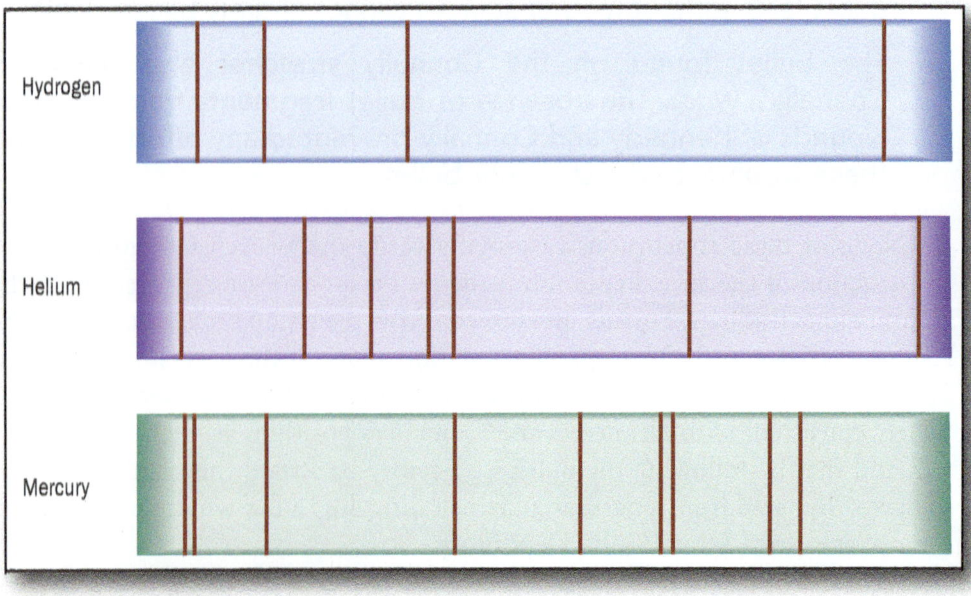

FIGURE 13–4 Some characteristic emission spectra.

This spectrum is in essence a "fingerprint" of an element and offers a practical method of identification. Sodium vapor, for example, always shows the same line spectrum, which differs from the spectrum of all other elements.

Atomic Structure

Any proposed theory that attempts to explain the origin of emission spectra must relate to the fundamental structure of the element—the atom. Scientists now know that the atom is composed of even more elementary particles that are collectively known as *subatomic particles*. The most important subatomic particles are the **proton**, **electron**, and **neutron**.

The properties of the proton, neutron, and electron are summarized in the following table:

Particle	Symbol	Relative Mass	Electrical Charge
Proton	P	1	+
Neutron	n	1	0
Electron	e	1/1837	–

As you can see, the masses of the proton and neutron are each about 1,837 times the mass of an electron. The proton has a positive electrical charge; the electron has a negative charge equal in magnitude to that of the proton; and the neutron is a neutral particle with neither a positive nor a negative charge.

A popular descriptive model of the atom, and the one that will be adopted for the purpose of this discussion, pictures an atom as consisting of electrons orbiting a central **nucleus** composed of protons and neutrons—an image that is analogous to our solar system, in which the planets revolve around the sun (see Figure 13–5).[1] To maintain a zero net electrical charge, the number of protons in the nucleus must always equal the number of electrons in orbit around the nucleus.

proton
A positively charged particle that is one of the basic structures in the nucleus of an atom

electron
A negatively charged particle that is one of the fundamental structural units of the atom

neutron
A particle with no electrical charge that is one of the basic structures in the nucleus of an atom

nucleus
The core of an atom, consisting of protons and neutrons

FIGURE 13–5
A popular model of the atom likens the electrons to planets orbiting the "sun" of the nucleus.
Montego/Shutterstock

With this knowledge, we can describe the atomic structure of the elements. For example, hydrogen has a nucleus consisting of one proton and no neutrons, and it has one orbiting electron. Helium has a nucleus comprising two protons and two neutrons, with two electrons in orbit around the nucleus (see Figure 13–6).

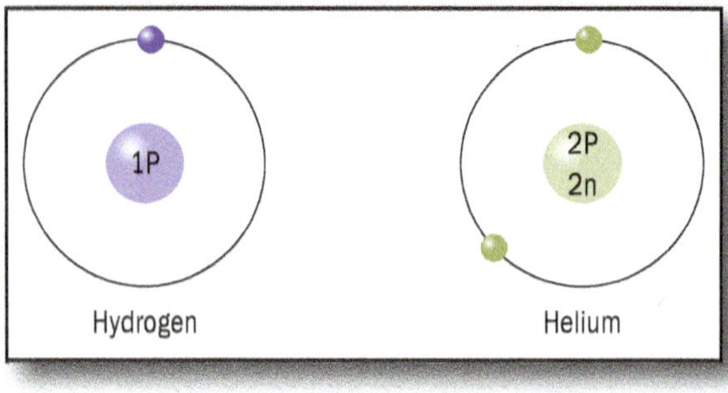

FIGURE 13–6
The atomic structures of hydrogen and helium.

The behavior and properties that distinguish one element from another must be related to the differences in the atomic structure of each element. One such distinction is that each element possesses a different number of protons. This number is called the **atomic number** of the element. As we look back at the periodic table on page 155, we see that the elements are numbered consecutively. Those numbers represent the atomic number or number of protons associated with each element. An element is therefore a collection of atoms that all have the same number of protons. Thus, each atom of hydrogen has one and only one proton, each atom of helium has 2 protons, each atom of silver has 47 protons, and each atom of lead has 82 protons in its nucleus.

atomic number
The number of protons in the nucleus of an atom

Isotopes and Radioactivity

Although the atoms of a single element must have the same number of protons, nothing prevents them from having different numbers of neutrons. The total number of protons and neutrons in a nucleus is known as the **atomic mass** number. Atoms with the same number of protons but differing solely in the number of neutrons are called **isotopes**.

For example, hydrogen consists of three isotopes: ordinary hydrogen, which has one proton and no neutrons in its nucleus, and two other isotopes called deuterium and tritium. Deuterium (or heavy hydrogen) also has one proton, but contains one neutron as well. Tritium has one proton and two neutrons in its nucleus.

Therefore, all the isotopes of hydrogen have an atomic number of 1 but differ in their atomic mass numbers. Hydrogen has an atomic mass of 1, deuterium a mass of 2, and tritium a mass of 3. The atomic structures of these isotopes are shown in Figure 13–7.

atomic mass
The sum of the number of protons and neutrons in the nucleus of an atom

isotope
An atom differing from another atom of the same element in the number of neutrons it has in its nucleus

FIGURE 13–7
Isotopes of hydrogen.

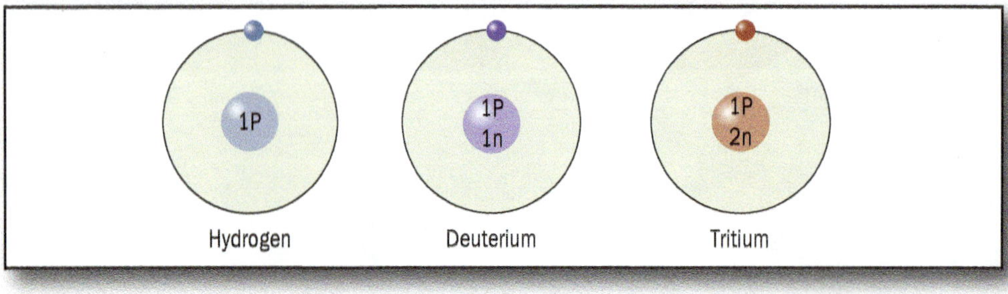

Case Files

Death by Radiation Poisoning

In November, 2006, Alexander V. Litvinenko lay at death's door step in a London hospital. He was in excruciating pain and had symptoms which included hair loss, the inability to make blood cells, and gastrointestinal distress. His organs slowly failed as he lingered for three weeks before dying. British investigators soon confirmed that Litvinenko died from the intake of polonium-210, a radioactive element, in what appeared to be its first use as a murder weapon (see the figure).

Litvinenko's death almost immediately set off an international uproar. Litvinenko, a former KGB operative, had become a vocal critic of the Russian spy agency FSB, the domestic successor to the KGB. In 2000, he fled to London, where he was granted asylum. Litvinenko continued to voice his criticisms of the Russian spy agency and also became highly critical of Russia's president, Vladimir Putin. Just before his death, he was believed to have compiled an incriminating report regarding the activities of senior Kremlin officials on behalf of a British company looking to invest millions in a project in Russia.

Suspicions immediately fell onto Andrei Lugovoi and Dmitri Kovtun, business associates of Mr. Litvinenko. Lugovoi was himself a former KGB officer. On the day he fell ill, Litvinenko met Lugovoi and Kovtun at the Pine Bar of the Millennium Hotel in London. At the meeting, Mr. Litvinenko drank tea out of a teapot later found to be highly radioactive. British officials have accused Lugovoi of poisoning Litvinenko. The precise nature of the evidence against him has not been made clear. However, investigators have linked him and Mr. Kovtun to a trail of polonium-210 radioactivity stretching from hotel rooms, restaurants, bars, and offices in London to Hamburg, Germany, and to British Airways planes that had flown to Moscow. Each man has denied killing Mr. Litvinenko.

Polonium-210 is highly radioactive and very toxic. By weight, it is about 250 million times as toxic as cyanide, so a particle the size of a dust particle could be fatal. It emits a radioactive ray known as an alpha particle. This form of radiation cannot penetrate the skin, so polonium-210 is effective as a poison can only be effective if it is swallowed, breathed in, or injected. The particles disperse through the body and first destroy fast-growing cells, like those in bone marrow, blood, hair, and the digestive tract. That would be consistent with Mr. Litvinenko's symptoms. There is no antidote for polonium poisoning.

Polonium does have industrial uses and is produced by commercial or institutional nuclear reactors. Polonium-210 has been found to be ideal for making antistatic devices that remove dust from film and lenses, as well as paper and textile plants. Its non-body-penetrating rays produce an electric charge on nearby air. Bits of dust with static attract the charged air, which neutralizes them. Once free of static, the dust is easy to blow or brush away. Manufacturers of such antistatic devices take great pains to make the polonium hard to remove from their products.

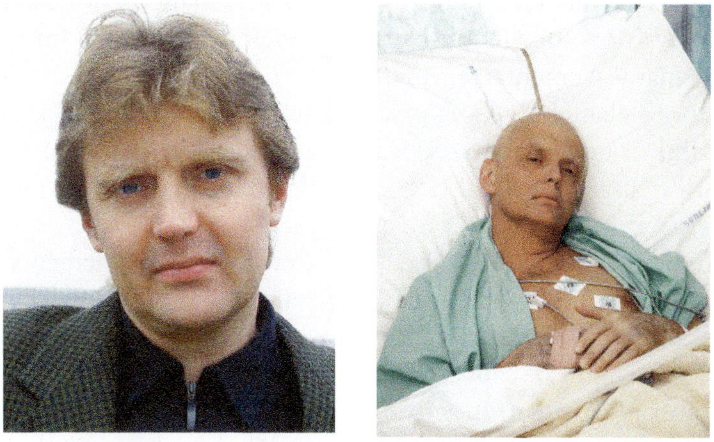

Alexander Litvinenko, former KGB agent, before and after he became sick.
ALISTAIR FULLER/AP Images (left) and Natasja Weitsz/Getty Images News/Getty Images (right)

Like hydrogen, most elements have two or more isotopes. Tin, for example, has 10 isotopes. Many of these isotopes are quite stable, and the isotopes of any one element have indistinguishable properties. Others, however, are not as stable and decompose over time by a process known as radioactive decay. **Radioactivity** is the emission of high-energy subatomic particles that accompanies the spontaneous disintegration of unstable nuclei.

radioactivity
The emission of high-energy subatomic particles that accompanies the spontaneous disintegration of unstable nuclei

Radioactivity is composed of three types of radiation: **alpha particle**, **beta particle**, and **gamma rays**. Alpha are helium atoms stripped of their orbiting electrons; thus, they are positively charged particles. Each alpha ray particle has a mass approximately four times that of a hydrogen atom. Beta are electrons, and gamma rays are a form of electromagnetic radiation similar to X-rays (discussed in Chapter 5), but of a higher frequency and energy. Fortunately, most naturally occurring isotopes are not radioactive, and those that are—radium, uranium, and thorium—are found in such small quantities in the earth's crust that their radioactivity presents no hazard to human survival.

alpha particle
Radiation composed of helium atoms minus their orbiting electrons

beta particle
Radiation composed of electrons

gamma ray
A high-energy form of electromagnetic radiation

When an atom is bombarded with neutrons, some neutrons are captured to form new isotopes. This is what happens in a nuclear reactor. A nuclear reactor is simply a source of neutrons that bombard the atoms of a specimen, thereby creating radioactive isotopes. The nucleus of an atom that has captured one or more neutrons is said to be activated, and it often begins to decompose immediately, emitting radioactivity.

Neutron Activation Analysis

Forensic chemists can characterize the trace elements in a specimen by bombarding it with neutrons and measuring the energy of the gamma rays emitted by the activated isotopes. The gamma rays of each element are associated with characteristic energy values, and thus exhibit unique levels of energy. This technique, known as neutron activation analysis, is depicted in Figure 13–8. Once an element has been identified, its concentration can be measured by the intensity of its gamma-ray radiation; the intensity of the radiation is directly proportional to the concentration of the element in a specimen.

The major advantage of neutron activation analysis is that it provides a nondestructive method for identifying and quantitating trace elements. A median detection sensitivity of one-billionth of a gram (1 nanogram) makes neutron activation analysis one of the most sensitive methods available for quantitative detection of many elements. Further, neutron activation can simultaneously analyze 20 to 30 elements. A major drawback to the technique is its expense. Only a handful of crime laboratories have access to a nuclear reactor; in addition, sophisticated analyzers are needed to detect and discriminate gamma-ray emissions.

Neutron activation has been used to characterize trace elements in metals, drugs, paint, soil, gunpowder residues, and hair. A typical illustration of its application occurred during the investigation of a theft of copper telegraphic wires in Canada. Four lengths of copper wire (A_1, A_2, A_3, A_4) found at the scene of the theft were compared by neutron activation with a length of copper wire (B) seized at a scrap yard and suspected of being stolen. All were bare, single-strand wire with the same general physical appearance and a diameter of 0.28 centimeter.

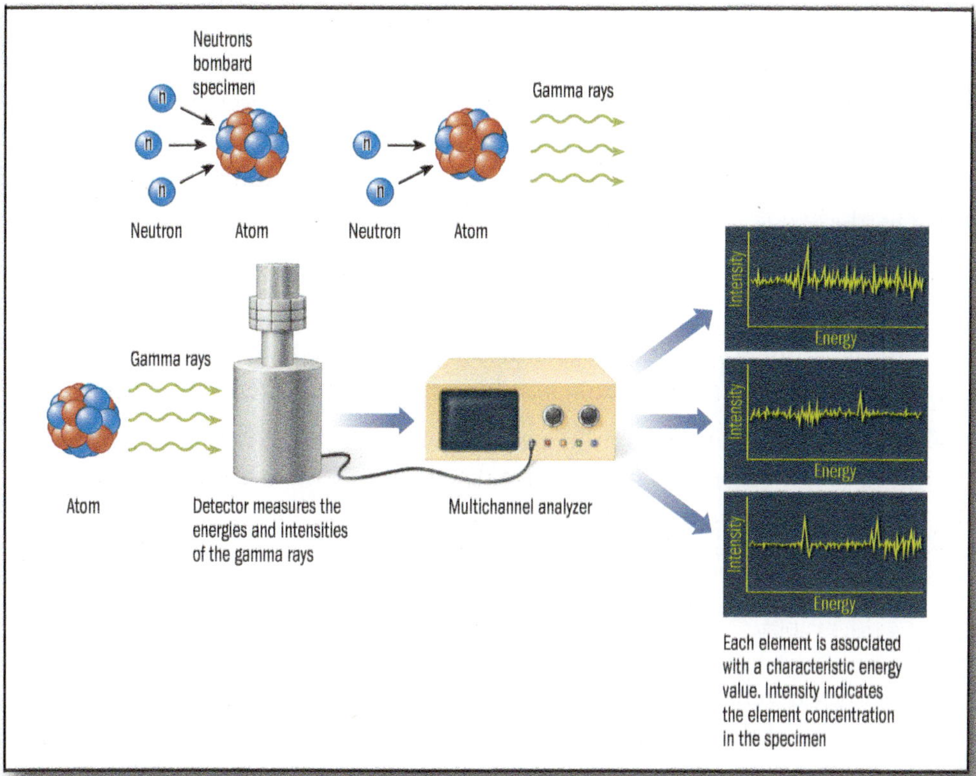

FIGURE 13–8 The neutron activation process requires the capture of a neutron by the nucleus of an atom. The new atom is now radioactive and emits gamma rays. A detector permits identification of the radioactive atoms present by measuring the energies and intensities of the gamma rays emitted.

Prior experiments had revealed that significant variations could be expected in the concentration levels of the trace elements selenium, gold, antimony, and silver for wires originating from different sources. Analysts compared these elements present in the wire involved in the theft. After exposing the wires to neutrons in a nuclear reactor, neutron activation analysis revealed a match between A_1 and B that was well within experimental error (see Table 13–2). The findings suggested a common origin of the control and suspect wires.

Table 13–2
Concentration of Trace Elements in Copper Wire

	Selenium	Gold	Antimony	Silver
Control Wire				
A_1	2.4	0.047	0.16	12.7
A_2	3.5	0.064	0.27	17.2
A_3	2.6	0.050	0.20	13.3
A_4	1.9	0.034	0.21	12.6
Suspect Wire				
B	2.3	0.042	0.15	13.0

Note: Average concentration measured in parts per million.

Source: Reprinted by permission of ASTM International from R. K. H. Chan, "Identification of Single-Stranded Copper Wires by Nondestructive Neutron Activation Analysis," Journal of Forensic Sciences 17 (1972): 93.

Quick Review

- Trace elements are small quantities of elements present in concentrations of less than 1 percent. They provide "invisible" markers that may establish the source of a material or provide additional points for comparison.
- The three most important subatomic particles are the proton, neutron, and electron. The proton has a positive electrical charge, the neutron has no electrical charge, and the electron has a negative electrical charge.
- Atomic number indicates the number of protons in the nucleus of an atom. Atomic mass refers to the total number of protons and neutrons in a nucleus.
- An isotope is an atom differing from other atoms of the same element in the number of neutrons in its nucleus.
- Radioactivity is the emission of high-energy subatomic particles that accompanies the spontaneous disintegration of the nuclei of unstable isotopes. The three types of radiation are alpha particles, beta particles, and gamma rays.
- In neutron activation analysis, a sample is bombarded with neutrons and the energy of the gamma rays emitted by the activated isotopes is measured. The gamma rays of each element are associated with characteristic energy values that helps identify the specific element that produces them.

Forensic Examination of Paint

Our environment contains millions of objects whose surfaces are painted. Thus paint, in one form or another, is one of the most prevalent types of physical evidence received by the crime laboratory.

Paint as physical evidence is perhaps most frequently encountered in hit-and-run and burglary cases. For example, a chip of dried paint or a paint smear may be transferred to the clothing of a hit-and-run victim on impact with an automobile, or paint smears could be transferred onto a tool during a burglary. Obviously, in many situations a transfer of paint from one surface to another could impart an object with an identifiable forensic characteristic.

In most circumstances, the criminalist must compare two or more paints to establish their common origin. For example, such a comparison may associate an individual or a vehicle with the crime site. However, the criminalist need not be confined to comparisons alone. Crime laboratories often help identify the color, make, and model of an automobile by examining small quantities of paint recovered at an accident scene. Such requests, normally made in hit-and-run cases, can lead to the apprehension of the responsible vehicle.

Composition of Paint

Paint is composed of a binder and pigments, as well as other additives, all dissolved or dispersed in a suitable solvent. Pigments impart color and hiding (or opacity) to paint and are usually mixtures of different inorganic and organic compounds added to the paint by the manufacturer to produce specific colors and properties. The binder is a polymeric substance that provides the support medium for the pigments and additives. After paint has been applied to a surface, the solvent evaporates, leaving behind a hard polymeric binder and any pigments that are suspended in it.

One of the most common types of paint examined in the crime laboratory is finishes from automobiles. Manufacturers apply a variety of coatings to the body of an automobile; this adds significant diversity to automobile paint and contributes to the forensic significance of automobile paint comparisons. The automotive finishing system for steel usually consists of at least four organic coatings:

- **Electrocoat Primer.** The first layer applied to the steel body of a car is the electrocoat primer. The primer, consisting of epoxy-based resins, is electroplated onto the steel body of the automobile to provide corrosion resistance. The resulting coating is uniform in appearance and thickness. The color of these primers ranges from black to gray.
- **Primer Surfacer.** Originally responsible for corrosion control, the surfacer usually follows the electrocoat layer and is applied before the basecoat. Primer surfacers are epoxy-modified polyesters or urethanes. The function of this layer is to completely smooth out and hide any seams or imperfections, because the basecoat will be applied on this surface. This layer is highly pigmented. Color pigments are used to minimize color contrast between primer and topcoats. For example, a light-gray primer may be used under pastel shades of a colored topcoat; a red oxide may be used under a dark-colored topcoat.
- **Basecoat.** The next layer of paint on a car is the basecoat or colorcoat. This layer provides the color and aesthetics of the finish and represents the "eye appeal" of the finished automobile. The integrity of this layer depends on its ability to resist weather, UV radiation, and acid rain. Most commonly, an acrylic-based polymer comprises the binder system of basecoats. Interestingly, the choice of automotive pigments is dictated by toxic and environmental concerns. Thus, the use of lead, chrome, and other heavy-metal pigments has been abandoned in favor of organicbased pigments. There is also a growing trend toward pearl luster or mica pigments. Mica pigments are coated with layers of metal oxide to generate interference colors. Also, the addition of aluminum flakes to automotive paint imparts a metallic look to the paint's finish.
- **Clearcoat.** An unpigmented clearcoat is applied to improve gloss, durability, and appearance. Most clearcoats are acrylic based, but polyurethane clearcoats are increasing in popularity. These topcoats provide outstanding etch resistance and appearance.

Microscopic Examination of Paint

The microscope has traditionally been and remains the most important instrument for locating and comparing paint specimens. Considering the thousands of paint colors and shades, it is quite understandable why color, more than any other property, imparts paint with its most distinctive forensic characteristics. Questioned and known specimens are best compared side by side under a stereoscopic microscope for color, surface texture, and color layer sequence (see Figure 13–9).

FIGURE 13–9
A stereoscopic microscope comparison of two automotive paints The questioned paint on the left has a layer structure consistent with the control paint on the right. *Courtesy Leica Microsystems, Inc., Buffalo, N.Y.*, www.leica-microsystems.com

The importance of layer structure for evaluating the evidential significance of paint evidence cannot be overemphasized. When paint specimens possess colored layers that match in number and sequence of colors, the examiner can begin to relate the paints to a common origin. How many layers must be matched before the criminalist can conclude that the paints come from the same source? There is no one accepted criterion. Much depends on the uniqueness of each layer's color and texture, as well as the frequency with which the particular combination of colors under investigation is observed. Because no books or journals have compiled this type of information, the criminalist is left to their own experience and knowledge when making this decision.

Unfortunately, most paint specimens do not have a layer structure of sufficient complexity to allow them to be individualized to a single source (see Figure 13–10). However, the diverse chemical composition of modern paints provides additional points of comparison between specimens. Specifically, a thorough comparison of paint must include a chemical analysis of the paint's pigments, its binder composition, or both.

FIGURE 13–10 Red paint chips peeling off a wall revealing underlying layers. *Jack Hollingsworth/ Photodisc/Getty Images*

Analytical Techniques Used in Paint Comparison

The wide variation in binder formulations in automobile finishes provides significant information. More important, paint manufacturers make automobile finishes in hundreds of varieties; this knowledge is most helpful to the criminalist who is trying to associate a paint chip with one car as distinguished from the thousands of similar models that have been produced in any one year. For instance, there are more than a hundred automobile production plants in the United States and Canada. Each can use one paint supplier for a particular color or vary suppliers during a model year. Although a paint supplier must maintain strict quality control over a paint's color, the batch formulation of any paint binder can vary, depending on the availability and cost of basic ingredients.

Characterization of Paint Binders An important extension of the application of gas chromatography to forensic science is the technique of **pyrolysis** *gas chromatography*. Many solid materials commonly encountered as physical evidence—for example, paint chips, fibers, and plastics—cannot be readily dissolved in a solvent for injection into the gas chromatograph. Thus, under normal conditions these substances cannot be subjected to gas chromatographic analysis. However, materials such as these can be heated, or pyrolyzed, to high temperatures (500–1,000°C) so that they will decompose into numerous gaseous products. Pyrolyzers permit these gaseous products to enter the carrier gas stream, where they flow into and through the GC column. The pyrolyzed material can then be characterized by the pattern produced by its chromatogram, or pyrogram.

pyrolysis The decomposition of organic matter by heat

Pyrolysis gas chromatography is particularly invaluable for distinguishing most paint formulations. In this process, paint chips as small as 20 micrograms are decomposed by heat into numerous gaseous products and are sent through a gas chromatograph.

As shown in Figure 13–11, the polymer chain is decomposed by a heated filament, and the resultant products are swept into and through a gas chromatograph column. The separated decomposition products of the polymer emerge and are recorded. The pattern of this chromatogram or "pyrogram" distinguishes one polymer from another. The result is a pyrogram that is sufficiently detailed to reflect the chemical makeup of the binder. Figure 13–12 illustrates how the patterns produced by paint pyrograms can differentiate acrylic enamel paints removed from two different automobiles.

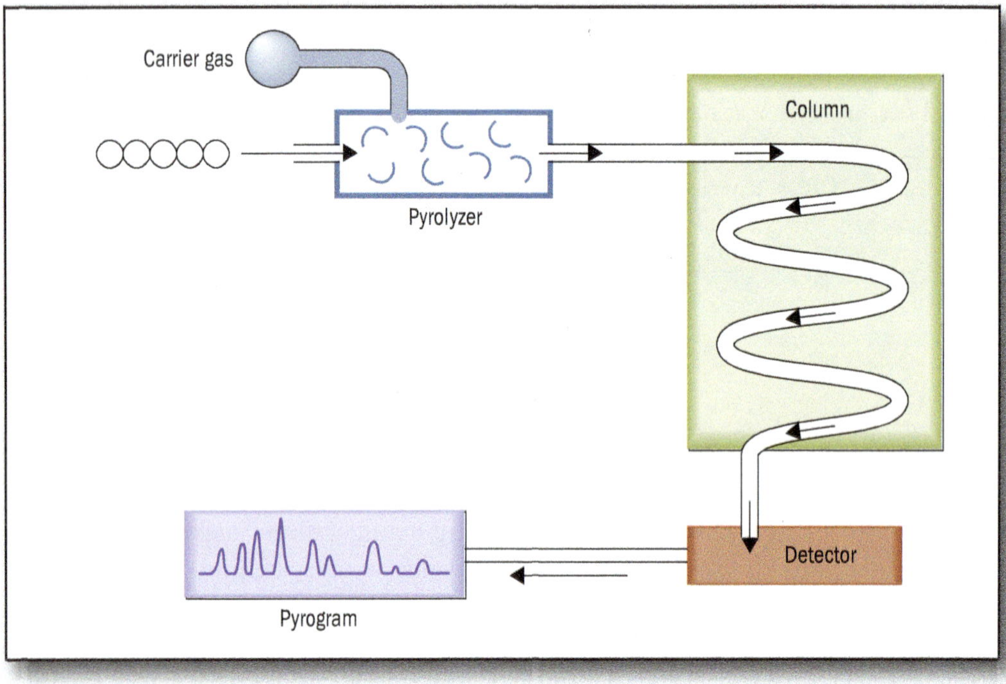

FIGURE 13–11 Schematic diagram of pyrolysis gas chromatography.

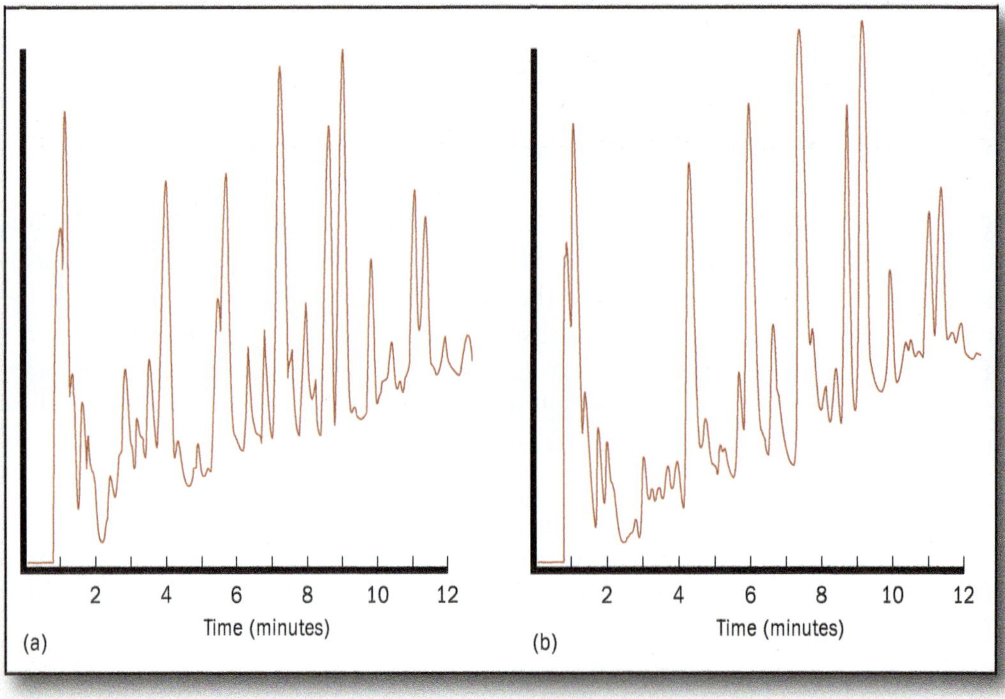

FIGURE 13–12 Paint pyrograms of acrylic enamel paints. (a) Paint from a Ford model and (b) paint from a Chrysler model.

Infrared spectrophotometry is still another analytical technique that provides information about the binder composition of paint.[2] Binders selectively absorb infrared radiation to yield a spectrum that is highly characteristic of a paint specimen.

Characterization of Paint Pigments The elements that constitute the inorganic pigments of paints can be identified by a variety of techniques: emission spectroscopy, neutron activation analysis, and X-ray spectroscopy. The emission spectrograph for example, can simultaneously detect 15 to 20 elements in most automobile paints. Some of these elements are relatively common to all paints and have little forensic value; others are less frequently encountered and provide excellent points of comparison between paint specimens.

We saw in Chapter 6 that organic compounds can be characterized by selective absorption of ultraviolet, visible, or infrared radiation. Equally significant to the forensic chemist is the knowledge that elements also selectively emit and absorb light. These observations form the basis of emission spectroscopy.

This spectrum is in essence a "fingerprint" of an element and offers a practical method of identification for liquid media such as paint. Sodium vapor, for example, always shows the same line spectrum, which differs from the spectrum of all other elements.

Inductively Coupled Plasma Emission Spectrometry (ICP) Recently, inductively coupled plasma (ICP) emission spectrometry has supplanted carbon arc emission spectroscopy for most applications. Like emission spectroscopy, ICP identifies and measures elements through light energy emitted by excited atoms. However, instead of using an electrical arc, the atoms are excited by placing the sample in a hot plasma torch. The torch is designed as three concentric quartz tubes through which argon gas flows (see Figure 13–13). A radio-frequency (RF) coil that carries a current is wrapped around the tubes. The RF current creates an intense magnetic field.

The process begins when a high-voltage spark is applied to the argon gas flowing through the torch. This strips some electrons from their argon atoms. These electrons are then caught and accelerated in the magnetic field such that they collide with other argon atoms, stripping off still more electrons. The collision of electrons and argon atoms continues in a chain reaction, breaking down the

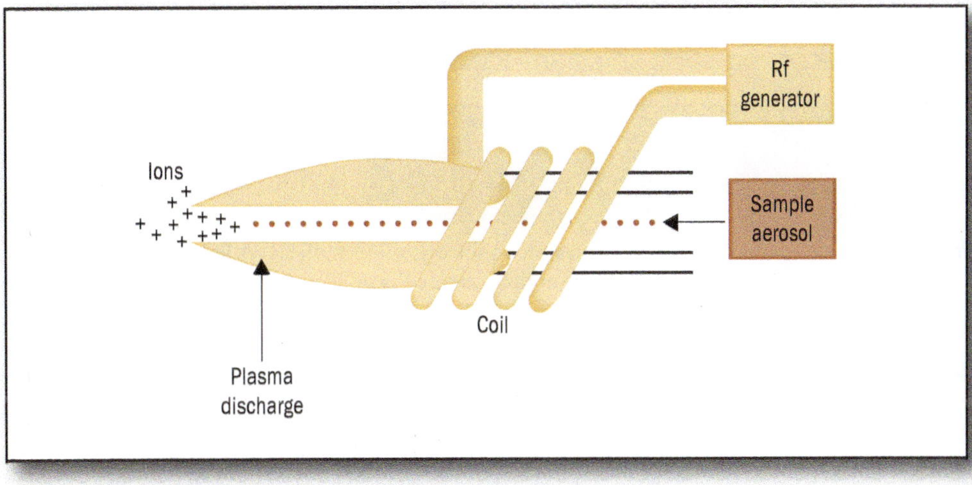

FIGURE 13–13 The creation of charged particles in the touch of an ICP discharge.

Inside the Science

The Carbon Arc Emission Spectrograph

An emission spectrograph is an instrument used to obtain and record the line spectra of elements. This instrument requires a means for vaporizing and exciting the atoms of elements so that they emit light, a means for separating this light into its component frequencies, and a means of recording the resultant spectrum. A simple emission spectrograph is depicted in Figure 1.

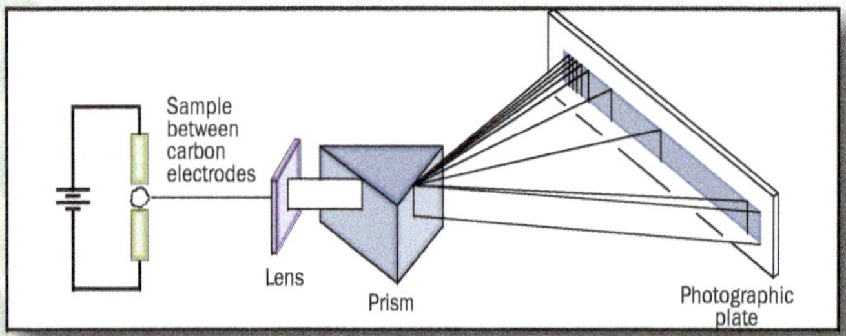

FIGURE 1 Parts of a simple emission spectrograph.

The specimen under investigation is inserted between two carbon electrodes through which a direct current arc is passed. The arc produces enough heat to vaporize and excite the specimen's atoms. A lens collects the light emitted by the excited atoms and focuses it onto a prism that disperses it into component frequencies. The separated frequencies of light are then directed toward a photographic plate, where they are recorded as line images. Normally, a specimen consists of numerous elements; hence, the typical emission spectrum contains many lines.

Each element in the spectrum can be identified when it is compared to a standard chart that shows the position of the principal spectral lines of all the elements. However, forensic analysis usually requires a rapid comparison of the elemental composition of two or more specimens. This is easily accomplished by comparing the emission spectra line for line, as illustrated in Figure 2, in which the emission spectra of two paint chips are shown to be comparable.

FIGURE 2 A comparison of paint chips 1 and 2 by emission spectrographic analysis. A line for line comparison shows that the paints have the same elemental composition. *Courtesy Richard Saferstein Ph.D.*

Inside the Science

The Origin of Emission Spectra

To explain the origin of atomic spectra, we must focus on the **electron orbitals** of the atom. As electrons move around the nucleus, they are confined to a path from which they cannot stray. This orbital path is associated with a definite amount of energy and is therefore called an *energy level*. Each element has its own set of characteristic energy levels at varying distances from the nucleus. Some levels are occupied by electrons; others are empty.

An atom is in its most stable state when all of its electrons are positioned in their lowest possible energy orbitals. When an atom absorbs energy, such as heat or light, its electrons are pushed into higher-energy orbitals. In this condition, the atom is in an **excited state**. However, because energy levels have fixed values, only a definite amount of energy can be absorbed in moving an electron from one level to another. This is an important observation; atoms absorb only a definite value of energy, and all other energy values are excluded, as shown in the figure.

A specific frequency of light is required to cause this transition, and its energy must correspond to the exact energy difference between the two orbitals involved. This energy difference is expressed by the relationship $E = hf$, where E represents the energy difference between the two orbitals, f is the frequency of absorbed light, and h is a universal constant called Planck's constant. Any energy value that is more or less than this difference will not produce the transition. Hence, an element is selective in the frequency of light it will absorb, and this selectivity is determined by the electron energy levels each element possesses.

In the same manner, if atoms are exposed to intense heat, enough energy will be generated to push electrons into unoccupied higher-energy orbitals. Normally the electron does not remain in this excited state for long, and it quickly falls back to its original energy level. As the electron falls back, it releases energy. An emission spectrum shows that this energy loss comes about in the form of light emission (see the figure). The frequency of light emitted is again determined by the relationship $E = hf$, where E is the energy difference between the upper and lower energy levels and f is the frequency of emitted light. Because each element has its own characteristic set of energy levels, each emits a unique set of frequency values. The emission spectrum thus provides a "picture" of the energy levels that surround the nucleus of each element.

Inside the Science (CONTINUED)

(a) The absorption of light by an atom, causing an electron to jump into a higher orbital, (b) The emission of light by an atom, caused by an electron falling back to a lower orbital.

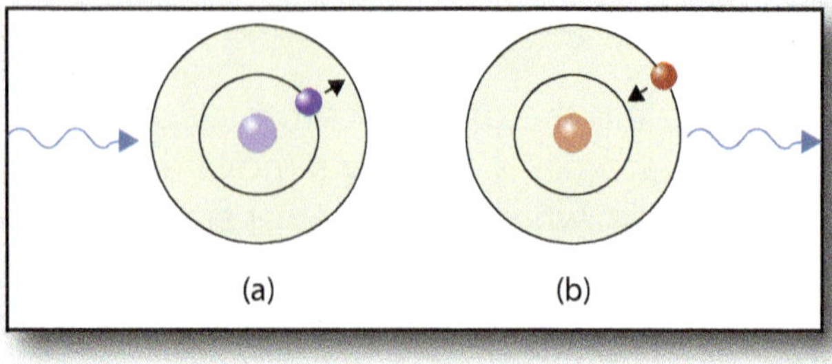

Thus, we see that as far as atoms are concerned, energy is a two-way street. Energy can be put into the atom at the same time that energy is given off; what goes in must come out. The chemist can study the atom using either approach.

electron orbital
The path of electrons as they move around the nuclei of atoms, each orbital is associated with a particular electronic energy level

excited state
The state in which an atom absorbs energy and an electron moves from a lower to a higher energy level

gas into argon atoms, argon ions, and electrons, forming an *inductively coupled plasma discharge*. The discharge is sustained by RF energy that is continuously transferred to it from the coil.

The plasma discharge acts like a very intense continuous flame generating temperatures in the range of 7,000–10,000°C. The sample, in aerosol form, is then introduced into the hot plasma, where it collides with the energetic argon electrons and generates charged particles (ions). The ions emit light of characteristic wavelengths that correspond to the identity of the elements in the sample.

The Significance of Paint Evidence

Once a paint comparison is completed, the task of assessing the significance of the finding begins. How certain can one be that two similar paints came from the same surface? For instance, a casual observer sees countless identically colored automobiles on our roads and streets. If this is the case, what value is a comparison of a paint chip from a hit-and-run scene to paint removed from a suspect car?

From previous discussions it should be apparent that far more is involved in paint comparison than matching surface paint colors. Paint layers beneath a surface layer offer valuable points of comparison. Furthermore, forensic analysts can detect subtle differences in paint binder formulations, as well as major or minor differences in the elemental composition of paint. Obviously, these properties cannot be discerned by the naked eye.

The significance of a paint comparison was convincingly demonstrated from data gathered at the Centre of Forensic Sciences, Toronto, Canada.[3] Paint chips randomly taken from 260 vehicles located in a local wreck yard were compared by color, layer structure, and, when required, by infrared spectroscopy. All were distinguishable except for one pair. In statistical terms, these results signify that

Inside the Science

ICP Analysis of Bullets

Mutilated bullets often are not suitable for traditional microscopic comparisons against an exemplar test-fired bullet. In such situations, ICP has been used to obtain an elemental profile of the questioned bullet fragment for comparison against an unfired bullet generally found in the possession of the suspect.

For a number of years forensic scientists have been aware of significant compositional differences among lead sources for the manufacture of lead-based bullets. Knowledge of these differences can be valuable when comparing bullets that are too mutilated to analyze with a microscope. Compositional differences in trace elements are typically reflected in the copper, arsenic, silver, antimony, bismuth, cadmium, and tin profiles of lead bullets.

When two or more bullets have comparable elemental compositions, evidence of their similarity may be offered in a court of law. In this respect, the comparison of lead bullets faces the same quandary as most common types of class physical evidence. The forensic analyst must convince a jury that the test results are meaningful to a criminal inquiry in the absence of any supporting statistical or probability data. Furthermore, creating meaningful databases to define the statistical significance of bullets compared by elemental profiles is currently an unrealistic undertaking.

Nevertheless, the significant diversity of bullet lead compositions in our population, like other class evidence such as fibers, hairs, paint, plastics, and glass, makes their chance occurrence at a crime scene and subsequent link to a defendant highly unlikely. However, care must be taken to avoid giving the impression that elemental profiles constitute a definitive match. Given the millions of bullets produced each year, one cannot conclusively rule out the possibility of a coincidental match with a non-case-related bullet.

if a crime-scene paint sample and a paint standard/reference sample removed from a suspect car compare by the previously discussed tests, the odds against the crimescene paint originating from another randomly chosen vehicle are approximately 33,000 to one. Obviously, this type of evidence is bound to forge a strong link between the suspect car and the crime scene.

After automotive paints, architectural paint comparisons are the most common paint analyses required of forensic laboratories. A large-scale population study of architectural paints collected throughout North America has been conducted for their forensic value.[4] Intercomparisons of nearly 960 randomly collected paints by visual, microscopic, and infrared technology resulted in a 99.99 percent differentiation of the paints, thus demonstrating the high diversity of architectural paints in our environment and their meaningful forensic value as class evidence.[4]

Color charts for automobile finishes are available from various paint manufacturers and refinishers (see Figure 13–14). Starting with the 1974 model year, the

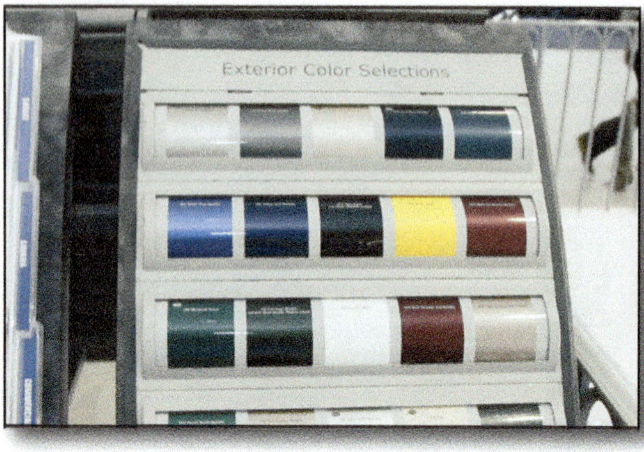

FIGURE 13–14 Automotive color chart of various car models. *Damian Dovarganes/AP Images*

Law Enforcement Standards Laboratory at the National Institute of Standards and Technology collected and disseminated to crime laboratories auto paint color samples from U.S. domestic passenger cars. This collection was distributed by Collaborative Testing Services, McLean, Virginia, through 1991. Since 1975, the Royal Canadian Mounted Police Forensic Laboratories have been systematically gathering color and chemical information on automotive paints. This computerized database, known as PDQ (Paint Data Query), allows an analyst to obtain information on paints related to automobile make, model, and year. The database contains such parameters as automotive paint layer colors, primer colors, and binder composition. A number of U.S. laboratories have access to PDQ.[5] Also, some laboratories maintain an in-house collection of automobile paints associated with various makes and models.

Collection and Preservation of Paint Evidence

As has already been noted, paint chips are most likely to be found on or near people or objects involved in hit-and-run incidents. The recovery of loose paint chips from a garment or from the road surface must be done with the utmost care to keep the paint chip intact. Paint chips may be picked up with a tweezers or scooped up with a piece of paper. Paper druggist folds and glass or plastic vials make excellent containers for paint. If the paint is smeared on or embedded in garments or objects, the investigator should not attempt to remove it; instead, it is best to package the whole item carefully and send it to the laboratory for examination.

When a transfer of paint occurs in hit-and-run situations (such as to the clothing of a pedestrian victim), uncontaminated standard/reference paint must always be collected from an undamaged area of the vehicle for comparison in the laboratory. The collected paint must be close to the area of the car that is suspected of being in contact with the victim. This is necessary because other portions of the car may have faded or been repainted.

Standard/reference samples are always removed so as to include all the paint layers down to the bare metal. This is best accomplished by removing a painted section with a clean scalpel or knife blade. Samples ¼ inch square are sufficient for laboratory examination. Each paint sample should be separately packaged and marked with the exact location of its recovery.

When a cross-transfer of paint occurs between two vehicles, again all of the layers, including the foreign as well as the underlying original paints, must be removed from each vehicle. A standard/reference sample from an adjacent undamaged area of each vehicle must also be taken in such cases. Carefully wipe the blade of any knife or scraping tool used before collecting each sample, to avoid cross-contamination of paints.

Tools used to enter buildings or safes often contain traces of paints as well as other substances such as wood and safe insulation. Care must be taken not to lose this type of trace evidence. The scene investigator should not try to remove the paint; instead, the investigator should package the tool for laboratory examination. Standard/reference paint should be collected from all surfaces suspected of having been in contact with the tool. Again, all layers of paint must be included in the sample.

When the tool has left its impression on a surface, standard/reference paint is collected from an uncontaminated area adjacent to the impression. No attempt should be made to collect the paint from the impression itself. If this is done, the impression may be permanently altered and its evidential value lost.

Quick Review

- Paint spread onto a surface dries into a hard film that is best described as consisting of pigments and additives suspended in the binder.
- Questioned and known paint specimens are best compared side by side under a stereoscopic microscope for color, surface texture, and color layer sequence.
- Pyrolysis gas chromatography and infrared spectrophotometry are used to distinguish most paint binder formulations.
- Emission spectroscopy and inductively coupled plasma are techniques available for determining the elemental composition of paint pigments.
- PDQ (Paint Data Query) is a computerized database that allows an analyst to obtain information on paints related to automobile make, model, and year.

Forensic Analysis of Soil

The term soil has many definitions; however, for forensic purposes, soil may be thought of as any disintegrated surface material, natural or artificial, that lies on or near the earth's surface. Therefore, forensic examination of soil not only is concerned with analysis of naturally occurring rocks, minerals, vegetation, and animal matter; it also encompasses detection of such manufactured objects as glass, paint chips, asphalt, brick fragments, and cinders, whose presence may impart soil with characteristics that make it unique to a particular location. When this material is collected accidentally or deliberately in a manner that associates it with a crime under investigation, it becomes valuable physical evidence.[6]

Case Files

The Predator

September in Arizona is hot and dry, much like the rest of the year—but September 1984 was a little different. Unusually heavy rains fell for two days, which must have seemed fitting for the friends and family of 8-year-old Vicki Lynn Hoskinson. Vicki went missing on September 17 of that year, and her disappearance was investigated as a kidnapping. A schoolteacher who knew Vicki remembered a suspicious vehicle loitering near the school that day, and he happened to jot down the license plate number. This crucial tip led police to 28-year-old Frank Atwood, recently paroled from a California prison. Police soon learned that Atwood had been convicted for committing sex offenses and for kidnapping a boy. This galvanized the investigators, who realized Vicki could be at the mercy of a dangerous and perverse man.

The only evidence the police had to work with was Vicki's bike, which was found abandoned in the middle of the street a few blocks from her home. Police found scrapes from her bike pedal on the underside of the gravel pan on Atwood's car, as well as pink paint apparently transferred from Vicki's bike to Atwood's front bumper. The police believed that Atwood deliberately struck Vicki while she was riding her bicycle, knocking her to the ground.

The pink paint on Atwood's bumper was first looked at microscopically and then examined by pyrolysis gas chromatography. This technique provides investigators with a "fingerprint" pattern of the paint sample, enabling them to compare this paint to any other paint evidence. In this case, the pink paint on Atwood's bumper matched the paint from Vicki's bicycle.

Vicki's skeletal remains were discovered in the desert, several miles away from her home, in the spring of 1985. Positive identification was made using dental records, but investigators wanted to see if the remains could help them determine how long she had been dead. Atwood was jailed on an unrelated charge three days after Vicki disappeared, so the approximate date of death was very important to proving his guilt.

Investigators found adipocere, a white, fatty residue produced during decomposition, inside Vicki's skull. This provided evidence that moisture was present around Vicki's body after her death, which did not make sense considering she was found in the Arizona desert! A check of the weather revealed that there had been an unusual amount of rainfall at only one time since Vicki was last seen alive: a mere 48 hours after her disappearance. This put Vicki's death squarely within Frank Atwood's three-day window of opportunity between her disappearance and his arrest. Frank Atwood was sentenced to death in 1987 for the murder of Vicki Lynn Hoskinson. Frank Atwood was executed on June 8, 2022.

The Significance of Soil Evidence

The value of soil as evidence rests with its prevalence at crime scenes and its transferability between the scene and the criminal. Thus, soil or dried mud found adhering to a suspect's clothing or shoes or to an automobile, when compared to soil samples collected at the crime site, may link a suspect or object to the crime scene. As with most types of physical evidence, forensic soil analysis is comparative in nature; soil found in the possession of the suspect must be carefully collected to be compared to soil samplings from the crime scene and its vicinity.

However, one should not rule out the value of soil even if the site of the crime has not been ascertained. For instance, small amounts of soil may be found on a person or object far from the actual site of a crime. A geologist who knows the local geology may be able to use geological maps to direct police to the general vicinity where the soil was originally picked up and the crime committed.

Forensic Examination of Soil

Most soils can be differentiated by their gross appearance. A side-by-side visual comparison of the color and texture of soil specimens is easy to perform and provides a sensitive property for distinguishing soils that originate from different locations. Soil is darker when it is wet; therefore, color comparisons must always be made when all the samples are dried under identical laboratory conditions. It is estimated that there are nearly 1,100 distinguishable soil colors; hence, color offers a logical first step in a forensic soil comparison (see Figure 13–15).

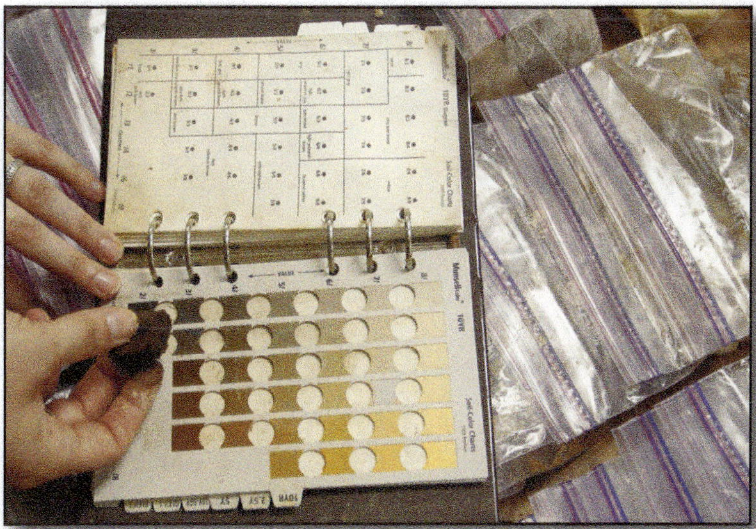

FIGURE 13–15 A technician classifies a soil sample based on a soil color chart.
Tampa Bay Times/ ZUMA Press Inc./Alamy Stock Photo

Microscopic Examination of Soil Low-power microscopic examination of soil reveals the presence of plant and animal materials as well as of artificial debris. Further high-power microscopic examination helps characterize minerals and rocks in earth materials. Although this approach to forensic soil identification requires the expertise of an investigator trained in geology, it can provide the most varied and significant points of comparison between soil samples. Only by carefully examining and comparing the minerals and rocks naturally present in soil can one take advantage of the large number of variations between soils and thus add to the evidential value of a positive comparison.[7]

mineral
A naturally occurring crystalline solid

A **mineral** is a naturally occurring crystal, and like any other crystal, its physical properties—for example, color, geometric shape, density, and refractive index or birefringence—are useful for identification. More than 2,200 minerals exist; however, most are so rare that forensic geologists usually encounter only about 20 of the more common ones. Rocks are composed of a combination of minerals and therefore exist in thousands of varieties on the earth's surface. They are usually identified by characterizing their mineral content and grain size (see Figure 13–16).

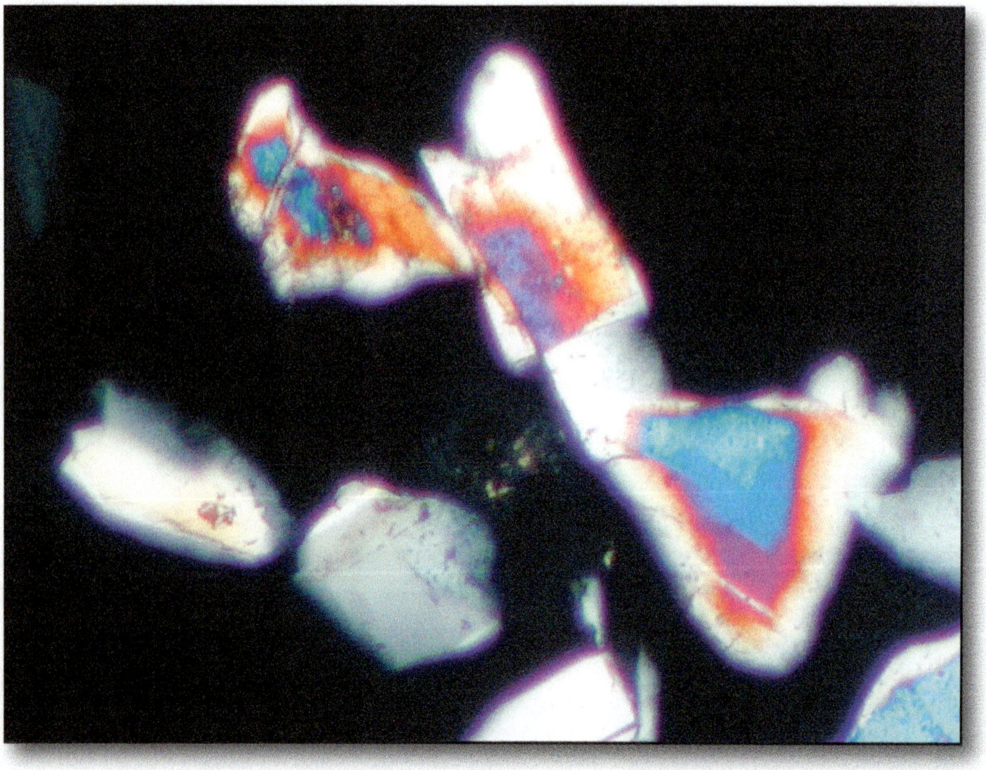

FIGURE 13–16
A mineral viewed under a microscope. *Courtesy Chris Palenik, Ph. D., Microtrace, Elgin, IL., www.microtracescientific.com*

Considering the vast variety of minerals and rocks and the possible presence of artificial debris in soil, the forensic geologist is presented with many points of comparison between two or more specimens. The number of comparative points and their frequency of occurrence must be considered before concluding similarity between specimens and judging the probability of common origin.

Rocks and minerals not only are present in earth materials but also are also used to manufacture a wide variety of industrial and commercial products. For example, the tools and garments of an individual suspected of breaking into a safe often contain traces of safe insulation. Safe insulation may be made from a wide combination of mineral mixtures that provide significant points of identification. Similarly, building materials such as brick, plaster, and concrete blocks are combinations of minerals and rocks that can easily be recognized and compared microscopically to similar minerals found on the breaking-and-entering suspect.

Variations in Soil

The ultimate forensic value of soil evidence depends on its variation at the crime scene. If, for example, soil is indistinguishable for miles surrounding the location of a crime, it will have limited value in associating soil found on the suspect with that particular site. Significant conclusions relating a suspect to a particular location through a soil comparison may be made when variations in soil composition occur every 10–100 feet from the crime site. However, even when such variations do exist, the forensic geologist usually cannot individualize soil to any one location unless an unusual combination of rare minerals, rocks, or artificial debris can be located.

No statistically valid forensic studies have examined the variability of soil evidence. A study conducted in southern Ontario, Canada, seems to indicate that soil in that part of Canada shows extensive diversity. It estimated a probability of less than 1 in 50 of finding two soils that are indistinguishable in both color and mineral properties but originate in two different locations separated by a distance of 1,000 feet. Based on these preliminary results, similar diversity may be expected in the northern United States, Canada, northern Europe, and eastern Europe. However, such probability values can only generally indicate the variation of soil within these geographical areas. Each crime scene must be evaluated separately to establish its own soil variation probabilities.

Collection and Preservation of Soil Evidence

When gathering soil specimens, the evidence collector must give primary consideration to establishing the variation of soil at the crime-scene area. For this reason, standard/reference soils should be collected at various intervals within a 100-foot radius of the crime scene, as well as at the site of the crime, for comparison to the questioned soil. Soil specimens also should be collected at all possible alibi locations that the suspect may claim.

All specimens gathered should be representative of the soil that was removed by the suspect. In most cases, only the top layer of soil is picked up during the commission of a crime. Thus, standard/reference specimens must be removed from the surface without digging too deeply into the unrepresentative subsurface layers. Approximately a tablespoon or two of soil is all the laboratory needs for a thorough comparative analysis. All specimens collected should be packaged in individual containers, such as plastic vials. Each vial should be marked to indicate the location at which the sampling was made.

Soil found on a suspect must be carefully preserved for analysis. If it is found adhering to an object, as in the case of soil on a shoe, the investigator must not remove it. Instead, each object should be individually wrapped in paper, with the soil intact, and transmitted to the laboratory. Similarly, loose soil adhering to garments should not be removed; these items should be carefully and individually wrapped in paper bags and sent to the laboratory for analysis. Care must be taken that particles that may accidentally fall off the garment during transportation will remain in the paper bag.

When a lump of soil is found, it should be collected and preserved intact. For example, an automobile tends to collect and build up layers of soil under the fenders, body, and so on. The impact of an automobile with another object may jar some of this soil loose. Once the suspect car has been apprehended, a comparison of the soil left at the scene with soil remaining on the automobile may help establish that the car was present at the accident scene. In these situations, separate samples are collected from under all the fender and frame areas of the vehicle; care is taken to remove the soil in lump form in order to preserve the order in which the soil adhered to the car. Undoubtedly, during the normal use of an automobile, soil will be picked up from numerous locations over a period of months and years. This layering effect may impart soil with greater variation, and hence greater evidential value, than that normally associated with loose soil.

The prevalence of soil in our environment makes it a common type of physical evidence at crime scenes. Proper collection and preservation by the criminal investigator will help ensure that a proper scientific examination can support investigative conclusions placing a suspect or object at the crime scene. Equally important is that soil, like other types of physical evidence, when properly collected and examined may exonerate the innocent from involvement in a crime.

Quick Review

- A side-by-side visual comparison of the color and texture of soil specimens provides a way to distinguish soils that originate from different locations.
- Minerals are naturally occurring crystalline solids found in soil. Their physical properties—for example, color, geometric shape, density, and refractive index or birefringence—are useful for characterizing soils.

Case Files

Soil: The Silent Witness

Alice Redmond was reported missing by her husband on a Monday night in 1983. Police learned that she had been seen with a co-worker, Mark Miller, after work that evening. When police questioned Miller, he stated that the two just "drove around" after work and then she dropped him off at home. Despite his statement, Miller was the prime suspect because he had a criminal record for burglary and theft.

Alice's car was recovered in town the following morning. The wheel wells were thickly coated in mud, which investigators hoped might provide a good lead. These hopes were dampened when police learned that Alice and her husband had attended a motorcycle race on Sunday, where her car was driven through deep mud.

After careful scrutiny, analysts found two colors of soil on the undercarriage of Alice's car. The thickest soil was brown; on top of the brown layer was a reddish soil that looked unlike anything in the county.

Investigators hoped the reddish soil, which had to have been deposited sometime after the Sunday night motorcycle event and before the vehicle was discovered on Tuesday morning, could link the vehicle to the location of Alice Redmond.

An interview with Mark Miller's sister provided a break in the case. She told police that Mark had visited her on Monday evening. During that visit, he confessed that he had driven Alice in her car across the Alabama state line into Georgia, killed her, and buried her in a remote location. Now that investigators had a better idea where to look for Alice, forensic analysts took soil samples that would prove or disprove Miller's sister's story.

Each field sample was dried and compared for color and texture by eye and stereomicroscopy to the reddish-colored soil gathered from the car. Next, soils that compared to the car were passed through a series of mesh filters, each of a finer gauge than the last. In this way, the components of the soil samples were physically separated by size. Finally, each fraction was analyzed and compared for mineral composition with the aid of a polarizing light microscope.

Only samples collected from areas across the Alabama state line near the suspected dump site were consistent with the topmost reddish soil recovered from Alice's car. This finding supported Miller's sister's story and was instrumental in Mark Miller's being charged with murder and kidnapping. After pleading guilty, the defendant led the authorities to where he had buried the body. The burial site was within a half mile of the location where forensic analysts had collected a soil sample consistent with the soil removed from Alice's vehicle.

Source: T. J. Hopen, "The Value of Soil Evidence," in M. M. Houck, ed., Trace Evidence Analysis: More Cases in Mute Witnesses, M. M. Houck, ed. (Elsevier Academic Press, Burlington, Mass.: Elsevier Academic Press, 2004), pp. 105–122.

Inside the Science

Nuclear Forensics

Nuclear forensics has emerged as a critical profession on the forefront in the war on terrorism. Nuclear forensic scientists are responsible for developing the means to analyze nuclear materials recovered from either the capture of intercepted intact nuclear materials or from post-explosion debris created as a result of a nuclear explosion. Nuclear forensics is becoming an increasingly important tool in the fight against illegal smuggling and trafficking of radiological and nuclear materials. In the United States, Lawrence Livermore National Laboratory and with seven other Department of Energy (DOE) national laboratories have been tasked by the Federal Bureau of Investigation and the Department of Homeland Security with developing the nation's technical forensics capability for nuclear and radiological materials. Organizations such as the European Commission's Institute for Transuranium Elements, located in Karlsruhe, Germany, have extended nuclear forensic capabilities onto an international scale.

View of a nuclear power plant. *Larry Lee Photography/Corbis/Getty Images*

A major focus of nuclear forensics is identifying signatures, which are the physical, chemical, and isotopic characteristics that distinguish one nuclear or radiological material from another. An example is stolen containers of uranium diverted during one of the mining, milling, conversion, enrichment, or fuel fabrication steps used to convert uranium ore to enriched fuel for nuclear power plants.

Inside the Science (CONTINUED)

Researchers analyze the material's chemical and isotopic composition, which includes measuring the amounts of trace elements as well as the ratio of parent isotopes to daughter isotopes. These measurements help to determine the source location and sample's age. They also examine the material's morphological characteristics such as shape, size, and texture. Analytical methods include electron microscopy, X-ray diffraction, and mass spectrometry. In addition, as a sample is moved from place to place, it picks up trace evidence such as pollen, hairs, fibers, and plant DNA, and finger prints. These so-called route materials may provide information about who has handled a sample and the path it has traveled.

Contracts with major U.S. uranium fuel suppliers have provided researchers with samples and manufacturing data. Forensic scientists are also seeking to obtain samples of uranium products worldwide to analyze the products' isotopic and trace-element content, grain size, and microstructure. Nations with nuclear capabilities are beginning to share information about their nuclear fuel processes and materials. The development of databases is essential to the nuclear forensic scientist's mission of identifying the origin of nuclear materials intercepted in the black market or associated with a terrorist event.

Chapter Review

- Trace elements are small quantities of elements present in concentrations of less than 1 percent. They provide "invisible" markers that may establish the source of a material or provide additional points for comparison.
- The three most important subatomic particles are the proton, neutron, and electron. The proton has a positive electrical charge, the neutron has no electrical charge, and the electron has a negative electrical charge.
- Atomic number indicates the number of protons in the nucleus of an atom. Atomic mass refers to the total number of protons and neutrons in a nucleus.
- An isotope is an atom differing from other atoms of the same element in the number of neutrons in its nucleus.
- Radioactivity is the emission of high-energy subatomic particles that accompanies the spontaneous disintegration of the nuclei of unstable isotopes. The three types of radiation are alpha particle rays, beta particle rays, and gamma rays.
- In neutron activation analysis, a sample is bombarded with neutrons and the energy of the gamma rays emitted by the activated isotopes is measured. The gamma rays of each element are associated with characteristic energy values that helps identify the specific element that produces them.
- Paint spread onto a surface dries into a hard film that is best described as consisting of pigments and additives suspended in the binder.
- Questioned and known paint specimens are best compared side by side under a stereoscopic microscope for color, surface texture, and color layer sequence.
- Pyrolysis gas chromatography and infrared spectrophotometry are used to distinguish most paint binder formulations.
- Emission spectroscopy and inductively coupled plasma are techniques available for determining the elemental composition of paint pigments.
- PDQ (Paint Data Query) is a computerized database that allows an analyst to obtain information on paints related to automobile make, model, and year.
- Aside-by-side visual comparison of the color and texture of soil specimens provides a way to distinguish soils that originate from different locations.
- Minerals are naturally occurring crystalline solids found in soil. Their physical properties—for example, color, geometric shape, density, and refractive index or birefringence—are useful for characterizing soils.

Review Questions

1. The nucleus of an atom contains
   a. neutrons.
   b. neutrons and electrons.
   c. protons and electrons.
   d. protons and neutrons.

2. Atoms having the same atomic number but different atomic masses are called
   a. isobars.
   b. isotopes.
   c. isotherms.
   d. isomers.

3. One of the most common types of paint examined in the crime laboratory involves
   a. finishes from fingernail polish on a victim's hands.
   b. finishes emanating from automobiles.
   c. finishes from the "bluing" of a firearm.
   d. finishes from furniture varnish at a crime scene.

4. A computerized database that allows an analyst to obtain information relating to an automobile's make, model, and year is
   a. CODIS.
   b. AFIS.
   c. PCR.
   d. PDQ.

5. Forensic soil comparison includes the comparison of color because
   a. color is an individual characteristic.
   b. soil color allows the examiner to determine the location of its origin.
   c. there are nearly 1,100 distinguishable soil colors so it offers a logical first step in soil comparison.
   d. color in soil is unique to a location.

6. True or False: "Trace elements" are important for the criminalist because they provide "invisible" markers that may establish the source of a material or at least provide additional points for comparison.

7. True or False: The major advantage of neutron activation analysis is that it provides a nondestructive method for identifying and quantifying trace elements.

8. True or False: Pyrolysis gas chromatography has proven to be an invaluable technique for distinguishing most paint pigments.

9. True or False: At the crime laboratory, a criminalist can determine the exact make and model of a car in question based on paint chips left behind at a crime scene.

10. True or False: When soil is found on a suspect's garments or shoes, the investigator should remove the soil by gently scraping the soil into an airtight container for later analysis.

11. What are trace elements?

12. Forensic analysis of the bullets recovered after the assassination of President John F. Kennedy focused on the concentration of what two trace elements?

13. Which of the following conclusions can be drawn from the forensic analysis of the bullets recovered after the assassination of President John F. Kennedy?
    a. The analysis absolutely verified the findings of the Warren Commission.
    b. The analysis cast doubt on the findings of the Warren Commission.
    c. The analysis generally supported the findings of the Warren Commission.
    d. The analysis absolutely disproved the findings of the Warren Commission.

14. Name the three most important subatomic particles and their electrical charges, and indicate where each is located in the atom.

15. How does atomic number differ from atomic mass?

16. What is radioactivity? What are the three types of radiation?

17. Briefly describe the process of neutron activation analysis.

18. In what types of criminal cases is paint evidence most frequently encountered?

19. Describe the basic composition of paint. What component of paint evaporates after paint is applied to a surface?

20. What characteristics does a criminalist look for when comparing paint chips under a microscope? Which of these characteristics is most important in evaluating the significance of paint evidence?

21. What technique has largely supplanted carbon arc emission spectroscopy for most applications? What is the main difference between the two techniques?

22. Why must paint collected from a vehicle involved in a hit-and-run accident be taken from the area of the car suspected of being in contact with the victim?

23. How should the investigator handle the collection of trace paint evidence left on a tool?

24. What is the first step in a forensic soil comparison?

25. What is a mineral? How are minerals useful in forensic soil analysis?

26. From what areas should standard/reference soils be collected when gathering soil evidence?

Learn the Vocabulary

Use the information from this chapter along with your prior knowledge and experience to learn the key terms. Choose from the following activities:

1. Make flashcards with a term on the front and its definition on the back. Use the flashcards to quiz a partner on correctly pronouncing and defining the terms.

2. Make an audio or video recording of each term and its definition.

3. Work with a partner to write a sentence for each term. Read the sentences out loud to a partner or share them with the class.

4. Write a list of new terms and language you have encountered in this chapter. Reuse the terms to write sentences and then read the sentences out loud to a partner. Reuse the terms again by writing a short speech. The speech might explain the terms or teach a lesson using the terms. Give the speech to a partner or to the class.

Application and Critical Thinking

1. Using the periodic table shown in Chapter 5, determine the atomic numbers for antimony, tin, barium, and lead. Which do you think has the largest atomic mass? Which do you think has the smallest atomic mass? Explain your answers.

2. You are investigating a hit-and-run accident and have identified a suspect vehicle. Describe, using details and specifics, how you would collect paint to determine whether the suspect vehicle was involved in the accident. Be sure to indicate the tools you would use and the steps you would take to prevent cross-contamination.

3. Criminalist Jared Heath responds to the scene of an assault on an unpaved lane in a rural neighborhood. Rain had fallen steadily the night before, making the area quite muddy. A suspect with very muddy shoes was apprehended nearby, but claimed to have picked up the mud either from his garden or from the unpaved parking lot of a local restaurant. Jared uses a spade to remove several samples of soil, each about 2 inches deep, from the immediate crime scene, and places each in a separate plastic vial. He collects the muddy shoes and wraps them in plastic as well. At the laboratory, he unpackages the soil samples and examines them carefully, one at a time. He then analyzes the soil on the shoes to see if it matches the soil from the crime scene. What mistakes, if any, did Jared make in his investigation?

Laboratory Experiments

This activity requires the use of the following practices of science and engineering:

- Asking questions and defining problems
- Planning and carrying out investigations
- Analyzing and interpreting data
- Obtaining, evaluating, and communicating information

This activity consists of the following crosscutting concepts:

- **Patterns**—Patterns are evident in the properties of soil that can be used to compare them.
- **Cause and effect**—The transfer of soil resulting from the hit and run can be used to determine if the automobiles were in contact with the accident scene.

Before you begin, read the scenario and steps out loud with a partner. Ask for clarification as necessary to make sure you understand the experiment.

The Scenario

A middle-aged woman, later identified as Ana Purna, was found dead on the shoulder of Route 66, obviously the victim of a hit and run accident. Investigators took samples of soil and fragments of glass from the scene. Deputies found two likely suspects (Nicole Waverly and Jason Evans) who could be placed on the highway at the estimated time of the accident. The following dried soil samples were submitted to your crime lab:

A. Soil taken from the clothes of the victim
B. Soil from the accident site
C. Soil scraped from under the right front fender of suspect Nicole Waverly's red Mazda CX5
D. Soil scraped from under the right front fender of suspect Jason Evans's Lexus IS250
E. Soil removed from Purna's yard

PROCEDURE

You need to perform the following analysis, complete the data table, and determine if either of the suspects were involved in the hit and run accident. While conducting your field investigation, be sure to use appropriate safety equipment and follow safety practices as outlined in Texas Education Agency-approved safety standards. Write a paragraph summarizing your findings. Be sure to use clear, basic, vocabulary as well as forensics key terms.

1. Examine each of the soil samples with a magnifying glass or under the stereomicroscope. Note the presence of unusual material or vegetation such as leaves, roots, pine needles, hair, metal specks, fibers, building materials, trash, or the like. Such objects can provide valuable clues.
2. Direct ultraviolet light on the soil samples. Note what particles, if any, fluoresce. Certain minerals fluoresce (e.g., fluorite, some calcites, willemite) as well as many manufactured articles such as fibers and plastics.
3. Pass a magnet through the soil samples to collect and identify any iron objects.
4. The color of a soil is generally related to the presence of particular minerals or organic matter. For example, red soils tend to have highly oxidized iron (that is, rust); black soils tend to contain organic matter, or humus. Wet soil is usually darker than dry soil. The soil must be dried (usually at 100oC for an hour) before making any color comparisons. Your samples have already been dried.

	Sample A	*Sample B*	*Sample C*	*Sample D*	*Sample E*
Physical Appearance					
Ultraviolet Light					
Magnet					
Color					
pH					

5. Certain soils may be especially acidic or basic. Those high in limestone ($CaCO_3$), for example, will be basic. Soils containing sulfides (S_{2-}) and sulfates ($SO\ _{2-}$) are usually acidic. Measure the pH by placing a small amount of sample in a test tube, add about a centimeter of distilled water, shake, let the mixture settle, note any discoloration of the water, then test with pH paper.

Endnotes

1. Actually, the electrons are moving so rapidly around the nucleus that they are best visualized as an electron cloud spread out over the surface of the atom.
2. P. G. Rodgers et al., "The Classification of Automobile Paint by Diamond Window Infrared Spectrophotometry, Part I: Binders and Pigments," *Canadian Society of Forensic Science Journal* 9 (1976): 1; T. J. Allen, "Paint Sample Presentation for Fourier Transform Infrared Microscopy," *Vibration Spectroscopy* 3 (1992): 217.
3. G. Edmondstone, J. Hellman, K. Legate, G. L. Vardy, and E. Lindsay, "An Assessment of the Evidential Value of Automotive Paint Comparisons," *Canadian Society of Forensic Science Journal* 37 (2004): 147.
4. D. W. Wright et al., "Analysis and Documentation of Architectural Paint Samples via a Population Study," *Forensic Science International* 209 (2011): 86.
5. J. L. Buckle et al., "PDQ—Paint Data Queries: The History and Technology behind the Development of the Royal Canadian Mounted Police Laboratory Services Automotive Paint Database," *Canadian Society of Forensic Science Journal* 30 (1997): 199. An excellent discussion of the PDQ database is also available in A. Beveridge, T. Fung, and D. MacDougall, "Use of Infrared Spectroscopy for the Characterisation of Paint Fragments," in B. Caddy, ed., *Forensic Examination of Glass and Paint* (New York: Taylor & Francis, 2001), pp. 222–233.
6. E. P. Junger, "Assessing the Unique Characteristics of Close-Proximity Soil Samples: Just How Useful Is Soil Evidence?" *Journal of Forensic Sciences* 41 (1996), 27.
7. W. J. Graves, "A Mineralogical Soil Classification Technique for the Forensic Scientist," *Journal of Forensic Sciences* 24 (1979): 323; M. J. McVicar and W. J. Graves, "The Forensic Comparison of Soil by Automated Scanning Electron Microscopy," *Canadian Society of Forensic Science Journal* 30 (1997): 241.

Forensic Aspects of Fire and Explosion Investigation

14

Sheila Fitzgerald/Shutterstock

Key Terms

- accelerant
- combustion
- endothermic reaction
- energy
- exothermic reaction
- flammable range
- flash point
- glowing combustion
- heat of combustion
- hydrocarbon
- ignition temperature
- modus operandi
- oxidation
- pyrolysis
- spontaneous combustion

Fire at the Ghost Ship

On December 2, 2016, at approximately 11:20 p.m. a fire broke out in a former warehouse that had been converted into an artist collective with living spaces known as the Ghost Ship. The warehouse was only permitted for industrial purposes while residential and entertainment uses were illegal. At the time the fire broke out, the warehouse was hosting a concert featuring artists from the house music record label 100 percent Silk. It was estimated that between 80 and 100 people were in attendance at the concert, and 36 people died. The cause of death was determined to be smoke inhalation. It remains the deadliest fire in Oakland, California history.

As part of the investigation into the cause of the fire, a report was issued by the Bureau of Alcohol, Tobacco and Firearms, the Alameda County Arson Task Force and the Oakland Fire Department. Investigators could not determine a cause for the deadly blaze due to the extreme damage. It was later revealed in documents released by the city of Oakland that several city agencies, including the police department, visited the Ghost Ship warehouse 35 times in the months and years leading up to the deadly fire.

Ghost ship master tenant Derick Almena and his assistant, Max Harris, are arrested and charged with involuntary manslaughter. Prosecutors alleged Almena and Harris knew well in advance of the fire hazards and took no steps to address them. In July 2018, prosecutors reach a plea deal with Almena and Harris. In exchange for a no contest plea to 36 counts of involuntary manslaughter, Almena would receive 9 years in prison and Harris 6 years. Family members of the victims are upset about the plea deal and the judge overseeing the case rejects the plea deal for the defendants failure to accept full responsibility for the fire.

In April 2019, Almena and Harris stand trial for their role in the deaths of 36 people in the Ghost Ship fire. On September 5, 2019, the jury acquitted Max Harris of all charges related to the fire, but was not able to reach a unanimous verdict as to the guilt or innocence of Derick Almena. With the jury hopelessly deadlocked at 10-2 in favor of conviction, the judge declared a mistrial. In a second trial, Derick Almena was sentenced to 12 years for the Ghost Ship fire. However, because of time already served while awaiting trial, he is serving the remainder of his sentence at home with an ankle monitor rather than returning to jail.

Learning Objectives

After studying this chapter, you should be able to:

14.1 Explain the chemical reactions that initiate and sustain fire

14.2 Explain the three mechanisms of heat transfer

14.3 Recognize how to locate the fire's origin along with telltale signs of an accelerant-initiated fire

14.4 Describe how to collect physical evidence at the scene of a suspected arson

14.5 Describe laboratory procedures used to detect and identify hydrocarbon residues

National Science Content Standards

Scientific Inquiry

Physical Science

Science in Personal and Social Perspective

Forensic Investigation of Arson

Arson often presents complex and difficult circumstances to investigate. Normally these incidents are committed at the convenience of a perpetrator who has thoroughly planned the criminal act and has left the crime scene long before any official investigation is launched. Furthermore, proving commission of the offense is more difficult because of the extensive destruction that frequently dominates the crime scene. The contribution of the criminalist is only one aspect of a comprehensive and difficult investigative process that must establish a motive, the **modus operandi**, and a suspect.

modus operandi
An offender's pattern of operation

The criminalist's function is limited; usually the criminalist is expected only to detect and identify relevant chemical materials collected at the scene and to reconstruct and identify igniters. Although a chemist can identify trace amounts of gasoline or kerosene in debris, no scientific test can determine whether an arsonist has used a pile of rubbish or paper to start a fire. Furthermore, a fire can have many accidental causes, including faulty wiring, overheated electric motors, improperly cleaned and regulated heating systems, and cigarette smoking—which usually leave no chemical traces. Thus, the final determination of the cause of a fire must consider numerous factors and requires an extensive on-site investigation. The ultimate determination must be made by an investigator whose training and knowledge have been augmented by the practical experiences of fire investigation. A trained fire investigator relies on experimental and engineering designs to analyze and evaluate the available evidence to come to a logical conclusion.

The Chemistry of Fire

Humankind's early search to explain the physical concepts underlying the behavior of matter always bestowed a central and fundamental role on fire. To ancient Greek philosophers, fire was one of the four basic elements from which all matter was derived. The medieval alchemist thought of fire as an instrument of transformation, capable of changing one element into another. One ancient recipe expresses its mystical power as follows: "Now the substance of cinnabar is such that the more it is heated, the more exquisite are its sublimations. Cinnabar will become mercury, and passing through a series of other sublimations, it is again turned into cinnabar, and thus it enables man to enjoy eternal life."

Today, we know of fire not as an element of matter but as a transformation process during which oxygen is united with some other substance to produce noticeable quantities of heat and light (a flame). Therefore, any insight into why and how a fire is initiated and sustained must begin with the knowledge of the fundamental chemical reaction of fire—**oxidation**.

oxidation
The combination of oxygen with other substances to produce new substances

Oxidation

In a simple description of oxidation, oxygen combines with other substances to produce new products. Thus, we may write the chemical equation for the burning of methane gas, a major component of natural gas, as follows:

CH_4	+	$2O_2$	$\longrightarrow$	CO_2	+	$2H_2O$
methane		oxygen	yields	carbon dioxide		water

However, not all oxidation proceeds in the manner that one associates with fire. For example, oxygen combines with many metals to form oxides. Thus, iron forms a red-brown iron oxide, or rust, as follows (see Figure 14–1):

$4Fe$	+	$3O_2$	$\longrightarrow$	$2Fe_2O_3$
iron		oxygen	yields	iron oxide

FIGURE 14–1
Rust forming on iron is an example of oxidation.
Wallenrock/Shutterstock

Yet chemical equations do not give us a complete insight into the oxidation process. We must consider other factors to understand all of the implications of oxidation or, for that matter, any other chemical reaction. Methane burns when it unites with oxygen, but merely mixing methane and oxygen does not produce a fire. Nor, for example, does gasoline burn when it is simply exposed to air. However, lighting a match in the presence of any one of these fuel–air mixtures (assuming proper proportions) produces an instant fire.

What are the reasons behind these differences? Why do some oxidations proceed with the outward appearances that we associate with a fire, while others do not? Why do we need a match to initiate some oxidations, but others proceed at room temperature? The explanation lies in a fundamental but abstract concept—energy.

Energy

energy
The ability or potential of a system or material to do work

Energy can be defined as the ability or potential of a system or material to do work. Energy takes many forms, such as heat energy, electrical energy, mechanical energy, nuclear energy, light energy, and chemical energy. For example, when methane is burned, the stored chemical energy in methane is converted to energy in the form of heat and light. This heat may be used to boil water or to provide high-pressure steam to turn a turbine. This is an example of converting chemical energy to heat energy to mechanical energy. The turbine can then be used to generate electricity, transforming mechanical energy to electrical energy. Electrical energy may then be used to turn a motor. In other words, energy can enable work to be done; heat is energy.

The quantity of heat from a chemical reaction comes from the breaking and formation of chemical bonds. Methane is a molecule composed of one carbon atom bonded with four hydrogen atoms:

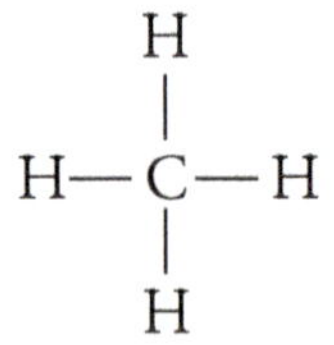

An oxygen molecule forms when two atoms of the element oxygen bond:

$$O=O$$

In chemical changes, atoms are not lost but merely redistributed during the chemical reaction; thus, the products of methane's oxidation will be carbon dioxide:

$$O=C=O$$

and water:

$$H{-}O{-}H$$

This rearrangement, however, means that the chemical bonds holding the atoms together must be broken and new bonds formed. We now have arrived at a fundamental observation in our dissection of a chemical reaction—that molecules must absorb energy to break apart their chemical bonds, and that they liberate energy when their bonds are re-formed.

The amount of energy needed to break a bond and the amount of energy liberated when a bond is formed are characteristic of the type of chemical bond involved. Hence, a chemical reaction involves a change in energy content; energy is going in and energy is given off. The quantities of energies involved are different for each reaction and are determined by the participants in the chemical reaction.

combustion
Rapid combination of oxygen with another substance, accompanied by production of noticeable heat and light

exothermic reaction
A chemical transformation in which heat energy is liberated

heat of combustion
The heat liberated during combustion

Combustion

All oxidation reactions, including the **combustion** of methane, are examples of reactions in which more energy is liberated than is required to break the chemical bonds between atoms. Such reactions are said to be **exothermic**. The excess energy is liberated as heat, and often as light, and is known as the **heat of combustion**. Table 14–1 summarizes the heats of combustion of some important fuels in fire investigation.

Although we will not be concerned with them, some reactions require more energy than they eventually liberate. These reactions are known as **endothermic reactions.**

endothermic reaction
A chemical transformation in which heat energy is absorbed from the surroundings

Table 14–1
Heats of Combustion of Fuels

Fuel	Heat of Combustion[a]
Crude oil	19,650 Btu/gal
Diesel fuel	19,550 Btu/lb
Gasoline	19,250 Btu/lb
Methane	995 Btu/cu ft
Natural gas	128–1,868 Btu/cu ft
Octane	121,300 Btu/gal
Wood	7,500 Btu/lb
Coal, bituminous	11,000–14,000 Btu/lb
Anthracite	13,351 Btu/lb

[a]A BTU (British thermal unit) is defined as the quantity of heat required to raise the temperature of 1 pound of water 1°F at or near its point of maximum density.

Source: John D. DeHaan, Kirk's Fire Investigation, 2nd ed. Upper Saddle River, N.J.: Prentice Hall, 1983.

Thus, all reactions require an energy input to start them. We can think of this requirement as an invisible energy barrier between the reactants and the products of a reaction (see Figure 14–2). The higher this barrier, the more energy required to initiate the reaction. Where does this initial energy come from? There are many sources of energy; however, for the purpose of this discussion we need to look at only one—heat.

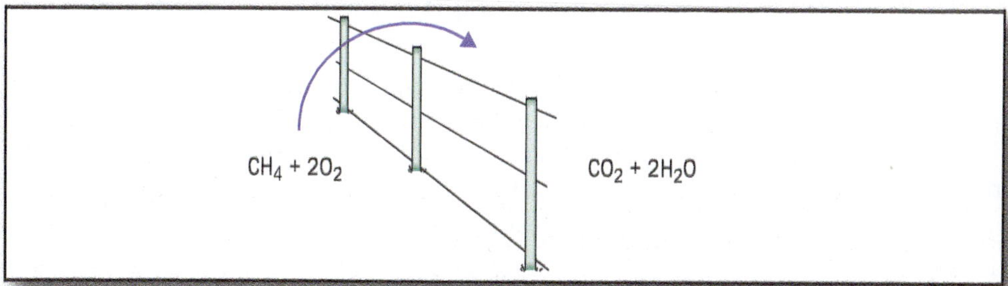

FIGURE 14–2
An energy barrier must be hurdled before reactants such as methane and oxygen can combine with one another to form the products of carbon dioxide and water.

Heat The energy barrier in the conversion of iron to rust is relatively small, and it can be surmounted with the help of heat energy in the surrounding environment at normal outdoor temperatures. Not so for methane or gasoline; these energy barriers are quite high, and a high temperature must be applied to start the oxidation of these fuels. Hence, before any fire can result, the temperature of these fuels must be raised enough to exceed the energy barrier. Table 14–2 shows that this temperature, known as the **ignition temperature**, is quite high for common fuels.

ignition temperature
The minimum temperature at which a fuel spontaneously ignites

Once combustion starts, enough heat is liberated to keep the reaction going by itself. In essence, the fire becomes a chain reaction, absorbing a portion of its own liberated heat to generate even more heat. The fire burns until either the oxygen or the fuel is exhausted.

Normally, a lighted match provides a convenient igniter of fuels. However, the fire investigator must also consider other potential sources of ignition—for example, electrical discharges, sparks, and chemicals—while reconstructing the initiation of a fire. All of these sources have temperatures higher than the ignition temperature of most fuels.

Table 14–2

Ignition Temperatures of Some Common Fuels

Fuel	Ignition Temperature, °F
Acetone	869
Benzene	928
Fuel oil #2	495
Gasoline (low octane)	536
Kerosene (fuel oil #1)	410
n-Octane	428
Petroleum ether	550
Turpentine	488

Source: John D. DeHaan, Kirk's Fire Investigation, 4th ed. Upper Saddle River, N.J.: Prentice Hall, 1997.

Speed of Reaction Although the liberation of energy explains many important features of oxidation, it does not explain all characteristics of the reaction. Obviously, although all oxidations liberate energy, not all are accompanied by a flame; witness the oxidation of iron to rust. Therefore, one other important consideration will make our understanding of oxidation and fire complete: the rate or speed at which the reaction takes place.

A chemical reaction, such as oxidation, takes place when molecules combine or collide with one another. The faster the molecules move, the greater the number of collisions between them and the faster the rate of reaction. Many factors influence the rate of these collisions. In our description of fire and oxidation, we consider only two: the physical state of the fuel and the temperature.

Physical state of fuel. A fuel achieves a reaction rate with oxygen sufficient to produce a flame only when it is in the gaseous state, because only in this state can molecules collide frequently enough to support a flaming fire. This remains true whether the fuel is a solid such as wood, paper, cloth, or plastic, or a liquid such as gasoline or kerosene.

For example, the conversion of iron to rust proceeds slowly because the iron atoms cannot achieve a gaseous state. The combination of oxygen with iron is thus restricted to the surface area of the metal exposed to air, a limitation that severely reduces the rate of reaction. On the other hand, the reaction of methane and oxygen proceeds rapidly because all the reactants are in the gaseous state. The speed of the reaction is reflected by the production of noticeable quantities of heat and light (a flame).

Fuel temperature. How, then, does a liquid or solid maintain a gaseous reaction? In the case of a liquid fuel, the temperature must be high enough to vaporize the fuel. The vapor that forms burns when it mixes with oxygen and combusts as a flame. The **flash point** is the lowest temperature at which a liquid gives off sufficient vapor to form a mixture with air that will support combustion. Once the flash point is reached, the fuel can be ignited by some outside source of temperature to start a fire. The ignition temperature of a fuel is always considerably higher than the flash point. For example, gasoline has a flash point of –50°F; however, an ignition temperature of 495°F is needed to start a gasoline fire.

flash point
The minimum temperature at which a liquid fuel produces enough vapor to burn

With a solid fuel such as wood, the process of generating vapor is more complex. A solid fuel burns only when exposed to heat intense enough to decompose the solid into gaseous products. This chemical breakdown of solid material is known as **pyrolysis**. The gaseous products of pyrolysis combine with oxygen to produce a fire (see Figure 14–3). Here again, fire can be described as a chain reaction. A match or other source of heat initiates the pyrolysis of the solid fuel, the gaseous products react with oxygen in the air to produce heat and light, and this heat in turn pyrolyzes more solid fuel into volatile gases.

pyrolysis
The decomposition of solid organic matter by heat

FIGURE 14–3
Intense heat causes solid fuels such as wood to decompose into gaseous products, a process called *pyrolysis*. *LiveMan/Shutterstock*

Typically, the rate of a chemical reaction increases when the temperature is raised. The magnitude of the increase varies from one reaction to another and also from one temperature range to another. For most reactions, a 10°C (18°F) rise in temperature doubles or triples the reaction rate. This observation explains in part why burning is so rapid. As the fire spreads, it raises the temperature of the fuel–air mixture, thus increasing the rate of reaction; this in turn generates more heat, again increasing the rate of reaction. Only when the fuel or oxygen is depleted does this vicious cycle come to a halt.

The Fuel–Air Mix As we have seen from our discussion about gaseous fuel, air (oxygen) and sufficient heat are the basic ingredients of a flaming fire. There is also one other consideration—the gas fuel–air mix. A mixture of gaseous fuel and air burns only if its composition lies within certain limits. If the fuel concentration is too low (lean) or too great (rich), combustion does not occur. The concentration range between the upper and lower limits is called the **flammable range**. For example, the flammable range for gasoline is 1.3–6.0 percent. Thus, in order for a gasoline–air mix to burn, gasoline must make up at least 1.3 percent, and no more than 6 percent, of the mixture.

flammable range The entire range of possible gas or vapor fuel concentrations in air that are capable of burning

Glowing Combustion Although a flaming fire can be supported only by a gaseous fuel, in some instances a fuel can burn without a flame. Witness a burning cigarette or the red glow of hot charcoals (see Figure 14–4). These are examples of **glowing combustion** or smoldering. Here combustion occurs on the surface of a solid fuel in the absence of heat high enough to pyrolyze the fuel. Interestingly, this phenomenon generally ensues long after the flames have gone out. Wood, for example, tends to burn with a flame until all of its pyrolyzable components have been expended; however, wood's carbonaceous residue continues to smolder long after the flame has extinguished itself.

glowing combustion Combustion on the surface of a solid fuel in the absence of heat high enough to pyrolyze the fuel

FIGURE 14–4 Red-hot charcoals are an example of glowing combustion.
Lev Kropotov/Shutterstock

Spontaneous Combustion One interesting phenomenon often invoked by arson suspects as the cause of a fire is **spontaneous combustion**. Actually, the conditions under which spontaneous combustion can develop are rather limited and rarely account for the cause of a fire. Spontaneous combustion is the result of a natural heat-producing process in poorly ventilated containers or areas. For example, hay stored in barns provides an excellent growing medium for bacteria whose activities generate heat. If the hay is not properly ventilated, the heat builds to a level that supports other types of heat-producing chemical reactions in the hay. Eventually, as the heat rises, the ignition temperature of hay is reached, spontaneously setting off a fire.

spontaneous combustion A fire caused by a natural heat-producing process in the presence of sufficient air and fuel

Another example of spontaneous combustion involves the ignition of improperly ventilated containers containing rags soaked with certain types of highly unsaturated oils, such as linseed oil. Heat can build up to the point of ignition as a result of a slow heat-producing chemical oxidation between the air and the oil. Of course, storage conditions must encourage the accumulation of the heat over a prolonged period of time. However, spontaneous combustion does not occur with hydrocarbon lubricating oils, and it is not expected to occur with most household fats and oils.

In summary, three requirements must be satisfied to initiate and sustain combustion:

1. **A fuel must be present.**
2. **Oxygen must be available in sufficient quantity to combine with the fuel.**
3. **Heat must be applied to initiate the combustion, and sufficient heat must be generated to sustain the reaction.**

Heat Transfer

Consider how a structural fire begins. The typical scenario starts with heat ignition at a single location. It may be an arsonist lighting a gasoline-soaked rag, a malfunctioning electric appliance sparking, or an individual falling asleep while smoking a cigarette in bed. How does a flame initially confined to a single location spread to engulf an entire structure? Understanding the anatomy of a fire begins with comprehending how heat travels through a burning structure.

The previous section stressed the importance of the role of heat in generating sufficient fuel vapors to support combustion, as well as the requirement that the heat source be hot enough to ignite the fuel's vapor. Once the fire begins, the heat generated by the fuel's reaction with air is fed back into the fuel–air mix to keep the chemical reaction going.

As a fire progresses, the heat created by the combustion process tends to move from a high-temperature region to one at a lower temperature. Understanding heat transfer from one location to another is important for reconstructing the origin of a fire, as well as for understanding why and how fire spreads through a structure. The three mechanisms of heat transfer are conduction, radiation, and convection.

Conduction Movement of heat through a solid object is caused by a process called conduction, in which electrons and atoms within the heated object collide with one another. Heat always travels from hot areas of a solid to cold ones by conduction. Solids whose atoms or molecules have loosely held electrons are good conductors of heat. Metals have the most loosely held electrons and are therefore excellent conductors of heat. Thus, when you insert one end of a metal object into an open flame, the entire object quickly becomes hot to the touch.

Materials that have electrons firmly attached to their molecules are poor conductors of heat. Poor conductors are called insulators. Wood is a good insulator; for that reason, metal objects that are subject to intense heat (such as skillets and saucepans) often have wooden handles (see Figure 14–5).

FIGURE 14–5
The wooden handle on this saucepan is a poor conductor of heat.
Somporn Naiyo/Alamy Stock Photo

In reconstructing a fire scene, it's important to keep in mind that heat may be transported through metals such as beams, nails, fasteners, bolts, and other good conductors to a location far from the initial heat source. Any fuel in contact with the conductor may be ignited, creating a new fire location. On the other hand, the conductivity of wood, plastic, and paper is very low, meaning that heat emanating from these surfaces does not spread well and does not cause ignitions far from the initial heat source.

Radiation *Radiation* is the transfer of heat energy from a heated surface to a cooler surface by electromagnetic radiation. A hot surface emits electromagnetic radiation of various wavelengths, and in a fire scene the electromagnetic radiation moves in a straight line from one surface to another. Radiant heat plays a key role in understanding how fire spreads throughout a structure. For example, all surfaces that face the fire are exposed to radiant heat and burst into flames when the surface reaches their ignition temperature. In very large fires, nearby structures and vehicles are often ignited at a distance by radiant heat.

Convection *Convection* is the transfer of heat energy by movement of molecules within a liquid or gas. Water being heated on a stove illustrates the concept of convection. As the water molecules on the bottom of the pot move faster, they spread apart and become less dense, causing them to move upward. Denser, cooler water molecules then migrate to the bottom of the pot. In this way, convection currents keep the fluid stirred up as warmer fluid moves away from the heat source and cooler fluid moves toward the heat source. Likewise, warm air expands, becoming less dense and causing it to rise and move toward the cooler surrounding air.

In a structural fire, the gaseous hot products of combustion expand, and convection moves the hot gases to the upper portions of the structure (see Figure 14–6). The convected hot gases become a source of heat, radiating heat energy downward onto all the surfaces below them. The hot surfaces of the exposed objects often pyrolyze or break down, releasing gaseous molecules. The phenomenon known as *flashover* occurs when all the combustible fuels simultaneously ignite, engulfing the entire structure in flame.

FIGURE 14–6
Convection causes flames to rise to the upper floor of a burning structure.
EvMedvedeva/Shutterstock

Quick Review

- Oxidation is the combination of oxygen with other substances to produce new substances.
- Combustion is rapid combination of oxygen with another substance, accompanied by production of noticeable heat and light.
- In an exothermic reaction, heat energy is liberated. In an endothermic reaction, heat energy is absorbed from the surroundings.
- Pyrolysis is the chemical breakdown of solid organic matter by heat. The gaseous products of pyrolysis combine with oxygen to produce a fire.
- Spontaneous combustion is fire caused by a natural heat-producing process in the presence of sufficient air and fuel.
- To initiate and sustain combustion, (1) a fuel must be present, (2) oxygen must be available in sufficient quantity to combine with the fuel, and (3) heat must be applied to initiate the combustion, and sufficient heat must be generated to sustain the reaction.
- The three mechanisms of heat transfer are conduction, radiation, and convection. Conduction is the movement of heat through a solid object. Radiation is the transfer of heat energy from a heated surface to a cooler surface by electromagnetic radiation. Convection is the transfer of heat energy by the movement of molecules within a liquid or gas.

Searching the Fire Scene

accelerant
Any material used to start or sustain a fire

The arson investigator should begin examining a fire scene for signs of arson as soon as the fire has been extinguished. Most arsons are started with petroleum-based **accelerants** such as gasoline or kerosene. Thus, the presence of containers capable of holding an accelerant arouse suspicions of arson. Discovery of an ignition device ranging in sophistication from a candle to a time-delay device is another indication of possible arson. A common telltale sign of arson is an irregularly shaped pattern on a floor or on the ground (see Figure 14–7) resulting from pouring an accelerant onto the surface. In addition to these visual indicators, investigators should look for signs of breaking and entering and theft, and they should begin interviewing any eyewitnesses to the fire.

FIGURE 14–7
Irregularly shaped pattern on the ground resulting from a poured ignitable liquid. *Courtesy Franklin County Crime Scene Unit, North Carolina*

Timelines of Investigation

Time constantly works against the arson investigator. Any accelerant residues that remain after a fire is extinguished may evaporate within a few days or even hours. Furthermore, safety and health conditions may necessitate that cleanup and salvage operations begin as quickly as possible. Once this occurs, a meaningful investigation of the fire scene is impossible. Accelerants in soil and vegetation can be rapidly degraded by bacterial action. Freezing samples containing soil or vegetation is an effective way to prevent this degradation.

The need to begin an *immediate* investigation of the circumstances surrounding a fire even takes precedence over the requirement to obtain a search warrant to enter and search the premises. The Supreme Court, explaining its position on this issue, stated in part:

> *... Fire officials are charged not only with extinguishing fires, but with finding their causes. Prompt determination of the fire's origin may be necessary to prevent its recurrence, as through the detection of continuing dangers such as faulty wiring or a defective furnace. Immediate investigation may also be necessary to preserve evidence from intentional or accidental destruction. And, of course, the sooner the officials complete their duties, the less will be their subsequent interference with the privacy and the recovery efforts of the victims. For these reasons, officials need no warrant to remain in a building for a reasonable time to investigate the cause of a blaze after it has been extinguished. And if the warrantless entry to put out the fire and determine its cause is constitutional, the warrantless seizure of evidence while inspecting the premises for these purposes also is constitutional....*

In determining what constitutes a reasonable time to investigate, appropriate recognition must be given to the exigencies that confront officials serving under these conditions, as well as to individuals' reasonable expectations of privacy.[1]

Locating the Fire's Origin

A search of the fire scene must focus on finding the fire's origin, which will prove most productive in any search for an accelerant or ignition device. In searching for a fire's specific point of origin, the investigator may uncover telltale signs of arson such as evidence of separate and unconnected fires or the use of "streamers" to spread the fire from one area to another. For example, the arsonist may have spread a trail of gasoline or paper to cause the fire to move rapidly from one room to another.

There are no fast and simple rules for identifying a fire's origin. Normally a fire tends to move upward, and thus the probable origin is most likely closest to the lowest point that shows the most intense characteristics of burning. Sometimes as the fire burns upward, a V-shaped pattern forms against a vertical wall, as shown in Figure 14–8. Because flammable liquids always flow to the lowest point, more severe burning found on the floor than on the ceiling may indicate the presence of an accelerant. If a flammable liquid was used, charring is expected to be more intense on the bottom of furniture, shelves, and other items rather than the top.

However, many factors can contribute to the deviation of a fire from normal behavior. Using burn patterns, such as depth of char, a V-shaped pattern, or low intense burn area, as an indicator of a fire's origin can prove to be misleading; particularly when a structural fire burns beyond flashover to full-room involvement. In these situations, air flow currents through the burning room can become a dominant factor in creating burn patterns.

Prevailing drafts and winds; secondary fires due to collapsing floors and roofs; the physical arrangement of the burning structure; stairways and elevator shafts; holes in the floor, wall, or roof; and the effects of the firefighter in suppressing the fire are all factors that the fire investigator must consider before determining conclusive findings regarding a fire's origin.

Once located, the point of origin should be protected to permit careful investigation. As at any crime scene, nothing should be touched or moved before notes, sketches, and photographs are taken. An examination must also be made

FIGURE 14–8 Typical V patterns illustrating upward movement of the fire. *David Schalliol/Moment/Getty Images*

for possible accidental causes, as well as for evidence of arson. The most common materials used by an arsonist to ensure the rapid spread and intensity of a fire are gasoline and kerosene or, for that matter, any volatile flammable liquid.

Searching for Accelerants

Fortunately, only under the most ideal conditions will combustible liquids be entirely consumed during a fire. When the liquid is poured over a large area, a portion of it will likely seep into a porous surface, such as cracks in the floor, upholstery, rags, plaster, wallboards, or carpet. Enough of the liquid may remain unchanged to permit its detection in the crime laboratory. In addition, when a fire is extinguished with water, the evaporation rate of volatile fluids may be slowed, because water cools and covers materials through which the combustible liquid may have soaked. Fortunately, water does not interfere with laboratory methods used to detect and characterize flammable liquid residues.

The search for traces of flammable liquid residues may be aided by the use of a sensitive portable vapor detector or "sniffer" (see Figure 14–9). This device can rapidly screen suspect materials for volatile residues by sucking in the air surrounding the questioned sample. The air is passed over a heated filament; if a combustible vapor is present, it oxidizes and immediately increases the temperature of the filament. The rise in filament temperature is then registered as a deflection on the detector's meter.

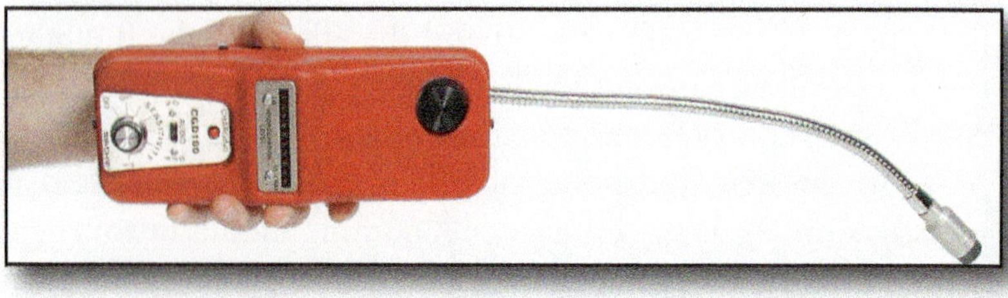

FIGURE 14–9 Portable hydrocarbon detector. *Courtesy Sirchie Finger Print Laboratories, Inc., Youngsville, N.C.,* www.sirchie.com

Of course, such a device is not a conclusive test for a flammable vapor, but it is an excellent screening device for checking suspect samples at the fire scene. Another approach is to use dogs that have been trained to recognize the odor of hydrocarbon accelerants.

Collection and Preservation of Arson Evidence

Two to three quarts of ash and soot debris must be collected at the point of origin of a fire when arson is suspected. The collection should include all porous materials and all other substances thought likely to contain flammable residues. These include such things as wood flooring, rugs, upholstery, and rags.

Packaging and Preservation of Evidence

Specimens should be packaged immediately in airtight containers so possible residues are not lost through evaporation. New, clean paint cans with friction lids are good containers because they are low cost, airtight, unbreakable, and available in a variety of sizes (see Figure 14–10). Wide-mouthed glass jars are also useful for packaging suspect specimens, provided that they have airtight lids. Cans and jars should be filled one-half to two-thirds full, leaving an air space in the container above the debris.

FIGURE 14–10 Various sizes of paint cans suitable for collecting debris at fire scenes. *Courtesy Sirchie Finger Print Laboratories, Inc., Youngsville, N.C., www.sirchie.com*

Large bulky samples should be cut to size at the scene as needed so that they will fit into available containers. Plastic polyethylene bags are not suitable for packaging specimens because they react with hydrocarbons and permit volatile hydrocarbon vapors to be depleted. Fluids found in open bottles or cans must be collected and sealed. Even when such containers appear empty, the investigator is wise to seal and preserve them in case they contain trace amounts of liquids or vapors.

Substrate Control

The collection of all materials suspected of containing volatile liquids must be accompanied by a thorough sampling of similar but uncontaminated control specimens from another area of the fire scene. This is known as *substrate control*. For example, if an investigator collects carpeting at the point of origin, the investigator must sample the same carpet from another part of the room, where it can be reasonably assumed that no flammable substance was placed.

In the laboratory, the criminalist checks the substrate control to be sure that it is free of any flammables. This procedure reduces the possibility (and subsequent argument) that the carpet was exposed to a flammable liquid such as a cleaning solution during normal maintenance. In addition, laboratory tests on the unburned control material may help analyze the breakdown products from the material's exposure to intense heat during the fire. Common materials such as plastic floor tiles, carpet, linoleum, and adhesives can produce volatile hydrocarbons when they are burned. These breakdown products can sometimes be mistaken for an accelerant.

Igniters and Other Evidence

The scene should also be thoroughly searched for igniters. The most common igniter is a match. Normally the match is completely consumed during a fire and is impossible to locate. However, there have been cases in which, by force of habit, matches have been extinguished and tossed aside only to be recovered later by the investigator. This evidence may prove valuable if the criminalist can fit the match to a book found in the possession of a suspect.

Arsonists can construct many other types of devices to start a fire. These include burning cigarettes, matches, firearms, ammunition, a mechanical match striker, electrical sparking devices, and a "Molotov cocktail"—a glass bottle containing flammable liquid with a cloth rag stuffed into it and lit as a fuse. Relatively complex mechanical devices are much more likely to survive the fire for later discovery. The broken glass and wick of the Molotov cocktail, if recovered, must be preserved as well.

One important piece of evidence is the clothing of the suspected perpetrator. If this individual is arrested within a few hours of initiating the fire, residual quantities of the accelerant may still be present in the clothing. As we will see in the next section, the forensic laboratory can detect extremely small quantities of accelerants, making the examination of a suspect's clothing a feasible investigative approach. Each item of clothing should be placed in a separate airtight container, preferably a new, clean paint can.

Quick Review

- Telltale signs of arson include evidence of separate and unconnected fires, the use of "streamers" to spread the fire from one area to another, and evidence of severe burning found on the floor as opposed to the ceiling of a structure.
- Other common signs of arson at a fire scene are the presence of accelerants and the discovery of an ignition device.
- Porous materials at the suspected point of origin of a fire should be collected and stored in airtight containers.

Analysis of Flammable Residues

Criminalists are nearly unanimous in judging the gas chromatograph to be the most sensitive and reliable instrument for detecting and characterizing flammable residues. Most arsons are initiated by petroleum distillates, such as gasoline and kerosene, that are composed of a complex mixture of **hydrocarbons**. The gas chromatograph separates the hydrocarbon components of these liquids and produces a chromatographic pattern characteristic of a particular petroleum product.

hydrocarbon
Any compound consisting of only carbon and hydrogen

The Headspace Technique

Before accelerant residues can be analyzed, they first must be recovered from the debris collected at the scene. The easiest way to recover accelerant residues from fire-scene debris is to heat the airtight container in which the sample is sent to the laboratory. When the container is heated, any volatile residue in the debris is driven off and trapped in the container's enclosed airspace. The vapor or headspace is then removed with a syringe, as shown in Figure 14–11.

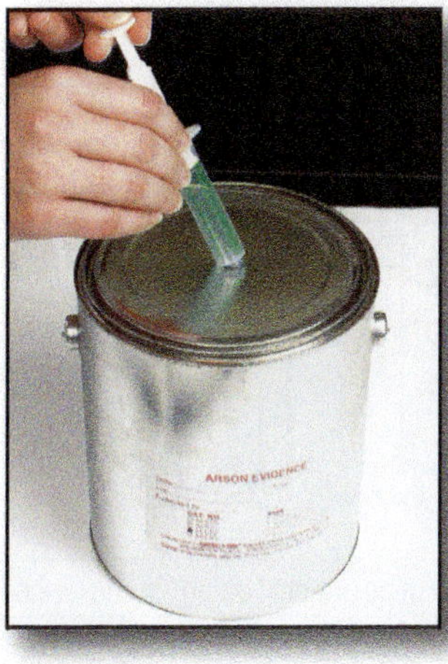

FIGURE 14–11
Removal of vapor from an enclosed container for gas chromatographic analysis.
Courtesy Richard Saferstein, Ph.D.

When the vapor is injected into the gas chromatograph, it is separated into its components, and each peak is recorded on the chromatogram. One way of classifying ignitable liquids is by their boiling point range, which is related to the number of carbon molecules that are present in the mixture of hydrocarbons comprising an ignitable fuel. A common classification system characterizes ignitable liquids based on their boiling-point range, which is related to the number of carbon molecules, as light, medium, and heavy petroleum distillates. Figure 14–12 illustrates examples of chromatograms for light, medium, and heavy petroleum distillates.

FIGURE 14–12 Chromatograms of ignitable liquids. (a) light petroleum distillate (b) medium petroleum distillate (c) heavy petroleum distillate.

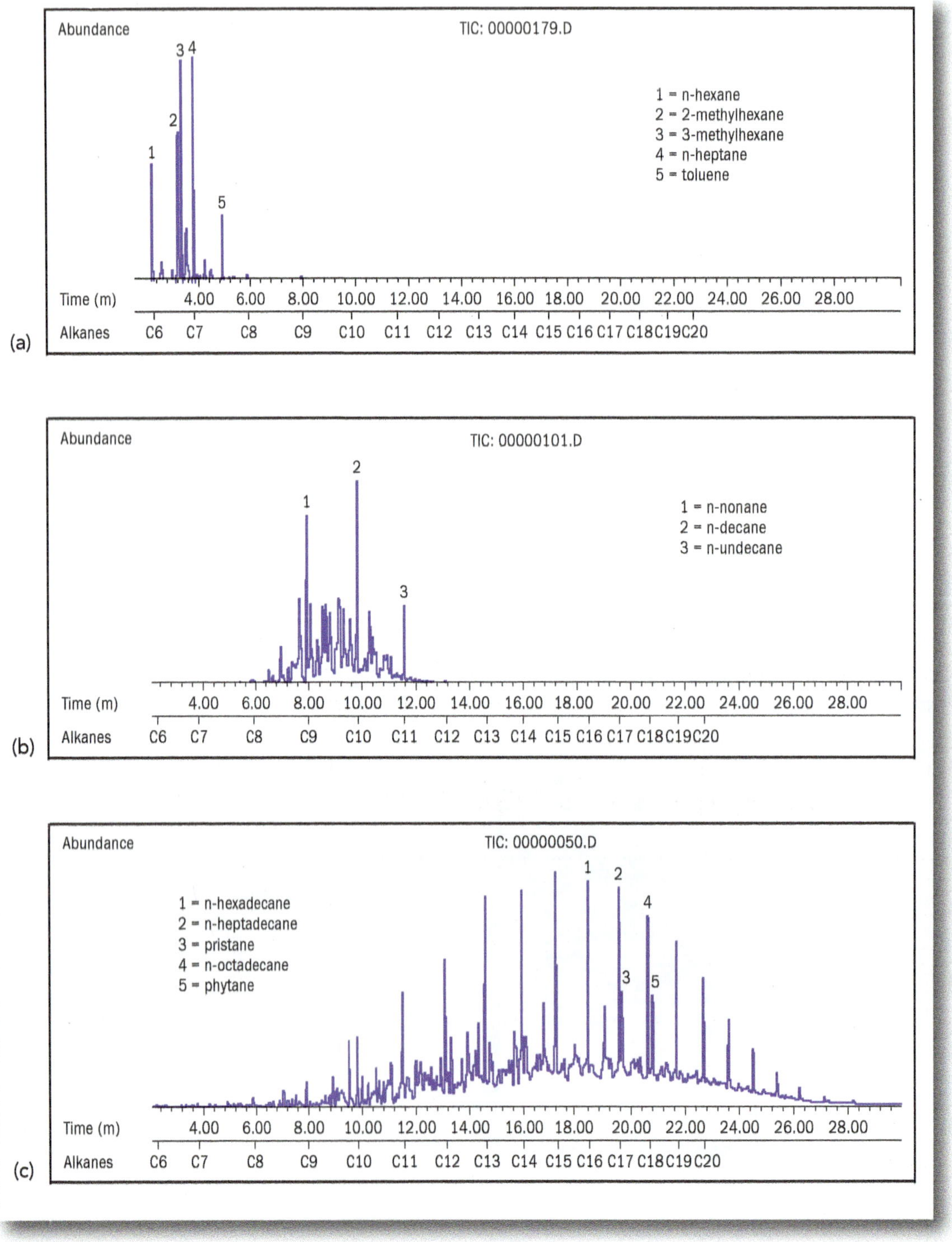

The identity of the volatile residue is determined when the pattern of the resultant chromatogram is compared to patterns produced by known petroleum products. For example, in Figure 14–13, a gas chromatographic analysis of debris recovered from a fire site shows a chromatogram similar to a known gasoline standard, thus proving the presence of gasoline.

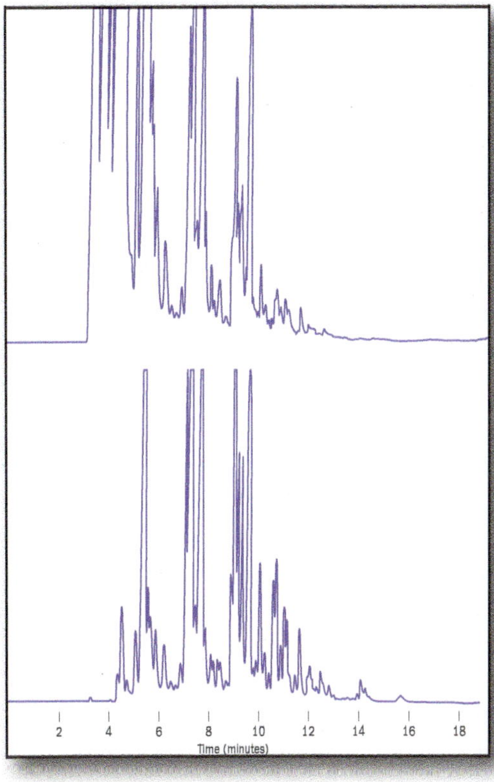

FIGURE 14–13
(top) Gas chromatograph of vapor from a genuine gasoline sample. (bottom) Gas chromatograph of vapor from debris recovered at a fire site. Note the similarity of the known gasoline to vapor removed from the debris.

Vapor Concentration

One major disadvantage of the headspace technique is that the size of the syringe limits the volume of vapor that can be removed from the container and injected into the gas chromatograph. To overcome this deficiency, many crime laboratories augment the headspace technique with a method called *vapor concentration*. One setup for this analysis is shown in Figure 14–14.

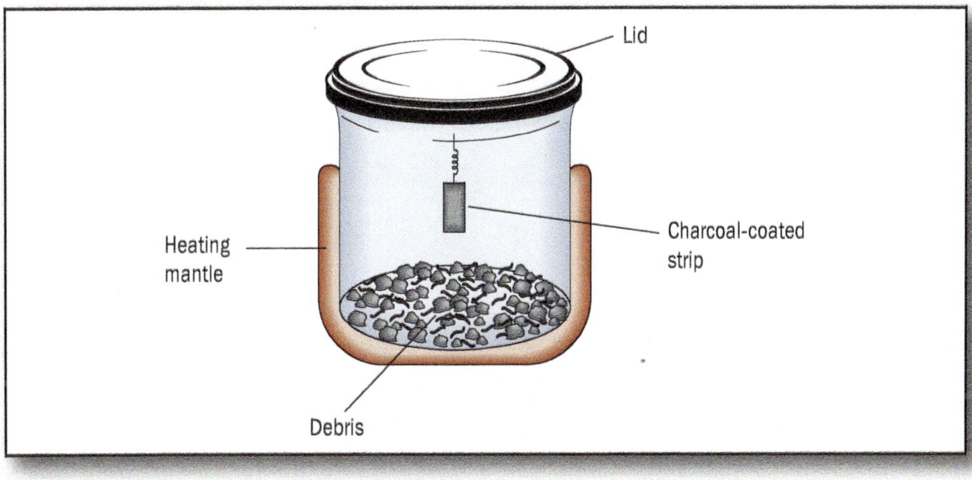

FIGURE 14–14
Apparatus for accelerant recovery by vapor concentration. The vapor in the enclosed container is exposed to charcoal, a chemical absorbent, where it is trapped for later analysis.

A charcoal-coated strip, similar to that used in environmental monitoring badges, is placed within the container holding the debris that has been collected from the fire scene.[2] The container is then heated to about 60°C for about one hour. At this temperature, a significant quantity of accelerant vaporizes into the container airspace. The charcoal absorbs the accelerant vapor with which it comes into contact. In this manner, over a short period of time, a significant quantity of the accelerant will be trapped and concentrated onto the charcoal strip.

Once the heating procedure is complete, the analyst removes the charcoal strip from the container and recovers the accelerant from the strip by washing it with a small volume of solvent (carbon disulfide). The solvent is then injected into the gas chromatograph for analysis. The major advantage of using vapor concentration with gas chromatography is its sensitivity. By absorbing the accelerant into a charcoal strip, the forensic analyst can increase the sensitivity of accelerant detection at least a hundredfold over that of the conventional headspace technique.

An examination of Figure 14–13 shows that identifying an accelerant such as gasoline by gas chromatography is an exercise in pattern recognition. Typically a forensic analyst compares the pattern generated by the sample to chromatograms from accelerant standards obtained under the same conditions. The pattern of gasoline, as with many other accelerants, can easily be placed in a searchable library. An invaluable reference known as "The Ignitable Liquids Reference Collection" or (ILRC) is found on the Internet at ilrc.ucf.edu. The ILRC is a useful collection showing chromatographic patterns for approximately 500 ignitable liquids.

Gas Chromatography/Mass Spectrometry

On occasion, discernible patterns are not obtainable by gas chromatography. This may be due to a combination of accelerants, or to the mixing of accelerant residue with heat-generated breakdown products of materials burning at the fire scene. Under such conditions, a gas chromatographic pattern can be difficult, if not impossible, to interpret. In these cases, gas chromatography combined with mass spectrometry (discussed in Chapter 6) has proven valuable for solving difficult problems in the detection of accelerant residues.

Complex chromatographic patterns can be simplified by passing the separated components emerging from the gas chromatographic column through a mass spectrometer. As each component enters the mass spectrometer, it is fragmented into a collection of ions. The analyst can then control which ions will be detected and which will go unnoticed. In essence, the mass spectrometer acts as a filter allowing the analyst to see only the peaks associated with the ions selected for a particular accelerant. In this manner, the chromatographic pattern can be simplified by eliminating extraneous peaks that may obliterate the pattern.[3] The process is illustrated in Figure 14–15.

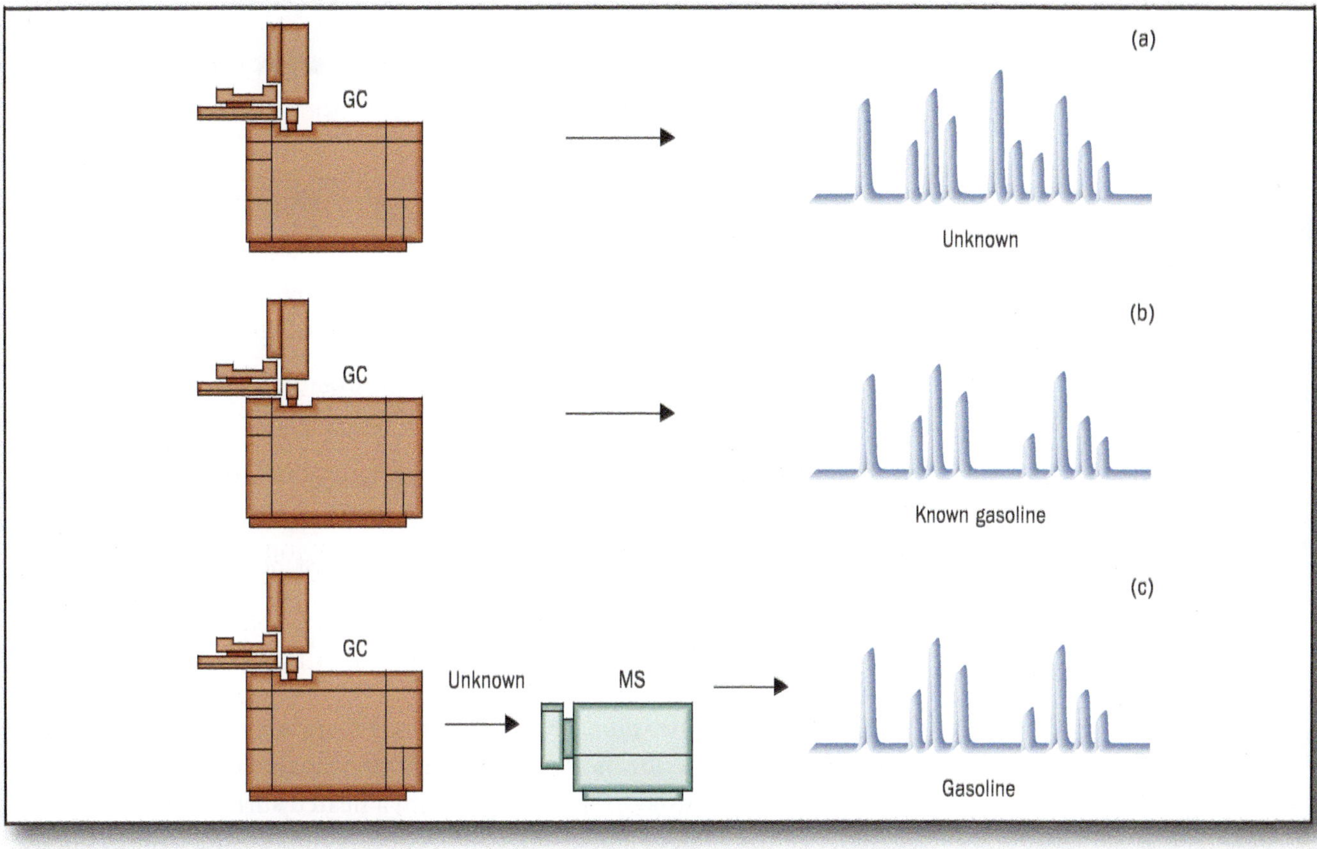

FIGURE 14–15 Chromatogram of a residue sample collected at a fire scene (a) shows a pattern somewhat like that of gasoline (b). However, a definitive conclusion that the unknown contained gasoline could be obtained only after extraneous peaks were eliminated from the unknown by the use of GC/MS (c).

Quick Review

- Most arsons are initiated by petroleum distillates such as gasoline and kerosene.
- The gas chromatograph is the most sensitive and reliable instrument for detecting and characterizing flammable residues. A gas chromatograph separates the hydrocarbon components and produces a chromatographic pattern characteristic of a particular petroleum product.
- By comparing select gas chromatographic peaks recovered from fire-scene debris to known flammable liquids, a forensic analyst may be able to identify the accelerant used to initiate the fire.
- Complex chromatographic patterns can be simplified by passing the separated components emerging from the gas chromatographic column through a mass spectrometer.

Chapter Review

- Oxidation is the combination of oxygen with other substances to produce new substances.
- Combustion is rapid combination of oxygen with another substance, accompanied by production of noticeable heat and light.
- In an exothermic reaction, heat energy is liberated. In an endothermic reaction, heat energy is absorbed from the surroundings.
- Pyrolysis is the chemical breakdown of solid organic matter by heat. The gaseous products of pyrolysis combine with oxygen to produce a fire.
- Spontaneous combustion is fire caused by a natural heat-producing process in the presence of sufficient air and fuel.
- To initiate and sustain combustion, (1) a fuel must be present, (2) oxygen must be available in sufficient quantity to combine with the fuel, and (3) heat must be applied to initiate the combustion, and sufficient heat must be generated to sustain the reaction.
- The three mechanisms of heat transfer are conduction, radiation, and convection. Conduction is the movement of heat through a solid object. Radiation is the transfer of heat energy from a heated surface to a cooler surface by electromagnetic radiation. Convection is the transfer of heat energy by the movement of molecules within a liquid or gas.
- Telltale signs of arson include evidence of separate and unconnected fires, the use of "streamers" to spread the fire from one area to another, and evidence of severe burning found on the floor as opposed to the ceiling of a structure.
- Other common signs of arson at a fire scene are the presence of accelerants and the discovery of an ignition device.
- Porous materials at the suspected point of origin of a fire should be collected and stored in airtight containers.
- Most arsons are initiated by petroleum distillates such as gasoline and kerosene.
- The gas chromatograph is the most sensitive and reliable instrument for detecting and characterizing flammable residues. A gas chromatograph separates the hydrocarbon components and produces a chromatographic pattern characteristic of a particular petroleum product.
- By comparing select gas chromatographic peaks recovered from fire-scene debris to known flammable liquids, a forensic analyst may be able to identify the accelerant used to initiate the fire.
- Complex chromatographic patterns can be simplified by passing the separated components emerging from the gas chromatographic column through a mass spectrometer.

Review Questions

1. If combustion is to be initiated and sustained, all of the following elements must be present except
   a. the presence of a fuel.
   b. the availability of oxygen.
   c. the application of heat.
   d. the elimination of carbon monoxide.

2. Most arson fires are started with
   a. lead-based paints.
   b. an oxidizing agent.
   c. petroleum-based accelerants.
   d. highly unsaturated oils.

3. The quantity of heat from a chemical reaction comes from
   a. the breaking and formation of chemical bonds.
   b. the presence of oxygen in the reaction.
   c. the emission of radiation.
   d. the composition of the fuel–air mix.

4. A search of the fire scene must focus on
   a. collecting trace evidence for later chemical analysis.
   b. estimating the value of monetary loss.
   c. ascertaining whether or not a crime has been committed.
   d. finding the fire's origin.

5. A gas chromatograph identifies an accelerant by a chromatogram's
   a. length.
   b. size.
   c. pattern.
   d. peak shape.

6. True or False: The criminalist plays a central role in the detection of arson and the subsequent apprehension of the suspected arsonist based on a determination as to the cause of a fire.

7. True or False: The flash point of a liquid fuel is the highest temperature at which a liquid gives off sufficient vapor to form a mixture with air that will support combustion.

8. True or False: Radiation is the transfer of heat energy by electromagnetic radiation.

9. True or False: Fire evidence can be contained in manila envelopes or plastic bags.

10. True or False: The most important instrument for identifying accelerants in the laboratory is the mass spectrometer.

11. What is oxidation?

12. Molecules must absorb energy to, and they liberate energy when.

13. What is combustion? Why is rusting not accompanied by combustion?
14. How does an exothermic reaction differ from an endothermic reaction?
15. A fire will burn until either of which two components is exhausted?
16. Name two factors that influence the speed of reaction of a fire.
17. In what physical state must a fuel exist in order to produce combustion when it reacts with oxygen? Why must it be in this state?
18. How high must the temperature of a liquid fuel be before the fuel will burn? What is the term for the lowest temperature at which this occurs?
19. What is pyrolysis? How does pyrolysis produce fire?
20. How does the fuel–air mix affect combustion?
21. Define and give two examples of glowing combustion.
22. Define *spontaneous combustion* and give two examples of conditions under which it may occur.
23. What three requirements must be satisfied to initiate and sustain combustion?
24. List and define the three mechanisms of heat transfer.
25. What is an insulator? Give an example of a good insulator.
26. List three common signs of arson at a fire scene.
27. Name three reasons why arson investigators must work quickly to collect evidence at a fire scene.
28. What are streamers? What does evidence of their presence at a fire scene suggest?
29. Why should fire evidence be packaged in airtight containers?
30. What instrument do most criminalists consider the most sensitive and reliable for detecting and characterizing flammable residues?
31. What is headspace? How is it recovered?
32. What is the main advantage of vapor concentration over the headspace technique?

Learn the Vocabulary

Use the information from this chapter along with your prior knowledge and experience to learn the key terms. Choose from the following activities:

1. Make flashcards with a term on the front and its definition on the back. Use the flashcards to quiz a partner on correctly pronouncing and defining the terms.
2. Make an audio or video recording of each term and its definition.
3. Work with a partner to write a sentence for each term. Read the sentences out loud to a partner or share them with the class.
4. Write a list of new terms and language you have encountered in this chapter. Reuse the terms to write sentences and then read the sentences out loud to a partner. Reuse the terms again by writing a short speech. The speech might explain the terms or teach a lesson using the terms. Give the speech to a partner or to the class.

Application and Critical Thinking

1. Indicate which method of heat transfer is most likely to be responsible for each of the following:
   a. ignition of papers in the room where a fire starts
   b. ignition of electrical wiring in a room adjoining the fire's point or origin
   c. ignition of roof timbers
   d. ignition of a neighboring house

2. It is late August in Houston, Texas, and you are investigating a fire that occurred at a facility that stores motor oils and other lubricating oils. A witness points out a man who allegedly ran from the structure about the same time that the fire started. You question the man, who turns out to be the owner of the facility. He tells you that he was checking his inventory when barrels of waste motor oil stored in an unventilated back room spontaneously burst into flames. The owner claims that the fire spread so rapidly that he had to flee the building before he could call 911. After speaking with several employees, you learn that the building has no air conditioning and that the oil had been stored for almost a year in the cramped back room. You also learn from a detective assisting on the case that the owner increased his insurance coverage on the facility within the past three months. Should you believe the owner's story, or should you suspect arson? Upon what do you base your conclusion?

3. Criminalist Mick Mickelson is collecting evidence from a fire scene. He gathers about a quart of ash and soot debris collected from several rooms surrounding the point of origin. He stored the debris in a new, clean paint can, filled about three-quarters full. Seeing several pieces of timber that he believes may contain accelerant residues, he cuts them and places them in airtight plastic bags. A short time later, a suspect is arrested and Mick searches him for any signs of an igniter or accelerants. He finds a cigarette lighter on the suspect and seizes it for evidence before turning the suspect over to the police. What mistakes, if any, did Mick make in collecting evidence?

4. Classify the following chromatograms of ignitable liquids below as low, medium, or high petroleum distillates. Refer to Figure 14–12.

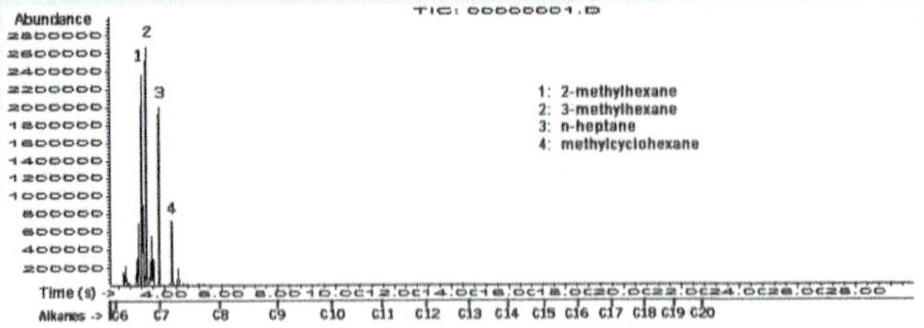

A. ____________________

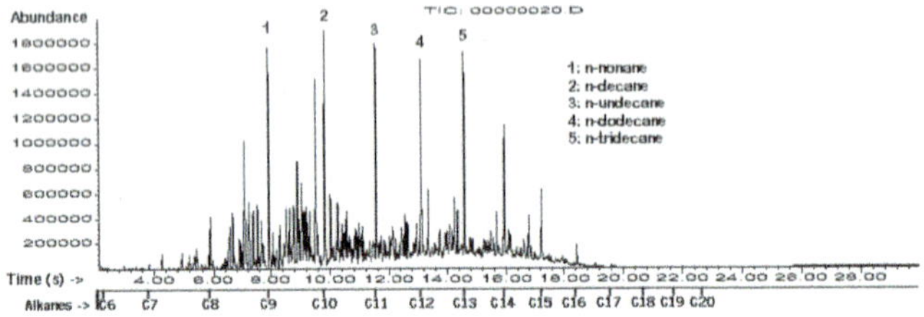

B. ____________________

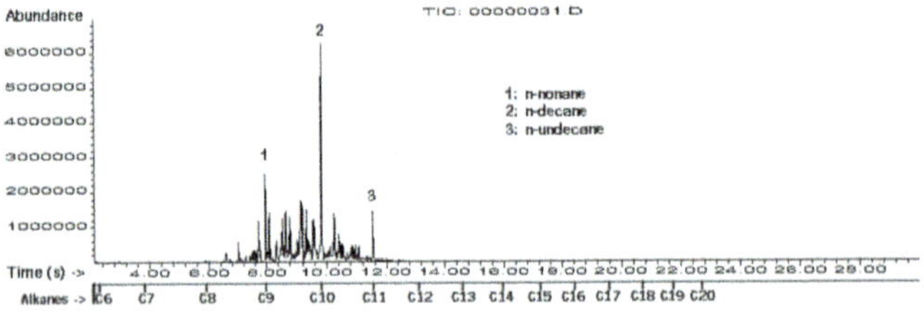

C. ____________________

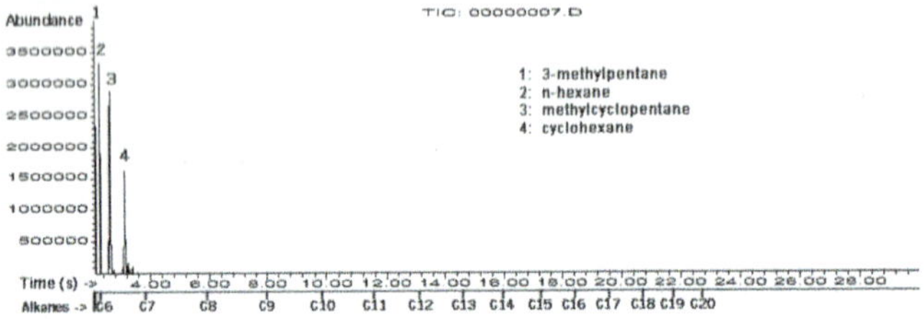

D. ____________________

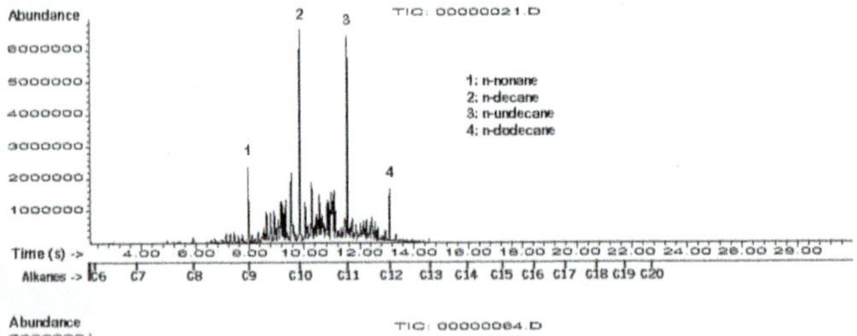

E. ____________________

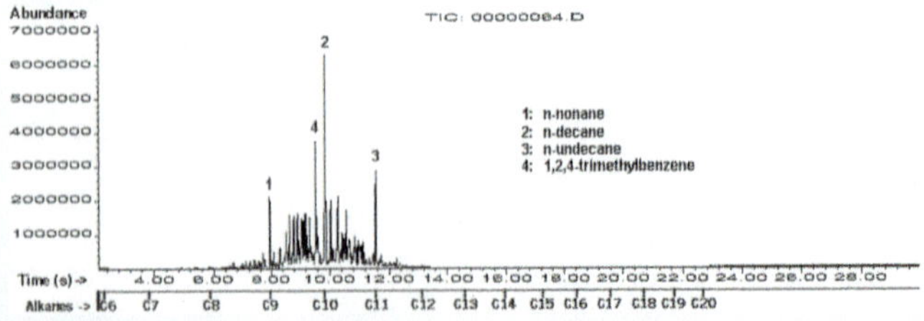

F. ____________________

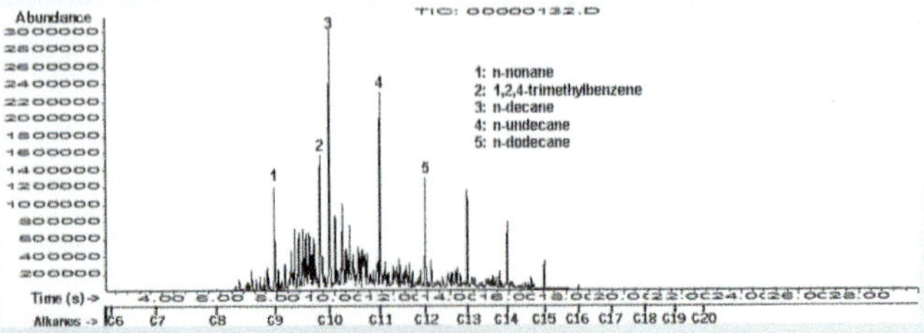

G. ____________________

Courtesy Richard Saferstein Ph.D.

Laboratory Experiments

This activity requires the use of the following practices of science and engineering:

- Analyzing and interpreting data
- Constructing explanations

This activity consists of the following crosscutting concepts:

- **Patterns**—The observed pattern peaks of the chromatograph can be used to determine the identity of an accelerant.
- **Cause and effect**—Burn patterns during a fire often leave a predictable pattern that can be used to reconstruct the cause of a fire and determine the fire's origin.

Before you begin, read the scenario and steps out loud with a partner. Ask for clarification as necessary to make sure you understand the experiment.

The Scenario

The Phoenix Fire Department is investigating an overnight fire that happened just before 10 PM in the 1100 block of Baseline Rd. When firefighters arrived, they found smoke coming out of the house. It took them about 25 minutes to put the fire out. According to witnesses, they saw three men run out of the house just before the fire started. Nobody was injured.

In order to determine if the fire was intentionally set, evaluate the following experimental evidence and consider the engineering design of the facility:

1. It appears that the fire was started in several locations: the kitchen, living room, and den.
2. There was more severe burning on the floor of each of the locations.
3. Samples of the wood flooring in each of the starting locations were collected and preserved.
4. Ash and soot debris was collected in each starting location.

PROCEDURE

In detail, describe how the evidence would be packaged and labeled.

After the laboratory analysis, you receive the following gas chromatograph from the flooring samples. What conclusions can you draw about the fire? Summarize your findings in paragraph form.

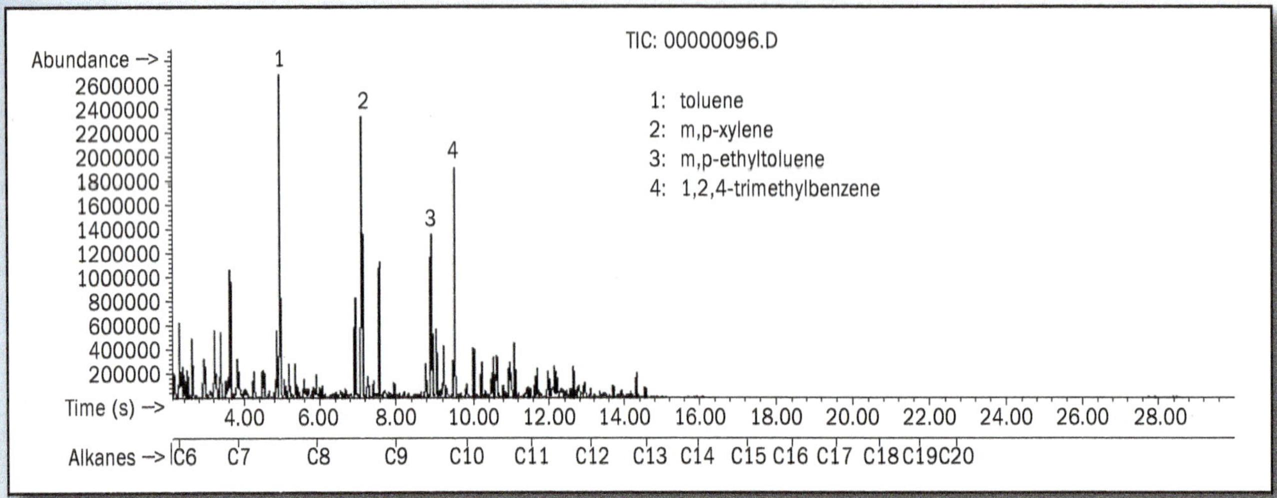

Courtesy Richard Saferstein, Ph.D.

Endnotes

1. Michigan v. Tyler, 436 U.S. 499 (1978).
2. R. T. Newman et al., "The Use of Activated Charcoal Strips for Fire Debris Extractions by Passive Diffusion. Part 1: The Effects of Time, Temperature, Strip Size, and Sample Concentration," *Journal of Forensic Sciences* 41 (1996): 361.
3. M. W. Gilbert, "The Use of Individual Extracted Ion Profiles versus Summed Extracted Ion Profiles in Fire Debris Analysis," *Journal of Forensic Sciences* 43 (1998): 871.

Forensic Investigation of Explosions

15

Naeblys/Shutterstock

Key Terms

black powder
deflagration
detonating cord
detonation
explosion
high explosive
low explosive
oxidizing agent
primary explosive
safety fuse
secondary explosive
smokeless powder (double-base)
smokeless powder (single-base)
X-ray diffraction

Beirut Explosion

On August 4, 2020, a large amount of ammonium nitrate stored at the port of the city of Beirut, Lebanon, exploded causing at least 200 deaths, 6,500 injuries, and billions of dollars property damage. A cargo of ammonium nitrate equivalent to approximately 1.1 kilotons of TNT had been stored in a warehouse without proper safety measures. The ship, *MV Rhosus*, and cargo were confiscated by Lebanese authorities six years prior. The cargo was preceded by a fire in the warehouse housing the *MV Rhosus*, but as of October 2020, the exact cause of the detonation is still under investigation.

There were reports that fireworks being stored in the warehouse were ignited by welding sparks from nearby workers, but it remains unclear whether ammunition or something else stored next to the ammonium nitrate might have been involved in worsening the warehouse fire and igniting the ammonium nitrate. Due to the extensive damage in the area of the blast, inspectors may never be able to say with certainty what triggered the explosion.

The Lebanese government declared a two-week state of emergency in response to the disaster. In its aftermath, protests erupted across Lebanon against the government for their failure to prevent the disaster.

Warehouses in the port were used to store explosives and chemicals including nitrates, common components of fertilizers and explosives. The failure to remove the materials from the warehouse and relocate them was attributed to mismanagement of the port, corruption of the government, and inaction of the flag registry's country and ship owner. An estimated 300,000 people were left homeless as a result of the blast.

The explosion was felt in Turkey, Syria, Israel, Palestine, and parts of Europe, and was heard in Cyprus, more than 160 miles away. It was detected by the United States Geological Survey as a seismic event of magnitude 3.3 on the Richter scale, and is considered one of the most powerful nonnuclear explosions in history.

Learning Objectives

After studying this chapter you should be able to:

15.1 Understand how explosives are classified

15.2 List some common commercial, homemade, and military explosives

15.3 Describe how to collect physical evidence at the scene of an explosion

15.4 Describe laboratory procedures used to detect and identify explosive residues

National Science Content Standards

Scientific Inquiry

Physical Science

Science and Technology

Science in Personal and Social Perspective

History and Nature of Science

Explosions and Explosives

The ready accessibility of potentially explosive laboratory chemicals, dynamite, and, in some countries, an assortment of military explosives has provided the criminal element of society with a lethal weapon. Unfortunately for society, explosives have become an attractive weapon to criminals bent on revenge, destruction of commercial operations, or just plain mischief.

Although politically motivated bombings have received considerable publicity worldwide, in the United States most bombing incidents are perpetrated by isolated individuals rather than by organized terrorists. These incidents typically involve homemade explosives and incendiary devices. The design of such weapons is limited only by the imagination and ingenuity of the bomber.

Like arson investigation, bomb investigation requires close cooperation of a group of highly specialized individuals trained and experienced in bomb disposal, bomb-site investigation, forensic analysis, and criminal investigation. The criminalist must detect and identify explosive chemicals recovered from the crime scene as well as identify the detonating mechanisms. This special responsibility concerns us for the remainder of this chapter.

The Chemistry of Explosions

explosion
A chemical or mechanical action caused by combustion, accompanied by creation of heat and rapid expansion of gases

Like fire, an **explosion** is the product of combustion accompanied by the creation of gases and heat. However, the distinguishing characteristic of an explosion is the rapid rate of the reaction. The sudden buildup of expanding gas pressure at the origin of the explosion produces the violent physical disruption of the surrounding environment.

Our previous discussion of the chemistry of fire referred only to oxidation reactions that rely on air as the sole source of oxygen. However, we need not restrict ourselves to this type of situation. For example, explosives are substances that undergo a rapid exothermic oxidation reaction, producing large quantities of gases. This sudden buildup of gas pressure constitutes an explosion. Detonation occurs so rapidly that oxygen in the air cannot participate in the reaction; thus, many explosives must have their own source of oxygen.

oxidizing agent
A substance that supplies oxygen to a chemical reaction

Chemicals that supply oxygen are known as **oxidizing agents**. One such agent is found in black powder, a *low explosive*, which is composed of a mixture of the following chemical ingredients:

75 percent potassium nitrate (KNO_3)

15 percent charcoal (C)

10 percent sulfur (S)

In this combination, oxygen-containing potassium nitrate acts as an oxidizing agent for the charcoal and sulfur fuels. As heat is applied to black powder, oxygen is liberated from potassium nitrate and simultaneously combines with charcoal and sulfur to produce heat and gases (symbolized by $\uparrow$), as represented in the following chemical equation:

$$\underset{\text{carbon}}{3C} + \underset{\text{sulfur}}{S} + \underset{\text{potassium nitrate}}{2KNO_3} \underset{\text{yields}}{\rightarrow}$$

Some explosives have their oxygen and fuel components combined within one molecule. For example, the chemical structure of nitroglycerin, the major constituent of dynamite, combines carbon, hydrogen, nitrogen, and oxygen:

$$\underset{\text{carbon dioxide}}{3CO_2\uparrow} + \underset{\text{nitrogen}}{N_2\uparrow} + \underset{\text{potassium sulfide}}{K_2S}$$

When nitroglycerin detonates, large quantities of energy are released as the molecule decomposes, and the oxygen recombines to produce large volumes of carbon dioxide, nitrogen, and water.

FIGURE 15–1 A violent explosion.
Jvdwolf/123RF

Consider, for example, the effect of confining an explosive charge to a relatively small, closed container. On detonation, the explosive almost instantaneously produces large volumes of gases that exert enormously high pressures on the interior walls of the container. In addition, the heat energy released by the explosion expands the gases, causing them to push on the walls with an even greater force. If we could observe the effects of an exploding lead pipe in slow motion, we would first see the pipe's walls stretch and balloon under pressures as high as several hundred tons per square inch. Finally, the walls would fragment and fly outward in all directions. This flying debris or shrapnel constitutes a great danger to life and limb in the immediate vicinity.

On release from confinement, the gaseous products of the explosion suddenly expand and compress layers of surrounding air as they move outward from the origin of the explosion. This blast effect, or outward rush of gases, at a rate that may be as high as 7,000 miles per hour creates an artificial gale that can overthrow walls, collapse roofs, and disturb any object in its path. If a bomb is sufficiently powerful, more serious damage will be inflicted by the blast effect than by fragmentation debris (see Figure 15–1).

Types of Explosives

The speed at which explosives decompose varies greatly from one to another and permits their classification as *high* and *low* explosives. In a low explosive, this speed is called the speed of **deflagration** (burning). It is characterized by very rapid oxidation that produces heat, light, and a subsonic pressure wave. In a high explosive, it is called the speed of detonation. **Detonation** refers to the creation of a supersonic shock wave within the explosive charge. This shock wave breaks the chemical bonds of the explosive charge, leading to the near instantaneous buildup of heat and gases.

deflagration
A very rapid oxidation reaction accompanied by the generation of a low-intensity pressure wave that can disrupt the surroundings

detonation
An extremely rapid oxidation reaction accompanied by a violent disruptive effect and an intense, highspeed shock wave

Low Explosives **Low explosives**, such as black and smokeless powders, decompose relatively slowly at rates up to 1,000 meters per second. Because of their slow burning rates, they produce a propelling or throwing action that makes them suitable as propellants for ammunition or skyrockets. However, the danger of this group of explosives must not be underestimated, because when any one of them is confined to a relatively small container, it can explode with a force as lethal as that of any known explosive.

low explosive
An explosive with a velocity of detonation less than 1,000 meters per second

FIGURE 15–2 **Samples of smokeless powders.** *Courtesy ATF Bureau of Alcohol, Tobacco, Firearms & Explosives*

black powder
Normally, a mixture of potassium nitrate, carbon, and sulfur in the ratio 75/15/10

safety fuse
A cord containing a core of black powder, used to carry a flame at a uniform rate to an explosive charge

smokeless powder (single-base)
An explosive consisting of nitrocellulose

smokeless powder (double-base)
An explosive consisting of a mixture of nitrocellulose and nitroglycerin

Black powder and smokeless powder. The most widely used explosives in the low-explosive group are black powder and smokeless powder. The popularity of these two explosives is enhanced by their accessibility to the public. Both are available in any gun store, and black powder can easily be made from ingredients purchased at any chemical supply house as well.

Black powder is a relatively stable mixture of potassium nitrate or sodium nitrate, charcoal, and sulfur. Unconfined, it merely burns; thus it commonly is used in safety fuses that carry a flame to an explosive charge. A **safety fuse** usually consists of black powder wrapped in a fabric or plastic casing. When ignited, a sufficient length of fuse will burn at a rate slow enough to allow a person adequate time to leave the site of the pending explosion. Black powder, like any other low explosive, becomes explosive and lethal only when it is confined.

The safest and most powerful low explosive is **smokeless powder**. This explosive usually consists of nitrated cotton or nitrocellulose (**single-base powder**) or nitroglycerin mixed with nitrocellulose (**double-base powder**). The powder is manufactured in a variety of grain sizes and shapes, depending on the desired application (see Figure 15–2).

Chlorate mixtures. The only ingredients required for a low explosive are fuel and a good oxidizing agent. The oxidizing agent potassium chlorate, for example, when mixed with sugar, produces a popular and accessible explosive mix. When confined to a small container—for example, a pipe—and ignited, this mixture can explode with a force equivalent to a stick of 40 percent dynamite.

Some other commonly encountered ingredients that may be combined with chlorate to produce an explosive are carbon, sulfur, starch, phosphorus, and magnesium filings. Chlorate mixtures may also be ignited by the heat generated from a chemical reaction. For instance, sufficient heat can be generated to initiate combustion when concentrated sulfuric acid comes in contact with a sugar–chlorate mix.

Gas–Air mixtures. Another form of low explosive is created when a considerable quantity of natural gas escapes into a confined area and mixes with a sufficient amount of air. If ignited, this mixture results in simultaneous combustion and sudden production of large volumes of gases and heat. In a building, walls are forced outward by the expanding gases, causing the roof to fall into the interior, and objects are thrown outward and scattered in erratic directions with no semblance of pattern.

Mixtures of air and a gaseous fuel explode or burn only within a limited concentration range. For example, the concentration limits for methane in air range from 5.3 to 13.9 percent. In the presence of too much air, the fuel becomes too diluted and does not ignite. On the other hand, if the fuel becomes too concentrated, ignition is prevented because there is not enough oxygen to support the combustion.

Mixtures at or near the upper concentration limit ("rich" mixtures) explode; however, some gas remains unconsumed because there is not enough oxygen to complete the combustion. As air rushes back into the origin of the explosion, it combines with the residual hot gas, producing a fire that is characterized by a whoosh sound. This fire is often more destructive than the explosion that preceded it. Mixtures near the lower end of the limit ("lean" mixtures) generally cause an explosion without accompanying damage due to fire.

High Explosives **High explosives** include dynamite, TNT, PETN, and RDX. They detonate almost instantaneously at rates of 1,000–8,500 meters per second, producing a smashing or shattering effect on their target. Unlike a low explosive, high explosives do not have to be containerized in order to amplify the damaging effect of their shock waves. High explosives are classified into two groups—primary and secondary explosives—based on their sensitivity to heat, shock, or friction.

high explosive
An explosive with a velocity of detonation greater than 1,000 meters per second

Primary explosives are ultrasensitive to heat, shock, or friction, and under normal conditions they detonate violently instead of burning. For this reason, they are used to detonate other explosives through a chain reaction and are often referred to as primers. Primary explosives provide the major ingredients of blasting caps and include lead azide, lead styphnate, and diazodinitrophenol (see Figure 15–3). Because of their extreme sensitivity, these explosives are rarely used as the main charge of a homemade bomb.

primary explosive
A high explosive that is easily detonated by heat, shock, or friction

Secondary explosives are relatively insensitive to heat, shock, or friction, and they normally burn rather than detonate when ignited in small quantities in open air. This group comprises most high explosives used for commercial and military blasting. Some common examples of secondary explosives are dynamite, TNT (trinitrotoluene), PETN (pentaerythritol tetranitrate), RDX (cyclotrimethylenetrinitramine), and tetryl (2,4,6-trinitrophenylmethylnitramine).

secondary explosive
A high explosive that is relatively insensitive to heat, shock, or friction

Dynamite. It is an irony of history that the prize most symbolic of humanity's search for peace—the Nobel Peace Prize—should bear the name of the developer of one of our most lethal discoveries—dynamite. In 1867, the Swedish chemist Alfred Nobel, searching for a method to desensitize nitroglycerin, found that when kieselguhr, a variety of diatomaceous earth, absorbed a large portion of nitroglycerin, it became far less sensitive but still retained its explosive force. Nobel later decided to use pulp as an absorbent because kieselguhr was a heat-absorbing material.

This so-called pulp dynamite was the beginning of what is now known as the straight dynamite series. These dynamites are used when a quick shattering action is desired. In addition to nitroglycerine and pulp, present-day straight dynamites also include sodium nitrate (which furnishes oxygen for complete combustion) and a small percentage of a stabilizer, such as calcium carbonate.

All straight dynamite is rated by strength; the strength rating is determined by the weight percentage of nitroglycerin in the formula. Thus, a 40 percent straight dynamite contains 40 percent nitroglycerin, a 60 percent grade contains 60 percent nitroglycerin, and so forth. However, the relative blasting power

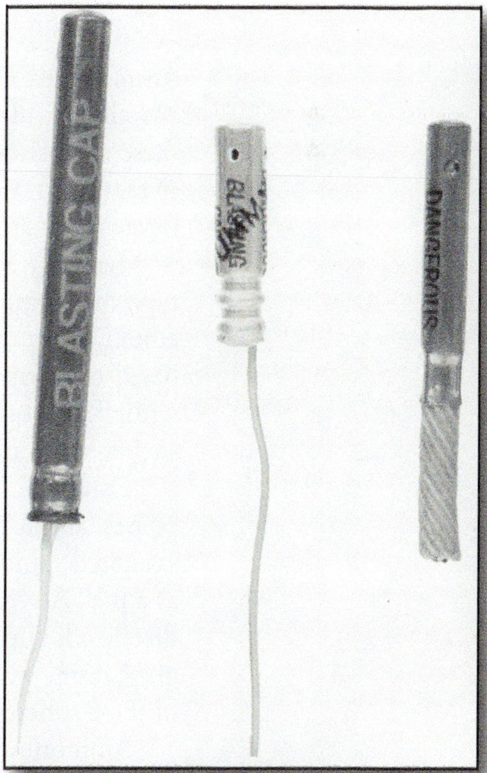

FIGURE 15–3 Blasting caps. The left and center caps are initiated by an electrical current; the right cap is initiated by a safety fuse.
Courtesy Richard Saferstein Ph.D.

FIGURE 15–4 Sticks of dynamite. *Yavapai County Sheriff's Office/AP Images*

of different strengths of dynamite is not directly proportional to their strength ratings. A 60 percent straight dynamite, rather than being three times as strong as a 20 percent, is only one and one-half times as strong (see Figure 15–4).

Ammonium nitrate explosives. In recent years, nitroglycerin-based dynamite has all but disappeared from the industrial explosives market. Commercially, these explosives have been replaced mainly by ammonium nitrate–based explosives, that is, water gels, emulsions, and ANFO explosives. These explosives mix oxygen-rich ammonium nitrate with a fuel to form a low-cost, stable explosive.

Typically, water gels have a consistency resembling that of set gelatin or gel-type toothpaste. They are characterized by their water-resistant nature and are employed for all types of blasting under wet conditions. These explosives are based on formulations of ammonium nitrate and sodium nitrate gelled with a natural polysaccharide such as guar gum. Commonly, a combustible material such as aluminum is mixed into the gel to serve as the explosive's fuel.

Emulsion explosives differ from gels in that they consist of two distinct phases, an oil phase and a water phase. In these emulsions, a droplet of a supersaturated solution of ammonium nitrate is surrounded by a hydrocarbon serving as a fuel. A typical emulsion consists of water, one or more inorganic nitrate oxidizers, oil, and emulsifying agents. Commonly, emulsions contain micron-sized glass, resin, or ceramic spheres known as microspheres or microballoons. The size of these spheres controls the explosive's sensitivity and detonation velocity.

Ammonium nitrate soaked in fuel oil is an explosive known as ANFO. Such commercial explosives are inexpensive and safe to handle and have found wide applications in blasting operations in the mining industry. Ammonium nitrate in the form of fertilizer makes a readily obtainable ingredient for homemade explosives. Indeed, in an incident related to the 1993 bombing of New York City's

World Trade Center, the FBI arrested five men during a raid on their hideout in New York City, where they were mixing a "witches' brew" of fuel oil and an ammonium nitrate–based fertilizer.

TATP. Triacetone triperoxide (TATP) is a homemade explosive that has been used as an improvised explosive by terrorist organizations in Israel and other Middle Eastern countries. It is prepared by reacting the common ingredients of acetone and hydrogen peroxide in the presence of an acid catalyst such as hydrochloric acid.

TATP is a friction- and impact-sensitive explosive that is extremely potent when confined in a container such as a pipe. The 2005 London transit bombings were caused by TATP-based explosives and provide ample evidence that terrorist cells have moved TATP outside the Middle East. A London bus destroyed by one of the TATP bombs is shown in Figure 15–5.

A plot to blow up ten international plane flights leaving Britain for the United States with a "liquid explosive" apparently involved plans to smuggle the peroxide-based TATP explosive onto the planes. This plot has prompted authorities to prohibit airline passengers from carrying liquids and gels onto planes.

Military high explosives. No discussion of high explosives would be complete without a mention of military high explosives. In many countries outside the United States, the accessibility of high explosives to terrorist organizations makes them common constituents of homemade bombs. RDX, the most popular and powerful military explosive, is often encountered in the form of a pliable plastic of doughlike consistency known as composition C–4 (a U.S. military designation).

TNT was produced and used on an enormous scale during World War II and may be considered the most important military bursting charge explosive. Alone or in combination with other explosives, it has found wide application in shells, bombs, grenades, demolition explosives, and propellant compositions. Interestingly, military "dynamite" contains no nitroglycerin but is composed of a mixture of RDX and TNT. Like other military explosives, TNT is rarely encountered in bombings in the United States.

FIGURE 15–5 A London bus destroyed by a TATP-based bomb. *Peter Macdiarmid/Getty Images News/Getty Images*

Case Files

Liquid Explosives

In 2006, security agencies in the United States and Great Britain uncovered a terrorist plot to use liquid explosives to destroy commercial airlines operating between the two countries. Of the hundreds of types of explosives, most are solid. Only about a dozen are liquid. But some of those liquid explosives can be readily purchased, and others can be made from hundreds of different kinds of chemicals that are not difficult to obtain. After the September 11 attacks, worries about solid explosives became the main concern. In 2001, Richard Reid was arrested for attempting to destroy an American Airlines flight flying out of Paris. Authorities later found a high explosive with a TATP (triacetone triperoxide) detonator hidden in the lining of his shoe. It is therefore not surprising that terrorists turned to liquids in this latest plot. A memo issued by federal security officials about the plot to blow up ten international planes highlighted a type of liquid explosive based on peroxide. The most common peroxide-based explosive is TATP, which can be used as a detonator or a primary explosive and has been used in terrorist-related bombings and by Palestinian suicide bombers.

In theory, scientists know how to detect peroxide-based explosives. The challenge is to design machines that can perform scans quickly and efficiently on thousands of passengers passing through airport security checks.

Current scanning machines at airports are designed to detect nitrogen-containing chemicals and are not designed to detect peroxide-containing explosive ingredients. Since 9/11, security experts have worried about the possibility of explosives in the form of liquids and gels getting onto airliners.

Not having the luxury of waiting for newly designed scanning devices capable of ferreting out dangerous liquids to be in place at airports, authorities decided to use a commonsense approach: to restrict the types and quantities of liquids that a passenger can carry onto a plane.

Liquids and gels discarded by airline passengers before boarding. *Stefano Paltera/ AP Images*

PETN is used by the military in TNT mixtures for small-caliber projectiles and grenades. Commercially, the chemical is used as the explosive core in a **detonating cord** or primacord. Instead of the slower-burning safety fuse, a detonating cord is often used to connect a series of explosive charges so that they will detonate simultaneously.

detonating cord A cordlike explosive containing a core of high-explosive material, usually PETN; also called primacord

Detonators. Unlike low explosives, bombs made of high explosives must be detonated by an initiating explosion. In most cases, detonators are blasting caps composed of copper or aluminum cases filled with lead azide as an initiating charge and PETN or RDX as a detonating charge. Blasting caps can be initiated by means of a burning safety fuse or by an electrical current.

Homemade bombs camouflaged in packages, suitcases, and the like are usually initiated with an electrical blasting cap wired to a battery. An unlimited number of switching-mechanism designs have been devised for setting off these devices; clocks and mercury switches are favored. Bombers sometimes prefer to employ outside electrical sources. For instance, most automobile bombs are detonated when the ignition switch of a car is turned on.

Quick Review

- Explosives are substances that undergo a rapid oxidation reaction with the production of large quantities of gases. The sudden buildup of gas pressure leads to the explosion.
- The speed at which an explosive decomposes determines whether it is classified as a high or low explosive.
- The most widely used low explosives are black powder and smokeless powder. Common high explosives include ammonium nitrate–based explosives (water gels, emulsions, and ANFO explosives).
- Among the high explosives, primary explosives are ultrasensitive to heat, shock, and friction and provide the major ingredients found in blasting caps. Secondary explosives normally constitute the main charge of a high explosive.

Collection and Analysis of Evidence of Explosives

The most important step in the detection and analysis of explosive residues is the collection of appropriate samples from the explosion scene. Invariably, undetonated residues of the explosive remain at the site of the explosion. The detection and identification of these explosives in the laboratory depend on the bomb-scene investigator's skill and ability to recognize and sample the areas most likely to contain such materials.

Detecting and Recovering Evidence of Explosives

The most obvious characteristic of a high or contained low explosive is the presence of a crater at the origin of the blast. Once the crater has been located, all loose soil and other debris must immediately be removed from the interior of the hole and preserved for laboratory analysis. Other good sources of explosive residues are objects located near the origin of detonation. Wood, insulation, rubber, and other soft materials that are readily penetrated often collect traces of the explosive. However, nonporous objects near the blast must not be overlooked. For instance, residues can be found on the surfaces of metal objects near the site of an explosion. Material blown away from the blast's origin should also be recovered because it, too, may retain explosive residues.

The entire area must be systematically searched, with great care given to recovering any trace of a detonating mechanism or any other item foreign to the explosion site. Wire-mesh screens are best used for sifting through debris. All personnel involved in searching the bomb scene must take appropriate measures to avoid contaminating the scene, including dressing in disposable gloves, shoe covers, and overalls.

In pipe-bomb explosions, particles of the explosive are frequently found adhering to the pipe cap or to the pipe threads, as a result of either being impacted into the metal by the force of the explosion or being deposited in the threads during the construction of the bomb. One approach for screening objects for the presence of explosive residues in the field or the laboratory is the ion mobility spectrometer (IMS).[1] A portable IMS is shown in Figure 15–6.

This handheld detector uses a vacuum to collect explosive residues from suspect surfaces. Alternatively, the surface suspected of containing explosive residues is wiped down with a Teflon-coated fiberglass disc and the collected residues are then drawn into the spectrometer off the disc. Once in the IMS, the explosive residues are vaporized by the application of heat. These vaporized substances are

FIGURE 15–6 A portable ion mobility spectrometer used to rapidly detect and tentatively identify trace quantities of explosives. *Courtesy 2013 Morpho Detection, Inc. All rights reserved. Mobiletrace is a registered trademark of Morpho Detection, Inc.*

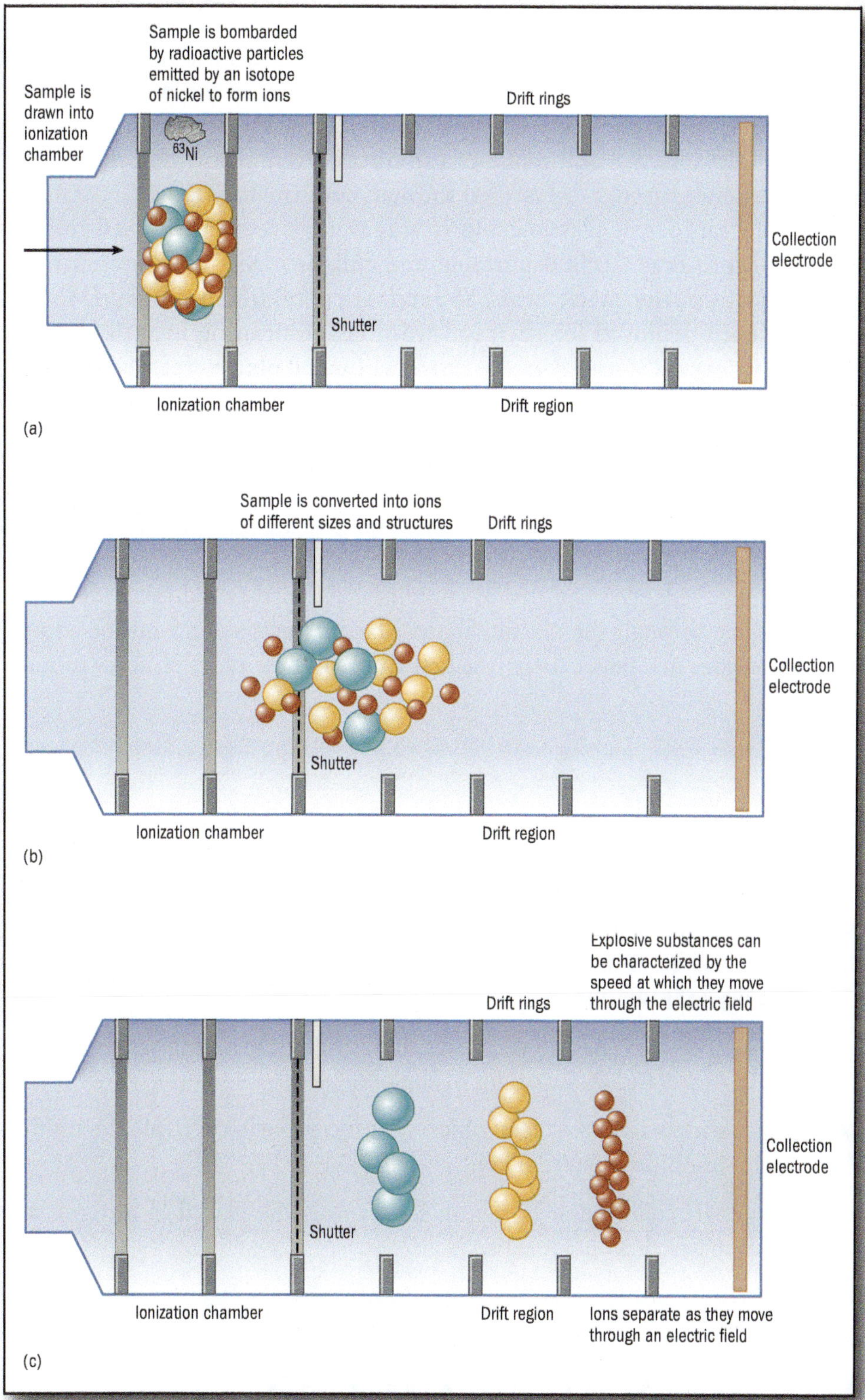

FIGURE 15–7 Schematic diagram of an ion mobility spectrometer. A sample is introduced into an ionization chamber, where bombardment with radioactive particles emitted by an isotope of nickel converts the sample to ions. The ions move into a drift region where ion separation occurs based on the speed of the ions as they move through an electric field.

exposed to a beam of electrons or beta particles emitted by radioactive nickel and converted into electrically charged molecules or ions. The ions are then allowed to move through a tube (drift region) under the influence of an electric field. A schematic diagram of an IMS is shown in Figure 15–7.

The preliminary identification of an explosive residue can be made by noting the time it takes the explosive to move through the tube. Because ions move at different speeds depending on their size and structure, they can be characterized by the speed at which they pass through the tube. Used as a screening tool, this method rapidly detects a full range of explosives, even at low detection levels. However, all results need to be verified through confirmatory tests.

The IMS can detect plastic explosives as well as commercial and military explosives. More than 10,000 portable and full-size IMS units are currently used at airport security checkpoints, and more than 50,000 handheld IMS analyzers have been deployed for chemical-weapons monitoring in various armed forces.

Collection and Packaging All materials collected for examination by the laboratory must be placed in airtight sealed containers and labeled with all pertinent information. Soil and other soft, loose materials are best stored in metal airtight containers such as clean paint cans. Debris and articles collected from different areas are to be packaged in separate airtight containers. Plastic bags should not be used to store evidence suspected of containing explosive residues. Some explosives can escape through the plastic. Sharp-edged objects should not be allowed to pierce the sides of a plastic bag. It is best to place these types of items in metal containers.

Quick Review

- The entire bomb site must be systematically searched to recover any trace of a detonating mechanism or any other item foreign to the explosion site. Objects located at or near the origin of the explosion must be collected for laboratory examination.
- The most obvious characteristic of a high or contained low explosive is the presence of a crater at the origin of the blast.
- A device widely used to screen objects for the presence of explosive residues is the ion mobility spectrometer.
- All materials collected at bombing scenes must be placed in airtight containers such as clean paint cans.

Analysis of Evidence Explosives

When the bomb-scene debris and other materials arrive at the laboratory, everything is first examined microscopically to detect particles of unconsumed explosive. Portions of the recovered debris and detonating mechanism, if found, are carefully viewed under a low-power stereoscopic microscope in a painstaking effort to locate particles of the explosive. Black powder and smokeless powder are relatively easy to

Table 15–1
Color Spot Tests for Common Explosives

Substance	Reagent Greiss[a]	Diphenylamine[b]	Alcoholic KOH[c]
Chlorate	No color	Blue	No color
Nitrate	Pink to red	Blue	No color
Nitrocellulose	Pink	Blue-black	No color
Nitroglycerin	Pink to red	Blue	No color
PETN	Pink to red	Blue	No color
RDX	Pink to red	Blue	No color
TNT	No color	No color	Red
Tetryl	Pink to red	Blue	Red-violet

[a]*Greiss reagent: Solution 1—Dissolve 1 g sulfanilic acid in 100 mL 30% acetic acid. Solution 2—Dissolve 0.5 g N-(1-napthyl) ethylenediamine in 100 mL methyl alcohol. Add solutions 1 and 2 and a few milligrams of zinc dust to the suspect extract.*
[b]*Diphenylamine reagent: Dissolve 1 g diphenylamine in 100 mL concentrated sulfuric acid*
[c]*Alcoholic KOH reagent: Dissolve 10 g of potassium hydroxide in 100 mL absolute alcohol*

locate in debris because of their characteristic shapes and colors (see Figure 15–2). However, dynamite and other high explosives present the microscopist with a much more difficult task and often must be detected by other means.

Following microscopic examination, the recovered debris is thoroughly rinsed with acetone. The high solubility of most explosives in acetone ensures their quick removal from the debris. When a water-gel explosive containing ammonium nitrate or a low explosive is suspected, the debris should be rinsed with water so that water-soluble substances (such as nitrates and chlorates) will be extracted. Table 15–1 lists a number of simple color tests the examiner can perform on the acetone and water extracts to screen for the presence of organic and inorganic explosives, respectively.

Screening and Confirmation Tests Once collected, the acetone extract is concentrated and analyzed using color spot tests, thin-layer chromatography (TLC), high-performance liquid chromatography (HPLC), and gas chromatography/mass spectrometry. The presence of an explosive is indicated by a well-defined spot on a TLC plate with an *Rf* value corresponding to that of a known explosive—for example, nitroglycerin, RDX, or PETN.

The high sensitivity of HPLC also makes it useful for analyzing trace evidence of explosives. HPLC operates at room temperature and hence does not cause explosives, many of which are temperature sensitive, to decompose during their analysis. When a water-gel explosive containing ammonium nitrate or a low explosive is suspected, the debris should be rinsed with water so that water-soluble substances (such as nitrates and chlorates) will be extracted.

When sufficient quantities of explosives are recoverable, confirmatory tests may be performed by either infrared spectrophotometry or **X-ray diffraction**. The former produces a unique "fingerprint" pattern for an organic explosive, as shown by the IR spectrum of RDX in Figure 15–8. The latter provides a unique diffraction pattern for inorganic substances such as potassium nitrate and potassium chlorate.

X-ray diffraction
An analytical technique for identifying crystalline materials

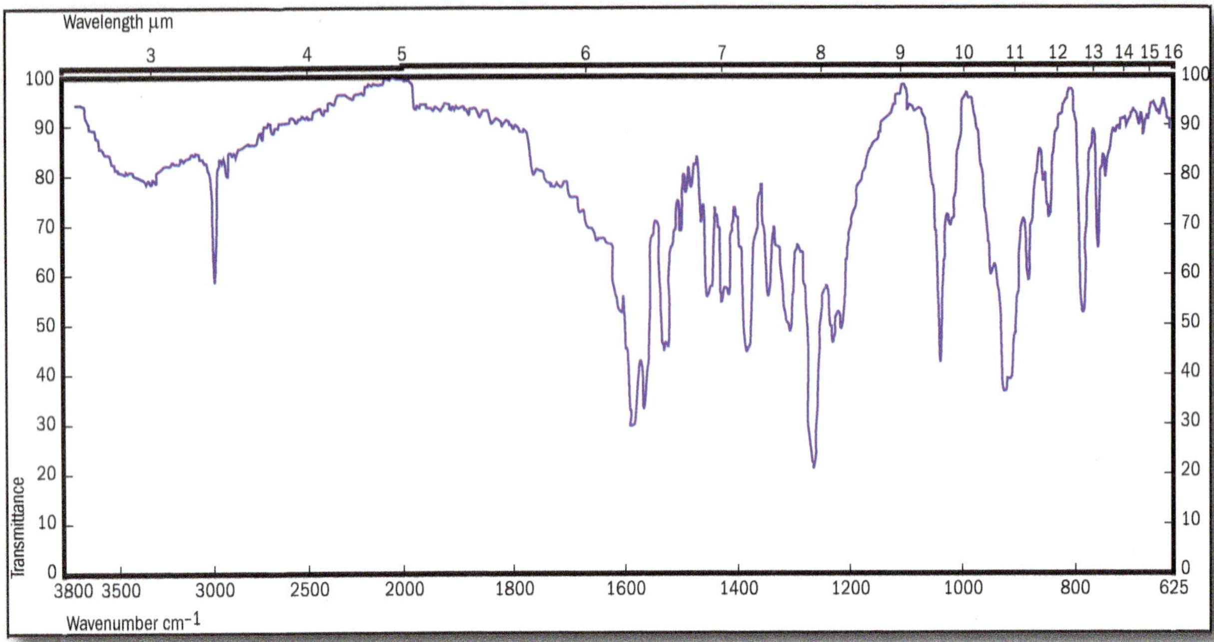

FIGURE 15–8 Infrared spectrum of RDX.

Inside the Science

X-Ray Diffraction

One way to elicit information about how elements have combined into a particular substance is to aim a beam of X-rays at the material and study how the X-rays interact with the atoms that compose the substance under investigation. This technique is known as X-ray diffraction.

X-ray diffraction can be applied only to the study of solid, crystalline materials—that is, solids with a definite and orderly arrangement of atoms. For example, sodium chloride (common table salt) is crystalline. Fortunately, many substances, including 95 percent of all inorganic compounds, are crystalline and are thus identifiable by X-ray diffraction analysis. The atoms in a crystal may be thought of as being composed of a series of parallel planes. As the X-rays penetrate the crystal, a portion of the beam is reflected by each of the atomic planes. As the reflected beams leave the crystal's planes, they combine with one another to form a series of light and dark bands known as a diffraction pattern. Every compound produces a unique diffraction pattern, thus giving analysts a means for "fingerprinting" compounds.

Figure 1 illustrates the X-ray diffraction process. Diffraction patterns for potassium nitrate and potassium chlorate, two common constituents of homemade explosives, are shown in Figure 2. Comparing a questioned specimen with a known X-ray pattern is a rapid and specific way to prove chemical identity.

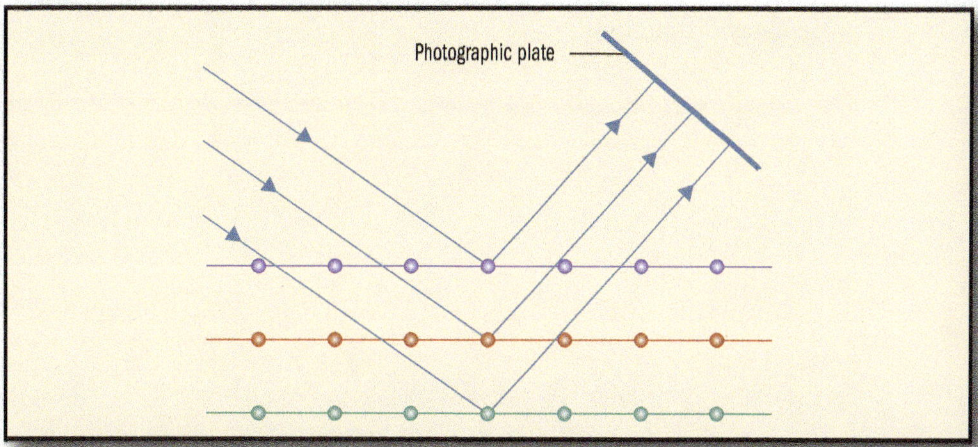

FIGURE 1 A beam of X-rays being reflected off the atomic planes of a crystal. The diffraction patterns that form are recorded on photographic film. These patterns are unique for each crystalline substance.

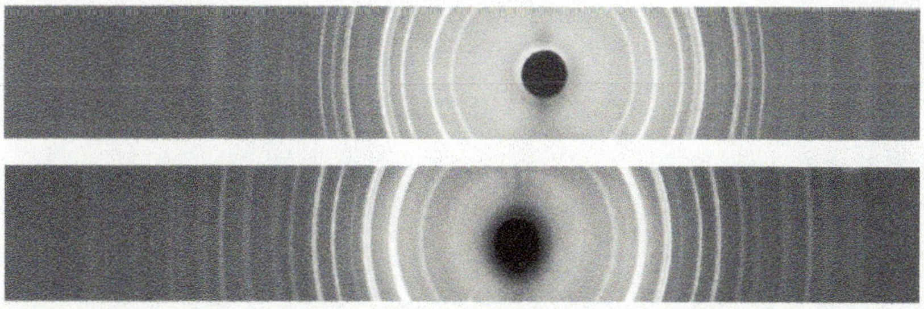

FIGURE 2 X-ray patterns for (a) potassium nitrate and (b) potassium chlorate. *Courtesy Richard Saferstein Ph.D.*

Quick Review

- Debris collected at explosion scenes is examined microscopically for unconsumed explosive particles.
- Recovered debris may be thoroughly rinsed with organic solvents and analyzed by testing procedures that include color spot tests, thin-layer chromatography, and gas chromatography/mass spectrometry.
- Unconsumed explosives are identified by either infrared spectrophotometry or X-ray diffraction.

Chapter Review

- Explosives are substances that undergo a rapid oxidation reaction with the production of large quantities of gases. The sudden buildup of gas pressure leads to the explosion.
- The speed at which an explosive decomposes determines whether it is classified as a high or low explosive.
- The most widely used low explosives are black powder and smokeless powder. Common high explosives include ammonium nitrate–based explosives (water gels, emulsions, and ANFO explosives).
- Among the high explosives, primary explosives are ultrasensitive to heat, shock, and friction and provide the major ingredients found in blasting caps. Secondary explosives normally constitute the main charge of a high explosive.
- The entire bomb site must be systematically searched to recover any trace of a detonating mechanism or any other item foreign to the explosion site. Objects located at or near the origin of the explosion must be collected for laboratory examination.
- The most obvious characteristic of a high or contained low explosive is the presence of a crater at the origin of the blast.
- A device widely used to screen objects for the presence of explosive residues is the ion mobility spectrometer.
- All materials collected at bombing scenes must be placed in airtight containers such as clean paint cans.
- Debris collected at explosion scenes is examined microscopically for unconsumed explosive particles.
- Recovered debris may be thoroughly rinsed with organic solvents and analyzed by testing procedures that include color spot tests, thin-layer chromatography, and gas chromatography/mass spectrometry.
- Unconsumed explosives are identified by either infrared spectrophotometry or X-ray diffraction.

Review Questions

1. Which type of explosive is classified into two groups, primary and secondary?
   a. low explosives
   b. medium explosives
   c. high explosives
   d. all explosives

2. Low explosives decompose at rates that vary up to
   a. 1,000 miles per hour.
   b. 1,000 meters per second.
   c. 1,000 meters per hour.
   d. 7,000 miles per hour.

3. In recent years, dynamite has been replaced commercially by
   a. PETN.
   b. nitroglycerin.
   c. TATP.
   d. ammonium nitrate–based explosives.

4. The single most important step in the detection and analysis of explosive residues is
   a. the collection of appropriate samples from the explosion scene.
   b. the determination of whether it is a low or a high explosive.
   c. the creation of a chain-of-custody form.
   d. locating the ignition source.

5. After bomb-scene debris has been examined microscopically, the next step is to
   a. examine the explosive using spectrophotometry.
   b. use X-ray diffraction to "fingerprint" the organic explosive.
   c. rinse the recovered debris with acetone to separate the debris from explosive material.
   d. view the detonating mechanism with a low-power stereoscopic microscope.

6. True or False: The speed at which explosives decompose varies greatly from one to another and permits their classification as high and low explosives.

7. True or False: The most widely used explosives in the low-explosive group are black powder and smokeless powder.

8. True or False: Military high explosives include RDX, TNT, and PETN.

9. True or False: Debris and particles collected from different areas at an explosion site are to be packaged in separate plastic containers.

10. True or False: The ion mobility spectrometer has become a widely used instrument for detecting the presence of explosive residues on objects.

11. What produces the violent physical disruption associated with an explosion?

12. What is the difference between deflagration and detonation?

13. What is the difference between low explosives and high explosives?

14. What is a safety fuse and what is it used for?

15. What ingredients are required to create a low explosive?

16. What are the two classes of high explosives? What is the difference between the two classes?

17. What are the main ingredients in straight dynamite? What other substances also are included in modern straight dynamites and what purpose do they serve?

18. What types of explosives largely have replaced dynamite for industrial uses? What are the advantages of these types of explosives?

19. What is ANFO?

20. Which is the most powerful and popular military explosive?

21. What is a detonator? What is the composition of most detonators?

22. What is the most obvious postexplosion characteristic of a high or contained low explosive?

23. How are soil and other soft loose materials collected at the scene of an explosion best stored?

24. What is the first procedure typically used to analyze bomb-scene debris that arrives in the laboratory?

25. List three procedures commonly used as screening tests for explosive residues.

26. Name two confirmatory tests for the presence of intact explosives located in debris.

27. What is a taggant and what is its purpose?

Learn the Vocabulary

Use the information from this chapter along with your prior knowledge and experience to learn the key terms. Choose from the following activities:

1. Make flashcards with a term on the front and its definition on the back. Use the flashcards to quiz a partner on correctly pronouncing and defining the terms.
2. Make an audio or video recording of each term and its definition.
3. Work with a partner to write a sentence for each term. Read the sentences out loud to a partner or share them with the class.
4. Write a list of new terms and language you have encountered in this chapter. Reuse the terms to write sentences and then read the sentences out loud to a partner. Reuse the terms again by writing a short speech. The speech might explain the terms or teach a lesson using the terms. Give the speech to a partner or to the class.

Application and Critical Thinking

1. The following pieces of evidence were found at separate explosion sites. For each item, indicate whether the explosion was more likely caused by low or high explosives, and explain your answer.
   a. lead azide residues
   b. nitrocellulose residues
   c. ammonium nitrate residues
   d. scraps of primacord
   e. potassium chlorate residues

2. Which color test or tests would you run first on a suspect sample to test for evidence of each of the following explosives? Explain your answers.
   a. tetryl
   b. TNT
   c. chlorate
   d. nitrocellulose

3. Criminalist Matt Weir is collecting evidence from the site of an explosion. Arriving on the scene, he immediately proceeds to look for the crater caused by the blast. After finding the crater, he picks through the debris at the site by hand, looking for evidence of detonators or foreign materials. Matt collects loose soil and debris from the immediate area, placing the smaller bits in paper folded into a druggist fold. Larger items he stores in plastic bags for transportation to the laboratory. What mistakes, if any, did Matt make in collecting and storing this evidence?

Laboratory Experiments

This activity requires the use of the following practices of science and engineering:

- Analyzing and interpreting data
- Constructing explanations

This activity consists of the following crosscutting concepts:

- **Patterns**—Patterns are evident in the analysis of bomb residue in a lab situation. These patterns in the results can lead to a conclusion of the composition of a bomb.

Before you begin, read the scenario and steps out loud with a partner. Ask for clarification as necessary to make sure you understand the experiment.

The Scenario

Jennifer Maddy, a Chico police spokeswoman, said officers were called to the 700 block of West 25th Street at about 11:30 PM Saturday to assist firefighters with items found at the scene of a vehicle fire. Firefighters determined that the fire was most likely the result of an explosion. A friend of the victim said that when she got into her car and started the ignition, the explosion occurred. A Chico police officer and bomb technician was called in to investigate. The bomb scene debris and other material were taken to the laboratory where you are to analyze the evidence.

PROCEDURE

Microscopic evidence indicates the presence of aluminum metal, which was verified by confirmatory tests.

The results of the color spot test are in the table below.

Debris Evidence	*Reagent Greiss*	*Diphenylamine*	*Alcoholic KOH*
A	Reddish pink	Blue	No color
B	Reddish pink	Blue	No color

The result of the thin-layer chromatography shows an Rf value of 0.21 which compared to the RDX standard that was run at the same time.

IR Spectrum shows that one of the components is RDX.

Summarize your findings based on the results of your analysis, including the bomb composition.

Endnotes

1. T. Keller et al., "Application of Ion Mobility Spectrometry in Cases of Forensic Interest," *Forensic Science/International/*161 (2006): 130.

Forensic Biometrics: Fingerprints and Facial Recognition

16

EPA European Pressphoto Agency b.v./US Department of Justice/ Alamy Stock Photo

Key Terms

anthropometry
arch
digital imaging
fluoresce
iodine fuming
latent fingerprint
livescan
loop
ninhydrin
Physical Developer
pixel
plastic print
portrait parlé
ridge characteristics (minutiae)
sublimation
superglue fuming
visible print
whorl

Boston Marathon Bombing

On April 15, 2013, two improvised explosive devices at the annual Boston Marathon killed three people and wounded 264 others.

Moments after the vicious Boston Marathon terrorist attack, investigations began in search of possible suspects. Local authorities and the FBI called the support of the public and in no time potential leads started pouring in. Over 2,000 reports and photos of the event were made. Dozens of investigators were tasked with scrutinizing these leads. Among the photographs sent to the FBI, Dzhokhar and Tamerlan Tsarnaev, two radicalized brothers, are seen in plain sight.

The bombs were discovered to be pressure cooker bombs similar to many IEDs found in other extremist terrorist attacks. Dzhokhar and Tamerlan left two bombs placed in a backpack and shopping bag among the onlookers of the Marathon. As they walked away, they remotely triggered the explosions.

The FBI set the task of finding the perpetrators to their most sophisticated facial recognition biometric system. Every image and video that was obtained during the event was scanned by a complex algorithm that would compare suspects against a repository of known terrorists and persons of interest. To the FBI and the public's surprise, the system failed. Out of the many images sent to law enforcement, a few had direct pictures of the Tsarnaev brothers. Both brothers had legally come into this country and had images in American immigration databases. Dzhokhar had a U.S. driver's license with an up-to-date picture. His brother Tamerlan was even the subject of a terrorism investigation and had many images in FBI databases. Despite all of these leads and a wealth of database information, the facial recognition system failed to match the Tsarnaevs' faces to identities.

The problem with facial recognition biometrics in 2013, and what still remains an issue today, is that software has a difficult time analyzing and identifying faces from grainy and low-resolution photographs. Tamerlan also could not be recognized due to the fact that he was wearing sunglasses, which can distort the results of the scan as well. Posture and light exposure can have an effect on this technology, and investigators learned this the hard way after the gruesome attack. Law enforcement eventually identified the Tsarnaev brothers using more conventional methods and apprehended them not long after. Facial recognition technology is evolving and continues to be updated and fine-tuned with the implementation of the Next Generation Information System.

Learning Objectives

After studying this chapter, you should be able to:

16.1 List and define the various forms of biometrics and their categories

16.2 Summarize the three fundamental principles of fingerprints

16.3 Explain the primary classification system of fingerprints

16.4 Describe the concept of an automated fingerprint identification system (AFIS)

16.5 Describe visible, plastic, and latent fingerprints and the techniques for developing latent fingerprints

16.6 Describe the proper procedures for preserving a developed latent fingerprint

16.7 Explain the scope of the FBI's Next Generation Identification System

National Science Content Standards

Scientific Inquiry

Physical Science

What Is Biometrics?

Biometrics is a cutting-edge form of access-control that accurately and efficiently identifies humans. This system uses an individual's biological and behavioural traits to grant access to an establishment or a computer network. An electronic device will require individuals to use a distinguishable feature of their body to positively identify themselves before they are able to bypass any type of access control system, such as a security system for a building. Once a match has been made, the person requesting access will be allowed on the premises by a simple electronic pulse that opens the door. If the sensor does not find a positive match within its database, it locks the attempted user out. The military has been using biometrics for an extended period of time to keep sensitive information secure, and to prevent espionage and theft with regard to valuable military assets. Law enforcement across the nation implements biometrics for record keeping and on-site access-control (i.e., only allowing staff on the premises).

Types of Forensic Biometrics

Now that we have established a basic definition of biometrics is, let's examine different types of biometric systems that exist. Many forms of biometric technologies are available today, but the majority of them can be split into two different groups.

Physiological Biometrics Fingerprints have been widely used to identify people for centuries. This form of biometric technology is the most cost-efficient and easy to install. With the addition of a fingerprint scanner on a door, an office can grant specific workers access to a sensitive area or a network of computers. Hand and vein scans within the hand work almost the same way. Humans have a unique set of physiological traits pertaining to the hand. When compared, the vein patterns that run through our hands are unique.

Retina and iris scans have been implemented in many government institutions as well as other areas with classified information. This technology is much more secure than the use of fingerprints due to the detailed map of a human eye. Using a tiny camera, a biometric system is able to take a picture of someone's eye and compare all of the intricate muscles and fibers that make it a unique set of characteristics. Eye scanning systems are expensive to install, yet necessary when trying to protect a valuable asset or piece of information.

Many forms of biometric technologies are available today, but the majority of them can be split into two different groups. The first group is ***physiological biometrics***, which contains: fingerprints, hand, iris, retina, and facial scans (see Figure 16–1). The second group is ***behavioral biometrics***, which is much less stable and includes handwriting, voice, keystroke, and gait recognition.

Facial scans are the least accurate form of biometrics currently available in the physiological category. The human face is just as unique as the eye or hand, but the technology available today has trouble distinguishing subtle traits from still images. This process becomes less accurate when using biometrics for identification purposes, such as picking a face out of a crowd and matching it to a database of wanted persons.

Fundamental Principles of Fingerprinting

Since the beginnings of criminal investigation, police have sought an infallible means of human identification. The first systematic attempt at personal identification was devised and introduced by a French police expert, Alphonse Bertillon, in 1883. The Bertillon system relied on a detailed description (**portrait parlé**) of the subject, combined with full-length and profile photographs and a system of precise body measurements known as **anthropometry**.

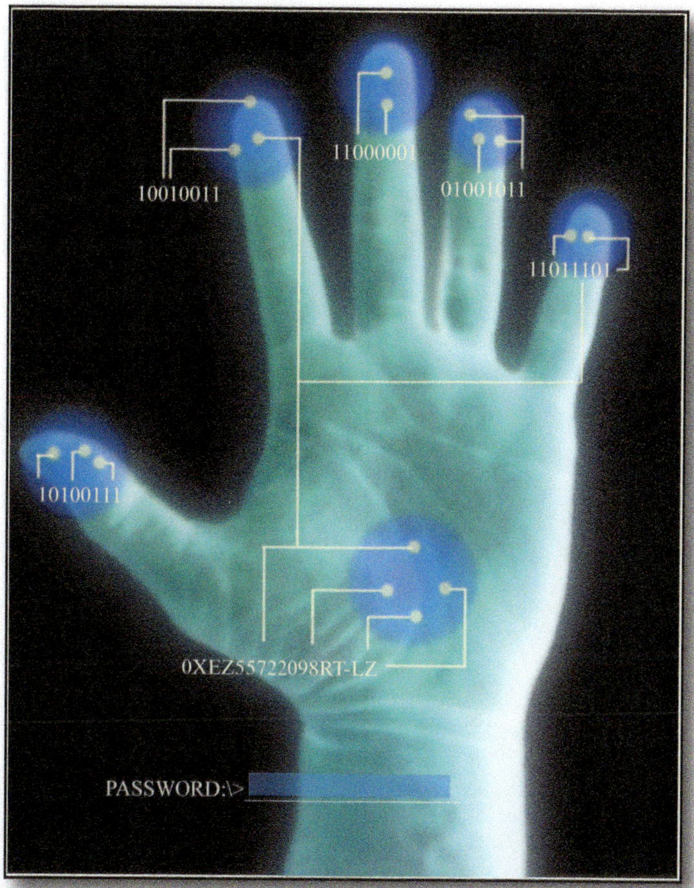

FIGURE 16–1 Biometric security scan of a human hand.
David Hernandez/Shutterstock

The use of anthropometry as a method of identification rested on the premise that the dimensions of the human bone system remained fixed from age 20 until death. Skeleton sizes were thought to be so extremely diverse that no two individuals could have exactly the same measurements. Bertillon recommended routine taking of 11 measurements of the human anatomy, including height, reach, width of head, and length of the left foot (see Figure 1–3).

For two decades, this system was considered the most accurate method of identification. But in the early years of the 20th century, police began to appreciate and accept a system of identification based on the classification of finger ridge patterns known as fingerprints. Today, the fingerprint is the pillar of modern criminal identification.

portrait parlé
A verbal description of a perpetrator's physical characteristics and dress provided by an eyewitness

anthropometry
A system of identification of individuals by measurement of parts of the body, developed by Alphonse Bertillon

Early Use of Fingerprints

The Chinese used fingerprints to sign legal documents as far back as three thousand years ago. Whether this practice was performed for ceremonial custom or as a means of personal identity remains a point of conjecture lost to history. In any case, the examples of fingerprinting in ancient history are ambiguous, and the few that exist did not contribute to the development of fingerprinting techniques as we know them today.

Several years before Bertillon began work on his system, William Herschel, an English civil servant stationed in India, started requiring natives to sign contracts with the imprint of their right hand, which was pressed against a stamp pad for the purpose. The motives for Herschel's requirement remain unclear; he may have envisioned fingerprinting as a means of personal identification or just as a form of the Hindu custom that a trace of bodily contact was more binding than a signature on a contract. In any case, he did not publish anything about his activities until after a Scottish physician, Henry Fauld, working in a hospital in Japan, published his views on the potential application of fingerprinting to personal identification.

In 1880, Fauld suggested that skin ridge patterns could be important for the identification of criminals. He told about a thief who left his fingerprint on a whitewashed wall, and how in comparing these prints with those of a suspect, he found that they were quite different. A few days later, another suspect was found whose fingerprints compared with those on the wall. When confronted with this evidence, the individual confessed to the crime.

Fauld was convinced that fingerprints furnished infallible proof of identification. He even offered to set up at his own expense a fingerprint bureau at Scotland Yard to test the practicality of the method. But his offer was rejected in favor of the Bertillon system. This decision was reversed less than two decades later.

Early Classification of Fingerprints

The extensive research into fingerprinting conducted by another Englishman, Francis Galton, provided the needed impetus that made police agencies aware of its potential application. In 1892, Galton published his classic textbook *Finger Prints*, the first book of its kind on the subject (see Figure 16–2). In his book, Galton discussed the anatomy of fingerprints and suggested methods for recording them. He also proposed assigning fingerprints to three pattern types—loops, arches, and whorls. Most important, the book demonstrated that no two prints are identical and that an individual's prints remain unchanged from year to year. At Galton's insistence, the British government adopted fingerprinting as a supplement to the Bertillon system.

The next step in the development of fingerprint technology was the creation of classification systems capable of filing thousands of prints in a logical and searchable sequence. Dr. Juan Vucetich, an Argentinian police officer fascinated by Galton's work, devised a workable concept in 1891. His classification system has been refined over the years and is still widely used today in most Spanish-speaking

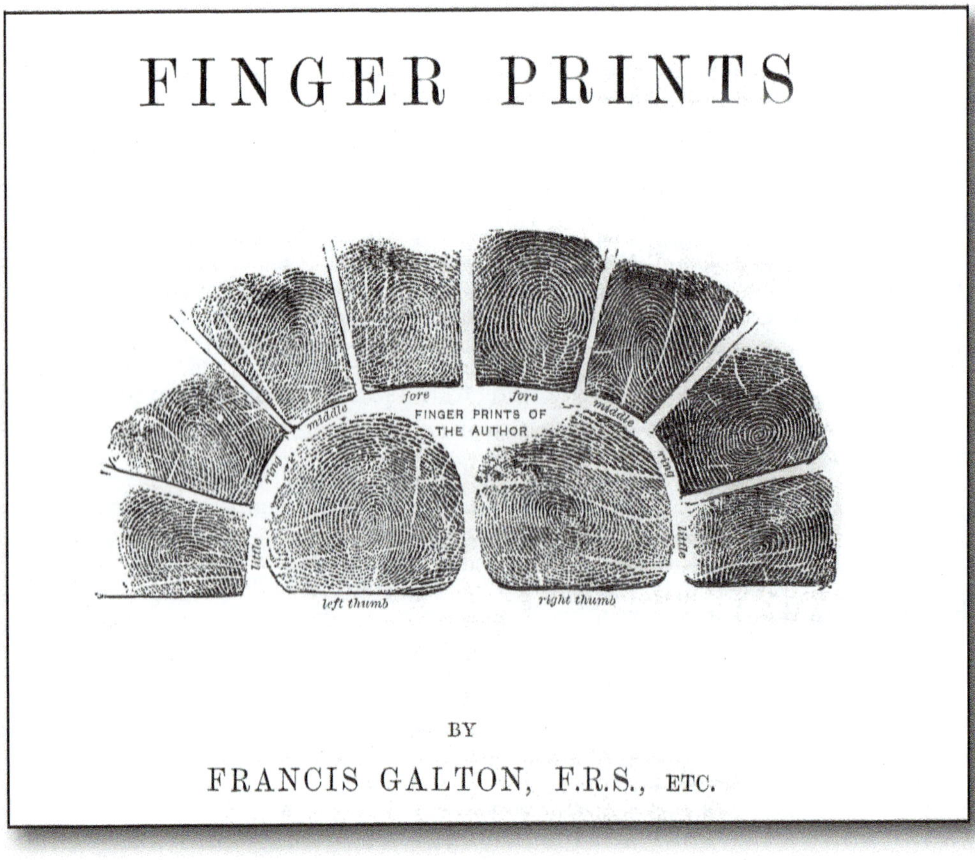

FIGURE 16–2 The cover of Sir Francis Galton's book *Finger Prints* which discusses the anatomy of fingerprints, how to record them, and classification by the three pattern types—loops, arches, and whorls. *Science History Images/Alamy Stock Photo*

FIGURE 16–3 Sir Edward Richard Henry, British pioneer of fingerprinting for identification purposes. *Reading Room 2020/Alamy Stock Photo*

countries. In 1897, another classification system was proposed by an Englishman, Sir Edward Richard Henry (see Figure 16–3). Four years later, Henry's system was adopted by Scotland Yard. Today, most English-speaking countries, including the United States, use some version of Henry's classification system to file fingerprints.

Adoption of Fingerprinting

Early in the 20th century, Bertillon's measurement system began to fall into disfavor. Its results were highly susceptible to error, particularly when the measurements were taken by people who were not thoroughly trained. The method was dealt its most severe and notable setback in 1903 when a convict, Will West, arrived at Fort Leavenworth prison. A routine check of the prison files startlingly revealed that a William West, already in the prison, could not be distinguished from the new prisoner by body measurements or even by photographs. In fact, the two men looked just like twins, and their measurements were practically the same. Subsequently, fingerprints of the prisoners clearly distinguished them.

In the United States, the first systematic and official use of fingerprints for personal identification was adopted by the New York City Civil Service Commission in 1901. The method was used for certifying all civil service applications. Several American police officials received instruction in fingerprint identification at the 1904 World's Fair in St. Louis from representatives of Scotland Yard. After the fair and the Will West incident, fingerprinting began to be used in earnest in all major cities of the United States. In 1924, the fingerprint records of the Bureau of Investigation and Leavenworth were merged to form the nucleus of the identification records of the new Federal Bureau of Investigation. The FBI has the largest collection of fingerprints in the world. By the beginning of World War I, England and practically all of Europe had adopted fingerprinting as their primary method of identifying criminals.

In 1999, the admissibility of fingerprint evidence was challenged in the case of *United States* v. *Byron C. Mitchell* in the Eastern District of Pennsylvania. The defendant's attorneys argued that fingerprints could not be proven unique under the guidelines cited in *Daubert* (see page 26). Government experts vigorously disputed this claim. After a four-and-a-half-day Daubert hearing, the judge upheld the admissibility of fingerprints as scientific evidence and ruled that (1) human friction ridges are unique and permanent and (2) human friction ridge skin arrangements are unique and permanent.

Fundamental Principles of Fingerprints

Since Galton's time, and as a result of his efforts, fingerprints have become an integral part of policing and forensic science. The principal reason for this is that fingerprints constitute a unique and unchanging means of personal identification. In fact, fingerprint analysts have formulated three basic principles of fingerprints that encompass these notions of the uniqueness and stability of fingerprint identification.

First Principle: Fingerprint Is an Individual Characteristic; No Two Fingers Have Yet Been Found to Possess Identical Ridge Chacteristics

The acceptance of fingerprint evidence by the courts has always been predicated on the assumption that no two individuals have identical fingerprints. Early fingerprint experts consistently referred to Galton's calculation, showing the possible existence of 64 billion different fingerprints, to support this contention. Later, researchers questioned the validity of Galton's figures and attempted to devise mathematical models to better approximate this value. However, no matter what mathematical model one refers to, the conclusions are always the same: the probability for the existence of two identical fingerprint patterns in the world's population is extremely small.

Not only is this principle supported by theoretical calculations, but just as important, it is verified by the millions of individuals who have had their prints classified during the past 110 years—no two have ever been found to be identical. The FBI has nearly 50 million fingerprint records in its computer database and has yet to find an identical image belonging to two different people.

ridge characteristics (minutiae)
Bifurcations, ending ridges, dots, short ridges, and enclosures/islands, which must match in two fingerprints in order for their common origin to be established

The individuality of a fingerprint is not determined by its general shape or pattern but by a careful study of its **ridge characteristics** (also known as **minutiae**), including bifurcations, ending ridges, dots, short ridges, and enclosures/islands. The identity, number, and relative location of characteristics such as those illustrated in Figure 16–4 impart individuality to a fingerprint. If two prints are to match, they must reveal characteristics that not only are identical but have the same relative location to one another in a print. In a judicial proceeding, a point- by-point comparison must be demonstrated by the expert,

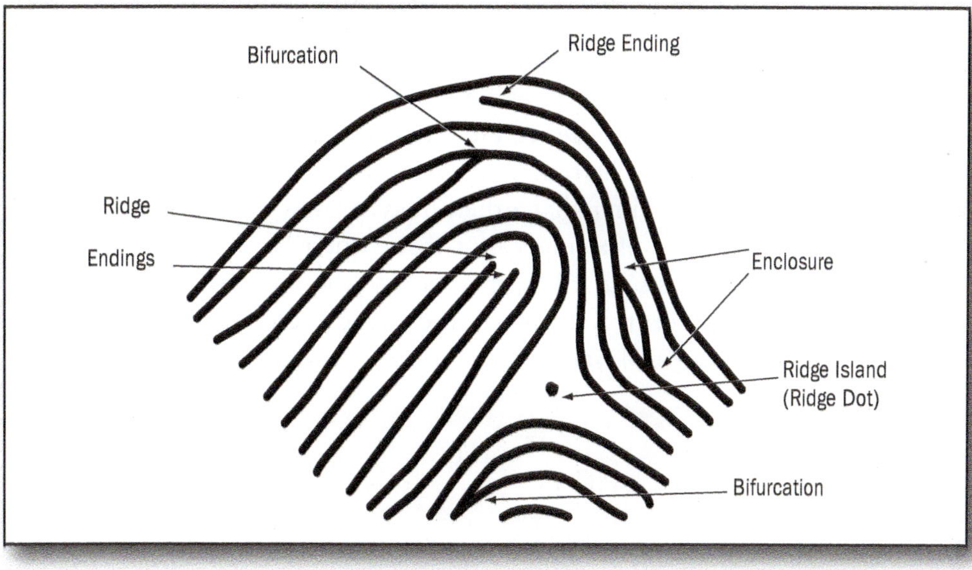

FIGURE 16–4 Fingerprint ridge characteristics. *Courtesy Sirchie Finger Print Laboratories, Inc., Youngsville, N.C., www.sirchie.com*

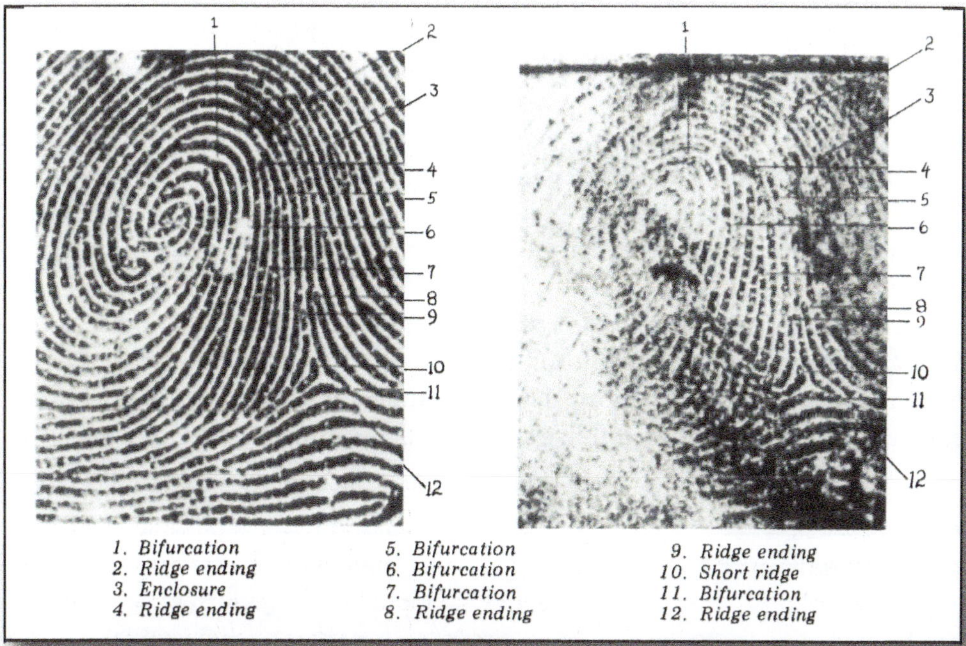

FIGURE 16–5 A fingerprint exhibit illustrating the matching ridge characteristics between the crimescene print and an inked impression of one of the suspect's fingers. *Courtesy Richard Saferstein, Ph.D.*

using charts similar to the one shown in Figure 16–5, in order to prove the identity of an individual.

An expert can easily compare the characteristics of the complete fingerprint; the average fingerprint has as many as 150 individual ridge characteristics. However, most prints recovered at crime scenes are partial impressions, showing only a segment of the entire print. Under these circumstances, the expert can compare only a small number of ridge characteristics from the recovered print to a known recorded print.

For years, experts have debated how many ridge comparisons are necessary to identify two fingerprints as the same. Numbers that range from 8 to 16 have been suggested as being sufficient to meet the criteria of individuality. However, the difficulty in establishing such a minimum is that no comprehensive statistical study has ever determined the frequency of occurrence of different ridge characteristics and their relative locations. Until such a study is undertaken and completed, no meaningful guidelines can be established for defining the uniqueness of a fingerprint.

In 1973, the International Association for Identification, after a three-year study of this question, concluded that "no valid basis exists for requiring a pre-determined minimum number of friction ridge characteristics which must be present in two impressions in order to establish positive identification." Hence, the final determination must be based on the experience and knowledge of the expert, with the understanding that others may profess honest differences of opinion on the uniqueness of a fingerprint if the question of minimal number of ridge characteristics exists. In 1995, members of the international fingerprint community at a conference in Israel issued the Ne'urim Declaration, which supported the 1973 International Association for Identification resolution.

Second Principle: A Fingerprint Remains Unchanged During an Individual's Lifetime

Fingerprints are a reproduction of friction skin ridges found on the palm side of the fingers and thumbs. Similar friction skin can also be found on the surface of the palms and soles of the feet. Apparently, these skin surfaces have been designed by nature to provide our bodies with a firmer grasp and a resistance to slippage. A visual inspection of friction skin reveals a series of lines corresponding to hills (ridges) and valleys (grooves). The shape and form of the skin ridges are what one sees as the black lines of an inked fingerprint impression.

Skin is composed of layers of cells. Those nearest the surface make up the outer portion of the skin known as the epidermis, and the inner skin is known as the *dermis*. A cross-section of skin (see Figure 16–6) reveals a boundary of cells separating the epidermis and dermis. The shape of this boundary, made up of *dermal papillae*, determines the form and pattern of the ridges on the surface of the skin. Once the dermal papillae develop in the human fetus, the ridge patterns remain unchanged throughout life except to enlarge during growth.

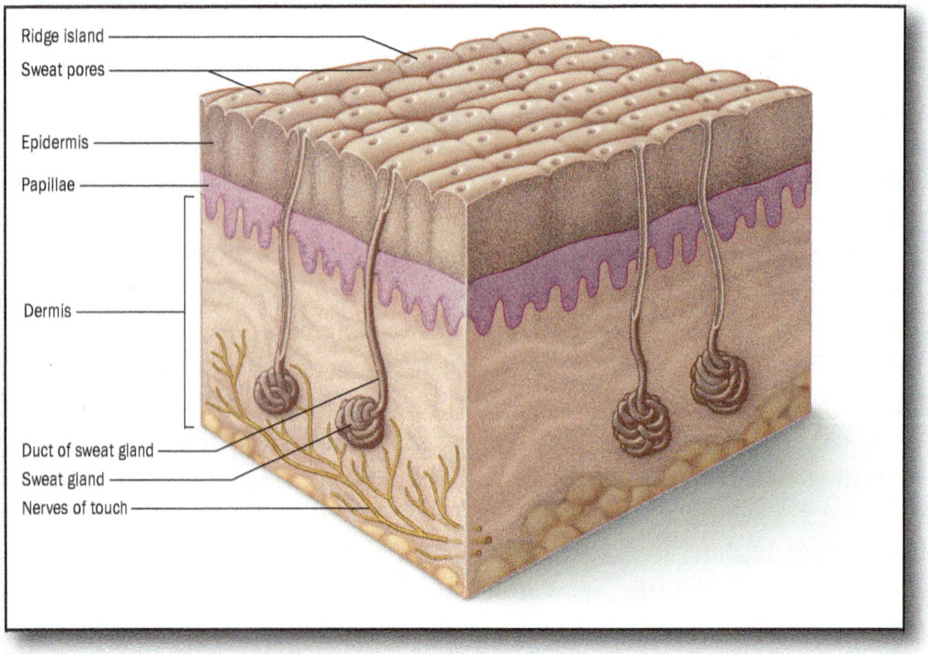

FIGURE 16–6 Cross-section of human skin.

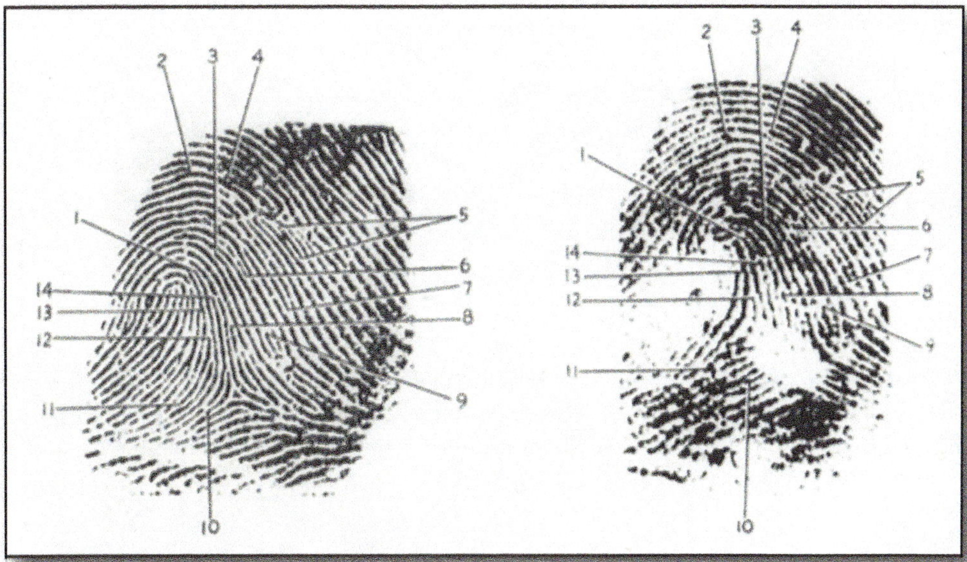

FIGURE 16–7 The right index finger impression of John Dillinger, before scarification on the left and afterward on the right. Comparison is proved by the 14 matching ridge characteristics. *Courtesy Sirchie Finger Print Laboratories, Youngsville, NC, www.sirchie.com*

Each skin ridge is populated by a single row of pores that are the openings for ducts leading from the sweat glands. Through these pores, perspiration is discharged and deposited on the surface of the skin. Once the finger touches a surface, perspiration, along with oils that may have been picked up by touching the hairy portions of the body, is transferred onto that surface, thereby leaving an impression of the finger's ridge pattern (a fingerprint). Prints deposited in this manner are invisible to the eye and are commonly referred to as **latent fingerprints.**

latent fingerprint
A fingerprint made by the deposit of oils and/or perspiration; it is invisible to the naked eye

Although it is impossible to change one's fingerprints, some criminals have tried to obscure them. If an injury reaches deeply enough into the skin and damages the dermal papillae, a permanent scar forms. However, for this to happen, such a wound would have to penetrate 1 to 2 millimeters beneath the skin's surface. Indeed, efforts at intentionally scarring the skin can only be self-defeating, for it is totally impossible to obliterate all of the ridge characteristics on the hand, and the presence of permanent scars merely provides new characteristics for identification.

Perhaps the most publicized attempt at obliteration was that of the notorious gangster John Dillinger, who tried to destroy his own fingerprints by applying a corrosive acid to them. Prints taken at the morgue after he was shot to death, compared with fingerprints recorded at the time of a previous arrest, proved that his efforts had been fruitless (see Figure 16–7).

Third Principle: Fingerprints Have General Ridge Patterns That Permit Them to Be Systematically Classified

All fingerprints are divided into three classes on the basis of their general pattern: **loops**, **whorls**, and **arches**. Sixty to 65 percent of the population have loops, 30 to 35 percent have whorls, and about 5 percent have arches. These three classes form the basis for all ten-finger classification systems presently in use.

loop
A class of fingerprints characterized by ridge lines that enter from one side of the pattern and curve around to exit from the same side of the pattern

whorl
A class of fingerprints that includes ridge patterns that are generally rounded or circular

arch
A class of fingerprints characterized by ridge lines that enter the print from one side and exit the other side

FIGURE 16–8 Loop-type pattern has ridges entering from one side of the print, recurving, and exiting from the same side. These patterns resemble a lake. *Courtesy Anthony Smith and Erika DiPalma*

A loop must have one or more ridges entering from one side of the print, recurving, and exiting from the same side. If the loop opens toward the little finger, it is called an "ulnar loop"; if it opens toward the thumb, it is a "radial loop." A typical loop pattern is illustrated in Figure 16–8.

Whorls are actually divided into four distinct groups, as shown in Figure 16–9: plain, central pocket loop, double loop, and accidental. A plain whorl and a central pocket loop have at least one ridge that makes a complete circuit. This ridge may be in the form of a spiral, oval, or any variant of a circle. As the name implies, the double loop is made up of two loops combined in one fingerprint. Any whorl classified as an accidental either contains two or more patterns (not including the plain arch) or is a pattern not covered by other categories. Hence, an accidental may consist of a combination of a loop and a plain whorl or a loop and a tented arch.

FIGURE 16–9 Whorl types. (a) Plain whorl. One or more ridges form a complete circle-like pattern around the center. These often look like a bull's eye. (b) Central pocket loop. Ridges form a loop pattern which go around the circle-like pattern in the center. These often look like a peacock eye. (c) Double loop whorl. Two separate loops are present in one impression. These often look like the symbol of the yin and the yang. (d) Accidental whorl. A mixture of two different types of patterns. These often look like an accident since several patterns are occupying the same space. *Courtesy Anthony Smith and Erika DiPalma*

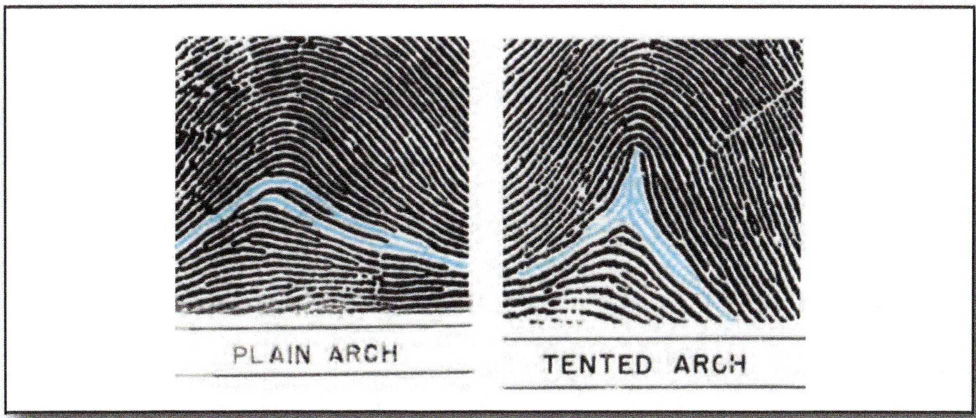

FIGURE 16–10 Arch patterns. (a) Plain arch. Ridges enter the impression, rise to the center, and exit the opposite side of the impression. These often resemble hills. (b) Tented arch. Ridges enter the impression, spike towards the center, and exit the opposite side of the impression. These often resemble a tent. *Courtesy Anthony Smith and Erika DiPalma*

Arches, the least common of the three general patterns, are subdivided into two distinct groups: plain arches and tented arches, as shown in Figure 16–10. The plain arch is the simplest of all fingerprint patterns; it is formed by ridges entering from one side of the print and exiting on the opposite side. Generally, these ridges tend to rise in the center of the print, forming a wavelike pattern. The tented arch is similar to the plain arch except that instead of rising smoothly at the center, there is a sharp upthrust or spike, or the ridges meet at an angle that is less than 90 degrees.[1]

The ACE-V Process

ACE-V is an acronym for the four-step process—analysis, comparison, evaluation, and verification—used to identify and individualize a fingerprint. The first step requires the examiner to identify any distortions associated with the friction ridges, as well as any external factors, such as surface or deposition factors or processing techniques, that may impinge on the print's appearance. If the examiner determines the latent print adequate, the examiner will declare the print to be of value for the comparison stage.

The comparison step requires the examiner to compare the questioned print to the known print at three levels. Level 1 looks at the general ridge flow and pattern configuration. Level 2 includes locating and comparing ridge characteristics, or minutiae. Level 2 details can individualize a print. Level 3 includes the examination and location of ridge pores, breaks, creases, scars, and other permanent minutiae. During the comparison phase, the examiner compares the latent print side by side with an exemplar print in their totality.

The evaluation stage requires one of three decisions to be arrived at. The decisions that can be reported are: identification (the latent print and exemplar came from the same source); exclusion (the latent print and exemplar did not come from the same source); inconclusive (one cannot determine that the latent print and exemplar came from the same source, or not, to a sufficiently strong level of certainty).

The final step in the process involves verification of the examiner's result. It requires an independent examination of the questioned and known prints by a second examiner. Ultimately, a consensus between the two examiners must be arrived at before a final conclusion is drawn.

Quick Review

- Fingerprints are a reproduction of friction skin ridges found on the palm side of the fingers and thumbs.
- The basic principles underlying the use of fingerprints in criminal investigations are as follows: (1) A fingerprint is an individual characteristic because no two fingers have yet been found to possess identical ridge characteristics; (2) a fingerprint remains unchanged during an individual's lifetime; and (3) fingerprints have general ridge patterns that permit them to be systematically classified.
- All fingerprints are divided into three classes on the basis of their general pattern: loops, whorls, and arches.
- The individuality of a fingerprint is determined not by its general shape or pattern, but by a careful study of its ridge characteristics. The expert must demonstrate a point-by-point comparison in order to prove the identity of an individual.
- When the finger touches a surface, perspiration and oils are transferred onto that surface, leaving a fingerprint. Prints deposited in this manner are invisible to the eye and are commonly referred to as latent or invisible fingerprints.

Classification of Fingerprints

The original Henry system, as adopted by Scotland Yard in 1901, converted ridge patterns on all ten fingers into a series of letters and numbers arranged in the form of a fraction. However, the system as it was originally designed could accommodate files of up to only 100,000 sets of prints. Thus, as collections grew in size, it became necessary to expand the capacity of the classification system. In the United States, the FBI, faced with the problem of filing ever-increasing numbers of prints, expanded its classification capacity by modifying and extending the original Henry system. These modifications are collectively known as the FBI system and are used by most agencies in the United States today. Although we will not discuss all of the different divisions of the FBI system, a description of just one part, the primary classification, will provide an interesting insight into the process of fingerprint classification.

The primary classification is part of the original Henry system and provides the first classification step in the FBI system. Using this classification alone, all of the fingerprint cards in the world could be divided into 1,024 groups. The first step in obtaining the primary classification is to pair up fingers, placing one finger in the numerator of a fraction, the other in the denominator. The fingers are paired in the following sequence:

$$\frac{\text{R. Index}}{\text{R. Thumb}} \quad \frac{\text{R. Ring}}{\text{R. Middle}} \quad \frac{\text{L. Thumb}}{\text{R. Little}} \quad \frac{\text{L. Middle}}{\text{L. Index}} \quad \frac{\text{L. Little}}{\text{L. Ring}}$$

The presence or absence of the whorl pattern is the basis for determination of the primary classification. If a whorl pattern is found on any finger of the first pair, it is assigned a value of 16; on the second pair, a value of 8; on the third pair, a value of 4; on the fourth pair, a value of 2; and on the last pair, a value of 1. Any finger with an arch or loop pattern is assigned a value of 0. Approximately 25 percent of the population falls into the 1/1 category; that is, all their fingers have either loops or arches.

After values for all ten fingers are obtained in this manner, they are totaled, and 1 is added to both the numerator and denominator. The fraction thus obtained is the primary classification. For example, if the right index and right middle fingers are whorls and all the others are loops, the primary classification is

$$\frac{16 + 0 + 0 + 0 + 0 + 1}{0 + 8 + 0 + 0 + 0 + 1} = \frac{17}{9}$$

A fingerprint classification system cannot in itself unequivocally identify an individual; it merely provides the fingerprint examiner with a number of candidates, all of whom have an indistinguishable set of prints in the system's file. The identification must always be made by a final visual comparison of the suspect print's and file print's ridge characteristics; only these features can impart individuality to a fingerprint. Although ridge patterns impart class characteristics to the print, the type and position of ridge characteristics give it its individual character.

Quick Review

- The primary classification is the first step in classifying fingerprints under the FBI system. The presence or absence of the whorl pattern is the basis for determination of the primary classification.

Automated Fingerprint Identification Systems

The Henry system and its subclassifications have proven to be a cumbersome system for storing, retrieving, and searching for fingerprints, particularly as fingerprint collections grow in size. Nevertheless, until the emergence of fingerprint computer technology, this manual approach was the only viable method for maintaining fingerprint collections. Since 1970, technological advances have made possible the classification and retrieval of fingerprints by computers. Automated Fingerprint Identification Systems (AFISs) have proliferated throughout the law enforcement community.

The premier model of all forensic database systems is the Integrated Automated Fingerprint Identification System (IAFIS), a national fingerprint and criminal history system maintained by the FBI. IAFIS, which first became operational in 1999, contains fingerprints and access to corresponding criminal history information for nearly seventy-five million subjects (or 750 million fingerprint images), which are submitted voluntarily to the FBI by state, local, and federal law enforcement agencies. In 2014, the IAFIS was effectively replaced and integrated into the Next Generation Identification (NGI) system. In the United States, each state has its own Automated Fingerprint Identification System (AFIS), which is linked to the FBI's NGI system. However, an AFIS can come in all sizes ranging from the FBI's NGI system to independent systems operated by cities, counties, and other agencies of local government (see Figure 16–11). Unfortunately, these

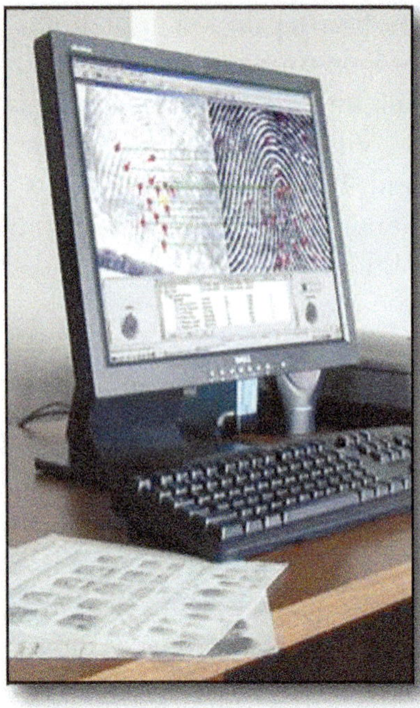

FIGURE 16–11 An AFIS system designed for use by local law enforcement agencies. *Courtesy AFIX Technologies Inc., Pittsburg, KS 66762, www.afix.net*

local systems often cannot be linked to the state's AFIS system because of differences in software configurations.

How AFIS Works

The heart of AFIS technology is the ability of a computer to scan and digitally encode fingerprints so that they can be subject to high-speed computer processing. The AFIS uses automatic scanning devices that convert the image of a fingerprint into digital minutiae that contain data showing ridges at their points of termination (ridge endings) and the branching of ridges into two ridges (bifurcations). The relative position and orientation of the minutiae are also determined, allowing the computer to store each fingerprint in the form of a digitally recorded geometric pattern.

The computer's search algorithm determines the degree of correlation between the location and relationship of the minutiae for both the search and file prints. In this manner, a computer can make thousands of fingerprint comparisons in a second. For example, a set of ten fingerprints can be searched against a file of 500,000 ten-finger prints (ten-prints) in about eight-tenths of a second. During the search for a match, the computer uses a scoring system that assigns prints to each of the criteria set by an operator. When the search is complete, the computer produces a list of file prints that have the closest correlation to the search prints. All of the selected prints are then examined by a fingerprint expert, who makes the final verification of the print's identity. Thus, the AFIS makes no final decisions on the identity of a fingerprint, leaving this function to the eyes of a trained examiner.

The speed and accuracy of ten-print processing by AFIS have made possible the search of single latent crime-scene fingerprints against an entire file's print collection. Before AFIS, police were usually restricted to comparing crimescene fingerprints against those of known suspects. The impact of the AFIS on no-suspect cases has been dramatic. In its first year of operation, San Francisco's AFIS computer conducted 5,514 latent fingerprint searches and achieved 1,001 identifications—a hit rate of 18 percent. This compares to the previous year's average of 8 percent for manual latent-print searches.

As an example of how an AFIS computer operates, one system has been designed to automatically filter out imperfections in a latent print, enhance its image, and create a graphic representation of the fingerprint's ridge endings and bifurcations and their direction. The print is then computer searched against file prints. The image of the latent print and a matching file print are then displayed side by side on a high-resolution video monitor, as shown in Figure 16–12. The matching latent and file prints are then verified and charted by a fingerprint examiner at a video workstation.

The stereotypical image of a booking officer rolling inked fingers onto a standard ten-print card for ultimate transmission to a database has, for the most part, been replaced with digital-capture devices (**livescan**) that eliminate ink and paper (see Figure 16–13). The livescan captures the image on each finger and the palms as they are lightly pressed against a glass platen. These livescan images can then be sent to the AFIS database electronically, so that within minutes the booking agency can enter the fingerprint record into the AFIS database and search the database for previous entries of the same individual.

livescan
An inkless device that captures the digital images of fingerprints and palm prints and electronically transmits the images to an AFIS

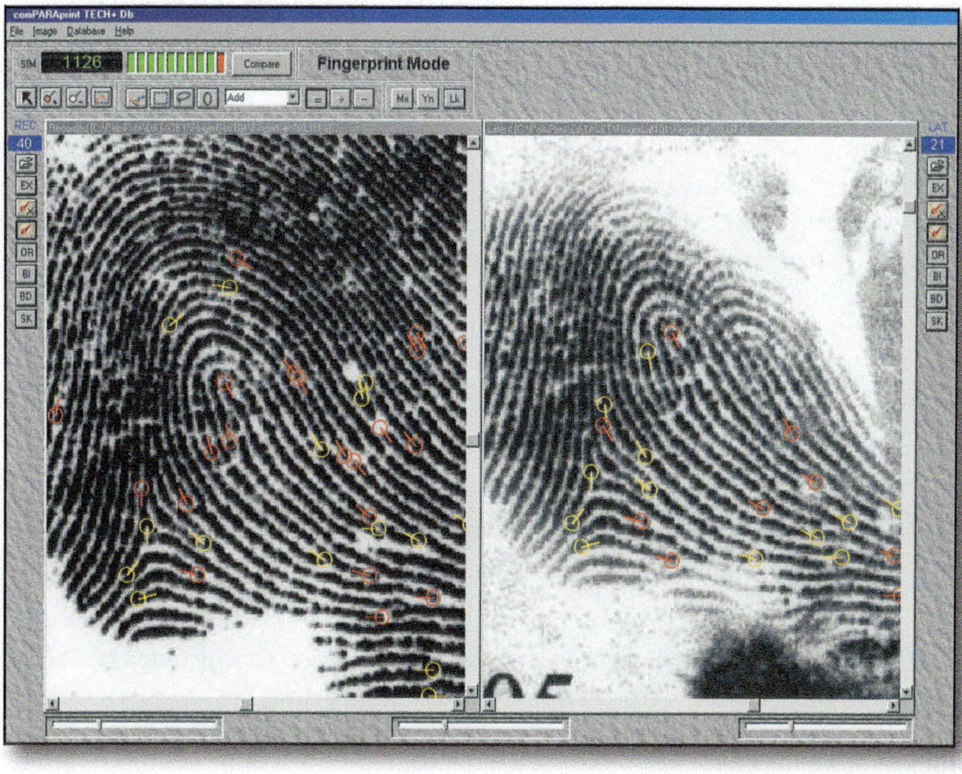

FIGURE 16–12 A side-by-side comparison of a latent print against a file fingerprint is conducted in seconds, and their similarity rating (SIM) is displayed on the upper-left portion of the screen.
Courtesy Sirchie Finger Print Laboratories, Inc., Youngsville, N.C., www.sirchie.com

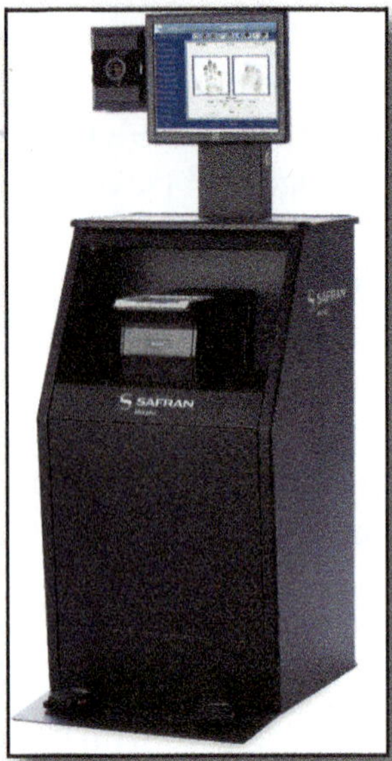

FIGURE 16–13 Livescan technology enables law enforcement personnel to print and compare a subject's fingerprints rapidly, without inking the fingerprints.
MorphoTrak, Inc

Considerations with AFIS

AFIS has fundamentally changed the way criminal investigators operate, allowing them to spend less time developing suspect lists and more time investigating the suspects generated by the computer. However, investigators must be cautioned against overreliance on a computer. Sometimes a latent print does not make a hit because of the poor quality of the file print. To avoid these potential problems, investigators must still print all known suspects in a case and manually search these prints against the crime-scene prints.

AFIS computers are available from several different suppliers. Each system scans fingerprint images and detects and records information about minutiae (ridge endings and bifurcations); however, they do not all incorporate exactly the same features, coordinate systems, or units of measure to record fingerprint information. These software incompatibilities often mean that although state systems can communicate with the FBI's IAFIS, they do not communicate with each other directly. Likewise, local and state systems frequently cannot share information with each other. Many of these technical problems will be resolved as more agencies follow transmission standards developed by the National Institute of Standards and Technology and the FBI.

Quick Review

- The FBI fingerprint database known as AFIS converts the image of a fingerprint into digital minutiae that contain data showing ridges at their points of termination (ridge endings) and their branching into two ridges (bifurcations).
- Livescan is an inkless device that captures digital images of fingerprints and palm prints and electronically transmits them to an AFIS.

Case Files

The Night Stalker

Richard Ramirez committed his first murder in June 1984. His victim was a 79-year-old woman who was stabbed repeatedly and sexually assaulted and then had her throat slashed. It would be eight months before Ramirez murdered again. In the spring, Ramirez began a murderous rampage that resulted in 13 additional killings and 5 rapes.

His modus operandi was to enter a home through an open window, shoot the male residents, and savagely rape his female victims. He scribed a pentagram and the words *Jack the Knife* on the wall of one of his victims, and another victim reported that he forced her to "swear to Satan" during the assault. His identity still unknown, the news media dubbed him the "Night Stalker." As the body count continued to rise, public hysteria and a media frenzy prevailed.

The break in the case came when the license plate of what seemed to be a suspicious car related to a sighting of the Night Stalker was reported to the police. The police determined that the car had been stolen and eventually located it, abandoned in a parking lot. After processing the car for prints, police found one usable partial fingerprint. This fingerprint was entered into the Los Angeles Police Department's brand-new AFIS computerized fingerprint system.

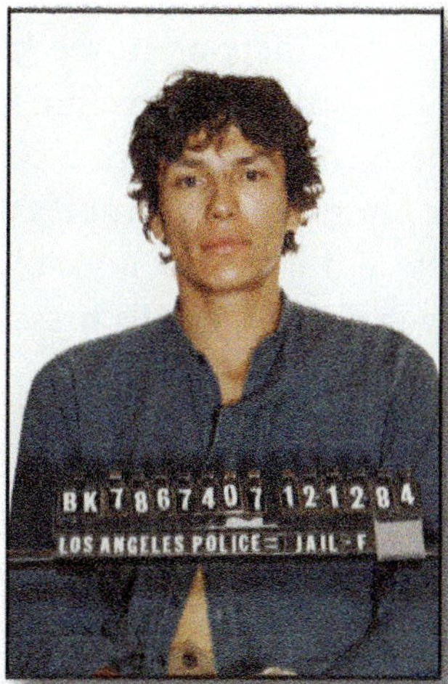

Richard Ramirez, the Night Stalker. *Bettmann/Getty Images*

Without AFIS, it would have taken a single technician, manually searching Los Angeles's 1.7 million print cards, 67 years to come up with the perpetrator's prints. Thanks to AFIS, it took only a few seconds to locate and identify them. The Night Stalker was identified as Richard Ramirez, who had been fingerprinted following a traffic violation some years before. Police searching the home of one of his friends found the gun used to commit the murders, and jewelry belonging to his victims was found in the possession of Ramirez's sister. Ramirez was convicted of murder and sentenced to death in 1989. Ramirez died of natural causes in 2013.

Methods of Detecting Fingerprints

Through common usage, the term *latent fingerprint* has come to be associated with any fingerprint impression discovered at a crime scene. Sometimes, however, prints found at the scene of a crime are quite visible to the eye, and the word latent is a misnomer.

visible print
A fingerprint made when the finger deposits a visible material such as ink, dirt, or blood onto a surface

plastic print
A fingerprint impressed in a soft surface

Actually, there are three kinds of crime-scene prints. **Visible prints**, also referred to as patent prints, are made by fingers touching a surface after the ridges have been in contact with a colored material such as blood, paint, grease, or ink; **plastic prints** are ridge impressions left on a soft material such as putty, wax, soap, or dust; and latent or invisible prints are impressions caused by the transfer of body perspiration or oils present on finger ridges to the surface of an object.

Locating Fingerprints

Locating visible or plastic prints at the crime scene normally presents little problem to the investigator, because these prints are usually distinct and visible to the eye. Locating latent or invisible prints is obviously much more difficult and requires the use of techniques to make the print visible. Although the investigator can choose from several methods for visualizing a latent print, the choice depends on the type of surface being examined.

Hard and nonabsorbent surfaces (such as glass, mirror, tile, and painted wood) require different development procedures from surfaces that are soft and porous (such as papers, cardboard, and cloth). Prints on the former are preferably developed by the application of a powder or treatment with superglue, whereas prints on the latter generally require treatment with one or more chemicals.

Sometimes the most difficult aspect of fingerprint examination is the location of prints. Recent advances in fingerprint technology have led to the development of an ultraviolet image converter for the purpose of detecting latent fingerprints. This device, called the Reflected Ultraviolet Imaging System (RUVIS), can locate prints on most nonabsorbent surfaces without the aid of chemical or powder treatments (see Figure 16–14).

FIGURE 16–14 A Reflected Ultraviolet Imaging System allows an investigator to directly view surfaces for the presence of untreated latent fingerprints. *Courtesy Sirchie Finger Print Laboratories, Inc., Youngsville, N.C., www.* sirchie.com

Case Files

The Mayfield Affair

On March 11, 2004, a series of ten explosions at four sites occurred on commuter trains traveling to or near the Atocha train station in Madrid, Spain. The death toll from these explosions was nearly 200, with more than 1,500 injured. On the day of the attack, a plastic bag was found in a van previously reported as stolen. The bag contained copper detonators like those used on the train bombs.

On March 17, the FBI received electronic images of latent fingerprints that were recovered from the plastic bag, and a search was initiated on the FBI's IAFIS. A senior fingerprint examiner encoded seven minutiae points from the high-resolution image of one suspect latent fingerprint and initiated an IAFIS search matching the print to Brandon Mayfield. Mayfield's prints were in the FBI's central database because they had been taken when he joined the military, where he served for eight years before being honorably discharged as a second lieutenant.

After a visual comparison of the suspect and file prints, the examiner concluded a "100 percent match." The identification was verified by a retired FBI fingerprint examiner with more than 30 years of experience who was working under contract with the bureau, as well as by a court-appointed independent fingerprint examiner (see the figure).

Mayfield, age 37, a Muslim convert, was arrested on May 6 on a material witness warrant. The U.S. Attorney's Office came up with a list of Mayfield's potential ties to Muslim terrorists, which they included in the affidavit they presented to the federal judge who ordered his arrest and detention. The document also said that although no travel records were found for Mayfield, "It is believed that Mayfield may have traveled under a false or fictitious name." On May 24, after the Spaniards had linked the print from the plastic bag to an Algerian national, Mayfield's case was thrown out. The FBI issued him a highly unusual official apology, and his ordeal became a stunning embarrassment to the U.S. government.

The impact of the Mayfield affair on fingerprint technology as currently practiced and the weight courts will assign to fingerprint matches remain open questions.

(a) Questioned print recovered in connection with the Madrid bombing Investigation
(b) File print of Brandon Mayfield. *(a) Courtesy U.S. Department of Justice, Washington, D.C.*

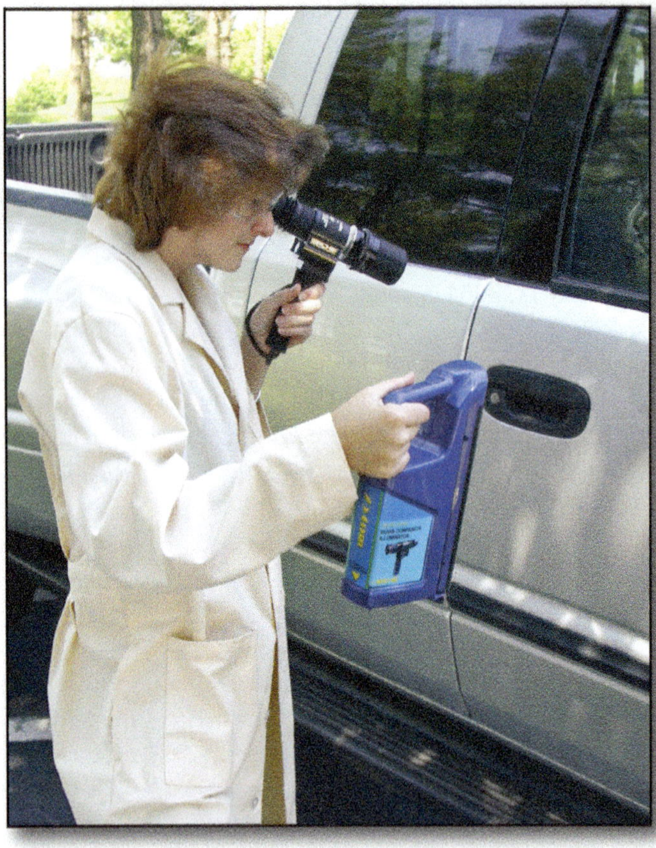

FIGURE 16–15 Using a Reflected Ultraviolet Imaging System with the aid of a UV lamp to search for latent fingerprints. *Courtesy Sirchie Finger Print Laboratories, Inc., Youngsville, N.C.,* www.sirchie.com

RUVIS detects the print in its natural state by aiming UV light at the surface suspected of containing prints. When the UV light strikes the fingerprint, the light is reflected back to the viewer, differentiating the print from its background surface. The transmitted UV light is then converted into visible light by an image intensifier. Once the print is located in this manner, the crime-scene investigator can develop it in the most appropriate fashion (see Figure 16–15).

Developing Latent Prints

Several techniques are available to the criminalist to develop latent prints on a variety of surfaces, including porous, nonporous, and adhesive surfaces. These include chemical methods such as powders and iodine fuming and the use of laser light.

Fingerprint Powders Fingerprint powders are commercially available in a variety of compositions and colors. These powders, when applied lightly to a nonporous surface with a camel's-hair or fiberglass brush, readily adhere to perspiration residues and/or deposits of body oils left on the surface (see Figure 16–16).

Experienced examiners find that gray and black powders are adequate for most latent-print work; the examiner selects the powder that affords the best color contrast with the surface being dusted. Hence, the gray powder, composed of an aluminum dust, is used on dark-colored surfaces. It is also applied to mirrors and metal surfaces that are polished to a mirrorlike finish, because these surfaces photograph as black. The black powder, composed basically of black carbon or charcoal, is applied to white or light-colored surfaces.

FIGURE 16–16 Developing a latent fingerprint on a surface by applying a fingerprint powder with a fiberglass brush. *Courtesy Sirchie Finger Print Laboratories, Inc., Youngsville, N.C., www.sirchie.com*

Other types of powders are available for developing latent prints. A magneticsensitive powder can be spread over a surface with a magnet in the form of a Magna Brush. A Magna Brush does not have any bristles to come in contact with the surface, so there is less chance that the print will be destroyed or damaged. The magnetic-sensitive powder comes in black and gray and is especially useful on such items as finished leather and rough plastics, on which the minute texture of the surface tends to hold particles of ordinary powder. Fluorescent powders are also used to develop latent fingerprints. These powders fluoresce under ultraviolet light. By photographing the fluorescence pattern of the developing print under UV light, it is possible to avoid having the color of the surface obscure the print.

iodine fuming
A technique for visualizing latent fingerprints by exposing them to iodine vapors

sublimation
A physical change from the solid directly into the gaseous state

Iodine Fuming Of the several chemical methods used for visualizing latent prints, **iodine fuming** is the oldest. Iodine is a solid crystal that, when heated, is transformed into a vapor without passing through a liquid phase; such a transformation is called **sublimation**. Most often, the suspect material is placed in an enclosed cabinet along with iodine crystals (see Figure 16–17). As the crystals are heated, the resultant vapors fill the chamber and combine with constituents of the latent print to make it visible.

Unfortunately, iodine prints are not permanent and begin to fade once the fuming process is stopped. Therefore, the examiner must photograph the prints immediately on development in order to retain a permanent record. Also, iodine-developed prints can be fixed with a 1 percent solution of starch in water, applied by spraying. The print turns blue and lasts for several weeks to several months.

The reasons why latent prints are visualized by iodine vapors are not yet fully understood. Many believe that the iodine fumes combine with fatty oils; however, there is also convincing evidence that the iodine may actually interact with residual water left on a print from perspiration.[2]

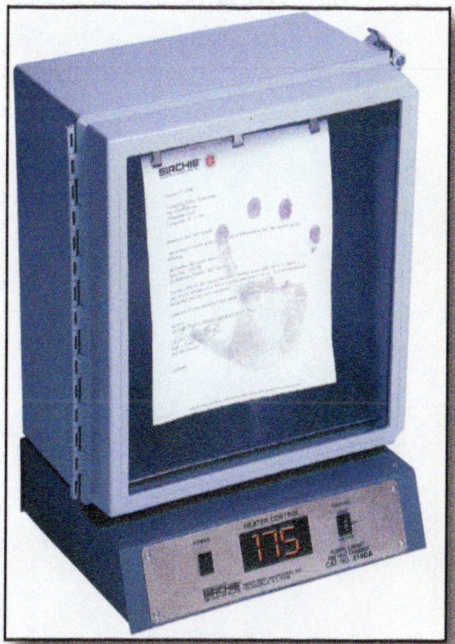

FIGURE 16–17 A heated fuming cabinet. *Courtesy Sirchie Finger Print Laboratories, Inc., Youngsville, N.C., www.sirchie.com*

ninhydrin
A chemical reagent used to develop latent fingerprints on porous materials by reacting with amino acids in perspiration

Ninhydrin Another chemical used for visualizing latent prints is **ninhydrin**. The development of latent prints with ninhydrin depends on its chemical reaction to form a purple-blue color with amino acids present in trace amounts in perspiration. Ninhydrin (triketohydrindene hydrate) is commonly sprayed onto the porous surface from an aerosol can. A solution is prepared by mixing the ninhydrin powder with a suitable solvent, such as acetone or ethyl alcohol; a 0.6 percent solution appears to be effective for most applications.

Generally, prints begin to appear within an hour or two after ninhydrin application; however, weaker prints may be visualized after 24 to 48 hours. The development can be hastened if the treated specimen is heated in an oven or on a hot plate at a temperature of 80–100°C. The ninhydrin method has developed latent prints on paper as old as 15 years.

Physical Developer
A silver nitrate–based reagent formulated to develop latent fingerprints on porous surfaces

Genetian Violet Another chemical used for visualizing latent prints is genetian or crystal violet. This method uses a chemical stain that reacts with oily and fatty surfaces that are typically found on the sticky surface of adhesive tape. For this method to work, the adhesive must be insoluble in water.

superglue fuming
A technique for visualizing latent fingerprints on nonporous surfaces by exposing them to cyanoacrylate vapors; named for the commercial product Super Glue

Physical Developer **Physical Developer** is a third chemical mixture used for visualizing latent prints. Physical Developer is a silver nitrate–based liquid reagent. This method has gained wide acceptance by fingerprint examiners, who have found it effective for visualizing latent prints that remain undetected by the previously described methods. Also, this technique is effective for developing latent fingerprints on porous articles that may have been wet at one time.

For most fingerprint examiners, the chemical method of choice is ninhydrin. Its extreme sensitivity and ease of application have all but eliminated the use of iodine for latent-print visualization. However, when ninhydrin fails, development with Physical Developer may provide identifiable results. Application of Physical Developer washes away any traces of proteins from an object's surface; hence, if one wishes to use all of the previously mentioned chemical development methods on the same surface, it is necessary to first fume with iodine, follow this treatment with ninhydrin, and then apply Physical Developer to the object.

FIGURE 16–18 **Superglue fuming a nonporous metallic surface in the search for latent fingerprints.** *Courtesy Sirchie Finger Print Laboratories, Inc., Youngsville, N.C., www.sirchie.com*

Superglue Fuming In the past, chemical treatment for fingerprint development was reserved for porous surfaces such as paper and cardboard. However, since 1982, a chemical technique known as **superglue fuming** has gained wide popularity for developing latent prints on nonporous surfaces such as metals, electrical tape, leather, and plastic bags.[3] See Figure 16–18.

Superglue is approximately 98–99 percent cyanoacrylate ester, a chemical that interacts with and visualizes a latent fingerprint. Cyanoacrylate ester fumes can be created when superglue is placed on absorbent cotton treated with sodium hydroxide. The fumes can also be created by heating the glue. The fumes and the evidential object are contained within an enclosed chamber for up to six hours. Development occurs when fumes from the glue adhere to the latent print, usually producing a white-appearing

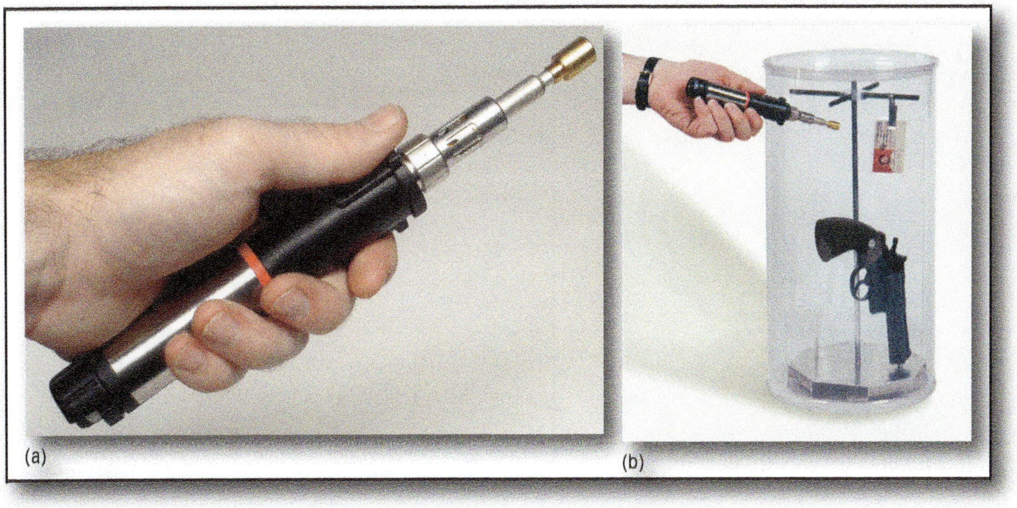

FIGURE 16–19 (a) A handheld fuming wand uses disposable cartridges containing cyanoacrylate. The wand is used to develop prints at the crime scene and (b) in the laboratory. *Courtesy Sirchie Finger Print Laboratories, Inc., Youngsville, N.C., www.sirchie.com*

latent print. A forensic scientist can use fingerprint powder to lift a latent print from a nonporous substrate after fuming with cyanoacrylate. Scientists then use powder and adhesive tape to collect the print. Interestingly, small enclosed areas, such as the interior of an automobile, have been successfully processed for latent prints with fumes from superglue.

Through the use of a small handheld wand, cyanoacrylate fuming is now easily done at a crime scene or in a laboratory setting. The wand heats a small cartridge containing cyanoacrylate. Once heated, the cyanoacrylate vaporizes, allowing the operator to direct the fumes onto the suspect area (see Figure 16–19).

Other Techniques for Visualization In recent years, researchers have explored a variety of new processes applicable to the visualization of latent fingerprints. However, for many years progress in this field was minimal. Fingerprint specialists traditionally relied on three chemical techniques—iodine, ninhydrin, and silver nitrate—to reveal a hidden fingerprint. Then, superglue fuming extended chemical development to prints deposited on nonporous surfaces.

Another hint of things to come emerged with the discovery that latent fingerprints could be visualized by exposure to laser light. This laser method took advantage of the fact that perspiration contains a variety of components that **fluoresce** when illuminated by laser light.

fluoresce
To emit visible light when exposed to light of a shorter wavelength

The next advancement in latent-fingerprint development occurred with the discovery that fingerprints could be treated with chemicals that would induce fluorescence when exposed to laser illumination. For example, application of zinc chloride after ninhydrin treatment or application of the dye rhodamine 6G after superglue fuming caused fluorescence and increased the sensitivity of detection on exposure to laser illumination. The discovery of numerous chemical developers for visualizing fingerprints through fluorescence quickly followed. This knowledge set the stage for the next advance in latent-fingerprint development—the *alternate light source.*

With the advent of chemically induced fluorescence, lasers were no longer needed to induce fingerprints to fluoresce through perspiration residues. Highintensity light sources or alternate light sources have proliferated and all but replaced laser lights (see Figure 16–20). High-intensity quartz halogen or

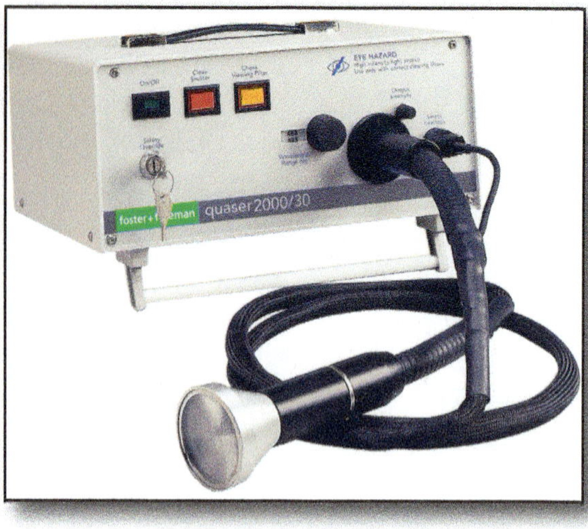

FIGURE 16–20 An alternate light source system incorporating a high-intensity light source. *Courtesy Foster & Freeman Limited, Worcestershire, U.K., www.fosterfreeman.co.uk*

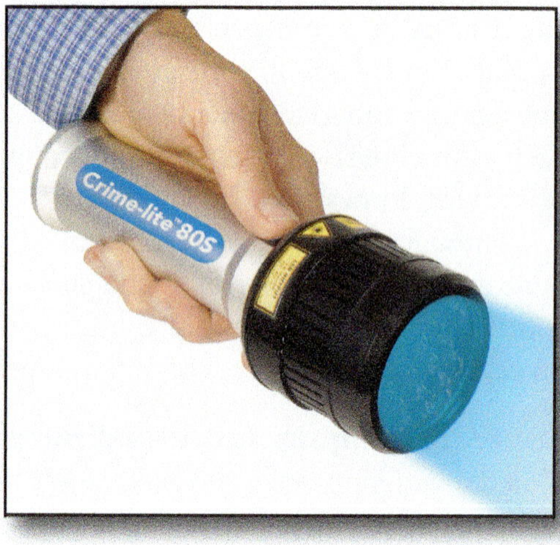

FIGURE 16–21 Lightweight handheld alternate light source that uses an LED light source. *Courtesy Foster & Freeman Limited, Worcestershire, U.K., www.fosterfreeman.co.uk*

xenonarc light sources can be focused on a suspect area through a fiber-optic cable. This light can be passed through several filters, giving the user more flexibility in selecting the wavelength of light to be aimed at the latent print. Alternatively, lightweight, portable alternate light sources that use light-emitting diodes (LEDs) are also commercially available (see Figure 16–21).

In most cases, these light sources have proven to be as effective as laser light in developing latent prints, and they are commercially available at costs significantly less than those of laser illuminators. Furthermore, these light sources are portable and can be readily taken to any crime scene.

A large number of chemical treatment processes are available to the fingerprint examiner, and the field is in a constant state of flux. Selection of an appropriate procedure is best left to technicians who have developed their skills through casework experience.

Newer chemical processes include a substitute for ninhydrin called DFO (1,8-diazafluoren-9-one). This chemical visualizes latent prints on porous materials when exposed to an alternate light source. DFO has been shown to develop 2.5 times more latent prints on paper than ninhydrin. A chemical called

Inside the Science

Fluorescence

Fluorescence occurs when a substance absorbs light and re-emits the light in wavelengths longer than that of the illuminating source. Importantly, substances that emit light or fluoresce are more readily seen either with the naked eye or through photography than are non-light-emitting materials. The high sensitivity of fluorescence serves as the underlying principle of many of the new chemical techniques used to visualize latent fingerprints.

The earliest use of fluorescence to visualize fingerprints came with the direct illumination of a fingerprint with argon–ion lasers. This laser type was chosen because its blue-green light output induced some of the perspiration components of a fingerprint to fluoresce (see the figure). The major drawback of this approach is that the perspiration components of a fingerprint are often present in quantities too minute to observe even with the aid of fluorescence.

The fingerprint examiner, wearing safety goggles containing optical filters, visually examines the specimen being exposed to the laser light. The filters absorb the laser light and permit the wavelengths at which latent-print residues fluoresce to pass through to the eyes of the wearer. The filter also protects the operator against eye damage from scattered or reflected laser light. Likewise, latent-print residue producing sufficient fluorescence can be photographed by placing this same filter across the lens of the camera. Examination of specimens and photography of the fluorescing latent prints are carried out in a darkened room.

Studies have demonstrated that common fingerprint-developing agents do not interfere with DNA-testing methods used for characterizing bloodstains.[4] Nonetheless, in cases involving items with material adhering to their surfaces and/or items that will require further laboratory examinations, fingerprint processing should not be performed at the crime scene. Rather, the items should be submitted to the laboratory, where they can be processed for fingerprints in conjunction with other necessary examinations.

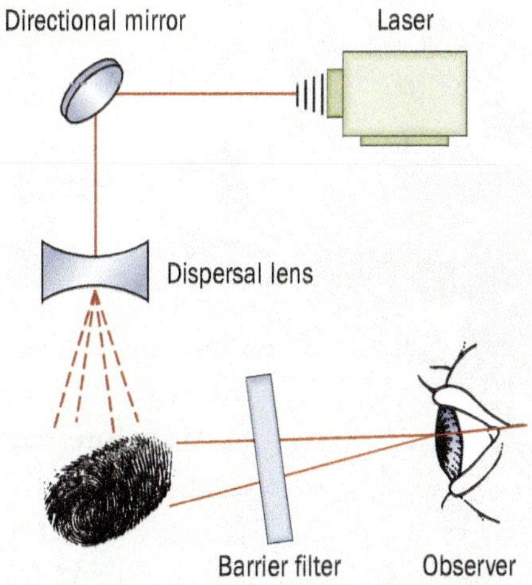

Schematic depicting latent-print detection with the aid of a laser. A fingerprint examiner, wearing safety goggles containing optical filters, examines the specimen being exposed to the laser light. The filter absorbs the laser light and permits the wavelengths at which latent-print residues fluoresce to pass through to the eyes of the examiner.
Courtesy Federal Bureau of Investigation, Washington, D.C.

1,2-indanedione is also emerging as a potential reagent for the development of latent fingerprints on porous surfaces. 1,2-indanedione gives both good initial color and strong fluorescence when reacted with amino acids derived from prints and thus has the potential to provide in one process what ninhydrin and DFO can do in two different steps. Dye combinations known as RAM, RAY, and MRM 10 when used in conjunction with superglue fuming have been effective in visualizing latent fingerprints by fluorescence.

Quick Review

- Visible prints are made when fingers touch a surface after the ridges have been in contact with a colored material such as blood, paint, grease, or ink.
- Plastic prints are ridge impressions left on a soft material, such as putty, wax, soap, or dust.
- Latent prints deposited on hard and nonabsorbent surfaces (such as glass, mirror, tile, and painted wood) are usually developed by the application of a powder, whereas prints on porous surfaces (such as papers and cardboard) generally require treatment with a chemical.
- Examiners use various chemical methods to visualize latent prints, such as iodine fuming, ninhydrin, and Physical Developer.
- Superglue fuming develops latent prints on nonporous surfaces.
- Latent fingerprints can also be treated with chemicals that induce fluorescence when exposed to a high-intensity light or an alternate light source.

Preservation of Developed Prints

Once the latent print has been visualized, it must be permanently preserved for future comparison and possible use in court as evidence. A photograph must be taken before any further attempts at preservation. Any camera equipped with a close-up lens will do; however, many investigators prefer to use a camera specially designed for fingerprint photography. Such a camera comes equipped with a fixed focus to take photographs on a 1:1 scale when the camera's open eye is held exactly flush against the print's surface (see Figure 16–22). In addition, photographs must be taken to provide an overall view of the print's location with respect to other evidential items at the crime scene.

Once photographs have been secured, one of two procedures is to be followed. If the object is small enough to be transported without destroying the print, it should be preserved in its entirety. The print should be covered with cellophane so it will be protected from damage. On the other hand, prints on large immovable objects that have been developed with a powder can best be preserved by "lifting." The most popular type of lifter is a broad adhesive tape similar to Scotch tape. Fingerprint powder is applied to the print, and the surface

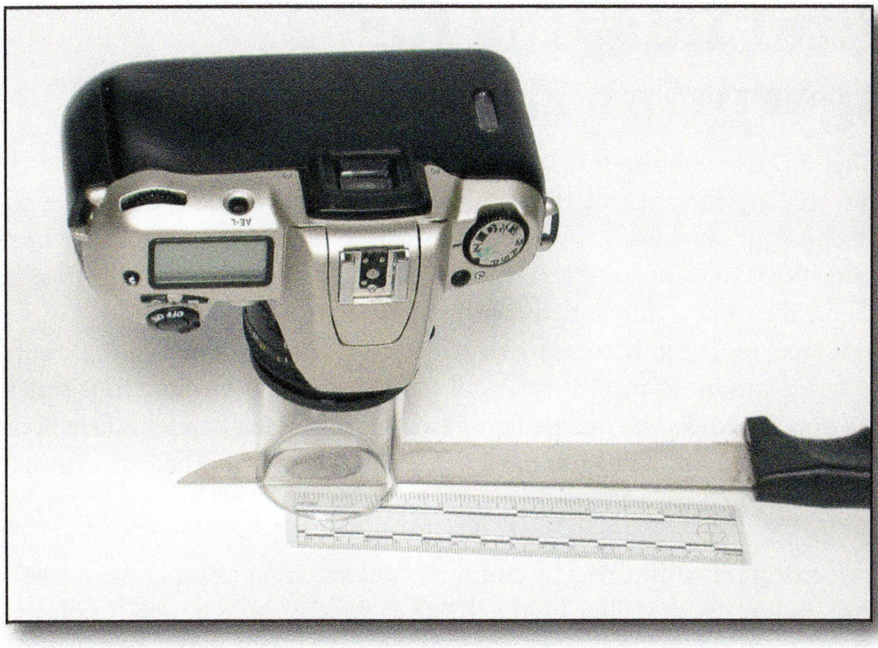

FIGURE 16–22 Camera fitted with an adapter designed to give an approximate 1:1 photograph of a fingerprint. *Courtesy Sirchie Finger Print Laboratories, Inc., Youngsville, N.C., www.sirchie.com*

containing the print is covered with the adhesive side of the tape. When the tape is pulled up, the powder is transferred to the tape. Then the tape is placed on a properly labeled card that provides a good background contrast with the powder.

A variation of this procedure is the use of an adhesive-backed clear plastic sheet attached to a colored cardboard backing. Before it is applied to the print, a celluloid separator is peeled from the plastic sheet to expose the adhesive lifting surface. The tape is then pressed evenly and firmly over the powdered print and pulled up (see Figure 16–23). The sheet containing the adhering powder is now pressed against the cardboard backing to provide a permanent record of the fingerprint. If a print is found on a nonporous surface, superglue fuming is a common procedure used to make a print visible. After the print is visible on the nonporous surface, fingerprint powders can then be applied. The print can then be lifted with adhesive for proper preservation.

FIGURE 16–23 "Lifting" a fingerprint. *Courtesy Sirchie Finger Print Laboratories, Inc., Youngsville, N.C., www.sirchie.com*

Digital Imaging for Fingerprint Enhancement

When fingerprints are lifted from a crime scene, they are not usually in perfect condition, making the analysis that much more difficult. Computers have advanced technology in most fields, and fingerprint identification has not been left behind. With the help of digital imaging software, fingerprints can now be enhanced for the most accurate and comprehensive analysis.

digital imaging
A process through which a picture is converted into a series of square electronic dots known as pixels; the picture is manipulated by computer software that changes the numerical value of each pixel

pixel
A square electronic dot that is used to compose a digital image

Digital imaging is the process by which a picture is converted into a digital file. The image produced from this digital file is composed of numerous square electronic dots called **pixels**. Images composed of only black and white elements are referred to as grayscale images. Each pixel is assigned a number according to its intensity. The grayscale image is made from the set of numbers to which a pixel may be assigned, ranging from 0 (black) to 255 (white). Once an image is digitally stored, it is manipulated by computer software that changes the numerical value of each pixel, thus altering the image as directed by the user. Resolution reveals the degree of detail that can be seen in an image. It is defined in terms of dimensions, such as 800 × 600 pixels. The larger the numbers, the more closely the digital image resembles the real-world image.

The input of pictures into a digital imaging system is usually done through the use of different types of equipment, such as scanners, digital cameras, and video cameras. After the picture is changed to its digital image, several methods can be employed to enhance the image. The overall brightness of an image, as well as the contrast between the image and the background, can be adjusted through contrast-enhancement methods. One approach used to enhance an image is *spatial filtering*. Several types of filters produce various effects. A low-pass filter is used to eliminate harsh edges by reducing the intensity difference between pixels. A second filter, the high-pass filter, operates by modifying a pixel's numerical value to exaggerate its intensity difference from that of its neighbor. The resulting effect increases the contrast of the edges, thus providing a high contrast between the elements and the background.

Frequency analysis, also referred to as *frequency Fourier transform* (FFT), is used to identify periodic or repetitive patterns such as lines or dots that interfere with the interpretation of the image. These patterns are diminished or eliminated to enhance the appearance of the image. Interestingly, the spacings between fingerprint ridges are themselves periodic. Therefore, the contribution of the fingerprint can be identified in FFT mode and then enhanced. Likewise, if ridges from overlapping prints are positioned in different directions, their corresponding frequency information is at different locations in FFT mode. The ridges of one latent print can then be enhanced while the ridges of the other are suppressed.

Color interferences can pose a problem when analyzing an image. For example, a latent fingerprint found on paper currency or a check may be difficult to analyze because of the distracting colored background. With the imaging software, the colored background can simply be removed to make the image stand out (see Figure 16–24). If the image itself is a particular color, such as a ninhydrin-developed print, the color can be isolated and enhanced to distinguish it from the background.

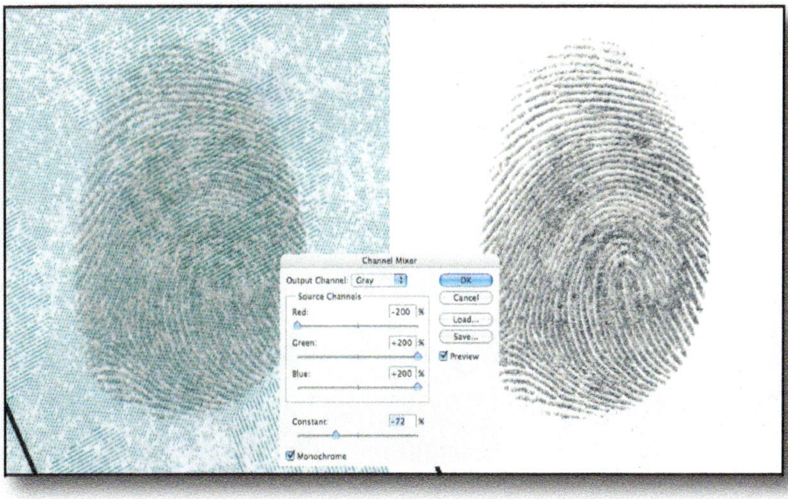

FIGURE 16–24 A fingerprint being enhanced in Adobe Photoshop. In this example, on the left is the original scan of an inked fingerprint on a check. On the right is the same image after using Adobe Photoshop's Channel Mixer to eliminate the green security background. *Courtesy Imaging Forensics, Fountain Valley, Calif., www.imagingforensics.com*

Digital imaging software also provides functions in which portions of the image can be examined individually. With a scaling and resizing tool, the user can select a part of an image and resize it for a closer look. This function operates much like a magnifying glass, helping the examiner view fine details of an image.

An important and useful tool, especially for fingerprint identification, is the compare function. This specialized feature places two images side by side and allows the examiner to chart the common features on both images simultaneously (see Figure 16–25). The zoom function is used in conjunction with the compare tool. As the examiner zooms into a portion of one image, the software automatically zooms into the second image for comparison.

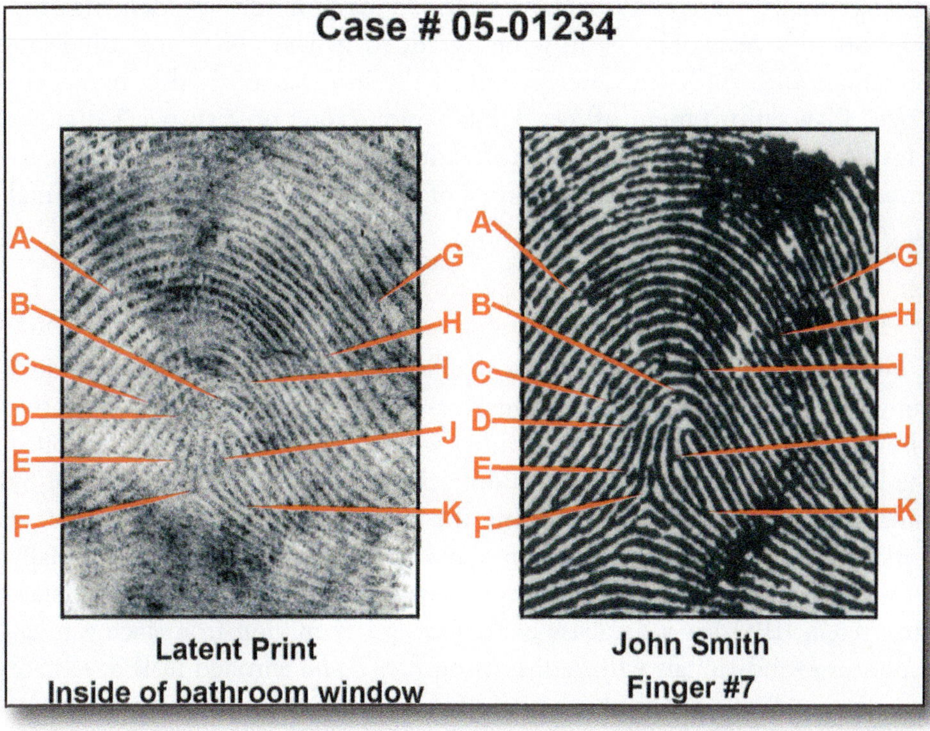

FIGURE 16–25 Current imaging software allows fingerprint analysts to prepare a fingerprint comparison chart. The fingerprint examiner can compare prints side by side and display important features that are consistent between the fingerprints. The time needed to create a display of this sort digitally is about 30 to 60 minutes. *Courtesy Imaging Forensics, Fountain View, Calif., www.imagingforensics.com*

Although digital imaging is undoubtedly an effective tool for enhancing and analyzing images, it is only as useful as the images it has to work with. If the details do not exist on the original images, the enhancement procedures are not going to work. The benefits of digital enhancement methods are apparent when weak images are made more distinguishable.

The Next Generation Identification System

The FBI has begun integrating biometric technology into their new identification system. As of September 2014, the FBI dropped the use of their Integrated Automated Fingerprint Identification System (IAFIS) for a brand new state-of-the-art identification system. The Next Generation Identification (NGI) system provides a broader selection of resources and information for law enforcement agencies across the country. This will potentially solve many cold cases and provide new leads for criminal investigations. The NGI is a necessary update for the FBI, as biometrics and new technologies have changed the way we fight crime and process individuals through the criminal justice system. Costing nearly $1.2 billion, the NGI will take a total of seven years before it will be complete.

According to the FBI, this new high-tech system will consist of seven identification focal points. The first new NGI increment will be known as the Rap Back service. The Rap Back service sends notifications to agencies throughout the United States pertaining to criminal activity of individuals that have already been processed through the system at one point (arrest, probation, parole). This system will be especially useful for agencies that use any form of criminal supervision, such as a GPS tracking anklet or a sex offender registry.

Scars, marks, and tattoos have been an easy mode of identification throughout the years. This is now being added to the NGI, providing law enforcement with in-depth descriptions of the individual's bodily marks. In practice, a law enforcement officer will be able to look up a suspect's distinctive tattoo to possibly establish identification. This can also be used in corrections, where tattoos provide a history of a felon's affiliations and criminal background.

The Interstate Photo System

The NGI system contains many new applications and programs to conduct thorough inquiries of suspected criminals, as well as generate new investigative leads. Much of the information contained within the NGI has been compiled by the FBI for many years, and new information is being provided to the FBI by criminal justice organizations throughout the United States. This expansive set of information is crucial for solving crimes and apprehending fugitives, especially due to the vast amount of law enforcement agencies that exist. The Interstate Photo System (IPS) has a database containing thirty million front-facing mugshot photos of individuals with tenprints on file. To be enrolled in IPS, all face photos must include a Tenprint submission of the individual (submission of all 10 fingerprints). The IPS database has two categories of photos: criminal identities (photos submitted as part of a lawful detention, an arrest, or incarceration) and civil identities (photos submitted for licensing, employment, security clearances, military service, volunteer service and immigration benefits). Over 80 percent of the photos in IPS are criminal.

The IPS users include the FBI and selected state and local law enforcement agencies, which can submit search requests to help identify an unknown person using, for example, a photo from a surveillance camera. The IPS is currently being used as an "investigative lead," and results are not to be interpreted as a positive identification of the suspected individual.

In addition to the IPS, the FBI has an internal unit called Facial Analysis, Comparison, and Evaluation (FACE) Services that provides facial recognition capabilities, among other things, to support active FBI investigations. FACE Services not only has access to IPS, but can also search or request to search databases owned by the Departments of State and Defense and sixteen states, which use their own facial recognition systems. Unlike IPS, which primarily contains criminal photos, FACE Services contains civil photos from state and federal government databases, such as visa applicant photos and selected states' driver's license photos. The total number of face photos available to FACE Services in all searchable repositories is over 411 million. FBI agents have requested almost 215,000 searches of external partners' databases. Of these requests, about thirty-six thousand have included searches on state driver's license databases. Any results returned as part of a FACE Services request is treated as an investigative lead only and is not considered a positive identification.

Additionally, the FBI has contracted the help of MorphoTrust to implement and maintain all facial recognition biometric systems. MorphoTrust is already responsible for biometric systems used in motor vehicle departments and other government establishments throughout the United States.

In April 2015, the FBI announced that facial recognition had become a fully active increment of the NGI system. Facial recognition works in tandem with the FBI's fingerprint system which holds over one hundred million records, and the FBI hopes to link facial recognition data with accompanying fingerprints, name, age, weight, and other essential characteristics.

Fingerprint and the National Palm Print System

The NGI makes use of fingerprint impressions with its brand new Tenprint System. All 10 fingerprints are acquired by police and these are added to a federal database of prints. The FBI makes use of a new algorithm that yields better results across its comparisons. With the dated IAFIS system, fingerprint search results were only 92 percent accurate, but the accuracy has been pumped up to 99 percent with NGI. The time required has also been reduced between acceptance of prints and positive identification.

Along with the addition of a new Tenprint system, the FBI has developed a new Latent print search algorithm. This will potentially solve many cold cases and provide new leads for criminal investigations. The latent print search will be much more accurate in finding matches and accessible to more than eighteen thousand law enforcement agencies across the country. The latent print search algorithm goes hand in hand with the FBI's new National Palm Print System (NPPS). This was previously unavailable to law enforcement and provided a major problem as many prints left at crime scenes are the offender's palm impression. With the addition of the NPPS and Tenprint system, many more prospective crimes will be solvable in the coming years.

In 2013, the FBI implemented the NPPS. The NPPS contains a database of millions of palm prints to be used in criminal investigations and processes. This works hand in hand with the Advanced Fingerprint Identification Technology (AFIT) system because now investigators can match fingerprints with any palm prints found at crime scenes. The NPPS will see great improvements in the search of latent fingerprint files from unsolved cold cases. Eighteen thousand law enforcement agencies will have access to this program and will be able to generate new leads for crimes that may have occurred decades ago and gone unsolved. All criminal and civil searches will now include data found in the Universal Latent File, which may generate hits for new fingerprints as they are added to the system. This will serve as a powerful investigative weapon and may provide closure for families of victims.

The FBI has also implemented the Repository of Individuals of Special Concern (RISC). RISC allows for officers to fingerprint individuals they come in contact with and quickly detect whether or not they are a threat. The system compares the suspect against a repository of sex offenders, wanted persons, suspected terrorists, and other high-risk individuals. This system may protect many officers from potential danger, as they will now have a better idea of whom they are dealing with. This application is completely mobile and can be accessed from a patrol vehicle or forward outpost using fingerprint data collected from two index fingers. Once a law enforcement officer has the fingerprints of an individual suspect, it can take as little as ten seconds for the RISC application to compare with all the 2.5 million prints contained within the database. Twenty-one states are currently participating in the RISC increment of the NGI, and the FBI hopes to eventually have the cooperation and assistance of all fifty states.

Searches within the Criminal Master File will now include everything found within the Unsolved Latent File. These new searches will lead law enforcement agencies to gain more hits when searching data that relates to unsolved cases. New algorithms and metrics make the infrastructure of the NGI more powerful and accurate. This will be a useful tool for law enforcement agencies across the country.

The combination of a mobile device and a fingerprint scanner has become a reality for law enforcement agencies. MorphoTrak, a leading developer of biometric technology, has developed a mobile fingerprint scanner the size of a cell phone. The law enforcement office will instruct a potential suspect to put a finger on the face of the device, and the print is transmitted to federal and state databases. The results are acquired very quickly and show information about the suspect including active warrants and identity. This can be very crucial, as potential suspects with open warrants may lie about their identity as a means of avoiding arrest.

Quick Review

- Once a latent print has been visualized, it must be permanently preserved for future comparison and for possible use as court evidence. A photograph must be taken before any further attempts at preservation are made.
- A common method for preserving prints developed with a powder is lifting the print with an adhesive tape.
- Digital imaging is a process through which a picture is converted into a series of square electronic dots known as pixels. Digital imaging can be used to enhance fingerprints.
- The FBI has begun integrating biometric technology into their new identification system.
- As of September 2014, the FBI dropped the use of their Integrated Automated Fingerprint Identification System (IAFIS) for a brand new state-of-the-art identification system.
- The Next Generation Identification (NGI) system provides a broader selection of resources, including updated fingerprint search algorithms; the Interstate Photo System; Facial Analysis, Comparison and Evaluation services; a National Palm Print System; and a Repository of Individuals of Special Concern (RISC) file.

Chapter Review

- List and define the various forms of biometrics and their categories.
- Most biometric technologies can be split into two different groups: physiological biometrics, which contains fingerprints, hand, iris, retina, and facial scans; and behavioral biometrics, which is much less stable and includes handwriting, voice, keystroke, and gait recognition.
- Explain the scope of the FBI's Next Generation Identification System
- Digital imaging is a process in which a picture is converted into a series of square electronic dots known as "pixels." Fingerprints can be enhanced with digital imaging.
- The NGI system provides a broader selection of resources and information for law enforcement agencies across the country. It will include a new Tenprint system, a new latent print search algorithm, the Rap Back service, and identifying information associated with specific scars, marks, and tattoos.
- Fingerprints are a reproduction of friction skin ridges found on the palm side of the fingers and thumbs.
- The basic principles underlying the use of fingerprints in criminal investigations are as follows: (1) A fingerprint is an individual characteristic because no two fingers have yet been found to possess identical ridge characteristics; (2) a fingerprint remains unchanged during an individual's lifetime; and (3) fingerprints have general ridge patterns that permit them to be systematically classified.
- All fingerprints are divided into three classes on the basis of their general pattern: loops, whorls, and arches.
- The individuality of a fingerprint is determined not by its general shape or pattern, but by a careful study of its ridge characteristics. The expert must demonstrate a point-by-point comparison in order to prove the identity of an individual.
- When the finger touches a surface, perspiration and oils are transferred onto that surface, leaving a fingerprint. Prints deposited in this manner are invisible to the eye and are commonly referred to as latent or invisible fingerprints.
- The primary classification is the first step in classifying fingerprints under the FBI system. The presence or absence of the whorl pattern is the basis for determination of the primary classification.
- The FBI fingerprint database known as AFIS converts the image of a fingerprint into digital minutiae that contain data showing ridges at their points of termination (ridge endings) and their branching into two ridges (bifurcations).
- Livescan is an inkless device that captures digital images of fingerprints and palm prints and electronically transmits them to an AFIS.
- Visible prints are made when fingers touch a surface after the ridges have been in contact with a colored material such as blood, paint, grease, or ink.
- Plastic prints are ridge impressions left on a soft material, such as putty, wax, soap, or dust.
- Latent prints deposited on hard and nonabsorbent surfaces (such as glass, mirror, tile, and painted wood) are usually developed by the application of

a powder, whereas prints on porous surfaces (such as papers and cardboard) generally require treatment with a chemical.

- Examiners use various chemical methods to visualize latent prints, such as iodine fuming, ninhydrin, and Physical Developer.
- Superglue fuming develops latent prints on nonporous surfaces.
- Latent fingerprints can also be treated with chemicals that induce fluorescence when exposed to a high-intensity light or an alternate light source.
- Once a latent print has been visualized, it must be permanently preserved for future comparison and for possible use as court evidence. A photograph must be taken before any further attempts at preservation are made.
- A common method for preserving prints developed with a powder is lifting the print with an adhesive tape.
- Digital imaging is a process through which a picture is converted into a series of square electronic dots known as pixels. Digital imaging can be used to enhance fingerprints.

Quick Lab: Fingerprinting

Materials:

Several sheets of blank paper
Fingerprinting pad (if available) or ink pad
Magnifying glass

Procedure:

Fingerprinting is perhaps the oldest method of scientific forensic identification. No two sets of same fingerprints are the same, yet they all exhibit characteristics that allow investigators to classify them for quicker identification. In this exercise, students should be matched up in teams of three. Each member of the group should roll the fingerprint of the right index finger of another group member onto a piece of paper. The paper should contain a print from every team member. Now, roll a print from each team member onto a separate blank piece of paper. Do NOT write the names of the team members on these papers. When the ink is dry, shuffle the papers and set two aside at random. Using the magnifying glass, each person on the team has 2 minutes to compare the unknown print to the named prints and determine who made it. Answer the following questions individually, then compare notes as a team and see if everyone agreed on the identity of the print. Be prepared to explain how you arrived at the identity of the print and answer questions from your team.

Follow-Up Questions:

1. **Who do you believe made the unknown print?**
2. **What is the general pattern of the print—arch, loop, or whorl?**
3. **Did everyone in the team agree on the identity of the unknown print?**

Review Questions

1. The International Association for Identification concluded that the minimum number of friction ridge characteristics which must be present in two impressions in order to establish positive identification is
   a. six.
   b. eight.
   c. ten.
   d. There is no minimum number.

2. For most fingerprint examiners, the chemical method of choice for visualizing latent prints is
   a. ninhydrin.
   b. iodine.
   c. silver nitrate.
   d. chlorate.

3. What was the original fingerprinting system adopted by Scotland Yard in 1901 which converted ridge patterns on all ten fingers into a series of letters and numbers arranged in the form of a fraction?
   a. Bertillon system
   b. Fauld system
   c. Henry system
   d. Galton system

4. Attempts at changing one's fingerprints by trying to obscure them has led to
   a. renewed efforts on the part of law enforcement to categorize the obliterated fingerprints.
   b. the creation of a new class of criminal who can avoid detection through currently available fingerprinting technology.
   c. the possibility of permanent scarring, which only provides new characteristics for identification because it is impossible to obliterate all ridge characteristics.
   d. self-injurious behavior that only results in the growth of new ridge characteristics on the fingertips.

5. A technique for visualizing latent fingerprints on nonporous surfaces by exposing them to cyanoacrylate vapors is
   a. sublimation.
   b. iodine fuming.
   c. fluorescing.
   d. superglue fuming.

6. True or False: The individuality of a fingerprint is determined by its general shape or pattern, and it has been empirically demonstrated that no two fingerprints are alike.

7. True or False: The most common type of fingerprint pattern is the arch pattern.

8. True or False: AFIS, the automated fingerprint identification system, makes it possible to search a single latent crime-scene fingerprint against an entire file's print collection.

9. True or False: Studies have recently demonstrated that common fingerprint developing agents interfere with DNA testing methods used for characterizing bloodstains, so DNA testing must precede fingerprint tests.

10. True or False: Digital imaging has become an effective tool for the enhancement and analysis of images because if the details do not exist on the original image, enhancement procedures are available to make the details clearer.

11. What are fingerprints?

12. What is the first fundamental principle of fingerprints?

13. What imparts individuality to a fingerprint?

14. What are ridge characteristics? What is another name for ridge characteristics?

15. What is the second fundamental principle of fingerprints?

16. What are dermal papillae and how are they related to fingerprints?

17. What is a latent fingerprint? Briefly describe how a latent fingerprint is formed.

18. Why is it pointless to try to obscure or obliterate one's fingerprints by scarring or otherwise damaging the skin?

19. What is the third fundamental principle of fingerprints?

20. Compare the three major fingerprint patterns of arches, loops, and whorls. Which is most common?

21. Which class of fingerprints includes ridge patterns that are generally rounded or circular?

22. Which type of fingerprint pattern has ridges entering from one side of the print, and exiting from the same side?

23. Which is the simplest of all fingerprint patterns, formed by ridges entering from one side of the print and exiting on the opposite side?

24. What is the primary classification? What is the basis for this classification?

25. What is an AFIS? What is the heart of AFIS technology?

26. When using AFIS, who makes the final verification of a print's identity?

27. What is livescan? What procedure has livescan largely replaced?

28. Name two main drawbacks to using AFIS.

29. Name the three kinds of crime-scene fingerprints.

30. How are prints from soft and porous surfaces preferably developed?

31. What is RUVIS and how does it work?

32. Name four common chemical methods for visualizing latent prints.

33. Explain how latent prints can be visualized when illuminated by laser light.

34. Name three reasons why alternate light sources have replaced lasers for visualizing latent prints.

35. What is the first thing that the criminalist must do after visualizing a print but before making any further attempts at preserving it?

36. Briefly describe how the criminalist should handle prints on small objects.

37. Describe the basic process used to "lift" a fingerprint. When should this procedure be used?

38. What is digital imaging? How is it useful for analyzing fingerprints?

39. Under what conditions is digital imaging not effective in enhancing latent fingerprints?

40. True or False: Over twenty-seven states are participating in the FBI's facial recognition program.

41. The __________system was implemented in 2014 to replace the existing fingerprint IAFIS.

42. The __________is a NGI database containing 23 million front-facing photos of individuals with ten prints on file.

43. The __________contains a NGI database of millions of palmprints to be used in criminal investigations and processes.

44. The AFIS system utilized complex matching algorithms that changed the fingerprint accuracy rating from __ to __ percent.

45. The FBI has implemented the __________system to quickly search a fingerprint database containing terrorists and other high risk individuals.

46. How can you distinguish between patent, plastic, and latent fingerprint impressions? Summarize your answer in writing.

Learn the Vocabulary

Use the information from this chapter along with your prior knowledge and experience to learn the key terms. Choose from the following activities:

1. Make flashcards with a term on the front and its definition on the back. Use the flashcards to quiz a partner on correctly pronouncing and defining the terms.
2. Make an audio or video recording of each term and its definition.
3. Work with a partner to write a sentence for each term. Read the sentences out loud to a partner or share them with the class.
4. Write a list of new terms and language you have encountered in this chapter. Reuse the terms to write sentences and then read the sentences out loud to a partner. Reuse the terms again by writing a short speech. The speech might explain the terms or teach a lesson using the terms. Give the speech to a partner or to the class.

Application and Critical Thinking

1. Classify each of the following prints as a loop, whorl, or arch.

(1). __________ (2). __________ (3). __________ (4). __________ (5). __________ (6). __________

2. Following is a description of the types of prints from the fingers of a criminal suspect. Using the FBI system, determine the primary classification of this individual.

Finger	Right Hand	Left Hand
Thumb	Whorl	Whorl
Index	Loop	Whorl
Middle	Whorl	Arch
Ring	Whorl	Whorl
Little	Arch	Whorl

3. While searching a murder scene, you find the following items that you believe may contain latent fingerprints. Indicate whether prints on each item should be developed using fingerprint powder or chemicals.

a. a leather sofa

b. a mirror

c. a painted wooden knife handle

d. blood-soaked newspapers

e. a revolver

4. Criminalist Frank Mortimer is using digital imaging to enhance latent fingerprints. Indicate which features of digital imaging he would most likely use for each of the following tasks:

a. isolating part of a print and enlarging it for closer examination

b. increasing the contrast between a print and the background surface on which it is located

c. examining two prints that overlap one another

5. The following are fingerprint patterns of three men and a woman with criminal records for robbery. Identify the following fingerprints according to the three groups and subgroups of fingerprints.

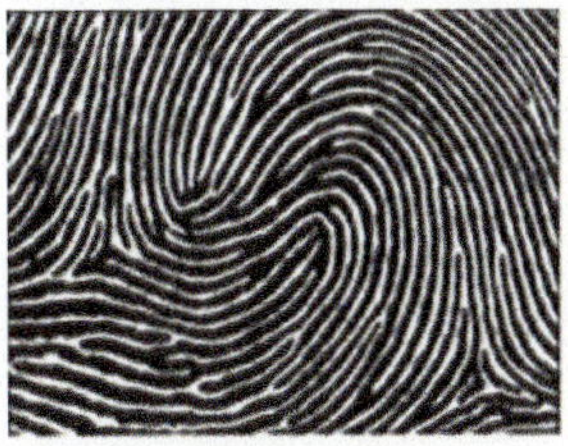

KJ

Ivan

Lisa

Charlie

Tented Arch__________ Double Loop Whorl__________ Plain Loop__________ Accidental Whorl__________

6. Identify the minutiae of fingerprints, including but not limited to bifurcations, ending ridges, dots, short ridges, and enclosures/islands. Count the number of bifurcations in the following print. Choose between 9, 11, and 13.

Number of bifurcations:

7. At the Museum of Culture Studies, a diary that belonged to Martin Luther King has been stolen and replaced by a fake. The only evidence is a fingerprint impression left by the thief on the fake diary. The police suspects four individuals who have had previous criminal records in similar crimes. Their fingerprints already exist in the police database. KJ, Ivan, Lisa, and Charlie are the four suspects. Carefully read the criminal's fingerprint impression and identify which if any suspect print matches with it.

Crime Scene Fingerprint

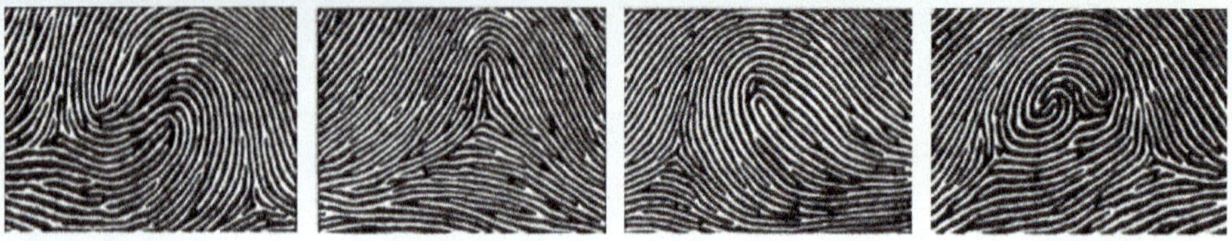

KJ Ivan Lisa Charlie

8. Explain the Integrated Automated Fingerprint Identification System (IAFIS). Describe the implications of the Next Generation Identification system and how it has changed forensic analysis of biometric evidence.

9. Perform and describe the procedures to develop and lift a latent print using cyanoacrylate and fingerprint powders on the following surfaces:

a. porous

b. adhesive surfaces

c. nonporous

Laboratory Experiments

This activity requires the use of the following practices of science and engineering:

- Asking questions and defining problems
- Planning and carrying out investigations
- Analyzing and interpreting data
- Using appropriate safety equipment and practices as outlined in Texas Education Agency-approved safety standards

This activity consists of the following crosscutting concept:

- Patterns—Fingerprints can be classified into categories based on their patterns, and can be individualized with various minutiae.

Before you begin, read the scenario and steps out loud with a partner. Ask for clarification as necessary to make sure you understand the experiment.

MATERIALS

Fingerprint brushes and powders
Lifting tapes and cards
Chemicals to develop prints, such as ninhydrin, crystal violet, and cyanoacrylate

The Scenario

A 911 call was made at 0615 hours from a hiker, who stated that he and his wife were walking across the park when they noticed a large object wrapped in plastic. They assumed it was garbage and went over to clean it up; as they got closer they noticed that it was a body. One went to the closest emergency phone (approximately 25 feet away) and called while the other stayed and watched over the body.

When officers arrived they could see that the body was of a dead male already in rigor. While looking around for evidence, they could see a wallet through the plastic, so they cut part of the plastic to gain access to the wallet. Other officers went to the address listed on his driver's license and found officers already there investigating a missing person called in by the victim's girlfriend, Jennifer Rivera. It was at this time that the officers realized both cases were related.

Both scenes were secured and your team was called out to investigate. The only evidence at the park was a single print on the plastic around the body.

PROCEDURE

1. Perform the best method for developing and lifting the print. Be sure to indicate if the print is on a porous or nonporous surface. Develop the print using the method you selected.
2. After developing the print, you must document the results via photography. You are given a copy of the print and 4 possible matches as determined by AFIS.
3. Indicate each print as being a loop, arch, or whorl.
4. Match a minimum of 5 minutiae between the crime scene print and the possible suspect. Be sure to identify the type of minutiae (such as bifurcations, ending ridges, dots, short ridges, and enclosures/islands) used to match the print to the suspect.
5. Determine which suspect matches the print.
6. A search of the victim's residence reveals some open mail scattered with unopened mail. This could be additional evidence for the case. Describe and perform the appropriate procedures to chemically develop latent prints on both the porous and adhesive surfaces of the envelopes. Be sure to indicate which chemicals will be used for each. Document the results via photography.

Endnotes

1. A tented arch is also any pattern that resembles a loop but lacks one of the essential requirements for classification as a loop.

2. M. A. Smrz et al., "Review of FBI Latent Print Unit Processes and Recommendations to Improve Practices and Quality," *Journal of Forensic Identification* 56 (2006): 402–433.
3. J. Almag, Y. Sasson, and A. Anati, "Chemical Reagents for the Development of Latent Fingerprints II: Controlled Addition of Water Vapor to Iodine Fumes— A Solution to the Aging Problem," *Journal of Forensic Sciences* 24 (1979): 431.
4. F. G. Kendall and B. W. Rehn, "Rapid Method of Superglue Fuming Application for the Development of Latent Fingerprints," *Journal of Forensic Sciences* 28 (1983): 777.
5. C. Roux et al., "A Further Study to Investigate the Effect of Fingerprint Enhancement Techniques on the DNA Analysis of Bloodstains," *Journal of Forensic Identification* 49 (1999): 357; C. J. Frégeau et al., "Fingerprint Enhancement Revisited and the Effects of Blood Enhancement Chemicals on Subsequent Profiler Plus™ Fluorescent Short Tandem Repeat DNA Analysis of Fresh and Aged Bloody Fingerprints," *Journal of Forensic Sciences* 45 (2000): 354; and P. Grubwieser et al., "Systematic Study on STR Profiling on Blood and Saliva Traces after Visualization of Fingerprints," *Journal of Forensic Sciences* 48 (2003): 733.

Firearms, Tool Marks, and Other Impressions

17

John Green/Cal Sport Media/Alamy Stock Photo

Key Terms

bore
breechblock
breechface
caliber
choke
distance determination
ejector
extractor
firearms identification
gauge
Greiss test
grooves
lands
rifling

The Case Against Aaron Hernandez

New England Patriots tight end, Aaron Hernandez, was arrested and charged with the murder of Odin Lloyd in the 2015 NFL offseason. The trial that followed made headlines, not just because of his status, but also because of what appeared to be overwhelming scientific evidence of his guilt. Two important pieces of the forensic puzzle were the footwear and tire tread impressions left behind at the scene where the victim's body was found shot in an industrial park.

Investigators identified consistency in the pattern, design, size, and manufacturer in one important footwear impression. According to the examiner, the impression matched a size 13 pair of Nike Air Jordan Retro 11 sneakers. Prosecutors introduced surveillance video of Hernandez at a gas station less than 90 minutes before the killing and at his home less than 10 minutes after wearing similar shoes. Nike consultant, Herbert Hedges, testified that in the footage Hernandez was wearing Nike Air Jordan Retro 11 sneakers even though investigators never recovered them.

Prosecutors alleged that Hernandez was driving a rented Nissan Altima when he and two accomplices picked up Lloyd and travelled to the industrial park to kill him. The tire tread examiner described the class characteristic examination performed on the tire impressions at the scene, comparing the "grooves," "sipes," and "wear bars." He first eliminated the Nissan's two front tires as the source of the tracks because the class characteristics didn't match. The two rear tires were of the same design as the one that left the track, so those were linked and the individual characteristics were compared to the evidence impression. Four stones wedged into the rear passenger-side tire helped investigators match it to the tire impression left at the scene.

Learning Objectives

After studying this chapter, you should be able to:

- 17.1 Describe the types of firearms and the techniques for rifling a barrel
- 17.2 Recognize the class and individual characteristics of bullets and cartridge cases
- 17.3 Discuss the various search systems developed for the FBI and ATF
- 17.4 Explain the procedure for determining how far a weapon was fired from a target
- 17.5 Identify the laboratory tests for determining whether an individual has fired a weapon
- 17.6 Discuss the procedures for collecting and preserving firearms evidence and restoring serial numbers
- 17.7 Explain the forensic significance of class and individual characteristics to the comparison of tool mark, footwear, and tire impressions
- 17.8 Discuss the preservation, lifting, casting, and comparison of impressions left at a crime scene

National Science Content Standards

Scientific Inquiry

History and Nature of Science

Just as natural variations in skin ridge patterns and characteristics provide a key to human identification, minute random markings on surfaces can impart individuality to inanimate objects. Structural variations and irregularities caused by scratches, nicks, breaks, and wear permit the criminalist to relate a bullet to a gun; a scratch or abrasion mark to a single tool; or a tire track to a particular automobile. Individualization, so vigorously pursued in all other areas of criminalistics, is frequently attainable in firearms and tool mark examination.

Although a portion of this chapter will be devoted to the comparison of surface features for the purposes of bullet identification, a complete description of the services and capabilities of the modern forensic firearms laboratory cannot be restricted to just this one subject, important as it may be. The high frequency of shooting cases means that the science of **firearms identification** must extend beyond mere comparison of bullets to include knowledge of the operation of all types of weapons, restoration of obliterated serial numbers on weapons, detection and characterization of gunpowder residues on garments and around wounds, estimation of muzzle-to-target distances, and detection of powder residues on hands. Each of these functions will be covered in this chapter.

firearms identification
A discipline mainly concerned with determining whether a bullet or cartridge was fired by a particular weapon

Types of Firearms

Generally, modern firearms can be divided into two categories: handguns and long guns. Handguns, or pistols, are firearms that are designed to be held and fired with one hand. The three most common types of handguns are single-shot handguns, revolvers, and semiautomatic pistols. All handguns can be classified as single-action or double-action firearms. Single-action firearms require the firing component to be manually cocked backward each time before the trigger is pulled in order to fire. Double-action firearms cock the firing component when the trigger is pulled and then reload the firing chamber after the round is fired.

Single-shot pistols can fire only one round, or shot, at a time. Each round must be manually loaded into the chamber before firing. The revolver features several firing chambers located within a revolving cylinder. As the revolver is fired, the cylinder can rotate clockwise or counterclockwise. Each firing chamber holds one cartridge, which is lined up with the barrel mechanically when the round is fired. The cartridge cases have to be manually ejected to reload the firing chambers. *Swing-out revolvers* feature a cylinder that swings out to the side of the weapon to be loaded (see Figure 17–1). Break-top revolvers are hinged so that both the barrel and the cylinder flip downward for loading. Solid-frame revolvers have no mechanism to uncover all the firing chambers at once. Instead, a small "gate" at the back of the gun allows one chamber to be loaded at a time; the cylinder is then rotated, and the next chamber is loaded with a cartridge.

FIGURE 17–1 A swing-out revolver features a cylinder that swings out to the side of the weapon to be loaded.
Ktsdesign/123RF

Semiautomatic pistols feature a removable magazine that is most often contained within the grip of the firearm. Once the magazine is loaded, the firing component is cocked by pulling the slide on the top of the gun rearward and then releasing it to load the

FIGURE 17–2 A semiautomatic pistol. *Dorling Kindersley/ Getty Images*

first round. The firing of the cartridge generates gases that are used to eject the cartridge case, cock the firing component, and load the next round. A semiautomatic pistol (see Figure 17–2) fires one shot per trigger pull.

An automatic firearm, such as a machine gun, fires as long as the trigger is pressed or until the ammunition is depleted. Long guns are either rifles or shotguns. Rifles and shotguns are designed to be fired while resting on the shoulder. The two principal differences between rifled firearms and shotguns are found in the ammunition and the barrel. Shotgun ammunition, called a shell, contains numerous ball-shaped projectiles, called slug. The barrel of a shotgun is smooth, without the grooves and lands found in rifles. A shotgun barrel can also be narrowed toward the muzzle in order to concentrate shot when fired. This narrowing of the barrel is called the choke of the shotgun. A shotgun may be single or double barreled. The two barrels of a double-barreled shotgun may be arranged horizontally (side by side) or vertically (one over another). The barrels may also have different choke diameters.

The various types of rifles and shotguns have different reloading mechanisms. The singleshot gun can chamber and fire only one round at a time. Just as with single-shot pistols, the round has to be loaded manually each time. Repeating long guns use a mechanical instrument of some sort to eject spent cartridge cases, load a new round, and cock the firing component after a round is fired. These include lever-action, pump or slide-action, bolt-action (see Figure 17–3), and semiautomatic (see Figure 17–4) long guns, the names of which refer to the loading mechanism used on each. Semiautomatic rifles use the force of the gas produced during firing to eject the spent cartridge case, load a new round, and cock the firing component. Semiautomatic firearms use a disconnector mechanism to fire one shot per trigger pull, whereas fully automatic firearms do not have such a mechanism and fire multiple consecutive shots with a single pull of the trigger.

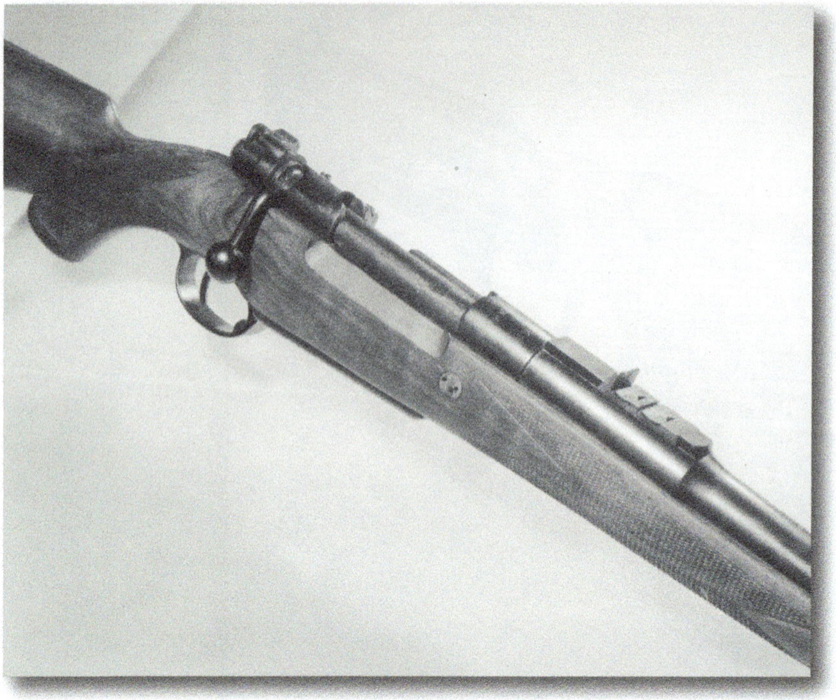

FIGURE 17–3 A bolt-action long gun uses the movement of a bolt mechanism to expel the spent cartridge case, load the next round, and cock the firing component. *George Pickow/Three Lions/Hulton Archive/Getty Images*

FIGURE 17–4 A semiautomatic long gun uses the energy from the firing reaction to expel the spent cartridge case, load the next round, and cock the firing component. *Tobkatrina/123RF*

Bullet and Cartridge Comparisons

The inner surface of the barrel of a gun leaves its markings on a bullet passing through it. These markings are peculiar to each gun. Hence, if one bullet found at the scene of a crime and another test-fired from a suspect's gun show the same markings, the suspect is linked to the crime. Because these inner surface markings, or striations, are so important for bullet comparison, it is important to know why and how they originate.

grooves
The cut or low-lying portions between the lands in a rifled bore

rifling
The spiral grooves formed in the bore of a firearm barrel that impart spin to the projectile when it is fired

bore
The interior of a firearm barrel

lands
The raised portion between the grooves in a rifled bore

The Gun Barrel

The gun barrel is produced from a solid bar of steel that has been hollowed out by drilling. The microscopic drill marks left on the barrel's inner surface are randomly irregular and in themselves impart a uniqueness to each barrel. However, the manufacture of a barrel requires the additional step of impressing its inner surface with spiral **grooves**, a step known as **rifling**. The surfaces of the original **bore** remaining between the grooves are called **lands** (see Figure 17–5).

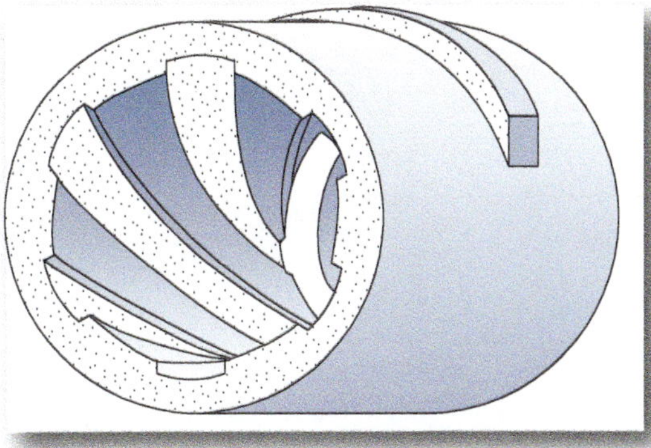

FIGURE 17–5 Interior view of a gun barrel, showing the presence of lands and grooves.

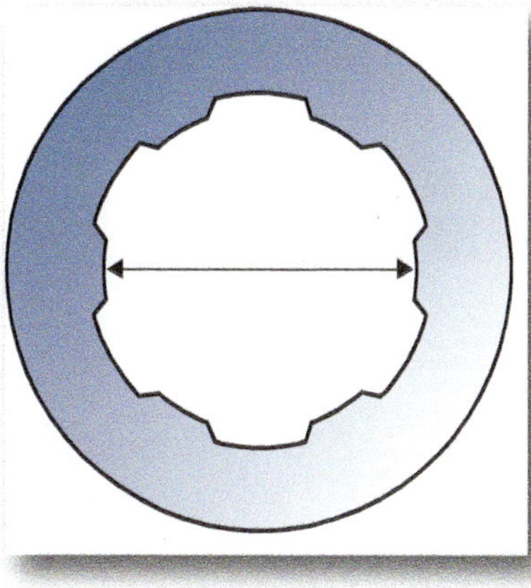

FIGURE 17–6 Cross-section of a barrel with six grooves. The diameter of the bore is the caliber. The diameter of the bore of a rifled firearm is usually expressed in hundredths of an inch or millimeters—for example, .22 caliber and 9 mm.

As a fired bullet travels through a barrel, it engages the rifling grooves. These grooves then guide the bullet through the barrel, giving it a rapid spin. This is done because a spinning bullet does not tumble end over end on leaving the barrel, but remains instead on a true and accurate course.

The diameter of the gun barrel, sketched in Figure 17–6, measured between opposite lands, is known as the **caliber** of the weapon. Caliber is normally recorded in hundredths of an inch or in millimeters—for example, .22 caliber and 9 mm. Actually, the term caliber, as it is commonly applied, is not an exact measurement of the barrel's diameter; for example, a .38-caliber weapon may actually have a bore diameter that ranges from 0.345 to 0.365 inch.

caliber
The diameter of the bore of a rifled firearm, usually expressed in hundredths of an inch or millimeters—for example, .22 caliber and 9 mm.

Rifling Methods Before 1940, barrels were rifled by having one or two grooves at a time cut into the surface with steel hook cutters. The cutting tool was rotated as it passed down the barrel, so that the final results were grooves spiraling to either the right or left. However, as the need for increased speed in weapons manufacture became apparent, newer techniques were developed that were far more suitable for the mass production of weapons.

The broach cutter, shown in Figure 17–7, consists of a series of concentric steel rings, with each ring slightly larger than the preceding one. As the broach passes through the barrel, it simultaneously cuts all grooves into the barrel at the required depth. The broach rotates as it passes through the barrel, giving the grooves their desired direction and rate of twist.

FIGURE 17–7 A segment of a broach cutter.
Ilan Amihai/ PhotoStock-Israel / Alamy Stock Photo

In contrast to the broach, the button process involves no cuttings. A steel plug or "button" impressed with the desired number of grooves is forced under extremely high pressure through the barrel. A single pass of the button down the barrel compresses the metal to create lands and grooves on the barrel walls that are negative forms of those on the button. The button rotates to produce the desired direction and rate of twist (see Figure 17–8).

Like the button process, the mandrel rifling hammer forging process involves no cutting of metal. A mandrel is a rod of hardened steel machined so its form is the reverse impression of the rifling it is intended to produce. The mandrel is inserted

FIGURE 17–8 (top) Cross-section of a .22-caliber rifled barrel. (bottom) A button used to produce the lands and grooves in the barrel. *Courtesy Richard Saferstein, Ph.D.*

into a slightly oversized bore, and the barrel is compressed with hammering or heavy rollers into the mandrel's form.

Every firearms manufacturer chooses a rifling process that is best suited to meet the production standards and requirements of its product. Once the choice is made, however, the class characteristics of the weapon's barrel remain consistent; each has the same number of lands and grooves, with the same approximate width and direction of twist. For example, .32 caliber Smith & Wesson revolvers have five lands and grooves twisting to the right. On the other hand, Colt .32-caliber revolvers exhibit six lands and grooves twisting to the left. Although these class characteristics permit the examiner to distinguish one type or brand name of weapon from another, they do not impart individuality to any one barrel; no class characteristic can do this.

If one could cut a barrel open lengthwise, careful examination of the interior would reveal the existence of fine lines, or *striations*, many running the length of the barrel's lands and grooves. These striations are impressed into the metal as the negatives of minute imperfections found on the rifling cutter's surface, or they are produced by minute chips of steel pushed against the barrel's inner surface by a moving broach cutter. The random distribution and irregularities of these markings are impossible to duplicate exactly in any two barrels. No two rifled barrels, even those manufactured in succession, have identical striation markings. These striations form the individual characteristics of the barrel.

Comparing Bullet Markings As the bullet passes through the barrel, its surface is impressed with the rifled markings of the barrel. The bullet emerges from the barrel carrying the impressions of the bore's interior surface (see Figure 17–9). Because there is no practical way to directly compare the markings on the fired bullet and those within a barrel, the examiner must obtain test bullets fired

FIGURE 17–9 A bullet is impressed with the rifling markings of the barrel when it emerges from the weapon. *Courtesy Richard Saferstein, Ph.D.*

through the suspect barrel for comparison. To prevent damage to the test bullet's markings and to facilitate the bullet's recovery, test firings are normally made into a recovery box filled with cotton or into a water tank (see Figure 17–10).

The number of lands and grooves, and their direction of twist, are obvious points of comparison during the initial stages of the examination. Any differences in these class characteristics immediately eliminate the possibility that both bullets traveled through the same barrel. A bullet with five lands and grooves could not possibly have been fired from a weapon of like caliber with six lands and grooves, nor could one having a right twist have come through a barrel impressed with a left twist. If both bullets carry the same class characteristics, the analyst must begin to match the striated markings on both bullets. This can be done only with the assistance of the comparison microscope (see Chapter 8).

Modern firearms identification began with the development and use of the comparison microscope. This instrument is the firearms examiner's most important tool. The test and evidence bullets are mounted on cylindrical adjustable holders beneath the objective lenses of the microscope, each pointing in the same direction (see Figure 17–11). Both bullets are observed simultaneously within the same field of view, and the examiner rotates one bullet until a well-defined land or groove comes into view.

Once the striation markings are located, the other bullet is rotated until a matching region is found. Not only must the lands and grooves of the test and evidence bullet have identical widths, but the longitudinal striations on each must coincide.

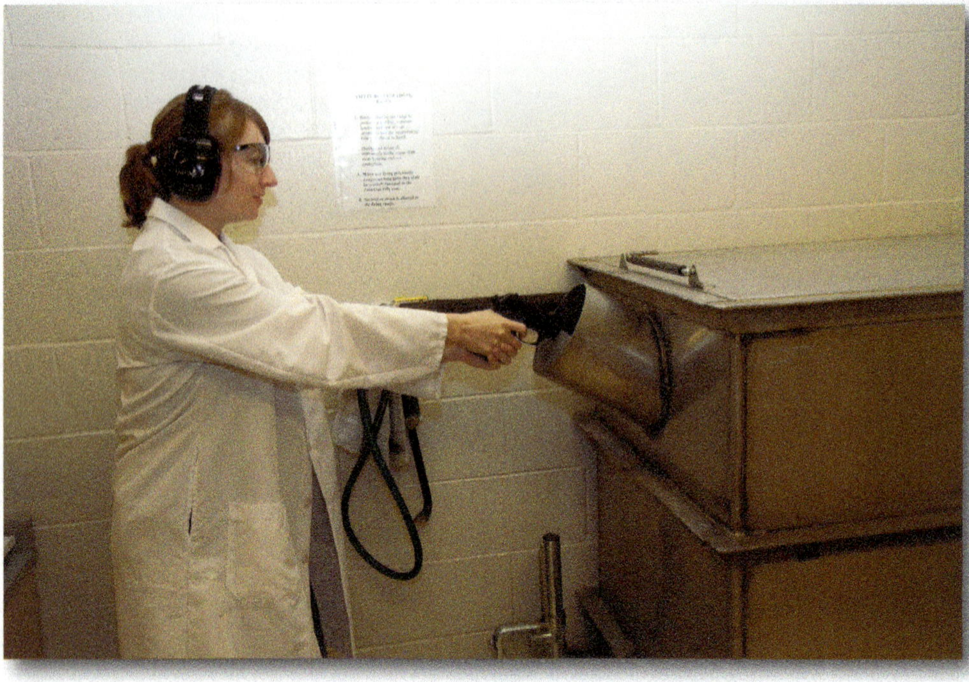

FIGURE 17–10 In ballistics testing, a suspect firearm is fired into a water tank. The bullet is slowed and stopped by the water, fished out undamaged, and compared to bullets from the crime scene. *Mikael Karlsson/Alamy Stock Photo*

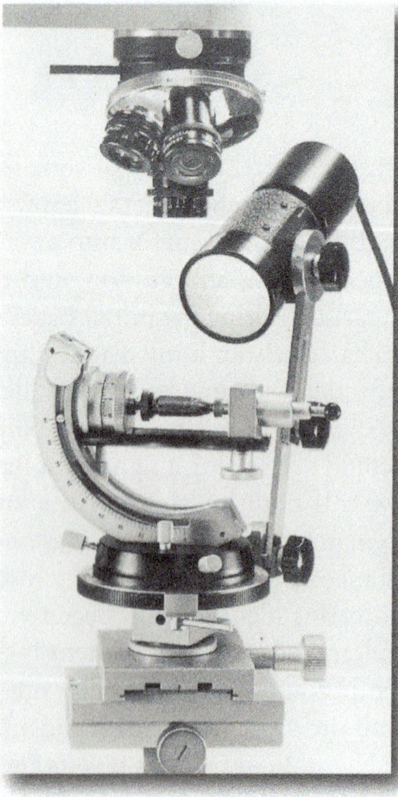

FIGURE 17–11 A bullet holder beneath the objective lens of a comparison microscope. *Courtesy Leica Microsystems, Buffalo, N.Y., www.leica-microsystems.com*

When a matching area is located, the two bullets are simultaneously rotated to obtain additional matching areas around the periphery of the bullets. Figure 17–12 shows a typical photomicrograph of a bullet match as viewed under a comparison microscope.

FIGURE 17–12 Photomicrograph of two bullets through a comparison microscope. The test bullet is on the right; the questioned bullet is on the left. *Courtesy Peter Diaczuk, John Jay College, City University of New York*

Considerations in Bullet Comparison Firearms examiners need to know the basic components of bullet cases to successfully carry out their comparison investigations. The basic components of bullet cases consist of a case with projectiles, gun powder, and a primer. Unfortunately, the firearms examiner rarely encounters a perfect match all around the bullet's periphery. The presence of grit and rust can alter the markings on bullets fired through the same barrel. More commonly, recovered evidence bullets become so mutilated and distorted on impact that they yield only a small area with intact markings.

Furthermore, striation markings on a barrel are not permanent structures; they are subject to continuing alteration through wear as succeeding bullets traverse the length of the barrel. Fortunately these changes are usually not dramatic and do not prevent the matching of two bullets fired by the same weapon. As with fingerprint comparison, there are no hard-and-fast rules governing the minimum number of points required for a bullet comparison. The final opinion must be based on the judgment, experience, and knowledge of the expert.

Frequently the firearms examiner receives a spent bullet without an accompanying suspect weapon and is asked to determine the caliber and possible make of the weapon. If a bullet appears not to have lost its metal, its weight may be one factor in determining its caliber. In some instances, the number of lands and grooves, the direction of twist, and the widths of lands and grooves are useful class characteristics for eliminating certain makes of weapons from consideration. For example, a bullet that has five lands and grooves and twists to the right could not come from a weapon manufactured by Colt, because Colts are not manufactured with these class characteristics.

Sometimes a bullet has rifling marks that set it apart from most other manufactured weapons, as in the case of Marlin rifles. These weapons are rifled by a technique known as *microgrooving* and may have 8 to 24 grooves impressed into their barrels; few other weapons are manufactured in this fashion. In this respect, the FBI maintains a record known as the General Rifling Characteristics File. This file contains listings of class characteristics, such as land and groove width dimensions, for known weapons. It is periodically updated and distributed to the law enforcement community to help identify rifled weapons from retrieved bullets.

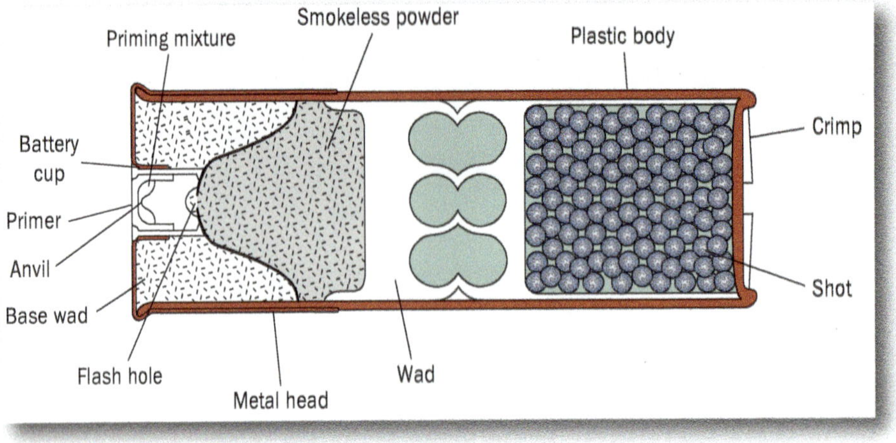

FIGURE 17–13 Cross-section of a loaded shotgun shell.

Unlike rifled firearms, a shotgun has a smooth barrel, so projectiles passing through a shotgun barrel are not impressed with any characteristic markings that can later be related back to the weapon. Shotguns generally fire small lead balls or pellets contained within a shotgun shell (see Figure 17–13). A paper or plastic wad pushes the pellets through the barrel on ignition of the cartridge's powder charge. By weighing and measuring the diameter of the shot recovered at a crime scene, the examiner can usually determine the size of shot used in the shell. The size and shape of the recovered wad may also reveal the gauge of the shotgun used and, in some instances, may indicate the manufacturer of the fired shell.

gauge
Size designation of a shotgun, originally the number of lead balls with the same diameter as the barrel that would make a pound

The diameter of the shotgun barrel is expressed by the term **gauge**.[1] The higher the gauge number, the smaller the barrel's diameter. For example, a 12-gauge shotgun has a bore diameter of 0.730 inch as contrasted to 0.670 inch for a 16-gauge shotgun. The exception to this rule is the .410-gauge shotgun, which refers to a barrel 0.41 inch in diameter.

Cartridge Cases

breechface
The rear part of a firearm barrel

The act of pulling a trigger releases the weapon's firing pin, causing it to strike the primer, which in turn ignites the powder. The expanding gases generated by the burning gunpowder propel the bullet forward through the barrel, simultaneously pushing the spent cartridge case or shell back with equal force against the **breechface**. As the bullet is marked by its passage through the barrel, the shell is also impressed with markings by its contact with the metal surfaces of the weapon's firing and loading mechanisms. As with bullets, these markings can be reproduced in test-fired cartridges to provide distinctive points of comparison for individualizing a spent shell to a rifled weapon or shotgun.

extractor
The mechanism in a firearm by which a cartridge or fired case is withdrawn from the chamber

ejector
The mechanism in a firearm that throws the cartridge or fired case from the firearm

The shape of the firing pin is impressed into the relatively soft metal of the primer on the cartridge case, revealing the minute distortions of the firing pin. These imperfections may be sufficiently random to individualize the pin impression to a single weapon. Similarly, the cartridge case, in its rearward thrust, is impressed with the surface markings of the breechface. The breechface, like any machined surface, is populated with random striation markings that become a highly distinctive signature for individualizing its surface.

Other distinctive markings that may appear on the shell as a result of metalto-metal contact are caused by the **extractor** and **ejector** mechanism and the magazine or clip, as well as by imperfections on the firing chamber walls. The

Case Files

Sacco and Vanzetti

Bettmann/Getty images

In 1920, two security guards were viciously gunned down by unidentified assailants. The security guards were transporting shoe factory payroll, nearly $16,000 in cash, at the time of the robbery-murder. Eyewitnesses described the assailants as "Italian-looking," one with a full handlebar mustache. The robbers had used two firearms, leaving behind three different brands of shells.

Two suspects were identified and arrested—Nicola Sacco and his friend, the amply mustachioed Bartolomeo Vanzetti. After denying owning any firearms, each was found to be in possession of a loaded pistol. In fact, Sacco's pistol was .32 caliber, the same caliber as the crime-scene bullets. In Sacco's pockets were found 23 bullets matching the brands of the empty shells found at the murder scene.

This case coincided with the "Red Scare," a politically turbulent time in post–World War I America. Citizens feared socialist zealots, and the media played up these emotions. Political maneuvering and the use of the media muddied the waters surrounding the case, and the fact that both suspects belonged to anarchist political groups that advocated revolutionary violence against the government only incited public animosity toward them. Sympathetic socialist organizations attempted to turn Sacco and Vanzetti into martyrs, calling their prosecution a "witch hunt."

The outcome of the trial ultimately depended on whether the prosecution could prove that Sacco's pistol fired the bullets that killed the two security guards. At trial, the ballistics experts testified that the bullets used were no longer in production and they could not find similar ammunition to use in test firings—aside from the unused cartridges found in Sacco's pockets. A forensics expert for the prosecution concluded that a visual examination showed that the bullets matched, leading the jury to return a verdict of guilty. Sacco and Vanzetti were sentenced to death.

Because of continued public protests, a committee was appointed in 1927 to review the case. Around this time, Calvin Goddard, at the Bureau of Forensic Ballistics in New York, perfected the comparison microscope for use in forensic firearms investigations. With this instrument, two bullets are viewed side by side to compare the striations imparted to a bullet's surface as it travels through the gun's barrel. The committee asked Goddard to examine the bullets in question. A test-fired bullet from Sacco's weapon was matched conclusively by Goddard to one of the crime-scene bullets. The fates of Sacco and Vanzetti were sealed, and they were put to death in 1927.

FIGURE 17–14 Comparison microscope photomicrograph showing a match between (a) firing pin impressions and (b) the breechface markings on two shells. *Courtesy Marmaduke St. John/Alamy Stock Photo; Ronald Welsh/Bureau of Forensic Services*

photomicrographs in Figure 17–14 reveal a comparison of the firing pin and breechface impressions on evidence and test-fired shells.

Firing pin, breechface, extractor, and ejector marks may also be impressed onto the surface of the brass portion of shells fired by a shotgun. These impressions provide points for individualizing the shell to a weapon that are just as valuable as cartridge cases discharged from a rifled firearm. Furthermore, in the absence of a suspect weapon, the size and shape of a firing pin impression and/or the position of ejector marks in relationship to extractor and other markings may provide some clue to the type or make of the weapon that fired the questioned shell, or at least eliminate a large number of possibilities.

Quick Review

- The manufacture of a gun barrel requires impressing its inner surface with spiral grooves, a step known as rifling. Rifling imparts spin to the projectile when it is fired, which keeps it on an accurate course.
- No two rifled barrels have identical striation markings. These striations form the individual characteristics of the barrel. The inner surface of the barrel of a gun leaves its striation markings on a bullet passing through it.
- The class characteristics of a rifled barrel include the number of lands and grooves and the width and direction of twist.
- The comparison microscope is a firearms examiner's most important tool because it allows two bullets to be observed and compared simultaneously.
- The firing pin, breechface, and ejector and extractor mechanism also offer a highly distinctive signature for individualization of cartridge cases.
- Unlike handguns, a shotgun is not rifled—it has a smooth barrel. Because of this, shotgun shells are not impressed with any characteristic markings that can be used to compare two shotgun shells to determine whether they were fired from the same weapon.

Automated Firearms Search Systems

The use of firearms, especially semiautomatic weapons, during the commission of a crime has significantly increased throughout the United States. Because of the expense of such firearms, the likelihood that a specific weapon will be used in multiple crimes has risen. The advent of computerized imaging technology has made possible the storage of bullet and cartridge surface characteristics in a manner analogous to the storage of automated fingerprint files. Using this concept, crime laboratories can be networked, allowing them to share information on bullets and cartridges retrieved from several jurisdictions.

Early Systems

The effort to build a national computerized database for firearms evidence in the United States had a rather confusing and inefficient start in the early 1990s. Two major federal law enforcement agencies, the FBI and the Bureau of Alcohol, Tobacco, Firearms and Explosives (ATF), offered the law enforcement community competing and incompatible computerized systems.

The automated search system developed for the FBI was known as DRUGFIRE. This system emphasized the examination of unique markings on the cartridge casings expended by the weapon. The specimen was analyzed through a microscope attached to a video camera. The magnification allowed for a close-up view to identify individual characteristics. The image was captured by a video camera, digitized, and stored in a database. Although DRUGFIRE emphasized cartridge-case imagery, the images of highly characteristic bullet striations could also be stored in a like manner for comparisons.

The *Integrated Ballistic Identification System (IBIS)*, developed for the Bureau of Alcohol, Tobacco, Firearms and Explosives, processed digital microscopic images of identifying features found on both expended bullets and cartridge casings. IBIS incorporated two software programs: Bulletproof, a bullet-analyzing module, and Brasscatcher, a cartridge-case-analyzing module. A schematic diagram of Bulletproof's operation is depicted in Figure 17–15.

NIBIN In 1999, members of the FBI and ATF joined forces to introduce the *National Integrated Ballistics Information Network (NIBIN)* program to the discipline of firearms examination. The new unified system incorporates both DRUGFIRE and IBIS technologies available in prior years. ATF has overall responsibility for the system sites, whereas the FBI is responsible for the communications network.

Agencies using the new NIBIN technology produce database files from bullets and cartridge casings retrieved from crime scenes or test fires from retrieved firearms. More than two hundred law enforcement agencies worldwide have adapted to this technology. The success of the system has been proven with more than 800,000 images compiled; nationwide, law enforcement agencies have connected more than 28,000 bullets and casings to more than one crime (see Figure 17–16).

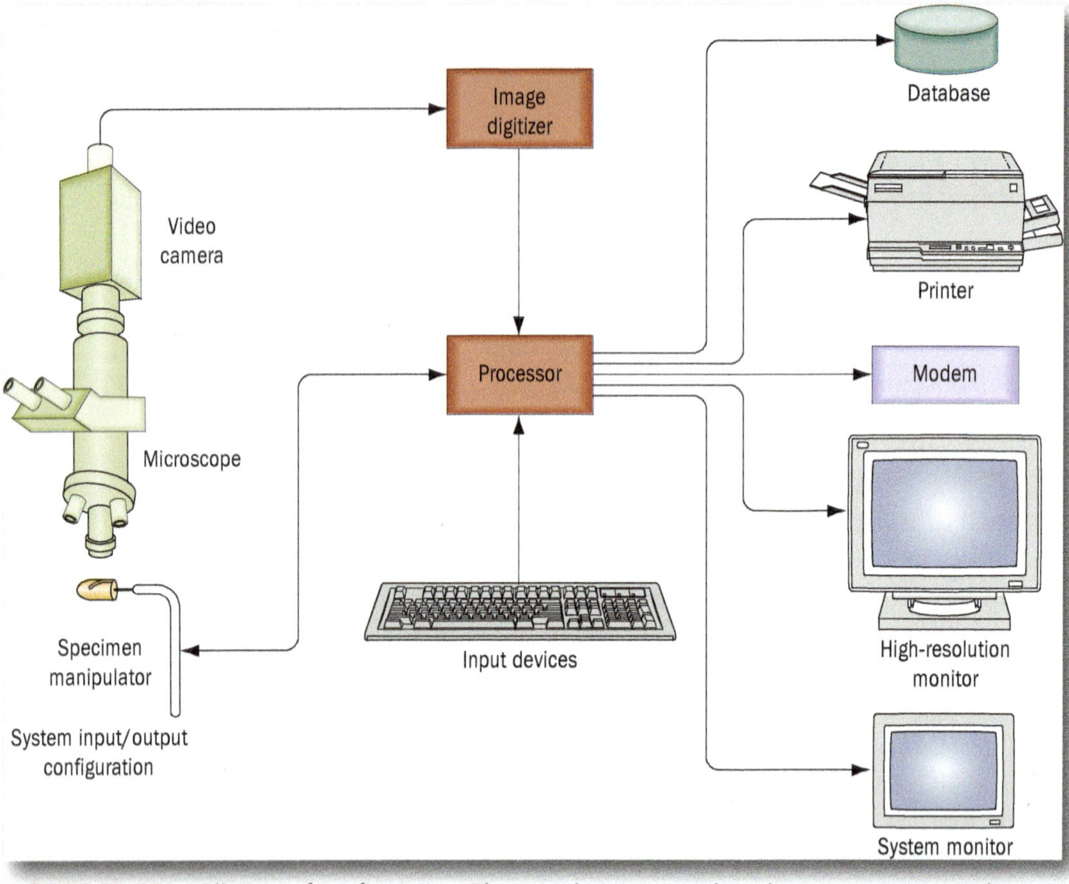

FIGURE 17–15 Bullet-proof configuration. The sample is mounted on the specimen manipulator and illuminated by the light source from a microscope. The image is captured by a video camera and digitized. The digital image is then stored in a database, available for retrieval and comparison. The search for a match includes analyzing the width of land and groove impressions along with both rifling and individual characteristics. The Brasscatcher software uses the same system configuration but emphasizes the analysis of expended cartridge casings rather than the expended bullets. *Courtesy Forensic Technology (WAI) Inc., Côte Quebec, Canada*

For example, in a recent case, a Houston security guard was shot and killed during a botched armed robbery. A bullet and .40-caliber Smith & Wesson cartridge casing were recovered and imaged into NIBIN. Earlier that day, a robbery-turned-double-homicide left two store clerks dead. Again, two bullets and two .40-caliber Smith & Wesson cartridge casings were recovered. Once they were processed into NIBIN, a correlation was found with the murder of the security officer and a separate aggravated robbery that occurred two weeks earlier. All three crimes were linked with a firearm believed to be a .40-caliber Smith & Wesson pistol.

Further investigation into the use of a victim's credit card helped police locate two suspects. In the possession of one suspect was a .40-caliber Smith & Wesson pistol. The gun was test-fired and imaged into NIBIN. The casing from the test-fired weapon matched the evidence obtained in the robbery and the aggravated robbery-homicides. A firearms examiner verified the associations by traditional comparisons. Before this computerized technology was developed, it would have taken years, or may have been impossible, to link all of these shootings to a single firearm.

In another example, the ATF laboratory in Rockville, Maryland, received 1,466 cartridge casings from the Ovcara mass burial site in Bosnia. After processing and imaging profiles for all casings, the examiners determined that 18 different firearms were used at the site. With the help of NIBIN technology and competent examiners, jurists were able to try and convict an individual for war crimes.

FIGURE 17–16 Bullets A, B, C, and D were acquired in the IBIS database at different times from different crime scenes. D is a fragmented bullet that had only three land impressions available for acquisition. Upon the entry of bullet D, IBIS found a potential matching candidate in the database: B. On the far right, bullet D is compared to bullet B using IBIS imaging software. Finally, a forensic firearms examiner using the actual evidence under a conventional comparison microscope will confirm the match between B and D. *Courtesy Forensic Technology (WAI) Inc., Côte St-luc, Quebec, Canada*

NIBIN serves only as a screening tool for firearms evidence. A computerized system does not replace the skills of the firearms examiner. NIBIN can screen hundreds of unsolved firearms cases and may narrow the possibilities to several firearms. However, the final comparison will be made by a forensic examiner through traditional microscopic methods.

Ballistic Fingerprinting

Participating crime laboratories in the United States are building databases of bullet and cartridge cases found at crime scenes and those fired in tests of guns seized from criminals. As these databases come online and prove their usefulness in solving crimes, law enforcement officials and the political community are scrutinizing the feasibility of scaling this concept up to create a system of *ballistic fingerprinting*.

This system would entail the capture and storage of appropriate markings on bullets and cartridges test-fired from handguns and rifles before they are sold to the public. Questions regarding who will be responsible for collecting the images and details of how will they be stored are but two of many issues to be determined. The concept of ballistic fingerprinting is an intriguing one for the law enforcement community and promises to be explored and debated intensely in the future.

Quick Review

- The advent of computerized imaging technology has made possible the storage of bullet and cartridge surface characteristics in a manner analogous to automated fingerprint files.
- Two automated firearms search systems are DRUGFIRE, developed by the FBI, and IBIS, developed by the ATF.
- NIBIN is the *National Integrated Ballistics Information Network*, a unified firearms search system that incorporates both DRUGFIRE and IBIS technologies.

Gunpowder Residues

Modern ammunition is propelled toward a target by the expanding gases created by the ignition of smokeless powder or nitrocellulose in a cartridge. Under ideal circumstances, all of the powder is consumed in the process and converted into the rapidly expanding gases. However, in practice the powder is never totally burned. When a firearm is discharged, unburned and partially burned particles of gunpowder in addition to smoke are propelled out of the barrel along with the bullet toward the target. If the muzzle of the weapon is sufficiently close, these products are deposited onto the target. The distribution of gunpowder particles and other discharge residues around the bullet hole permits a **distance determination**, an assessment of the distance from which a handgun or rifle was fired.

distance determination The process of determining the distance between the firearm and a target, usually based on the distribution of powder patterns or the spread of a shot pattern

Distance Determination

In incidents involving gunshot wounds, it is often necessary to determine the distance from which the weapon was fired. For example, in incidents involving a shooting death, the suspect often pleads self-defense as the motive for the attack. Such claims are fertile grounds for distance determinations, because finding the proximity of the people involved is necessary to establish the facts of the incident. Similarly, careful examination of the wounds of suicide victims usually reveals characteristics associated with a very close-range gunshot wound. The absence of such characteristics strongly indicates that the wound was not self-inflicted and signals the possibility of foul play.

The accuracy of a distance determination varies according to the circumstances of the case. When the investigator is unable to recover a suspect weapon, the best that the examiner can do is to state whether a shot could have been fired within some distance from the target. More exact opinions are possible only when the examiner has the suspect weapon in hand and has knowledge of the type of ammunition used in the shooting.

Handguns and Rifles The precise distance from which a handgun or rifle has been fired must be determined by carefully comparing the powder residue pattern on the victim's clothing or skin to test patterns made when the suspect weapon is fired at varying distances from a target. A white cloth or a fabric comparable to the victim's clothing may be used as a test target (see Figure 17–17). Because the spread and density of the residue pattern vary widely between weapons and ammunition, such a comparison is significant only when it is made with the suspect weapon and suspect ammunition, or with ammunition of the same type and make. By comparing the test and evidence patterns, the examiner may find enough similarity in shape and density on which to judge the distance from which the shot was fired.

Without the weapon, the examiner is restricted to looking for recognizable characteristics around the bullet hole. Such findings are at best approximations made as a result of general observations and the examiner's experience. However, some noticeable characteristics should be sought. For instance, when the weapon is held in contact with or less than 1 inch from the target, a heavy concentration of smokelike vaporous lead usually surrounds the bullet entrance hole. Often

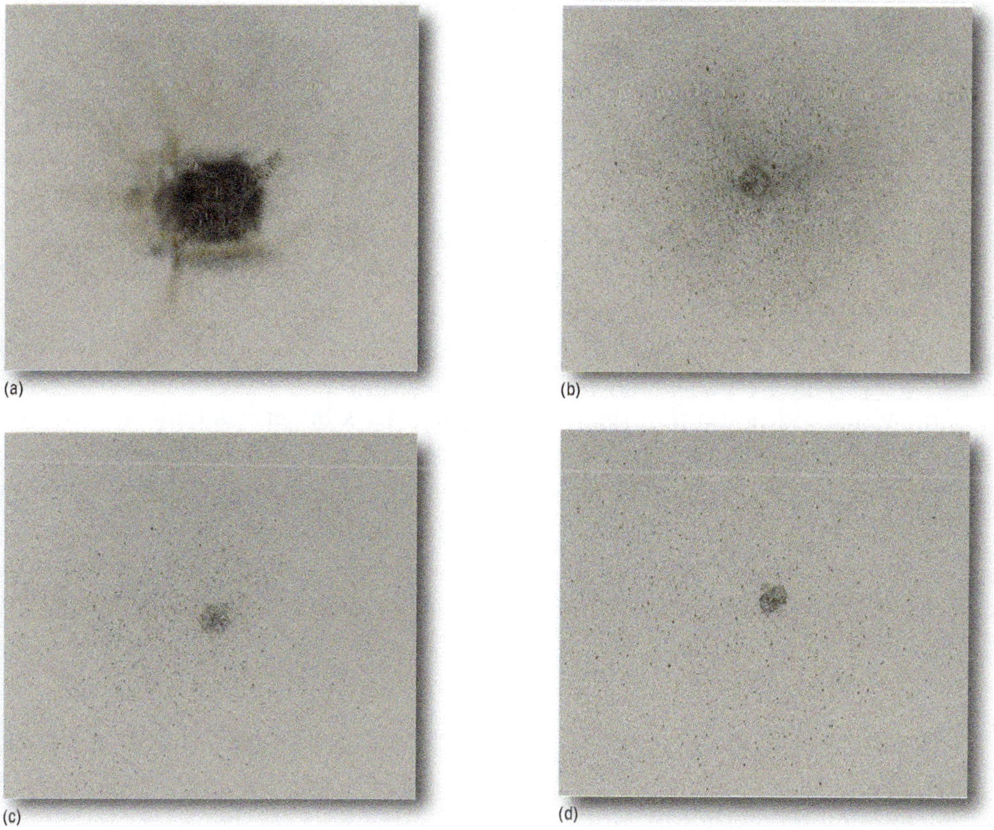
(a) (b) (c) (d)

FIGURE 17–17 Test powder patterns made with a Glock 9mm luger fired at the following distances: (a) contact, (b) 6 inches, (c) 12 inches, and (d) 18 inches. *Courtesy Michelle D. Miranda, MS, D-ABC*

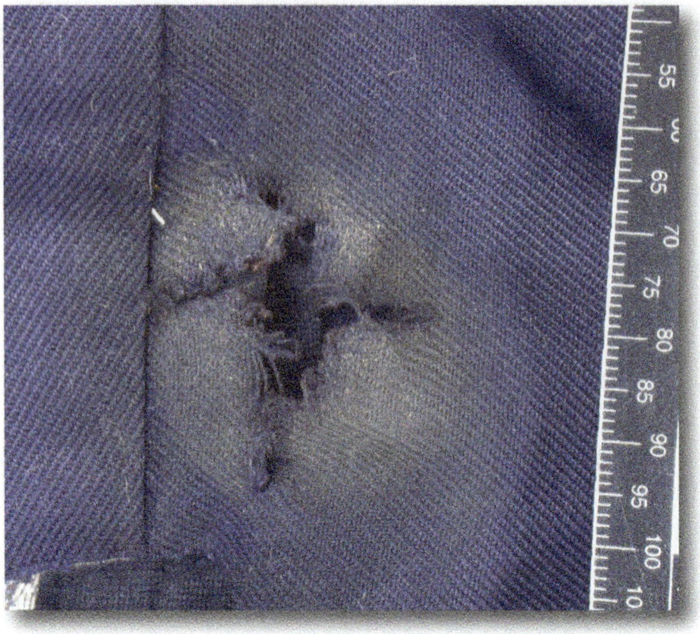

FIGURE 17–18
A contact shot. *Courtesy Michelle D. Miranda, MS, D-ABC*

loose fibers surrounding a contact hole show scorch marks from the flame discharge of the weapon, and some synthetic fibers may show signs of being melted as a result of the heat from the discharge. Furthermore, the blowback of muzzle gases may produce a stellate (star-shaped) tear pattern around the hole. Such a hole is invariably surrounded by a rim of a smokelike deposit of vaporous lead (see Figure 17–18).

A halo of vaporous lead (smoke) deposited around a bullet hole normally indicates a discharge 12–18 inches or less from the target. The presence of scattered specks of unburned and partially burned powder grains without any accompanying soot can often be observed at distances up to approximately 25 inches. Occasionally, however, scattered gunpowder particles are noted at a firing distance as far out as 36 inches. With ball powder ammunition, this distance may be extended to 6–8 feet.

Finally, a weapon that has been fired more than 3 feet from a target usually does not deposit any powder residues on the target's surface. In these cases, the only visual indication that the hole was made by a bullet is a dark ring, known as bullet wipe, around the perimeter of the entrance hole. Bullet wipe consists of a mixture of carbon, dirt, lubricant, primer residue, and lead wiped off the bullet's surface as it passes through the target. Again, in the absence of a suspect weapon, these observations are general guidelines for estimating target distances. Numerous factors—barrel length, caliber, type of ammunition, and type and condition of the weapon fired—influence the amount of gunpowder residue deposited on a target.

Shotguns The determination of firing distances involving shotguns must also be related to test firings performed with the suspect weapon, using the same type of ammunition known to be used in the crime. In the absence of a weapon, the muzzle-to-target distance can be estimated by measuring the spread of the discharged shot. With close-range shots varying in distance up to 4–5 feet, the shot charge enters the target as a concentrated mass, producing a hole somewhat larger than the bore of the barrel. As the distance increases, the pellets progressively separate and spread out. Generally speaking, the spread in the pattern made by a 12-gauge shotgun increases 1 inch for each yard of distance. Thus, a 10-inch pattern would be produced at approximately 10 yards.

Of course, this is only a rule of thumb; normally, many variables can affect the shot pattern. Other factors include the barrel length, the size and quantity of the pellets fired, the quantity of powder charge used to propel the pellets, and the choke of the gun under examination. **Choke** is the degree of constriction placed at the muzzle end of the barrel. The greater the choke, the narrower the shotgun pattern and the faster and farther the pellets will travel.

choke
An interior constriction placed at or near the muzzle end of a shotgun's barrel to control shot dispersion

Powder Residues on Garments

When garments or other evidence relevant to a shooting are received in the crime laboratory, the surfaces of all items are first examined microscopically for gunpowder residue. These particles may be identifiable by their characteristic colors, sizes, and shapes. However, the absence of visual indications does not preclude the possibility that gunpowder residue is present. Sometimes the lack of color contrast between the powder and garment or the presence of heavily encrusted deposits of blood can obscure the visual detection of gunpowder. Often, an infrared photograph of the suspect area overcomes the problem. Such a photograph may enhance the contrast, thus revealing vaporous lead and powder particles deposited around the hole (see Figure 17–19). In other situations, this may not help, and the analyst must use chemical tests to detect gunpowder residues.

Nitrites are one type of chemical product that results from the incomplete combustion of smokeless (nitrocellulose) powder. One test method for locating powder residues involves transferring particles embedded on the target surface to chemically treated gelatin-coated photographic paper. This procedure is known

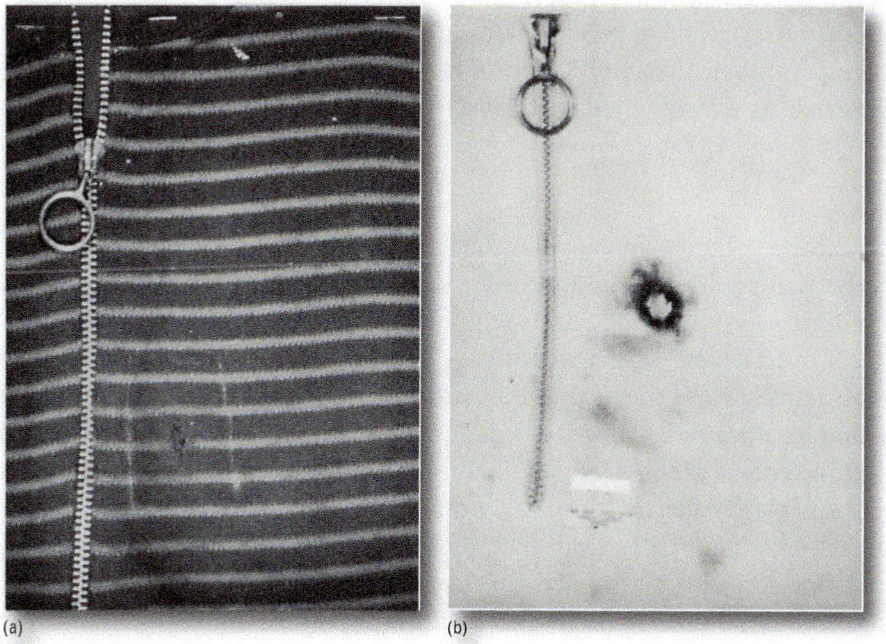

FIGURE 17–19 (a) A shirt bearing a powder stain, photographed under normal light. (b) Infrared photograph of the same shirt. *Courtesy Richard Saferstein, Ph.D.*

Greiss test
A chemical test used to develop patterns of gunpowder residues around bullet holes

as the **Greiss test**. The examiner presses the photographic paper onto the target with a hot iron; once the nitrite particles are on the paper, they are made easily visible by chemical treatment. In addition, comparing the developed nitrite pattern to nitrite patterns obtained from test firings at known distances can be useful in determining the shooting distance from the target. A second chemical test is then performed to detect any trace of lead residue around the bullet hole. The questioned surface is sprayed with a solution of sodium rhodizonate, followed by a series of oversprays with acid solutions. This treatment turns lead particles pink, followed by blue-violet.

Quick Review

- The distribution of gunpowder particles and other discharge residues around a bullet hole permits an assessment of the distance from which a handgun or rifle was fired.
- The precise distance is determined by carefully comparing the powder residue pattern on the victim's clothing to test patterns made when the suspect weapon is fired at varying distances from a target.
- The Greiss test is a chemical test used to develop patterns of gunpowder residues around bullet holes. It tests for the presence of nitrates.

Primer Residues on the Hands

The firing of a weapon not only propels residues toward the target, but also blows gunpowder and primer residues back toward the shooter (see Figure 17–20). As a result, traces of these residues are often deposited on the firing hand of the shooter, and their detection can provide valuable information as to whether an individual has recently fired a weapon.

Detecting Primer Residues

Early efforts at demonstrating powder residues on the hands centered on chemical tests that could detect unburned gunpowder or nitrates. For many years, the *dermal nitrate test* was popular. It required the application of hot paraffin or wax to the suspect's hand with a paintbrush. After drying into a solid crust, the paraffin was removed and tested with diphenylamine. A blue color indicated a positive reaction for nitrates. However, the dermal nitrate test has fallen into disfavor with law enforcement agencies, owing mainly to its lack of specificity. Common materials such as fertilizers, cosmetics, urine, and tobacco all give positive reactions that are indistinguishable from that obtained for gunpowder by this test.

Efforts to identify a shooter now center on the detection of primer residues deposited on the hand of a shooter at the time of firing. With the exception of most .22-caliber ammunition, primers currently manufactured contain a blend of lead styphnate, barium nitrate, and antimony sulfide. Residues from these

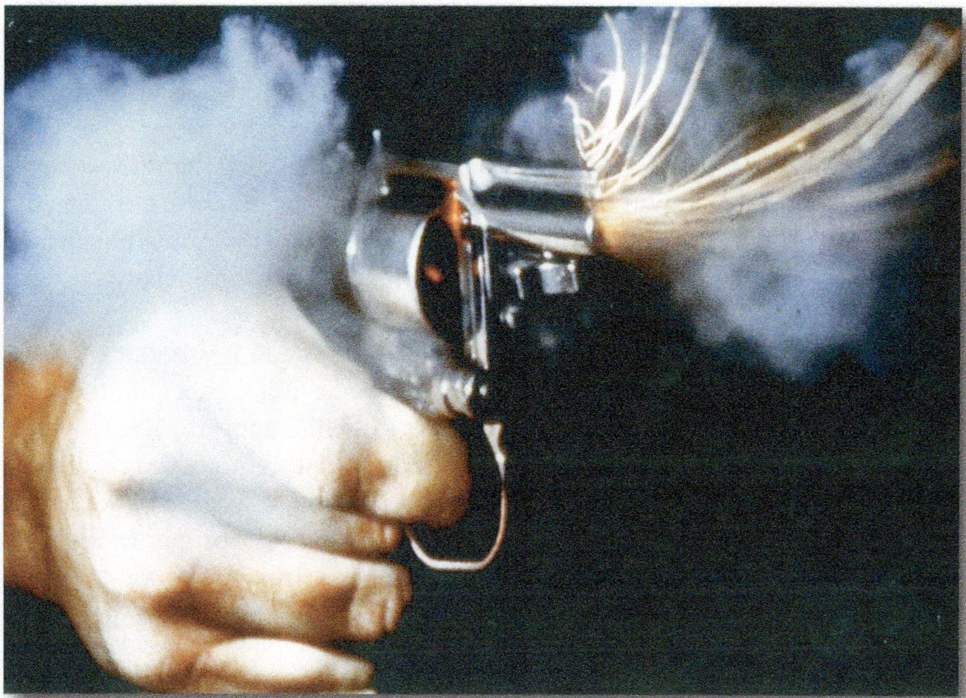

FIGURE 17–20 When a handgun is fired, gunpowder and primer residues are normally blown back toward the hand of the shooter. *Courtesy Forensic Technology Inc.*

materials are most likely to be deposited on the thumb web and the back of the firing hand of a shooter, because these areas are closest to gases escaping along the side or back of the gun during discharge. In addition, individuals who handle a gun without firing it may have primer residues deposited on the palm of the hand coming in contact with the weapon.

However, with the handling of a used firearm, the passage of time, and the resumption of normal activities following a shooting, gunshot residues from the back of the hand are frequently redistributed to other areas, including the palms. Therefore, it is not unusual to find higher levels of barium and antimony on the palms than on the backs of the hands of known shooters. Another possibility is the deposition of significant levels of barium and antimony on the hands of an individual who is near a firearm when it is discharged.

Tests for Primer Residues

Determination of whether a person has fired or handled a weapon or has been near a discharged firearm is normally made by measuring the presence and possibly the amount of barium and antimony on the relevant portions of the suspect's hands. A variety of materials and techniques are used for removing these residues. The most popular approach, and certainly the most convenient for the field investigator, is to apply an adhesive tape or adhesive to the hand's surface in order to remove any adhering residue particles.

Swabbing Another approach is to remove any residues present by swabbing both the firing and nonfiring hands with cotton that has been moistened with 5 percent nitric acid. The front and back of each hand are separately swabbed.

All four swabs, along with a moistened control, are then forwarded to the crime laboratory for analysis.

In any case, once the hands are treated for the collection of barium and antimony, the collection medium must be analyzed for the presence of these elements. High barium and antimony levels on the suspect's hand(s) strongly indicate that the person fired or handled a weapon or was near a firearm when it was discharged. Because these elements are normally present in small quantities (less than 10 micrograms) after a firing, only the most sensitive analytical techniques can be used to detect them.

Unfortunately, even though most specimens submitted for this type of analysis have been from individuals strongly suspected of having fired a gun, there has been a low rate of positive findings. The major difficulty appears to be the short time that primer residues remain on the hands. These residues are readily removed by intentional or unintentional washing, rubbing, or wiping of the hands. In fact, one study demonstrated that it is difficult to detect primer residues on cotton hand swabs taken as soon as two hours after firing a weapon.[2] Hence, some laboratories do not accept cotton hand swabs taken from living people six or more hours after a firing has occurred.

In cases that involve suicide victims, a higher rate of positives for the presence of gunshot residue is obtained when the hand swabbing is conducted before the person's body is moved or when the hands are protected by paper bags.[3] However, hand swabbing or the application of an adhesive cannot be used to detect firings with most 22-caliber rim-fire ammunition. Such ammunition may contain only barium or neither barium nor antimony in its primer composition.

SEM Testing Most laboratories that can detect gunshot residue require application of an adhesive to the shooter's hands. Microscopic primer and gunpowder particles on the adhesive are then found with a scanning electron microscope (SEM). Adhesive containing stubs are applied against the surface to be sampled by repeatedly dabbing (see Figure 17–21). A minimum of 20–30 dabs should be applied to the surface. The characteristic size and shape of these particles

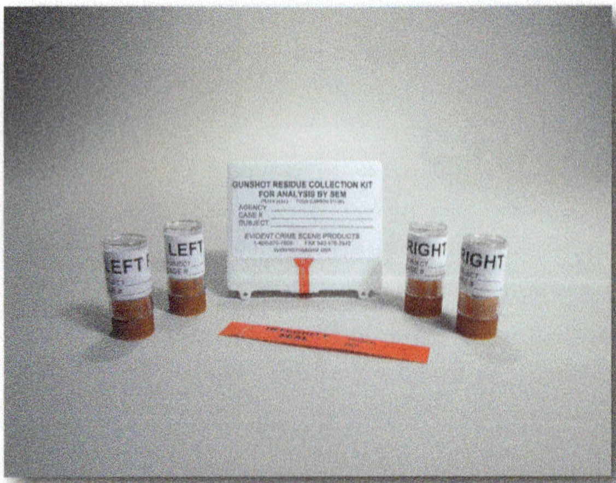

(a)

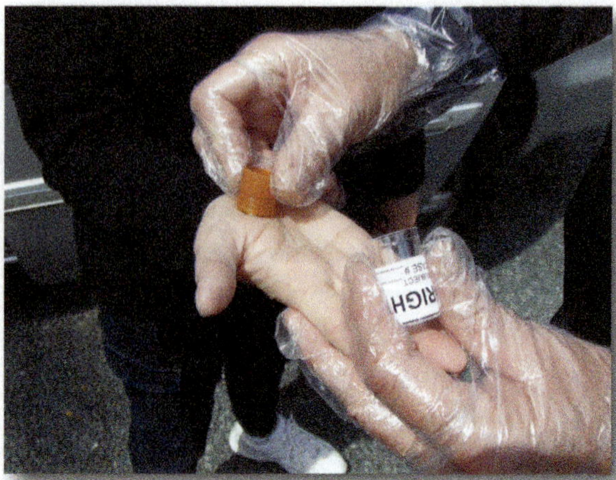

(b)

FIGURE 17–21 (a) Adhesive stubs used to sample a suspect's shooter's hands. (b) Sampling a suspect's hand for gunshot residue with an adhesive stub. *Courtesy Evident, Union Hall, VA 24176-4025, www.evidentcrimescene.com*

distinguishes them from other contaminants on the hands (see Figure 17–22). When the SEM is linked to an X-ray analyzer, an elemental analysis of the particles can be conducted. A finding of a select combination of elements (lead, barium, and antimony) confirms that the particles were indeed primer residue (see Figure 17–23).

The major advantage of the SEM approach for primer residue detection is its enhanced specificity over hand swabbing. The SEM characterizes primer particles by their size and shape as well as by their chemical composition. Unfortunately, the excessive operator time required to find and characterize gunshot residue has deterred this technique's use. The availability of automated particle search and identification systems for use with scanning electron microscopes may overcome this problem. Results of work performed with automated systems show it to be significantly faster than a manual approach for finding gunshot residue particles.

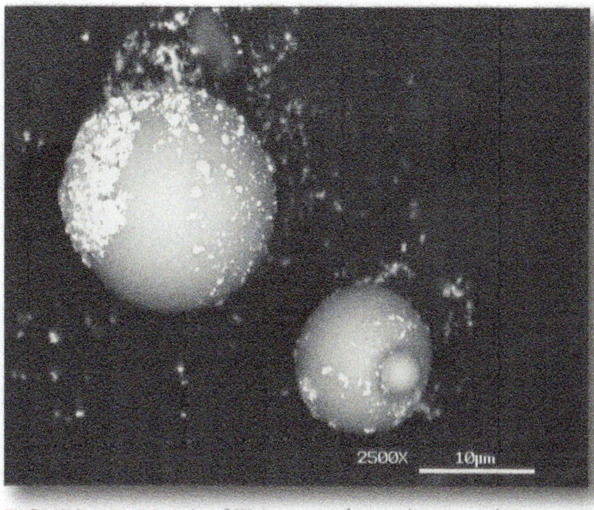

FIGURE 17–22 An SEM view of gunshot residue particles. *Courtesy JEOL USA*

Other Primer Residue Tests Neutron activation analysis and flameless atomic absorption spectrophotometry are analytical methods sensitive enough to detect barium and antimony in gunshot residues in hand swabs. However, the need for access to a neutron source, expensive counting equipment, and extensive regulatory requirements limit neutron activation analysis technology to a small number of crime laboratories. On the other hand, flameless atomic absorption spectrophotometry can be purchased at a cost well within the budgets of most crime laboratories, and a number of laboratories use this instrument to detect barium and antimony on a shooter's hands.

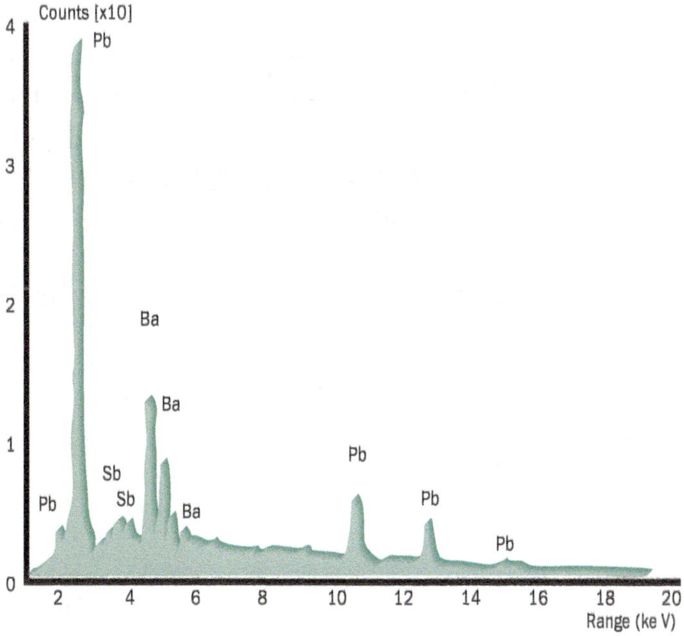

FIGURE 17–23 Spectrum showing the presence of lead, barium, and antimony in gunshot residue. *Courtesy Jeol USA Inc.*

Inside the Science

Atomic Absorption Spectrophotometry

When an atom is vaporized, it absorbs many of the same frequencies of light that it emits in an excited state. Selective absorption of light by atoms is the basis for a technique known as *atomic absorption spectrophotometry*. A simple atomic absorption spectrophotometer is illustrated in the figure.

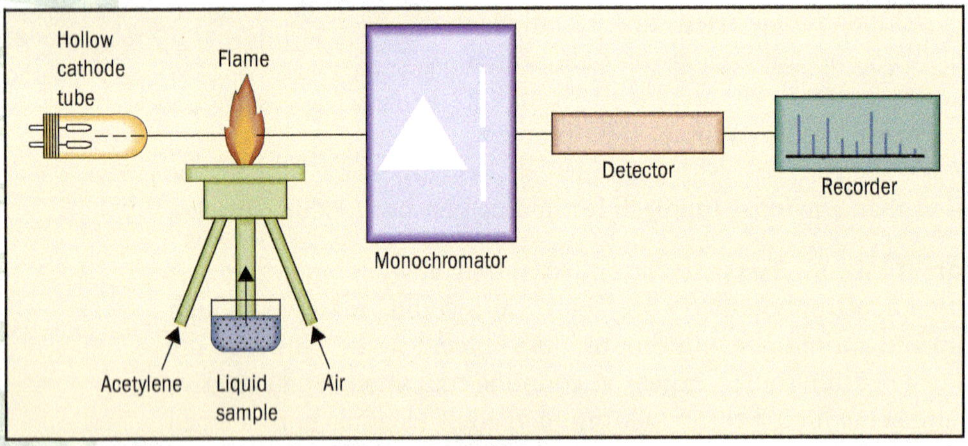

Parts of a simple flame atomic absorption spectrophotometer.

In atomic absorption spectrophotometry, the specimen is heated to a temperature that is hot enough to vaporize its atoms while leaving a substantial number of atoms in an unexcited state. Normally the specimen is inserted into an air-acetylene flame to achieve this temperature. The vaporized atoms are then exposed to radiation emitted from a light source. The technique achieves great specificity by using as its radiation source a discharge tube made of the same element being analyzed in the specimen.

When the discharge lamp is turned on, it emits only the frequencies of light that are present in the emission spectrum of the element. Likewise, the sample absorbs these frequencies only when it contains the same element. Therefore, to determine the presence of antimony in a specimen, the atomic absorption spectrophotometer must be fitted with a discharge lamp that is constructed of antimony. Under these conditions, the sample absorbs light only if it contains antimony.

Once the radiation has passed through the sample, a monochromator, consisting of a prism or a diffraction grating and a slit, isolates the desired radiation frequency and transmits it to a detector. The detector converts the light into an electrical signal, the intensity of which is recorded on a strip-chart recorder.

Atomic absorption spectrophotometry measures the absorption of light by the element of interest. The concentration of the absorbing element is directly proportional to the quantity of the light absorbed. The higher the concentration of the element, the more light is absorbed. For this reason, atomic absorption spectroscopy is most useful for accurately determining an element's concentration in a sample. Furthermore, the technique is sufficiently sensitive to find wide application in detecting and quantitating elements that are present in trace levels. However, one drawback is that the analyst can determine only one element at a time, each time having to select the proper lamp to match the element under investigation.

Although atomic absorption spectrophotometry has been used for chemical analysis since 1955, it has not yet found wide application for solving forensic problems. However, a modification in the design of the instrument may change this situation. By substituting a heated graphite furnace or a heated strip of metal (tantalum) for the flame, analysts have achieved a more efficient means of atomic volatilization and as a result have substantially increased the sensitivity of the technique. Many elements can now be detected at levels that approach one-trillionth of a gram.

The high sensitivity of "flameless" atomic absorption now equals or surpasses that of most known analytical procedures. Considering the relative simplicity and low cost of the technique, atomic absorption spectrophotometry has become an attractive method for detecting and measuring the smallest levels of trace elements in physical evidence.

As far as atoms are concerned, energy is a two-way street. Energy can be put into the atom at the same time that energy is given off; what goes in must come out. The chemist can study the atom using either approach. Atomic absorption spectrophotometry carefully measures the value and amount of light energy going into the atom; emission spectroscopy collects and measures the various light energies given off. The result is the same: atoms are identified by the existence of characteristic energy levels.

Quick Review

- Firing a weapon propels residues toward the target and blows gunpowder and primer residues back toward the shooter. Traces of these residues are often deposited on the firing hand of the shooter, providing valuable information as to whether an individual has recently fired a weapon.
- Examiners measure the amount of barium and antimony on the relevant portion of the suspect's hands or characterize the morphology of particles containing these elements to determine whether a person has fired or handled a weapon or was near a discharged firearm.

Serial Number Restoration

Today, many manufactured items, including automobile engine blocks and firearms, are impressed with a serial number for identification. Increasingly the criminalist must restore such a number when it has been removed or obliterated by grinding, rifling, or punching.

Serial numbers are usually stamped on a metal body or frame, or on a plate, with hard steel dies. These dies strike the metal surface with a force that allows each digit to sink into the metal at a prescribed depth. Serial numbers can be restored because the metal crystals in the stamped zone are placed under a permanent strain that extends a short distance beneath the original numbers. When a suitable etching agent is applied, the strained area dissolves faster than the unaltered metal, thus revealing the etched pattern in the form of the original numbers (see Figure 17–24). However, if the zone of strain has been removed, or if the area has been impressed with a different strain pattern, the number usually cannot be restored.

Before any treatment with the etching reagent, the obliterated surface must be thoroughly cleaned of dirt and oil and polished to a mirrorlike finish. The reagent is swabbed onto the surface with a cotton ball. The choice of etching reagent depends on the type of metal surface being worked on. A solution of hydrochloric acid (120 milliliters), copper chloride (90 grams), and water (100 milliliters) generally works well for steel surfaces.

FIGURE 17–24 Obliterated or altered serial numbers on firearms can be restored by analysts using chemical means. *Courtesy Federal Bureau of Investigation*

Collection and Preservation of Firearms Evidence

Firearms

The Hollywood technique of picking up a weapon by its barrel with a pencil or stick in order to protect fingerprints must be avoided. This practice only disturbs powder deposits, rust, or dirt lodged in the barrel, and consequently may alter the striation markings on test-fired bullets. If recovery of latent fingerprints is a primary concern, hold the weapon by the edge of the trigger guard or by the checkered portion of the grip, which usually does not retain identifiable fingerprints.

The most important consideration in handling a weapon is safety. Before any weapon is sent to the laboratory, all precautions must be taken to prevent an accidental discharge of a loaded weapon in transit. In most cases, it will be necessary to unload the weapon. If this is done, first a record should be made of the weapon's hammer and safety position; likewise, the location of all fired and unfired ammunition in the weapon must be recorded.

When a revolver is recovered, the chamber position should be indicated by a scratch mark on the cylinder where it aligns with the barrel. Each chamber is designated a number on a diagram, and as each cartridge or casing is removed, it should be marked to correspond to the number of its corresponding chamber in the diagram. Knowledge of the cylinder position of a cartridge casing may be useful for later determination of the sequence of events, particularly in shooting cases, when more than one shot was fired. Each round should be placed in a separate box or envelope. If the weapon is an automatic, the magazine must be removed and checked for prints and the chamber then emptied.

As with any other type of physical evidence recovered at a crime scene, firearms evidence must be marked for identification, and a chain of custody must be established. When a firearm is recovered, an identification tag should be attached to the trigger guard. The tag should include appropriate identifying data, including the weapon's serial number, make, and model, and the investigator's initials.

When a weapon is recovered from an underwater location, no effort should be made to dry or clean it. Instead, the firearm should be transported to the laboratory in a receptacle containing enough of the same water to keep it submerged. This procedure prevents rust from developing during transport.

Ammunition

Protection of class and individual markings on bullets and cartridge cases must be the primary concern of the field investigator. Thus, extreme caution is needed when removing a lodged bullet from a wall or other object. If the bullet's surface is accidentally scratched during this operation, valuable striation markings could be obliterated. It is best to free bullets from their target by carefully breaking away the surrounding support material while avoiding direct contact with the projectile.

Bullets, cartridge casings, and discharged shells from shotguns should just be placed in a container that is appropriately marked for identification. It is recommended that the investigator not directly mark these items with a scribe. In any case, the investigator must protect the bullet by wrapping it in tissue paper before placing it in a pillbox or an envelope for shipment to the crime laboratory. Minute traces of evidence such as paint and fibers may be adhering to the bullet; the investigator must take care to leave these trace materials intact.

When semiautomatic or automatic weapons have been fired, the ejection pattern of the casings can help establish the relationship of the suspect to their victim. For this reason, the investigator must note the exact location where a shell casing was recovered.

In incidents involving shotguns, any wads recovered are to be packaged and sent to the laboratory. An examination of the size and composition of the wad may reveal information about the type of ammunition used and the gauge of the shotgun.

Gunpowder Deposits

The clothing of a firearms victim must be carefully preserved to prevent damage or disruption to powder residues deposited around a bullet or shell hole. Cutting or tearing of clothing in the area of the holes must be avoided when removing the clothing. All wet clothing should be air-dried out of direct sunlight and then folded carefully so as not to disrupt the area around the bullet hole. Each item should be placed in a separate paper bag.

Trajectory Analysis of Bullet Strikes

In order to determine the trajectory of a bullet strike, the bullet hole must be examined. Bullets entering at a 90° angle will create a circular entry hole. Bullets entering at an angle will create an elliptical entry hole. The first step in calculating the bullet's trajectory is to determine the impact angle. Using basic trigonometry, the trajectory of the bullet can be calculated.

Quick Review

- Criminalists can restore serial numbers removed or obliterated by grinding, rifling, or punching.
- Because the metal crystals in the stamped zone are placed under a permanent strain that extends a short distance beneath the original numbers, the serial number can be restored through chemical etching.
- A suspect firearm should never be picked up by inserting an object into its barrel because this practice may alter the striation markings on test-fired bullets.
- Before unloading a suspect weapon, the weapon's hammer and safety position should be recorded, as well as the location of all fired and unfired ammunition in the weapon.
- Protection of class and individual markings on bullets and cartridge cases is the primary concern of the field investigator when recovering bullets and cartridge casings.
- Bullet strike trajectory can be determined by examining the shape of the entry. The angle of entry can be calculated using basic trigonometry.

Tool Marks

A tool mark is any impression, cut, gouge, or abrasion caused by a tool coming into contact with another object. Most often, tool marks are encountered at burglary scenes that involve forcible entry into a building or safe. Generally, these marks occur as indented impressions into a softer surface or as abrasion marks caused by the tool cutting or sliding against another object. Successful forensic analysis relies on careful recovery and documentation of the tool marks.

Comparing the Class of Tool Marks

Typically, an indented impression is left on the frame of a door or window as a result of the prying action of a screwdriver or crowbar. Careful examination of these impressions can reveal important class characteristics—that is, the size and shape of the tool. However, they rarely reveal any significant individual characteristics that could permit the examiner to individualize the mark to a single tool. Such characteristics, when they do exist, usually take the form of discernible random nicks and breaks that the tool has acquired through wear and use (Figure 17–25).

Just as the machined surfaces of a firearm are impressed with random striations during its manufacture, the edges of a pry bar, chisel, screwdriver, knife, or cutting tool likewise display a series of microscopic irregularities that look like ridges and valleys. Such markings are left as a result of the machining processes used to cut and finish tools. The shape and pattern of such minute imperfections are further modified by damage and wear during the life of the tool. Considering the variety of patterns that the hills and valleys can assume, it is highly unlikely that any two tools will be identical. Hence, these minute imperfections impart individuality to each tool.

If the edge of a tool is scraped against a softer surface, it may cut a series of striated lines that reflect the pattern of the tool's edge. Markings left in this manner are compared in the laboratory through a comparison microscope with test tool marks made from the suspect tool. The result can be a positive comparison, and hence a definitive association of the tool with the evidence mark, when a sufficient quantity of striations match between the evidence and test markings.

A major problem of tool mark comparisons is the difficulty in duplicating in the laboratory the tool mark left at the crime scene. A thorough comparison

FIGURE 17–25 A comparison of a tool mark with a suspect screwdriver. Note how the presence of nicks and breaks on the tool's edge helps individualize the tool to the mark. *Courtesy Richard Saferstein, Ph.D.*

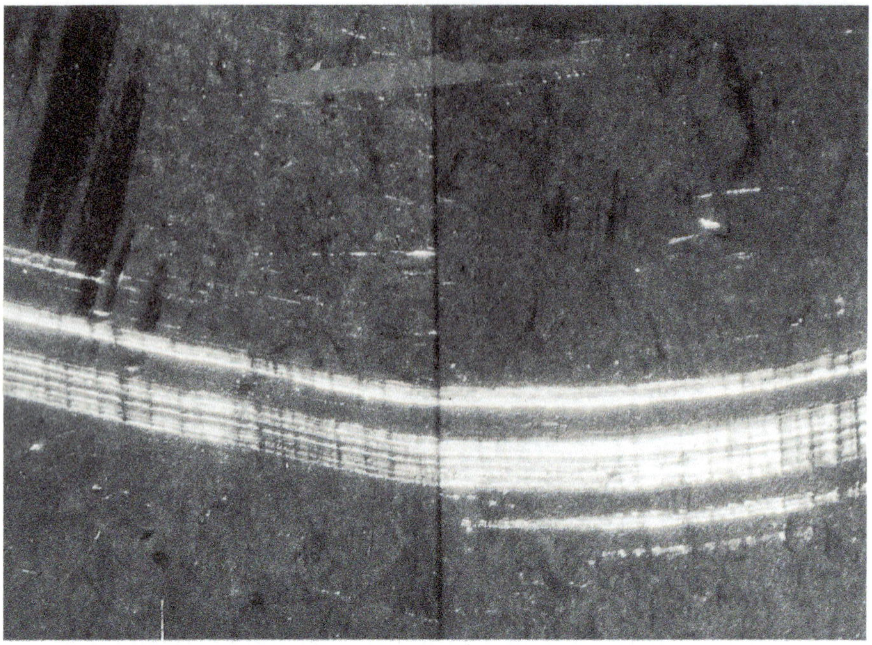

FIGURE 17–26 A photomicrograph of a tool mark comparison seen under a comparison microscope. *Courtesy Leica Microsystems, Buffalo, N.Y., www.leicamicrosystems.com*

requires preparing a series of test marks by applying the suspect tool at various angles and pressures to a soft metal surface (lead is commonly used). This approach gives the examiner ample opportunity to duplicate many of the details of the original evidence marking. A photomicrograph of a typical tool mark comparison is illustrated in Figure 17–26.

Collecting Tool Mark Evidence

Whenever practical, to perform recovery and documentation techniques on tool mark evidence, the entire object or the part of the object bearing a tool mark should be submitted to the crime laboratory for examination. When removal of the tool mark is impractical, the only recourse is photographing the marked area to scale and making a cast of the mark. Liquid silicone casting material is best for reproducing most of the fine details of the mark (see Figure 17–27). However, even under the best conditions, the clarity of many of the tool mark's minute details will be lost or obscured in a photograph or cast. Of course, this will reduce the chance of individualizing the mark to a single tool.

The crime-scene investigator must never attempt to fit the suspect tool into the tool mark. Any contact between the tool and the marked surface may alter the mark and will, at the least, raise serious questions about the integrity of the evidence. The suspect tool and mark must be packaged in separate containers, with every precaution taken to avoid contact between the tool or mark and another hard surface. Failure to protect the tool or mark from damage could result in the destruction of its individual characteristics.

Furthermore, the tool or its impression may contain valuable trace evidence. Chips of paint adhering to the mark or tool provide perhaps the best example of how the transfer of trace physical evidence can occur as a result of using a tool to gain forcible entry into a building. Obviously, the presence of trace evidence greatly

(a) (b)

FIGURE 17–27 **(a) Casting a tool mark impression with a silicone-based putty. (b) Impression alongside suspect tool.** *Courtesy Sirchie Finger Print Laboratories, Inc., Youngsville, N.C., www.sirchie.com*

enhances the evidential value of the tool or its mark and requires special care in handling and packaging the evidence to avoid losing or destroying these items.

Quick Review

- The presence of minute imperfections on a tool imparts individuality to that tool. The shape and pattern of such imperfections are further modified by damage and wear during the life of the tool.
- The comparison microscope is used to compare crime-scene tool marks with test impressions made with the suspect tool.

Other Impressions

From time to time, other types of impressions are left at a crime scene. This evidence may take the form of a shoe, tire, or fabric impression. Similar to tool marks, footwear impressions and tire tread impressions can show both class and individual characteristics about the type of shoe or tire at the scene of the crime. Class characteristics of footwear impressions are features made during the manufacturing of the shoe. For example, work boots made from the same company will all have the same design on the base of the shoe. Individual characteristics are unique to the footwear that occurs from wear and tear when an individual wears the shoes. For example, an individual characteristic of a footwear impression could be pebbles stuck in the treads of the shoe from walking through gravel. It may be as varied as a shoe impression left on a piece of paper at the scene of a burglary (Figure 17–28), a hit-and-run victim's garment that has come into violent contact with an automobile (Figure 17–29), or the impression of a bloody shoe print left on a floor or carpet at a homicide scene (Figure 17–30).

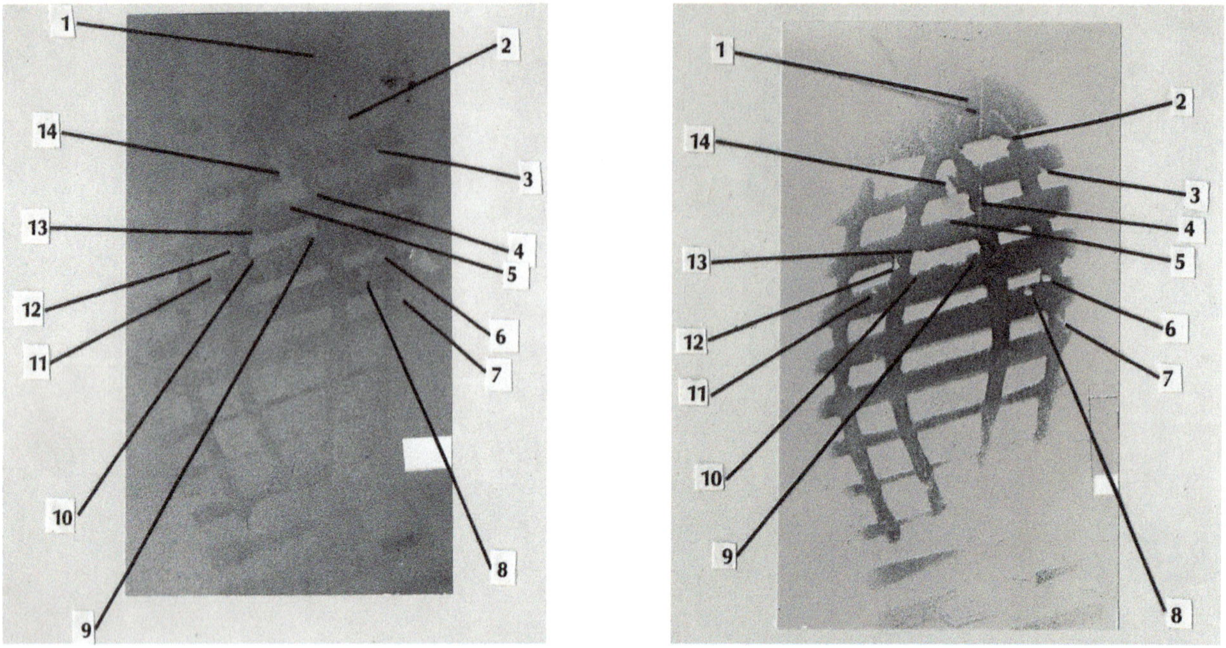

FIGURE 17–28 (a) Impression of shoe found at a crime scene. (b) Test impression made with suspect shoe. A sufficient number of points of comparison exist to support the conclusion that the suspect shoe left the impression at the crime scene. *Courtesy Richard Saferstein Ph.D.*

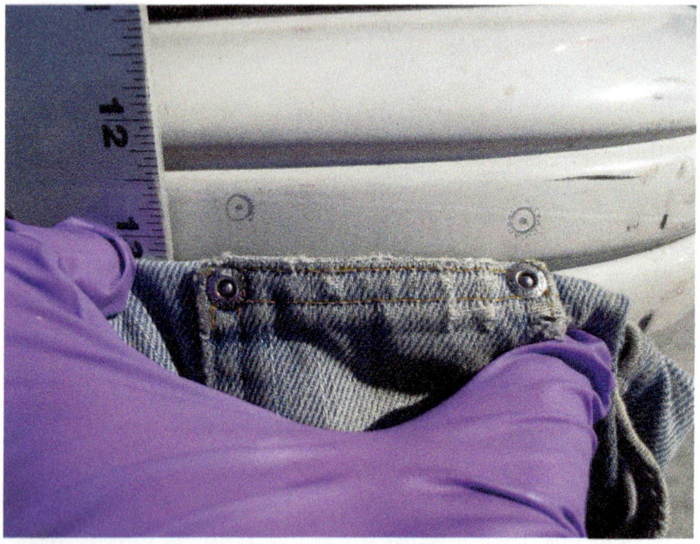

FIGURE 17–29 A pattern impression on a car bumper arising from a hit-and-run. Note rivets from the jeans are present in the impression. The writing from the rivet on the right side of the bumper is visible. *Courtesy Peter Diaczuk, John Jay College, City University of New York*

Preserving Impressions

The primary consideration in collecting impressions at the crime scene is the preservation of the impression or its reproduction for later examination in the crime laboratory. This is the first type of documentation of a footwear impression. Before any impression is moved or otherwise handled, it must be photographed (a scale should be included in the picture) to show all the observable details of the impression. If the impression cannot be taken to a laboratory, forensic scientists use the photograph to analyze the impression of footwear or tire treads for surface characteristics. Several shots should be taken directly over the

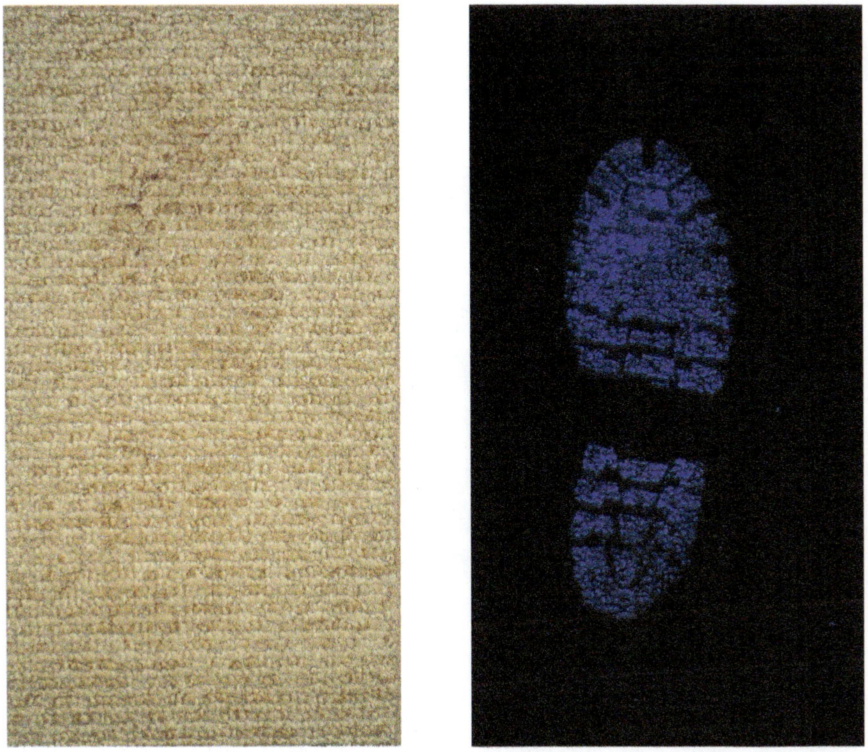

(a) (b)

FIGURE 17–30 (a) A section of a carpet under normal light showing a faint footprint in blood. (b) Same section of the carpet after spraying with luminal. *Courtesy Sirchie Finger Print Laboratories, Inc., Youngsville, N.C., www.sirchie.com*

impression as well as at various angles around the impression. Skillful use of side lighting for illumination will help highlight many ridge details that might otherwise remain obscured. Photographs should also be taken to show the position of the questioned impression in relation to the overall crime scene.

Although photography is an important first step in preserving an impression, it must be considered merely a backup procedure that is available to the examiner if the impression is damaged before reaching the crime laboratory. Naturally, the examiner prefers to receive the original impression for comparison to the suspect shoe, tire, garment, and so forth. In most cases when the impression is on a readily recoverable item, such as glass, paper, or floor tile, the evidence is easily transported intact to the laboratory.

Lifting Impressions

Forensic scientists need to analyze the surface characteristics of footwear impressions in a laboratory. These impressions can be left at the scene on many different types of surfaces. Impressions left on a soil surface would be best collected by casting. Other footwear impressions, like those on hard floors or latent footprints, are retrieved similarly to fingerprints. If an impression is encountered on a surface that cannot be submitted to the laboratory, the investigator may be able to preserve the print in a manner that is analogous to lifting a fingerprint. This is especially true of impressions made in light deposits of dust or dirt. A lifting material large enough to lift the entire impression should be used. Carefully place the lifting material over the entire impression. Use a fingerprint roller to eliminate any air pockets before lifting the impression off the surface.

FIGURE 17–31 Electrostatic lifting of a dust impression off a floor using an electrostatic unit. *Courtesy Sirchie Finger Print Laboratories, Inc., Youngsville, N.C., www.sirchie.com*

A more exotic approach to lifting and preserving dust impressions involves the use of a portable electrostatic lifting device. The principle is similar to that of creating an electrostatic charge on a comb and using the comb to lift small pieces of tissue paper. A sheet of mylar film is placed on top of the dust mark, and the film is pressed against the impression with the aid of a roller. The high-voltage electrode of the electrostatic unit is then placed in contact with the film while the unit's earth electrodes are placed against a metal plate (earth plate) (see Figure 17–31). A charge difference develops between the mylar film and the surface below the dust mark so that the dust is attached to the lifting film. In this manner, dust prints on chairs, walls, floors, and the like can be transferred to the mylar film. Floor surfaces up to 40 feet long can be covered with a mylar sheet and searched for dust impressions. The electrostatic lifting technique is particularly helpful in recovering barely visible dust prints on colored surfaces. Dust impressions can also be enhanced through chemical development (see Figure 17–32).

Casting Impressions

Shoe and tire marks impressed into soft earth at a crime scene are best recovered and documented by photography and casting. Class I dental stone, a form of gypsum, is widely recommended for making casts of shoe and tire impressions. A series of photographs clearly illustrating the steps in casting an impression are found in Figure 17–33. The cast should be allowed to air-dry for 24–48 hours before it is shipped to the forensic science laboratory for examination. Figure 17–34 illustrates a cast made from a shoe print in mud. The cast compares to the suspect shoe.

An aerosol product known as Snow Impression Wax is available for casting snow impressions. The recommended procedure is to spray three light coats of the wax at an interval of one to two minutes between layers, and then let it dry for ten minutes.

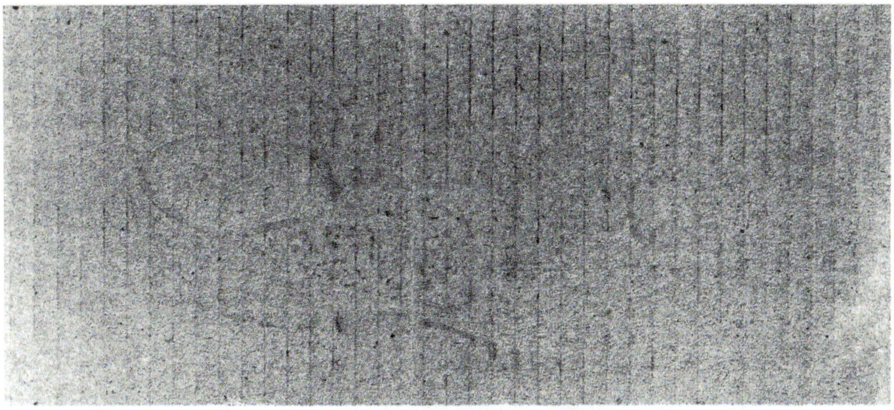

(a)

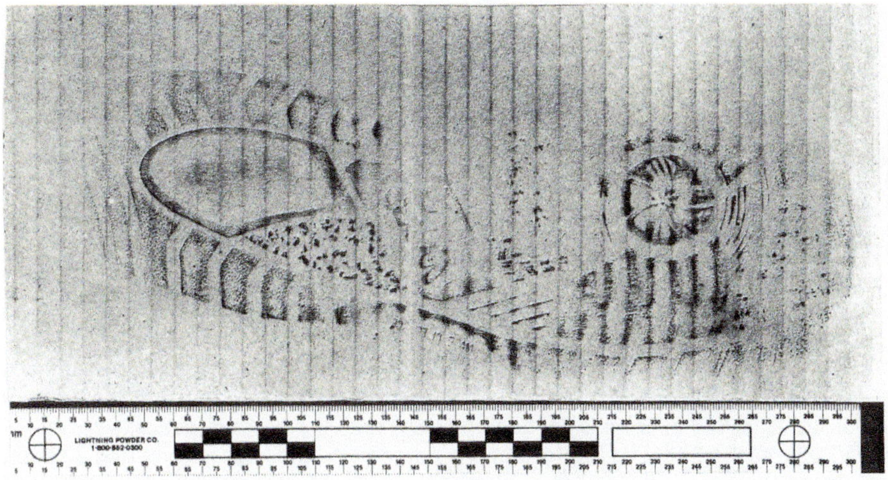

(b)

FIGURE 17–32 (a) A dust impression of a shoe print on cardboard before enhancement. (b) Shoe print after chemical enhancement with Bromophenol Blue and exposure to water vapor. *Courtesy Division of Identification and Forensic Science (DIFS), Israel Police Headquarters, Jerusalem, Israel*

A viscous mixture of Class I dental stone is then poured onto the wax-coated impression. After the casting material has hardened, the cast can be removed.

Several chemicals can be used to develop and enhance footwear impressions made with blood. In areas where a bloody footwear impression is very faint or where the subject has tracked through blood leaving a trail of bloody impressions, chemical enhancement can visualize latent or nearly invisible blood impressions.

Several blood enhancement chemicals have been examined for their impact on STR DNA typing. None of the chemicals examined had a deleterious effect, on a short-term basis, on the ability to carry out STR DNA typing on the blood.[4]

Comparing Impressions

Whatever the circumstances, the laboratory procedures used to examine any type of impression remain the same. Of course, a comparison is possible only when an item suspected of having made the impression is recovered. Test impressions may be necessary to compare the characteristics of the suspect item with the evidence impression.

FIGURE 17–33 To cast a footwear impression at a crime scene. (a) The impression is hardened using aerosol hairspray. (b) The correct amount of water is added to a known amount of dental stone. (c) The mixture is kneaded by hand until the desired (pancake batter-like) consistency is reached. (d) The dental stone is poured into the impression using a spoon as a medium to disperse the flow. (e) The impression is filled with dental stone and allowed to dry before removal. *Courtesy Sirchie Finger Print Laboratories, Inc., Youngsville, N.C., www.sirchie.com*

The evidential value of the impression is determined by the number of class and individual characteristics that the examiner finds. Agreement with respect to size, shape, or design may permit the conclusion that the impression could have been made by a particular shoe, tire, or garment; however, one cannot entirely exclude other possible sources from having the same class characteristics. More significant is the existence of individual characteristics arising out of wear, cuts, gouges, or other damage. A sufficient number or the uniqueness of such points of comparison support a finding that both the evidence and test impressions originated from one and only one source.

When tire tread impressions are left at a crime scene, the laboratory can examine the design of the impression and possibly determine the style and/or manufacturer of the tire. This may be particularly helpful to investigators when a suspect tire has not yet been located.

(a) (b) (c)

FIGURE 17–34 **(a) Shoe impression in mud. (b) Cast of shoe impression. (c) Shoe suspected of leaving muddy impression.** *Courtesy Sirchie Finger Print Laboratories, Inc., Youngs- ville, N.C., www.sirchie.com*

Not only can class and individual characteristics be obtained from the cast of the prints, but soil and organic plant material that was embedded into the footwear or tire tread can be recovered and analyzed, providing additional connections between a suspect, victim, and crime scene. Once a cast has been made of the impression, any soil or plant material attached to the print should be photographed and then transported intact to the lab. The soil should be allowed to dry, likely causing it to fall off of the cast, so it can be collected for analysis.

New computer software may help the forensic scientist compare shoe prints. For example, an automated shoe print identification system developed in England, called Shoeprint Image Capture and Retrieval (SICAR), incorporates multiple databases to search known and unknown footwear files for comparison against footwear specimens. Using the system, an impression from a crime scene can be compared to a reference database to find out what type of shoe caused the imprint. That same impression can also be searched in the suspect and crime databases to reveal whether that shoe print matches the shoes of a person who has been in custody or the shoe prints left behind at another crime scene. When matches are made during the searching process, the images are displayed side by side on the computer screen (see Figure 3–12).

Case Files

The O. J. Simpson Trial—Who Left the Impressions at the Crime Scene?

On the night of June 12, 1994, Nicole Brown—ex-wife of football star O. J. Simpson—and her friend Ron Goldman were brutally murdered on the grounds outside her home in Brentwood, California. O. J. Simpson was arrested for their murders, but professed his innocence. At the crime scene, investigators found bloody shoe impressions along the concrete walkway leading up to the front door of Brown's condominium. These shoe impressions were of extremely high quality and of intricate detail. The news media broadcast countless images of these bloody shoe prints on television, making it obvious to the killer that those shoes would surely link him to the crime.

Famed FBI shoe print examiner William J. Bodziak investigated the footwear evidence from the scene. His first task was to identify the brand of shoe that made the marks. Because the pattern was clear and distinct, with complete toe-to-heel detail, this seemed a simple task at first. Bodziak compared this pattern to the thousands of sole patterns in the FBI's database. None matched. He then went to his reference collection of books and trade show brochures, again with no success.

Bodziak's experience told him that these were expensive, Italian-made casual dress shoes with a sole made from synthetic material. Using this knowledge, he shopped the high-end stores for a similar tread pattern, but still was unable to identify the shoes. He then drew a composite sketch of the sole and faxed the image to law enforcement agencies and shoe manufacturers and distributors worldwide. The owner of the American distributing company for Bruno Magli shoes was the only one to respond.

Further exhaustive investigation revealed that these were extremely rare shoes. There were two styles of shoe bearing this exact sole design. They were available for only two years, and from a mere 40 stores in the United States and Puerto Rico. The Lorenzo style shoe had a bootlike upper that came to the ankle. The Lyon style shoe had the lower, more typical dress shoe cut. The impressions were made by a size 12 shoe, and it was later determined that only 299 pairs of size 12 with this tread pattern were sold in the United States.

Simpson flatly denied ever owning these shoes, adding that he would never wear anything so ugly. However, he was known to wear a size 12, and photographs taken almost nine months before the murders show Simpson wearing a pair of black leather Bruno Magli Lorenzo shoes. These shoes were available in several colors, so this narrows the number of shoes matching Simpson's pair of Lorenzos (this size, color, and style) sold in the United States to 29 pairs.

Proving that Simpson owned a pair of shoes that had the exact pattern found printed in blood at the crime scene was an essential component of the case, but it was not done in time to be used during the criminal prosecution. The photographs of Simpson in his Bruno Magli shoes were released after the culmination of the criminal trial, so the jury never heard the direct evidence that Simpson owned these shoes. However, this proved to be an important link uniting Simpson with the crime scene in the civil trial. Although O. J. Simpson was acquitted of the murders of Nicole Brown and Ron Goldman in the criminal trial, he was judged responsible for their murders in the civil court case.

Quick Review

- Shoe and tire marks impressed into soft earth at a crime scene are best preserved by photography and casting.
- The electrostatic lifting technique is particularly helpful in recovering barely visible dust prints on floor surfaces.
- In areas where a bloody footwear impression is very faint or where the subject has tracked through blood, leaving a trail of bloody impressions, chemical enhancement can visualize latent or nearly invisible blood impressions.

Chapter Review

- The manufacture of a gun barrel requires impressing its inner surface with spiral grooves, a step known as rifling. Rifling imparts spin to the projectile when it is fired, which keeps it on an accurate course.
- No two rifled barrels have identical striation markings. These striations form the individual characteristics of the barrel. The inner surface of the barrel of a gun leaves its striation markings on a bullet passing through it.
- The class characteristics of a rifled barrel include the number of lands and grooves and the width and direction of twist.
- The comparison microscope is a firearms examiner's most important tool because it allows two bullets to be observed and compared simultaneously.
- The firing pin, breechface, and ejector and extractor mechanism also offer a highly distinctive signature for individualization of cartridge cases.
- Unlike handguns, a shotgun is not rifled—it has a smooth barrel. Because of this, shotgun shells are not impressed with any characteristic markings that can be used to compare two shotgun shells to determine whether they were fired from the same weapon.
- The advent of computerized imaging technology has made possible the storage of bullet and cartridge surface characteristics in a manner analogous to automated fingerprint files.
- Two automated firearms search systems are DRUGFIRE, developed by the FBI, and IBIS, developed by the ATF.
- NIBIN is the *National Integrated Ballistics Information Network*, a unified firearms search system that incorporates both DRUGFIRE and IBIS technologies.
- The distribution of gunpowder particles and other discharge residues around a bullet hole permits an assessment of the distance from which a handgun or rifle was fired.
- The precise distance is determined by carefully comparing the powder residue pattern on the victim's clothing to test patterns made when the suspect weapon is fired at varying distances from a target.
- The Greiss test is a chemical test used to develop patterns of gunpowder residues around bullet holes. It tests for the presence of nitrates.
- Firing a weapon propels residues toward the target and blows gunpowder and primer residues back toward the shooter. Traces of these residues are often deposited on the firing hand of the shooter, providing valuable information as to whether an individual has recently fired a weapon.
- Examiners measure the amount of barium and antimony on the relevant portion of the suspect's hands or characterize the morphology of particles containing these elements to determine whether a person has fired or handled a weapon or was near a discharged firearm.
- Criminalists can restore serial numbers removed or obliterated by grinding, rifling, or punching.

- Because the metal crystals in the stamped zone are placed under a permanent strain that extends a short distance beneath the original numbers, the serial number can be restored through chemical etching.
- A suspect firearm should never be picked up by inserting an object into its barrel because this practice may alter the striation markings on test-fired bullets.
- Before unloading a suspect weapon, the weapon's hammer and safety position should be recorded, as well as the location of all fired and unfired ammunition in the weapon.
- Protection of class and individual markings on bullets and cartridge cases is the primary concern of the field investigator when recovering bullets and cartridge casings.
- The presence of minute imperfections on a tool imparts individuality to that tool. The shape and pattern of such imperfections are further modified by damage and wear during the life of the tool.
- The comparison microscope is used to compare crime-scene tool marks with test impressions made with the suspect tool.
- Shoe and tire marks impressed into soft earth at a crime scene are best preserved by photography and casting.
- The electrostatic lifting technique is particularly helpful in recovering barely visible dust prints on floor surfaces.
- In areas where a bloody footwear impression is very faint or where the subject has tracked through blood, leaving a trail of bloody impressions, chemical enhancement can visualize latent or nearly invisible blood impressions.

Quick Lab: Tool Marks

Materials:

Six used screwdrivers
Piece of wooden board or stiff cardboard
Jar of Play-doh or other easy-to-use modeling clay (one for each group)

Procedure:

Tool marks and other impressions left at a crime scene are most helpful when they exhibit individual characteristics—variations that distinguish one particular tool, tire, or shoe from all others of its type. In this exercise, students will identify tools based on their individual characteristics. Student should be matched up in teams of four or five. Place the screwdrivers on a table where they are accessible to all teams and mark each with an identifying number or letter. Each team places their modeling clay onto the board and molds it into a shape suitable to hold impressions. One team member takes the clay to the table and makes impressions in the clay with three of the screwdrivers, noting on a separate paper which screwdrivers made each impression. When all teams have finished making impressions, teams trade their impressions with another and note which team they trade with.

Each team now has 5 minutes to study the impressions and the screwdrivers and determine which ones made the impressions in the clay. The instructor should allow no more than two teams to study the screwdrivers at the same time.

Follow Up Questions:

1. **Which screwdrivers made the impressions in your clay? Consult with the team you traded with to determine if you correctly identified all three marks.**
2. **Were some of the marks easier to identify than the others? What factors made identification easier?**
3. **Tool marks are often found in harder substances such as wood. How would identifying tool marks in wood be different from identifying them in clay?**
4. **Analyze both the class and individual characteristics of the tool mark impressions created. Analyze the recovery and documentation of the characteristics of the surface.**

Review Questions

1. The raised areas between the grooves in a rifled bore are called
   a. rifling.
   b. lands.
   c. grooves.
   d. caliber.

2. The most important tool available to the firearms examiner in the identification of firearms is
   a. the comparison microscope.
   b. the digitizing videocamera.
   c. mylar film and a high-voltage electrode.
   d. laser illumination.

3. Tool marks are most often encountered at which type of crime scene?
   a. arson
   b. homicide
   c. sexual assault
   d. burglary

4. What is the primary consideration in collecting impressions at a crime scene?
   a. the preservation of the impression or its reproduction for later use
   b. maintaining the chain of evidence
   c. the impression's admissibility in a courtroom
   d. the obliteration of the impression based on weather conditions

5. In the examination of tool mark impressions, individuality of a tool can be ascertained by
   a. patterns of grooves and lands resulting from the machining process.
   b. patterns of hills and valleys as minute imperfections.
   c. patterns of striations created by the manufacturer to trademark the tool.
   d. patterns of stains from oil and other maintenance activities.
6. True or False: Unlike other areas of criminalistics, the goal of individualization is not yet a reality in firearm and tool mark examination.
7. True or False: Rifled barrels manufactured in succession will have identical striation markings.
8. True or False: As a shotgun shell passes through a barrel, its surface is impressed with striations from the lands and grooves of the barrel.
9. True or False: Barium and antimony specimens found in primer on a suspect's hand after firing a gun remain for a period of 36 hours and confirm the suspect's involvement in a crime approximately 75 percent of the time.
10. True or False: Other forms of impressions that may constitute evidence at a crime scene include shoe, tire, and fabric impressions.
11. What is rifling and what is its purpose?
12. What are grooves and lands?
13. What is caliber?
14. List the class characteristics of a rifled barrel.
15. What are striations? How are they produced?
16. How are striations useful for comparing bullets?
17. Name three factors that can prevent a firearms examiner from obtaining a perfect match of striation markings around the periphery of two bullets.
18. How does a shotgun barrel differ from the barrel of a handgun? How does this difference affect the ability of a firearms examiner to compare shotgun shells?
19. Describe the relationship between a shotgun's gauge and the diameter of its barrel.
20. Briefly describe the sequence of events that occur from the time the trigger of a firearm is pulled to the time the bullet is fired.
21. What parts of a firearm leave impressions on a cartridge case that constitute individual characteristics of that weapon?
22. What is NIBIN?
23. What is distance determination and upon what is it based?
24. Briefly describe how a firearms examiner makes a distance determination for a handgun or rifle shot.
25. What is bullet wipe? Of what does it consist?
26. How does a firearms examiner estimate a distance determination for a shooting involving a shotgun?

27. As the distance to the target of a shotgun blast increases, what do the pellets do?

28. What is choke? How is it related to the speed, range, and spread pattern of shotgun pellets?

29. What is the Greiss test? What substance is being tested for in this procedure?

30. Name and briefly describe two popular approaches for collecting gunshot residue from a suspect's hands.

31. What device is often used to locate primer and gunpowder particles by their characteristic size and shape?

32. What is the major advantage of the SEM approach for primer residue detection? What is its major drawback?

33. Why is it possible to restore an obliterated serial number? Name two situations in which an obliterated serial number usually cannot be restored.

34. Why should a suspect firearm never be picked up by inserting an object into its barrel?

35. How should a suspect weapon discovered underwater be transported to the laboratory and why?

36. What typically imparts individual characteristics to a tool mark?

37. How does the criminalist record a tool mark for comparison when removal of the original tool mark is impractical?

38. Why must the crime-scene investigator never attempt to fit a suspect tool into a tool mark?

39. Name two procedures used to preserve impressions that cannot be submitted to the laboratory.

40. What is SICAR and how is it useful in comparing impressions?

41. How can you identify the components and characteristics of bullet and cartridge cases?

42. What type of information is available through the *National Integrated Ballistics Information Network*? Write a summary of your findings.

Learn the Vocabulary

Use the information from this chapter along with your prior knowledge and experience to learn the key terms. Choose from the following activities:

1. Make flashcards with a term on the front and its definition on the back. Use the flashcards to quiz a partner on correctly pronouncing and defining the terms.

2. Make an audio or video recording of each term and its definition.

3. Work with a partner to write a sentence for each term. Read the sentences out loud to a partner or share them with the class.

4. Write a list of new terms and language you have encountered in this chapter. Reuse the terms to write sentences and then read the sentences out loud to a partner. Reuse the terms again by writing a short speech. The speech might explain the terms or teach a lesson using the terms. Give the speech to a partner or to the class.

Application and Critical Thinking

1. From each of the following descriptions of bullet holes, use general guidelines to estimate the distance from the shooter to the target.

a. a few widely scattered gunpowder particles with no soot around the entrance hole

b. a dark ring around the bullet hole, but no soot or gunpowder particles

c. a halo of soot surrounding the entrance hole along with scattered specks of powder grains

d. scorch marks and melted fibers surrounding the entrance hole

2. You are investigating a shooting involving a 12-gauge shotgun with a moderately high choke. The spread of the pattern made by the pellets measures 12 inches. In your opinion, which of the following is probably closest to the distance from the target to the shooter? Explain your answer and explain why the other answers are likely to be incorrect.

a. 18 yards

b. 12 yards

c. 6 yards

d. 30 yards

3. Criminalist Ben Baldanza is collecting evidence from the scene of a shooting. After locating the revolver suspected of firing the shots, Ben picks the gun up by the grip, unloads it, and places the ammunition in an envelope. He then attaches an identification tag to the grip. Searching the scene, Ben finds a bullet lodged in the wall. He uses pliers to grab the bullet and pull it from the wall, then inscribes the bullet with his initials and places it in an envelope. What mistakes, if any, did Ben make in collecting this evidence? If you were assisting on the crime scene investigation, how could you conduct and calculate a trajectory analysis of the bullet strike?

4. How would you go about collecting impressions in each of the following situations?

a. You discover a shoe print in dry dirt.

b. You discover a tool mark on a windowsill.

c. You discover tire marks in soft earth.

d. You discover a shoe print on a loose piece of tile.

e. You discover a very faint shoe print in dust on a colored linoleum floor.

5. Gunshot residue patterns (a) through (d) (contact, 1 inch, 6 inches, and 18 inches) from a 40-caliber pistol are shown in the figures. Match the firing distance to each pattern.

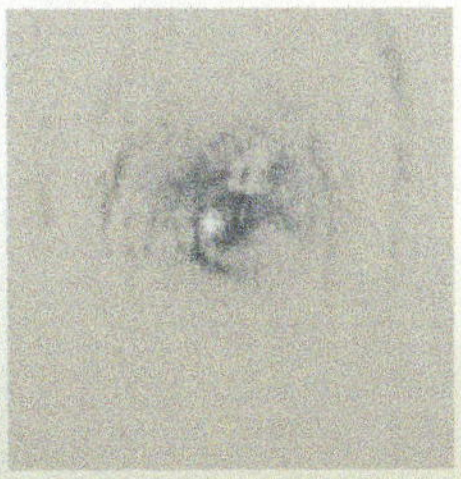

Richard Saferstein

(a)________________

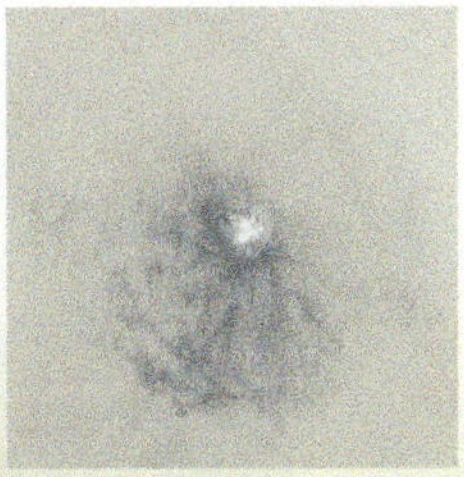

Richard Saferstein

(b)________________

Richard Saferstein

(c)________________

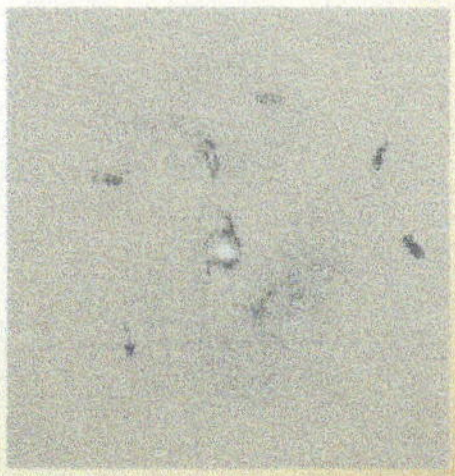

Richard Saferstein

(d)________________

6. A dead body is found lying in a field. Near the body investigators locate a tire impression. Describe how you would analyze the class and individual characteristics of the tread impression. Describe the recovery and documentation of soil and or organic plant material that may be a part of the surface characteristics.

7. Describe the mechanism of modern firearms, such as long guns and handguns.

8. Describe the composition of and method of analysis for gunshot residue and primer residue.

9. Working with a partner, create a poster or presentation that identifies and recognizes the type of information available through the *National Integrated Ballistics Information Network.*

Laboratory Experiments

This activity requires the use of the following practices of science and engineering:

- Asking questions and defining problems
- Planning and carrying out investigations
- Analyzing and interpreting data
- Obtaining, evaluating, and communicating information
- Using appropriate safety equipment and practices as outlined in Texas Education Agency-approved safety standards

This activity consists of the following crosscutting concept:

- **Patterns**—The patterns in shoe prints allow for the classification of prints which in turn allows for the identification of the footwear.

Before you begin, read the scenario and steps out loud with a partner. Ask for clarification as necessary to make sure you understand the experiment.

The Scenario

You respond to a reported robbery at a two-story townhouse on April 2, 2014 at 3:42 PM. When you show up on scene the first responding officer on the scene gives you the following information:

I received a call from the homeowner, Ricardo Gonzales, who has been out of town for the last week on business. He said when he arrived home, the front door was open. When he entered the home, he saw that the house had been ransacked. His TV, computer, and several firearms were missing. When he entered the kitchen, he noticed glass on the floor and that the window in the back of the house had been broken. He called 911. The scene was secured immediately upon arrival of the first responding officer. The residence was dusted for fingerprints and a shoe impression was found in the mud outside of the rear window sill. No other evidence was recovered.

This robbery was not an isolated case, as several of the homes in the area had been robbed. Your team will analyze the foot impression at the crime scene and compare using as many indicators as possible to tie one of the suspects' shoes to the impression.

Exercise 1: Making a Cast

MATERIALS

A shoebox containing the impression from the crime scene
Ruler
One small can of inexpensive aerosol hair spray
Large mixing bowl (or 1-gallon zip-top bag)
Paint stirrer or large, long-handled spoon
Five-pound bag of Plaster of Paris (or dental stone)
Water
Wire hangers, clipped into 3- to 4-inch pieces, or craft sticks

PROCEDURE

1. Holding the can of hairspray about 8 inches from the soil, apply an even layer to the impression using a sweeping motion. Be sure that even the deepest edges of the impression receive the same amount of coverage. More is not necessarily better—do not over apply.
2. Allow to stand for 10 minutes to allow the hairspray to dry.
3. Add ½ of the 5-pound bag of Plaster of Paris to a large mixing bowl.* Add enough water to the bowl to produce a pancake batter-like consistency. No lumps. (Again, the amount of water you need will vary with the ambient humidity and plaster manufacturer.) Alternately, you can measure the Plaster of Paris into a 1-gallon size plastic zip-top bag and add the water in increments. Seal the bag and mix by working back and forth with your fingers.
4. Pour the plaster into the impression, using a paint stirrer to broadcast the stream so it does not destroy the fine details of the impression. Pour to reach at least ½ inch in thickness.
5. Place the cut-up hangers or craft sticks horizontally and vertically on the surface of the web plaster. These will serve to reinforce the cast.
6. Prepare a second mix of plaster of Paris in the same way, and apply it over the first to a total thickness of at least 1 inch.
7. Label the wet plaster surface with the date and your name or group number before it dries.
8. Store it to dry completely overnight.

*Alternately, you can measure the plaster of Paris into a 1-gallon size plastic zip-top bag and add the water in increments. Seal the bag and mix by working back and forth with your fingers. Dental stone may be substituted for the plaster of Paris. This material can be pre-weighed and mixed with a measured amount of water according to the manufacturer's instructions.

Exercise 2: Cast/Footwear Comparison

PROCEDURE

1. Collect your cast prepared in Exercise 1. The first step for the documentation and analysis of surface characteristics is to take a photograph. Photograph your print from all angles. Once dried, carefully analyze your impression. Recover and document any soil or organic plant material collected on the cast.

2. Using a magnifying glass, examine the cast to make depth and details more important. Rotate the cast and examine all sides for unique features. Look for and analyze the following individual characteristics of the footwear impression: cuts that interrupt the pattern on the sole, wear or rub marks on the outer edge, or even apparent manufacturing defects. Also look at and analyze class characteristics of the impression, such as the general tread pattern of the shoe.
3. Your group will devise a numbering scheme that assigns the unique identifiers (try to have 10) to the cast such as a cut, gouge, tear, imbedded pebble, or pattern of wear which is unique to that shoe. List each item under its unique identifier and write a brief description in the data table. Be sure to include the cast dimensions as an identifier (length, width of heal, and width at widest point).
4. By comparing length, width, sole pattern, and unique features, you should be able to deduce which suspect's shoe is the most likely mold for the impression evidence found at the "crime scene." Outline your findings in your report using explicit examples.

Data Table Part 1

Identifiers	*Description of Cast*
1	
2	
3	
4	
5	
6	
7	
8	
9	
10	

Data Table Part 2

Identifiers (same as above)	*Description of Shoe #1*	*Description of Shoe #2*	*Description of Shoe #3*
1			
2			
3			
4			
5			
6			
7			
8			
9			
10			

Endnotes

1. Originally, the number of lead balls with the same diameter as the barrel that would make a pound. For example, a 20-gauge shotgun has an inside diameter equal to the diameter of a lead ball that weighs 1/20 of a pound.

2. J. W. Kilty, "Activity after Shooting and Its Effect on the Retention of Primer Residues," *Journal of Forensic Sciences* 20 (1975): 219.

3. G. E. Reed et al., "Analysis of Gunshot Residue Test Results in 112 Suicides," *Journal of Forensic Sciences* 35 (1990): 62.

4. C. J. Frégeau et al., "Fingerprint Enhancement Revisited and the Effects of Blood Enhancement Chemicals on Subsequent Profiler Plus™ Fluorescent Short Tandem Repeat DNA Analysis of Fresh and Aged Bloody Fingerprints," *Journal of Forensic Sciences* 45 (2000): 354.

Document Examination

18

Pat Sullivan/AP/Shutterstock

PICKING UP THE SCENT

Trail heats up in missing wife mystery

Cops shake the dust off '82 Durst case

WIFE MISSING: 1000 REWARD

Courtesy New York Daily News/ Getty Images

Key Terms

- charred document
- erasure
- exemplar
- indented writings
- infrared luminescence
- natural variations
- obliteration
- questioned document

The Life and Deaths of Robert Durst

Robert Durst married Kathleen McCormack in April of 1973. The two lived busy lives in Manhattan, where Kathie was a medical student and Robert worked for his family's real estate business. By all accounts, their relationship started as a loving one, but after nearly 10 years of marriage, the union was in decline. Kathie was treated at a Bronx hospital for facial bruises and told a friend that Robert had beat her. One night after dinner with friends in January 1982, Kathie went missing. She was never seen or heard from again.

After the disappearance of his wife, many people—including the police—focused on Durst as a suspect. He often sought solace and counsel from his close friend, Susan Berman. Berman acted as a representative for Durst, making statements on his behalf to the news media following Kathie's disappearance and even providing police with an alibi for his whereabouts on the night she went missing. Some 20 years later, Berman had fallen on hard times. Police believed that she knew too much about the disappearance of Kathie, and began extorting Durst, who began providing her with checks for large sums of money. On December 24, 2000, Susan Berman was found murdered execution-style in her home in California. Durst is known to have been in Northern California days before Berman was killed, and to have flown from San Francisco to New York the night before Berman's body was discovered.

Police discovered the crime after they received an anonymous letter written to the Beverly Hills Police Department alerting them to a "cadaver" at Berman's address. The handwriting in the letter was compared to other documents authored by Durst, including letters mailed to Berman. A curious misspelling of the word "Beverley" appeared in both the "cadaver" letter to police and letters known to be authored by Durst. Handwriting experts, as well as Durst himself, have acknowledged the similarity between his handwriting and the handwriting on the letter to the Beverly Hills Police. Robert Durst was arrested and indicted on charges of first degree murder related to the death of Susan Berman in March of 2015 based in part on the handwriting analysis performed on the "cadaver" letter. Robert Durst was convicted of murdering Susan Berman in 2021 and sentenced to life in prison. He died January 10, 2022.

Learning Objectives

After studying this chapter, you should be able to:

- 18.1 Define the term *questioned document*
- 18.2 Explain the factors considered when comparing variations in handwriting and the challenges associated with it
- 18.3 List some important guidelines for the collection of known writings for comparison to a questioned document
- 18.4 Recognize some of the class and individual characteristics of printers and photocopiers
- 18.5 List some of the techniques document examiners use to uncover alterations, erasures, and obliterations
- 18.6 Summarize some of the other problems faced by document examiners

National Science Content Standards

Scientific Inquiry

Science and Technology

Science in Personal and Social Perspective

History and Nature of Science

The Document Examiner

Ordinarily, the work of the document examiner involves examining handwriting and typescript to ascertain the source or authenticity of a questioned document. However, document examination is not restricted to a mere visual comparison of words and letters. The document examiner must know how to use microscopy, photography, and even such analytical methods as chromatography to uncover all efforts, both brazen and subtle, to change the content or meaning of a document.

Alterations of documents through overwriting, erasures, or the more obvious crossing out of words must be recognized and characterized as efforts to alter or obscure the original meaning of a document. The document examiner identifies such efforts and recovers the original contents of the writing. An examiner may even reconstruct writing on charred or burned papers, or uncover the meaning of indented writings found on a paper pad after the top sheet has been removed.

questioned document
Any document about which some issue has been raised or that is the subject of an investigation

Any object that contains handwritten or typewritten markings whose source or authenticity is in doubt may be referred to as a **questioned document.** Such a broad definition covers all of the written and printed materials we normally encounter in our daily activities. Letters, checks, driver's licenses, contracts, wills, voter registrations, passports, petitions, and even lottery tickets are commonly examined in crime laboratories. However, we need not restrict our examples to paper documents. Questioned documents may include writings or other markings found on walls, windows, doors, or any other objects.

Document examiners possess no mystical powers or scientific formulas for identifying the authors of writings. They apply knowledge gathered through years of training and experience to recognize and compare the individual characteristics of questioned and known authentic writings. For this purpose, gathering documents of known authorship or origin is critical to the outcome of the examination. Collecting known writings may entail considerable time and effort and may be further hampered by uncooperative or missing witnesses. However, the uniqueness of handwriting makes this type of physical evidence, like fingerprints, one of few definitive individual characteristics available to the investigator, a fact that certainly justifies an extensive investigative effort.

Handwriting Comparisons

Document experts continually testify that no two individuals write exactly alike. This is not to say that there cannot be marked resemblances between two individuals' handwritings, because many factors make up the total character of a person's writing.

General Style

Perhaps the most obvious feature of handwriting to the layperson is its general style. As children, we all learn to write by attempting to copy letters that match a standard form or style shown to us by our teachers. The style of writing acquired by the learner is that which is fashionable for the particular time and locale. In the United States, for example, the two most widely used systems are the Palmer method, first introduced in 1880, and the Zaner-Bloser method, introduced in

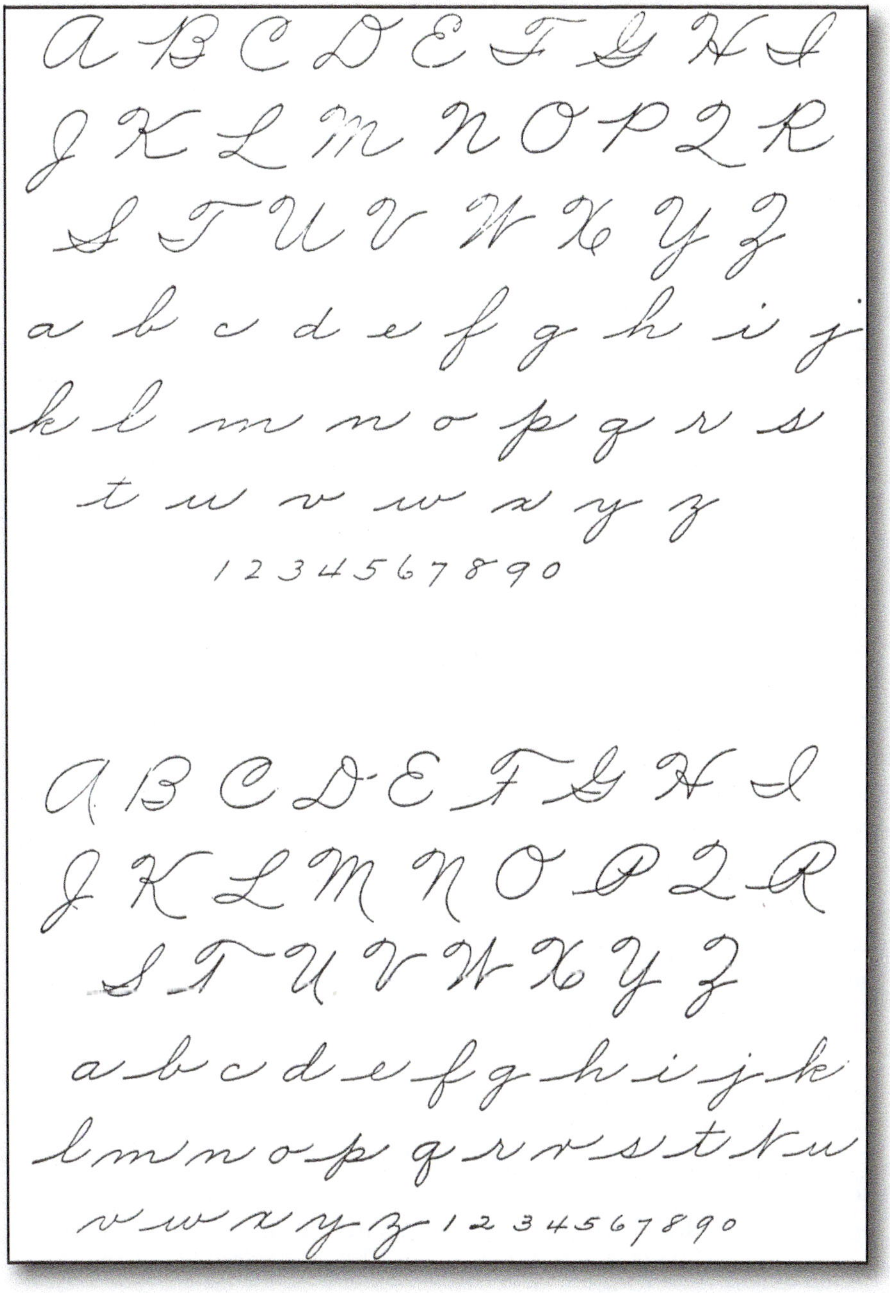

FIGURE 18–1 (top) Example of Zaner-Bloser handwriting; (bottom) Example of Palmer handwriting. *Courtesy Robert J. Phillips, Document Examiner, Audubon, New Jersey*

1895 (see Figure 18–1). To some extent, both of these systems are taught in nearly all 50 states.

The early stages of learning and practicing handwriting are characterized by a conscious effort by the student to copy standard letter forms. Many pupils in a handwriting class tend at first to have writing styles that are similar to one another, with minor differences attributable to skill in copying. However, as initial writing skills improve, a child normally reaches the stage at which the nerve and motor responses associated with the act of writing become subconscious. The individual's writing now begins to take on innumerable habitual shapes and patterns that distinguish it from all others. The document examiner looks for these unique writing traits.

Variations in Handwriting

The unconscious handwriting of two different individuals can never be identical. Individual variations associated with mechanical, physical, and mental functions make it extremely unlikely that all of these factors can be exactly reproduced by any two people. Thus, variations are expected in angularity, slope, speed, pressure, letter and word spacings, relative dimensions of letters, connections, pen movement, writing skill, and finger dexterity.

Furthermore, many other factors besides pure handwriting characteristics should be considered. The arrangement of the writing on the paper may be as distinctive as the writing itself. Margins, spacings, crowding, insertions, and alignment are all results of personal habits. Spelling, punctuation, phraseology, and grammar can be personal and, if so, combine to individualize the writer.

In a problem involving the authorship of handwriting, all characteristics of both the known and questioned documents must be considered and compared. Dissimilarities between the two writings strongly indicate two writers, unless these differences can logically be accounted for by the facts surrounding the preparation of the documents. Because any single characteristic, even the most distinctive one, may be found in the handwriting of other individuals, no single handwriting characteristic can by itself be taken as the basis for a positive comparison. The final conclusion must be based on a sufficient number of common characteristics between the known and questioned writings to effectively preclude their having originated from two different sources.

What constitutes a sufficient number of personal characteristics? Here again, there are no hard-and-fast rules for making such a determination. The expert examiner can make this judgment only in the context of each particular case.

Challenges to Handwriting Comparison

When the examiner receives a reasonable amount of known handwriting for comparison, sufficient evidence to determine the source of a questioned document is usually easy to find. Frequently, however, circumstances prevent a positive conclusion or permit only the expression of a qualified opinion. Such situations usually develop when an insufficient number of known writings are available for comparison. Although nothing may be found that definitely points to the questioned and known handwriting being of different origin, not enough personal characteristics may be present in the known writings that are consistent with the questioned materials.

Difficulties may also arise when the examiner receives questioned writings containing only a few words, all deliberately written in a crude, unnatural form or all very carefully written and thought out so as to disguise the writer's natural style—a situation usually encountered with threatening or obscene letters. It is extremely difficult to compare handwriting that has been very carefully prepared to a document written with such little thought for structural details that it contains only the subconscious writing habits of the writer. However, although one's writing habits may be relatively easy to change for a few words or sentences, maintaining such an effort grows more difficult with each additional word.

When an adequate amount of writing is available, the attempt at total disguise may fail. This was illustrated by Clifford Irving's attempt to forge letters in the name of the late industrialist Howard Hughes in order to obtain lucrative publishing contracts for Hughes's life story. Figure 18–2 shows forged signatures of Howard Hughes along with Clifford Irving's known writings. By comparing these signatures, document examiner R. A. Cabbane of the U.S. Postal Inspection Service detected many examples of Irving's personal characteristics in the forged signatures.

For example, note the formation of the letter *r* in the word *Howard* on lines 1 and 3, as compared with the composite on line 6. Observe the manner in which the terminal stroke of the letter *r* tends to terminate with a little curve at the baseline of Irving's writing and the forgery. Notice the way the bridge of the *w* drops in line 1 and also in line 6. Also, observe the similarity in the formation of the letter *g* as it appears on line 1 as compared with the second signature on line 5.

The document examiner must also be aware that writing habits may be altered beyond recognition by the influence of drugs or alcohol. Under these circumstances, it may be impossible to obtain known writings of a suspect written under conditions comparable to those at the time the questioned document was prepared.

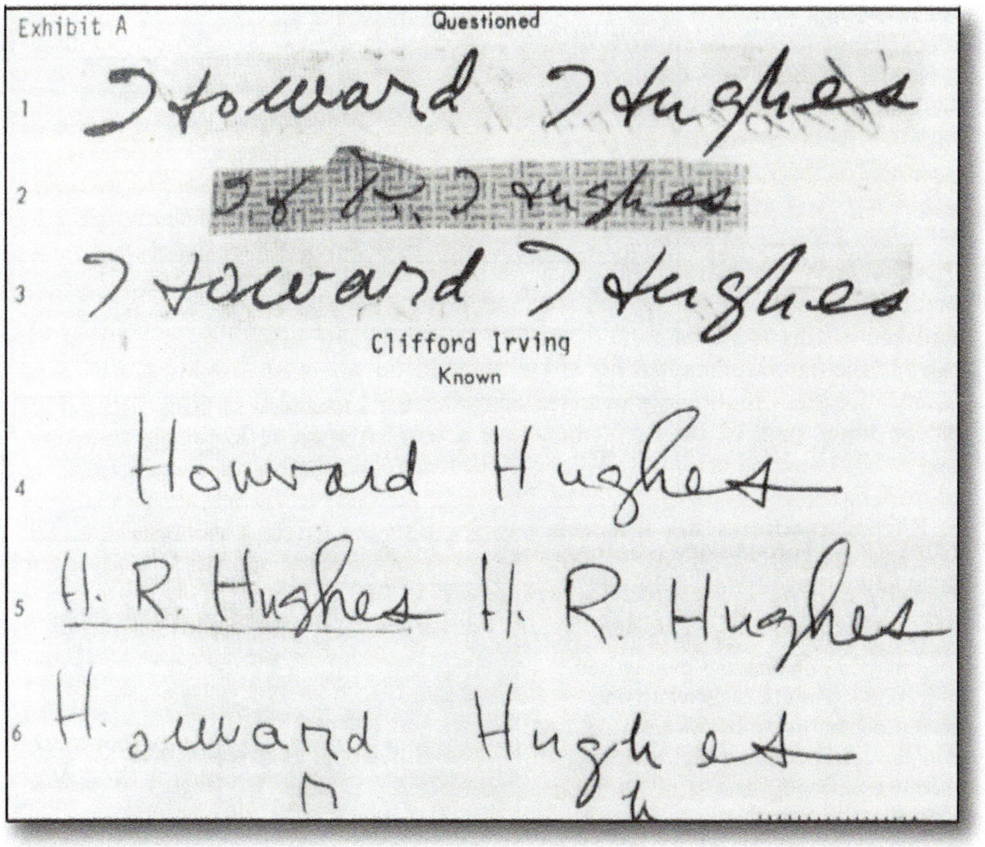

FIGURE 18–2 Forged signatures of Howard Hughes and examples of Clifford Irving's writing.
Courtesy American Society for Testing and Materials from the Journal of Forensic Sciences, copyright 1975

Collection of Handwriting Exemplars

exemplar
An authentic sample used for comparison purposes, such as handwriting

Collection of an adequate number of known writings (**exemplars**) is critical for determining the outcome of a comparison. Generally, known writings of the suspect furnished to the examiner should be as similar as possible to the questioned document. This is especially true with respect to the writing implement and paper. Styles and habits may be somewhat altered if a person switches from a pencil to a ballpoint pen or to a fountain pen. The way the paper is ruled, or the fact that it is unruled, may also affect the handwriting of a person who has become particularly accustomed to one type or the other. Known writings should also contain some of the words and combinations of letters present in the questioned document.

natural variations
Normal deviations found between repeated specimens of an individual's handwriting

The known writings must be adequate in number to show the examiner the range of **natural variations** in a suspect's writing characteristics. No two specimens of writing prepared by one person are ever identical in every detail. Variation is an inherent part of natural writing. In fact, a signature forged by tracing an authentic signature can often be detected even if the original and tracing coincide exactly, because no one ever signs two signatures exactly alike (see Figures 18–3 and 18–4).

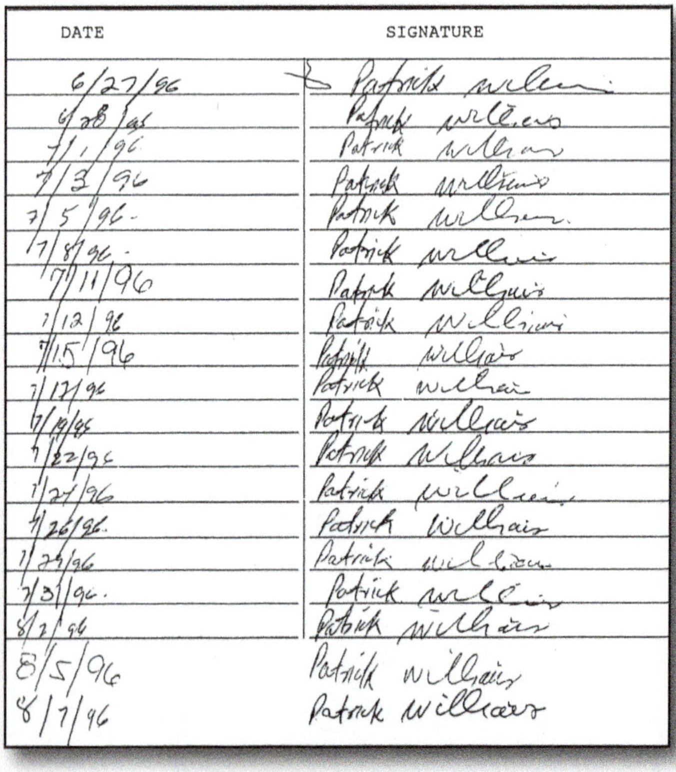

FIGURE 18–3 No two specimens of writing from one person are ever identical. *Courtesy Robert J. Phillips, Document Examiner, Audubon, New Jersey*

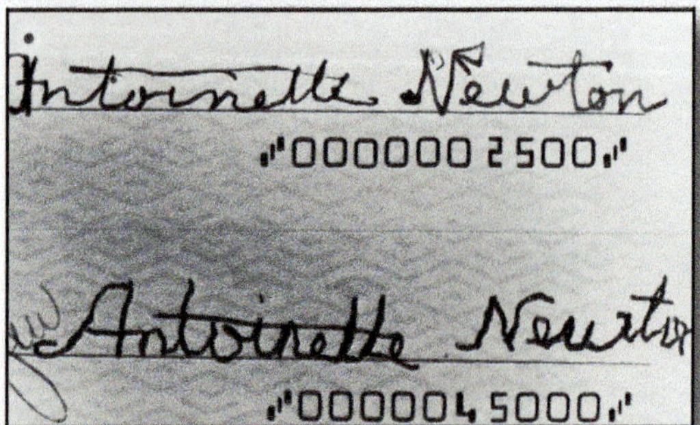

FIGURE 18–4 Forged signature. The top signature is genuine. The bottom signature is a simulated forgery. The adaption to the space is incorrect; there are numerous pen lifts and over-writes. The forged signature is awkward and drawn. *Courtesy Robin D. Williams, MFS, MS, D-BFDE, Omni Document Examinations, Green Bay, WI, rwilliams@docexamination.com*

Case Files

Hitler's Diaries

In 1981, a spectacular manuscript attributed to Adolf Hitler was disclosed by the brother of an East German general. These documents included Hitler's 27-volume diary and an unknown third volume of his autobiography, *Mein Kampf*. The existence of these works was both culturally and politically significant to the millions who were affected by World War II.

Authentication of the diaries was undertaken by two world-renowned experts, one Swiss and one American. Both declared that the handwritten manuscripts were identical to the known samples of Adolf Hitler's handwriting that they were given. Bidding wars began for publishing rights, and a major national newspaper in the United States won with a price near $4 million.

The publishing company that originally released the documents to the world market undertook its own investigation, which ultimately revealed a clever but devious plot. The paper on which the diaries were written contained a whitener that didn't exist until 1954, long after Hitler committed suicide. The manuscript binding threads contained viscose and polyester, neither of which was available until after World War II. Further, the inks used in the manuscript were all inconsistent with those in use during the year these pages were allegedly written.

Moreover, the exemplars sent to the Swiss and American experts as purportedly known examples of Hitler's handwriting were actually from the same source as the diaries. Thus, the experts were justified in proclaiming that the documents were authentic because they *were* written by the same hand—it just wasn't Hitler's. Chemical analysis of the inks later determined that the "Hitler diaries" were in fact less than one year old—spectacular, but fake!

Many sources are available to the investigator for establishing the authenticity of the writings of a suspect. An important consideration in selecting sample writings is the age of the genuine document relative to the questioned one. It is important to try to find standards that date closely in time to the questioned document. For most typical adults, basic writing changes are comparatively slow. Therefore, material written within two or three years of the disputed writing is usually satisfactory for comparison; as the age difference between the genuine and unknown specimens becomes greater, the standard tends to become less representative.

Despite the many potential sources of handwriting exemplars, obtaining an adequate set of collected standards may be difficult or impossible. In these situations, handwriting may have to be obtained voluntarily or under court order from the suspect. Ample case law supports the constitutionality of taking handwriting specimens. In *Gilbert v. California*,[1] the Supreme Court upheld the taking of handwriting exemplars before the appointment of counsel. The Court also reasoned that handwriting samples are identifying physical characteristics that lie outside the protection privileges of the Fifth Amendment. Furthermore, in *United States v. Mara*,[2] the Supreme Court ruled that taking a handwriting sample did not constitute an unreasonable search and seizure of a person and hence did not violate Fourth Amendment rights.

As opposed to nonrequested specimens (written without the thought that they may someday be used in a police investigation), requested writing samples may be consciously altered by the writer. However, the investigator can take certain steps to minimize attempts at deception. The requirement of several pages of writing normally provides enough material that is free of nervousness or attempts at deliberate disguise for a valid comparison. In addition, the writing of dictation yields exemplars that best represent the suspect's subconscious style and characteristics.

Other steps that can be taken to minimize a conscious writing effort, as well as to ensure conditions approximating those of the questioned writing, can be summarized as follows:

1. **The writer should be allowed to write sitting comfortably at a desk or table and without distraction.**
2. **The suspect should not under any conditions be shown the questioned document or be told how to spell certain words or what punctuation to use.**
3. **The suspect should be furnished a pen and paper similar to those used in the questioned document.**
4. **The dictated text should be the same as the contents of the questioned document, or at least should contain many of the same words, phrases, and letter combinations found in the document. In handprinting cases, the suspect must not be told whether to use uppercase (capital) or lowercase (small) lettering. If, after writing several pages, the writer fails to use the desired type of lettering, the writer can then be instructed to include it. Altogether, the text must be no shorter than a page.**
5. **Dictation of the text should take place at least three times. If the writer is trying to disguise the writing, noticeable variations should appear among the three repetitions. Discovering this, the investigator must insist on continued repetitive dictation of the text.**
6. **Signature exemplars can best be obtained when the suspect is required to combine other writings with a signature. For example, instead of compiling a set of signatures alone, the writer might be asked to fill out completely 20 to 30 separate checks or receipts, each of which includes a signature.**
7. **Before requested exemplars are taken from the suspect, a document examiner should be consulted and shown the questioned specimens.**

Quick Review

- Any object with handwriting or print whose source or authenticity is in doubt may be referred to as a questioned document.
- Document examiners gather documents of known authorship or origin and compare them to the individual characteristics of questioned writings.
- Collecting an adequate number of known writings is critical for determining the outcome of a handwriting comparison. Known writing should contain some of the words and combinations of letters in the questioned document.
- The unconscious handwriting of two different individuals can never be identical. However, the writing style of an individual may be altered beyond recognition by the influence of drugs or alcohol.

Typescript Comparisons

The document examiner analyzes not only handwritten documents, but digital ones as well. Document-creating machines include a wide variety of devices; some examples are computer printers, photocopiers, fax machines, and typewriters.

Photocopier, Printer, and Fax Examination

With the emergence of digital technology, document examiners are confronted with a new array of machines capable of creating documents subject to alteration or fraudulent use. Personal computers use daisy wheel, dot-matrix, ink-jet, and laser printers. More and more, the document examiner encounters problems involving these machines, which often produce typed copies that have only inconspicuous defects.

In the cases of photocopiers, fax machines, and computer printers, an examiner may need to identify the make and model of a machine that may have been used in printing a document. Alternatively, the examiner may need to compare a questioned document with test samples printed from a suspect machine. Typically, the examiner generates approximately ten samples through each machine to obtain a sufficient representation of a machine's characteristics. A side-by-side comparison is then made between the questioned document and the printed exemplars to compare markings produced by the machine.

Photocopiers Transitory defect marks originating from random debris on the glass platen, inner cover, or mechanical portions of a copier produce images. These images are often irregularly shaped and sometimes form distinctive patterns. Thus, they become points of comparison as the document examiner attempts to link the document to suspect copiers. The gradual change, shift, or duplication of these marks may help the examiner date the document.

Fax Machines Fax machines print a header known as the *transmitting terminal identifier (TTI)* at the top of each fax page. For the document examiner, the TTI is an very important point of comparison (see Figure 18–5). The header

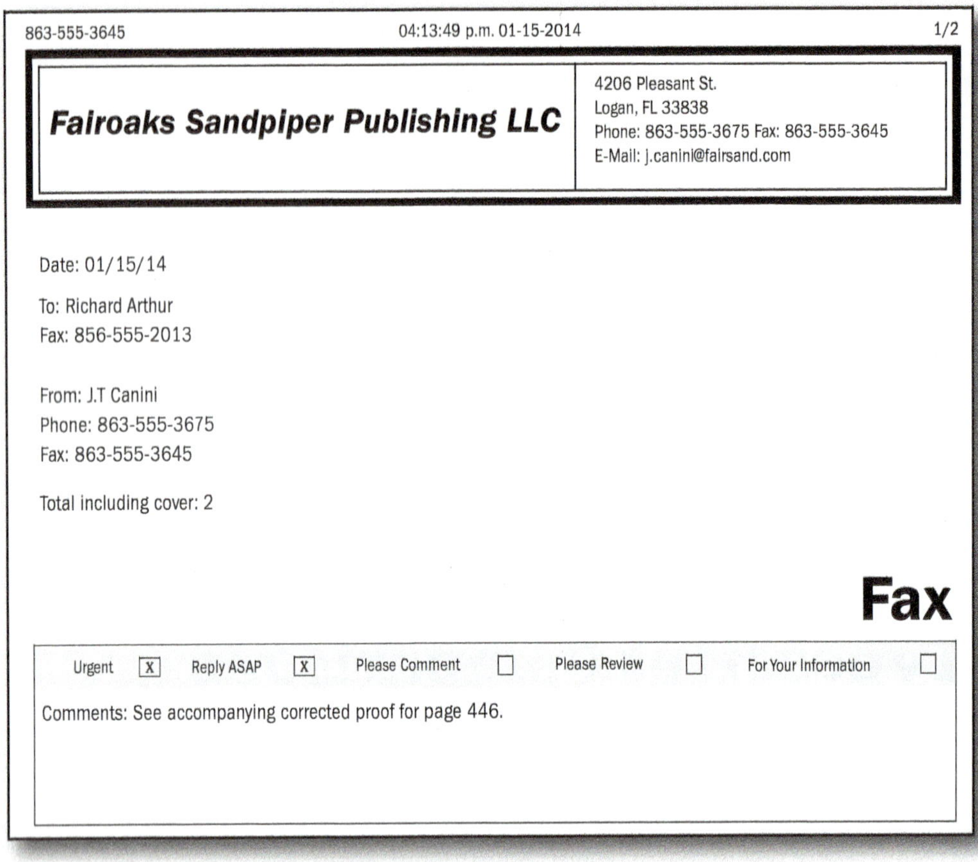
863-555-3645 04:13:49 p.m. 01-15-2014 1/2

Fairoaks Sandpiper Publishing LLC

4206 Pleasant St.
Logan, FL 33838
Phone: 863-555-3675 Fax: 863-555-3645
E-Mail: j.canini@fairsand.com

Date: 01/15/14

To: Richard Arthur
Fax: 856-555-2013

From: J.T Canini
Phone: 863-555-3675
Fax: 863-555-3645

Total including cover: 2

Fax

Urgent [X] Reply ASAP [X] Please Comment [ ] Please Review [ ] For Your Information [ ]

Comments: See accompanying corrected proof for page 446.

FIGURE 18–5 A fax page showing a transmitting terminal identifier (TTI).

and exemplars sent to the Swiss and American experts as purportedly the document's text should have different type styles. TTIs can be fraudulently prepared and placed in the appropriate position on a fax copy. However, a microscopic examination of the TTI's print quickly reveals significant characteristics that distinguish it from a genuine TTI.

In determining the fax machine's model type, the examiner usually begins by analyzing the TTI type style. The fonts of that line are determined by the sending machine. The number of characters, their style, and their position in the header are best evaluated through a collection of TTI fonts organized into a useful database. One such database is maintained by the American Society of Questioned Document Examiners.

Computer Printers The determination of what model of computer printer has been used requires extensive analysis of the specific printer technology and type of ink used. Visual and microscopic techniques help determine the technology and toner used. Generally printers are categorized as *impact* and *nonimpact* printers by the mechanism of their toner application. Nonimpact printers, such as ink-jet and laser printers, and impact printers, such as thermal and dot-matrix printers, all have characteristic ways of printing documents. Character shapes, toner differentiation, and toner application methods are easily determined with a low-power microscope and help the examiner narrow the possibilities of model type.

In analyzing computer printouts and faxes, examiners use the same approach for comparing the markings on a questioned document to exemplar documents generated by a suspect machine. These markings include all possible transitory patterns arising from debris and other extraneous materials. When the suspect machine is not available, the examiner may need to analyze the document's class characteristics to identify the make and model of the machine. It is important to identify the printing technology, the type of paper, the type of toner or ink used, the chemical composition of the toner, and the type of toner-to-paper fusing method used in producing the document.

Examination of the toner usually involves microscopic analysis to characterize its surface morphology, followed by identification of the inorganic and organic components of the toner. These results separate model types into categories based on their mechanical and printing characteristics. Typically, document examiners access databases to help identify the model type of machine used to prepare a questioned document. The resulting list of possibilities produced by the database hopefully reduces the number of potential machines to a manageable number. Obviously, once a suspect machine is identified, the examiner must perform a side-by-side comparison of questioned and exemplar printouts, as described previously.

Quick Review

- The examiner compares the individual type character's style, shape, and size to a complete reference collection of past and present typefaces.
- Use of a printing device results in wear and damage to the machine's moving parts in a way that is both random and irregular, thereby imparting individual characteristics to it.
- Transitory defect marks originating from random debris on the glass platen, inner cover, or mechanical portions of a copier produce irregularly shaped images that may serve as points of comparison.
- A TTI, or transmitting terminal identifier, is a header at the top of each page of a fax document. It is useful in document comparison because it serves as a way to distinguish between a real and a fraudulently prepared fax document.

Alterations, Erasures, and Obliterations

Documents are often altered or changed after preparation, to hide their original intent or to perpetrate a forgery. Documents can be changed in several ways, and for each way, the application of a special discovery technique is necessary.

One of the most common ways to alter a document is to try to erase parts of it, using an India rubber eraser, sandpaper, a razor blade, or a knife to remove writing or type by abrading or scratching the paper's surface. All such attempts at erasure disturb the upper fibers of the paper. These changes are apparent when the suspect area is examined under a microscope using direct light or by allowing the light to strike the paper obliquely from one side (side lighting; see Figure 18–6). Although microscopy may reveal whether an **erasure** has been made, it does not necessarily indicate the original letters or words present. Sometimes so much of the paper has been removed that identifying the original contents is impossible.

erasure
The removal of writing, typewriting, or printing from a document, normally accomplished by either chemical means or an abrasive instrument

FIGURE 18–6 Erasure in a checkbook deposit stub revealed by photography with the use of oblique lighting. The amount of cash deposited was changed and the difference was pocketed by the bookkeeper. *Courtesy Robin D. Williams, MFS, MS, D-BFDE, Omni Document Examinations, Green Bay, WI, rwilliams@* docexamination.com

In addition to abrading the paper, the perpetrator may also obliterate words with chemicals. In this case, strong oxidizing agents are placed over the ink, producing a colorless reaction product. Although such an attempt may not be noticeable to the naked eye, examination under the microscope reveals discoloration on the treated area of the paper. Sometimes examination of the document under ultraviolet or infrared lighting reveals the chemically treated portion of the paper. Interestingly, examination of documents under ultraviolet light may also reveal fluorescent ink markings that go unnoticed in room light, as seen in Figure 18–7.

infrared luminescence
A property exhibited by some dyes that emit infrared light when exposed to blue-green light

Some inks, when exposed to blue-green light, absorb the radiation and reradiate infrared light. This phenomenon is known as **infrared luminescence**. Thus, alteration of a document with ink differing from the original can sometimes be detected by illuminating the document with blue-green light and using infrared-sensitive film to record the light emanating from the document's surface. In this fashion, any differences in the luminescent properties of the inks are observed. Infrared luminescence has also revealed writing that has been erased. Such writings may be recorded by invisible residues of the original ink that remain embedded in the paper even after an erasure.

Another important application of infrared photography arises from the observation that inks differ in their ability to absorb infrared light. Thus, illuminating a document with infrared light and recording the light reflected off the document's surface with infrared-sensitive film enables the examiner to differentiate inks of a dissimilar chemical composition (see Figure 18–8).

FIGURE 18–7 (a) A twenty-dollar bill as it appears under room light. (b) The bill illuminated with ultraviolet light reveals ink writing. *Courtesy Sirchie Finger Print Laboratories, Inc., Youngsville, N.C., www. sirchie.com*

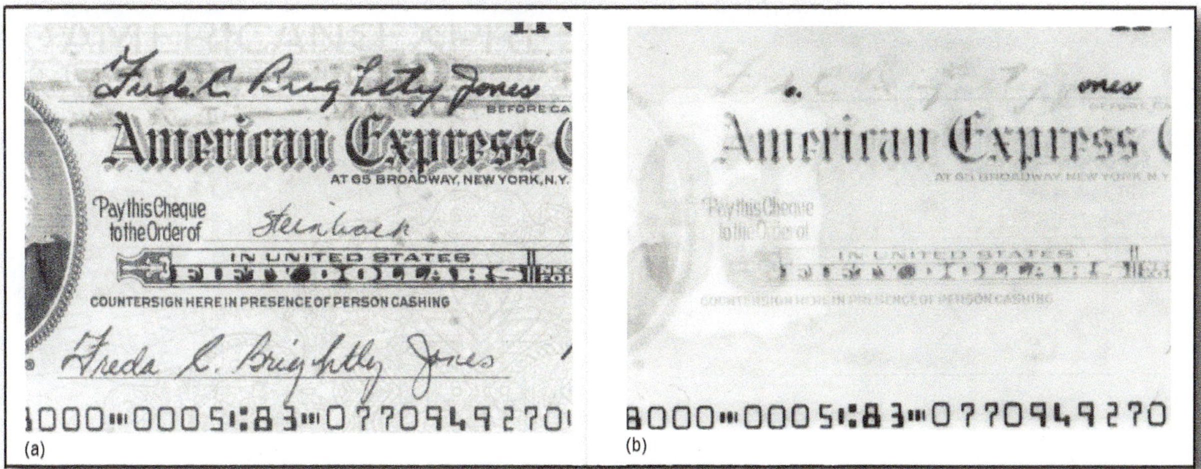

FIGURE 18–8 (a) This photograph, taken under normal illumination, shows the owner of an American Express check to be "Freda C. Brightly Jones." Actually, this signature was altered. The check initially bore the signature "Fred C. Brightly Jr." (b) This photograph taken under infrared illumination, using infrared-sensitive film, clearly shows that the check was altered by adding *a* to Fred and *ones* to *Jr*. The ink used to commit these changes is distinguishable because it absorbs infrared light, whereas the original ink does not. *Courtesy Richard Saferstein, Ph.D.*

obliteration
Blotting out or smearing over writing or printing to make the original unreadable

Intentional **obliteration** of writing by overwriting or crossing out is seldom used for fraudulent purposes because of its obviousness. Nevertheless, such cases may be encountered in all types of documents. Success at permanently hiding the original writing depends on the material used to cover the writing. If it is done with the same ink as was used to write the original material, recovery will be difficult if not impossible. However, if the two inks are of a different chemical composition, photography with infrared-sensitive film may reveal the original writing. Infrared radiation may pass through the upper layer of writing while being absorbed by the underlying area (see Figure 18–9).

Close examination of a questioned document sometimes reveals staple holes, crossing strokes or strokes across folds of perforations in the paper that are not in a sequence that is consistent with the natural preparation of the document. Again, these differences can be shown by microscopic or photographic scrutiny (see Figure 18–10).

charred document
Any document that has become darkened and brittle through exposure to fire or excessive heat

Infrared photography sometimes reveals the contents of a document that has been accidentally or purposely charred in a fire. Another way to decipher **charred documents** involves reflecting light off the paper's surface at different angles in order to contrast the writing against the charred background (see Figure 18–11).

FIGURE 18–9 An order number was obliterated with magic marker to cover up a theft. Infrared photography was used to penetrate the covering ink to reveal the original writing. *Courtesy Robin D. Williams, MFS, MS, D-BFDE, Omni Document Examinations, Green Bay, WI, rwilliams@docexamination.com*

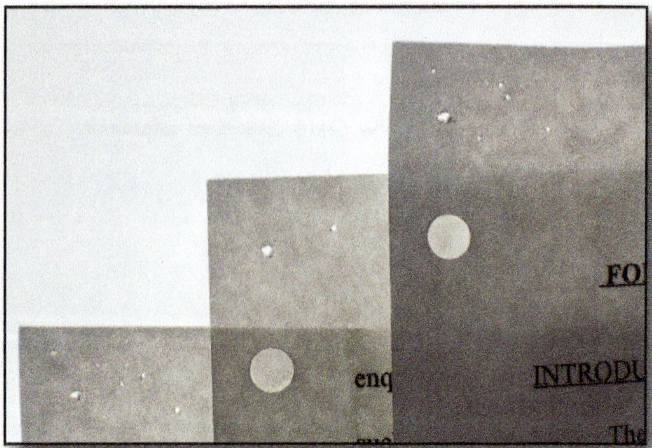

FIGURE 18–10 Staple hole exam. The middle sheet was substituted based on the evidence of staple hole pattern differences. *Courtesy Jacqueline A Joseph, D-BFDE, Handwriting and Document Forensics, Portland, OR. www.jjhandwriting.com*

FIGURE 18–11 Decipherment of charred papers seized in the raid of a suspected bookmaking establishment. The charred documents were photographed with reflected light.
Courtesy Richard Saferstein, Ph.D.

Digital image processing is the method by which the visual quality of digital pictures is improved or enhanced, resulting in digital data recovery. *Digitizing* is the process by which the image is stored in memory. This is commonly done by scanning an image with a flatbed scanner or a digital camera and converting the image by computer into an array of digital intensity values called *pixels*, or picture elements. Once the image has been digitized, an image-editing program such as Adobe Photoshop is used to adjust the image. An image may be enhanced through lightening, darkening, and color and contrast controls. Examples of how the technology is applied to forensic document examination are shown in Figures 18–12 and 18–13.

FIGURE 18–12 This composite demonstrates the various changes that can be applied to a digitized image in order to reveal information that has been obscured. Using a photo editor (Adobe Photoshop), the original was duplicated and pasted as a second layer. Colors were changed in selected areas of the image using the "screen" and "exclusion" options. "Replace color" allows the user to choose a specific color or range of colors and lighten, darken, or change the hue of the colors selected. "Level" and "curves" tools can adjust the lightest and darkest color ranges and optimize contrast, highlights, and shadow detail of the image for additional clarity. *Courtesy Bob Garrett, IDMAN Forensics*

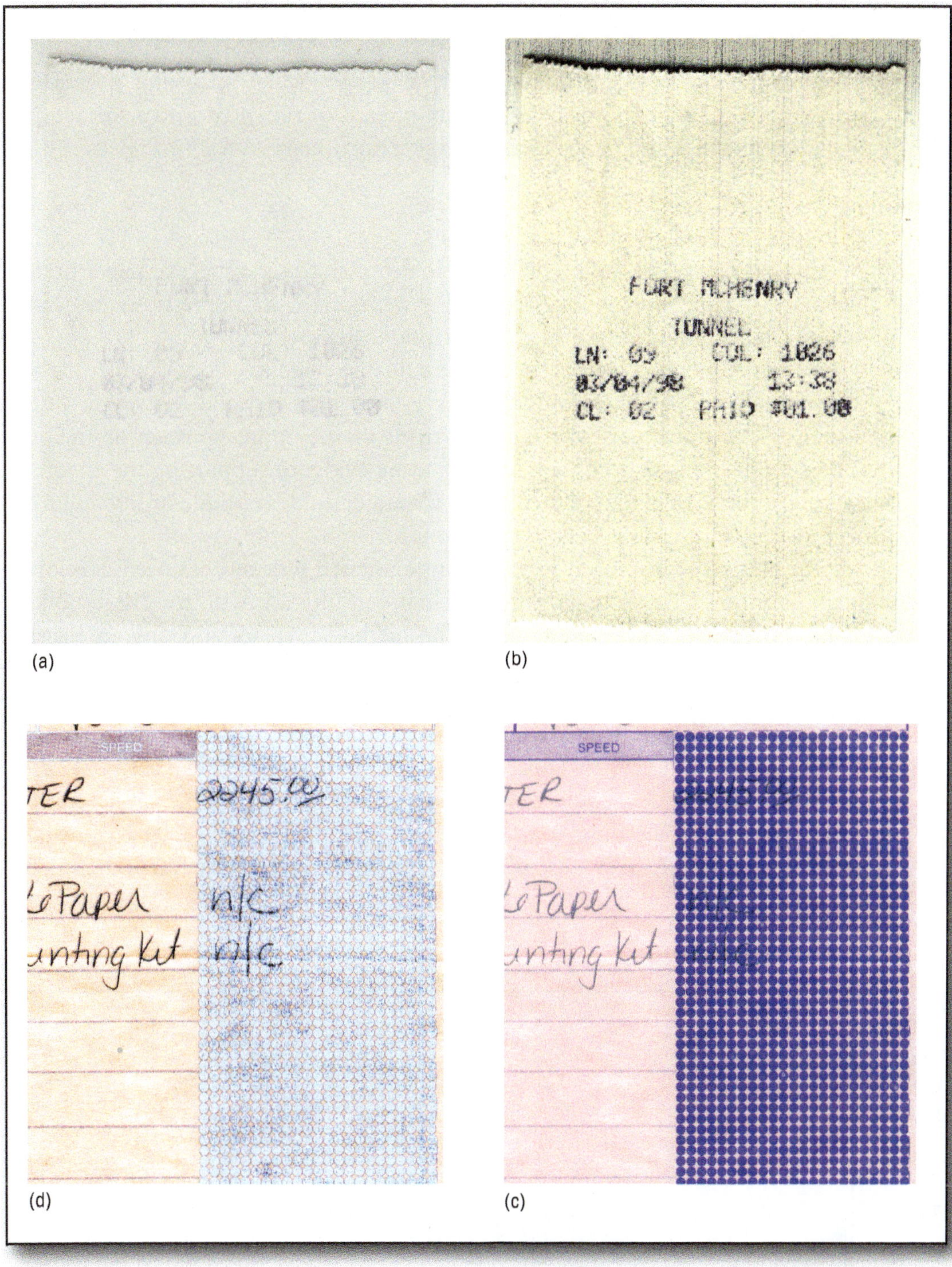

FIGURE 18–13 (a) Receipts have been used in investigations to establish a victim's whereabouts, provide suspects with alibis, and substantiate a host of personal conduct. Unfortunately, due to wear, age, or poor printing at the register, the receipt is often unreadable. This can be corrected using photo-editing software. In this example, the original toll receipt was scanned at the highest color resolution, which allows more than 16 million colors to be reproduced. The image was then manipulated, revealing the printed details, by adjusting the lightest and darkest levels and the color content of the image. (b) Invoices may contain details about a transaction that are important to an investigation. The copy that ships with the merchandise may have that information blocked out. This information may be recovered using digital imaging. The left figure shows the original shipping ticket. The right figure shows the information revealed after replacing the color of the blocking pattern. *Courtesy Bob Garrett, IDMAN Forensics*

Other Document Problems

Document examiners encounter other challenges when analyzing questioned documents, including visualizing writing pressed or indented into a surface and analyzing the inks and paper used in suspect documents.

Indented Writings

indented writings
Impressions left on paper positioned under a piece of paper that has been written on

Indented writings are the partially visible depressions on a sheet of paper underneath the one on which the visible writing was done. Such depressions are caused by the application of pressure on the writing instrument.

Indented writings have proved to be valuable evidence. For example, the top sheet of a bookmaker's records may have been removed and destroyed, but it still may be possible to determine the writing by the impressions left on the pad. These impressions may contain incriminating evidence supporting the charge of illegal gambling activities. When paper is studied under oblique or side lighting, its indented impressions are often readable.

An innovative approach to visualizing indented writings has been developed at the London College of Printing in close consultation with the Metropolitan Police Forensic Science Laboratory.[3] The method involves applying an electrostatic charge to the surface of a polymer film that has been placed in contact with a questioned document, as shown in Figure 18–14. Indented impressions on the document are revealed by applying a toner powder to the charged film. For many documents examined by this process, clearly readable images have been produced from impressions that could not be seen or were barely visible under normal illumination. An instrument that develops indented writings by electrostatic detection is commercially available and is routinely used by document examiners.

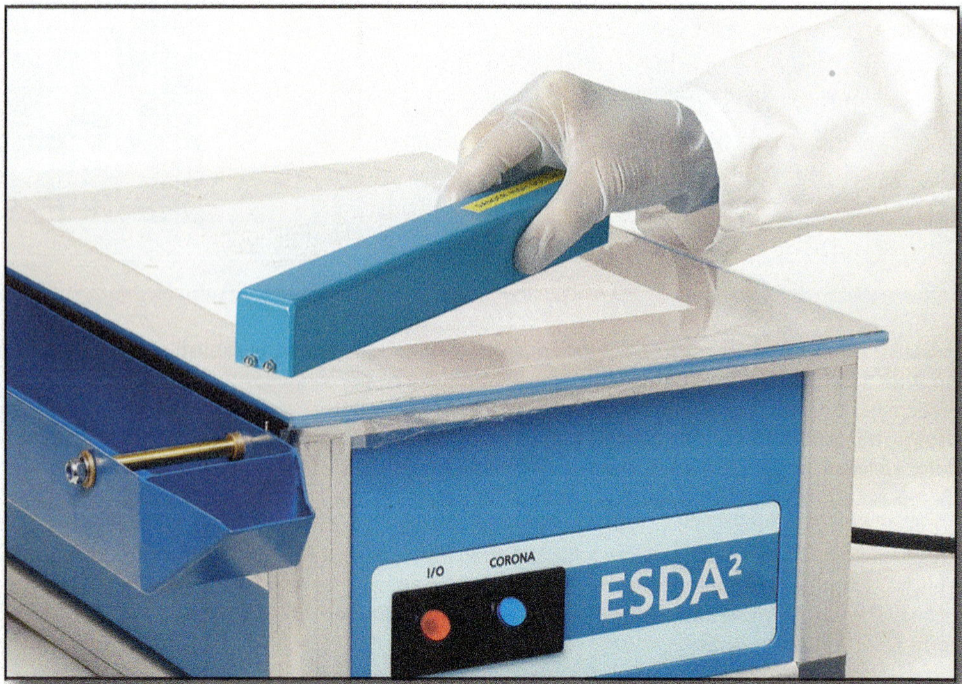

FIGURE 18–14 An electostatic detection apparatus (ESDA) works by applying an electrostatic charge to a document suspected of containing indented writings. The indentations are then visualized by the application of charge-sensitive toner. *Courtesy Foster & Freeman Limited, Worcestershire, U.K., www.fosterfreeman.co.uk*

Ink and Paper Comparisons

A study of the chemical composition of writing ink present on documents may verify whether known and questioned documents were prepared by the same pen. A nondestructive approach to comparing ink lines is accomplished with a visible microspectrophotometer.[4] A case example illustrating the application of this approach to ink analysis appears in Figure 8–12. Thin-layer chromatography is also suitable for ink comparisons. Most commercial inks, especially ballpoint inks, are actually mixtures of several organic dyes. These dyes can be separated on a properly developed thin-layer chromatographic plate. The separation pattern of the component dyes is distinctly different for inks with different dye compositions and thus provides many points of comparison between a known and a questioned ink.

Ink can be removed from paper with a hypodermic needle with a blunted point to punch out a small sample from a written line. About ten plugs or microdots of ink are sufficient for chromatographic analysis. U.S. Secret Service and the Internal Revenue Service jointly maintain the U.S. International Ink Library. This collection includes more than 8,500 inks, which date back to the 1920s. Each year new pen and ink formulations are added to the reference collection. These inks have been systematically cataloged according to dye patterns developed by thin-layer chromatography (TLC; see Figure 18–15). On several occasions, this approach has been used to prove that a document has been fraudulently backdated. For example, in one instance, it was possible to establish that a document dated 1958 was backdated because a dye identified in the questioned ink had not been synthesized until 1959.

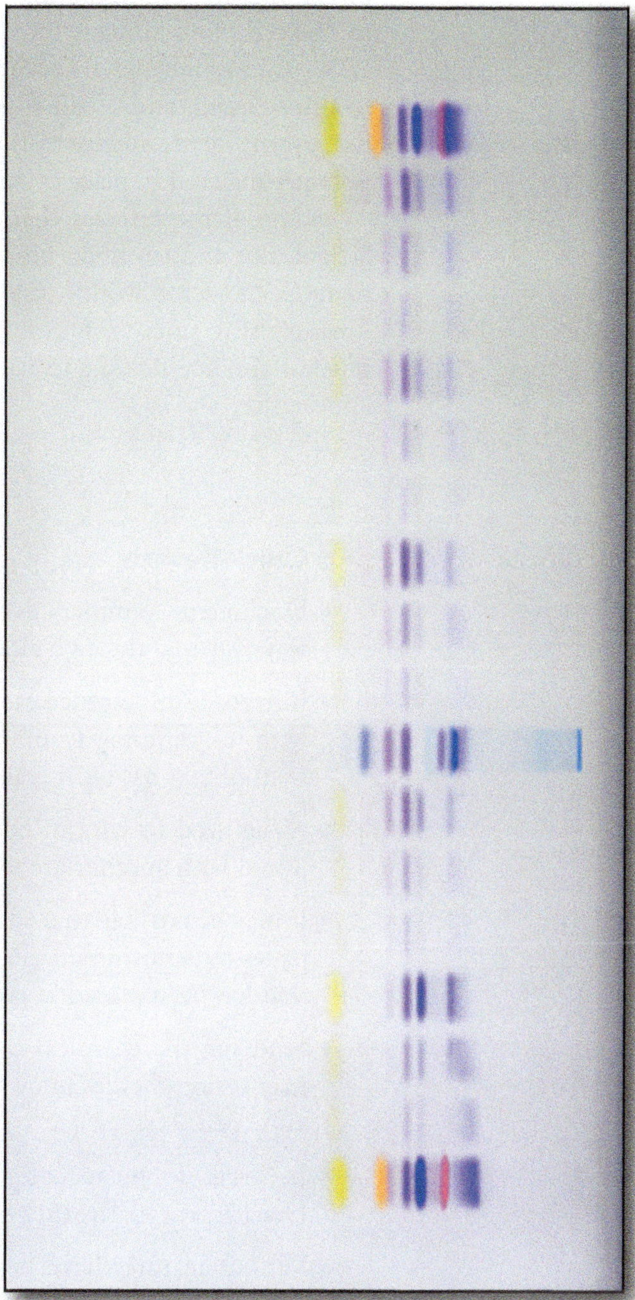

FIGURE 18–15 Chart demonstrating different TLC patterns of blue ballpoint inks. *Courtesy US Secret Service Laboratory, Washington, D.C.*

To further aid forensic chemists in ink-dating matters, several ink manufacturers, at the request of the U.S. Treasury Department, voluntarily tag their inks during the manufacturing process. The tagging program allows inks to be dated to the exact year of manufacture by changing the tags annually.

Another area of inquiry for the document examiner is the paper on which a document is written or printed. Paper is often made from cellulose fibers found in wood and fibers recovered from recycled paper products. The most common features associated with a paper examination are general appearance, color, weight, and watermarks. Other areas of examination include fiber identification and the characterization of additives, fillers, and pigments present in the paper product.

Counterfeiting Currency

One of the oldest of all known crimes is counterfeiting. Since the 19th century when banks issued their own currency, people have attempted to create counterfeit money. Various techniques have been used through the years by counterfeiters, such as offset printing to color copiers, computers, scanner, and inkjet printers. The U.S. Department of Treasury has made changes to currency to try to make it more difficult to counterfeit. Some of the features incorporated into U.S. currency include 3D security ribbon, color-shifting ink, portrait watermark, security thread, and a bell in the inkwell. Other countries use features such as fluorescent fibers, microscript, holograms, embossed ink, currency is printed on a polymer material in place of paper, infrared images, and pearl ink.

Some of the features that can help identify counterfeit currency include a lack of blue and red fibers embedded in the paper; blurred, broken, or indistinct margins and scroll work; variations in ink color between the serial numbers and Treasury seal; uneven, blunt, or broken sawtooth points around Treasury seal; portrait appears lifeless and flat; or the background being too dark or mottled in appearance.

Quick Review

- Document examiners deal with evidence that has been changed in several ways, such as through alterations, erasures, and obliterations.
- Infrared luminescence can be used to detect alterations to a document made with ink differing from the original. Infrared luminescence can also reveal writing that has been erased.
- A digitized image can be lightened, darkened, and color and contrast adjusted with appropriate software.
- It may be possible to read indented writing—the impressions left on a paper pad—by applying an electrostatic charge to the surface of a polymer film that has been placed in contact with a questioned document.
- Studying the chemical composition of writing ink present on documents may verify whether known and questioned documents were prepared by the same pen.
- In order to make counterfeiting U.S. money more difficult, the U.S. Department of Treasury has made changes to currency.
- Various features have been added to money in other countries to make counterfeiting currency more difficult.

Chapter Review

- Any object with handwriting or print whose source or authenticity is in doubt may be referred to as a questioned document.
- Document examiners gather documents of known authorship or origin and compare them to the individual characteristics of questioned writings.
- Collecting an adequate number of known writings is critical for determining the outcome of a handwriting comparison. Known writing should contain some of the words and combinations of letters in the questioned document.
- The unconscious handwriting of two different individuals can never be identical. However, the writing style of an individual may be altered beyond recognition by the influence of drugs or alcohol.
- The examiner compares the individual type character's style, shape, and size to a complete reference collection of past and present typefaces.
- Use of a printing device results in wear and damage to the machine's moving parts in a way that is both random and irregular, thereby imparting individual characteristics to it.
- Transitory defect marks originating from random debris on the glass platen, inner cover, or mechanical portions of a copier produce irregularly shaped images that may serve as points of comparison.
- A TTI, or transmitting terminal identifier, is a header at the top of each page of a fax document. It is useful in document comparison because it serves as a way to distinguish between a real and a fraudulently prepared fax document.
- Variations in vertical and horizontal alignment and perpendicular misalignment of characters, as well as defects in each typeface, are valuable for proving the identity of a typewriter.
- Document examiners deal with evidence that has been changed in several ways, such as through alterations, erasures, and obliterations.
- Infrared luminescence can be used to detect alterations to a document made with ink differing from the original. Infrared luminescence can also reveal writing that has been erased.
- A digitized image can be lightened, darkened, and color and contrast adjusted with appropriate software.
- It may be possible to read indented writing—the impressions left on a paper pad—by applying an electrostatic charge to the surface of a polymer film that has been placed in contact with a questioned document.
- Studying the chemical composition of writing ink present on documents may verify whether known and questioned documents were prepared by the same pen.

Quick Lab: Handwriting Comparison

Materials:

Piece of paper with exemplar signature (prepared by instructor)
Notebook paper
Scissors
Tape

Procedure:

Handwriting comparison is a basic tool in the arsenal of investigators looking into cases involving possible forgery. In this exercise, divided into teams of three or four. Each team member will sign their name on the same piece of paper. Your instructor will then give each student a piece of paper containing a signature. Working alone for 2 minutes, each student will attempt to copy the signature on a blank piece of paper. When the time has expired, use the scissors to cut your finished "forgery" from the piece of paper and fold it so no one can see it. One member of each team will collect the scraps of paper containing the team's "forgeries," mix them, then unfold and tape each to the piece of paper containing the tam members' signatures. Mark the forgeries with a letter (A, B, C, etc.) to identify each. Each member will now have 3 minutes to perform handwriting comparisons by comparing the exemplar signatures to the "forgeries." Team members will analyze characteristics such as letter, line, and format, and determine which team member created each "forgery." Write your answers on a separate piece of paper. When all members have finished, each team member reveals which forgery they created. Check your answers to see how many you guessed correctly.

Review Questions

1. Critical to the examination of questioned documents is
   a. the preservation of the questioned document in its original form.
   b. the recognition that only handwritten or typewritten documents are suitable for examination.
   c. the gathering of documents of known authorship.
   d. the knowledge and understanding on the part of the document examiner of the current scientific formula to apply to the examination.

2. A nondestructive approach to comparing ink lines is accomplished with
   a. a comparison microscope.
   b. a visible microspectrophotometer.
   c. infrared spectroscopyluminescence.
   d. digitizing.

3. Chemical erasures of words or numbers may be revealed by examination of handwriting under
   a. visible light.
   b. X-ray fluorescence.
   c. UV light.
   d. infrared luminescence.

4. Handwriting containing inks of different chemical compositions may be distinguished by photography with
   a. UV light.
   b. visible light.
   c. IR light.
   d. all of the above

5. The age of an ink may be determined by utilizing an ink library based on analysis by
   a. gas chromatography.
   b. mass spectrometry.
   c. thin-layer chromatography.
   d. infrared spectrophotometry.

6. True or False: Because people sign their names the same way every time, if an original signature was traced over, the handwriting examiner would have to conclude that the tracing was produced by the signer of the original.

7. True or False: The age difference between genuine and unknown specimens is not an important factor in a handwriting comparison.

8. True or False: A single handwriting characteristic can by itself be taken as a basis for a positive comparison.

9. True or False: Random wear and damages to a printer impart it with class characteristics.

10. True or False: A TTI, or transmitting terminal identifier, is a header at the top of each page of a fax document.

11. What is a questioned document?

12. Name at least five characteristics of handwriting in which one might expect to encounter variations between individuals.

13. Name at least five factors besides handwriting characteristics that can impart individual variations to writing.

14. Describe two situations in which a document examiner may be prevented from coming to a positive conclusion about a questioned document.

15. What are exemplars? Why are they important for document examination?

16. What are natural variations? How can they be useful for detecting forgeries?

17. When comparing sample writing to a suspect document, the age difference between the documents should be no more than
    a. 6 to 12 months.
    b. 12 to 18 months.
    c. two to three years.
    d. five to seven years.
18. What constitutional principle did the Supreme Court address in *Gilbert v. California*, and how did the court rule?
19. What constitutional principle did the Supreme Court address in the case of *United States v. Mara, and how did the court rule?*
20. What distinguishing marks serve as points of comparison on documents produced by a photocopier?
21. What is a TTI? Name two ways in which it is useful in document comparison.
22. Name the two general categories of printers and list two examples of each type.
23. Name five important characteristics of a printer, photocopier, or fax machine that a document examiner must identify when analyzing a document in a situation in which the suspect machine is not available.
24. What two processes typically are involved in examination of toner?
25. What two questions are document examiners most often asked about typewriters?
26. What individual characteristics of a typewriter are valuable for proving its identity?
27. Name two ways in which infrared luminescence is useful in examining documents.
28. Describe two methods for visualizing indented writing.
29. Name two analytical techniques used to analyze writing inks.
30. List the most common features associated with a paper examination.

Learn the Vocabulary

Use the information from this chapter along with your prior knowledge and experience to learn the key terms. Choose from the following activities:

1. Make flashcards with a term on the front and its definition on the back. Use the flashcards to quiz a partner on correctly pronouncing and defining the terms.
2. Make an audio or video recording of each term and its definition.
3. Work with a partner to write a sentence for each term. Read the sentences out loud to a partner or share them with the class.
4. Write a list of new terms and language you have encountered in this chapter. Reuse the terms to write sentences and then read the sentences out loud to a partner. Reuse the terms again by writing a short speech. The speech might explain the terms or teach a lesson using the terms. Give the speech to a partner or to the class.

Application and Critical Thinking

1. Criminalist Julie Sandel is investigating a series of threatening notes written in pencil and sent to a local politician. A suspect is arrested, and Julie directs the suspect to prepare writing samples to compare to the writing on the notes. She has the suspect sit at a desk in an empty office and gives him a pen and a piece of paper. She begins to read one of the notes and asks the suspect to write the words she dictates. After reading about half a page, she stops, then dictates the same part of the note a second time for the suspect. At one point, the suspect indicates that he does not know how to spell one of the words, so Julie spells it for him. After completing the task, Julie takes the original notes and the dictated writing from the suspect to a document examiner. What mistakes, if any, did Julie make?

2. In each of the following situations, indicate how you would go about recovering original writing that is not visible to the naked eye.
   a. The original words have been obliterated with a different ink than was used to compose the original.
   b. The original words have been obliterated by chemical erasure.
   c. The original writing was made with fluorescent ink.
   d. The original documents have been charred or burned.

3. You have been asked to determine whether a handwritten will, supposedly prepared 30 years ago, is authentic or a modern forgery. What aspects of the document would you examine to make this determination? Explain how you would use thin-layer chromatography to help you come to your conclusion.

4. Research different types of examinations performed on digital and physical evidence in a forensic laboratory. In a short presentation, explain your findings.

5. Investigate the security features incorporated in the U.S. and foreign currency to prevent counterfeiting. Describe which of these features is the most effective in preventing counterfeiting.

Laboratory Experiments

This activity requires the use of the following practices of science and engineering:

- Asking questions and defining problems
- Planning and carrying out investigations
- Analyzing and interpreting data
- Constructing explanations
- Obtaining, evaluating, and communicating information
- Using appropriate safety equipment and practices as outlined in Texas Education Agency-approved safety standards

This activity consists of the following crosscutting concept:

- **Patterns**—The patterns in characteristics of handwriting allow for the identification of the author.

Before you begin, read the scenario and steps out loud with a partner. Ask for clarification as necessary to make sure you understand the experiment.

The Scenario

A letter was found posted on a wall in the student union at the University of Arizona near where a statue of Wilbur the Wildcat should be. The note read as follows:

> I am in possession of
> Wilbur. Unless you provide
> $10,000.00 within one week,
> I will destroy the statue. I
> will contact you with further
> instructions in two days.

Handwriting samples from 5 individuals seen near the student union during the time Wilbur went missing have been provided. Study the ransom note and the suspects' handwriting samples to determine if any of the suspects could have written the ransom note and taken Wilbur. You will summarize your findings and submit data tables to support your conclusion.

DATA FOR HANDWRITING COMPARISON

Complete one data table for the ransom note and each of the suspect samples.

Characteristic #	*Yes*	*No*	*Comments (and measurements in mm, if required)*
1. Is line quality smooth?			
2. Are words and margins equally spaced?			Margins: Words:
3. Is the ratio of small letters to capital letters consistent? What is the ratio?			
4. Is the writing continuous?			
5. Are letters connected between capitals and lowercase letters?			
6. Are letter formations complete?			(Be specific, which letters?)
7. Is all of the writing in cursive?			(Be specific, which words?)
8. Is the pen pressure the same throughout?			
9. Do all the letters slant to the right?			
10. Are all the letters written on the line?			
11. Are there fancy curls or loops?			(Which letters?)
12. Are all i's and t's dotted and crossed (top, middle, or not)?			i's t's

Endnotes

1. 388 U.S. 263 (1967).
2. 410 U.S. 19 (1973).
3. D. M. Ellen, D. J. Foster, and D. J. Morantz, "The Use of Electrostatic Imaging in the Detection of Indented Impressions," *Forensic Science International* 15 (1980): 53.
4. P. W. Pfefferli, "Application of Microspectrophotometry in Document Examination," *Forensic Science International* 23 (1983): 129.

Computer Forensics

Andrew W. Donofrio

19

Getty Images New/Getty Images

Key Terms

- bit
- bookmark
- byte
- central processing unit (CPU)
- cluster
- cookies
- file slack
- firewall
- hacking
- hard disk drive (HDD)
- hardware
- Internet cache
- Internet history
- latent data
- Message Digest 5 (MD5)/Secure Hash Algorithm (SHA)
- motherboard
- operating system (OS)
- partition
- RAM slack
- random-access memory (RAM)
- sector
- software
- swap file
- temporary files
- unallocated space
- visible data

The BTK Killer

Dennis Rader was arrested in February 2005 and charged with committing ten murders since 1974 in the Wichita, Kansas area. The killer, whose nickname stands for "bind, torture, kill," hadn't murdered since 1991, but he resurfaced in early 2004 by sending a letter to a local newspaper taking credit for a 1986 slaying. Included with the letter were a photocopy of the victim's driver's license and three photos of her body. The BTK killer was back to his old habit of taunting the police.

Three months later, another letter surfaced. This letter detailed some of the events surrounding BTK's first murder victims. In 1974, he strangled Joseph and Julie Otero along with two of their children. Shortly after those murders, BTK sent a letter to a local newspaper in which he gave himself the name BTK. In December 2004, a package found in a park contained the driver's license of another BTK victim along with a doll covered with a plastic bag, its hands bound with pantyhose.

The major break in the case came when BTK sent a message on a computer disk to a local TV station. "Erased" information on the disk was recovered and restored by forensic computer specialists, and the disk was traced to the Christ Lutheran Church in Wichita. The disk was then quickly linked to Dennis Rader, the church council president. The long odyssey of the BTK killer was finally over.

Learning Objectives

After studying this chapter you should be able to:

19.1 Describe the hardware and software components of a computer

19.2 Explain how data is stored and retrieved from a hard disk drive

19.3 Describe the proper procedure for preserving computer evidence at a crime scene

19.4 Explain the analysis of visible and latent data

19.5 Relate various areas found on the computer where a user's Internet activities can be investigated

19.6 Explain the role of IP in forensic investigation of Internet communications

19.7 Discuss the techniques used to investigate unauthorized computer intrusion

19.8 Describe the types of services offered by modern mobile devices, such as cell phones, and the potential investigative value they have

National Science Content Standards

Scientific Inquiry

Science and Technology

History and Nature of Science

Since the 1990s, few fields have progressed as rapidly as computer technology. Computers are no longer a luxury, nor are they in the hands of just a select few. Technology and electronic data are a part of everyday life and permeate all aspects of society. Consequently, computers have become increasingly important as sources of evidence in an ever-widening spectrum of criminal activities.

Investigators frequently encounter computers and other digital devices in all types of cases. As homicide investigators sift for clues, they may inquire whether the method for a murder was researched on the Internet, whether signs of an extramarital affair can be found in e-mail or remnants of instant messages (which may provide motive for a spouse killing or murder for hire), or whether an obsessed stalker threatened the victim before a murder. Arson investigators want to know whether financial records on a computer may provide a motive in an arsonfor-profit fire. A burglary investigation would certainly be aided if law enforcement determined that the proceeds from a theft were being sold online—perhaps through eBay or a similar online auction site.

Accessibility to children and the perception of anonymity has given sexual predators a way to seek out child victims online. The vulnerability of computers to hacker attacks is a constant reminder of security issues surrounding digitally stored data. Finally, the fact that computers control most of our critical infrastructure makes technology an appetizing target for would-be terrorists.

Computer forensics involves preserving, acquiring, extracting, analyzing, and interpreting computer data. Although this is a simple definition, it gets a bit more complicated. Part of this complication arises from technology itself. More and more devices are capable of storing electronic data: cell phones, personal digital assistants (PDAs), iPods®, digital cameras, flash memory cards, smart cards, jump drives, and many others. Each method for extracting data from these devices presents unique challenges. However, sound forensic practices apply to all these devices. The most logical place to start to examine these practices is with the most common form of electronic data: the personal computer.

Andrew W. Donofrio is a lieutenant with the prosecutor's office in Bergen County, New Jersey, and is a leading computer forensics examiner for Bergen County, with more than 20 years' experience in law enforcement. He has conducted more than five hundred forensic examinations of computer evidence and frequently lectures on the subject throughout the state, as well as teaching multiday courses on computer forensics and investigative topics at police academies and colleges in New Jersey. Detective Sergeant Donofrio writes regularly on Internet-related and computer forensics issues for a number of law enforcement publications and has appeared as a guest expert on Internet-related stories on MSNBC.

From Input to Output: How Does the Computer Work?

Hardware versus Software

Before we get into the nuts and bolts of computers, we must establish the important distinction between **hardware** and software. Hardware comprises the physical components of the computer: the computer chassis, monitor, keyboard, mouse, hard disk drive, random-access memory (RAM), central processing unit (CPU), and so on (see Figure 19–1). The list is much more extensive but, generally speaking, if it is a computer component or peripheral that you can see, feel, and touch, it is hardware.

hardware
The physical components of a computer: case, keyboard, monitor, motherboard, RAM, HDD, mouse, and so on; generally speaking, if it is a computer component you can touch, it is hardware

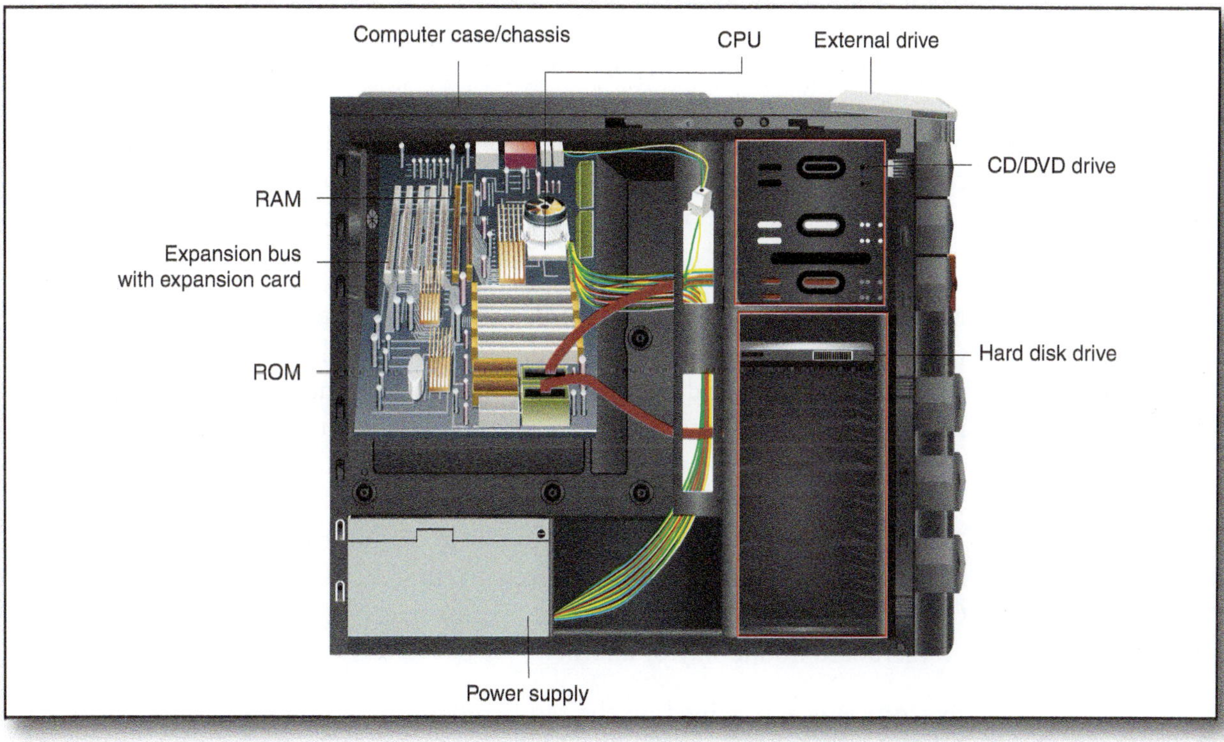

FIGURE 19–1 Cutaway diagram of a personal computer. *Courtesy Richard Saferstein, Ph.D.*

Software, conversely, is a set of instructions compiled into a program that performs a particular task. Software consists of programs and applications that carry out a set of instructions on the hardware. Operating systems (Windows®, Mac® OS, Linux®, Unix®), word-processing programs (Microsoft® Word, WordPerfect®), web-browsing applications (Internet Explorer®, Safari®, Firefox®), and accounting applications (Quicken®, QuickBooks®, Microsoft Money) are all examples of software.

software
A set of instructions compiled into a program that performs a particular task

It is important not to confuse software with the physical media that it comes on. When you buy an application such as Microsoft Office, it comes on a compact disc (CD). The CD containing this suite of applications is typically referred to as software, but this is technically wrong. The CD is external computer media that contains the software; it is a container for and a medium to load the set of instructions onto the hard disk drive (the hardware).

Hardware Components

Computer Case/Chassis The case is the physical box holding the fixed internal computer components in place. Cases come in many shapes and sizes: a full upright tower chassis, a slim desktop model sitting on the desktop, or an all-in-one monitor/computer case like the iMac. For our purposes, the term *system unit* is probably most appropriate when describing a chassis seized as evidence. The term *system unit* accurately references the chassis, including the motherboard and other internal components.

Power Supply The term *power supply* is actually a misnomer, because it doesn't actually supply power—the power company does that. Rather, a computer's power supply converts power from the wall outlet to a usable format for the computer and its components. Different power supplies have different wattage ratings. The use, or more specifically the components, of the computer dictate the appropriate power supply.

motherboard
The main system board of a computer (and many other electronic devices) that delivers power, data, and instructions to the computer's components

Motherboard The main circuit board in a computer (or other electronic devices) is referred to as the **motherboard**. Motherboards contain sockets for chips (such as the CPU and ROM, discussed shortly) and slots for add-on cards. Examples of add-on cards are a video card to connect the computer to the monitor, a network card or modem to connect to an internal network or the Internet, and a sound card to connect to speakers.

Sockets on the motherboard typically accept things such as random-access memory (RAM) or the central processing unit (CPU). The keyboard, mouse, CD-ROM drives, monitor, and other peripherals or components connect to the motherboard in one way or another.

System Bus Contained on the motherboard, the system bus is a vast complex network of wires that carries data from one hardware device to another. This network is analogous to a complex highway. Data is sent along the bus in the form of ones and zeros (or, more appropriately stated, as electrical impulses representing an "on" or "off" state—this two-state computing is also known as binary computing).

central processing unit (CPU)
The part of the computer that processes information and performs computations

Central Processing Unit (CPU) The **central processing unit (CPU)**, also referred to as a processor, is the brain of the computer; it is the part of the computer that actually computes. It is the main (and typically the largest) chip that plugs into a socket on the motherboard. Basically, all operations performed by the computer are run through the CPU. The CPU carries out the program steps to perform the requested task, such as opening and working in a Microsoft Word document or performing advanced mathematical algorithms.

Read-Only Memory (ROM) This rather generic term describes special chips on the motherboard. ROM chips store programs called *firmware*, used to start the boot process (in which the computer starts up before the system is fully functioning) and configure a computer's components. This technology is referred to as the BIOS, for *basic input-output system*. The operation of the BIOS is relevant to several computer forensics procedures, particularly the boot sequence. As will become clear later, it is important not to boot the actual computer under investigation to the original hard disk drive. This would cause changes to the data, thus compromising the integrity of evidence. The BIOS allows investigators to control the boot process to some degree.

Random-Access Memory (RAM) **Random-access memory (RAM)** stores software programs and instructions while the computer is turned on; it takes the physical form of chips that plug into the motherboard. Most of the data on a computer is stored on the hard disk drive (HDD). However, if the computer had to access the HDD each time it wanted data, it would run slowly and inefficiently. Instead, the computer, aware that it may need certain data at a moment's notice, stores the data in RAM. This takes the burden off the computer's processor and hard disk drive (HDD). RAM is referred to as *volatile memory* because it is not permanent; its contents undergo constant change and are forever lost once power is taken away from the computer.

random-access memory (RAM) The volatile memory of the computer; when power is turned off, its contents are lost; programs and instructions are loaded into RAM while they are in use

Hard Disk Drive (HDD) Generally speaking, the **hard disk drive (HDD)** is the primary component of storage in the personal computer (see Figure 19–2). It typically stores the operating system, programs, and data files created by the user, such as documents, spreadsheets, accounting information, or a company database. Unlike RAM, the HDD is permanent storage and retains its information even after the power is turned off. HDDs work off a controller that is typically part of the motherboard, but sometimes takes the form of an add-on (expansion) card plugged into the motherboard.

hard disk drive (HDD) Typically the main storage location within the computer, consisting of magnetic platters contained in a case

FIGURE 19–2 An inside view of the platter and read/write head of a hard disk drive. *Courtesy STILLFX/Shutterstock*

Input Devices Input devices are used to get data into the computer or to give the computer instructions. Input devices constitute part of the "user" side of the computer. Examples include the keyboard, mouse, joystick, and scanner.

Output Devices Output devices are equipment through which data is obtained from the computer. Output devices are also part of the "user" side of the computer and provide the results of the user's tasks. They include the monitor, printer, and speakers.

Quick Review

- Computer forensics involves preserving, acquiring, extracting, and interpreting computer data.
- Software programs are applications that carry out a set of instructions.
- The central processing unit (CPU) is the brain of the computer—the main chip responsible for doing the actual computing.
- The motherboard is the main circuit board within a computer.
- Read-only memory (ROM) chips store programs that control the boot (startup) process and configure a computer's components.
- Random-access memory (RAM) is volatile memory, which is lost when power is turned off. Programs are loaded into RAM because of its faster read speed than accessing the program on the hard drive constantly.
- The hard disk drive (HDD) is typically the primary location of data storage within the computer.

Putting It All Together

A person approaches the computer, sits down, and presses the power button. The power supply wakes up and delivers power to the motherboard and all of the hardware connected to the computer. At this point the flash ROM chip on the motherboard (the one that contains the BIOS) conducts a power-on self test (POST) to make sure everything is working properly.

The flash ROM also polls the motherboard to check the hardware that is attached and follows its programmed boot order, thus determining from what device it should boot. Typically the boot device is the HDD, but it can also be a CD, or USB drive. If it is the HDD, the HDD is then sent control. It locates the first sector of its disk (known as the master boot record), determines its layout (partition[s]), and boots an operating system (Windows, Mac OS, Linux, Unix). The person is then presented with a computer work environment, commonly referred to as a desktop.

Now ready to work, the user double-clicks an icon on the desktop, such as a Microsoft Word shortcut, to open the program and begin to type a document. The CPU processes this request, locates the Microsoft Word program on the HDD (using a predefined map of the drive called a *file system table*), carries out the programming instructions associated with the application, loads Microsoft

Word into RAM via the system bus, and sends the output to the monitor by way of the video controller, which is either located on or attached to the motherboard.

The user then begins to type, placing the data from the keyboard into RAM. At the end, the user may print the document or simply save it to the HDD for later retrieval. If printed, the data is taken from RAM, processed by the CPU, placed in a format suitable for printing, and sent through the system bus to the external port where the printer is connected. If the document is saved, the data is taken from RAM, processed by the CPU, passed to the HDD controller (IDE, SCSI, or SATA) by way of the system bus, and written to a portion of the HDD. The HDD's file system table is updated so it knows where to retrieve that data later. In actuality, the boot process is more complex than the way it has been described here and requires the forensic examiner to possess an in-depth knowledge of its process.

The preceding example illustrates how three components perform most of the work: the CPU, RAM, and system bus. The example can get even more complicated as the user opens more applications and performs multiple tasks simultaneously (multitasking). Several tasks can be loaded into RAM at once, and the CPU is capable of juggling them all. This allows for the multitasking environment and the ability to switch back and forth between applications. All of this is orchestrated by the operating system and is written in the language of the computer—ones and zeros. The only detail missing, and one that is important from a forensic standpoint, is a better understanding of how data is stored on the hard disk drive (see Figure 19–2).

Storing and Retrieving Data

As mentioned earlier, most of the data in a computer is stored on the hard disk drive (HDD). However, before beginning to understand how data is stored on the HDD, it is first important to understand the role of the **operating system (OS)**. An OS is the bridge between the human user and the computer's electronic components. It provides the user with a working environment and facilitates interaction with the system's components. Each OS supports certain types of file systems that store data in different ways, but some support the methods of others.

operating system (OS)
Software that allows the user to interact with the hardware and manages the file system and applications

Formatting and Partitioning the HDD

Generally speaking, before an OS can store data on a HDD, the HDD must first be formatted, or prepared to accept the data in its current form. Before the HDD can be formatted, a partition must be defined. A **partition** is nothing more than a contiguous (adjacent) set of blocks (physical areas on the HDD in which data can be stored) that are defined and treated as an independent disk. Thus, a hard disk drive can hold several partitions, making a single HDD appear as several disks.

partition
A contiguous set of blocks that are defined and treated as an independent disk

Partitioning a drive can be thought of as dividing a container that begins as nothing more than four sides with empty space on the inside. Imagine that we then cut a hole in the front of the container and place inside two drawers containing the hardware to open and close the drawers. We have just created a two-drawer filing cabinet and defined each drawer as a contiguous block of storage.

A partitioning program then defines the partitions that will later hold the data on the HDD. Just as the style, size, and shape of a filing cabinet drawer can vary, so too can partitions.

After a hard drive is partitioned, it is typically formatted. The formatting process *initializes* portions of the HDD, so that they can store data, and creates the structure of the file system. There are various types of file systems—methods for storing and organizing computer files and data so they are easier to locate and access. Each has a different way of storing, retrieving, and allocating data.

At the conclusion of these steps, we say that the drive is *logically defined.* The term logically is used because no real divisions are made. If you were to crack open the HDD before or after partitioning and formatting, to the naked eye the platters would look the same.

Mapping the HDD

sector
The smallest unit of data addressable by a hard disk drive, generally consisting of 512 bytes

byte
A group of eight bits

bit
Short for *binary digit*; taking the form of either a one or a zero, it is the smallest unit of information on a machine

cluster
A group of sectors in multiples of two; typically the minimum space allocated to a file

As shown in Figure 19–3, disks are logically divided into sectors, clusters, tracks, and cylinders. A **sector** is the smallest unit of data that a hard drive can address; sectors are typically 512 bytes in size (a **byte** is eight bits; a **bit** is a single one or zero).[1] A **cluster** usually is the minimum space allocated to a file. Clusters are groups of sectors; their size is defined by the file system, but they are always in multiples of two. A cluster, therefore, consists of two, four, six, or eight sectors, and so on. (With modern file systems, the user can exercise some control over the number of sectors per cluster.) Tracks are concentric circles that are defined around the platter. Cylinders are groups of tracks that reside directly above and below each other.

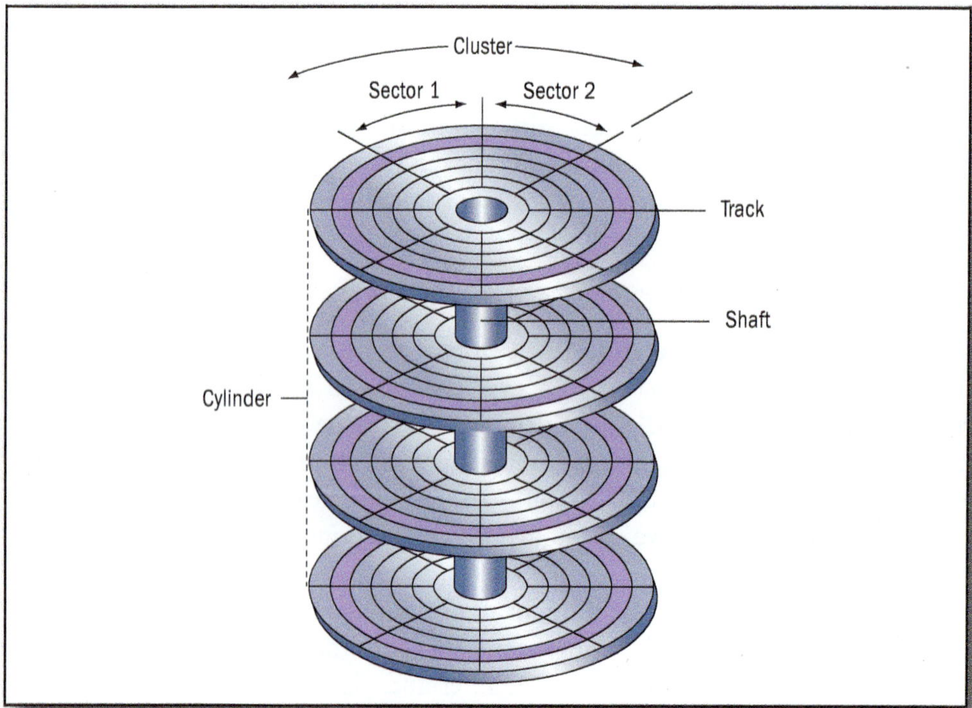

FIGURE 19–3 Parts in a hard disk drive.

Additionally, the HDD has a file system table (map) of the layout of the defined space in that partition. FAT file systems use a *file allocation table* (which is where the acronym *FAT* comes from) to track the location of files and folders (data) on the HDD, whereas NTFS file systems use, among other things, a *master file table (MFT)*. Each file system table tracks data in different ways, and forensic computer examiners should be versed in the technical nuances of the HDDs they examine. It is sufficient for our purposes here, however, to merely visualize the file system table as a map to where the data is located. This map numbers sectors, clusters, tracks, and cylinders to keep track of the data.

One way to envision a partition and file system is as a room full of safe-deposit boxes. The room itself symbolizes the entire partition, and the boxes with their contents represent clusters of data. In order to determine who rented which box, and subsequently where each depositor's property is, a central database is needed. This is especially true if a person rented two boxes located in opposite ends of the room. A similar situation often arises when storing data on a computer, because not all the data in a particular file or program is always stored in contiguous (adjacent) clusters on the HDD. Thus, a database tracking the locations of the safe-deposit boxes is much like a file system table tracking the location of data within the clusters.

This example is also useful to understand the concept of reformatting a HDD, which involves changing the way the disk stores data. If the database managing the locations of the safe-deposit boxes were wiped out, the property in them would still remain; we just wouldn't know what was where. So too with the hard disk drive. If a user were to wipe the file system table clean—for example, by reformatting it—the data itself would not be gone. Both the database tracking the locations of the safe-deposit boxes and the file system table tracking the location of the data in the cluster are maps—not the actual contents. (Exceptions exist with some file systems, such as an NTFS file system, which stores data for very small files right in its file system table, known as the master file table.)

Other Common Storage Devices

Although the HDD is the most common storage device for the personal computer, many others exist. Methods for storing data and the layout of that data can vary from device to device. A CD-ROM, for example, uses a different technology and format for writing data than a USB thumb drive. Fortunately, regardless of the differences among devices, the same basic forensic principles apply for acquiring the data. Common storage devices include the following:

CDs and DVDs Compact discs (CDs) and digital video discs (DVDs) are two of the most common forms of storing all sorts of external data, including music, video, and data files. Both types of media consist of plastic discs with an aluminum layer containing the data that is read by a beam of laser light in the CD/DVD reader. Different CDs are encoded in different ways, which makes forensic examination of such discs difficult at times.

USB Thumb Drives and Smart Media Cards These devices, which can store a large amount of data, are known as *solid-state storage devices* because they have no moving parts. Smart media cards are typically found in digital cameras and PDAs, whereas USB thumb drives come in many shapes, sizes, and storage capacities.

Tapes Tapes come in many different formats and storage capacities. Each typically comes with its own hardware reader and sometimes a proprietary application to read and write its contents. Tapes and thumb drives are typically used for backup purposes and consequently have great forensic potential.

Network Interface Card (NIC) Very rarely do we find a computer today that doesn't have a NIC. Whether they are on a local network or the Internet, when computers need to communicate with each other, they typically do so through a NIC. NICs come in many different forms: add-on cards that plug into the motherboard, hard-wired devices on the motherboard, add-on cards (PCMCIA) for laptops, and universal serial bus (USB) plug-in cards, to name a few. Some are wired cards, meaning they need a physical wired connection to participate on the network, and others are wireless, meaning they receive their data via radio waves.

Quick Review

- The computer's operating system (OS) is the bridge between the human user and the computer's electronic components. It provides the user with a working environment and facilitates interaction with the system's components.
- Formatting is the process of preparing a hard disk drive to store and retrieve data in its current form.
- A sector is the smallest unit of data that a hard drive can address. A cluster usually is the minimum space allocated to a file. Clusters are groups of sectors.
- A FAT is a file allocation table. It tracks the location of files and folders on the hard disk drive.

Processing the Electronic Crime Scene

Processing the electronic crime scene has a lot in common with processing a traditional crime scene. The investigator must first ensure that the proper legal requirements (search warrant, consent, and so on) have been met so that the scene can be searched and the evidence seized. The investigator should then devise a plan of approach based on the facts of the case and the physical location.

Documenting the Scene

The scene should be documented in as much detail as possible before disturbing any evidence and before the investigator lays a finger on any computer components. Of course there are circumstances in which an investigator may have to act quickly and pull a plug before documenting the scene, such as when data is in the process of being deleted.

Crime-scene documentation is accomplished through two actions: sketching and photographing. The electronic crime scene is no different. The scene should be sketched in a floor plan fashion (see Figure 19–4) and then overall photographs of the location taken. In a case in which several computers are connected together in a network, a technical network sketch should also be included if possible.

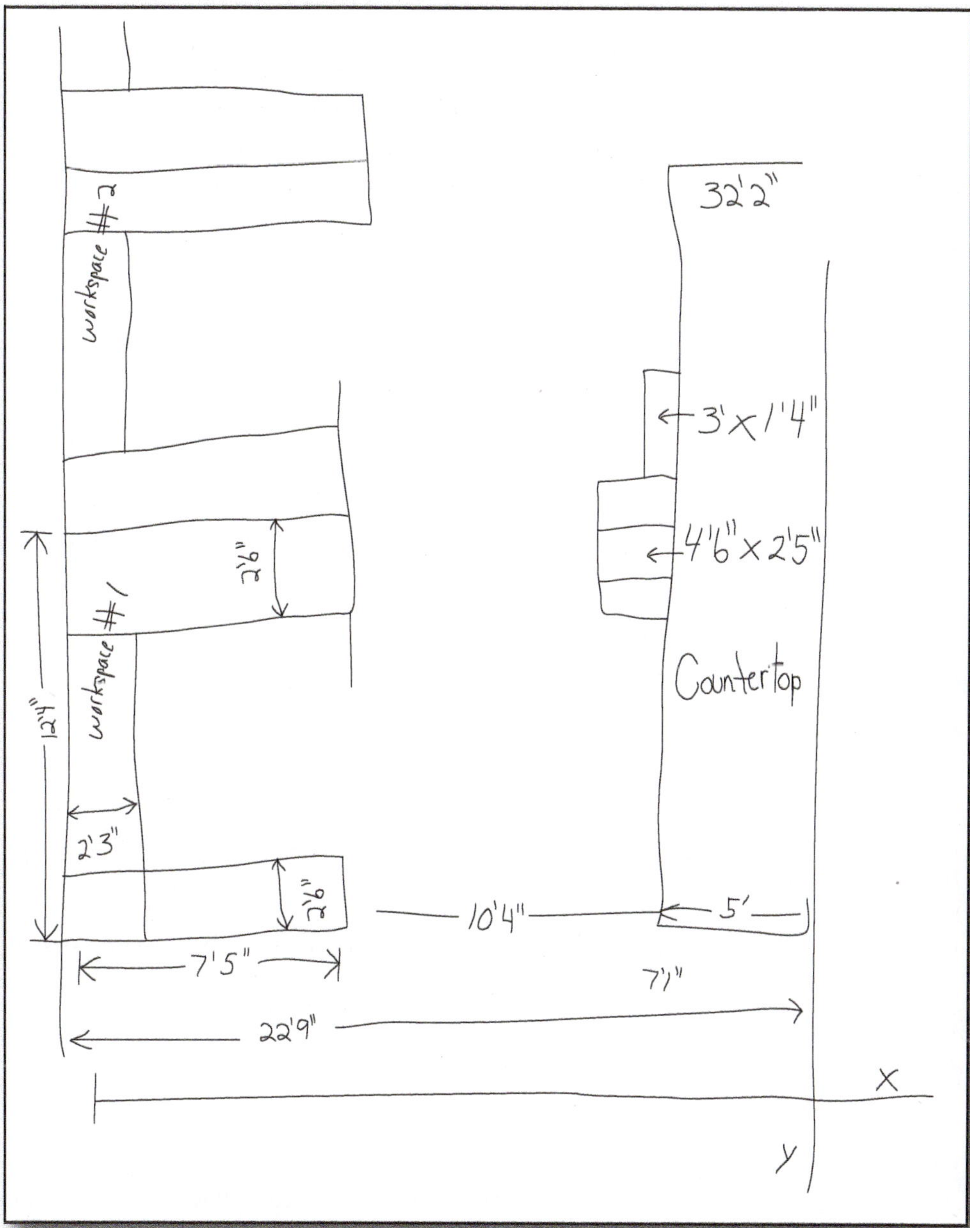

FIGURE 19–4 Rough sketch made at a crime scene, with necessary measurements included.

After investigators photograph the overall layout, close-up photographs should be shot. A close-up photograph of any running computer monitor should be taken. All the connections to the main system unit, such as peripheral devices (keyboard, monitor, speakers, mouse, and so on), should be photographed. If necessary, system units should be moved delicately and carefully to facilitate the connections photograph (see Figure 19–5). Close-up photographs of equipment serial numbers should be taken if practical.

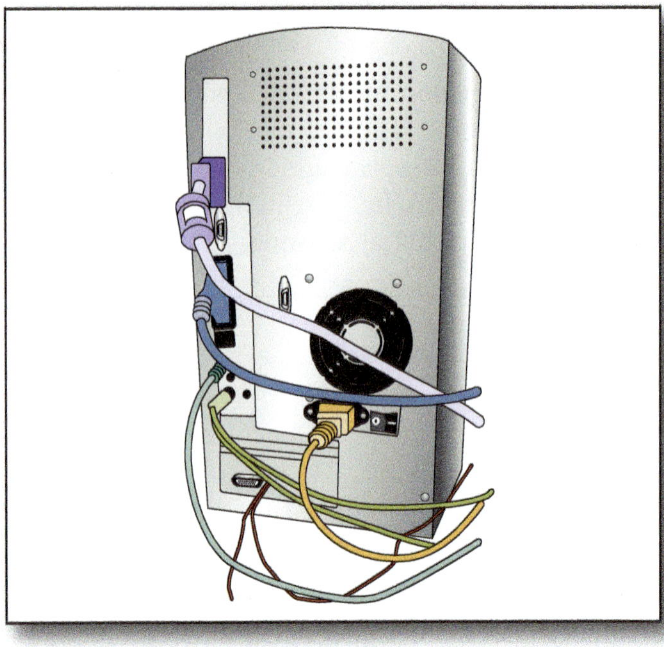

FIGURE 19–5 Back of a computer showing all connections.

At this point, investigators must decide whether to perform a live acquisition of the data, perform a system shutdown (as in the case of server equipment), pull the plug from the back of the computer,[2] or a combination thereof. Several factors influence this decision. For example, if encryption is being used and pulling the plug will encrypt the data, rendering it unreadable without a password or key, pulling the plug would not be prudent. Similarly, any data that exists in RAM and has not been saved to the HDD will be lost if power to the system is discontinued. Regardless of how investigators decide to proceed, the equipment most likely will be seized. Exceptions exist in the corporate environment, where servers are fundamental to business operations.

After the photographs and sketches are complete, but before disconnecting the peripherals from the computer, a label should be placed on the cord of each peripheral, with a corresponding label placed on the port to which it is connected. A numbering scheme should be devised to identify the system unit if several computers are at the scene (Figure 19–6). The combination of sketching, photographing, and labeling should adequately document the scene, prevent confusion of which component went with which system unit, and facilitate reconstruction if necessary for lab or courtroom purposes.

Forensic Image Acquisition

Once a computer has been seized, the data it contains must be obtained for analysis. The number of electronic items that potentially store evidentiary data are too vast to cover in this section. The hard disk drive will be used as an example, but the same "best practices" principles apply for other electronic devices as well.

The goal in obtaining data from a HDD is to do so without altering even one bit of data. Thus, throughout the entire process, the forensic computer examiner must use the least intrusive method to retrieve data. Because booting a HDD to its operating system changes many files and could potentially destroy evidentiary data, obtaining data is generally accomplished by removing the HDD from the system and placing it in a laboratory forensic computer so that a *forensic image*

FIGURE 19–6 Back of a computer with each component correlated with its port.

can be created. However, the BIOS of the seized computer sometimes interprets the geometry of the HDD differently than the forensic computer does. *Geometry* refers to the functional dimensions of a drive, including the number of heads, cylinders, and sectors per track. In these instances, the image of the HDD must be obtained using the seized computer.

Regardless of the computer in which the HDD is placed, the examiner must ensure that, when creating the forensic image, the drive to be analyzed is in a "write-blocked" read-only state in which no new data can be added to the drive. Furthermore, the examiner needs to be able to prove that the forensic image they obtained includes every bit of data and caused no changes (writes) to the HDD.

To this end, a sort of fingerprint of the drive is taken before and after imaging through the use of a **Message Digest 5 (MD5)/Secure Hash Algorithm (SHA)**, or similar algorithm. Before imaging the drive, the algorithm is run, and a 32-character alphanumeric string is produced based on the drive's contents. The algorithm is then run against the resulting forensic image. If nothing changed, the same alphanumeric string is produced, thus demonstrating that the image is all-inclusive of the original contents and that nothing was altered in the process.

Message Digest 5 (MD5)/Secure Hash Algorithm (SHA)
A software algorithm used to "fingerprint" a file or contents of a disk in order to verify the integrity of data; in forensic analysis it is typically used to verify that an acquired image of suspect data was not altered during the process of imaging

A forensic image of the data on a hard disk drive (or any type of storage medium) is merely an exact duplicate of the entire contents of that medium. In other words, all portions of a hard disk drive—even blank portions—are copied from the first bit (one or zero) to the last. Why would investigators want to copy what appears to be blank or unused portions of the HDD? The answer is simple: to preserve *latent data*, discussed later in the chapter. Data exists in areas of the drive that are, generally speaking, unknown and inaccessible to most end users. This data can be valuable as evidence. Therefore, a forensic image—one that copies every single bit of information on the drive—is necessary.[3] A forensic image differs from a backup or standard copy in that it takes the entire contents, not only data the operating system is aware of.

Many forensic software packages come equipped with a method to obtain a forensic image. The most popular software forensic tools—EnCase®, Forensic

Toolkit® (FTK), Forensic Autopsy (Linux-based freeware), and SMART (Linuxbased software by ASR Data)—all include a method to obtain a forensic image. All produce self-contained image files that can then be interpreted and analyzed. They also allow image compression to conserve storage. The fact that self-contained, compressed files are the result of forensic imaging allows many images from different cases to be stored on the same forensic storage drive. This makes case management and storage much easier (see Figure 19–7).

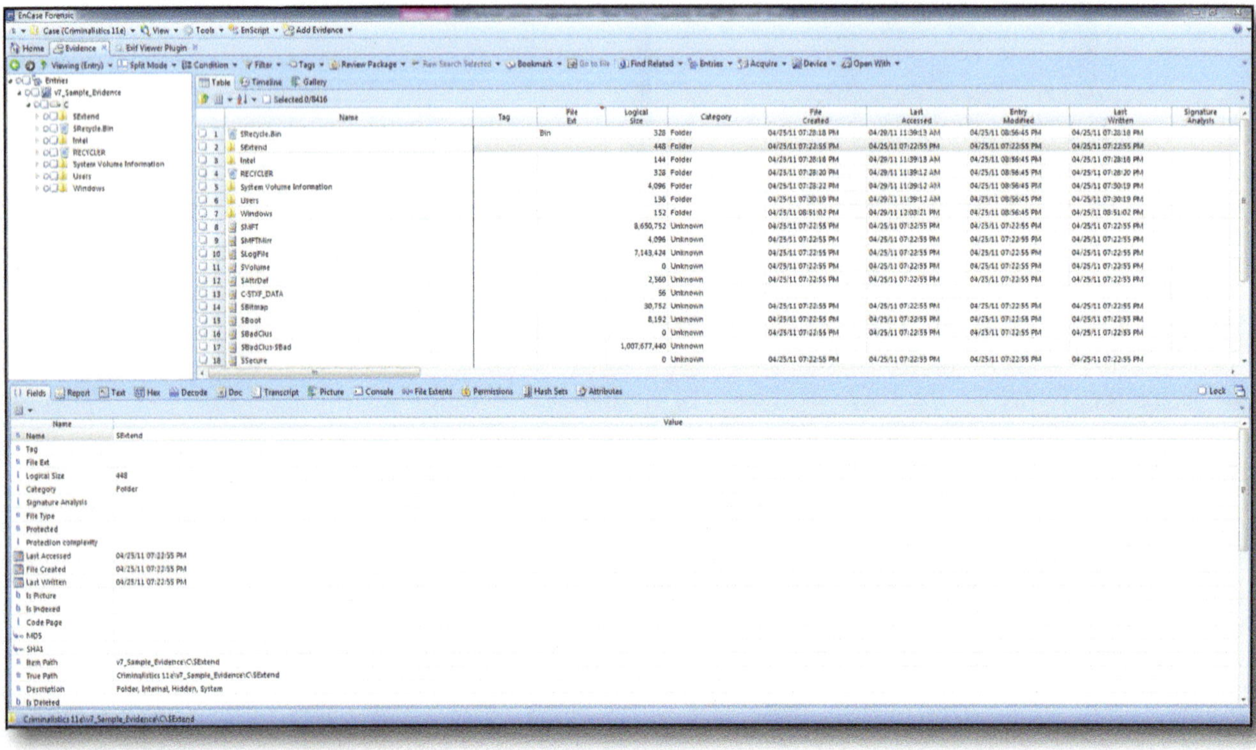

FIGURE 19–7 Screen shot of EnCase Software. Used by permission by EnCase, Guidance Software, Inc.

Quick Review

- Aspects of a computer that should be photographed close up at an electronic crime scene include (1) the screen of any running computer monitor, (2) all the connections to the main system unit, such as peripheral devices (keyboard, monitor, speakers, mouse, and so on), and (3) equipment serial numbers.
- Two situations in which an investigator would not unplug a computer at an electronic crime scene are (1) if encryption is being used and pulling the plug will encrypt the data, rendering it unreadable without a password or key, and (2) if data exists in RAM that has not been saved to the HDD, and will thus be lost if power to the system is discontinued.
- The primary goal in obtaining data from a HDD is to do so without altering even one bit of data. To this end, a Message Digest 5 (MD5)/Secure Hash Algorithm (SHA) takes a "fingerprint" of a hard disk drive (HDD) before and after forensic imaging.

Analysis of Electronic Data

Analysis of electronic data is virtually limitless and bound only to the level of skill of the examiner. The more familiar an examiner is with computers, operating systems, application software, data storage, and a host of other disciplines, the more prepared the examiner will be to look for evidentiary data.

Because computers are vast and complex, discussing each area, file, directory, log, or computer process that could potentially contain evidentiary data is beyond the scope of one chapter—and may be beyond the scope of an entire book. What follows are some of the more common areas of analysis. While reading this section, reflect on your own knowledge of computers and consider what other data might be of evidentiary value and where it might be found.

Visible Data

The category of **visible data** includes all information that the operating system is presently aware of and thus is readily accessible to the user. Here we present several common types of visible data considered in many investigations. This list is by no means exhaustive and can include any information that has value as evidence.

visible data
All data that the operating system is presently aware of and thus is readily accessible to the user

Data/Work Product Files One place to find evidence is in documents or files produced by the suspect. This category is extremely broad and can include data from just about any software program. Microsoft Word and WordPerfect wordprocessing programs typically produce text-based files such as typed documents and correspondence. These programs, and a host of other word-processing programs, have replaced the typewriter. They are common sources of evidence in criminal cases, particularly those involving white-collar crime.

Also relevant in white-collar crime and similar financial investigations are any data related to personal and business finance. Programs such as QuickBooks and Peachtree accounting packages can run the entire financial portion of a small to midsize business. Similarly, personal bank account records in the computer are often managed with personal finance software such as Microsoft Money and Quicken. Moreover, criminals sometimes use these programs as well as spreadsheet applications to track bank accounts stolen from unsuspecting victims. Forensic computer examiners should familiarize themselves with these programs, the ways in which they store data, and methods for extracting and reading the data.

Advances in printer technology have made high-quality color printing both affordable and common in many homes. Although this is a huge benefit for home office workers and those interested in graphic arts, the technology has been used for criminal gain. Counterfeiting and check and document fraud are easily perpetrated by most home computer users. All that is required is a decent ink-jet printer and a scanner. Including the computer, a criminal could set up a counterfeiting operation for less than $1,500. Examiners must learn the graphics and photo-editing applications used for such nefarious purposes. Being able to recognize the data produced by these applications and knowing how to display the images is key to identifying the evidence.

Page file
A file or defined space on the HDD to which data is written, or swapped, to free RAM for applications that are in use

Page File Data When an application is running, the program and the data being accessed are loaded into RAM. A computer's RAM can read data much faster than the hard disk drive, which is why the programs are loaded here. RAM, however, has its limits. Some computers have more than a gigabyte or two. Regardless of the amount, though, most operating systems (Microsoft Windows™, Linux, and so on) are programmed to conserve RAM when possible. This is where the *Page file* comes in. The operating system attempts to keep only data and applications that are presently being used in RAM. Other applications that were started, but are currently waiting for user attention, may be swapped out of RAM and written to the Page file on the hard disk drive.[4]

For example, a manager of a retail store may want to type a quarterly report based on sales. The manager starts Microsoft Word and begins his report. Needing to incorporate sales figure data from a particular spreadsheet, he opens Microsoft Excel®. Depending on what is running on the computer, the original Word document may be swapped from RAM to the swap space on the HDD to free up space for Excel. As the manager goes back and forth between the programs (and maybe checks his e-mail in between), this swapping continues. Data that is swapped back and forth is sometimes left behind in the memory space. Even as this area is constantly changed, some of the data is orphaned in unallocated space, an area of the HDD discussed later in this chapter.

Page file can be defined as a particular file or even a separate HDD partition, depending on the operating system and file system type. Data in the memory space can be read by examining the HDD through forensic software or a utility that provides a binary view, such as Norton™ Disk Editor or WinHex (see Figure 19–8).

FIGURE 19–8 As the user switches between applications and performs multiple tasks, data is swapped back and forth between RAM and the computer's hard drive. This area on the hard drive is referred to as swap space.

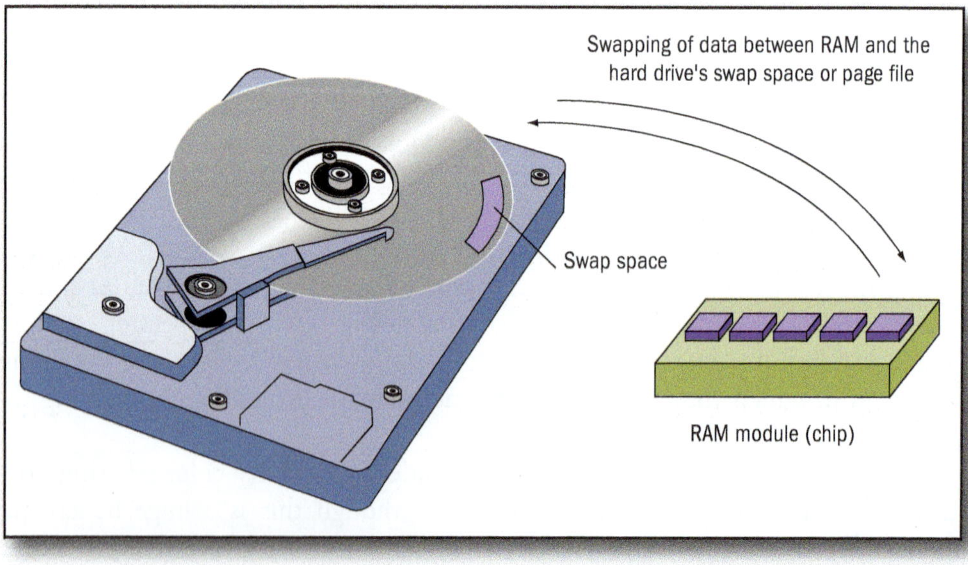

temporary files
A copy of the current file that the computer saves in memory automatically and updates periodically

Temporary Files Any user who has suffered a sudden loss of power in the middle of typing a document can attest to the value of a **temporary file**. Most programs automatically save a copy of the file being worked on in a temporary file. After typing a document, working on a spreadsheet, or working on a slide presentation, the user can save the changes, thus promoting the temporary copy to an actual file. This is done as a sort of backup on the fly. If the computer experiences a sudden loss of power or other catastrophic failure, the temporary file can be recovered, limiting the amount of data lost. The loss is limited because the temporary file is not updated in real time. Rather, it is updated periodically (typically defaulted to every ten minutes in most programs), depending on the application's settings.

Temporary files can sometimes be recovered during a forensic examination. Some of the data that may have been orphaned from a previous version may be recoverable, if not the complete file. This is true even when a document has been typed and printed, but never saved. The creation of the temporary file makes it possible for some of this "unsaved" data to be recovered during analysis.

Another type of temporary file valuable to the computer investigator is the *print spool file*. When a print job is sent to the printer, a spooling process delays the sending of the data so the application can continue to work while the printing takes place in the background. To facilitate this, a temporary print spool file is created; this file typically includes the data to be printed and information specific to the printer. There are different methods for accomplishing this, and thus the files created as a result of this process vary. It is sometimes possible to view the data in a readable format from the files created during the spooling process.

Latent Data

The term **latent data** includes data that are blocked (not necessarily intentionally) from a user's view. It includes areas of files and disks that typically are not apparent to the computer user, but contain data nonetheless. Latent data are one of the reasons a forensic image of the media is created. If a standard copy were all that was produced, only the logical data (that which the operating system is aware of) would be captured. Getting every bit of data ensures that potentially valuable evidence in latent data is not missed.

latent data
Areas of files and disks that are typically not apparent to the computer user (and often not to the operating system), but contain data nonetheless

Once the all-inclusive forensic image is produced, how are the latent data viewed? Utilities that allow a user to examine a hard disk drive on a binary level (ones and zeros) are the answer. Applications such as Norton Disk Editor and WinHex provide this type of access to a hard disk drive or other computer media. Thus, these applications, sometimes also referred to as hex editors (for the hexadecimal shorthand of computer language), allow all data to be read on the binary level independent of the operating system's file system table. Utilities such as these can write to the media under examination, thus changing data. Consequently, a software or hardware write-blocker should be used.

A more common option in data forensics is to use specialized forensic examination software. EnCase and Forensic Toolkit for Windows and SMART and Forensic Autopsy for Linux are examples of forensic software. Each allows a search for evidence on the binary level and provides automated tools for performing common forensic processing techniques. Examiners should be cautious, however, about relying too heavily on automated tools. Merely using an automated tool without understanding what is happening in the background and why evidentiary data may exist in particular locations would severely impede the ability to testify to the findings.

Slack Space Slack space is empty space on a hard disk drive created because of the way the HDD stores files. Recall that although the smallest unit of data measure is one bit (either a one or a zero), a HDD cannot address or deal with such a small unit. In fact, not even a byte (eight bits) can be addressed. Rather, the smallest unit of addressable space by a HDD is the sector. HDDs typically group sectors in 512-byte increments, whereas CD-ROMs allocate 2,048 bytes per sector.

If the minimum addressable unit of the HDD is 512 bytes, what happens if the file is only 100 bytes? In this instance there are 412 bytes of slack space. It does not end here, however, because of the minimum cluster requirement. As you may recall, clusters are groups of sectors used to store files and folders. The cluster is the minimum storage unit defined and used by the logical partition. It is because of the minimum addressable sector of the HDD and the minimum unit of storage requirement of the volume that we have slack space.

Minimum cluster allocation must be defined in a sector multiple of two. Thus, a cluster must be a minimum of two, four, six, or eight sectors, and so on. Returning to our initial example of the 100-byte file, suppose an HDD has a two-sectors-percluster volume requirement. This means that the HDD will isolate a minimum of two 512-byte sectors (a total of 1,024 bytes) of storage space for that 100-byte file. The remaining 924 bytes would be slack space (see Figure 19–9).

FIGURE 19–9 Slack space illustrated in a two-sector cluster. Cluster sizes are typically greater than two sectors, but two sectors are displayed here for simplicity.

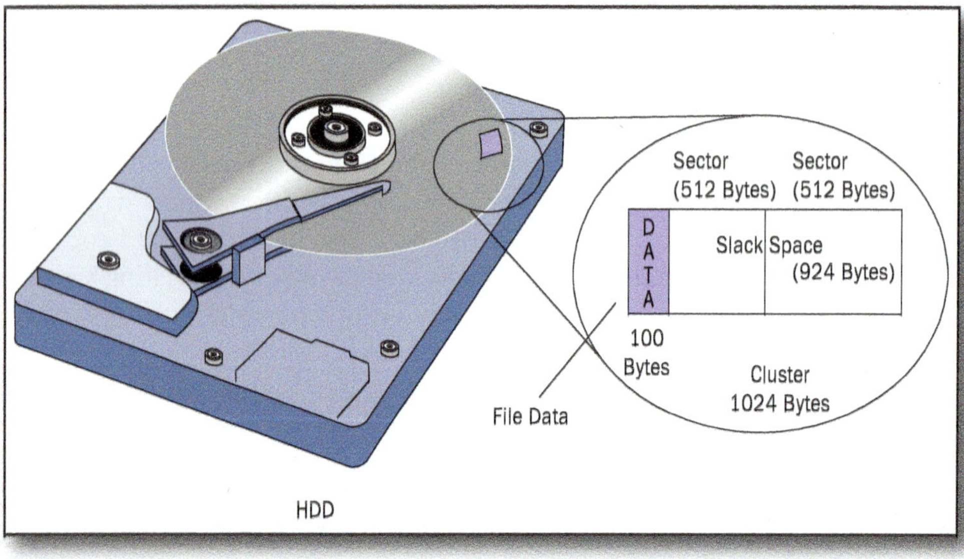

To illustrate this point, let us expand on the previous example of safe-deposit boxes. The bank offers safe-deposit boxes of a particular size. This is the equivalent of the HDD's clusters. A person wanting to place only a deed to a house in the box gets the same size box as a person who wants to stuff it full of cash. The former would have empty space should they desire to place additional items in the box. This empty space is the equivalent of slack space. But what if the box becomes full and the person needs more space? That person must then get a second box. Similarly, if a file grows to fill one cluster and beyond, a second cluster (and subsequent clusters as needed) is allocated. The remaining space in the second cluster is slack space. This continues as more and more clusters are allocated depending on file size and file growth.

RAM slack
The area beginning at the end of the logical file and terminating at the end of that sector; in some older operating systems this area is padded with information in RAM

file slack
The area that begins at the end of the last sector that contains logical data and terminates at the end of the cluster

This example is a bit of an oversimplification because there are actually two types of slack space: RAM slack and file slack. **RAM slack** occupies the space from where the actual (logical) data portion of the file ends to where the first allocated sector in the cluster terminates. **File slack**, therefore, occupies the remaining space of the cluster.

Let us go back to the 100-byte file with the two-sectors-per-cluster minimum requirement. Following the end of the logical data (the end of the 100 bytes), the remaining 412 bytes of that sector is RAM slack; the additional 512 bytes completing the cluster is then file slack. See Figure 19–10 for a visual depiction. The question now becomes: What can I expect to find in slack space, and why is this important? The answer: junk—valuable junk.

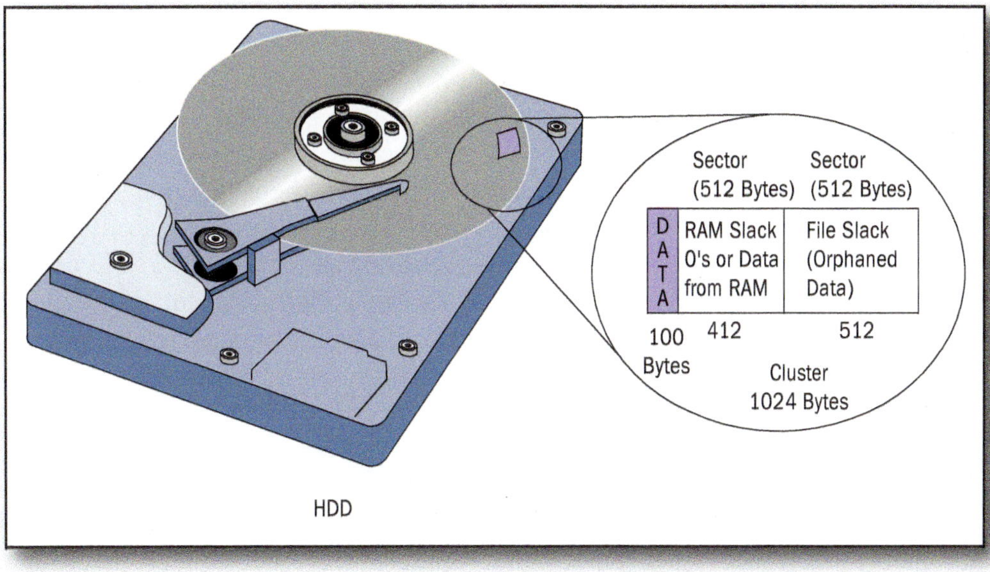

FIGURE 19–10 File slack.

RAM slack is a concept that was more relevant in older operating systems. Remember that the minimum amount of space the HDD can address is the 512-byte sector. Therefore, if the file size is only 100 bytes, the remaining space must be padded. Some operating systems pad this area with data contained in RAM. This could include web pages, passwords, data files, or other data that existed in RAM when the file was written. Modern Windows operating systems pad this space with zeros, but some examinations may still yield valuable data in this area.

File slack, on the other hand, can contain a lot of old, orphaned data. To illustrate this point, let's take the 100-byte file example a bit further. Let's say that before the 100-byte file was written to the HDD, occupying one cluster (two sectors totaling 1024 bytes), a 1,000-byte file occupied this space but was deleted by the user. When a file is "deleted" the data still remains behind, so it is probably a safe bet that data from the original 1,000-byte file remains in the slack space of the new 100-byte file now occupying this cluster. This is just one example of why data exists in file slack and why it may be valuable as evidence.

In one final attempt to illustrate this point, let us again build on our safe-deposit box analogy. Suppose a person rents two safe-deposit boxes, each box representing a sector and the two combined representing a cluster. If that person places the deed to his house in the first box, the remaining space in that box would be analogous to RAM slack. The space in the second box would be the equivalent of file slack. The only difference is that unlike the empty spaces of the safe-deposit box, the slack space of the file most likely contains data that may be valuable as evidence.

The data contained in RAM and file slack is not really the concern of the operating system. As far as the OS is concerned, this space is empty and therefore ready to be used. Until that happens, however, an examination with one of the aforementioned tools will allow a look into these areas, thus revealing the orphaned data. The same is true for unallocated space.

unallocated space
The area of the HDD that the operating system (file system table) sees as empty (containing no logical files) and ready for data; simply stated, it is the unused portion of the HDD, but is not necessarily empty

Unallocated Space Latent evidentiary data also resides in **unallocated space**. What is unallocated space, how does data get in there, and what is done to access this space?

If we have an 80-GB hard drive and only half of the hard drive is filled with data, then the other half, or 40 GB, is unallocated space (see Figure 19–11). Returning to our safe-deposit box analogy, if the entire bank of safe-deposit boxes contains 100 boxes, but only 50 are currently in use, then the other 50 would be the equivalent of unallocated space. The HDD's unallocated space typically contains a lot of useful data. The constant shuffling of files on the HDD causes data to become orphaned in unallocated space as the logical portion of the file is rewritten to other places. Some examples of how data is orphaned may help.

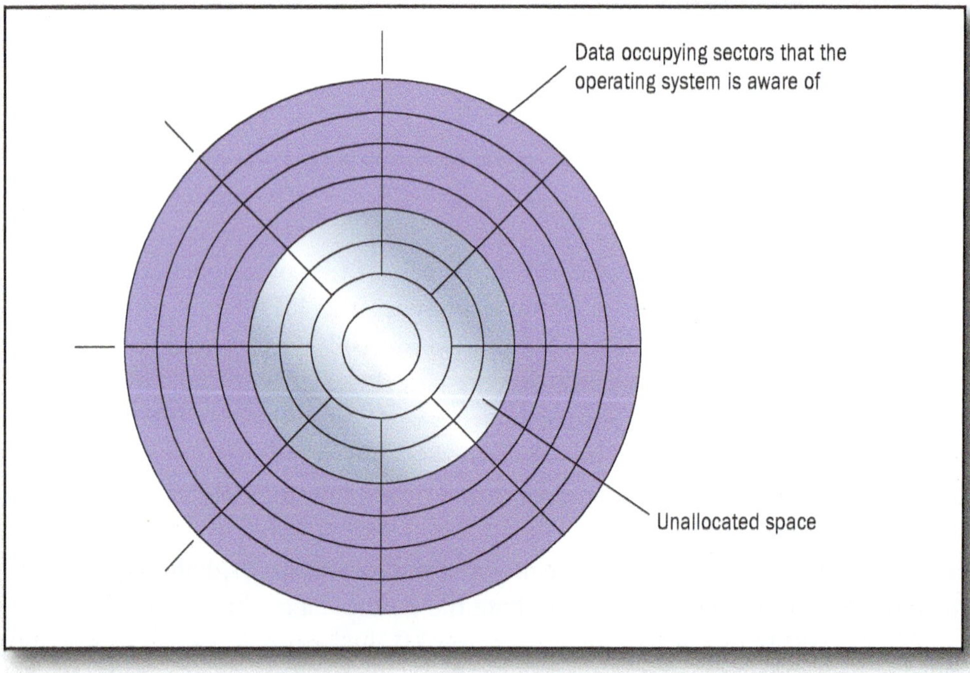

FIGURE 19–11 Simplistic view of a hard drive platter demonstrating the concept of unallocated space.

Defragmenting. Defragmenting a HDD involves moving noncontiguous data back together. Remember that the HDD has minimum space reservation requirements. Again, if a file requires only 100 bytes of space, the operating system may allocate much more than that for use. If the file grows past what has been allocated for it, another cluster is required. If, however, a different file occupies the next cluster in line, then the operating system will have to find another place for that first file on the drive. In this scenario, the file is said to be *fragmented* because data for the same file is contained in noncontiguous clusters. In the case of the HDD, the shuffling of files causes data to be orphaned in unallocated space.

Ultimately, fragmentation of numerous files can degrade the performance of a HDD, causing the read/write heads to have to traverse the platters to locate the data. Defragmenting the HDD takes noncontiguous data and rearranges it so it is in contiguous clusters. Building yet again on our safe-deposit box analogy, if a renter eventually needs to store more property than his original box can hold, the bank will rent him a second box. If, however, all the boxes around his are occupied and the only free one is in another section of the room, then his property is "fragmented." The bank would have to "defrag" the safe-deposit boxes to get the property of users with more than one box into adjacent boxes.

Swap file/swap space. Recall that a computer uses the HDD to maximize the amount of RAM by constantly swapping data in and out of RAM to a predetermined location on the HDD, thus freeing valuable RAM. The constant read and write operations of RAM cause a constant change in the page file or memory space. Data can become orphaned in unallocated space from this constant swapping to and from the HDD.

Deleted files. The deletion of files is another way that data becomes orphaned in unallocated space. Data from deleted files can manifest itself in different ways during a forensic examination. The actions that occur when a file is deleted vary among file systems. However, generally speaking, the data is not gone. For example, consider what happens when a user or program deletes a file in a Windows operating system with a FAT file system. When a file is deleted, the first character in the file's directory entry (its name) is replaced with the Greek letter sigma. When the sigma replaces the first character, the file is no longer viewable through conventional methods and the operating system views the space previously occupied by the file as available. The data, however, is still there.

This example doesn't account for the actions of the Windows Recycle Bin. When the Windows operating system is set up to merely place the deleted file in the Recycle Bin, the original directory entry is deleted and one is created in the Recycle folder for that particular user. The new Recycle folder entry is linked to another file, the *info* or *info2* file, which includes some additional data, such as the location of the file before its deletion in case the user wishes to restore it to that location. Detailed discussions of the function of the Recycle Bin are beyond the scope of this chapter, but suffice it to say that even when the Recycle Bin is emptied, the data usually remains behind until overwritten. Moreover, Windows NTFS partitions and Linux EXT partitions handle deleted files differently, but in both cases data typically remains.

What if a new file writes data to the location of the original file? Generally speaking, the data is gone. This is, of course, unless the new file only partially overwrites the original. In this instance we return to the unallocated-space orphaned-data scenario. If a file that occupied two clusters is deleted and a new file overwrites one of the clusters, then the data in the second cluster is orphaned in unallocated space. Of course, yet a third file can overwrite the second cluster entirely, but until then the data remains in unallocated space.

Let us once again look to our safe-deposit box analogy. If, for example, the owner of two safe-deposit boxes stopped renting them, the bank would list them as available. If the owner didn't clean them out, the contents would remain unchanged. If a new owner rented one of the boxes, the contents from the former owner would be replaced with the new owner's possessions. The second box would therefore still contain orphaned contents from its previous owner. The contents would remain in this "unallocated box" space until another renter occupies it.

Quick Review

- The types of computer evidence can be grouped under two major subheadings: visible and latent data.
- Visible data is data that the operating system is aware of and thus is easily accessible to the user. It includes any type of user-created data, such as word-processing documents, spreadsheets, accounting records, databases, and pictures.
- Temporary files created by programs as a sort of backup on the fly can prove valuable as evidence. Data in the swap space (used to conserve the valuable RAM within the computer system) can also yield evidentiary data.
- Latent data is data that the operating system is not aware of. The constant shuffling of data through deletion, defragmentation, swapping, and so on is one of the ways data is stored in latent areas.
- Latent data can exist in both RAM slack and file slack. RAM slack is the area from the end of the logical file to the end of the sector. File slack is the remaining area from the end of the final sector containing data to the end of the cluster.
- Latent data may be found in unallocated space—space on a HDD that the operating system sees as empty and ready for data.
- When a user deletes files the data typically remains behind, so deleted files are another source of latent data.

Forensic Analysis of Internet Data

It's important from the investigative standpoint to be familiar with the evidence left behind from a user's Internet activity. A forensic examination of a computer system reveals quite a bit of data about a user's Internet activity. The data described next would be accessed and examined using the forensic techniques outlined in the previous sections of this chapter.

Internet Cache

Evidence of web browsing typically exists in abundance on the user's computer. Most web browsers (Internet Explorer, Safari, and Firefox) use a caching system to expedite web browsing and make it more efficient. This was particularly true in the days of dial-up Internet access. When a user accesses a website, such as the *New York Times* home page, the data is fed from that server (in this example the *New York Times*), via the Internet service provider, over whatever type of connection the user has, to the user's computer. If that computer is accessing the Internet via a dial-up connection, the transfer of the *New York Times* home page may take a while, because the data transfer rate and capabilities (bandwidth) of the telephone system is limited. Even with the high-speed access of a DSL line or cable connection, conservation of bandwidth is always a consideration. Taking that into account, web browsers store (cache) portions of the pages visited onto the local hard disk drive. This way, if the page is revisited, portions of it can be reconstructed more quickly from this saved data, rather than having to pull it yet again from the Internet and use precious bandwidth.

This **Internet cache** is a potential source of evidence for the computer investigator. Portions of, and in some cases, entire visited web pages can be reconstructed. Even if deleted, these cached files can often be recovered (see the section on deleted data). Investigators must know how to search for this data within the particular web browser used by a suspect.

Internet cache
Portions of visited web pages placed on the local hard disk drive to facilitate quicker retrieval once revisited

Internet Cookies

Cookies provide another area where potential evidence can be found. To appreciate the value of cookies you must first understand how they get onto the computer and their intended purpose. **Cookies** are placed on the local hard disk drive by websites the user has visited, if the user's web browser (such as Internet Explorer) is set to allow this to happen. Internet Explorer places cookies in a dedicated directory. The website uses cookies to track certain information about its visitors, such as the history of visits, purchasing habits, passwords, and personal information used to recognize the user for later visits.

cookies
Files placed on a computer from a visited website; they are used to track visits and usage of that site

Consider a user who registers for an account at the Barnes and Noble bookstore website, and then returns to the same site from the same computer a few days later. The site will then display "Welcome, *Your User Name*." This data is retrieved from the cookie file placed on the user's hard disk drive by the website during the initial visit and registration with the site.

It is helpful to think of cookies almost like a Caller ID for websites. The site recognizes and retrieves information about the visitor, as when a salesperson recognizes the caller from a Caller ID display and quickly pulls the client's file. Cookie files can be a valuable source of evidence. In Internet Explorer, they take the form of plain text files, which can typically be opened with a standard text viewer or word-processing program, revealing part of the data. The existence of the files themselves, regardless of the information contained within, can be of evidentiary value to show a history of web visits.

A typical cookie may resemble the following: rsaferstein@forensicscience.txt. From this we can surmise that someone using the local computer login *rsaferstein* accessed the forensic science website. It is possible that the cookie was placed there by an annoying pop-up ad, but considered against other evidence in the computer data, the presence of this cookie may be of corroborative value.

Internet History

Most web browsers track the history of web page visits for the computer user. This is probably done merely for a matter of convenience. Like the "recent calls" list on a cell phone, the **Internet history** provides an accounting of sites most recently visited, with some storing weeks' worth of visits. Users can go back and access sites they recently visited just by going through the browser's history. Most web browsers store this information in one particular file; Internet Explorer uses the *index.dat* file. On a Microsoft Windows system, an *index.dat* file is created for each login user name on the computer.

Internet history
An accounting of websites visited; different browsers store this information in different ways

The history file can be located and read with most popular computer forensic software packages. It displays the uniform resource locator (URL) of each website, along with the date and time the site was accessed. An investigation involving Internet use almost always includes an examination of Internet history data.

In some respects, the term "*Internet history*" is wrong because it doesn't encompass all of its functions. Several browsers—Internet Explorer, for one—store other valuable evidence independent of Internet access. It is not uncommon to see files accessed over a network listed in the history. Similarly, files accessed on external media, such as CDs or thumb drives, may also appear in the history. Regardless, the Internet history data is a valuable source of evidence worthy of examination (see Figure 19–12).

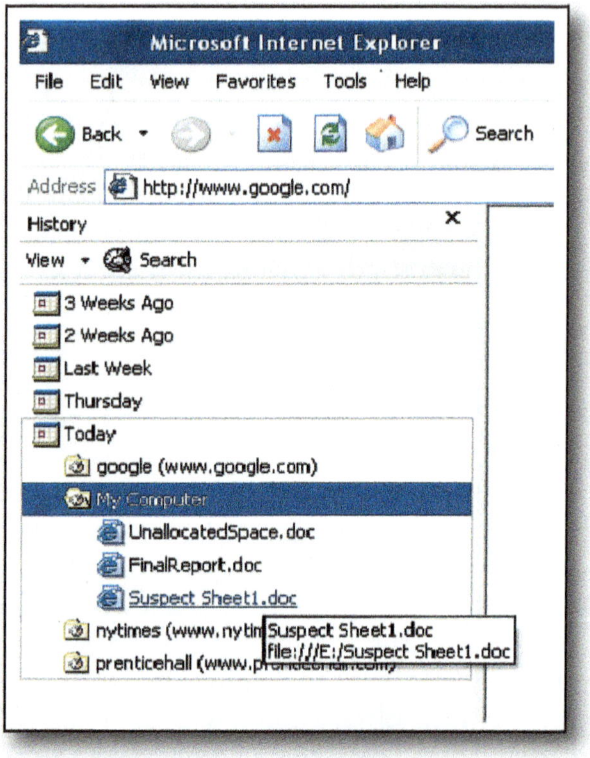

FIGURE 19–12 The Internet history displays more than just web browsing activity. Here we see Microsoft Word documents and a picture accessed on the current day. *Microsoft Corporation*

Bookmarks and Favorite Places

bookmark
A feature that enables the user to designate favorite sites for fast and easy access

Another way users can access websites quickly is to store them in their **bookmarks** or "Favorites." Like a preset radio station, web browsers allow users to bookmark websites for future visits (see Figure 19–13). A lot can be learned from a user's bookmarked sites. You may learn what online news a person is interested in or what type of hobbies they have. You may also see that person's favorite child pornography or computer hacking sites bookmarked.

In Internet Explorer the favorites are kept in a folder with link (shortcut) files to a particular URL. They can be organized in subfolders or grouped by type. The same is true for the Firefox web browser, except that Firefox bookmarks are stored in a document created in hypertext markup language (HTML), the same language interpreted by web browsers themselves.

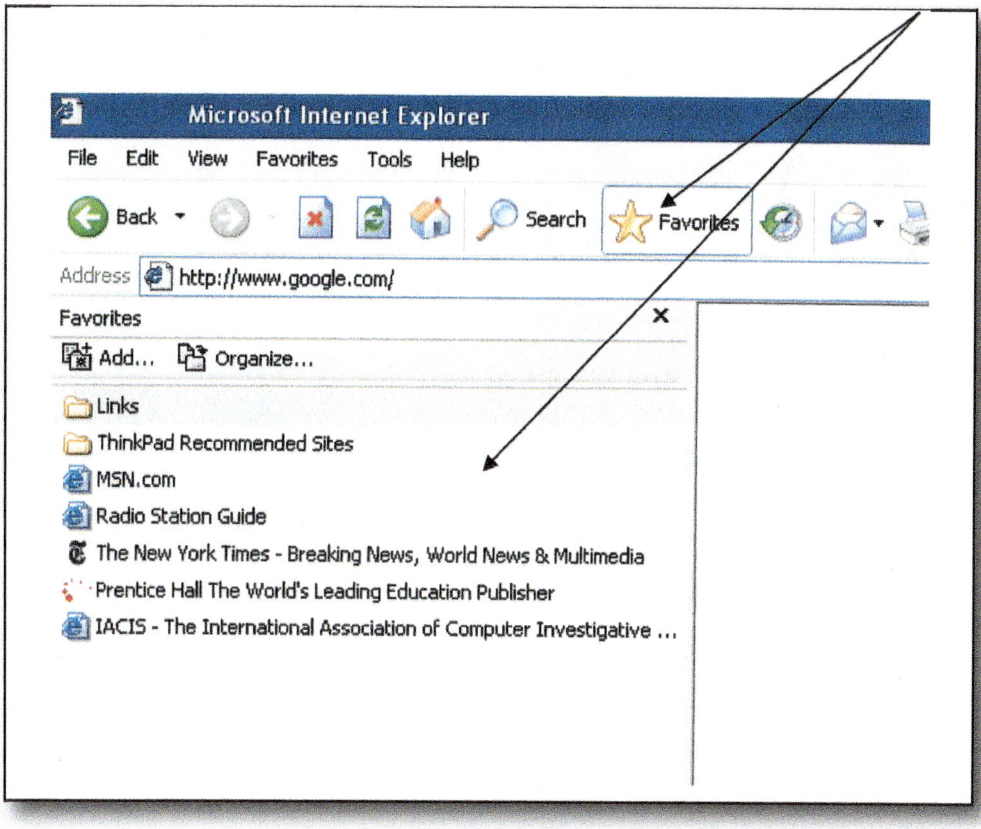

FIGURE 19–13 Bookmarks or favorite places can be saved for quick access in most web browsers. *Microsoft Corporation*

Forensic Investigation of Internet Communications

Computer investigations often begin with or are centered on Internet communication. Whether it is a chat conversation among many people, an instant message conversation between two individuals, or the back-and-forth of an e-mail exchange, human communication has long been a source of evidentiary material. Regardless of the type, investigators are typically interested in communication.

Role of the IP

With all of the computer manufacturers and software developers, some rules are necessary for computers to communicate on a global network. Just as any human language needs rules for people to communicate successfully, so does the language of computers. Computers that participate on the Internet, therefore, must be provided with an address known as an Internet protocol (IP) address from the Internet service provider to which they connect.

IP addresses take the form ###.###.###.###, in which, generally speaking, ### can be any number from 0 to 255. A typical IP address might look like this: 66.94.234.13. Not only do these IP addresses provide the means by which data can be routed to the appropriate location, but they also provide the means by which most Internet investigations are conducted (see Figure 19–14). Thus, the IP address may lead to the identity of a real person. If an IP address is the link to the identity of a real person, then it is quite obviously valuable for identifying someone on the Internet.

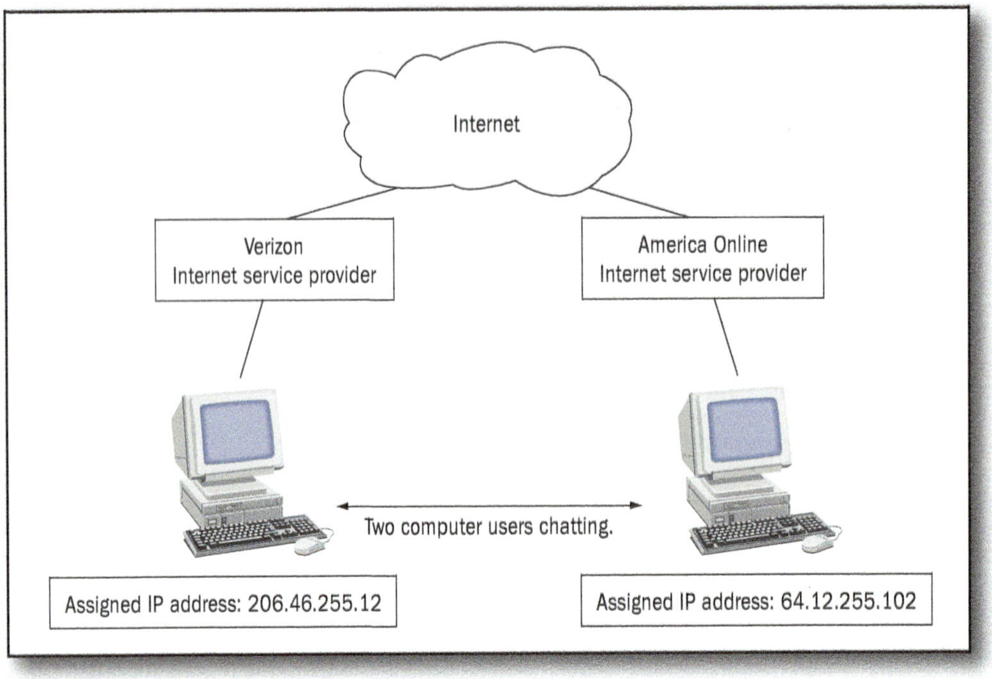

FIGURE 19–14 Two computers communicating by sending data to each other's IP address via the Internet. An IP address is assigned to each computer by its respective Internet service providers.

To illustrate, let's assume that a user of the Internet, fictitiously named John Smith, connects to the Internet from his home by way of a Verizon™ DSL connection. Verizon in this case would be responsible for providing Smith with his IP address. Verizon was issued a bank of IP addresses to service its customers from a regulatory body designed to track the usage of IP addresses (obviously so no two were used at the same time).

Suppose that Smith, while connected to the Internet, decides to threaten an ex-girlfriend by sending her an e-mail telling her he is going to kill her. That e-mail must first pass through Smith's Internet service provider's computers (in this case, Verizon) on its way to its destination—Smith's ex-girlfriend. The e-mail would be stamped by the servers that it passes through, and this stamp would include the IP address given to Smith by Verizon for his session on the Internet.

An investigator responsible for tracking that e-mail would locate the originating IP address stamped in the e-mail header. That IP address could be researched using one of many Internet sites (www.samspade.org, www.arin.net) to determine which Internet service provider was given this IP as part of a block to service its customers. The investigator then files a subpoena with the Internet service provider (Verizon) asking which of its customers was using that IP address on that date and time.

IP addresses are located in different places for different methods of Internet communications. E-mail has the IP address in the header portion of the mail. This may not be readily apparent and may require a bit of configuration to reveal. Each e-mail client is different and needs to be evaluated on a case-by-case basis. For an instant message or chat session, the particular provider (the one providing the chat mechanism—AOL®, Yahoo!®, and so on) would be contacted to provide the user's IP address.

E-Mail, Chat, and Instant Messaging

E-mail can be read by a number of *clients* or software programs. Two of the most popular ways to access, read, and store e-mail in today's Internet environment, however, are Microsoft Outlook® and an Internet browser. Some people even use a combination of the two.

If an e-mail account is linked through Microsoft Outlook, then the e-mail is stored in a compound file (a file with several layers). Typically, a compound file exists for received (inbox), sent, and deleted e-mail. Users can also create new categories (shown as folders in Outlook) and categorize saved e-mail there. Most computer forensic software applications can view (mount) these compound files so that the e-mail can be seen, including any file attachments. These files can also be imported into a clean copy (one not attached to an account) of Microsoft Outlook and the e-mail viewed there. Investigators must also be aware that in a computer network environment, the user's Outlook files may not reside on the user's workstation computer, but rather on a central mail or file server.

Most accounts offer the ability to access e-mail through a web-based interface as well. This way, users can access their e-mail remotely from other computers. For e-mail accessed through a web browser, the information presented earlier on Internet-based evidence applies. The web interface converts the e-mail into a document suitable for reading in a web browser. Consequently, web-based e-mail is often found in the Internet cache. This is particularly true of free Internet e-mail providers such as Hotmail and Yahoo!.

Much of the evidence from Internet communication is also derived from chat and instant message technology. This is particularly true in the world of child sexual exploitation over the Internet. Various technologies provide chat and instant message services. Most chat and instant message conversations are not saved by the parties involved. Although most of the software does allow for conversation archiving, it is typically turned off by default. Therefore, conversations of this nature typically exist in the volatile memory space of random-access memory (RAM).

Recall that RAM is termed volatile because it holds data only if it has power. Unplugging the computer will cause the data located in RAM to be lost. If, however, chat or instant message conversations are relevant as evidence and the computer was turned off, thus erasing the data in RAM, all may not be lost. Remember that there is an interaction between the computer system's RAM and the hard disk drive. RAM is a commodity, and therefore the computer's operating system makes an effort to conserve it as best as possible. This is done by swapping/paging that information back and forth into the swap space/paging file. Therefore, remnants of chat conversations are often found in the swap space/paging file during a forensic examination of the hard disk drive. These remnants, however, are typically fragmented, disconnected, and incomplete. Therefore, if the chat or instant message is still present on the screen (and thus probably still in RAM) the investigator needs a method by which to preserve and collect it.

A detailed discussion of capturing volatile data from RAM is beyond the scope of this chapter. Suffice it to say that many commercial forensic software packages can capture this data. Similarly, Linux-based tools can accomplish this as well. The examiner may even be able to export the data remotely to another device. Regardless of the method, the data needs to be acquired.

Furthermore, many programs such as AOL Instant Messenger®, Yahoo! Messenger, and mIRC® (Internet Relay Chat) create files regarding the rooms or channels a user chatted in or the screen names with which a user sent instant messages. Each application needs to be researched and the computer forensic examination guided by an understanding of how it functions.

hacking
Has various meanings, but is frequently used as a slang term for an unauthorized computer or network intrusion

Unauthorized computer intrusion, more commonly referred to as **hacking**, is the concern of every computer administrator. Hackers penetrate computer systems for a number of reasons. Sometimes the motive is corporate espionage; other times it is merely for bragging rights within the hacker community. Most commonly, though, a rogue or disgruntled employee with some knowledge of the computer network is looking to cause damage. Whatever the motivation, corporate America frequently turns to law enforcement to investigate and prosecute these cases.

Generally speaking, when investigating an unauthorized computer intrusion, investigators concentrate their efforts in three locations: *log files, volatile memory*, and *network traffic*. Logs typically document the IP address of the computer that made the connection. Logs can be located in several locations on a computer network. Most servers on the Internet track connections made to them through the use of logs. Additionally the router (the device responsible for directing data) may contain log files detailing connections.

firewall
Hardware or software designed to protect intrusions into an Internet network

Similarly, devices known as **firewalls** may contain log files listing computers that were allowed access to the network or an individual system. Firewalls are devices (taking the form of either hardware or software) that permit only requested traffic to enter a computer system (or, more appropriately, a network). In other words, if a user didn't send out a request for Internet traffic from a specific system, the firewall should block its entry. If the log files captured the IP address of the intruder, then revealing the user behind the IP is the same process as for e-mail. Investigating a computer intrusion, however, does get a bit more complicated than this.

Frequently, in cases of unlawful access to a computer network, the perpetrator attempts to cover the tracks of their IP address. In these instances, advanced investigative techniques may be necessary to discover the hacker's true identity. When an intrusion is in progress, the investigator may have to capture volatile data (data in RAM). The data in RAM at the time of an intrusion may provide valuable clues into the identity of the intruder, or at least the intruder's method of attack. As in the case of an instant message or chat conversation, the data in RAM needs to be acquired.

Another standard tactic for investigating intrusion cases is to document all programs installed and running on a system, in order to discover malicious software installed by the perpetrator to facilitate entry. The investigator uses specialized software to document running processes, registry entries, open ports, and any installed files.

Additionally, the investigator may want to capture live network traffic as part of the evidence collection and investigation process. Traffic that travels the network does so in the form of data packets. In addition to data, these packets also contain source and destination IP addresses. If the attack requires two-way communication, as in the case of a hacker stealing data, then data needs to be transmitted back to the hacker's computer using the destination IP address. Once this is learned, the investigation can focus on that system. Moreover, the type of data that is being transmitted on the network may be a clue as to what type of attack is being launched, whether any important data is being stolen, or what types of malicious software, if any, are involved in the attack.

Quick Review

- An investigator tracking the origin of an e-mail seeks out the sender's IP address in the e-mail's header. Chat and instant messages are typically located in a computer's random-access memory (RAM).
- Tracking the origin of unauthorized computer intrusions (hacking) requires investigating a computer's log file, RAM, and network traffic.
- A firewall is a device designed to protect against intrusions into a computer network.

Chapter Review

- Computer forensics involves preserving, acquiring, extracting, and interpreting computer data.
- Software programs are applications that carry out a set of instructions.
- The central processing unit (CPU) is the brain of the computer—the main chip responsible for doing the actual computing.
- The motherboard is the main circuit board within a computer.
- Read-only memory (ROM) chips store programs that control the boot (startup) process and configure a computer's components.
- Random-access memory (RAM) is volatile memory, which is lost when power is turned off. Programs are loaded into RAM because of its faster read speed.
- The hard disk drive (HDD) is typically the primary location of data storage within the computer.
- The computer's operating system (OS) is the bridge between the human user and the computer's electronic components. It provides the user with a working environment and facilitates interaction with the system's components.
- Formatting is the process of preparing a hard disk drive to store and retrieve data in its current form.
- A sector is the smallest unit of data that a hard drive can address. A cluster usually is the minimum space allocated to a file. Clusters are groups of sectors.
- A FAT is a file allocation table. It tracks the location of files and folders on the hard disk drive.
- Aspects of a computer that should be photographed close up at an electronic crime scene include (1) the screen of any running computer monitor, (2) all the connections to the main system unit, such as peripheral devices (keyboard, monitor, speakers, mouse, and so on), and (3) equipment serial numbers.
- Two situations in which an investigator would not unplug a computer at an electronic crime scene are (1) if encryption is being used and pulling the plug will encrypt the data, rendering it unreadable without a password or key, and (2) if data exists in RAM that has not been saved to the HDD, and will thus be lost if power to the system is discontinued.
- The primary goal in obtaining data from a HDD is to do so without altering even one bit of data. To this end, a Message Digest 5 (MD5)/Secure Hash Algorithm (SHA) takes a "fingerprint" of a hard disk drive (HDD) before and after forensic imaging.
- The types of computer evidence can be grouped under two major subheadings: visible and latent data.

- Visible data is data that the operating system is aware of and thus is easily accessible to the user. It includes any type of user-created data, such as word-processing documents, spreadsheets, accounting records, databases, and pictures.
- Temporary files created by programs as a sort of backup on the fly can prove valuable as evidence. Data in the swap space (used to conserve the valuable RAM within the computer system) can also yield evidentiary data.
- Latent data is data that the operating system is not aware of. The constant shuffling of data through deletion, defragmentation, swapping, and so on is one of the ways data is stored in latent areas.
- Latent data can exist in both RAM slack and file slack. RAM slack is the area from the end of the logical file to the end of the sector. File slack is the remaining area from the end of the final sector containing data to the end of the cluster.
- Latent data may be found in unallocated space—space on a HDD that the operating system sees as empty and ready for data.
- When a user deletes files, the data typically remains behind, so deleted files are another source of latent data.
- An investigator tracking the origin of an e-mail seeks out the sender's IP address in the e-mail's header. Chat and instant messages are typically located in a computer's random-access memory (RAM).
- Tracking the origin of unauthorized computer intrusions (hacking) requires investigating a computer's log file, RAM, and network traffic.
- A firewall is a device designed to protect against intrusions into a computer network.

Review Questions

1. Which of the following is not a utility for quickly accessing web pages a user has visited previously?
   a. bookmarks
   b. Internet history
   c. a search engine
   d. a favorites

2. Which of the following carries data from one hardware device to another?
   a. system bus
   b. central processing unit (CPU)
   c. random-access memory (RAM)
   d. network interface card (NIC)

3. The volatile memory of the computer is known as
   a. ROM.
   b. BIOS.
   c. RAM.
   d. the CPU.

4. Areas of files and disks that are not apparent to the user, and sometimes not even to the operating system, are called
   a. missing data.
   b. latent data.
   c. exceptional data.
   d. hidden data.

5. An investigator responsible for tracking e-mail communication would begin by locating
   a. the computer that originated the questioned e-mail.
   b. the receiving IP address.
   c. the originating IP address.
   d. the ISP customer database assigned an IP address.

6. True or False: Processing the electronic crime scene has very little in common with processing a traditional crime scene and requires that the investigator take a substantially different approach.

7. True or False: The goal in obtaining data from a HDD is to do so without altering any data.

8. True or False: If so motivated, most home computer users could easily perpetrate counterfeiting and check and document fraud using a decent ink-jet printer and a scanner.

9. True or False: Because web browsers typically do not use a caching system, evidence of web browsing is not typically available to the forensic examiner.

10. What is the difference between hardware and software?

11. What are CDs, DVDs, and other containers for programs that are loaded into the memory of a computer called?

12. What is a computer's motherboard?

13. What type of memory stores software programs and instructions while the computer is turned on? Because it is not permanent, and its contents are forever lost once power is taken away from the computer, what is it also called?

14. What is firmware? What is another name for firmware? Why is it important to forensic computer investigation?

15. What is the role of the computer's operating system (OS)?

16. What is formatting? Why must a hard disk drive be formatted?

17. What are sectors and clusters? How are they related to one another?

18. What is the smallest unit of information on a computer? What do eight of them constitute?

19. What is a FAT and what purpose does it serve?

20. Name two situations in which an investigator would not immediately unplug a computer at an electronic crime scene.

21. What is the purpose of a Message Digest 5 (MD5)/Secure Hash Algorithm (SHA)? Why would a forensic computer examiner run such an algorithm?

22. Why would investigators want to copy blank or unused portions of the HDD?

23. What is swap space?

24. In which of the following places would a computer forensic investigator not look for latent data?
    a. RAM slack
    b. file slack
    c. unallocated space
    d. temporary files

25. What is slack space?

26. What is fragmentation? What effect does fragmentation have on a hard disk drive (HDD)?

27. What is an IP address? What form do addresses typically take?

28. What is a human-readable name assigned to an IP address called?

29. What is a browser?

30. What is the purpose of an Internet cache?

31. List three places where a forensic computer examiner might look to determine what websites a computer user has visited recently.

32. Why is an IP important to forensic computer examination?

33. What is hacking? Who most commonly engages in hacking, and for what purpose?

34. What is a firewall and how does it work?

Learn the Vocabulary

Use the information from this chapter along with your prior knowledge and experience to learn the key terms. Choose from the following activities:

1. Make flashcards with a term on the front and its definition on the back. Use the flashcards to quiz a partner on correctly pronouncing and defining the terms.
2. Make an audio or video recording of each term and its definition.
3. Work with a partner to write a sentence for each term. Read the sentences out loud to a partner or share them with the class.
4. Write a list of new terms and language you have encountered in this chapter. Reuse the terms to write sentences and then read the sentences out loud to a partner. Reuse the terms again by writing a short speech. The speech might explain the terms or teach a lesson using the terms. Give the speech to a partner or to the class.

Application and Critical Thinking

1. If a file system defines a cluster as six sectors, how many bits of information can be stored on each cluster? Explain your answer.
2. Criminalist Tom Parauda is investigating the scene of a crime involving a computer. After he arrives, he photographs the overall scene and takes close-up shots of all the connections to the single computer involved, as well as photos of the serial numbers of the computer and all peripheral devices. Tom then labels the cord to each peripheral device, then disconnects them from the computer. After making sure that all data in RAM has been saved to the hard disk drive, he unplugs the computer from the wall. What mistakes, if any, did Tom make?
3. You are investigating a case in which an accountant is accused of keeping fraudulent books for a firm. Upon examining his computer, you notice that the suspect uses two different accounting programs that are capable of reading the same types of files. Given this information, where would you probably begin to search for latent data on the computer and why?
4. You are examining two computers to determine the IP address from which several threatening e-mails were sent. The first computer uses Microsoft Outlook as an e-mail client, and the second uses a web-based e-mail client. Where would you probably look first for the IP addresses in each of these computers?

Endnotes

1. One million bytes is referred to as a megabyte (MB); 1 billion bytes is termed a gigabyte (GB).
2. Pulling the plug should always be done by removing the plug that is connected to the back of the computer. If a plug is removed from the wall, it may be unknowingly connected to a battery backup (UPS). Under these circumstances, the UPS may cause an alert to the system and keep the unit powered on.
3. In this instance, *bit* is both metaphorical and literal. Every bit of information is needed, so we must get it all. So too every bit, as in the smallest unit of data storage—a one or a zero—must be imaged.
4. Actually, the more appropriate term is probably *paging* as opposed to swapping. This is because entire programs are typically not swapped in and out of memory to the swap space; rather, *pages* of memory are placed there.This activity requires the use of the following practices of science and engineering:

Mobile Device Forensics

Peter Stephenson

20

Charles Krupa/AP/Shutterstock

Key Terms

- analog
- architecture
- broadband
- CDMA (Code Division Multiple Access)
- file system
- geolocation
- GSM (Global System for Mobile Communication)
- GPS (Global Positioning System)
- logical extraction
- operating system (OS)
- physical extraction
- SIM (Subscriber Identification Module) card
- SD (Secure Digital) card
- SMS (Short Message Service)
- Wi-Fi

Michelle Carter: The Texting Suicide Case

On Sunday, July 13, 2014, 18-year-old Conrad Roy was found dead of carbon monoxide poisoning. When police began investigating the circumstances surrounding his death, they uncovered a series of text messages between Roy and his long distance girlfriend, Michelle Carter. From at least July 6 to July 12, 2014, Carter counseled him to overcome his doubts, devised a plan to run a combustion engine within his truck in order to poison him with carbon monoxide, and by directed him to go back in his truck after he exited it, when he became frightened that the plan was working.

She told him that he was "strong" enough to execute the suicide plan and that he would be happy once he was dead. In a text to Conrad, she wrote that he would go "straight to heaven guided by God." Conrad replied: "And I will be happy again." Carter wrote: "Yes, you will, (smiley face)...there is no way you can fail ... You're strong ... I love you to the moon and back and deeper than the ocean and higher than the pines, too, babe, forever and always. It's painless and quick." Below are excerpts of text messages exchanged between Michelle Carter and Conrad Roy on the day that he died:

CONRAD: Like, why am I so hesitant lately. Like two weeks ago I was willing to try everything and now I'm worse, really bad, and I'm LOL not following through. It's eating me inside.

CARTER: You're so hesitant because you keeping over thinking it and keep pushing it off. You just need to do it, Conrad. The more you push it off, the more it will eat at you. You're ready and prepared. All you have to do is turn the generator on and you will be free and happy. No more pushing it off. No more waiting.

CONRAD: You're right.

CONRAD: I don't know. I'm freaking out again. I'm over thinking.

CARTER: I thought you wanted to do this. This time is right and you're ready. You just need to do it. You can't keep living this way. You just need to do it like you did the last time and not think about it and just do it, babe. You can't keep doing this every day.

CONRAD: I do want to but I'm like freaking for my family I guess. I don't know.

Learning Objectives

After studying this chapter, you should be able to:

20.1 Identify the types of computing devices categorized as "mobile devices"

20.2 Explain the forensic challenges in examining mobile devices in comparison with personal computers

20.3 Explain the two procedures used to extract useful data from mobile devices

20.4 Explain the architecture of mobile devices that provides forensically valuable artifacts

20.5 Describe the types of evidence that can be found on mobile devices and how it is recovered

20.6 Explain how mobile devices fit into a digital investigation

CARTER: Conrad, I told you I'll take care of them. Everyone will take care of them to make sure they won't be alone and people will help them get through it. We talked about this and they will be okay and accept it. People who commit suicide don't think this much. They just could do it.

CONRAD: I'm ready.

CARTER: Good because it's time, babe. You know that. When you get back from them (sic) beach you've gotta go do it. You're ready. You're determined. It's the best time to do it.

CONRAD: Okay, I will.

CARTER: Are you back?

CONRAD: No more thinking.

CARTER: Yes. No more thinking. You need to just do it. No more waiting.

CONRAD: I don't know. I'm stressing.

CARTER: You're fine. It's gonna be okay. You just gotta do it, babe. You can't think about it.

CONRAD: Okay. Okay. I got this.

CARTER: Yes, you do. I believe in you. Did you delete the messages?

CONRAD: Yes. But you're going to keep messaging me.

CARTER: I will until you turn on the generator.

Michelle Carter was charged and convicted of involuntary manslaughter in the death of Conrad Roy and served 11 months in prison.

Of all of the areas of digital forensics, mobile device forensics may be the most complicated. It is complicated for several reasons. First, there is a huge number—growing daily—of mobile devices. Second, these devices often have little in common, even those from the same manufacturer. Additionally, the proliferation of mobile devices as substitutes for full-size or laptop computers is significant and increasing rapidly. Finally, these devices are an amalgam of radio and computing technologies and may in some cases be treated differently under the law. The forensics involved certainly is complicated by this paradigm.

Mobile devices began as an outgrowth of ship-to-shore radios in World War II. Additionally, handheld radio transceivers, or walkie-talkies, were available then, and they evolved into mobile phones for cars in the 1940s. The real explosion in mobile devices came much later. Before that, we saw the Motorola handheld phone debuting in 1973. From that point on, advances in cellular technology enabled the mobile phone boom that followed. Early mobile phone systems were **analog** (1G). They were followed by digital networks (2G).

When mobile **broadband** networks (3G) arrived on the scene in Japan in 2001, the mobile device landscape changed forever. Now it was possible to do much more than talk on a cell phone. Now the cell phone had the possibility of behaving like a small computer and could transfer data—at Internet speeds. The smartphone was born. When native IP networks arrived (4G), the smartphone became a node on the Internet just like any other computing device. With the proliferation of **Wi-Fi** networking, smartphones evolved into tablets, and products such as the iPad® became viable substitutions for small computers.

analog
The traditional method of modulating radio signals so that they can carry information

broadband
Describes a communication channel that can provide higher-speed data communication than a standard telephone circuit

Wi-Fi
A term describing a wireless local area network

The Mobile Device Neighborhood: What Makes a Mobile Device "Mobile"?

We start with the notion of cellular systems. A *cellular system* is a network of relatively short-distance transceivers that are spaced strategically so that low-power transmitters can reach the phones in their coverage areas and the very-low-power transmitters in the cell phones can reach the cell tower. Since the 1960s the concepts of *handoff* and *frequency reuse* allowed users to move between cells without dropping a call. Usually. As we will see, 1G networks suffered from this problem, much to the consternation of their users. Figure 20–1 shows the layout of cell towers and their coverage areas.

Dr. Peter Stephenson, a cyber-criminologist and educator with fifty years of technology experience, has written, edited, or contributed to eighteen books and several hundred articles in major national and international trade, technical, and scientific publications. He is an associate professor teaching network attack and defense, digital forensics, and cyber investigation on both the graduate and undergraduate levels at Norwich University in Vermont, is the chief information security officer for the university, and is the director of the Center for Advanced Computing and Digital Forensics. He holds one of the first PhDs in the world in digital investigation and was one of the first recipients of the prestigious Certified Cyber Forensic Professional designation from (ISC). In addition, he holds the CISSP and CISM designations and is a member of the American Academy of Forensic Sciences and the Vidocq Society. He holds a master's degree in diplomacy with a concentration in terrorism.

FIGURE 20–1
Cell tower geographic configuration. *Courtesy Richard Saferstein, Ph.D.*

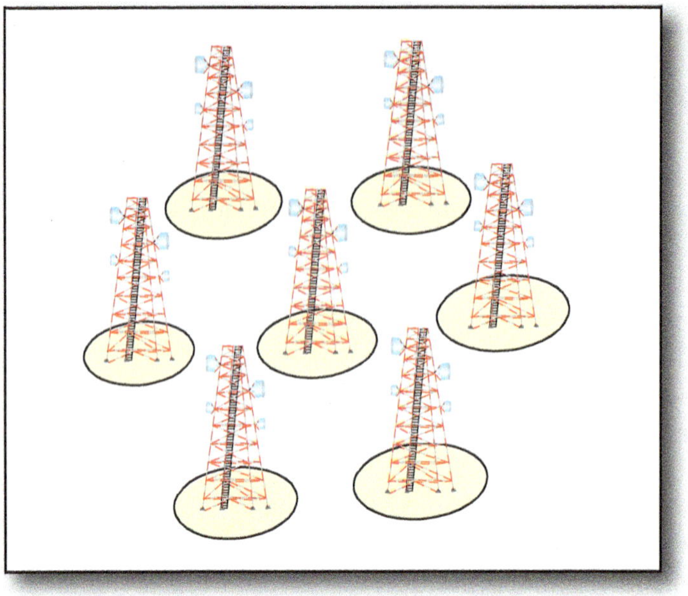

Although there is overlap as indicated by the red circles, the transmitting patterns are hexagonal because of the transmission patterns of the tower's antenna array.

Returning to the early (1G) analog phones, we have several issues that laid the groundwork for modern mobile devices. Some of these are still with us and can bedevil the forensic examiner. 1G networks are analog. That means that they behave in exactly the same way as older radio stations behave. A mobile device is made up of a computer and one or more radios. The computer may be quite primitive and the radios may communicate with a network of some type. Since cellular networks are much different from Wi-Fi networks, devices that communicate with both need two radios. 1G devices usually could communicate only with the cellular system.

GSM (Global System for Mobile Communication)
A set of standards for second-generation cellular networks

CDMA (Code Division Multiple Access)
A spread-spectrum technology for cellular networks

Digital (2G) cellular networks appeared in the 1990s using two standards: **GSM (Global System for Mobile Communications)** and **CDMA (Code Division Multiple Access)**. This new generation moved phones into the small, handheld form, and, because they were digital, the new networks opened the door for practical data communications and the beginning of what was referred to as "feature phones." These phones had more features than simply being able to make and receive phone calls, hence the name. Feature phones could send and receive SMS (text) messages, synchronize with e-mail, and provide other features that were heading the mobile device genre in the direction of smartphones and tablets.

Communications were slow, however, and such things as surfing the web and transferring photos were not particularly practical. Smaller radios were, however, practical because of the increased density of cell tower installations, and that saved battery life, further reducing the physical size of the phones.

Mobile broadband (3G) data opened the floodgates for mobile communications and, thus, the explosion of mobile devices on the market. The architectural functionality that distinguishes 2G from 3G is that 2G systems were circuit switched and 3G systems are packet switched. Without delving too deeply into the technical differences between the two, we may say that, fundamentally, circuit switching is similar to landline telephone systems and packet switching is like the Internet. Indeed, one of the benefits of packet switching is the ability to connect more readily to the Internet.

The advent of packet-switched mobile phone networks allowed virtually any kind of data to be accessed by the mobile device, and the smartphone was born. Moving photos over the network, streaming video and television, video chat, and other advanced services now could be supported. 3G was launched in Japan in 2001.

Native IP (4G) networks differ technologically from 3G networks in that they access the Internet directly, increasing speed and bandwidth dramatically. With 4G, virtually any form of communication possible with a PC was now possible from a smartphone or tablet. The age of the mobile device was now in full swing.

Quick Review

- A "cellular system" is a network of relatively short-distance transceivers that are spaced strategically so that low-power transmitters can reach the phones in their coverage areas and the very-low-power transmitters in the cell phones can reach the cell.
- Digital (2G) cellular networks moved phones into the small, handheld form, and, because they were digital, the new networks opened the door for practical data communications and the beginning of what was referred to as "feature phones."
- The architectural functionality that distinguishes 2G from 3G is that 2G systems were circuit switched and 3G systems are packet switched.
- The advent of packet-switched mobile phone networks allowed virtually any kind of data to be accessed by the mobile device, and the smart phone was born.
- Native IP (4G) networks differ technologically from 3G networks in that they access the Internet directly, increasing speed and bandwidth dramatically.

Forensic Challenges: Mobile Devices as Small Computers—Sort Of

Modern mobile devices are, in many respects, little more than small PCs. However, they do have some unique aspects that complicate the digital forensic process a bit. Let's begin by going back to those 1G devices. They still exist in some parts of the world and even with die-hard users in the United States in regions where 1G networks still exist. While those are extremely rare, the more common occurrence is getting a 1G phone tied to a 15-year-old crime. The problem with these old devices is that they do not have the capabilities that we are used to with a computer. Virtually none have what we would classify as an **operating system (OS)**, and about the most you can hope to get off them is the phone directory. Some may keep a call log, but the number of calls retained is fairly small. Anything that has been deleted is no longer available. Unlike many operating systems, we are unable to recover deleted material.

operating system (OS) A custom-designed program that controls the components of mobile devices and facilitates how they function

When we get to 2G, the ballgame changes considerably. These phones still exist, mostly with people who are satisfied with their limited capabilities and who live in a 2G network area. The capabilities of these phones, while considerably fewer than smartphones, are well ahead of 1G phones, and most of the 2G devices have real operating systems. Mostly these operating systems are custom designed for the phones—such as Palm® and BlackBerry®—but in some cases they are the forerunners of smartphone operating systems.

Some 2G phones, such as feature phones, come closer to smartphones than others. However, the ability to recover deleted messages, for example, varies significantly from phone model to model. However, just because a phone has an OS does not mean that the OS is the same as a PC's. While the OS performs many of the same functions as that of a PC, it does not necessarily follow that it performs them in the same way or, in many cases, even as well. The most popular operating systems for mobile devices—including phones and tablets—are Apple iOS, Google Android®, and Microsoft Windows Phone OS. Primitive versions of these were widely available for 2G devices.

Some had other features, but the higher the bandwidth requirement, the poorer these devices performed. From a forensic perspective it is a toss-up as to how much data you will be able to extract from a 2G device. More important, during this time the manufacturers of mobile devices started to go wild with new product releases. The problem with that was—and still is—that even if two phones looked a lot alike, had the same functionality, and used the same OS, there was no guarantee that they were the same, and if your mobile device forensic tool did not have a driver specifically designed for that particular phone, you were out of luck forensically. This continues to be the single most difficult issue plaguing mobile device forensics: it is nearly impossible to stay current with the available mobile device models.

architecture
The basic components of a mobile device

3G and 4G phones are the closest in **architecture** and design to a PC. They behave the same way—especially 4G devices—and they have the ability to download and install applications ("apps") the same as any PC or Mac. They are the same in architectural issues such as processors and file systems, but the nature of those architectures are different between mobile devices and computers. This is because mobile devices have special requirements such as multiple radios (Wi-Fi and 4G, for example), size restrictions, and storage space. Apps are both a boon and a devilment to the forensic examiner. Apps each have their own specific operating parameters, and how they communicate with the outside—*if* they communicate—is inconsistent from app to app. Apple has taken major steps to standardize the development of apps for its iPad and iPhone®, but Android apps are far less constrained. Furthermore, just because an app is written for the iOS operating system does not mean that it runs equally well or behaves the same on iPad and iPhone. In that regard, Android apps are a bit less forgiving.

geolocation
Assessment of the actual geographical location of a mobile device

For the forensic analyst, understanding what is running on the mobile device under examination is a key issue, and one that is nontrivial to figure out. Once the examiner understands what is running, it is equally difficult to figure out what the app is doing and how it is interacting with the user and the outside. One interesting aspect of mobile device forensics is **geolocation**. Some devices and many apps report out the geographical location of the device (see Figure 20–2). That can make it much easier to track the owner's movements.

FIGURE 20–2
Representation of geolocation using a mobile device. *Andre Nantel/ Shutterstock*

Additionally, mobile devices that offer geolocation include not just smartphones and tablets, but we also can analyze **GPS (Global Positioning System)** devices such as Garmin® or TomTom®. However, special tools and drivers are required, as with other mobile devices, and GPS units are becoming as prolific and ever-changing as smartphones and tablets.

GPS (Global Positioning System)
A system for determining position by comparing radio signals from several satellites

Another forensic challenge with phones, particularly, is the chipset used to build the phone. This adds another variable to the mix along with OS, model, manufacturer, and apps. The chipset is the hardware that makes the device work. Today there is quite a large number of chipsets, and the most difficult to deal with are knock-offs of American chips manufactured in China. Not all of these behave the same as American chips, and special driver sets for mobile device forensic tools are required.

The big problem, of course, is that even though some of these devices are really small computers with computer-like operating systems, they usually cannot be examined using typical computer forensic tools. Each device has its own quirks, and each device needs special connectors and special device drivers on the tool that is examining it to decipher what is on the device's storage. Device storage also takes several forms, such as onboard nonvolatile memory and mini SD cards that add storage in a modern smartphone or tablet. This does not include devices that predate the computer-like architectures. They need special drivers and connectors as well, but the amount of information that can be gleaned from them is much less because the amount of information they store is much less.

Quick Review

- The most popular operating systems for mobile devices—including phones and tablets—are Apple iOS, Google Android, and Microsoft Windows Phone OS.
- 3G and 4G phones are the closest in architecture and design to a PC. They behave the same way—especially 4G devices—and they have the ability to download and install applications ("apps") the same as any PC or Mac.
- One interesting aspect of mobile device forensics is geolocation. The GPS can locate the user's activities and, when used with a timeline, can place the user in the vicinity of a crime. That can make it much easier to track the owner's movements.
- Each device has its own quirks, and each device needs special connectors and special device drivers on the tool that is examining it to decipher what is on the device's storage. Device storage also takes several forms, such as onboard nonvolatile memory and mini-SD cards that add storage in a modern smartphone or tablet.

Extracting Useful Data: The Differences in Various Types of Mobile Devices

When working a mobile device, the investigator has several sources of information available. Probably the most useful is web searching. Doing a search on the phone model often reveals a wealth of information such as other investigators' experiences, battery charging techniques, what can and cannot be recovered if it was deleted, and so on. Some phones and tablets are fairly modern and rather straightforward. With those, connection to the forensic tool and extraction are fairly simple to do. But some devices—both very old and very new—are not so obvious. Some research before connecting is important. All mobile devices should be kept in a Faraday bag or box. This prevents changes from being made to the device remotely. These changes might be initiated by the owner of the device, such as a remote wipe to preserve a picture of innocence by destroying evidence, or unintentional, such as changes made by the device's carrier that could overwrite evidence. The efficacy of Faraday enclosures has been debated by experts, but the consensus still is to use them.

So, the examiner's first step is to determine what they are working with. Is it a very old feature phone, a typical iPad or iPhone, or a state-of-the-art, just-announced-last-week smartphone? The examiner must do a little research, select a tool, and then make the next decision: physical or logical extraction (or both)? Just as with computers, **physical extraction** is the best bet. Physical forensic images are bit-by-bit copies of the file system, including deleted data. **Logical extraction** is a snapshot of the file system showing what the file system wants the user to see. Here the examiner gets the same view that the user gets.

physical extraction
A duplicate of data located on a mobile device

logical extraction
A snapshot of the file system of a mobile device

Some tools, such as Cellebrite's UFED Touch, are quite clear about which devices support physical extraction. For a device that supports physical extraction, that is the way to image the device. Logical extractions are useful only when the physical option is not available because of the device itself. On some cell phones, an exchange of text messages may hold evidence in a murder. However, one side of the exchange may be missing—obviously someone deleted it—and because of the architecture of the phone, retrieving the deleted messages is not possible. The only solution is to acquire the other phone in the conversation and extract it from that phone. If that phone is not available, the examiner is left with a tantalizing snippet that may or may not include evidence.

Tools such as UFED and MPE+ greatly simplify analysis. And these are not the only available tools for mobile device analysis. Others include Parabin's product, Device Seizure, a forensic tool that started mobile device analysis and an excellent tool by Oxygen. When selecting a tool for mobile device forensics, one should look at the field and pick more than one, in the same way that most digital forensic labs use more than one computer forensic tool.

Mobile device forensic analysis can provide an overlay to physical evidence and timelines as well as computer forensic timelines to give a clearer picture of the events preceding and following a crime event. However, the efficacy with which the examiner can gather this information depends a lot on the generation of mobile device being tested. Technological capabilities vary widely from first- through fourth-generation devices, though third-and fourth-generation devices tend to have a lot of power.

When you analyze the device be sure to follow the recommendations of the tool you are using for analysis. Do an Internet search to learn as much about the device make and model as you can before attempting acquisition and analysis if you are unfamiliar with the specific device.

YouTube has a remarkable number of videos that show detailed step-by-step procedures for device disassembly. Special tools may be required. Working on a carpeted surface often raises the danger of static electricity. That can damage the chips in a device. It may be wise to wear a grounded antistatic wristband when working on a mobile device, especially if disassembly will be required.

When the examiner has identified the device and the procedures to extract its data, the next step is to run the tools and take a forensic image. Examiners make it a practice to run the imager twice, taking one of the images and treating it as evidence. The other is the working image. This ensures that if necessary, the examiner can make another copy of the evidence original as a new work copy if the working copy is inadvertently damaged.

Logical images are fairly fast for most devices, depending on how much memory the device has and how much of the memory is full. Physical images can take a very long time to make because on a large storage device, even if the storage is not full, the imager must look at the entire memory footprint, not just the part that has something save to it. The examiner should decide, based on what can be done for the particular device, whether to obtain a physical or logical extraction or both. The logical extraction is fairly fast, and one may want to examine it for obvious evidence while the tool is making a physical image of the target.

It's important that the examiner select the proper connector—or "pigtail"—for the device from their tool kit.

Inside the Science

Analysis of Submerged Devices

Many people go to great lengths to conceal their mobile activity. Some even go as far as to physically destroying their phone altogether by submerging it in oil, drain cleaner, fuel, or some other flammable liquid. However, digital evidence can be very resilient, and digital forensics practitioners equipped with the right tools and knowledge can still find valuable information from a damaged smartphone.

A team of scientists from VTO Labs, which included interns from Marshall University's Forensic Science Graduate Program, studied the effects of submerging smartphones in different liquids. In their study, they submerged 49 phones, of the same model and operating system, in a variety of oil-based, flammable, and clandestine chemical liquids for seven days. Their results showed that a full forensic image could still be extracted from all of them—even when circuit board components were severely corroded.

Even in the most severe case of physical damage—the drain cleaner trial—the team was able to image all 3,909,091,328 bytes of data on the phones, which is the same exact amount imaged before the phones were submerged. Trials of all other liquids yielded the same results. The other liquids were vegetable oil, new motor oil, used motor oil, regular gasoline, diesel fuel, kerosene heating oil, ammonia, and hydrogen peroxide.

Due to the damage to each phone's exterior and circuit board, researchers had to remove and repair the memory chips from the drain-cleaner-drenched devices, and take the data directly from the chips, a technique known as "chip-off." But before this, the devices were cleaned using specific methods—the methods that they determined to be most effective.

The lab had previously tested to see how digital evidence fared after devices were submerged in various types of water—including fresh water, salt water, tap water, and chlorinated pool water—for varying amounts of time. In the previous study, the team had managed to extract data from phones submerged for as long as three years. They found that salt water was actually more damaging to phones than oil-based liquids, gasoline, or drain cleaner.

Quick Review

- All mobile devices should be kept in a "Faraday" bag or box. This prevents changes from being made to the device remotely.
- Physical forensic images are bit-by-bit copies of the file system, including deleted data. Logical extraction is a snapshot of the file system showing what the file system wants the user to see.
- Mobile device forensic analysis can provide an overlay to physical evidence and timelines as well as computer forensic timelines to give a clearer picture of the events preceding and following a crime event.
- Examiners make it a practice to run the forensic image twice, taking one of the images and treating it as evidence.
- The examiner should decide, based on what can be done with the particular device, whether to obtain a physical or logical extraction or both.

Mobile Device Architecture: What Is Inside the Device and What Is It Used For?

All mobile devices have an architecture. Just like computers, the architecture defines the basic components of the device. How and where is data stored? What kind of processor is used? What does the file system look like? Is there a formal file system? These and other questions form the basis of the analysis. Extraction is affected only if there is additional storage—such as an extra plug-in SD card—that must be analyzed. Most analysis tools acquire these add-in storage modules, but it is helpful to know which so data won't be missed.

SIMs and SDs

SD (Secure Digital) cards are storage expansion cards that many mobile devices can accept. The SD card adds memory for storing such things as photos and music. SD cards are *nonvolatile*, meaning that even if the power is turned off on the device, you won't lose your favorite tunes or your pictures of Great Aunt Susie just as she was going down the water slide at the local swimming pool (see Figure 20–3).

SIM (Subscriber Identification Module) cards are different. Each SIM has an international mobile subscriber identity (IMSI) number that associates the phone with the subscriber's mobile network. In many cases you can keep all of your subscriber information when you change mobile

FIGURE 20–3
GB Micro SD Card (right) and adapter. *Krisda chamchuen/Alamy Stock Photo*

SD (Secure Digital) card
A storage expansion card for a mobile device

SIM (Subscriber Identification Module) card
The card that is inserted into a mobile device that identifies the user account to the network, handles authentication, and provides storage for basic user data and network information

phones simply by switching the SIM to the new phone. SIMs may also store text messages and other user data as well as the user's phone book and the phone number of the device. Not all mobile devices use SIMs. The information on the SIM may be stored in the device itself. However, for forensic purposes, the examiner must be able to access that information because it is a way to identify the target device unambiguously during a forensic analysis.

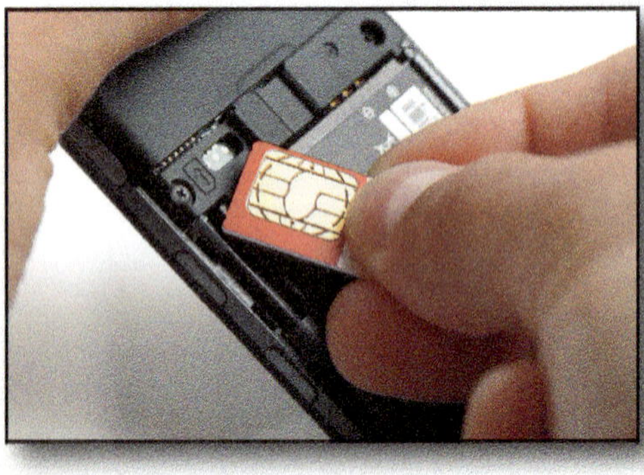

FIGURE 20–4 A SIM inserted in a cell phone. *Brian Jackson/ Alamy Stock Photo*

Also on each SIM is the integrated circuit card identifier (ICCID). That number also is printed on the SIM. The ICCID contains the issuer identification number (IIN), the individual account identification, and a check digit. The forensic image of the mobile device is acquired and the tool will extract and record the ICCID. That is usually the way that the examiner identifies the phone, but if it is a SIM and not embedded in the mobile device, the examiner must bear in mind that they are seeing only the SIM and not necessarily the device itself. Figure 20–4 shows the SIM ready for placement in the cell phone. Note that this information is embedded in the tablet, not in an external card.

It often is desirable to clone the SIM in much the same way as one would take a physical image of the mobile device or a computer. By cloning the SIM, the investigator retains a perfect copy for evidentiary purposes. Some mobile device forensics vendors, such as AccessData, developer of the MPE+, provide forensic SIMs for use as targets to which the investigator may clone the evidence SIM for preservation.

In addition to memory, the typical mobile device contains a digital signal processor, a microprocessor, a radio frequency transmitter/receiver, audio components, and a power supply that takes battery power to run the device as well as providing the ability to charge the battery. Thus, the mobile device is a simple radio transceiver with a few digital twists and turns that make it more like a computer in some regards.

This hybrid construction—a radio and a computer—mean that forensic investigators must focus on the laws that cover both. In some regards, thinking from the computer side, they must be concerned about such things as privacy. From the radio side, there also are protections of which to be aware that vary greatly from jurisdiction to jurisdiction.

File Systems—or Not

file system
A software mechanism that defines the way files are named, stored, organized, and accessed

To say that every mobile device has a **file system** is both correct and incorrect. Certainly there needs to be a way to keep track of such things as phone books, but in some cases the examiner is simply looking at a flat file database. A *flat file database* is a single file, usually human-readable, that contains some collection of data, in this case names and phone numbers. The earliest, most primitive mobile devices had such a file structure.

When we think of a file system, though, we usually think of some organized method of collecting files of various types and keeping track of them. In computers we have file tables that track both the files in the file system and their physical location on the computer's hard drive. A real file system must be able to do those two things: track the logical location of the files on the device and tie them to the files' physical locations. More modern mobile devices do both of those things, sometimes with a proprietary file system and occasionally with a more open one.

Tracking the location of files, both physically and logically, is done with some sort of database. Sometimes the database is proprietary and tied to the file structure itself, and sometimes it is a standard database such as SQL Lite (SQLite). In either case the analysis tool must know what file system structure the device uses, and it must know how to address files on the device both physically and logically. Extracting file information from such a data structure is critical to the analysis of mobile devices just as it is on computers. And the extraction and analysis are quite similar.

One important aspect of file system analysis is accessibility. Can the examiner access deleted files, for example? Is there some indication, as is the case with the Windows operating system on a computer (in the link or .lnk files), that a file existed even if it has been deleted? The answer, unfortunately, is "maybe." As shown in Table 20–1, different types of operating systems have different types of file systems, some more recoverable than others.

Table 20–1
Example Mobile Device File Systems (Not Exhaustive)

BlackBerry	SQLite or MS exFAT
Android	Ext4, YAFFS, or vendor proprietary
iOS	HFSX

It may not be possible to recover deleted file items such as e-mails, texts, and photos from a mobile device. For example, BlackBerrys cannot be recovered directly. To recover deleted files, the examiner must go to the backup—usually on a PC somewhere—and recover from that. One can also recover from the BlackBerry server if available. Recovery of deleted items from iOS devices is somewhat easier, and tools such as Cellebrite's UFED can do that with a physical extraction. Androids are usually recoverable as well since they are, at heart, a form of the Linux operating system.

Analyzing Mobile Devices: Finding Forensically Valuable Artifacts

The detailed analysis of mobile devices is a book in itself. There are so many different devices (between 50,000 and 100,000 by some counts) and multiple generations of devices that a full analysis addressing all of these possibilities is not attainable in a single chapter. As well, a detailed understanding of computer and mobile device architecture is necessary to grasp many of the finer points of analysis. It suffices here to discuss the process and give some examples that are at opposite ends of the mobile device spectrum. We have introduced the Apple iPad and the Samsung SCH R350.

Let's start with our Samsung phone. The R350 is a feature phone popular around 2009. It had a lot of features but some, such as its music player, required an optional memory card. Overall, the R350 was a very good stepping-stone to today's smartphones if you deployed all of its features. The file system is EFS (encrypting file system). By looking at the Project Tree on the UFED extraction report it's possible to find out if the user browsed the web and the history of web browsing. Further, by browsing down the Project Tree on the UFED extraction report, one may find folders associated with a phone book and calender.

The iPhone is replete with data for one's perusal. The calendar is referred to as *Calendar* rather than *task list*. That is because *Calendar* is the phone's terminology and the UFED simply uses it for consistency.

SMS (Short Message Service)
A cellular network facility that allows users to send and receive text messages

In addition to the calendar we can follow the project tree in the UFED report and see several other resources that our iPhone gives up to our analysis. There is a call log with details, chats including deleted messages, contacts including those who were recently contacted, cookies, **SMS (Short Message Service)** messages (texts), voice mails, and more. This is a veritable treasure trove of data on the phone's user and the user's behavior. But there is much more if we look a bit deeper. For example, it would be nice to take all of the phone's activities and place them on a timeline. Digital forensic investigations depend on timelines for their success. Indeed, when overlaid on the timelines of a physical crime, the timelines from mobile devices and computers provide an excellent yardstick by which to measure the play of events surrounding the crime itself. Because of the vast amount of data the iPhone 4S smartphone offers, we can create just such a timeline.

The UFED also provides us with the ability to analyze phone activity by caller, giving us incoming, outgoing, missed, and SMS (text) calls.

Another useful type of mobile device is the GPS. GPS can locate the user's activities and, when used with a timeline, can place the user in the vicinity of a crime. Timelines are the meat and potatoes of digital forensic investigations. Because computers and mobile devices have a fairly accurate clock, examiners can match the activities on these devices to physical crime activities to do a precrime, pericrime (during the crime event), and postcrime analysis of a suspect's behavior.

For example, with the Garmin® nuvi® 40 GPS, the UFED provides a timeline, but since this is a GPS, the timeline is associated with a specific location. In addition to the timeline graph, there are specific entries noting where the GPS was located at a particular latitude and Longitude at a particular time.

Following the track of locations using Google Maps and these coordinates is a straightforward way to track the progress of the GPS during some particular period of time. To make the tracking easier, the GPS provides a list of journeys, each with the coordinates of waypoints over the course of the trip. That can be correlated back to the timeline view for devices that have been used heavily, simplifying the tracking of the GPS on specific dates of interest.

Additionally, the GPS provides a list of favorite destinations with their coordinates.

The FBI v. Apple: The Encryption Dispute

Case File

In December, 2015, Syed Rizwan Farook and his wife Tashfeen Malik killed 14 people and injured 22 others in a mass shooting and attempted bombing attack at the Inland Regional Center in San Bernardino, California. The couple were killed in a shootout with the police shortly after their attack. Farook's work phone, an Apple iPhone 5C, was recovered intact. The phone had been locked with a four digit password. Two-month effort by the FBI to unlock the phone failed due to the phone's advanced encryption system. As a result, the FBI asked Apple, Inc. to create a version of the iPhone operating system to allow the FBI to disable certain security features. Apple declined citing its policy of never undermining the security systems of it products. The FBI then proceeded to have a court order issued requiring that Apple comply with the government's request. However, two months later, the FBI withdrew the request stating that it had unlocked the phone with the help of a third party.

The use of mobile devices are playing an increasing role in criminal investigations. In addition to the San Bernardino case, Apple has received at least 10 different requests from federal courts to extract data like contacts, photos and calls from locked iPhones. Apple has consistently objected to these requests taking the position that to make access easier for law enforcement will also make it easier for criminal hackers to access personal data placed on their customer's phone.

The standoff between Apple and the FBI is at the center of an escalating battle between technology companies that are encrypting data in order to safeguard their customers' privacy and government's desires to access encrypted data to secure criminal investigative information.

Quick Review

- Secure Digital (SD) cards are storage expansion cards that many mobile devices can accept. The SD card adds memory for storing such things as photos and music. SD cards are nonvolatile.
- Subscriber Identification Module (SIM) cards have an international mobile subscriber identity (IMSI) number that associates the phone with the subscriber's mobile network.
- Each SIM is the integrated circuit card identifier (ICCID). The ICCID contains the issuer identification number (IIN), the individual account identification, and a check digit.
- In addition to memory, the typical mobile device contains a digital signal processor, a microprocessor, a radio frequency transmitter/receiver, audio components, and a power supply that takes battery power to run the device as well as providing the ability to charge the battery.

Hybrid Crime Assessment: Fitting the Mobile Device into the Digital Forensic Investigation

Now that we've gotten a look at how digital forensics is performed on mobile devices, it would be useful to fit that process into the investigative process as a whole. In any investigation chain of evidence is very important. *Chain of evidence* (as opposed to *chain of custody*, which describes access to evidence) describes the events and concomitant evidence that make up the events of the crime.

There are two types of chains of evidence. *Temporal chains* show events in the order in time in which they occurred. This commonly is called a timeline and is the easiest way for laypeople, such as triers of fact, to visualize a crime, especially a complicated one. *Causal chains* of evidence describe the events of a crime in terms of cause and effect. The links in the chain are the pieces of evidence, and they are tied together based upon how one link affects one or more other links. We could say that our first link causes the next link to occur, which in turn causes two other links—events—to happen, and so on, until the events of the crime fully describe the crime itself. For our purposes—and to keep this manageable—we will stick with the temporal chain. This is where a technique called *hybrid crime assessment* enters the picture.

Hybrid crime assessment is a technique that investigators can use when faced with a physical crime—such as murder, rape, or robbery—that has a digital element to it: a computer, a cell phone, or some other mobile device. The idea is to tie all of these elements together into a single crime scene and use the timelines to build a picture and describe the events, and supporting evidence, of the crime. This works very well. So well, in fact, that it has been used in at least one case to describe the last few minutes of a shooting victim's life when the closest physical event that could be corroborated occurred more than a half hour before the shooting and had nothing to do with the crime itself.

In that regard, it is important to recognize that there are events in the timeline that have little or nothing to do with the actual crime but can act as markers—rather like quarter-hour chimes on a clock that simply happen but do not play any role beyond marking time. In the case of the shooting, the activities of the victim on his computer ended moments before he was murdered. Given the physical crime scene and the position of the body, the activities surrounding his use of the computer were very important. They placed, with little question, the time of the event within an eight-minute period, which corroborated a witness account. They also allowed some level of description of the events immediately following the event, since the computer was turned back on after it had been shut down briefly.

This is a simplistic example of correlating the events of a physical crime with a timeline of events on a computer. In this case, the period of inactivity on the computer could be measured, and that timeline—very accurate since it was based upon the computer's clock—could be laid against the physical timeline to fill in gaps and corroborate witness accounts. When we add the dimension of a mobile device, which works in essentially the same manner as the computer for our purposes, we have a layered view of the timeline of an event and the pre-event and post-event elements/evidence that make up the layers. The mobile device is one of the layers and, in most cases, the most accurate one.

Mobile devices may be synchronized to the network clock. That means that the network provider—such as AT&T®, Verizon, or T-Mobile®—is sending the clock signal to the mobile device. Since the network provider is likely to use a time standard such as the U.S. Naval Observatory to obtain time signals, it is likely that the mobile device is extremely accurate as a yardstick for measuring when events happen. How the user behaves with the mobile device can be measured precisely against that yardstick.

Mobile devices, like most computers, perform various types of housekeeping. That means that the device is constantly doing things on its own to maintain its own operation. For example, it must continue to stay in contact with the mobile network (or the Wi-Fi network if it is set up to connect to one). It must make sure that all of its internal functions are working properly, so it is constantly testing parts of itself. When the device is in use, it does more. When it is in standby, it does less. When it is off, obviously, it does none. There are variations on that as well. For example, if you put the device in "airplane mode," the radio is off but the rest of the computer (in the phone) still is working. By checking logs, when they are present, we can see when the radio was turned off or on.

Inside the Science

What Is a StingRay?

A StingRay is an international mobile subscriber identity catcher (IMSI) device designed in 2013 by the Harris Corporation. This technology is used by military, intelligence, and law enforcement agencies across the world for cellphone surveillance. It allows these agencies to intercept mobile phone traffic, send fake text messages, inject malware, and track the movements of mobile phone users. StingRays act as cell site simulators that collect data from users by mirroring cell towers. The StingRay equipment has two operating modes: active and passive. In the active mode the device simply mimics the behavior of a cell tower in order to force all nearby mobile phones and other cellular data devices to connect to it. In the passive mode, however, the StingRay actively interacts with cellular devices and performs functions like data extraction and location tracking. The StingRay family of devices can be mounted in vehicles, on airplanes, helicopters, and drones. Hand-carried versions are referred to under the trade name KingFish. Currently over 70 state and local law enforcement agencies in 24 states along with 13 federal agencies across the United States use StingRay technology to monitor mobile phone users.

Case File

Carpenter v. United States

Carpenter v. United States was a landmark United States Supreme Court case that involved the process of obtaining call logs and cell site location information (CSLI) from cellphone companies as part of a police investigation. Prior to *Carpenter*, government entities could obtain cellphone location records by court order by claiming the information was required as part of an investigation. Law enforcement agencies could compel a third party entity (cell phone provider) to provide those records because they kept them in the normal course of their business in the cell phone industry.

In *Carpenter*, the Supreme Court determined what type of legal authorization is required by a government entity in order to compel third-party wireless service providers to turn over historical records containing the physical locations of cellphones. Attorneys for Carpenter argued that the authorization should be a search warrant. Attorneys for the United States argued that the authorization should be a court order for disclosure. Requirements to obtain an "order for disclosure" are less stringent than the requirements to obtain a warrant.

An order for disclosure is a type of court order typically instructing a third party to turn over information in its possession related to an investigation. To obtain an order for disclosure from a court, a governmental entity must provide specific and articulable facts showing that there are reasonable grounds to believe that the information sought is relevant and material to an ongoing criminal investigation. Government entities are not required to show probable cause to obtain an order for disclosure.

Under the Fourth Amendment to the United States Constitution, people have the right to be free from unreasonable searches of their "persons, houses, papers, and effects" unless the government entity obtains a search warrant supported by probable cause. Probable cause to search exists when facts and circumstances known to the government entity provide the basis for a reasonable person to believe that a crime was committed and the items to be searched are relevant to the crime.

The Court held, in a 5–4 decision authored by Chief Justice Roberts, that obtaining the records without a warrant supported by probable cause and signed by a judge or magistrate violated the Fourth Amendment to the United States Constitution. After *Carpenter*, government entities must obtain a warrant in order to access these records for domestic criminal investigations.

As we have seen, the amount of information that we can get from a mobile device varies greatly with the device. Unfortunately, most books that deal with mobile device forensics assume that the digital investigator will invariably encounter a fairly recent device. That means that the oldest devices are likely to be feature phones. That, as it turns out, is not a practical position to take. Feature phones of widely varying capabilities are generally available, and there are mobile phones that are little more than that: mobile phones. These and older phones often have very little to offer in the way of establishing timelines but should be examined in any event.

However, when a mobile device has information to give up, it can be extremely useful. For example, when a mobile device is set to use Wi-Fi, it will recognize any Wi-Fi network in its range. It may not be able to join the network because of the security settings on the Wi-Fi access point, but the mobile device will see it and note that it exists. When that happens, the device takes note of the network and logs it. This is a function of today's mobile devices such as smartphones, of course. The iPhone 4S can look at its timeline, expand it to get maximum resolution, and see, for example, that at 15:42:30 on 6/12/2012 the phone was in a Best Buy store. We see that because the timeline shows that it joined the Best Buy network at that time.

Taken with other evidence, such as GPS information, witness interviews, or the individual's own timeline as reported during an interview/interrogation, this becomes a good corroborator and may, in some cases, provide an alibi as well as it might otherwise place the individual near the crime scene. If one opts to take a bit more complicated look, the mobile device can be examined for everything happening on it at a particular time. Then the investigator can put all of the pieces together from both the mobile device logs—which are quite precise—and the physical evidence, which may not be as precise.

Quick Review

- "Temporal" chains of evidence show events in the order in time in which they occurred. This is commonly called a timeline.
- "Causal" chains of evidence describe the events of a crime in terms of cause and effect. The links in the chain are the pieces of evidence, and they are tied together based upon how one link affects one or more other links.
- "Hybrid crime assessment" is a technique that investigators can use when faced with a physical crime—such as murder, rape, or robbery—that has a digital element to it—a computer, a cell phone, or some other mobile device.
- The amount of information that one can get from a mobile device varies greatly with the device.

Chapter Review

- A "cellular system" is a network of relatively short-distance transceivers that are spaced strategically so that low-power transmitters can reach the phones in their coverage areas and the very-low-power transmitters in the cell phones can reach the cell.
- Digital (2G) cellular networks moved phones into the small, handheld form, and, because they were digital, the new networks opened the door for practical data communications and the beginning of what was referred to as "feature phones."
- The architectural functionality that distinguishes 2G from 3G is that 2G systems were circuit switched and 3G systems are packet switched.
- The advent of packet-switched mobile phone networks allowed virtually any kind of data to be accessed by the mobile device, and, so, the smart phone was born.
- Native IP (4G) networks differ technologically from 3G networks in that they access the Internet directly, increasing speed and bandwidth dramatically.
- The most popular operating systems for mobile devices—including phones and tablets—are Apple iOS, Google Android, and Microsoft Windows Phone OS.
- 3G and 4G phones are the closest in architecture and design to a PC. They behave the same way—especially 4G devices—and they have the ability to download and install applications ("apps") the same as any PC or Mac.
- One interesting aspect of mobile device forensics is geolocation. The GPS can locate the user's activities and, when used with a timeline, can place the user in the vicinity of a crime. That can make it much easier to track the owner's movements.
- Each device has its own quirks, and each device needs special connectors and special device drivers on the tool that is examining it to decipher what is on the device's storage. Device storage also takes several forms, such as on-board nonvolatile memory and mini-SD cards that add storage in a modern smartphone or tablet.
- All mobile devices should be kept in a "Faraday" bag or box. This prevents changes from being made to the device remotely.
- Physical forensic images are bit-by-bit copies of the file system, including deleted data. Logical extraction is a snapshot of the file system showing what the file system wants the user to see.
- Mobile device forensic analysis can provide an overlay to physical evidence, and timelines (as well as computer forensic timelines) can give a clearer picture of the events preceding and following a crime event.
- Examiners make it a practice to run the forensic image twice, taking one of the images and treating it as evidence.
- The examiner should decide, based on what can be done with the particular device, whether to obtain a physical or logical extraction or both.

- Secure Digital (SD) cards are storage expansion cards that many mobile devices can accept. The SD card adds memory for storing such things as photos and music. SD cards are nonvolatile.
- Subscriber Identification Module (SIM) cards have an international mobile subscriber identity (IMSI) number that associates the phone with the subscriber's mobile network.
- Each SIM is the integrated circuit card identifier (ICCID). The ICCID contains the issuer identification number (IIN), the individual account identification, and a check digit.
- In addition to memory, the typical mobile device contains a digital signal processor, a microprocessor, a radio frequency transmitter/receiver, audio components, and a power supply that takes battery power to run the device as well as providing the ability to charge the battery.
- "Temporal" chains of evidence show events in the order in time in which they occurred. This is commonly called a timeline.
- "Causal" chains of evidence describe the events of a crime in terms of cause and effect. The links in the chain are the pieces of evidence, and they are tied together based upon how one link affects one or more other links.
- "Hybrid crime assessment" is a technique that investigators can use when faced with a physical crime—such as murder, rape, or robbery—that has a digital element to it: a computer, a cell phone, or some other mobile device.
- The amount of information that one can get from a mobile device varies greatly with the device.

Review Questions

1. Early mobile phone systems were followed by digital ________ networks.
2. True or False: The architectural functionality that distinguishes 2G from 3G is that 2G systems were circuit switched and 3G systems are packet switched.
3. True or False: One of the benefits of packet switching is the ability to connect more readily to the Internet.
4. It's (easy, difficult) to stay current with the available mobile device models.
5. Apple has taken major steps to standardize the development of apps for its ________ and ________.
6. Some devices and many apps report the ________ of the device. That can make it much easier to track the owner's movements.
7. When working on a mobile device, the investigator has several sources of information available. Probably the most useful source of information available to an investigator is ________.
8. An examiner should decide whether to obtain a(n) ________ extraction or ________ extraction or both of a mobile device.

9. True or False: If the examiner has a device that supports physical extraction, that is the way to image the device. Logical extractions are useful only when the physical option is not available because of the device itself.

10. True or False: Logical extractions are bit-by-bit copies of the file system, including deleted data.

11. Examiners make it a practice to run an extracted image (once, twice).

12. The ________ extraction is fairly fast, and one may want to examine it for obvious evidence while a tool is making a physical image of the target.

13. Just like computers, the ________ defines the basic components of the mobile device.

14. ________ are storage expansion cards that many mobile devices can accept.

15. True or False: SD cards are *nonvolatile*, meaning that even if the power is turned off on the device, you won't lose your music or photos.

16. In many cases a user can keep their subscriber information when changing mobile phones by simply switching the ________ card to the new phone.

17. It often is desirable to ________ the SIM in much the same way as one would take a physical image of the mobile device or a computer in order to retain a copy for evidentiary purposes.

18. True or False: It's always possible to recover deleted file items such as e-mails, texts, and photos from a mobile device.

19. ________ describes the events and concomitant evidence that make up the events of the crime.

20. ________ chains of evidence show events in the order in time in which they occurred.

21. ________ chains of evidence describe the events of a crime in terms of cause and effect.

22. ________ crime assessment attempts to tie elements of a crime together into a single crime scene and use the timelines to build a picture and describe the events and supporting evidence of the crime.

23. It is likely that the mobile device is extremely accurate as a yardstick for measuring when events happen, as the device may be synchronized to a(n) ________ clock.

24. When a mobile device is set to use ________, it will recognize any ________ network in its range.

25. True or False: Mobile device forensic analysis can provide an overlay to physical evidence and timelines, as well as computer forensic timelines, to give a clearer picture of the events preceding and following a crime event.

Learn the Vocabulary

Use the information from this chapter along with your prior knowledge and experience to learn the key terms. Choose from the following activities:

1. Make flashcards with a term on the front and its definition on the back. Use the flashcards to quiz a partner on correctly pronouncing and defining the terms.
2. Make an audio or video recording of each term and its definition.
3. Work with a partner to write a sentence for each term. Read the sentences out loud to a partner or share them with the class.
4. Write a list of new terms and language you have encountered in this chapter. Reuse the terms to write sentences and then read the sentences out loud to a partner. Reuse the terms again by writing a short speech. The speech might explain the terms or teach a lesson using the terms. Give the speech to a partner or to the class.

Application and Critical Thinking

1. What precautions should the examiner take when seizing/analyzing a live, turned-on mobile device?
2. Differentiate chain of evidence and chain of custody and give examples of both in the context of an investigation where mobile devices play an important part. Share your response with a partner. Compare your responses with theirs.
3. How can law enforcement make use of the locations of cell phone towers?
4. How are today's generations of mobile devices different from and the same as personal computers?
5. What are SIMs and SD cards and why does a mobile device need them? Do all mobile devices have one or both of these? If not, what substitutes?
6. What is the IMSI, where might it reside, and what is it used for? How would the digital forensic investigator use it?
7. If there is a GPS capability on a smartphone, how might the investigator make use of it? Is it useful for correlation? What kind and how would such correlation be accomplished?

Further References

Ayers, R., et al., Cell phones Forensic Tools: An Overview and Analysis Update, http://csrc.nist.gov/publications/nistir/nistir-7387.pdf

Digital Evidence and Forensics, 2010, http://www.nij.gov/topics/forensics/evidence/digital/welcome.htm

Electronic Crime Scene Investigation: A Guide for First Responders, 2nd ed., 2008, https://www.ncjrs.gov/pdffiles1/nij/219941.pdf

Jansen, W. and R. Ayers, *Guidelines on Cell Phone Forensics*, http://csrc.nist.gov/publications/nistpubs/800-101/SP800-101.pdf

Careers in Forensic Science

21

Dr. Jurgen Scriba/Science Source

In previous chapters we have encountered some of the many disciplines within forensic science pathology, toxicology, anthropology, archaeology, and entomology, to name a few. However, the profession of forensic science encompasses an even wider range of activities, with work taking place in the field, the laboratory, and the courtroom. A wealth of information and resources on the topic of forensic science and its specialties are available to be researched and explored in museums, libraries, and online platforms. Professional organizations and private companies often utilize mentors employed in science, technology, engineering, and mathematics (STEM) fields. In this chapter, we will explore some of the various careers within forensic science and discuss the kind of training necessary to pursue each.

Disciplines in Forensic Science

The wide variety of activities comprising forensic science is reflected in the structure of the American Academy of Forensic Sciences (AAFS), the largest forensic science organization in the world. Each of the five thousand scientists who are members of the AAFS belongs to one of the academy's 11 divisions: Criminalistics, Digital & Multimedia Sciences, Engineering Sciences, General, Jurisprudence, Odontology, Pathology/Biology, Physical Anthropology, Psychiatry/Behavioral Science, Questioned Documents, and Toxicology. Each career in forensic science requires various college courses, licensure, certifications, and physical and mental capabilities. Physical requirements may include the ability to get out to the scene and crouch down for an extended time to examine and collect evidence. Mental capabilities include the ability to think logically and independently, attention to detail, and the ability to work under pressure to meet a deadline.

Criminalistics

As we have seen throughout this text, the criminalist works primarily with physical evidence—analyzing, comparing, identifying, and interpreting items that may help solve a crime. The criminalist applies techniques from the physical and natural sciences to the examination of physical evidence. However, examining, testing, and interpreting results are only part of the responsibilities of the criminalist. The criminalist must also effectively communicate technical details and scientific findings to nonspecialist members of the law enforcement community.

The extent and depth of versatility expected of the forensic scientist are usually determined by the size of the crime laboratory's staff. Scientists in smaller laboratories are often expected to be generalists, performing a wide variety of tasks in order to fulfill the varied objectives of the laboratory. Their counterparts in larger facilities enjoy the luxury of working in specialized areas, relying on a teamwork approach to provide the spectrum of scientific skills needed for the comparison or identification of physical evidence. Currently there is a huge demand for forensic DNA analysts. Generally these positions require a graduate degree in biochemistry or forensic science.

In addition to their technical responsibilities, the newly trained criminalist must discover and master the role of the expert witness. A good courtroom demeanor and the ability to communicate thoughts and ideas in clear, concise terms are absolutely essential if the scientist's examination and conclusions are to be properly and effectively presented at a hearing or in court.

Education and Training The minimum educational requirement is a bachelor's degree in a science such as chemistry, biology, physics, or molecular biology. Coursework must include mathematics, chemistry, and/or biology. Many government criminalist positions have either local or national civil service educational and work experience requirements. Interested job applicants should inquire with appropriate civil service authorities to determine these requirements for a particular jurisdiction. The criminalist also participates in continuing education courses throughout their career to remain current with the ever-changing technology and procedures in the field. The American Board of Criminalistics (www.criminalistics.com) offers certification for trained criminalists who pass a rigorous examination.

Career Opportunities Criminalists work in a wide array of settings, including forensic laboratories in local, regional, and state police and sheriff's departments and in district attorneys' offices. They may also serve federal agencies such as the U.S. military; Drug Enforcement Administration (DEA); Alcohol, Tobacco, Firearms and Explosives (ATF); Federal Bureau of Investigation (FBI); Department of Justice (DOJ); U.S. Postal Service (USPS); Secret Service (SS); Central Intelligence Agency (CIA); and U.S. Fish and Wildlife Service. Trained criminalists also may teach at community colleges and universities.

Digital and Media Sciences

Digital data are collected routinely in most investigations. Digital & Multimedia Sciences professionals perform and conduct digital forensic examinations. More likely than not, someone involved in a crime operated a computer, used a mobile phone/device, or accessed the Internet. Digital scientists are knowledgeable in the skills required to extract data from these devices. Additionally, digital audio, photography, and video recording devices are nearly everywhere; most businesses and many local and state governments have security cameras/devices that can yield a surprising amount of photos/video.

Education and Training Candidates for an apprenticeship program in digital and multimedia sciences should possess a minimum of a bachelor's degree, preferably in computer science, information technology, or engineering. In the United States, there are undergraduate degree programs with emphasis in digital forensics, computer forensics, and media forensics. There are also graduate degree and graduate certificate programs in these fields. Digital forensic examiners also may have various types of certifications.

Career Opportunities Digital forensic examiners are employed in both the public and private sector. Many large police organizations, as well as most state and federal law enforcement agencies, generally employ digital forensic experts. Scientists who specialize in the field of digital & multimedia sciences work or practice in the following forensic areas: computer-related crime investigator; computer specialist; image analyst/examiner; audio analyst/examiner; video analyst/examiner; speech scientist; and facial identification/biometrics.

Engineering Sciences

Forensic engineers apply principles of engineering to resolve legal issues, primarily in civil, rather than criminal, cases. However, a forensic engineer may also assist in criminal matters or cases involving violation of government regulations. These areas of activity cover investigations into structural failure; reconstruction of automobile accidents; exploring the causes of fires or explosions; and evaluation of construction, manufacturing, or maintenance procedures. The forensic engineer's work applies in personal-injury suits, construction claims, contract or warranty disputes, patent or copyright infringements, and criminal and regulatory matters. Certification in the forensic engineering sciences is available from the International Institute of Forensic Engineering Sciences (www.iifes.org).

Education and Training The minimum educational requirement for a forensic engineer is a bachelor's degree in engineering or an allied science. However, an advanced degree is desirable, as is registration in a professional engineering society. Because of the wide range of engineering disciplines, a forensic engineer should have professional experience in the particular field(s) of engineering that are the subject of their investigation.

Career Opportunities Although some forensic engineers work for sizable corporations or government agencies, most are either self-employed or employed by small firms. Some forensic engineers hold a full-time job in another field (such as engineering consulting or teaching) and perform forensic services on a part-time or contract basis.

Odontology

As we saw in previous chapters, forensic odontology applies the principles of dental science to law. This includes such activities as identifying human remains through dental analysis as well as comparing bite marks, using both physical and biological dental evidence. Forensic odontologists frequently are called upon after catastrophic events involving large numbers of casualties for which identification from other physical remains is difficult or impossible. Bite mark analysis may involve collecting saliva for DNA profiling and matching.

Education and Training The forensic odontologist must of course possess a Doctor of Dental Science (D.D.S.) degree, but this basic education by itself is not sufficient to function in this field. Courses in forensic science and medicolegal death investigation are strongly recommended. Additional courses and advanced training in approved forensic techniques and procedures are required to prepare a forensic odontologist to conduct proper crime-scene investigations. Several professional organizations, including the American Board of Forensic Odontology, the American Society of Forensic Odontology, the New York Society of Forensic Dentistry, and the New York County Dental Society, offer advanced training in this field. The American Board of Forensic Odontology (www.abfo.org) serves as a highly regarded credentialing body for forensic dentists.

Career Opportunities Most forensic odontologists are engaged in private dental practice. They typically have either formal appointments or consulting relationships with coroners, medical examiners' offices, state and local government agencies, and branches of the military. They also often offer private consultations for insurance companies and law firms. Qualified forensic odontologists frequently provide expert testimony in criminal and civil cases involving personal injuries, worker's compensation, malpractice suits, potential violations of the dentist–patient relationship, and identification of bite marks in criminal cases. Once a commitment is made to enter this field, the dental investigator needs to be current in the most accurate methods available, be aware of ethical values and conflicts, and possess the dedication to render assistance in a timely and professional manner.

Pathology/Biology

Forensic pathology is the application of the principles of pathology—and medicine in general—to the legal process. Forensic pathologists perform autopsies to determine the causes of an individual's death and the circumstances surrounding the death. Understanding the circumstances of a person's death allows the pathologist to determine whether the death was the result of natural causes, an accident, suicide, or homicide. Forensic pathologists also identify remains using medical information, dental records, and other peculiarities of an individual. Forensic pathologists help public health officials recognize and control epidemic diseases. Clinical forensic pathologists study patterns of injury in living people to help law investigations into crimes such as child and elder abuse. The principal professional organization for forensic pathologists is the National Association of Medical Examiners (www.thename.org).

Education and Training Forensic pathologists undergo the same training and education as any medical doctor: four years of college, followed by four years of medical school, and then a residency in pathology. After college, medical school, internship, and residency, an additional one- or two-year fellowship in forensic pathology must be completed. This is then followed by certification in pathology or one of its subspecialties from the American Board of Pathology.

Career Opportunities Forensic pathologists are usually employed by city, county, or state medical examiners' offices and hospitals. Some federal government agencies, such as the Centers for Disease Control (CDC) and the Armed Forces Institute of Pathology (AFIP), also employ forensic pathologists.

Physical Anthropology

Forensic anthropologists are trained in the study of human skeletal biology, and they typically use this training to identify deceased individuals when skeletal remains are the only useful evidence. They also determine whether recovered bones are human or non-human and determine sex based on skeletal structure. Forensic pathologists may estimate the cause and time of death when remains are unsuitable for examination by a pathologist.

Education and Training Forensic anthropologists usually earn a Ph.D. in anthropology with an emphasis on human osteology (the study of bones) and anatomy. Practical experience in forensic anthropology typically is required before a court accepts an individual as an expert witness in the field. The American Board of Forensic Anthropology (ABFA) (www.theabfa.org) provides professional certification for experts in the field of forensic anthropology. To earn certification, an individual must submit case reports for review that demonstrate practical experience in the field. The applicant must also pass both a written and a practical examination.

Career Opportunities Most forensic anthropologists work out of laboratories at major research institutions or universities. In addition, various state and local medical examiner offices use forensic anthropologists as medical investigators or administrators. The federal government employs forensic anthropologists at the U.S. Army Central Identification Laboratory and the Armed Forces Institute of Pathology. State and federal law enforcement agencies also hire physical anthropologists to act as special agents and laboratory personnel. The largest group of forensic anthropologists works for the military in the U.S. Army's Central Identification Laboratory in Hawaii. A major professional organization in this field is the American Association of Physical Anthropologists (www.physanth.org).

Psychiatry/Behavioral Science

The forensic psychologist or psychiatrist deals with issues of human behavior and mental illness as they relate to matters of civil and criminal law. This involves such activities as determining an individual's competence to stand trial, testify, or waive legal representation, and determining whether mental illness mitigates a suspect's responsibility for their actions. For example, the forensic psychiatrist often gives an opinion about whether a defendant may be innocent by reason of mental illness or defect. Other areas in which such expertise is applied includes cases of involuntary psychiatric hospitalization, a patient's right to refuse treatment, and competency to participate in do-not-resuscitate decisions. Forensic psychology includes psychological evaluation and expert testimony regarding such criminal forensic issues as trial competency and forensic behavioral analysis.

Education and Training Forensic psychiatrists are medical doctors who complete a course of study similar to that of forensic pathologists. This includes four years of college followed by medical school and residency training in psychiatry. Some forensic psychiatrists take an additional year or two of postresidency training in psychiatry and the law, and many continue independent study and on-the-job training after completing their formal education. The American Board of Psychiatry and Neurology offers certification in forensic psychiatry for those who pass special examinations. Forensic psychologists usually obtain a Ph.D. degree and are licensed by a state board, and may be board certified by the American Board of Professional Psychology (www.abpp.org).

Career Opportunities As with other professionals in the psychiatric profession, forensic psychiatrists and forensic psychologists may be employed in private practice, by hospitals, or by city, county, state, or federal government agencies. They often work in a prison or state hospital setting. The major professional organization for forensic psychiatrists is the American Academy of Psychiatry and the Law (www.aapl.org).

Questioned Documents

As we saw in Chapter 18, the document examiner analyzes questioned documents and related material, such as ink, paper, toner from a copier or fax, and computer printers. Detecting forged documents and counterfeit currency is a large part of the document examiner's work.

Education and Training Document examiners should possess at least a bachelor's degree, preferably in a scientific field. Several colleges and universities offer questioned-document or related courses as part of criminal justice, forensic science, or criminalistics degree programs, although no document examination degree programs currently exist. In addition to a college education, the prospective document examiner must complete an apprenticeship program lasting approximately two years under the direct supervision of a full member or fellow of the questioned-documents section of the AAFS, a member of the American Society of Questioned Document Examiners, or a diplomate of the American Board of Forensic Document Examiners (www.abfde.org).

Career Opportunities Forensic document examiners are often consultants working in private practice, but many large police organizations, as well as most state and federal law enforcement agencies, also employ forensic document experts.

Toxicology

Forensic toxicology deals with the effects of drugs and chemicals on the human body and the application of that knowledge to questions of law. Forensic toxicologists perform a wide range of activities, including determining whether an individual has consumed illegal drugs and identifying the substances involved. Forensic toxicologists are often called upon to determine whether an individual was driving under the influence of alcohol or drugs. They may be involved in postmortem toxicology, which involves determining the contribution drugs or chemicals make to the circumstances of a death. The forensic toxicologist investigates cases as varied as animal poisoning, the use of drugs in sexual assault, and drug use and doping in human and animal sports. Currently, many forensic toxicologists are employed by private laboratories engaged in conducting workplace drug testing and forensic urine testing on employees of organizations and industries associated with the public's safety or engaged in performing hazardous work.

Education and Training Training for a career in forensic toxicology requires a bachelor's degree in a physical science, ideally including a solid background in chemistry and coursework in pharmacology. Several colleges and universities offer graduate coursework in forensic toxicology at the master's or Ph.D. level. Membership in the forensic toxicology section of AAFS requires at least one year of additional experience and further requirements of scholarly work or advanced study. The American Board of Forensic Toxicology and the Forensic Toxicologist Certification Board offer professional certification to scientists with work experience in forensic toxicology. A major forensic toxicology organization is the Society of Forensic Toxicologists (www.soft-tox.org).

Career Opportunities Many forensic toxicologists work in police laboratories, medical examiners' offices, and workplace drug-testing laboratories. Other career opportunities for forensic toxicologists exist in hospitals, universities, corporations, and agencies that monitor drug use in sports.

College Courses in Forensic Science

Degree Programs in Forensic Science

It is something of an irony that although most of the careers discussed previously require at least a bachelor's degree, few colleges and universities in the United States offer undergraduate or graduate degrees in forensic science. Institutions offering Forensic Science Education Programs Accreditation Commission (FEPAC) accredited undergraduate and graduate programs include the following:

Albany State University, Albany, GA: Bachelor of Science Degree in Forensic Science

University of Alabama at Birmingham, Birmingham, AL: Master of Science in Forensic Science

Alfred State—Suny College of Technology, Alfred, NY: Bachelor of Science Degree in Forensic Science Technology

Arcadia University, Glenside, PA: Master of Science Degree in Forensic Science

Boston University School of Medicine: Master of Science in Biomedical Forensic Science

Buffalo State SUNY, Buffalo, NY: Bachelor of Science Degree in Forensic Chemistry

Cedar Crest College, Allentown, PA: Bachelor of Science Degree in Forensic Science and Bachelor of Science in Genetic Engineering, Concentration in Forensic Science and Master of Science Degree in Forensic Science

Duquesne University, Pittsburg, PA: Master of Science Degree in Forensic Science

Eastern Kentucky University, Richmond, KY: Bachelor of Science Degree in Forensic Science (Biology and Chemistry Options) and Bachelor of Science Degree in Digital Forensics and Cybersecurity

Fayetteville State University, Fayetteville, NC: Bachelor of Science Degree in Forensic Biology Concentration

University of Central Florida, Orlando, FL: Bachelor of Science Degree in Forensic Science

Florida International University, Miami, FL: Accreditation for both the Certificate Programs in conjunction with the Bachelor of Science Degree in a natural science such as chemistry or biology and the Master of Science in Forensic Science

The George Washington University, Washington, DC: Master of Science Degree in Concentration in Forensic Chemistry, Molecular Biology, Forensic Toxicology

University of Illinois at Chicago, Chicago, IL: Master of Science Degree in Forensic Science

Indiana University Purdue University, Indianapolis, IN: Bachelor of Science Degree in Forensic Science and Investigative Sciences and Master of Science Degree in Forensic Science and Investigative Sciences

John Jay College of Criminal Justice, New York, NY: Master of Science Degree in Forensic Science

Laurentian University, Sudbury, ON: Bachelor of Science in Forensic Science (Single Specialization) and Honors B.Sc. Degree in Forensic Science and Chemistry (Combined Specialization)

Liberty University, Lynchburg, VA: Bachelor of Science Degree in Forensic Science

Loyola University, Chicago, IL: Bachelor of Science Degree in Forensic Science

Madonna University, Livonia, MI: Bachelor of Science Degree in Emphasis in Digital Evidence

Marshall University, Huntington, WV: Master of Forensic Science Degree, Emphasis in Digital Evidence

Michigan State University, East Lansing, MI: Master of Science Degree, Forensic Chemistry Track

University of Mississippi, University, MS: Bachelor of Science Degree in Forensic Chemistry

University of New Haven, West Haven, CT: Bachelor of Science Degree in Forensic Science and Master of Science Degree in Forensic Science

University of North Texas, Denton, TX: Bachelor of Science Degree in Biochemistry, Biology, and Chemistry

Ohio University, Athens, OH: Bachelor of Science Degree in Forensic Chemistry

Oklahoma State University, Tulsa, OK: Master of Science Degree in Forensic Science, Forensic Toxicology and Forensic Biology Tracks

University of Central Oklahoma, Edmond, OK: Bachelor of Science Degree in Forensic Science - Chemistry & Molecular Biology and Bachelor of Science Degree in Forensic Science - Digital Forensics

University of Ontario Institute of Technology, Oshawa, Ontario: Bachelor of Science (Honours) in Forensic Science

The Pennsylvania State University, University Park, PA: Bachelor of Science Degree in Forensic Science and Master of Professional Studies in Forensic Science

Sam Houston State University, Huntsville, TX: Master of Science Degree in Forensic Science

Syracuse University, Syracuse, NY: Master of Science Degree in Forensic Science

University of Tampa, Tampa, FL: Bachelor of Science Degree in Forensic Science

Texas A&M University, College Station, TX: Bachelor of Science Degree in Forensic and Investigative Sciences

Towson University, Towson, MD: Bachelor of Science in Forensic Chemistry and Master of Science Degree in Forensic Science

Virginia Commonwealth University, Richmond, VA: Bachelor of Science Degree in Forensic Science and the Master of Science Degree in Forensic Science Programs

West Chester University of Pennsylvania, West Chester, PA: Bachelor of Science Degree in Forensic and Toxicological Chemistry

West Virginia University, Morgantown, WV: Bachelor of Science Degree in Forensic and Investigative Science Program and Master of Science Degree in Forensic and Investigative Sciences

This is not to suggest, however, that course and degree programs in forensic sciences are not available at other colleges and universities. In fact, more than a hundred schools offer associate's, bachelor's, and master's degrees in some area of forensic science, and others provide seminars and courses in forensic specialties such as DNA typing and profiling. Although most of these programs have not been accredited by the FEPAC, most working forensic scientists obtained their education and degrees from non-FEPAC-accredited schools. A complete list of undergraduate and graduate programs in forensic science in the United States and abroad can be found at the AAFS website (www.aafs.org).

A small number of schools have initiated programs and courses in computer forensics. Interested students should search out schools on the AAFS website for these programs.

Required College Courses

Certain natural science courses are required for any student in forensic science. Unlike other criminal justice professionals, a forensic scientist requires a foundation in chemistry, biology, physics, and mathematics. The minimum general core requirements recommended for undergraduate forensic science programs include the following:

- General chemistry I and II and lab for science majors (eight credit hours)
- Organic chemistry I and II and lab (eight credit hours)
- Biology I and II for science majors (four to eight credit hours; classes with laboratory components are preferable, if available)
- Physics I and II for science majors and lab (eight credit hours)
- Calculus (three credit hours)
- Statistics for science majors (three credit hours)

An undergraduate degree in forensic science is expected to be an interdisciplinary degree that includes substantial laboratory work and an emphasis on advanced coursework in chemistry or biology. Students can use these additional courses to begin to specialize along a forensic science discipline track, such as forensic biology or forensic chemistry. Specialized science courses may be selected from any of the following:

- Biochemistry
- Molecular biology
- Genetics
- Population genetics
- Inorganic chemistry
- Analytical/quantitative chemistry
- Physical chemistry
- Instrumental analysis
- Cell biology
- Pharmacology
- Calculus II
- Microbiology

As you can see, the path to becoming a forensic scientist is neither brief nor easy. It begins with rigorous college coursework, typically followed by additional training and certification in the specific area of forensic science one chooses to pursue. It involves lifelong learning and the dedication to seek out the newest and best methods to gather, analyze, and interpret data needed to solve all types of crimes.

Despite these hurdles, forensic science is a fascinating profession that challenges one's scientific knowledge, powers of observation, critical-thinking skills, deductive and inductive reasoning, and ability to think creatively and imagine various possible scenarios when presented with facts and physical evidence. Hopefully, this text will provide both a spark of interest and the basic information that will spur you to take the first steps toward pursuing a career in this challenging and ever-evolving field.

Quick Review

- Forensic science encompasses a wide range of activities with different kinds of training necessary for each.
- The American Academy of Forensic Science (AAFS) is the largest forensic science organization in the world.
- There are 11 divisions within the AAFS: Criminalistics, Digital & Multimedia Sciences, Engineering Sciences, General, Jurisprudence, Odontology, Pathology/Biology, Physical Anthropology, Psychiatry/Behavioral Science, Questioned Documents, and Toxicology.

Chapter Review

- Forensic science encompasses a wide range of activities with different kinds of training necessary for each.
- The American Academy of Forensic Science (AAFS) is the largest forensic science organization in the world.
- There are 11 divisions within the AAFS: Criminalistics, Digital & Multimedia Sciences, Engineering Sciences, General, Jurisprudence, Odontology, Pathology/Biology, Physical Anthropology, Psychiatry/Behavioral Science, Questioned Documents, and Toxicology.
- Criminalists primarily work with physical evidence applying techniques from the physical and natural sciences in their examination.
- Criminalists need a minimum of a bachelor's degree in a science and work in a wide array of settings.
- Professionals working in digital and multimedia sciences perform and conduct digital forensic examinations.
- To enter an apprenticeship program, a digital and media science professional must have a minimum of a bachelor's degree in computer science, information technology, or engineering.
- Digital forensic examiners can be employed in the public or private sector.
- Forensic engineers apply principles of engineering to resolve legal issues.
- A forensic engineer needs a minimum of a bachelor's degree in engineering or an allied science, with advanced degrees the more desirable.
- Forensic engineers typically work for large corporations or government agencies.
- Odonlotogists apply the principles of dental science to law, typically being asked to identify human remains through dental analysis.
- Forensic odontologists must possess a Doctor of Dental Science (D.D.S) degree with additional courses and advanced training in approved forensic techniques and procedures.
- Most forensic odontologists work in a private dental practice with consulting relationships with coroners, medical examiners' offices, state and local government agencies, and branches of the military.
- Forensic pathology applies the principles of pathology to legal processes.
- Forensic pathologists obtain the same education and training as medical doctors: four years of college followed by four years of medical school with a residency in pathology. In addition, a one- or two-year fellowship in forensic pathology must be completed.
- Forensic pathologists typically work for a city, county, or state medical examiners' office and hospitals.

- Forensic anthropologists use their training in the study of human skeletal biology to identify deceased individuals when skeletal remains are the only useful evidence.
- Forensic anthropologists typically earn a Ph.D. in anthropology with an emphasis on human osteology and anatomy. Professional certification is available through the American Board of Forensic Anthropology (ABFA)
- Most forensic anthropologists work from laboratories at major research institutions or universities. Many state and local medical examiner offices use forensic anthropologists as medical investigators or administrators.
- Forensic psychologists or psychiatrists deal with issues of human behavior and mental illness related to matters of civil and criminal law.
- Forensic psychiatrists complete four years of college followed by medical school and residency training in psychiatry.
- Forensic psychologists obtain a Ph.D. degree and are licensed by a state board.
- Forensic psychiatrists and psychologists may be employed in private practice or by hospitals, or various levels of governmental agencies.
- Document examiners analyze questioned documents and related materials.
- Document examiners should possess a minimum of a bachelor's degree in a scientific field, in addition to completing an apprenticeship program under the direct supervision of a full member or fellow of the questioned-documents section of the AAFS.
- Forensic document examiners are often consultants working in private practice, but can be employed by state and federal law enforcement agencies.
- Forensic toxicology deals with the effects of drugs and chemicals on the human body and the application of that knowledge to questions of law.
- A forensic toxicologist must have a minimum of a bachelor's degree in physical science, ideally with a solid background in chemistry and coursework in pharmacology.
- Most forensic toxicologists work in police laboratories, medical examiner's offices, and workplace drug-testing laboratories.
- Most forensic careers require at least a bachelor's degree, but few colleges and universities in the U.S. offer undergraduate or graduate degrees in forensic science.
- The minimum general core requirements include one or more courses in biology, chemistry, organic chemistry, physics, calculus, and statistic
- Becoming a forensic scientists requires rigorous college coursework, and usually additional training and certification in the specific area chosen.

Review Questions

1. What is the minimum level of education required by most disciplines in forensic science?

2. What is the world's largest forensic science organization?

3. True or False: You can earn a bachelor's degree in forensic science at any college or university in the United states.

4. True or False: Forensic science requires critical-thinking skills, and the ability to think creatively.

Application and Critical Thinking

1. Research two professions within forensic science disciplines you have learned about in this course. Identify and analyze how these professions have changed and developed over time, starting with early historical developments and tracing the professions through to current advancements in the fields today. Differentiate the functions and responsibilities of a professional working in these two fields.

2. Choose a specialty in forensic science to explore and describe. To gather your information you should research and explore resources such as libraries, professional organizations, online platforms, and a professional employed in the area you chose to research. Be sure to include relevant discipline-specific information:
   a. Describe the collegiate course requirements to work in this field.
   b. Describe the type of licensure or certifications that can be obtained and/or required.
   c. Describe the physical and mental capabilities required.
   d. Present your specialty to the class.

3. Select three or four forensic disciplines that interest you. Conduct additional research on these careers. Differentiate the roles, responsibilities, and functions of each of these professionals working in the criminal justice system. Careers you may want to choose from include forensic scientists, crime scene investigators, criminologists, court systems personnel, and medicolegal death investigators. Create a short written summary of your findings.

Index

C

G

N

T

U

V